Organic Reactions

ADVISORY BOARD

John E. Baldwin	James A. Marshall
Peter Beak	Jerrold Meinwald
George A Boswell, Jr.	Scott J. Miller
Engelbert Ciganek	Larry E. Overman
Dennis Curran	Leo A. Paquette
Samuel Danishefsky	Gary H. Posner
Huw M. L. Davies	T. V. RajanBabu
John Fried	Hans J. Reich
Jacquelyn Gervay-Hague	James H. Rigby
Heinz W. Gschwend	William R. Roush
Stephen Hanessian	Scott D. Rychnovsky
Richard F. Heck	Martin Semmelhack
Louis Hegedus	Charles Sih
Herbert O. House	Amos B. Smith, III
Robert C. Kelly	Barry M. Trost
Andrew S. Kende	Milán Uskokovic
Laura Kiessling	James D. White
Steven V. Ley	

FORMER MEMBERS OF THE BOARD
NOW DECEASED

Roger Adams	Louis F. Fieser
Homer Adkins	Ralph F. Hirschmann
Werner E. Bachmann	John R. Johnson
A. H. Blatt	Robert M. Joyce
Virgil Boekelheide	Willy Leimgruber
Theodore L. Cairns	Frank C. McGrew
Arthur C. Cope	Blaine C. McKusick
Donald J. Cram	Carl Niemann
David Y. Curtin	Harold R. Snyder
William G. Dauben	Boris Weinstein

Organic Reactions

VOLUME 76

EDITORIAL BOARD

SCOTT E. DENMARK, *Editor-in-Chief*

DALE BOGER
ANDRÉ B. CHARETTE
VITTORIO FARINA
JEFFREY S. JOHNSON
MICHAEL J. MARTINELLI
STUART W. MCCOMBIE

JOHN MONTGOMERY
ANDREW J. PHILLIPS
TOMISLAV ROVIS
STEVEN M. WEINREB
PETER WIPF

ROBERT BITTMAN, *Secretary*
Queens College of The City University of New York, Flushing, New York

JEFFERY B. PRESS, *Secretary*
Press Consulting Partners, Brewster, New York

LINDA S. PRESS, *Editorial Coordinator*

DANIELLE SOENEN, *Editorial Assistant*

ENGELBERT CIGANEK, *Editorial Advisor*

ASSOCIATE EDITORS

CAROLE J. R. BATAILLE
SANDOR CACCHI
JEAN-MARC CAMPAGNE
TIMOTHY J. DONOHOE
GIANCARLO FABRIZI
ANNE GAUCHER
ANTONELLA GOGGIAMANI
PAOLO INNOCENTI
SYLVAIN MARQUE
DAMIEN PRIM

A JOHN WILEY & SONS, INC., PUBLICATION

Published by John Wiley & Sons, Inc., Hoboken, New Jersey

Copyright © 2012 by Organic Reactions, Inc. All rights reserved.

Published simultaneously in Canada.

No part of this publication may be reproduced, stored in a retrieval system, or transmitted in any form or by any means, electronic, mechanical, photocopying, recording, scanning, or otherwise, except as permitted under Section 107 or 108 of the 1976 United States Copyright Act, without either the prior written permission of the Publisher, or authorization through payment of the appropriate per-copy fee to the Copyright Clearance Center, Inc., 222 Rosewood Drive, Danvers, MA 01923, (978) 750-8400, fax (978) 750-4470, or on the web at www.copyright.com. Requests for permission need to be made jointly to both the publisher, John Wiley & Sons, Inc., and the copyright holder, Organic Reactions, Inc. Requests to John Wiley & Sons, Inc., for permissions should be addressed to the Permissions Department, John Wiley & Sons, Inc., 111 River Street, Hoboken, NJ 07030, (201) 748-6011, fax (201) 748-6008, or online at http://www.wiley.com/go/permission.

Limit of Liability/Disclaimer of Warranty: While the publisher and author have used their best efforts in preparing this book, they make no representations or warranties with respect to the accuracy or completeness of the contents of this book and specifically disclaim any implied warranties of merchantability or fitness for a particular purpose. No warranty may be created or extended by sales representatives or written sales materials. The advice and strategies contained herein may not be suitable for your situation. You should consult with a professional where appropriate. Neither the publisher nor author shall be liable for any loss of profit or any other commercial damages, including but not limited to special, incidental, consequential, or other damages.

For general information on our other products and services or for technical support, please contact our Customer Care Department within the United States at (800) 762-2974, outside the United States at (317) 572-3993 or fax (317) 572-4002.

Wiley also publishes its books in a variety of electronic formats. Some content that appears in print may not be available in electronic formats. For more information about Wiley products, visit our web site at www.wiley.com.

Library of Congress Catalog Card Number: 42-20265
ISBN: 978-0-470-63843-9

Printed in the United States of America

10 9 8 7 6 5 4 3 2 1

INTRODUCTION TO THE SERIES
ROGER ADAMS, 1942

In the course of nearly every program of research in organic chemistry, the investigator finds it necessary to use several of the better-known synthetic reactions. To discover the optimum conditions for the application of even the most familiar one to a compound not previously subjected to the reaction often requires an extensive search of the literature; even then a series of experiments may be necessary. When the results of the investigation are published, the synthesis, which may have required months of work, is usually described without comment. The background of knowledge and experience gained in the literature search and experimentation is thus lost to those who subsequently have occasion to apply the general method. The student of preparative organic chemistry faces similar difficulties. The textbooks and laboratory manuals furnish numerous examples of the application of various syntheses, but only rarely do they convey an accurate conception of the scope and usefulness of the processes.

For many years American organic chemists have discussed these problems. The plan of compiling critical discussions of the more important reactions thus was evolved. The volumes of *Organic Reactions* are collections of chapters each devoted to a single reaction, or a definite phase of a reaction, of wide applicability. The authors have had experience with the processes surveyed. The subjects are presented from the preparative viewpoint, and particular attention is given to limitations, interfering influences, effects of structure, and the selection of experimental techniques. Each chapter includes several detailed procedures illustrating the significant modifications of the method. Most of these procedures have been found satisfactory by the author or one of the editors, but unlike those in *Organic Syntheses*, they have not been subjected to careful testing in two or more laboratories. Each chapter contains tables that include all the examples of the reaction under consideration that the author has been able to find. It is inevitable, however, that in the search of the literature some examples will be missed, especially when the reaction is used as one step in an extended synthesis. Nevertheless, the investigator will be able to use the tables and their accompanying bibliographies in place of most or all of the literature search so often required. Because of the systematic arrangement of the material in the chapters and the entries in the tables, users of the books will be able to find information desired by reference to the table of contents of the appropriate chapter. In the interest of economy, the entries in the indices have been kept to a minimum, and, in particular, the compounds listed in the tables are not repeated in the indices.

The success of this publication, which will appear periodically, depends upon the cooperation of organic chemists and their willingness to devote time and effort to the preparation of the chapters. They have manifested their interest already by the almost unanimous acceptance of invitations to contribute to the work. The editors will welcome their continued interest and their suggestions for improvements in *Organic Reactions*.

INTRODUCTION TO THE SERIES
SCOTT E. DENMARK, 2008

In the intervening years since "The Chief" wrote this introduction to the second of his publishing creations, much in the world of chemistry has changed. In particular, the last decade has witnessed a revolution in the generation, dissemination, and availability of the chemical literature with the advent of electronic publication and abstracting services. Although the exponential growth in the chemical literature was one of the motivations for the creation of *Organic Reactions*, Adams could never have anticipated the impact of electronic access to the literature. Yet, as often happens with visionary advances, the value of this critical resource is now even greater than at its inception.

From 1942 to the 1980's the challenge that *Organic Reactions* successfully addressed was the difficulty in compiling an authoritative summary of a preparatively useful organic reaction from the primary literature. Practitioners interested in executing such a reaction (or simply learning about the features, advantages, and limitations of this process) would have a valuable resource to guide their experimentation. As abstracting services, in particular *Chemical Abstracts* and later *Beilstein*, entered the electronic age, the challenge for the practitioner was no longer to locate all of the literature on the subject. However, *Organic Reactions* chapters are much more than a surfeit of primary references; they constitute a distillation of this avalanche of information into the knowledge needed to correctly implement a reaction. It is in this capacity, namely to provide focused, scholarly, and comprehensive overviews of a given transformation, that *Organic Reactions* takes on even greater significance for the practice of chemical experimentation in the 21st century.

Adams' description of the content of the intended chapters is still remarkably relevant today. The development of new chemical reactions over the past decades has greatly accelerated and has embraced more sophisticated reagents derived from elements representing all reaches of the Periodic Table. Accordingly, the successful implementation of these transformations requires more stringent adherence to important experimental details and conditions. The suitability of a given reaction for an unknown application is best judged from the informed vantage point provided by precedent and guidelines offered by a knowledgeable author.

As Adams clearly understood, the ultimate success of the enterprise depends on the willingness of organic chemists to devote their time and efforts to the preparation of chapters. The fact that, at the dawn of the 21st century, the series continues to thrive is fitting testimony to those chemists whose contributions serve as the foundation of this edifice. Chemists who are considering the preparation of a manuscript for submission to *Organic Reactions* are urged to contact the Editor-in-Chief.

PREFACE TO VOLUME 76

The impact of transition-metal catalysis in synthetic organic chemistry can hardly be overstated. Three of the last ten Nobel Prizes in Chemistry have been awarded for the discovery and development of transition-metal-catalyzed reactions that fundamentally changed the practice of organic synthesis (2001: oxidation/reduction (Knowles, Noyori and Sharpless); 2005: olefin metathesis (Chauvin, Grubbs, Schrock); 2010: cross coupling (Heck, Negishi, Suzuki)). In fact, 14 of the 56 chapters published in *Organic Reactions* Volumes 50–75 are dedicated to the myriad manifestations of this family of increasingly important chemical transformations. It is thus no surprise that each of the three newest family members contained in this volume describe transition-metal-catalyzed reactions.

The first chapter by Timothy J. Donohoe, Carole J. R. Bataille, and Paolo Innocenti describes a versatile modification of the osmium-catalyzed dihydroxylation of alkenes. The ability to introduce vicinal hydroxy groups with high predictability and selectivity is a powerful and enabling transform in organic synthesis (see Noe, Letavic, Snow, and McCombie in Volume 66). Whereas the directing effect of hydrogen-bonding groups on epoxidation has been known for over 50 years, the corresponding control of dihydroxylation has only recently been harnessed in a synthetically practical fashion. Pioneering studies from the Donohoe laboratories have demonstrated that hydrogen-bonding functional groups (alcohols, secondary amides, and carbamates) serve as powerful directing moieties for the introduction of vicinal diols on proximal (allylic and homoallylic) double bonds. This mode of diastereoselective oxidation has allowed the synthesis of densely functionalized building blocks as well as enabling the late-stage introduction of key functional groups on advanced intermediates in complex molecule synthesis.

The second chapter by Damien Prim, Sylvain Marque, Anne Gaucher, and Jean-Marc Campagne covers the transition-metal-catalyzed α-arylation of enolates. The formation of carbon-carbon bonds facilitated by the α-hydrogen acidity of carbonyl compounds is among the most commonly employed reactions in organic synthesis. Classic chemical transformations and their modern variants (alkylation, aldol, Claisen, and Michael reactions) are reliable workhorse processes. However, because of the inability to employ aromatic electrophiles, the formation of C-aryl bonds is notably absent from this list. Once only the purview of stoichiometric transformations involving arylbismuth and aryllead reagents, the introduction of palladium, nickel, or copper catalysis in combination with standard enolate chemistry now enables the use of common aryl electrophiles (halides and triflates) to engage in direct bond formation. The scope of this transformation is broad and allows arylation of enolates derived from ketones, aldehydes, esters, amides, nitriles, and active methylene compounds. An important variant also incorporates the use of enoxysilane derivatives of ketones and esters. As is so often the case in transition-metal-catalyzed reactions, the choice of ligands and conditions is critical. Prim, Marque, Gaucher, and Campagne provide expert guidance for selection of these variables and show that with chiral phosphine ligands, even enantioselective C-aryl bond formation can be achieved. The usefulness of this process has been demonstrated by its application as a strategy level transform in total synthesis endeavors.

The third chapter by Sandro Cacchi, Giancarlo Fabrizi, and Antonella Goggiamani reviews the application of transition metal catalysis for the synthesis of one of the most important classes of heterocyclic compounds, namely indoles. The indole subunit is ubiquitous in alkaloid natural products as well as many signaling molecules, bioactive pharmaceutical agents, and fragrances. Because of its prevalence, many general methods for the preparation of indoles have been developed dating back to the Fischer indole synthesis reported in 1883. Indole syntheses comprise no fewer than 10 name reactions involving different precursors and bond constructions. Once again, the advent of palladium catalysis has enabled the multifaceted construction of substituted indoles from three basic building blocks: aromatic halides or anilines, alkynes, and amines. Cacchi, Fabrizi, and Goggiamani thoroughly detail the versatility of this family of reactions and illustrate the manifold strategies that can be used to create any of the four bonds comprising the pyrrole subunit of the indole moiety. Moreover, strategies that allow the simultaneous introduction of substituents on C(2) or C(3) of the pyrrole ring are described.

It is appropriate here to acknowledge the expert assistance of the entire editorial board, in particular Stuart McCombie (Donohoe, et al.) and Dale Boger (Prim et al. and Cacchi et al.), who shepherded the three chapters in this volume. The contributions of the authors, editors, and the publisher were expertly coordinated by the responsible secretary, Jeffery B. Press. In addition, the *Organic Reactions* enterprise could not maintain the quality of production without the dedicated efforts of its editorial staff, Dr. Linda S. Press and Dr. Danielle Soenen. Insofar as the essence of *Organic Reactions* chapters resides in the massive tables of examples, the authors' and editorial coordinators' painstaking efforts are highly prized.

SCOTT E. DENMARK
Urbana, Illinois

CONTENTS

CHAPTER   PAGE

1. HYDROGEN-BONDING-MEDIATED DIRECTED OSMIUM DIHYDROXYLATION
   Timothy J. Donohoe, Carole J. R. Bataille, and Paolo Innocenti ... 1

2. TRANSITION-METAL-CATALYZED α-ARYLATION OF ENOLATES
   *Damien Prim, Sylvain Marque, Anne Gaucher,
   and Jean-Marc Campagne* 49

3. INDOLES VIA PALLADIUM-CATALYZED CYCLIZATION
   Sandro Cacchi, Giancarlo Fabrizi, and Antonella Goggiamani 281

CUMULATIVE CHAPTER TITLES BY VOLUME 535

AUTHOR INDEX, VOLUMES 1–76 551

CHAPTER AND TOPIC INDEX, VOLUMES 1–76 557

CHAPTER 1

HYDROGEN-BONDING-MEDIATED DIRECTED OSMIUM DIHYDROXYLATION

Timothy J. Donohoe, Carole J. R. Bataille, and Paolo Innocenti

Department of Chemistry, University of Oxford, Chemistry Research Laboratory, Oxford, OX1 3TA, UK

CONTENTS

	Page
Introduction	2
Mechanism and Stereochemistry	4
Scope and Limitations	7
Nature of the Amine	7
Nature of the Directing Group	8
Steric Effects	9
Nature of the Substrate	11
Allylic versus Homoallylic Substrates	11
Conformational Factors Determined by the Alkene Substitution Pattern	12
Site Selectivity of the Directed Dihydroxylation Reaction	13
Alternative Directing Groups	13
Comparison with other Methods	14
Experimental Conditions	15
Experimental Procedures	15
General Procedure for Stoichiometric Dihydroxylation	15
$OsO_4 \cdot TMEDA$	15
Isolation Procedures for 0.50 mmol of Substrate	15
Sodium Sulfite	15
Acidic Methanol	16
Ethylenediamine	16
($1R^*,2S^*,3S^*$)-Cyclohexane-1,2,3-triol [Directed Dihydroxylation of an Allylic Cyclic Alcohol Using $OsO_4 \cdot TMEDA$]	16
2,2,2-Trichloro-N-[($1R^*,2R^*,3S^*$)-2,3-dihydroxycyclohexyl]acetamide [Directed Dihydroxylation of an N-Allylic Cyclic Amide Using $OsO_4 \cdot TMEDA$]	16
($1R^*,2S^*,3S^*$)-Cyclopentane-1,2,3-triol [Directed Dihydroxylation of an Allylic Cyclic Alcohol Using $OsO_4 \cdot TMEDA$]	17
($2R^*,3R^*,4S^*,5S^*$)-2-(Acetoxymethyl)tetrahydro-$2H$-pyran-3,4,5-triyl Triacetate [Directed Dihydroxylation of an Allylic Cyclic Alcohol Using $OsO_4 \cdot TMEDA$ and Subsequent Peracetylation]	17

---

timothy.donohoe@chem.ox.ac.uk
Organic Reactions, Vol. 76, Edited by Scott E. Denmark et al.
© 2012 Organic Reactions, Inc. Published 2012 by John Wiley & Sons, Inc.

Tricyclic Tetraol [Directed Dihydroxylation of an Exocyclic Allylic Alcohol Using OsO$_4$·TMEDA] 18
(2S_P^*,3S^*,4R^*,5S^*,6R^*)-2-Phenylmethoxy-6-(hydroxymethyl)-3,4,5-trihydroxy-1,2-oxaphosphorinane-2-oxide) [Directed Dihydroxylation of an Allylic Cyclic Alcohol Using OsO$_4$·TMEDA] 19
Osmate Ester of (3aR^*,4S^*,5S^*,6R^*,7R^*,7aR^*)-3-Benzyl-4-benzyloxy-5,6,7-trihydroxyhexahydrobenzo[d]oxazol-2(3H)-one [Preparation of an Osmate Ester Using OsO$_4$·TMEDA] 19
(2R^*,3R^*,4R^*)-2-Hydroxymethyl-2-(4-methoxybenzyl)tetrahydrofuran-3, 4-diol [Directed Dihydroxylation of a Homoallylic Alcohol Using Catalytic OsO$_4$•Quinuclidine and NMO] 20
(2R^*,3R^*,4R^*)-2-Hydroxymethyl-2-(4-methoxybenzyl)tetrahydrofuran-3, 4-diol [Directed Dihydroxylation of a Homoallylic Alcohol Using Catalytic OsO$_4$ and TMO] 20
(2R^*,3R^*,4R^*)-2-Hydroxymethyl-2-(4-methoxybenzyl)tetrahydrofuran-3, 4-diol [Directed Dihydroxylation of a Homoallylic Alcohol Using Catalytic OsO$_4$, TMO, and Polymer-Bound DABCO] 21
2,2,2-Trichloro-N-[(1R^*,2R^*,3S^*,5S^*)-2,3-dihydroxy-5-isopropyl-2-methylcyclohexyl]acetamide [Directed Dihydroxylation of an Allylic Cyclic Amide Using Catalytic OsO$_4$ and QNO] 21
(2R^*,3R^*,4S^*,5S^*,6S^*)-Methyl 3,4,5-Trihydroxy-6-methoxytetrahydro-2H-pyran-2-carboxylate [Directed Dihydroxylation of an Allylic Cyclic Alcohol Using OsO$_4$·Pyridine] 22
(2R^*,3S^*,4S^*,5S^*,6S^*)-2-{2-[(2S^*,3S^*,6R^*)-3-Acetoxy-6-methoxy-3,6-dihydro-2H-pyran-2-yl]ethyl}-3,4,5-triacetoxy-6-methoxytetrahydropyran [Directed Dihydroxylation Using OsO$_4$ and a Chiral Amine] 22
TABULAR SURVEY 23
   Table 1. Directed Dihydroxylation of Allylic Cyclic Alcohols 24
   Table 2. Directed Dihydroxylation of Acyclic and Exocyclic Allylic Alcohols 30
   Table 3. Directed Dihydroxylation of Homoallylic Cyclic Alcohols . . . 34
   Table 4. Directed Dihydroxylation of Homoallylic Exocyclic Alcohols . . . 36
   Table 5. Directed Dihydroxylation of N-Allylic Amine Derivatives . . . 37
   Table 6. Directed Dihydroxylation of N-Homoallylic Cyclic Amides . . . 45
REFERENCES 47

INTRODUCTION

This review focuses on the dihydroxylation of alkenes using osmium tetroxide (OsO$_4$) that is directed by alcohols and amine derivatives through hydrogen bonding between the substrate and the oxidant.

Discussion focuses on the different types of directing groups that are viable. The outcome from directed dihydroxylation of all the major classes of alkenes, including cyclic and acyclic substrates and varied alkene substitution patterns, is also addressed (Eqs. 1 and 2).[1]

HYDROGEN-BONDING-MEDIATED DIRECTED OSMIUM DIHYDROXYLATION

$$\begin{array}{c}\text{OH}\\ R^1\diagdown\!\!\!\diagup\\ R^2\quad R^3\end{array}\xrightarrow[\text{then cleavage of osmate ester}]{\text{OsO}_4,\ \text{TMEDA},\ \text{CH}_2\text{Cl}_2,\ -78°}\begin{array}{c}\text{OH}\\ R^1\!\!\diagdown\!\!\!\diagup\!\text{OH}\\ R^2\diagdown\!\!\!\diagup\!\text{OH}\\ R^3\end{array}+\begin{array}{c}\text{OH}\\ R^1\!\!\diagdown\!\!\!\diagup\!\text{OH}\\ R^2\diagdown\!\!\!\diagup\!\text{OH}\\ R^3\end{array}\quad\text{(Eq. 1)}$$

highly *syn* selective

$$\begin{array}{c}\text{NHCOCF}_3\\ R^1\diagdown\!\!\!\diagup\\ R^2\diagdown\!\!\!\diagup\\ R^3\end{array}\xrightarrow[\text{2. Ac}_2\text{O, py}]{\text{1. OsO}_4,\ \text{TMEDA},\ \text{CH}_2\text{Cl}_2,\ -78°}\begin{array}{c}\text{NHCOCF}_3\\ R^1\\ R^2\\ R^3\diagdown\!\!\!\diagup\!\text{OH}\\ \text{OH}\end{array}+\begin{array}{c}\text{NHCOCF}_3\\ R^1\\ R^2\\ R^3\diagdown\!\!\!\diagup\!\text{OH}\\ \text{OH}\end{array}\quad\text{(Eq. 2)}$$

highly *syn* selective

The mechanism section outlines the different reactivity patterns that various ligands can impart onto the osmium oxidant, together with the importance of choosing a solvent that encourages hydrogen bonding. The influence that the directing group has on *syn* selectivity is also discussed, in both the context of its position in space with respect to the alkene, and the relationship between the pKa of the acidic proton and *syn* selectivity.

Criegee first reported the controlled oxidation of alkenes using stoichiometric amounts of OsO_4,[2] and later expanded upon those original observations by noting that pyridine acts as a ligand for osmium and accelerates the dihydroxylation process.[3] Osmium tetroxide has since established itself as the reagent of choice for the *syn*-dihydroxylation of olefins, primarily because of its inertness toward other functional groups and lack of over-oxidation products.[4]

Researchers from the UpJohn company reported a convenient and reliable procedure for dihydroxylation that involved substoichiometric amounts of OsO_4 (typically 5 mol %) and *N*-methymorpholine-*N*-oxide (NMO) as a stoichiometric co-oxidant. This landmark paper defined a procedure that has since enjoyed widespread use.[5]

Observations as to the outcome from the dihydroxylation of chiral substrates were given a basis by Kishi, who reported that *anti* selectivity is generally attained during the oxidation of a wide range of allylic alcohols and protected derivatives thereof.[6,7,8,9] This mode of reactivity, whereby the heteroatom compels oxidation to occur on the opposite face of the alkene (most easily envisaged in cyclic systems) has proven to be very reliable with few exceptions reported. In fact, the high level of *anti* selectivity that is observed in such dihydroxylations has led to a problem: how to overturn this bias and obtain dihydroxylation on the same face as the directing group? Because the facial bias of the substrate (particularly allylic alcohols) is so strong, and often cyclic *cis*-alkenes are involved, it is frequently not possible to use the impressive asymmetric dihydroxylation system developed by Sharpless to control the diastereoselective dihydroxylation of a chiral substrate.[10,11] Therefore, the notion of a heteroatom-directed dihydroxylation becomes an interesting and useful proposition; and as such, the method discussed here forms an excellent counterpart to that described by Kishi.

Remarkably, only a few other synthetic methods are known that accomplish the direct addition of a diol unit or a protected diol unit across an alkene while controlling the stereochemical course of the process. In fact, in addition to oxidation with high-valent metal oxo species, only iodine/silver acetate, the Woodward modification of the Prevost reaction,[12] will add two oxygen atoms in a *syn* fashion across an alkene. While this reaction has not enjoyed widespread use in the chemistry community it is discussed in some detail in the "Comparison with Other Methods" section.

MECHANISM AND STEREOCHEMISTRY

The hydrogen bond accepting ability of OsO_4 is enhanced upon complexation by amines. This behavior can be explained simply by the coordination of a Lewis base to the metal center, which leads to increased electron density on the oxo-ligands. Equations 3 and 4 compare the differences of reactivity between the OsO_4–NMO and the OsO_4•TMEDA complexes.

$$\text{(Eq. 3)}$$

$$\text{(Eq. 4)}$$

Corey showed, through low-temperature X-ray crystallographic analysis, that chiral 1,2-diamines form unique bidentate complexes with OsO_4.[13,14] These findings suggest that an OsO_4•diamine system should benefit from the bidentate nature of the ligand, which would exert an enhanced donor effect on the metal and also on the oxo-ligands. Spectroscopic analysis of the complex, formed at low temperature between OsO_4 and TMEDA (N,N,N',N'-tetramethyl-1,2-ethanediamine), has been carried out. ^{1}H NMR spectra of a 1:1 mixture of OsO_4 and TMEDA reveal the presence of a single, symmetrical compound. Low temperature IR spectroscopy studies indicate a reduction in the Os=O bond order as one traverses the series OsO_4, OsO_4•monodentate amine, OsO_4•chelating-diamine. These findings support the hypothesis that the increase in *syn* selectivity in directed dihydroxylation, following the order $OsO_4 < OsO_4$•monodentate amine $< OsO_4$•chelating-diamine, arises from an augmentation in hydrogen bond forming ability.[15]

The importance of hydrogen-bonding is further substantiated by the dihydroxylation of methyl ether **1** (R = Me) (Eq. 5) and *N*-methyl trichloroacetamide **2** (R = Me) (Eq. 6).[15] The absence of a hydrogen bond donor in these substrates has a pivotal influence on the stereochemical outcome of the reaction: the *anti* isomer is obtained as the major product in both cases. Also, it is noteworthy that these dihydroxylation reactions are significantly slower than the oxidation of the parent alcohol or trichloroacetamide.

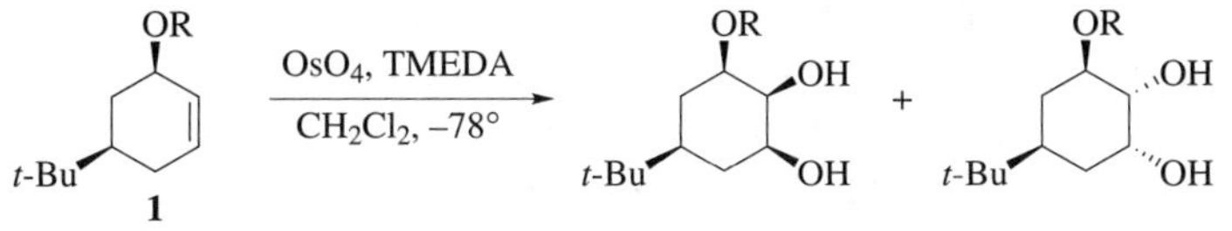

[Eq. 5 scheme: Compound **1** (t-Bu, OR-substituted cyclohexene) → OsO₄, TMEDA, CH₂Cl₂, −78 °C → two diol products]

R	dr
Me	(65%) 1:25
H	(91%) 24:1

(Eq. 5)

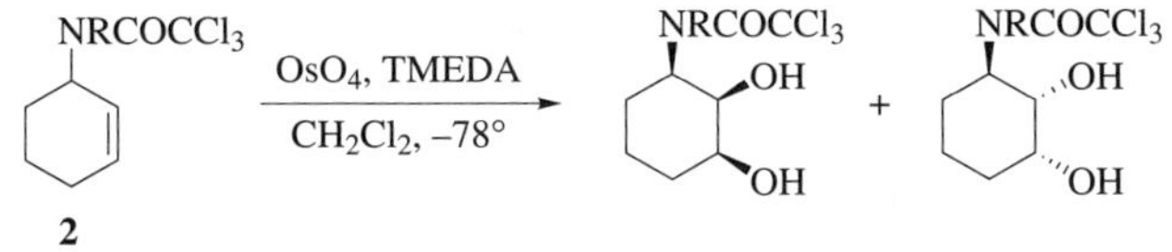

[Eq. 6 scheme: Compound **2** (NRCOCCl₃ cyclohexene) → OsO₄, TMEDA, CH₂Cl₂, −78 °C → two diol products]

R	dr
Me	(78%) 1:10
H	(99%) >25:1

(Eq. 6)

Further studies established that the OsO₄·TMEDA complex reacts through a hydrogen bond between the substrate and an oxo ligand (see **A**, Fig. 1), rather than a non-ligated amino group of TMEDA (see **B**, Fig. 1).

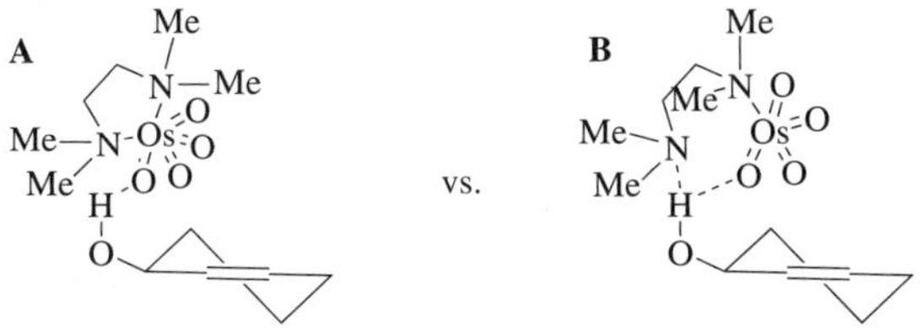

Figure 1. Possible hydrogen-bonding between the substrate and the OsO₄·TMEDA complex.

The results for the dihydroxylation of alcohol **3** in the presence of several bifunctional analogues of TMEDA are shown in Eq. 7. It is noteworthy that all of the amines fail to match the *syn* selectivity observed with TMEDA.[15]

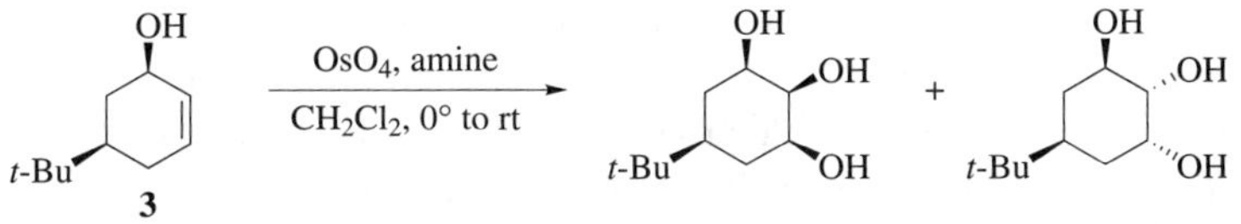

[Scheme: Compound **3** (t-Bu, OH-substituted cyclohexene) → OsO₄, amine, CH₂Cl₂, 0 °C to rt → two diol products]

Amine	dr
none	1:1.7
Me₃N	1.2:1
Me₂N(CH₂)₂OMe	1.1:1
Me₂N(CH₂)₂OH	1.1:1
Me₂NCH₂NMe₂	1.6:1
Me₂N(CH₂)₂NMe₂	4.6:1
Me₂N(CH₂)₃NMe₂	2.8:1

(Eq. 7)

High levels of *syn* stereoselectivity are achieved for chelating amines only;[15] if the hypothetical model **B** were correct, it is expected that amines with pendant oxygen functionality should be able to form a hydrogen bond to the substrate and hence direct the dihydroxylation to some degree. Clearly this is not the case, as the level of selectivity in these reactions is comparable to those found using a simple monodentate amine such as Me_3N. These studies provide further evidence for the existence and reaction of a chelated OsO_4·TMEDA complex. Model **A** is, therefore, to be considered the reacting species.[15]

More information on the OsO_4·TMEDA system can be gathered by a closer analysis of the osmate esters produced, which are quite stable and can be easily purified. The X-ray crystal structure of *syn*-osmate ester **4**, obtained from the corresponding alkene and OsO_4·TMEDA, clearly shows the chelating nature of the diamine ligand (Fig. 2).[15]

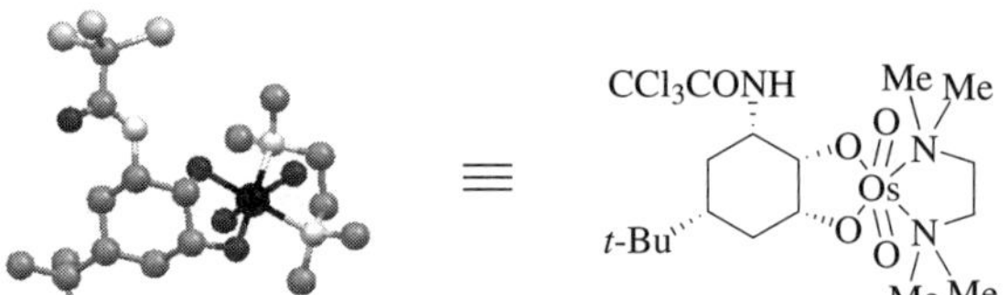

Figure 2. X-ray crystal structure of *syn*-osmate ester **4**. Hydrogen atoms have been omitted for clarity.

Another feature of the dihydroxylation reaction using OsO_4 in the presence of amines is the increased reactivity of the reagents towards alkenes. On the basis of literature data, approximate relative rate values for olefin oxidation with OsO_4, OsO_4·quinuclidine, and OsO_4·TMEDA are 1, 100, and 10,000 respectively.[13,16] The use of TMEDA as an additive generates an extraordinarily powerful dihydroxylating system, which is able to react with alkenes even at $-78°$. Under the same conditions, both OsO_4 and OsO_4·quinuclidine are essentially inert. This unique feature of the complex has enabled wide use in different dihydroxylation reactions where standard protocols are found to be ineffective.[17,18]

A disadvantage of the OsO_4·TMEDA system is the requirement for stoichiometric amounts of transition metal due to the inability of the resulting osmate(VI) ester to undergo either direct hydrolysis or in situ oxidation to a more easily hydrolyzed Os(VIII) species. By switching to monodentate amines such as quinuclidine, introduced as its *N*-oxide (QNO), the reactivity and hydrogen-bonding ability of the osmium complex decrease but the dihydroxylation reaction can be carried out with a substoichiometric amount of metal.[19] As the reaction progresses and QNO is reduced, OsO_4 can bind to the released quinuclidine and oxidize the alkene preferentially in a *syn* fashion. The resulting osmate ester is then able to undergo fast oxidation with more QNO, and subsequent hydrolysis (there is no need for the addition of water, as QNO is normally used as a monohydrate) releases the product and regenerates the catalytic species, as shown in Scheme 1.

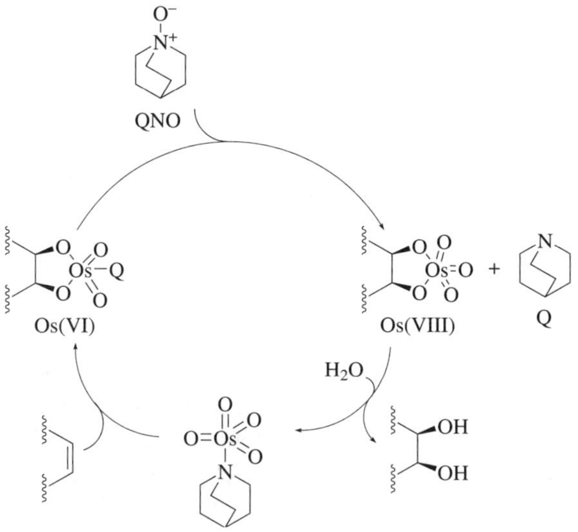

Scheme 1

SCOPE AND LIMITATIONS

Although the osmium(VIII) dihydroxylation reaction can be influenced by a number of factors (electronic effects, steric effects, etc.), this chapter focuses on reactions wherein hydrogen-bonding effects are important. The presence of a directing group (usually an amide or alcohol) in either the allylic or homoallylic position combined with a complex of osmium tetroxide with an amine (generally $OsO_4 \cdot TMEDA$) can allow *syn* stereoselectivity and site selectively in the oxidation of a double bond.

Success of the hydrogen-bonding-mediated directed dihydroxylation depends upon a few essential elements. The level of stereoselectivity attained can be widely variable depending upon the geometry, substitution pattern, and position of the alkene relative to the directing group, and other steric or stereoelectronic factors.

Nature of the Amine

The weaker directing effect of the $OsO_4 \cdot$quinuclidine complex results in moderate levels of diastereoselectivity with allylic alcohols. Better results are obtained when trichloroacetamides are used as the directing element. Good levels of *syn* selectivity can be attained with trichloroacetamides due to the enhanced hydrogen bond forming ability of these acidic species, which allows a stronger interaction between the osmium complex and the substrate (Eq. 8).[15,19,20] Protocols that are catalytic in OsO_4[19] are less selective than the stoichiometric method[15,20] but do provide significant levels of *syn* selectivity, with the QNO system being slightly superior to the Me_3NO (TMO) system.

(Eq. 8)

Conditions		dr
OsO$_4$, TMEDA, CH$_2$Cl$_2$, −78° to rt	(99%)	>25:1
OsO$_4$ (5 mol %), QNO•H$_2$O (1.3 equiv), CH$_2$Cl$_2$, rt	(86%)	4.3:1
OsO$_4$ (cat), Me$_3$NO, CH$_2$Cl$_2$	(93%)	3:1
OsO$_4$ (cat), NMO, Me$_2$CO/H$_2$O, rt	(98%)	1:3.2

The use of a monodentate amine also represents a distinct advantage when the dihydroxylation of hindered allylic trichloroacetamides is required. Because of the smaller size of the OsO$_4$•quinuclidine complex compared to the OsO$_4$•TMEDA system, increased levels of selectivity are obtained in the directed oxidation of sterically demanding substrates.[19] Replacing QNO•H$_2$O, which needs to be prepared beforehand, with commercially available Me$_3$NO•2H$_2$O makes the dihydroxylation process easier to perform while maintaining good levels of *syn* selectivity.[19]

Nature of the Directing Group

The dihydroxylation can be directed if an alcohol or secondary amide group is present within reasonable proximity of the alkene. In general, suitably activated amide derivatives are prone to higher *syn* selectivity than their alcohol counterparts (Eqs. 9 and 10). The enhanced acidity of the trichloroacetamide and trifluoroacetamide relative to that of the corresponding alcohol (pKa values are approximately 11.2, 10.7, and 15 respectively) means that hydrogen-bonding to the OsO$_4$•TMEDA reagent is more effective, resulting in a higher *syn* selectivity. Oxidation of amide derivatives bearing less acidic proton donors (Me$_3$CONHR, *t*-BuOCONHR) afford only moderate *syn* selectivities.[15] The more acidic sulfonamides are not as selective, a result that is probably due to their greater steric bulk. Good levels of *syn* selectivity can be attained with substrates bearing amide directing groups using the hydrogen-bonding conditions catalytic in OsO$_4$ (QNO•H$_2$O, CH$_2$Cl$_2$).[19,21]

(Eq. 9)

Conditions		dr
OsO$_4$, TMEDA, CH$_2$Cl$_2$, −78° to rt	(91%)	25:1
OsO$_4$ (5 mol %), QNO•H$_2$O (1.3 equiv), CH$_2$Cl$_2$, rt	(—)	1.2:1
OsO$_4$ (cat), NMO, Me$_2$CO/H$_2$O, rt	(91%)	1:4

HYDROGEN-BONDING-MEDIATED DIRECTED OSMIUM DIHYDROXYLATION

(Eq. 10)

Conditions		dr
OsO$_4$, TMEDA, CH$_2$Cl$_2$, −78° to rt	(96%)	24:1
OsO$_4$ (cat), Me$_3$NO, CH$_2$Cl$_2$	(81%)	6:1
OsO$_4$ (5 mol %), QNO•H$_2$O (1.3 equiv), CH$_2$Cl$_2$, rt	(77%)	13:1
OsO$_4$ (cat), NMO, Me$_2$CO/H$_2$O, rt	(86%)	1:1.6

An additional hydroxy group in the vicinity of the allylic hydroxy group can reduce the selectivity of the hydroxylation. Equations 11 and 12 illustrate this effect.[15,21]

(76%) 7:1 dr (Eq. 11)

(67%) 2:1 dr (Eq. 12)

Steric Effects

Adverse steric effects can, of course, affect the *syn* selectivity dramatically. The bulk of the OsO$_4$•TMEDA complex hampers its ability to oxidize the hindered faces of alkenes. 1-Amino-2-cyclohexene derivatives and 2-cyclohexenols give the best *syn* selectivity when the donor group is in an equatorial position. When a conformationally locked substrate contains a pseudoaxially disposed directing group, the *syn* selectivity is poor (Eq. 13),[15,20] because hydrogen-bonding of the large oxometal species is discouraged by sterics (Fig. 3). As was mentioned previously, the *syn* selectivity of dihydroxylation of hindered allylic trichloroacetamides is increased when TMEDA, a bidentate ligand, is replaced by quinuclidine, a monodentate ligand. Even though the OsO$_4$•quinuclidine complex displays weaker inherent hydrogen bond accepting ability, the reduced steric bulk provides moderate *syn* selectivity in this system.

(Eq. 13)

Conditions		dr
OsO$_4$, TMEDA, CH$_2$Cl$_2$, −78° to rt	(97%)	1.7:1
OsO$_4$, quinuclidine, CH$_2$Cl$_2$, −78°	(66%)	4.9:1

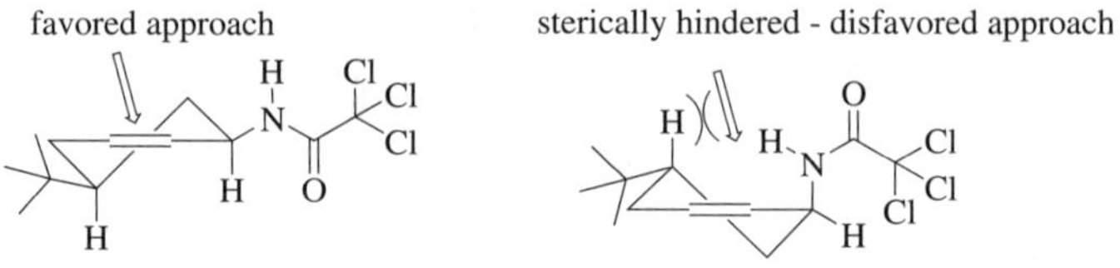

Figure 3. Steric effects in a conformationally locked substrate.

The same lack of *syn* selectivity is also observed with pseudo-axially biased alcohol **5** (Eqs. 14 and 15).[15,21]

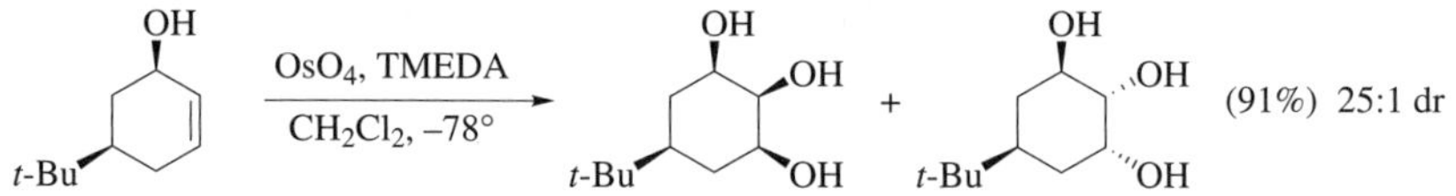

(91%) 25:1 dr

(Eq. 14)

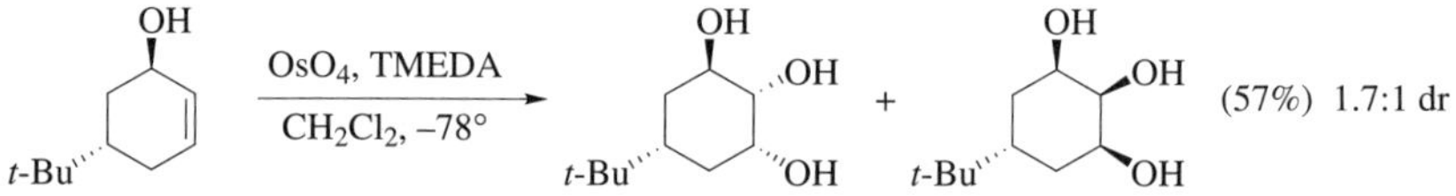

(57%) 1.7:1 dr

(Eq. 15)

Clearly, the directing functionality must also be placed in a position where it can interact freely with the osmium complex. In contrast to allylic substrates, the directing group in homoallylic substrates needs to be in an axial position to deliver the oxidant intramolecularly. For example, poor selectivity is observed when 4-trichloroacetamido-1-cyclohexene is oxidized, probably because the bulky amide group has to adopt an unfavored axial position in order to deliver the oxidant (Eq. 16).[22] However, when the trichloroacetamide is replaced by the smaller and more acidic trifluoro derivative (approximate pKa of $Cl_3C(O)CNHR = 11.2$ and $F_3C(O)CNHR = 10.7$), the oxidation proceeds with excellent *syn* selectivity (Eq. 16).[22]

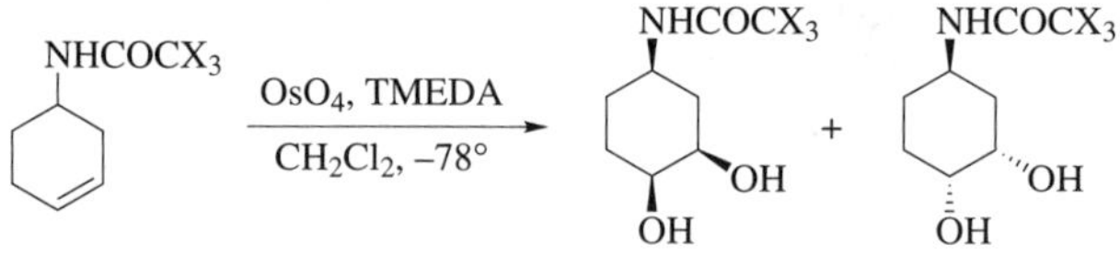

X		dr
Cl	(92%)	1.2:1
F	(98%)	>20:1

(Eq. 16)

It is noteworthy that in both the trichloroacetamide and trifluoroacetamide cases, the *syn* selectivity is greatly affected by the presence of an alkyl group on the carbon bearing the amide functionality (Eq. 17).[22] In this example, the directing group may be able to adopt the preferred axial conformation, but cannot point the N–H group towards the alkene without encountering steric hindrance from the geminal alkyl group, which leads to dramatic reduction of *syn* selectivity.[22,23,24]

$$\text{(Eq. 17)}$$

X	dr
Cl (92%)	1:5
F (67%)	3:1

The requirement for an axial directing group would also explain the difference in selectivity between substrate **7** (which must always have one hydroxy group in an axial position) and substrate **6** (Eqs. 18 and 19).[22,25]

(92%) 3:1 dr

$$\text{(Eq. 18)}$$

(82%) 12.4:1 dr

$$\text{(Eq. 19)}$$

With the five-membered ring homoallylic alcohol **8**, the dihydroxylation using OsO$_4$·TMEDA proceeds unselectively (Eq. 20). The exocyclic methylene side-chain may be sufficiently bulky to interfere with effective directed dihydroxylation. This hypothesis is supported by the results of substrate **9**, wherein an alkyl substituent has been introduced to block the face of the alkene opposite to the hydroxymethyl group. As expected, the directed dihydroxylation then proceeds well and with good *syn* selectivity. Furthermore, the directed dihydroxylation of substrate **10** confirms this rationale as the *p*-methoxybenzyl group completely blocks *anti* attack and therefore excellent *syn* selectivity is obtained.[22,25]

$$\text{(Eq. 20)}$$

	R		dr
8	H	(55%)	1:1
9	Me	(83%)	6:1
10	*p*-MeOC$_6$H$_4$CH$_2$	(82%)	99:1

Nature of the Substrate

Allylic versus Homoallylic Substrates. As a rule, allylic substrates lead to better stereoselectivity than homoallylic substrates. The main reason for this is that the directing group in the latter is now positioned further away from the double bond where it cannot influence the approach of the oxidant as easily. In cyclic homoallylic systems, it is more difficult for the hydrogen bond donor group to adopt a position that allows the osmium complex to attack the double bond in a *syn* selective fashion whilst participating in hydrogen-bonding. This issue has been already detailed in the "Steric Effects" section. Eqs. 21 and 22 directly compare examples of allylic and homoallylic alcohols.[22,25]

R		dr
OH	(98%)	9:1
NHCOCCl$_3$	(99%)	24:1

(Eq. 21)

R		dr
OH	(92%)	3:1
NHCOCCl$_3$	(92%)	1.2:1

(Eq. 22)

Conformational Factors Determined by the Alkene Substitution Pattern. In cyclic systems, the rigidity of the structure and consequent steric effects lead to high levels of *syn* selectivity. In acyclic systems, the alkene substitution pattern is crucial to obtaining high *syn* stereoselectivity. It is interesting to note that the *syn* selectivity increases dramatically in acyclic systems when the double bond bears a *cis* substituent, as in substrate **11** (Eqs. 23 and 24).[24,26]

(79%) 25:1 dr

(Eq. 23)

(84%) 3:1 dr

(Eq. 24)

Within each type of alkene, the levels of *syn* selectivity reported in the literature for directed epoxidation with peracid (most notably *m*-CPBA) are similar to those observed for directed dihydroxylation. It is suggested that, in the transition structure, the dihedral angle between the C–O and the C=C is most favorable at approximately 120°. The two possible transition structures are distinguished by the difference in A$^{[1,3]}$ strain between the R group and the R$_{cis}$ substituent

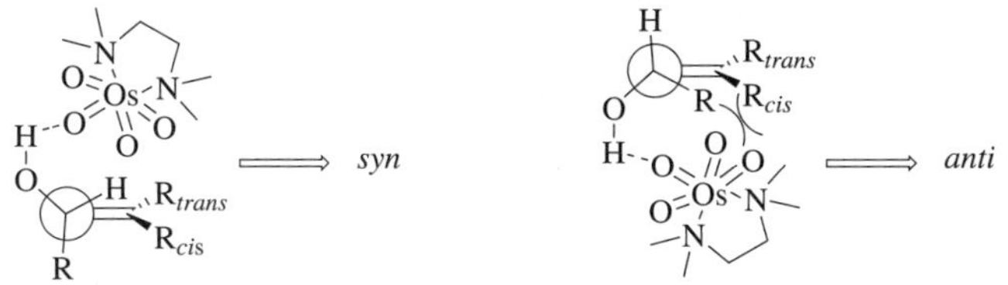

Figure 4. The two possible transition structures for directed dihydroxylation.

and explains why a large group in the R_{cis} position leads to higher levels of stereocontrol than the same group in the R_{trans} position (Fig. 4).[24]

Site Selectivity of the Directed Dihydroxylation Reaction. The directed dihydroxylation also expresses high site selectivity. Treatment of geraniol (**12**) with the $OsO_4 \cdot$TMEDA complex leads to highly selective oxidation of the 2,3-alkene (Eq. 25).[21] When the same substrate is oxidized under Sharpless asymmetric dihydroxylation conditions, the site selectivity is reversed, and oxidation of the most electron-rich double bond is observed.[27]

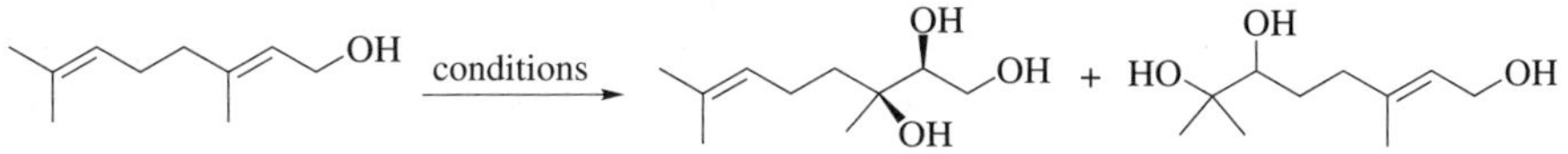

Conditions		I:II
OsO_4, TMEDA, CH_2Cl_2, –78° to rt	(74%)	>25:1
OsO_4 (1 mol %), $(DHQD)_2$-PHAL (5 mol %), $K_3Fe(CN)_6$, K_2CO_3, $MeSO_2NH_2$, t-BuOH/H_2O, 0°	(89%)	1:>49

(Eq. 25)

Alternative Directing Groups

Dihydroxylation reactions of allylic alcohols normally give the *anti* product under standard osmium tetroxide oxidation conditions.[6,7] However, scattered reports in the literature suggest that the natural steric bias of certain substrates may be overcome when heteroatomic substituents such as sulfoximines and nitro groups are present within the molecule and a reagent–substrate interaction is postulated to occur. Sulfoximine-directed dihydroxylation of alkene **13**, followed by desulfurization affords triol **14** as a single diastereomer (Eq. 26).[28] Osmium tetroxide oxidation of cyclopentene **15** unexpectedly gives all-*syn* product **16** (Eq. 27).[29] Although association of OsO_4 with the nitrosulfone side-chain is suggested to account for this selectivity,[29] the results from oxidizing a number of simpler analogs do not support a substrate–oxidant association and are interpreted in terms of substrate conformation.[30]

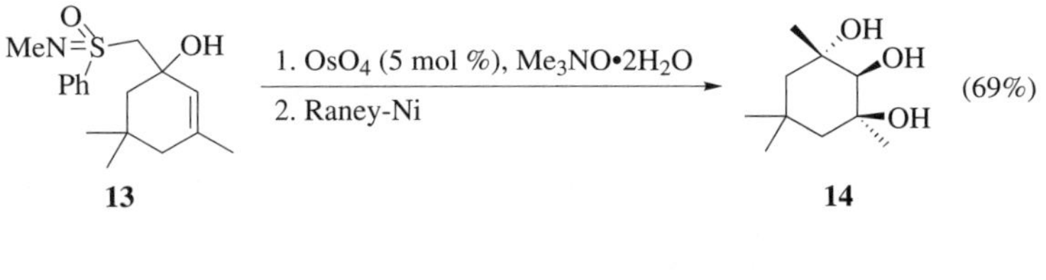

(Eq. 26)

<table>
<tr><td></td><td>AcO⟶[NO₂,SO₂Ph] cyclopentene</td><td>OsO₄ (cat), NMO →</td><td>AcO⟶[NO₂,SO₂Ph,HO,OH] cyclopentane (66%)</td><td>(Eq. 27)</td></tr>
<tr><td></td><td>15</td><td></td><td>16</td><td></td></tr>
</table>

Although interesting, these findings are of limited utility because they cannot be easily interpreted, rationalized, and extended; whereas hydrogen-bonding may come into play in some cases, steric effects are sometimes sufficient to account for the configuration of the products. On the contrary, the $OsO_4 \cdot TMEDA$ system relies unequivocally on the hydrogen-bonding ability of the metal complex and shows broad applicability over a large number of allylic alcohol and amine derivatives and enhanced reactivity towards alkenes even at very low temperature.

COMPARISON WITH OTHER METHODS

It is noteworthy that the modified Woodward alkene oxidation,[12] which involves the reaction of the alkene with AgOAc and I_2 in HOAc, followed by the addition of H_2O, affords moderate levels of *syn* selectivity in the *cis*-dihydroxylation of some allylic alcohols (Eqs. 28 and 29).[31] Selectivities depend upon the alkene substituents and configuration and the size of the *O*-protecting group, and are generally modest. The *syn* selectivity reflects attack of the iodonium ion on the face of the alkene that is opposite to the –OR group, followed by neighboring group attack within the initially formed β-acetoxy iodocompound. The reversal of stereoselectivity when the same protocol is applied to the free alcohol is attributed to hydrogen-bonding between the –OH and the electrophile.

Ph–CH=CH–CH(OR)–CH(CH₃)₂
1. I_2, AgOAc, AcOH; H_2O, 90°
2. DIBALH, THF, –78° (R = Piv)
3. Ac_2O, DMAP, py
→ Ph–[OAc,OAc,OAc] **I** + Ph–[OAc,OAc,OAc] **II**

R		I:II
H	(—)	1:12
Ac	(71%)	2.5:1
Piv	(65%)	4:1

(Eq. 28)

Bn–CH₂–CH=CH–CH(OR)–CH(CH₃)₂
1. I_2, AgOAc, AcOH; H_2O, 90°
2. DIBALH, THF, –78° (R = Piv)
3. Ac_2O, DMAP, py
→ Bn–[OAc,OAc,OAc] **I** + Bn–[OAc,OAc,OAc] **II**

R		I:II
H	(75%)	1:2.6
Ac	(90%)	14.3:1
Piv	(85%)	12.4:1

(Eq. 29)

Alternative, direct oxidations of an alkene to a *syn*-diol have been reported in the literature; we restricted our search to reactions of prochiral substrates possessing a stereogenic center in the allylic position. Although stereocontrolled reactions involving other high-valent metal oxidants are known, no coordination-induced directing effect has been described. For example, ruthenium(VIII)-promoted dihydroxylation leads to *anti* selectivity with respect to the original stereocenter (Eq. 30),[32] and stereocontrolled permanganate-mediated oxidation of a steroidal enone is presumably sterically directed away from the angular methyl group (Eq. 31).[33]

(Eq. 30)

(Eq. 31)

EXPERIMENTAL CONDITIONS

The osmium-mediated dihydroxylation reaction is carried out under an inert atmosphere such as argon or nitrogen and the solvents (CH_2Cl_2, acetone, THF) must be anhydrous. *Osmium tetroxide is toxic, volatile, and sublimes quite easily; it should therefore be handled in a well-ventilated fume-hood. The aqueous layers from the osmium-mediated reactions and any other waste materials should be disposed of properly.*

EXPERIMENTAL PROCEDURES

General Procedure for Stoichiometric Dihydroxylation

OsO_4·TMEDA.[15] To a solution of substrate (0.50 mmol) and TMEDA (0.55 mmol) in CH_2Cl_2 precooled to −78° was added a solution of OsO_4 (0.53 mmol) in CH_2Cl_2 (∼1 mL). The solution turned deep red and then brown-black. It was stirred until the reaction was complete (TLC analysis, ca. 1 h) before being allowed to warm to rt.

Isolation Procedures for 0.50 mmol of Substrate

Sodium Sulfite.[15] After completion of the oxidation, the solvent was removed under vacuum and replaced with THF (10 mL) and sodium sulfite (aq saturated solution, 10 mL). This mixture was heated at reflux for 3 h and the work-up completed as indicated.

Acidic Methanol.[15] After completion of the oxidation, the solution was concentrated under vacuum and the resulting residue was dissolved in MeOH (10 mL) before addition of HCl (concentrated, ~5 drops). The solution was stirred for 2 h, concentrated under vacuum, and the product isolated as indicated.

Ethylenediamine.[15] After completion of the oxidation, ethylenediamine (5.0 equiv) was added to the crude reaction mixture at rt and the resulting solution was stirred for 48 h during which time a brown precipitate formed. The solution was then concentrated under vacuum and the product isolated as indicated.

(1R*,2S*,3S*)-Cyclohexane-1,2,3-triol [Directed Dihydroxylation of an Allylic Cyclic Alcohol Using OsO$_4$·TMEDA].[15] 2-Cyclohexene-1-ol (50 mg, 0.51 mmol) was oxidized with OsO$_4$·TMEDA using the sodium sulfite work-up. The crude reaction mixture was then concentrated under vacuum to afford a grey powder; EtOH (30 mL) was added and the suspension stirred at rt for 1 h. Filtration of the resulting suspension through Celite and concentration of the filtrate under vacuum gave a colorless solid (80 mg). Purification by column chromatography (SiO$_2$, EtOAc/petroleum ether 7:1) afforded the title compound as an inseparable mixture of isomers (66 mg, 98%, syn/anti 9:1): IR (film) 3192, 2927 cm^{-1}; ^{1}H NMR (300 MHz, D$_2$O) δ 3.81 (t, J = 2.6 Hz, 1H), 3.52 (ddd, J = 10.0, 4.6, 2.6 Hz, 2H), 1.80–1.00 (m, 6H); ^{13}C NMR (75 MHz, D$_2$O) δ 72.6, 70.3, 26.3, 18.8; CIMS (m/z): [M + NH$_4$]$^+$ 150 (100); CI (m/z): [M + NH$_4$]$^+$ calcd for C$_6$H$_{16}$NO$_3$, 150.1130; found, 150.1128.

2,2,2-Trichloro-N-[(1R*,2R*,3S*)-2,3-dihydroxycyclohexyl]acetamide [Directed Dihydroxylation of an N-Allylic Cyclic Amide Using OsO$_4$·TMEDA].[15] 2,2,2-Trichloro-N-(cyclohex-2-enyl)acetamide (100 mg, 0.412 mmol) was oxidized with OsO$_4$·TMEDA using the methanolic work-up; the resulting orange mixture was purified by column chromatography (SiO$_2$, petroleum ether/EtOAc 1.5:1) to yield the title product (111 mg, 99%) as a colorless oil: IR (film) 3407, 2942, 1698, 1515 cm^{-1}; ^{1}H NMR (300 MHz, CDCl$_3$) δ 7.86 (br s, 1H), 4.06–3.86 (m, 3H), 3.70–3.00 (m, 2H), 1.84–1.58 (m, 5H), 1.46–1.32 (m, 1H); ^{13}C NMR (75 MHz, CDCl$_3$) δ 161.8, 92.6, 70.8, 70.3, 52.1, 28.4, 26.1, 18.2; CIMS (m/z): 295 (91), [M + NH$_4$]$^+$ 293 (100); CI (m/z): [M + H]$^+$ calcd for C$_8$H$_{13}$NO$_3$Cl$_3$, 275.9961; found, 275.9966.

(1R*, 2S*, 3S*)-Cyclopentane-1,2,3-triol [Directed Dihydroxylation of an Allylic Cyclic Alcohol Using OsO$_4$·TMEDA].[15] Cyclopent-2-enol (100 mg, 1.19 mmol) was oxidized with OsO$_4$·TMEDA using the ethylenediamine work-up. The residue was redissolved by sonication in a mixture of EtOH (7.5 mL) and EtOAc (40 mL); the resulting solution was filtered through Celite and concentrated under vacuum. Column chromatography (SiO$_2$, EtOAc) afforded the title compound as a clear oil (66 mg, 76%, syn/anti 7:1, by ^{1}H NMR). (1R*, 2S*, 3S*)-Cyclopentane-1,2,3-triol was obtained by repeated chromatography: IR (neat) 3365, 2962, 2926 cm^{-1}; ^{1}H NMR (300 MHz, CD$_3$OD) δ 4.09–4.02 (m, 2H), 3.82 (t, $J = 5$ Hz, 1H), 1.95–1.76 (m, 4H); ^{13}C NMR (75 MHz, CD$_3$OD) δ 71.9, 69.9, 27.1; CIMS (m/z): [M + NH$_4$]$^+$ 154 (100), 90(40); CI (m/z): [M + NH$_4$]$^+$ calcd for C$_5$H$_{14}$NO$_3$, 136.0974; found, 136.0979.

(2R*, 3R*, 4S*, 5S*)-2-(Acetoxymethyl)tetrahydro-2H-pyran-3,4,5-triyl Triacetate [Directed Dihydroxylation of an Allylic Cyclic Alcohol Using OsO$_4$·TMEDA and Subsequent Peracetylation].[15] (2R*, 3S*)-2-(Hydroxymethyl)-3,6-dihydro-2H-pyran-3-ol (100 mg, 0.771 mmol) was oxidized with OsO$_4$·TMEDA using the sodium sulfite work-up. The aqueous mixture was concentrated under vacuum to a grey solid, which was powdered before the sequential addition of pyridine (10 mL), Ac$_2$O (5 mL) and DMAP (cat). The resulting black suspension was stirred at rt under an atmosphere of nitrogen for 48 h; Et$_2$O (100 mL) was then added and the mixture filtered through Celite (washing further with Et$_2$O (200 mL)). The filtrate was washed with HCl (aq solution, 2M, 100 mL), NaHCO$_3$ (aq saturated solution, 100 mL) and brine (100 mL). The organic extracts were dried (MgSO$_4$) and concentrated under vacuum to afford a light-brown oil (201 mg) as a mixture of isomers (syn/anti 6:1 by ^{1}H NMR). Purification by column chromatography (SiO$_2$, petroleum ether/EtOAc 6:1) gave the title product (161 mg, 63%) as a colorless oil: [α]$^{27}_D$ + 7.5 (c 0.2, CHCl$_3$); IR (film) 2996, 1747 cm^{-1}; ^{1}H NMR (300 MHz, CDCl$_3$) δ 5.60 (t, $J = 2.6$ Hz, 1H), 4.95 (ddd, $J = 10.0, 5.5, 2.6$ Hz, 1H), 4.84 (ddd, $J = 10.0, 5.5, 2.6$ Hz, 1H), 4.14–4.10 (m, 2H), 3.88–3.78 (m, 2H), 3.60 (t, $J = 10.0$ Hz, 1H), 2.10 (s, 3H), 2.02 (s, 3H), 1.94 (s, 3H); ^{13}C NMR (75 MHz, CDCl$_3$) δ 170.69, 169.89, 169.30, 169.11, 71.71, 67.77, 66.40, 66.27, 63.45, 62.49, 20.70 (2) and 20.55 (2); CIMS (m/z): [M + NH$_4$]$^+$ 350(10), 249(100); CI (m/z): [M + NH$_4$]$^+$ calcd for C$_{14}$H$_{24}$NO$_9$, 350.1451; found, 350.1454.

Tricyclic Tetraol [Directed Dihydroxylation of an Exocyclic Allylic Alcohol Using OsO$_4$·TMEDA].[17]   A solution of alkene **17** (590 mg, 0.842 mmol) in CH$_2$Cl$_2$ (32.4 mL) was cooled to −78° and treated sequentially with TMEDA (0.32 mL, 2.1 mmol) and OsO$_4$ (531 mg, 2.09 mmol). The reaction mixture was stirred at this temperature for 2 h, allowed to warm to rt over 15 min, and concentrated under vacuum. The residue was taken up in THF (80 mL), acetone (40 mL), and water (40 mL), treated with sodium bisulfite (7.5 g), and stirred for 3 h. Water (100 mL) and EtOAc (100 mL) were then added, the aqueous layer was extracted with EtOAc (2 × 50 mL), and the combined organic phases were concentrated under vacuum. THF (80 mL), acetone (40 mL), water (40 mL), and sodium bisulfite (4.0 g) were added, and the mixture was stirred at rt for 20 h. The resultant mixture was filtered through a pad of Celite and the residue rinsed with EtOAc (3 × 150 mL). The layers were separated, the aqueous phase was extracted with EtOAc (2 × 50 mL), the combined organic phases were evaporated, and the residue was purified by column chromatography (SiO$_2$, hexanes/EtOAc 1.2:1) to give the title product as a colorless oil (460 mg, 72%). The spectroscopic properties of the tricyclic tetraol were identical to those previously reported:[34] [α]^{20}$_D$ +5.2 (c 0.56, CHCl$_3$); IR (neat) 3470, 1719, 1706, 1514 cm^{-1}; ^{1}H NMR (400 MHz, CDCl$_3$) δ 7.94 (d, J = 7.2 Hz, 2H), 7.54–7.52 (m, 1H), 7.44–7.40 (m, 2H), 7.26 (d, J = 8.5 Hz, 2H), 6.81 (d, J = 8.5 Hz, 2H), 5.71 (d, J = 5.8 Hz, 1H), 4.79 (d, J = 4.0 Hz, 1H), 4.55 (dd, J = 4.3, 11.6 Hz, 1H), 4.43 (d, J = 10.5 Hz, 1H), 4.06 (d, J = 10.6 Hz, 1H), 3.99 (s, 1H), 3.87 (br s, 1H), 3.82 (d, J = 5.8 Hz, 1H), 3.74 (s, 3H), 3.56 (d, J = 10.2 Hz, 1H), 3.43–3.41 (m, 1H), 3.36 (s, 1H), 3.16–3.06 (m, 1H), 2.77 (s, 1H), 2.77–2.72 (m, 1H), 2.36–2.26 (m, 2H), 2.18–2.16 (m, 1H), 1.94–1.93 (m, 1H), 1.88–1.80 (m, 1H), 1.80–1.70 (m, 1H), 1.30 (s, 3H), 1.03 (s, 3H), 0.87 (s, 3H), 0.78 (s, 9H), 0.01 (s, 3H), −0.06 (s, 3H); ^{13}C NMR (75 MHz, CDCl$_3$) δ 212.3, 208.9, 165.2, 159.4, 152.8, 133.8, 129.8, 129.5, 128.98, 128.9, 113.8, 84.2, 81.6, 75.2, 74.0, 73.5, 71.5, 67.2, 62.9, 58.7, 55.2, 51.3, 42.1, 38.9, 38.0, 32.9, 31.0, 29.6, 25.7, 22.7, 18.2, 10.0, −2.1, −4.2; ES HRMS (m/z): [M + Na$^+$] calcd for C$_{40}$H$_{56}$O$_{11}$SiNa, 763.3490; found, 763.3432.

(2S_P^*,3S^*,4R^*,5S^*,6R^*)-2-Phenylmethoxy-6-(hydroxymethyl)-3,4,5-trihydroxy-1,2-oxaphosphorinane-2-oxide) [Directed Dihydroxylation of an Allylic Cyclic Alcohol Using OsO$_4$·TMEDA].[35]   To a solution of OsO$_4$

(43 mg, 0.17 mmol) in CH_2Cl_2 (0.6 mL) at $-78°$ was added TMEDA (22 mg, 0.19 mmol) followed by the alcohol **18** (68 mg, 0.13 mmol) in CH_2Cl_2 (1.0 mL). The reaction mixture was stirred for 3 h at $-78°$, warmed to rt, and stirred for 15 min. The solution was concentrated under vacuum to give the crude osmate ester, which was dissolved in MeOH (1 mL) and treated with citric acid (40 mg, 0.21 mmol) for 24 h. The solution was concentrated under vacuum; the residue was dissolved in a small amount of MeOH, and filtered through silica gel (EtOAc/MeOH 9:1). The crude product was dissolved in MeOH (1 mL), treated with a catalytic amount of TsOH·H_2O, and stirred for 8 h. The solution was then concentrated under vacuum and the crude product was purified by column chromatography (SiO_2, EtOAc/MeOH 9:1) to afford the title compound (28 mg, 70%): ^{1}H NMR (400 MHz, $CDCl_3$) δ 7.44–7.31 (m, 5H), 5.21–5.11 (m, 2H), 4.55–4.50 (m, 1H), 4.24 (dt, $J = 33.7, 2.7$ Hz, 1H), 4.01 (dd, $J = 9.8, 3.4$ Hz, 1H), 3.90 (ddd, $J = 12.5, 4.4, 2.9$ Hz, 1H), 3.75 (dd, $J = 9.8, 2.1$ Hz, 1H); ^{13}C NMR (100 MHz, $CDCl_3$) δ 137.7 (d, $J_{CP} = 6.4$ Hz), 129.7, 129.6, 129.2, 78.4 (d, $J_{CP} = 4.4$ Hz), 75.5 (d), 71.4, 69.6 (dt, $J_{CP} = 6.4$ Hz), 69.0 (d), 67.7 (d, $J_{CP} = 144.5$ Hz), 62.8 (dt, $J_{CP} = 8.0$ Hz); ^{31}P NMR δ 24.5; HRMS-FAB (m/z): $[M + H]^+$ calcd for $C_{38}H_{36}O_8P$, 651.2148; found, 651.2131.

Osmate ester of (3aR^*,4S^*,5S^*,6R^*,7R^*,7aR^*)-3-Benzyl-4-benzyloxy-5,6,7-trihydroxyhexahydrobenzo[d]oxazol-2(3H)-one [Preparation of an Osmate Ester Using OsO_4·TMEDA].[36] A solution of OsO_4 (140 mg, 0.551 mmol) in CH_2Cl_2 (0.7 mL) was added to a solution of (3aS^*, 4S^*, 5S^*,7aS^*)-3-benzyl-4-(benzyloxy)-5-hydroxy-3,3a,4,5-tetrahydrobenzo[d]oxazol-2(7aH)-one (184 mg, 0.532 mmol) and TMEDA (87.0 μL, 0.580 mmol) in CH_2Cl_2 at $-78°$ and the reaction mixture was stirred for 2 h. The solution was allowed to warm to rt, concentrated onto silica and the crude material was purified by column chromatography (SiO_2, CH_2Cl_2/MeOH 19:1) to afford the title compound as a brown foam (379 mg, 100%): ^{1}H NMR (400 MHz, $CDCl_3$) δ 7.36–7.18 (m, 10H), 4.90 (dd, $J = 8.8, 4.8$ Hz, 1H), 4.84 (d, $J = 15.2$ Hz, 1H), 4.75 (d, $J = 11.6$ Hz, 1H), 4.68 (t, $J = 5.6$ Hz, 1H), 4.56 (dd, $J = 11.2, 5.6$ Hz, 1H), 4.53 (m, 1H), 4.48 (d, $J = 12.0$ Hz, 1H), 4.01 (app quart, 1H), 4.01 (app quart, 1H), 3.93 (d, $J = 15.2$ Hz, 1H), 3.76, (d, $J = 2.8$ Hz, 1H), 3.16–3.08 (m, 4H), 2.93 (s, 3H), 2.90 (s, 3H), 2.89 (s, 3H), 2.84 (s, 3H); ^{13}C NMR (100 MHz, $CDCl_3$) δ 158.7, 138.0, 136.4, 128.6, 128.4, 128.1, 128.0, 127.7, 127.6, 87.9, 82.1, 75.2, 74.6, 72.9, 68.0, 64.7, 64.3, 53.6, 52.6, 52.3, 52.0, 51.6, 46.8; ESI$^+$ (m/z): [M + MeCN + NH_4]$^+$ 724 (100); ESI$^+$ (m/z): [M + MeCN + NH_4]$^+$ calcd for $C_{27}H_{38}N_3O_8Os$, 724.2274; found, 724.2278.

(2R*,3R*,4R*)-2-Hydroxymethyl-2-(4-methoxybenzyl)tetrahydrofuran-3,4 -diol [Directed Dihydroxylation of a Homoallylic Alcohol Using Catalytic OsO$_4$·Quinuclidine and NMO].[37] 4-Methylmorpholine-N-oxide (240 mg, 2.01 mmol) was added to a stirred solution of 2-hydroxymethyl-2-(4-methoxybenzyl)-2,5-dihydrofuran (150 mg, 0.681 mmol) in acetone (20 mL) and water (5 mL) at rt, followed by quinuclidine (5 mg, 7 mol %) and OsO$_4$ (5 mg, 3 mol %). The reaction mixture was stirred overnight. Acetone was removed under vacuum before the addition of EtOAc (20 mL) and brine (20 mL). The organic layer was dried (MgSO$_4$) and concentrated under vacuum to give the crude product as a mixture of diastereomers (*syn/anti* 2.1:1, by HPLC). Purification by column chromatography (SiO$_2$, CH$_2$Cl$_2$/i-PrOH 19:1) gave (2R*,3S*,4S*)-2-hydroxymethyl-2-(4-methoxybenzyl)tetrahydrofuran-3,4-diol (41 mg, 24%) as a crystalline solid, mp 103–105°, and (2R*, 3R*, 4R*)-2-(hydroxymethyl)-2-(4-methoxybenzyl)tetrahydrofuran-3,4-diol (89 mg, 52%) as a crystalline solid, mp 93–95°. Analytical data for the major isomer: R$_f$ (CH$_2$Cl$_2$/i-PrOH 95:5) 0.26; IR 3232, 1249 cm^{-1}; ^{1}H NMR (400 MHz, CD$_3$OD) δ 7.20–7.16 (m, 2H), 6.85–6.81 (m, 2H), 4.04 (d, $J = 5.3$ Hz, 1H), 3.81 (dd, $J = 8.3, 4.7$ Hz, 1H), 3.76 (s, 3H), 3.71–3.61 (m, 3H), 3.52 (d, $J = 11.4$ Hz, 1H), 2.81 (d, $J = 13.9$ Hz, 1H), 2.71 (d, $J = 13.9$ Hz, 1H); ^{13}C NMR (100 MHz, CD$_3$OD) δ 158.9, 131.7, 129.1, 113.5, 85.4, 75.2, 71.9, 71.8, 64.3, 54.6, 40.2; ESI$^+$ (m/z): [M + Na$^+$] 277 (100); ESI$^+$ (m/z): [M + Na]$^+$ calcd for C$_{13}$H$_{18}$O$_5$Na, 277.1046; found, 277.1046; Anal. Calcd for C$_{13}$H$_{18}$O$_5$: C, 61.41; H, 7.14. Found: C, 61.37; H, 7.16.

(2R*,3R*,4R*)-2-Hydroxymethyl-2-(4-methoxybenzyl)tetrahydrofuran-3,4 -diol [Directed Dihydroxylation of a Homoallylic Alcohol Using Catalytic OsO$_4$ and TMO].[37] Trimethylamine-N-oxide dihydrate (5.7 g, 51 mmol) was added to a stirred solution of 2-hydroxymethyl-2-(4-methoxybenzyl)-2,5-dihydrofuran (3.78 g, 17.2 mmol) in CH$_2$Cl$_2$ (200 mL) at rt. OsO$_4$ (50 mg, 0.20 mmol) was then added and the mixture stirred overnight. Sodium sulfite (aq saturated solution, 20 mL) was added and the mixture stirred for 20 min. The organic layer was dried (MgSO$_4$) and concentrated under vacuum to give the crude product as a mixture of diastereomers (*syn/anti* 6.7:1, by HPLC). Purification by column chromatography (SiO$_2$, CH$_2$Cl$_2$/i-PrOH 19:1) gave the *anti*-triol (0.45 g, 10%) and the *syn*-triol (3.04 g, 70%).

OsO$_4$ (cat), Me$_3$NO•H$_2$O
polymer-bound DABCO, CH$_2$Cl$_2$, rt

(84%) 7.3:1 dr

(2R*,3R*,4R*)-2-Hydroxymethyl-2-(4-methoxybenzyl)tetrahydrofuran-3,4-diol [Directed Dihydroxylation of a Homoallylic Alcohol Using Catalytic OsO$_4$, TMO, and Polymer-Bound DABCO].[37] Polymer-bound 1,4-diazabicyclo[2.2.2]octane chloride (100 mg, 1% DVB, 100-200 mesh) was added to a solution of OsO$_4$ (26 mg) in cyclohexane (5 mL); the solvent was then evaporated, and the solid so obtained (100 mg, ~10 mol % OsO$_4$) was added to a solution of 2-hydroxymethyl-2-(4-methoxybenzyl)-2,5-dihydrofuran (100 mg, 0.451 mmol) in CH$_2$Cl$_2$ (15 mL) at rt. Trimethylamine-N-oxide dihydrate (150 mg, 1.40 mmol) was added and the mixture was shaken overnight; the polymer was then removed by filtration and the filtrate was concentrated under vacuum to give the crude product as a mixture of diastereomers (*syn*/*anti* 7.3:1, by HPLC). Purification by column chromatography (SiO$_2$, CH$_2$Cl$_2$/i-PrOH 19:1) gave the *anti*-triol (12 mg, 10%) and the *syn*-triol (85 mg, 74%).

1. OsO$_4$ (5 mol%),
   QNO•H$_2$O (1.3 equiv), CH$_2$Cl$_2$, rt
2. H$^+$, MeOH

(95%) 20:1 dr

19

2,2,2-Trichloro-N-((1R*,2R*,3S*,5S*)-2,3-dihydroxy-5-isopropyl-2-methylcyclohexyl)acetamide [Directed Dihydroxylation of an Allylic Cyclic Amide Using Catalytic OsO$_4$ and QNO].[19] Quinuclidine (1.00 g, 9.01 mmol) was dissolved in CH$_2$Cl$_2$ (20 mL) under nitrogen and cooled to −78° before the addition of recrystallised m-CPBA (1.94 g, 11.24 mmol) in one portion. The mixture was stirred for 30 min and then allowed to warm to rt. The crude reaction mixture was flushed through a column of silica gel using CH$_2$Cl$_2$ as eluent until all of the benzoic acid byproduct was removed, then the solvent gradient was increased (CH$_2$Cl$_2$/MeOH 2.3:1) to strip quinuclidine-N-oxide from the column. Concentration under vacuum produced a brown oil, which crystallised on standing under high vacuum conditions to produce QNO as an off-white solid (1.25 g, 95%) and this was stored under reduced pressure. KF analysis showed this hygroscopic material contained 11% water by weight (~QNO•H$_2$O), and on standing open to air this increased to 40% water by weight (~QNO•5H$_2$O).

Quinuclidine-N-oxide monohydrate (0.36 g, 2.18 mmol) was added in one portion to a stirred solution of trichloroacetamide **19** (0.50 g, 1.68 mmol) in

CH$_2$Cl$_2$ at rt. OsO$_4$ (0.02 g, 0.08 mmol) was then added and the reaction mixture was stirred until complete consumption of the starting amide was observed by TLC. MeOH (10 mL) and HCl (concd, 4 drops) were added and the resulting solution was stirred for 2 h and then concentrated under vacuum to afford a dark yellow, viscous oil. Purification by column chromatography (SiO$_2$, petroleum ether/Et$_2$O 1:4) yielded the title compound (0.53 g, 95%, *syn/anti* 20:1, by HPLC) as a colorless crystalline solid. The analytical data for the product was not reported in this reference.

(2R^*, 3R^*, 4S^*, 5S^*, 6S^*)-Methyl 3,4,5-Trihydroxy-6-methoxytetrahydro-2H-pyran-2-carboxylate [Directed Dihydroxylation of an Allylic Cyclic Alcohol Using OsO$_4$·Pyridine].[38] (2R^*, 3R^*, 6S^*)-Methyl 3-Hydroxy-6-methoxy-3,6-dihydro-2H-pyran-2-carboxylate (50 mg, 0.25 mmol) in pyridine (4 mL) was added to an OsO$_4$ (70 mg, 0.28 mmol, 1.1 equiv) solution in pyridine (0.5 mL). After 2 h at rt, the reaction was quenched with NaHSO$_3$ (aq saturated solution, 1 mL) and dry loaded onto SiO$_2$. Purification by column chromatography (SiO$_2$, CH$_2$Cl$_2$/MeOH 9:1) gave the title compound as a colorless oil (60 mg, 0.25 mmol, 100%): IR (neat) 3418, 2930, 1736, 1084, 734 cm^{-1}; ^{1}H NMR (300 MHz, CDCl$_3$) δ 4.56 (d, $J = 6.8$ Hz, 1H), 4.26–4.16 (m, 3H), 3.86 (dd, $J = 3.0, 9.0$ Hz, 1H), 3.49 (s, 3H), 3.47 (dd, $J = 3.4, 7.1$ Hz, 1H), 2.78 (br s, 3H), 1.26 (t, $J = 7.1$ Hz, 3H); ^{13}C NMR (75 MHz, CDCl$_3$) δ 170.8, 102.4, 73.5, 71.0, 69.7, 69.5, 62.2, 57.8, 14.4; EIMS (m/z): [M−H$_2$O] 218 (1), [M + H−MeOH] 205 (2), 71 (100). CIMS (m/z): [M + H]$^+$ 237 (1), [M + H−MeOH] 205 (18), 187 (100).

(2R^*, 3S^*, 4S^*, 5S^*, 6S^*)-2-{2-[(2S^*, 3S^*, 6R^*)-3-Acetoxy-6-methoxy-3,6-dihydro-2H-pyran-2-yl]ethyl}-3,4,5-triacetoxy-6-methoxytetrahydropyran [Directed Dihydroxylation Using OsO$_4$ and a Chiral Amine].[39] A solution of diamine **20** (33.9 mg, 0.071 mmol) in CH$_2$Cl$_2$ (0.5 mL) was added to a stirred solution of OsO$_4$ (20 mg, 0.071 mmol) in CH$_2$Cl$_2$ (1.5 mL). The yellow solution was cooled to −20°; (2R*, 3S*, 6R*)-2-(2-((2R*, 3R*, 6S*)-3-hydroxy-6-methoxy-3,6-dihydro-2H-pyran-2-yl)ethyl)-6-methoxy-3,6-dihydro-2H-pyran-3-ol (20 mg, 0.071 mmol) was added in one portion and the reaction mixture was stirred for 5 h, warmed to rt, stirred for 2 d and evaporated under vacuum. The residue was dissolved in THF/sodium sulfite (aq saturated solution, 1:1, 2 mL), refluxed for 2 h and evaporated under vacuum to give a crude product. Crude (2R*, 3R*, 4R*, 5R*, 6R*)-2-(2-((2R*, 3R*, 6S*)-3-hydroxy-6-methoxy-3,6-dihydro-2H-pyran-2-yl)ethyl)-6-methoxytetrahydro-2H-pyran-3,4,5-triol was dissolved in a mixture of Ac$_2$O (2 mL) and pyridine (1 mL), stirred for 5 h at rt, and evaporated under vacuum to give the crude product. Purification by column chromatography (petroleum ether/EtOAc 2.3:1) afforded the title product (13.2 mg, 39%) as a viscous colorless oil: [α]$_D^{20}$ +90.9 (c 0.033, CHCl$_3$); R$_f$ 0.31 (petroleum ether/EtOAc 2.3:1); IR (thin film) 2960, 2924, 2853, 1742, 1678, 1455, 1373, 1259, 1083 cm^{-1}; ^{1}H NMR (500 MHz, CDCl$_3$) δ 6.08 (ddd, J = 10.0, 5.5, 1.0 Hz, 1H), 6.00 (ddd, J = 10.0, 3.0, 0.4 Hz, 1H), 5.27 (t, J = 3.8 Hz, 4H), 5.21 (dd, J = 3.8, 1.2 Hz, 1H), 5.10 (dt, J = 3.8, 1.2 Hz, 1H), 4.91 (d, J = 3.0 Hz, 1H), 4.76 (d, J = 1.2 Hz, 1H), 4.02 (td, J = 9.3, 2.6 Hz, 1H), 3.97 (ddd, J = 9.3, 3.8, 1.0 Hz, 1H), 3.41 (s, 3H), 3.39 (s, 3H), 2.14 (s, 3H), 2.14 (s, 3H), 2.08 (s, 3H), 1.99 (s, 3H) and 1.77–1.68 (m, 4H); ^{13}C NMR (125 MHz, CDCl$_3$) δ 170.6, 170.4, 170.1, 169.6, 130.4, 125.9, 99.4, 95.2, 68.8, 68.3, 67.5, 66.0, 64.5, 55.7, 55.2, 27.1, 26.4, 21.0, 20.8, 20.7, 20.6; ESI$^+$ (m/z): [M + Na]$^+$ 511 (100); ESI$^+$ (m/z): [M + Na]$^+$ calcd for C$_{22}$H$_{32}$O$_{12}$Na, 511.1791; found, 511.1768.

TABULAR SURVEY

The literature has been covered through the end of September 2007. The tables are organized by substrate type. Entries in the tables are in order of increasing number of carbons. Protecting groups and O-methyl groups are excluded from the count. The symbol (—) indicates that no yield was reported and the symbol — indicates that no dr ($syn/anti$) was reported.

Abbreviations used in the tables are as follows:

ee	enantiomeric excess
eq	equivalents
QNO	quinuclidine N-oxide
TBDPS	$tert$-butyldiphenylsilyl

TABLE 1. DIRECTED DIHYDROXYLATION OF ALLYLIC CYCLIC ALCOHOLS

Substrate	Conditions	Product(s), Yield(s) (%), and dr (*syn:anti*)	Refs.
C5			
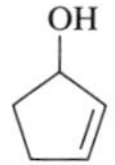 (cyclopentenol)	OsO4, TMEDA, CH2Cl2, −78° to rt; then NH2(CH2)2NH2	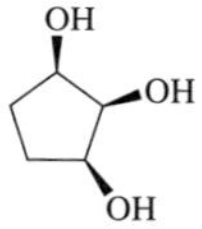 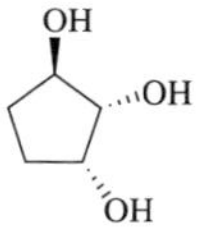 **I** + **II** (76), **I:II** = 7:1	21, 15
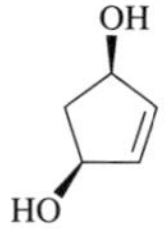 (cyclopentenediol)	1. OsO4, TMEDA, CH2Cl2, −78° to rt; then NH2(CH2)2NH2 2. Ac2O, py	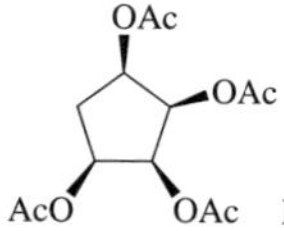 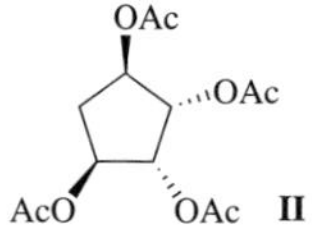 **I** + **II** (73), **I:II** = 25:1	15
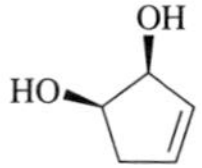 (cyclopentenediol isomer)	1. OsO4, TMEDA, CH2Cl2, −78° to rt; then NH2(CH2)2NH2 2. Ac2O, py	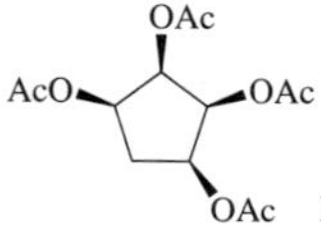 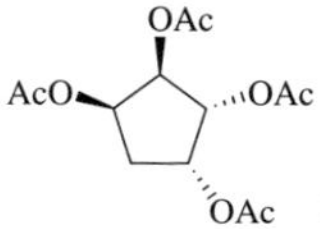 **I** + **II** (67), **I:II** = 2:1	15
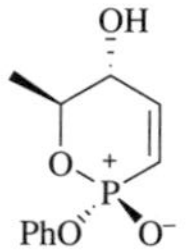 (phosphate sugar)	OsO4, TMEDA, CH2Cl2, −60° to rt; then citric acid, MeOH	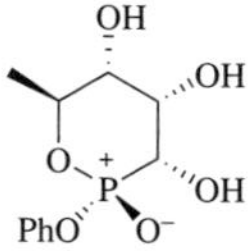 **I** 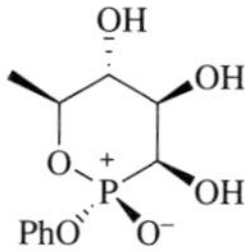 + **II** (70), **I:II** = 13.5:1	35
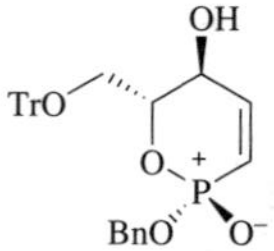 (BnO phosphate sugar)	OsO4, TMEDA, CH2Cl2, −60° to rt; then TsOH, MeOH	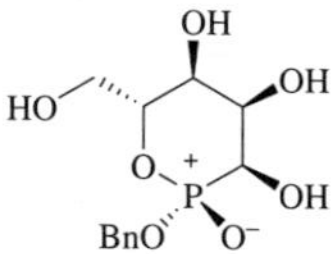 **I** 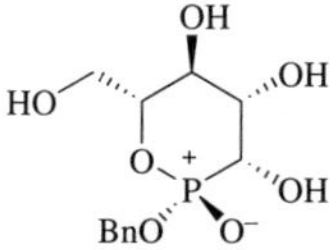 + **II** (70), **I:II** = 24:1	35
C6			
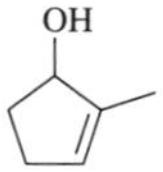 (methylcyclopentenol)	OsO4, TMEDA, CH2Cl2, −78° to rt; then NH2(CH2)2NH2	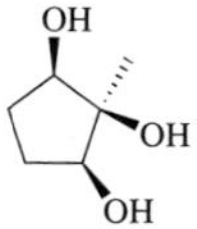 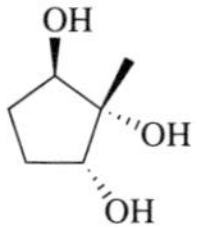 **I** + **II** (88), **I:II** = 25:1	15

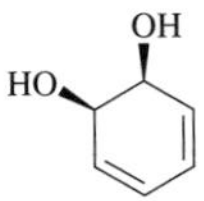

Substrate	Conditions	Products (I + II, ratio)	Refs.
(3,4-dihydroxycyclohexa-1,5-diene-like substrate)	OsO₄, TMEDA, CH₂Cl₂, −78° to rt; then Na₂SO₃	**I + II** (54), **I:II** = 16:1	21, 15
cyclohexenol	OsO₄, TMEDA, CH₂Cl₂, −78° to rt; then Na₂SO₃	**I + II** (98), **I:II** = 9:1	15
bromo-diol diene	OsO₄ (cat), Me₃NO·2H₂O, CH₂Cl₂	**I + II** (79), **I:II** = 4.6:1	21
bromo-diol diene	OsO₄, quinuclidine, CH₂Cl₂, −78°	**I + II** (40), **I:II** = 2:1	
MeO-bromo diene	OsO₄ (cat), Me₃NO·2H₂O, CH₂Cl₂	**I + II** (71), **I:II** = 2.2:1	40
pyranose diol	1. OsO₄, TMEDA, CH₂Cl₂, −78° to rt; then Na₂SO₃ 2. Ac₂O, py	**I + II** (63), **I:II** = 6:1	21, 15

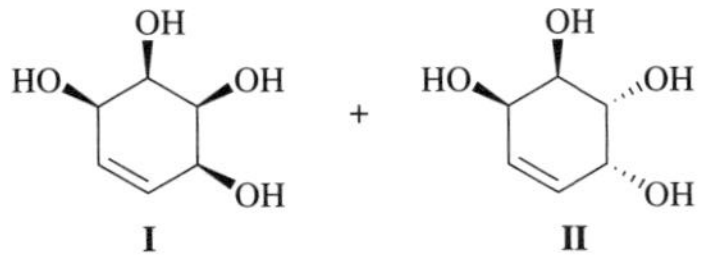

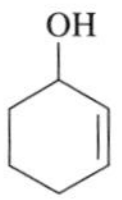

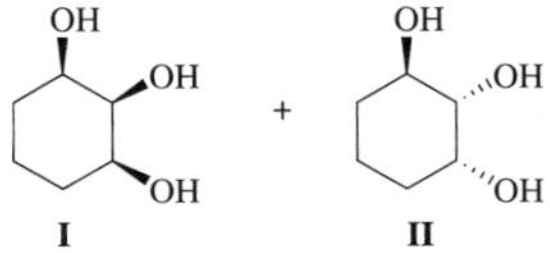

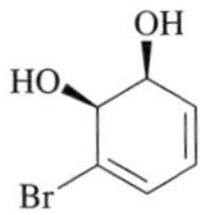

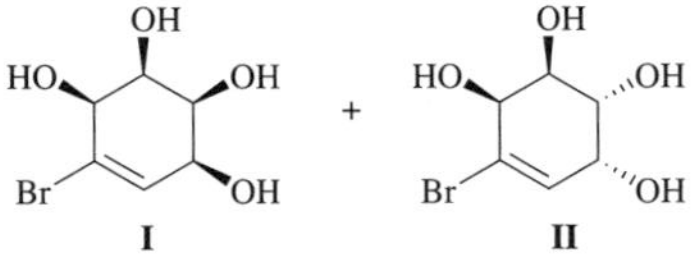

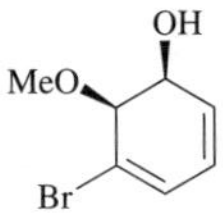

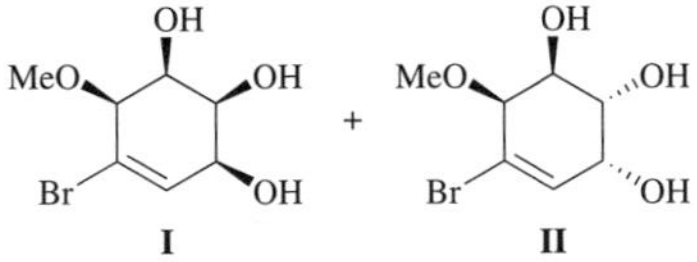

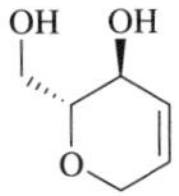

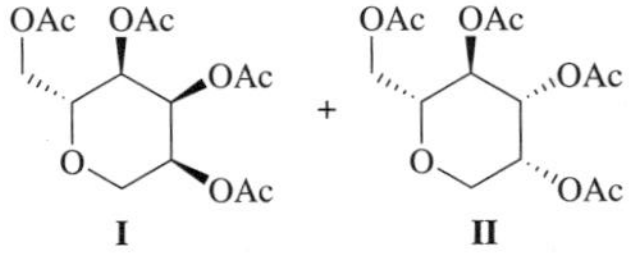

TABLE 1. DIRECTED DIHYDROXYLATION OF ALLYLIC CYCLIC ALCOHOLS (*Continued*)

Substrate	Conditions	Product(s), Yield(s) (%), and dr (*syn:anti*)	Refs.
C_6			
(MeO₂C, OH, OMe pyranose)	OsO₄•py, py, 0° to rt; then NaHSO₃	(100), 1 diastereomer	38
(BnN/BnO bicyclic with OH)	OsO₄, TMEDA, CH₂Cl₂, −78° to rt	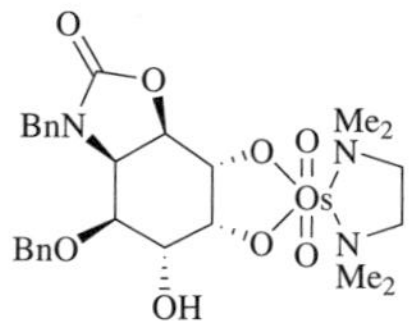 (100), 1 diastereomer	36
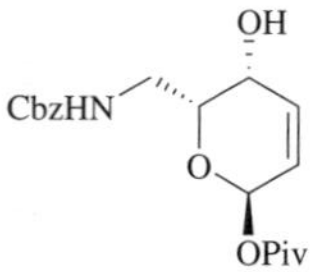 (CbzHN, OH, OPiv pyranose)	OsO₄, TMEDA, CH₂Cl₂, −78°; then NH₂(CH₂)₂NH₂	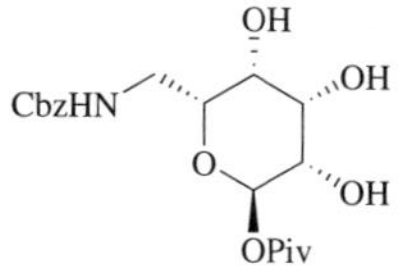 (47), 1 diastereomer	41
(TBSO, OH, OPiv pyranose)	OsO₄, TMEDA, CH₂Cl₂, −78° to rt; then Na₂SO₃	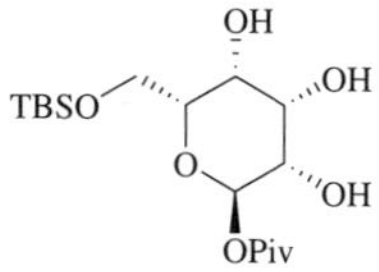 (80), 1 diastereomer	42, 20, 15
C_7			
(MeO₂C, Me, OH, OMe pyranose)	OsO₄•py, py, 0° to rt; then NaHSO₃	(60), 1 diastereomer	38

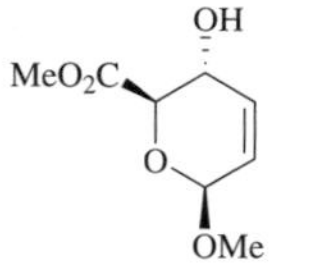

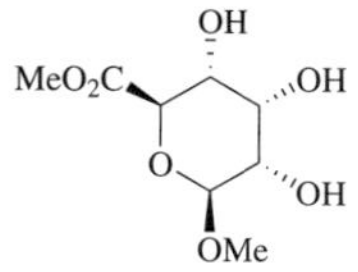

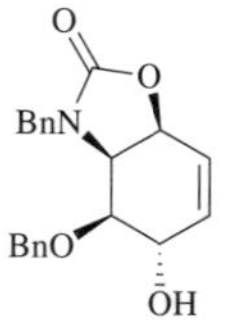

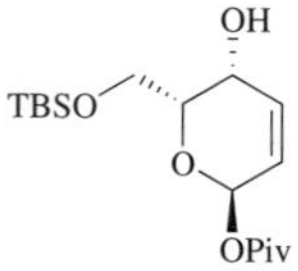

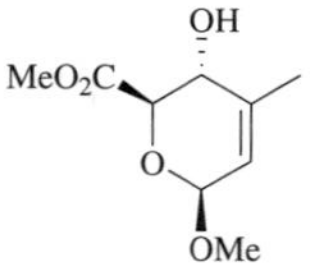

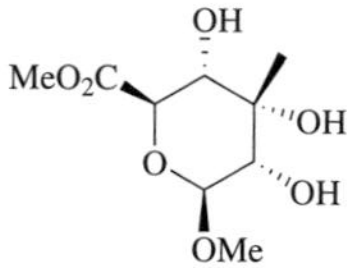

26

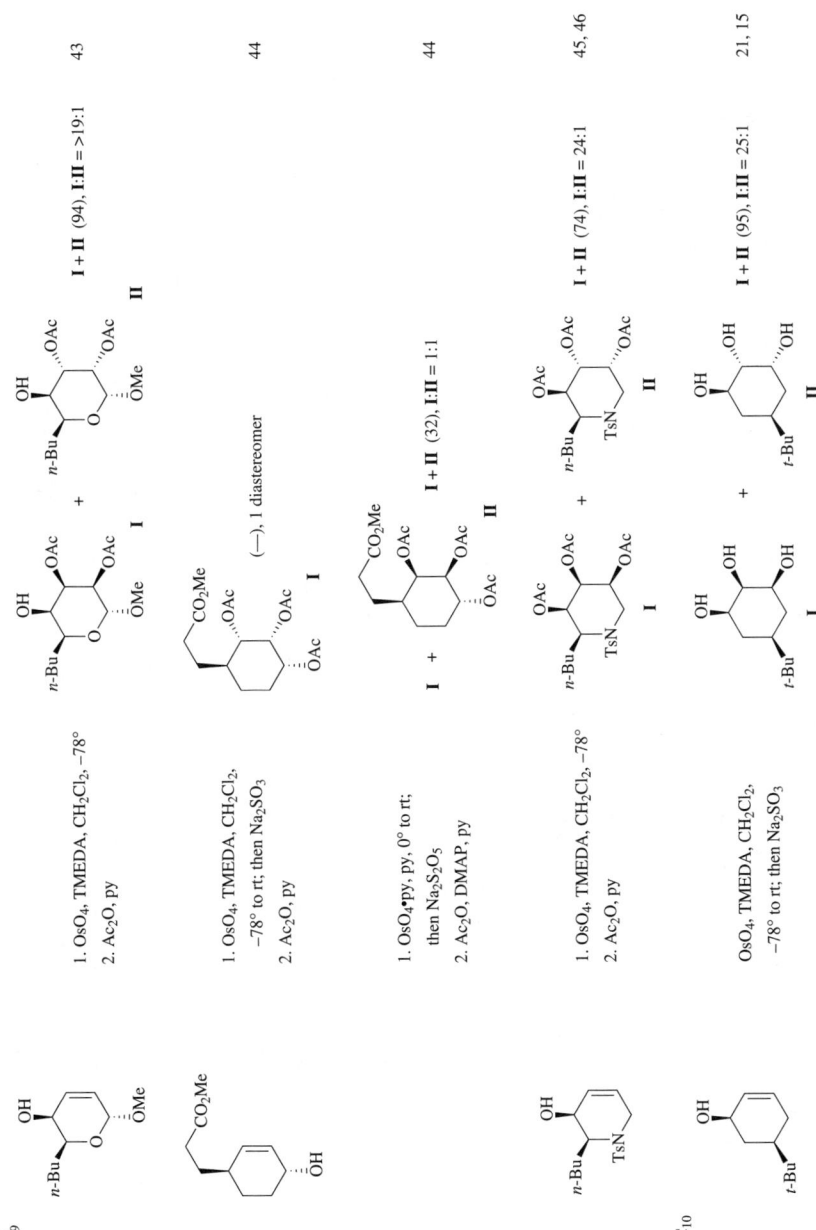

TABLE 1. DIRECTED DIHYDROXYLATION OF ALLYLIC CYCLIC ALCOHOLS (*Continued*)

Substrate	Conditions	Product(s), Yield(s) (%), and dr (*syn:anti*)	Refs.

C₁₀ substrate (t-Bu cyclohexenol), OsO₄, amine, CH₂Cl₂, 0° to rt; then Na₂SO₃

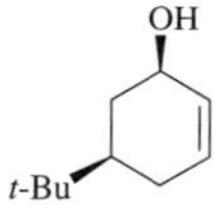

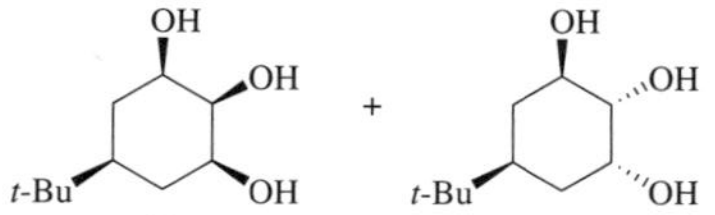

Amine	I:II
TMEDA	4.6:1
Me₃N	1.2:1
Me₂NCH₂NMe₂	1.6:1
Me₂N(CH₂)₂OH	1.1:1
Me₂N(CH₂)₂OMe	1.1:1
Me₂N(CH₃)₃NMe₂	2.8:1

Ref. 15

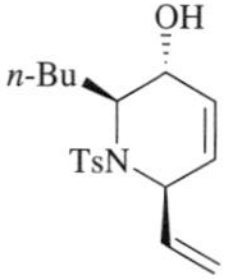

C₁₂:

1. OsO₄, TMEDA, CH₂Cl₂, −78°
2. Ac₂O, py

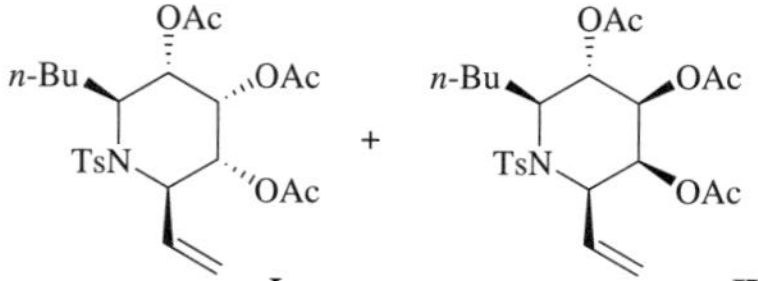

I + II (25), **I:II** = 3:1 — Refs. 45, 46

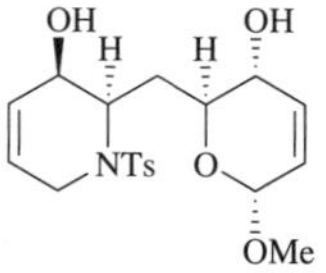

1. OsO₄, TMEDA, CH₂Cl₂, −78°
2. Ac₂O, py

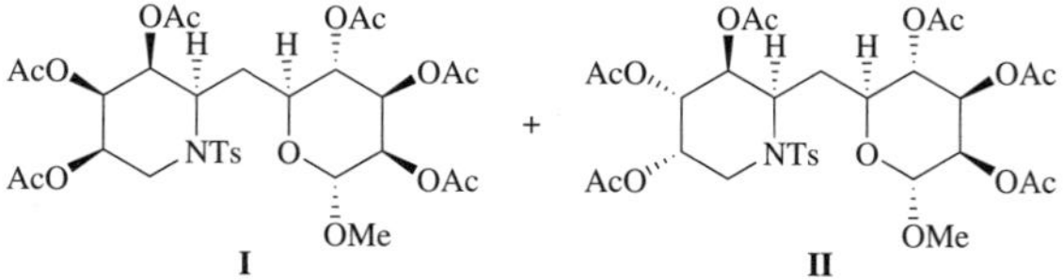

I + II (17), **I:II** = 24:1 — Refs. 45, 46

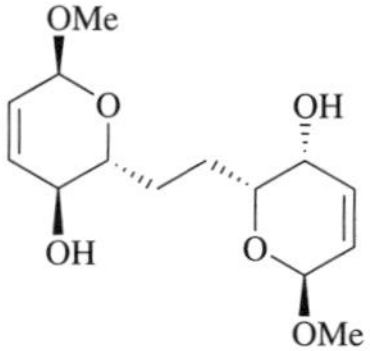

1. OsO₄, TMEDA, CH₂Cl₂, −78°
2. Ac₂O, py

I + II (83), **I:II** = 19:1 — Refs. 47, 48

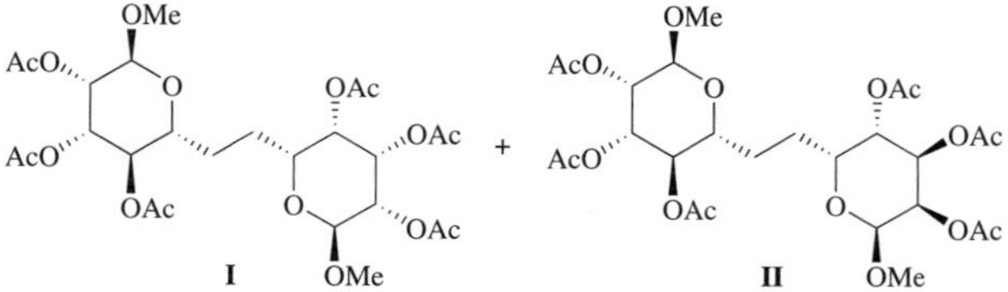

28

TABLE 2. DIRECTED DIHYDROXYLATION OF ACYCLIC AND EXOCYCLIC ALLYLIC ALCOHOLS

Substrate	Conditions	Product(s), Yield(s) (%), and dr (syn:anti)	Refs.
C₆			
 (pentadienyl alcohol)	OsO₄, TMEDA, CH₂Cl₂, −78° to rt; then HCl, MeOH	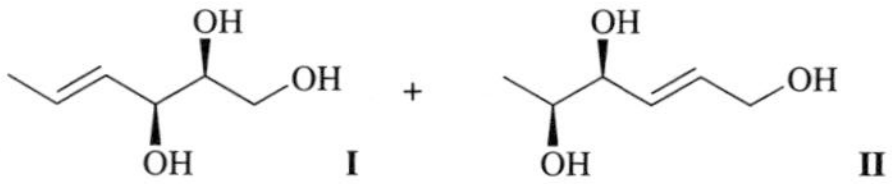 I + II (70), I:II = 5:1	21
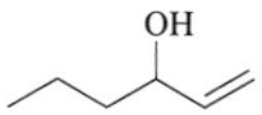 (hex-1-en-3-ol)	1. OsO₄, TMEDA, CH₂Cl₂, −78° to rt; then HCl, MeOH 2. Ac₂O, py	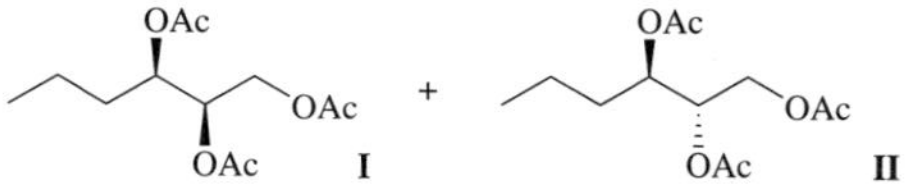 I + II (74), I:II = 5:1	50
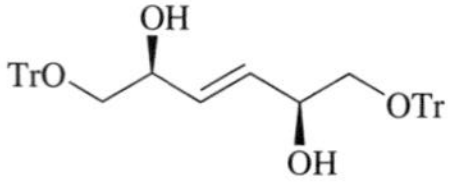 (TrO/OH allylic diol)	OsO₄, TMEDA, CH₂Cl₂, −78° to rt; then HCl, MeOH	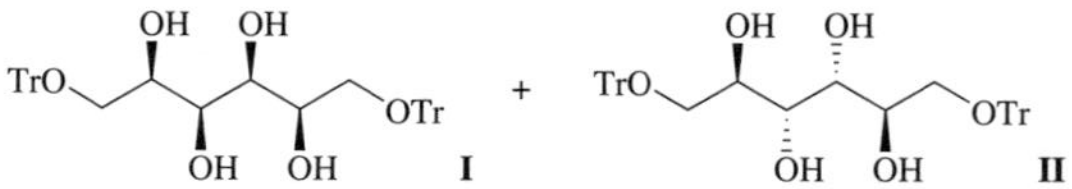 I + II (95), I:II = 1.1:1	51
C₉			
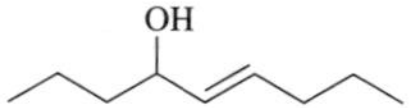 (non-3-en-5-ol)	1. OsO₄, TMEDA, CH₂Cl₂, −78° to rt; then HCl, MeOH 2. Ac₂O, py	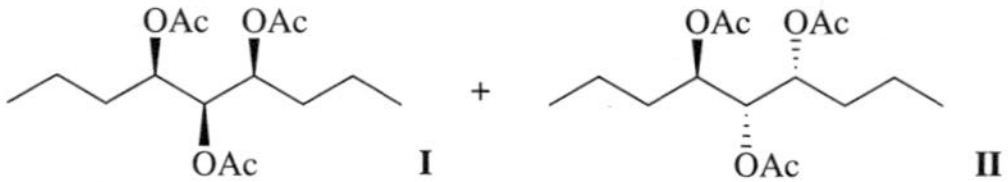 I + II (83), I:II = 3:1	50
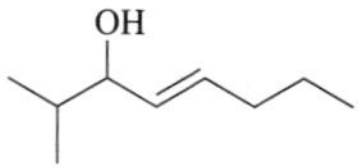 (2-methyl-oct-3-en-5-ol)	OsO₄, TMEDA, CH₂Cl₂, −78° to rt; then HCl, MeOH	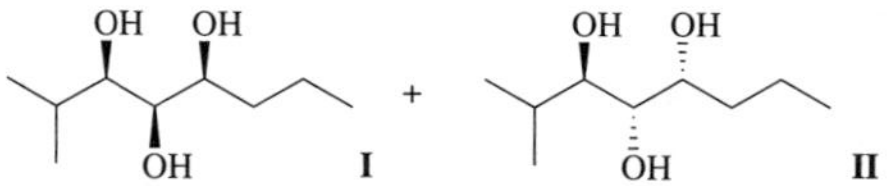 I + II (84), I:II = 3:1	50

31

TABLE 2. DIRECTED DIHYDROXYLATION OF ACYCLIC AND EXOCYCLIC ALLYLIC ALCOHOLS (*Continued*)

Substrate	Conditions	Product(s), Yield(s) (%), and dr (*syn:anti*)	Refs.

C_{10} — 1. OsO$_4$, TMEDA, CH$_2$Cl$_2$, −78° to rt; then HCl, MeOH; 2. Ac$_2$O, py

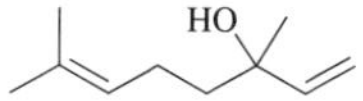

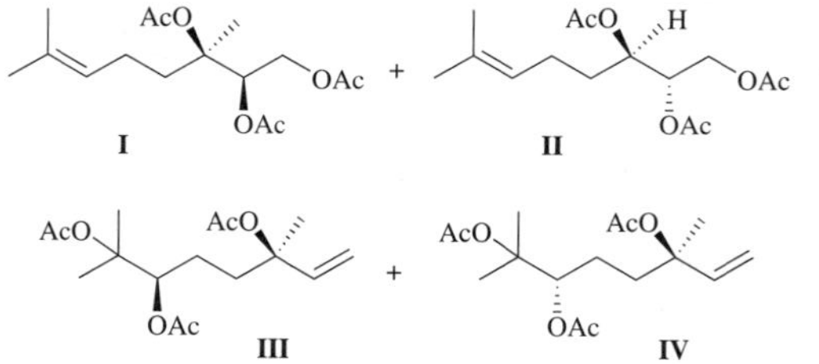

I + II	I:II	III + IV	III:IV
(70)	2.5:1	(6)	—

21

C_{14} — 1. OsO$_4$, TMEDA, CH$_2$Cl$_2$, −78° to rt; then HCl, MeOH; 2. Ac$_2$O, py

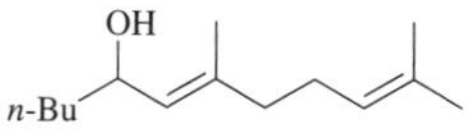

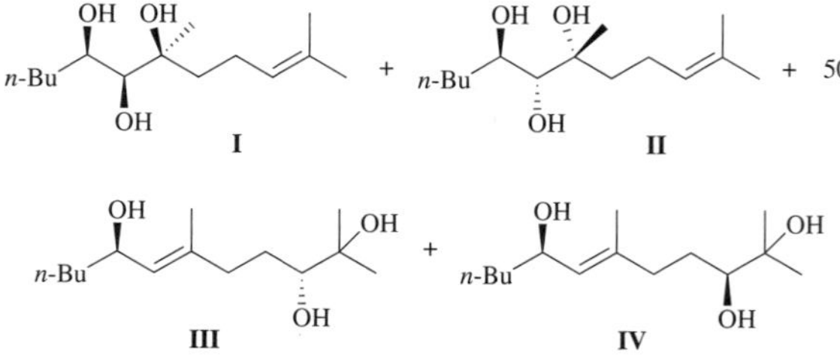

I + II	I:II	III + IV	III:IV
(70)	24:1	(<5)	—

50

C_{19} — OsO$_4$ (2.5 eq), TMEDA (2.1 eq), CH$_2$Cl$_2$, −78° to rt; then NaHSO$_3$, THF/acetone/H$_2$O (2:1:1)

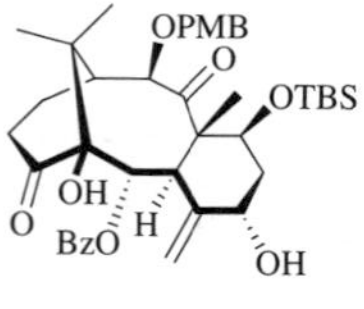

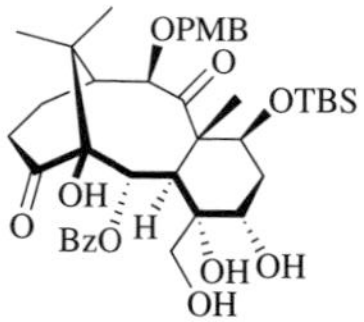

(72), 1 diastereomer

17

C$_{22}$

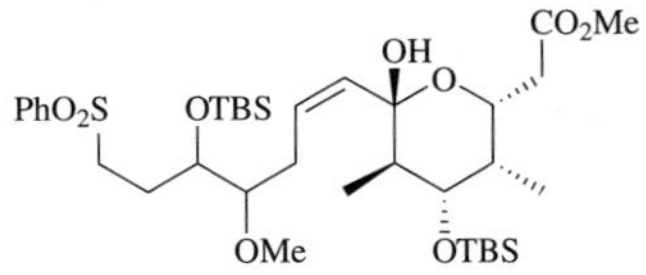

OsO_4 (1.0 eq), TMEDA (1.1 eq), CH_2Cl_2, –78° to rt; then $NaHSO_3$

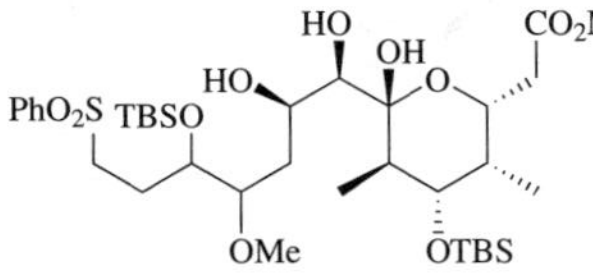

I +

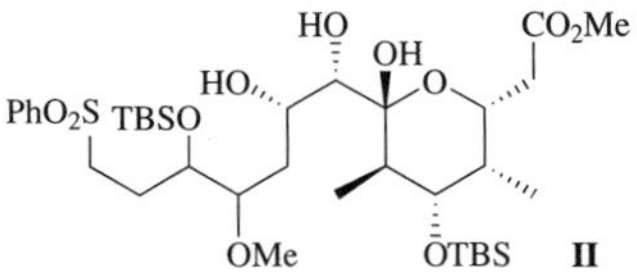

II

I + II (100), **I:II** = 9:1

52

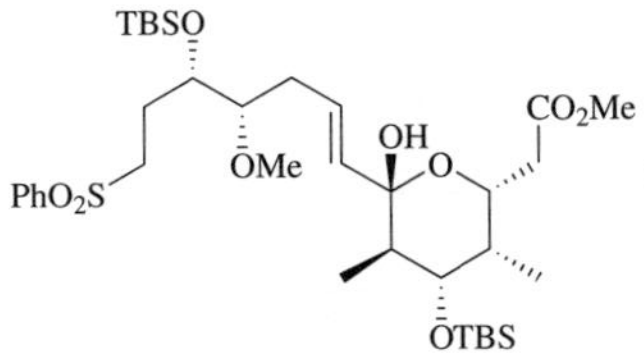

OsO_4 (1.0 eq), TMEDA (1.1 eq), CH_2Cl_2, –78° to rt; then $NaHSO_3$

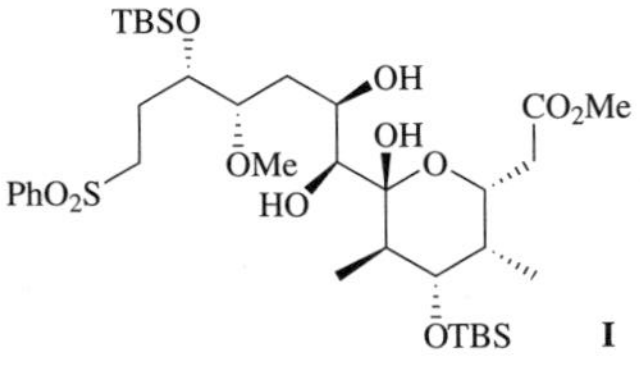

I +

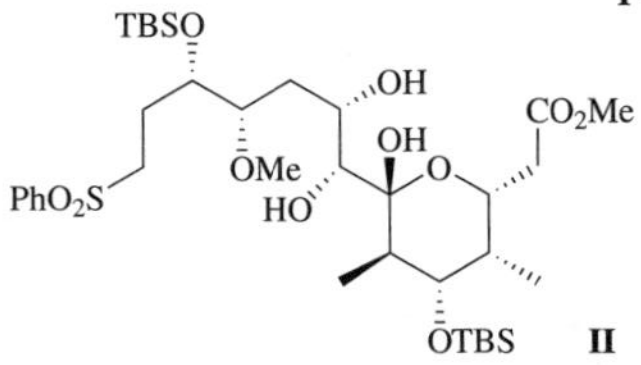

II

I + II (100), **I:II** = 5:1

52

TABLE 3. DIRECTED DIHYDROXYLATION OF HOMOALLYLIC CYCLIC ALCOHOLS

Substrate	Conditions	Product(s), Yield(s) (%), and dr (*syn:anti*)	Refs.
C₅			
(cyclopentenol, HO-)	1. OsO₄, TMEDA, CH₂Cl₂, −78° to rt; then HCl, MeOH 2. Ac₂O, py	**I** (AcO, OAc, OAc) + **II** (AcO, OAc, OAc) **I + II** (71), **I:II** = 25:1	25, 22
(dihydrofuran-CH₂OH)	1. OsO₄, TMEDA, CH₂Cl₂, −78° to rt; then HCl, MeOH 2. Ac₂O, py	**I** (AcO, O, OAc, OAc) + **II** (AcO, O, OAc, OAc) **I + II** (55), **I:II** = 1:1	25, 22
(methyl dihydrofuran-CH₂OH)	1. OsO₄, TMEDA, CH₂Cl₂, −78° to rt; then HCl, MeOH 2. Ac₂O, py	**I** (AcO, O, OAc, OAc) + **II** (AcO, O, OAc, OAc) **I + II** (83), **I:II** = 6:1	25, 22
C₆			
(cyclohexenol)	1. OsO₄, TMEDA, CH₂Cl₂, −78° to rt; then HCl, MeOH 2. Ac₂O, py	**I** (AcO, OAc, OAc) + **II** (AcO, OAc, OAc) **I + II** (92), **I:II** = 3:1	25, 22
(cyclohexenediol)	1. OsO₄, TMEDA, CH₂Cl₂, −78° to rt; then HCl, MeOH 2. Ac₂O, py	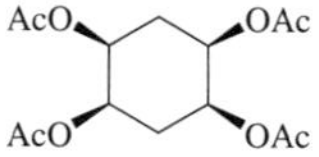 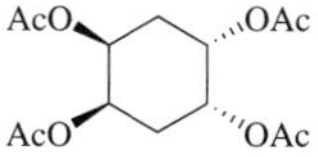 **I** (AcO, OAc, OAc) + **II** (AcO, OAc, OAc) **I + II** (55), **I:II** = 12.4:1	25, 22
(SMe, HO, AcHN cyclopentene)	OsO₄, TMEDA, CH₂Cl₂, −78° to rt	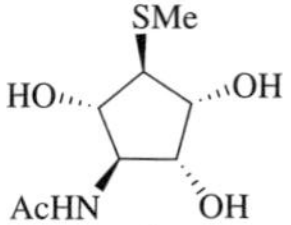 **I** 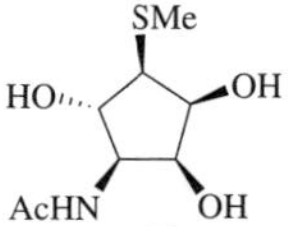 (SMe, HO, OH, AcHN) + **II** (SMe, HO, OH, AcHN) **I + II** (71), **I:II** = 6:1	53

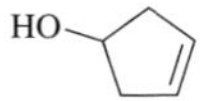

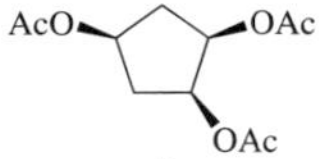

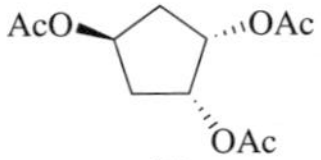

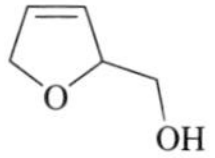

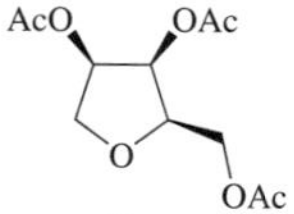

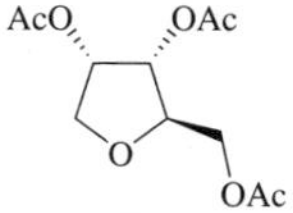

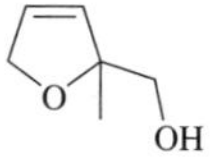

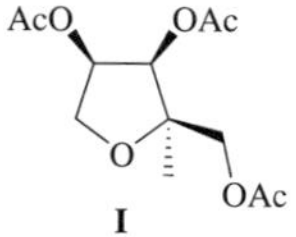

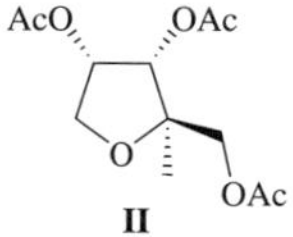

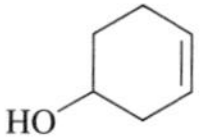

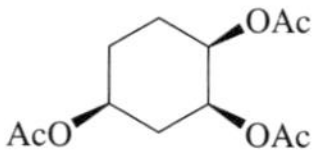

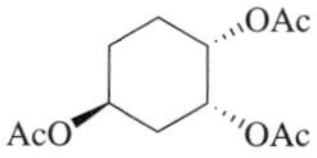

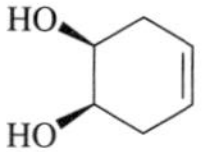

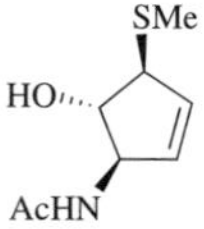

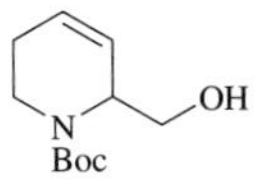

Substrate	Conditions	Products	Refs.
C12 (tetrahydropyridine-Boc with CH2OH)	OsO4, TMEDA, CH2Cl2, −78° to rt; then HCl, MeOH	I + II (86), I:II = 2:1	54
C12 (dihydrofuran with PMB, CH2OH)	OsO4, TMEDA, CH2Cl2, −78° to rt; then HCl, MeOH	I + II (82), I:II = 99:1	37
	OsO4 (cat), quinuclidine, acetone/H2O	I + II (76), I:II = 2.1:1	37
	OsO4 (cat), Me3NO·2H2O, CH2Cl2	I + II (80), I:II = 6.7:1	37
	OsO4 (cat), Me3NO·2H2O, polymer-bound DABCO, CH2Cl2	I + II (84), I:II = 7.3:1	37
C20 (SiPh2, allyl carbinol)	OsO4, TMEDA, CH2Cl2, −78° to rt	I + II (—), I:II = 1.2:1	55

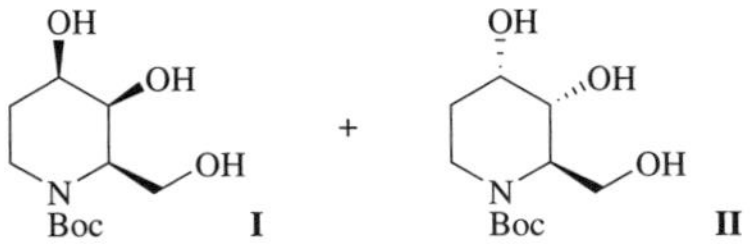

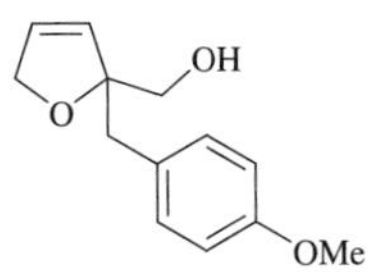

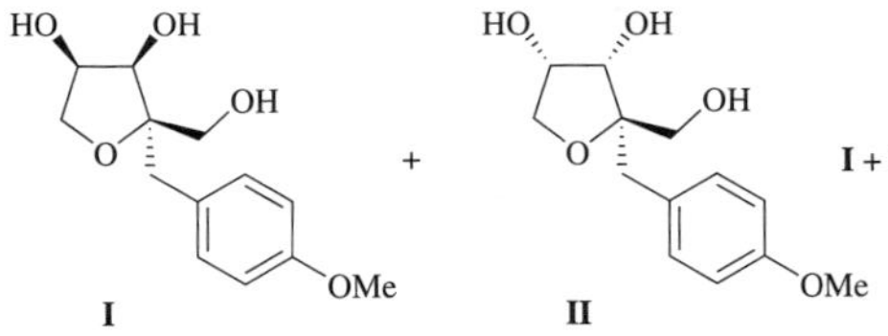

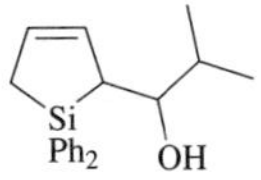

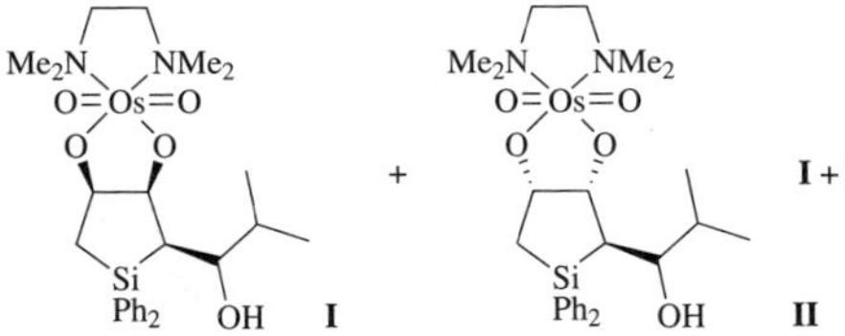

TABLE 4. DIRECTED DIHYDROXYLATION OF HOMOALLYLIC EXOCYCLIC ALCOHOLS

Substrate	Conditions	Product(s), Yield(s) (%), and dr (syn:anti)	Refs.
C7	OsO4, TMEDA, CH2Cl2, −78° to rt; then HCl, MeOH	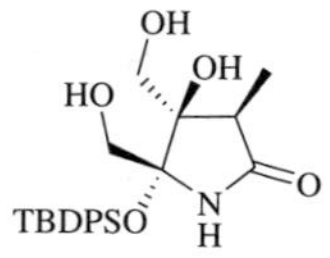 (99), 1 diastereomer	56
C8	OsO4, TMEDA, CH2Cl2, −78° to rt; then HCl, MeOH	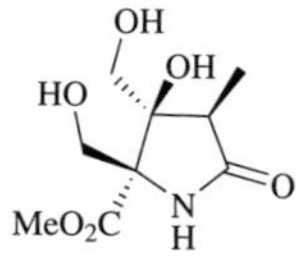 (35), 1 diastereomer	56

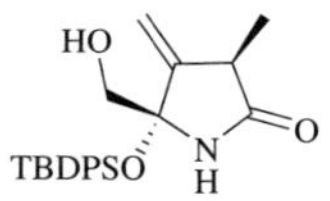

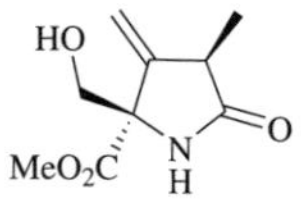

TABLE 5. DIRECTED DIHYDROXYLATION OF N-ALLYLIC AMINE DERIVATIVES

Substrate	Conditions	Product(s), Yield(s) (%), and dr (syn:anti)	Refs.
C$_5$ 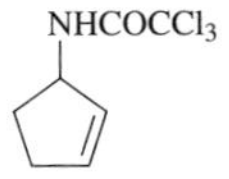 NHCOCCl$_3$	OsO$_4$, TMEDA, CH$_2$Cl$_2$, −78° to rt; then HCl, MeOH	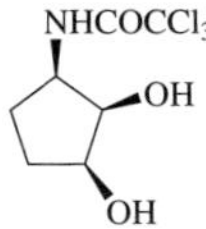 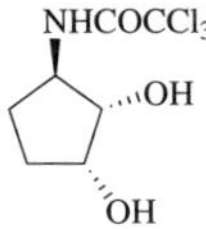 I + II (80), I:II = 24:1	20, 15
	OsO$_4$ (cat), Me$_3$NO•2H$_2$O (1.5 eq), CH$_2$Cl$_2$, rt; then HCl, MeOH	I + II (84), I:II = 7.8:1	19
	OsO$_4$ (0.05 eq), QNO•H$_2$O (1.3 eq), CH$_2$Cl$_2$, rt; then HCl, MeOH	I + II (69), I:II = 13:1	19
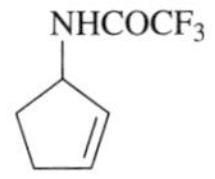 NHCOCF$_3$	OsO$_4$, TMEDA, CH$_2$Cl$_2$, −78° to rt; then aq Na$_2$SO$_3$, reflux	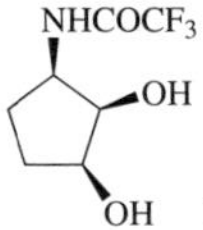 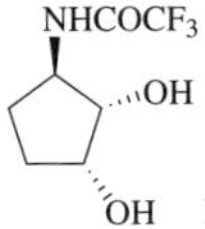 I + II (—), I:II = 25:1	15
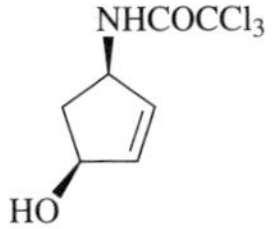 NHCOCCl$_3$ HO	1. OsO$_4$, TMEDA, CH$_2$Cl$_2$, −78° to rt; then HCl, MeOH 2. Ac$_2$O, py	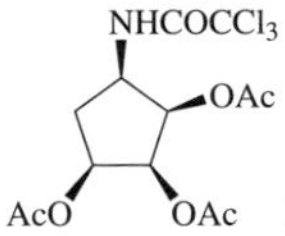 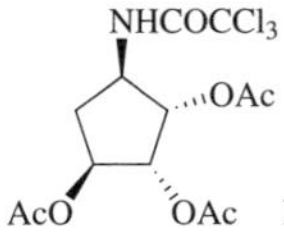 I + II (83), I:II = 17:1	15
 NHBoc	OsO$_4$, TMEDA, CH$_2$Cl$_2$, −78° to rt; then aq Na$_2$SO$_3$, reflux	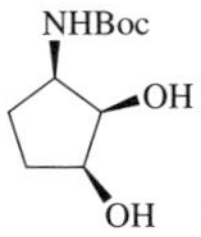 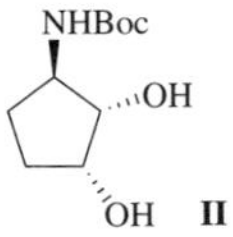 I + II (—), I:II = 9:1	15
C$_6$ 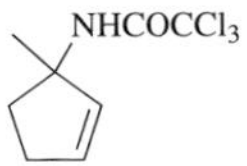 NHCOCCl$_3$	OsO$_4$, TMEDA, CH$_2$Cl$_2$, −78° to rt; then HCl, MeOH	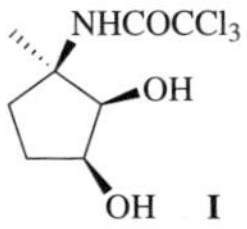 I 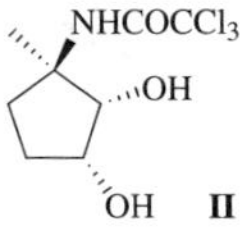 + II (81), I:II = 24:1	15

TABLE 5. DIRECTED DIHYDROXYLATION OF *N*-ALLYLIC AMINE DERIVATIVES (*Continued*)

Substrate	Conditions	Product(s), Yield(s) (%), and dr (*syn:anti*)	Refs.
C6 NHCOCCl3 (methylcyclopentene)	OsO4 (cat), Me3NO•2H2O (1.5 eq), CH2Cl2, rt; then HCl, MeOH	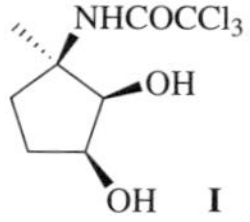 I 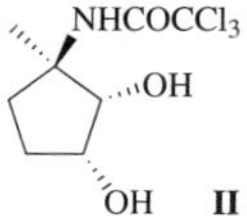 (NHCOCCl3, OH, OH cyclopentane) + II (NHCOCCl3, OH, OH cyclopentane) I + II (81), I:II = 9:1	19
	OsO4 (0.05 eq), QNO•H2O (1.3 eq), CH2Cl2, rt; then HCl, MeOH	I + II (69), I:II = 13:1	19
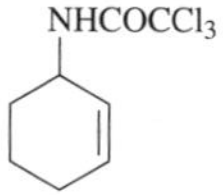 NHCOCCl3 (cyclohexene)	OsO4, TMEDA, CH2Cl2, −78° to rt; then HCl, MeOH	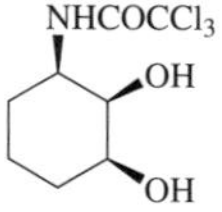 I 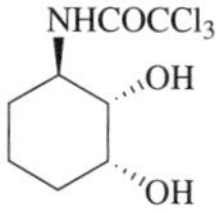 (NHCOCCl3, OH, OH cyclohexane) + II (NHCOCCl3, OH, OH cyclohexane) I + II (99), I:II = 24:1	20, 15
	OsO4 (cat), Me3NO•2H2O (1.5 eq), CH2Cl2, rt; then HCl, MeOH	I + II (93), I:II = 3:1	19
	OsO4 (0.05 eq), QNO•H2O (1.3 eq), CH2Cl2, rt; then HCl, MeOH	I + II (86), I:II = 4.3:1	19
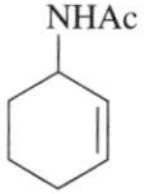 NHAc (cyclohexene)	OsO4, TMEDA, CH2Cl2, −78° to rt; then aq Na2SO3, reflux	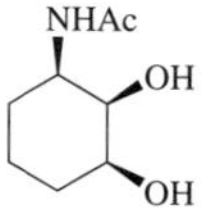 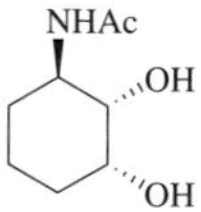 I (NHAc, OH, OH cyclohexane) + II (NHAc, OH, OH cyclohexane) I + II (—), I:II = 1.8:1	15
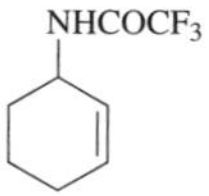 NHCOCF3 (cyclohexene)	OsO4, TMEDA, CH2Cl2, −78° to rt; then aq Na2SO3, reflux	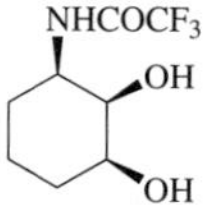 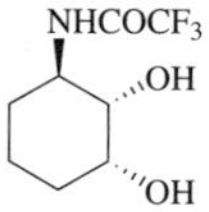 I (NHCOCF3, OH, OH cyclohexane) + II (NHCOCF3, OH, OH cyclohexane) I + II (—), I:II = 24:1	15

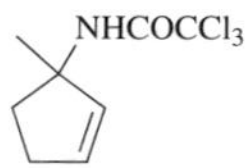

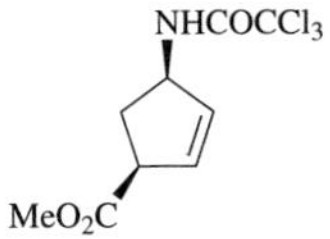

Substrate	Conditions	Products	Ref.

Cyclopentene substrate with NHCOCCl₃ and MeO₂C groups:

- OsO₄, TMEDA, CH₂Cl₂, −78° to rt; then HCl, MeOH → **I + II** (93), **I:II** = 5:1 — 20, 15
- OsO₄ (cat), Me₃NO·2H₂O (1.5 eq), CH₂Cl₂, rt; then HCl, MeOH → **I + II** (81), **I:II** = 1.5:1 — 19
- OsO₄ (0.05 eq), QNO·H₂O (1.3 eq), CH₂Cl₂, rt; then HCl, MeOH → **I + II** (69), **I:II** = 1.6:1 — 19

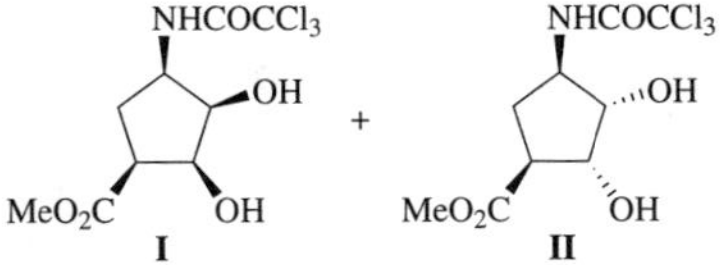

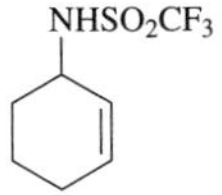

Cyclohexene substrate with NHSO₂CF₃:

- OsO₄, TMEDA, CH₂Cl₂, −78° to rt; then aq Na₂SO₃, reflux → **I + II** (—), **I:II** = 4.2:1 — 15

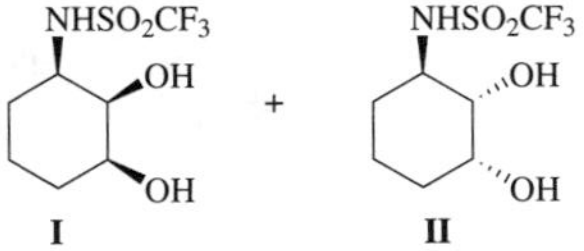

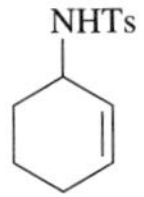

Cyclohexene substrate with NHTs:

- OsO₄, TMEDA, CH₂Cl₂, −78° to rt; then aq Na₂SO₃, reflux → **I + II** (—), **I:II** = 3.5:1 — 15

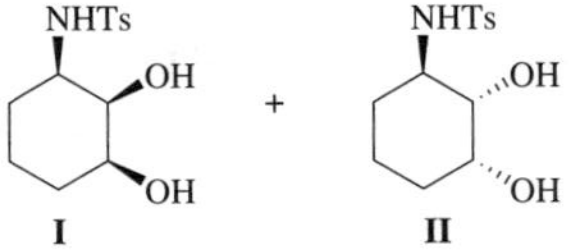

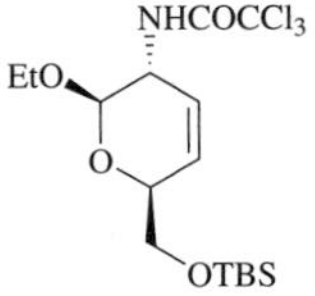

Dihydropyran substrate with NHCOCCl₃, EtO, OTBS:

- OsO₄ (cat), Me₃NO·2H₂O (1.5 eq), CH₂Cl₂, rt; then HCl, MeOH → **I + II** (80), **I:II** = 20:1 — 19
- OsO₄ (0.05 eq), QNO·H₂O (1.3 eq), CH₂Cl₂, rt; then HCl, MeOH → **I + II** (91), **I:II** = 20:1 — 19

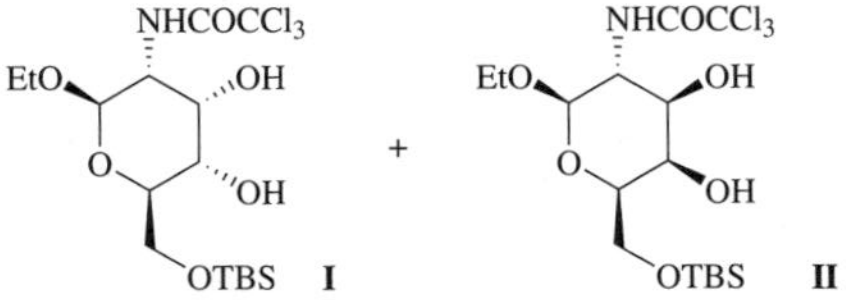

TABLE 5. DIRECTED DIHYDROXYLATION OF *N*-ALLYLIC AMINE DERIVATIVES (*Continued*)

Substrate	Conditions	Product(s), Yield(s) (%), and dr (*syn:anti*)	Refs.
C_6 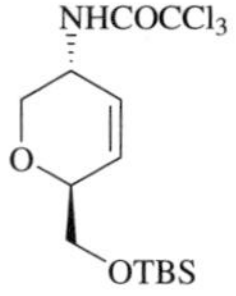 (NHCOCCl$_3$, OTBS allylic pyran)	1. OsO$_4$, TMEDA, CH$_2$Cl$_2$, −78° to rt; then HCl, MeOH 2. Ac$_2$O	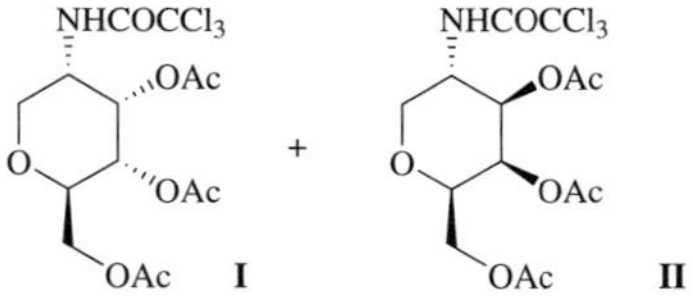 **I** (NHCOCCl$_3$, OAc, OAc, OAc) + **II** (diastereomer) **I + II** (78), **I:II** = 24:1	20
	OsO$_4$ (cat), Me$_3$NO•2H$_2$O (1.5 eq), CH$_2$Cl$_2$, rt; then HCl, MeOH	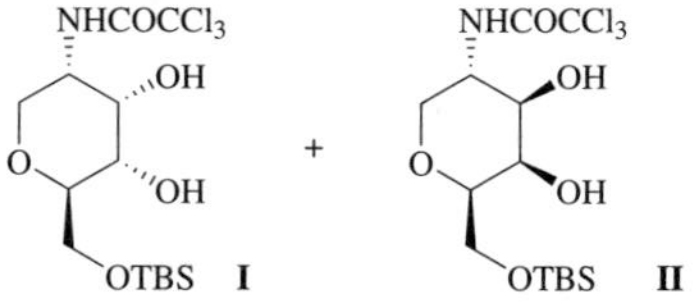 **I** (NHCOCCl$_3$, OH, OH, OTBS) + **II** **I + II** (85), **I:II** = 20:1	19
	OsO$_4$ (0.05 eq), QNO•H$_2$O (1.3 eq), CH$_2$Cl$_2$, rt; then HCl, MeOH	**I + II** (89), **I:II** = 25:1	19
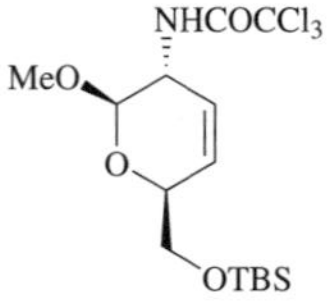 (NHCOCCl$_3$, MeO, OTBS pyran)	1. OsO$_4$, TMEDA, CH$_2$Cl$_2$, −78° to rt; then HCl, MeOH 2. Ac$_2$O	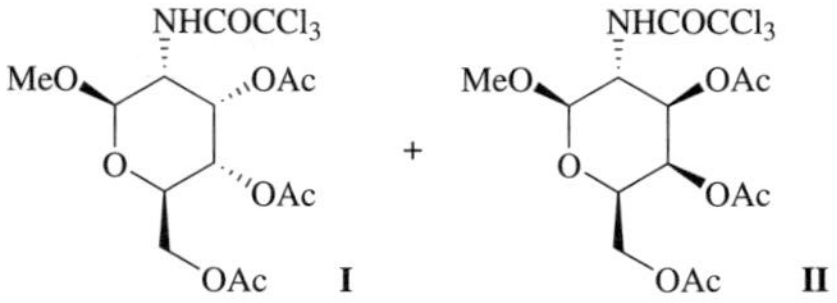 **I** (NHCOCCl$_3$, OAc, OAc, MeO, OAc) + **II** **I + II** (92), **I:II** = 4:1	57
	1. OsO$_4$, quinuclidine (1.1 eq), CH$_2$Cl$_2$, −78° to rt; then HCl, MeOH 2. Ac$_2$O	**I + II** (90), **I:II** = 24:1	57
	OsO$_4$ (cat), QNO•H$_2$O (1.3 eq), CH$_2$Cl$_2$, rt; then HCl, MeOH	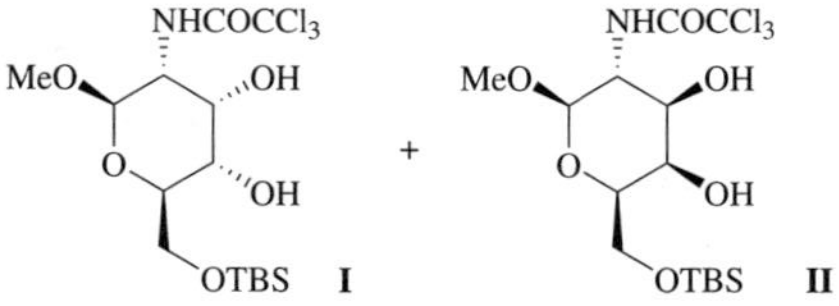 **I** (NHCOCCl$_3$, OH, OH, MeO, OTBS) + **II** **I + II** (80), **I:II** = 20:1	57

Substrate	Conditions	Products	Yield (I:II)	Ref
BnO—[dihydropyran with NHCOCF₃, OTBS]	OsO₄ (cat), Me₃NO·2H₂O (1.5 eq), CH₂Cl₂, rt; then HCl, MeOH	I: BnO—[pyran, NHCOCF₃, OAc, OAc, OAc] + II: [epimer]	**I + II** (87), **I:II** = 14:1	57
BnO—[dihydropyran with NHCOCF₃, OTBS]	1. OsO₄, TMEDA, CH₂Cl₂, −78° to rt; then HCl, MeOH 2. Ac₂O	I: BnO—[pyran, NHCOCF₃, OAc, OAc, OAc] + II	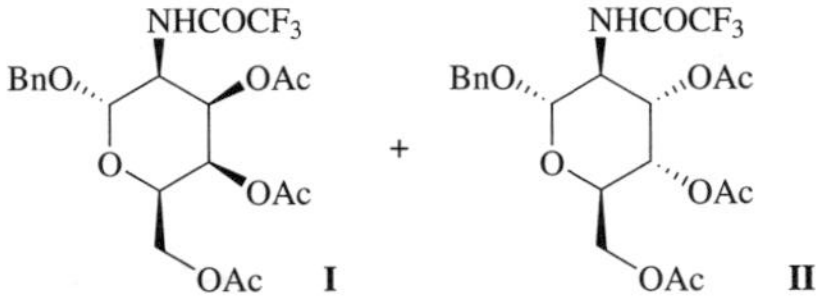 **I + II** (82), **I:II** = 25:1	15
BnO—[dihydropyran with NHCOCCl₃, OTBS]	1. OsO₄, TMEDA, CH₂Cl₂, −78° to rt; then HCl, MeOH 2. Ac₂O	I: BnO—[pyran, NHCOCCl₃, OAc, OAc, OAc] + II	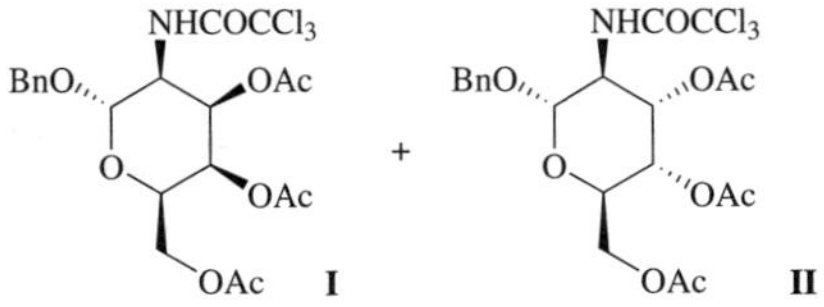 **I + II** (89), **I:II** = 25:1	57
OBn—[dihydropyran with NHCOCCl₃, TBSO]	OsO₄ (cat), Me₃NO·2H₂O (1.5 eq), CH₂Cl₂, rt; then HCl, MeOH	I: [pyran OBn, OH, OH, NHCOCCl₃, TBSO] + II	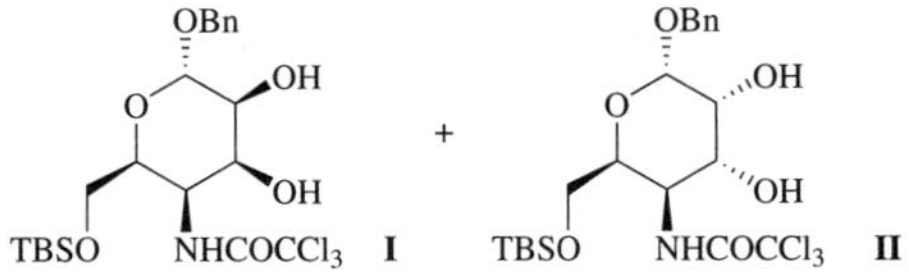 **I + II** (90), **I:II** = 2:1	19
OBn—[dihydropyran with NHCOCCl₃, TBSO]	OsO₄ (0.05 eq), QNO·H₂O (1.3 eq), CH₂Cl₂, rt; then HCl, MeOH	I + II	**I + II** (94), **I:II** = 2.4:1	19
C₇ [cycloheptenyl NHCOCCl₃]	1. OsO₄, TMEDA, CH₂Cl₂, −78° to rt; then HCl, MeOH 2. Ac₂O, py	I: [cycloheptane NHCOCCl₃, OAc, OAc] + II	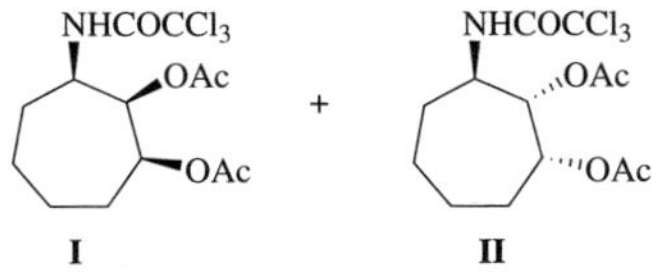 **I + II** (78), **I:II** = 25:1	15

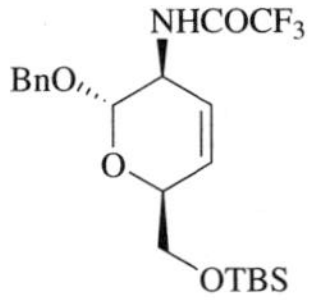

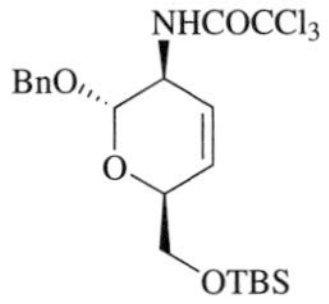

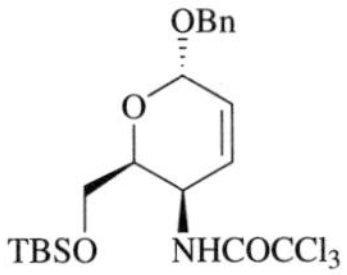

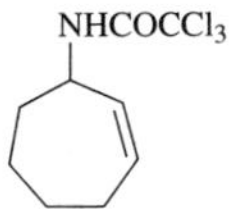

41

TABLE 5. DIRECTED DIHYDROXYLATION OF *N*-ALLYLIC AMINE DERIVATIVES (*Continued*)

Substrate	Conditions	Product(s), Yield(s) (%), and dr (*syn:anti*)	Refs.
C₇ NHCOCCl₃ (cycloheptenyl)	1. OsO₄ (cat), Me₃NO•2H₂O (1.5 eq), CH₂Cl₂, rt; then HCl, MeOH 2. Ac₂O, py	**I** (NHCOCCl₃, OAc, OAc) + **II** (NHCOCCl₃, OAc, OAc); **I + II** (87), **I:II** = 10:1	19
	1. OsO₄ (0.05 eq), QNO•H₂O (1.3 eq), CH₂Cl₂, rt; then HCl, MeOH 2. Ac₂O, py	**I + II** (82), **I:II** = 17:1	19
C₉ NHCOCCl₃ (gem-dimethyl cyclohexenyl)	OsO₄, TMEDA, CH₂Cl₂, –78° to rt; then HCl, MeOH	**I** (NHCOCCl₃, OH, OH) + **II** (NHCOCCl₃, OH, OH); **I + II** (81), **I:II** = 11:1	20, 15
	OsO₄ (cat), Me₃NO•2H₂O (1.5 eq), CH₂Cl₂, rt; then HCl, MeOH	**I + II** (79), **I:II** = 20:1	19
	OsO₄ (0.05 eq), QNO•H₂O (1.3 eq), CH₂Cl₂, rt; then HCl, MeOH		19
C₁₀ NHCOCCl₃ (t-Bu cyclohexenyl)	OsO₄, TMEDA, CH₂Cl₂, –78° to rt; then HCl, MeOH	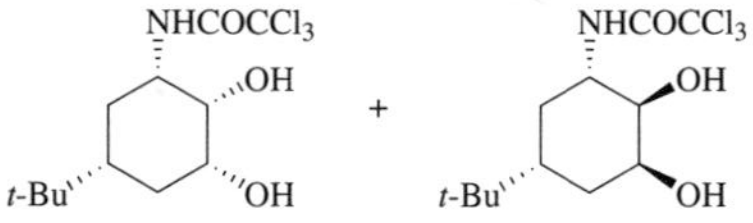 **I** (t-Bu, NHCOCCl₃, OH, OH) + **II** (t-Bu, NHCOCCl₃, OH, OH); **I + II** (96), **I:II** = 24:1	20, 15

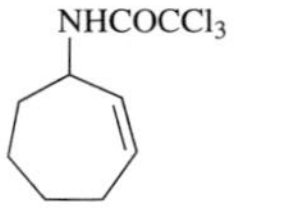

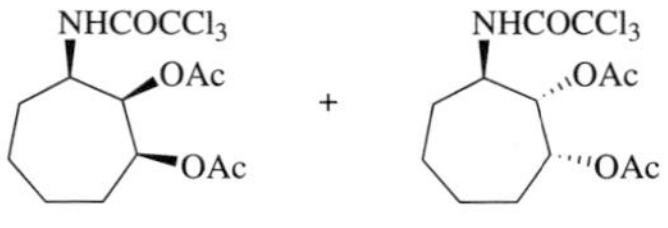

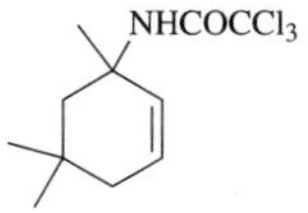

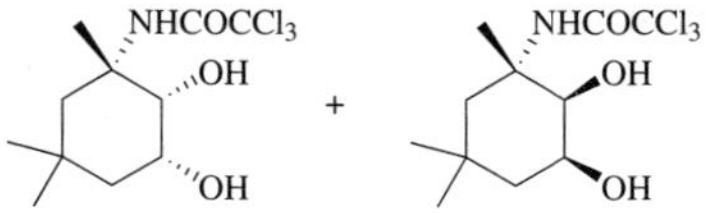

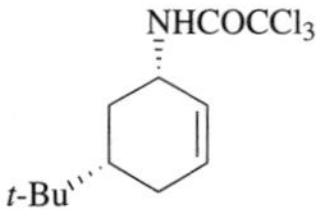

42

Substrate 1: (1S,4R)-4-tert-butyl-N-(cyclohex-2-en-1-yl)-2,2,2-trichloroacetamide

Products **I** and **II**:
- **I**: (1S,2S,3R,5R)-3-(2,2,2-trichloroacetamido)-5-tert-butylcyclohexane-1,2-diol (NHCOCCl₃, OH, OH cis; t-Bu)
- **II**: (1R,2R,3S,5R)-3-(2,2,2-trichloroacetamido)-5-tert-butylcyclohexane-1,2-diol

Conditions	Yield (%) and I:II ratio	Refs.
OsO₄ (cat), Me₃NO·2H₂O (1.5 eq), CH₂Cl₂, rt; then HCl, MeOH	**I + II** (81), **I:II** = 6:1	19
OsO₄ (0.05 eq), QNO·H₂O (1.3 eq), CH₂Cl₂, rt; then HCl, MeOH	**I + II** (77), **I:II** = 13:1	19
OsO₄, TMEDA, CH₂Cl₂, −78° to rt; then HCl, MeOH	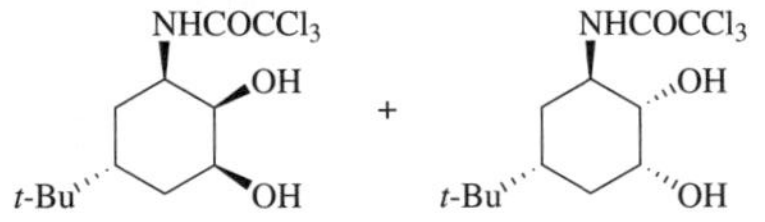 **I + II** (97), **I:II** = 1.7:1	20, 15
OsO₄, quinuclidine, CH₂Cl₂, −78°; then HCl, MeOH	**I + II** (66), **I:II** = 4.9:1	20

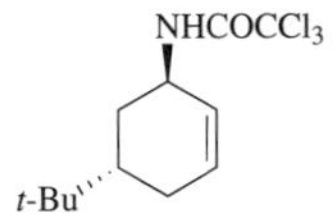

Substrate 2: (R)-N-(2-methyl-5-isopropylcyclohex-2-en-1-yl)-2,2,2-trichloroacetamide

Products **I** and **II** (2-methyl analogues with 5-isopropyl substituent):

Conditions	Yield (%) and I:II ratio	Refs.
OsO₄ (cat), Me₃NO·2H₂O (1.5 eq), CH₂Cl₂, rt; then HCl, MeOH	**I + II** (91), **I:II** = 1.6:1	19
OsO₄ (0.05 eq), QNO·H₂O (1.3 eq), CH₂Cl₂, rt; then HCl, MeOH	**I + II** (88), **I:II** = 2.1:1	19
OsO₄ (cat), Me₃NO·2H₂O (1.5 eq), CH₂Cl₂, rt; then HCl, MeOH	**I + II** (84), **I:II** = 13:1	19
OsO₄ (0.05 eq), QNO·H₂O (1.3 eq), CH₂Cl₂, rt; then HCl, MeOH	**I + II** (95), **I:II** = 20:1	19

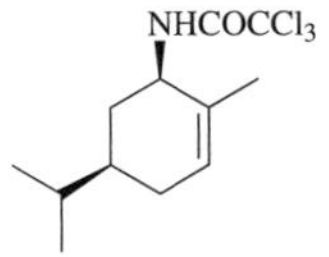

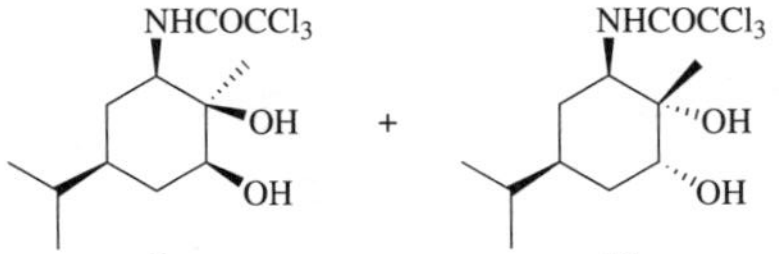

TABLE 5. DIRECTED DIHYDROXYLATION OF *N*-ALLYLIC AMINE DERIVATIVES (*Continued*)

Substrate	Conditions	Product(s), Yield(s) (%), and dr (*syn:anti*)	Refs.

C_{10} substrate with Cl$_3$COCHN group, OsO$_4$, TMEDA, CH$_2$Cl$_2$, –78° to rt; then HCl, MeOH

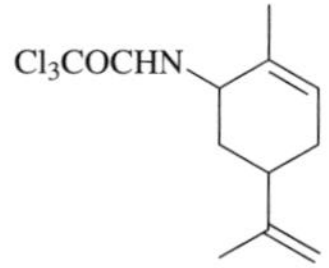

Products I + II (cyclohexane diols with Cl$_3$COCHN) + III + IV (cyclohexene diols with Cl$_3$COCHN)

I + II	I:II	III + IV	III:IV
(71)	25:1	(<3)	—

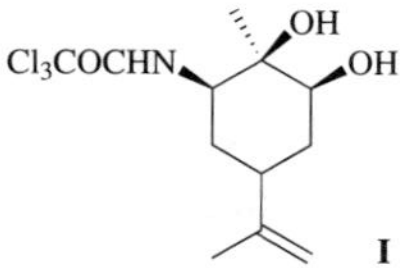

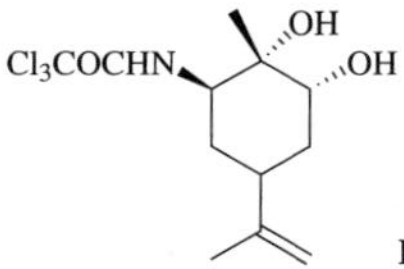

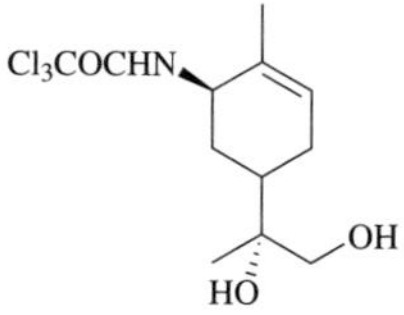

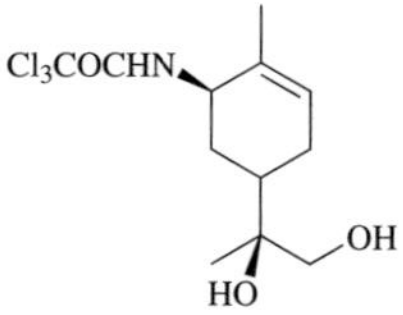

Ref. 58

C_{15} bicyclic NH-containing alkene substrate, OsO$_4$, TMEDA, CH$_2$Cl$_2$, –78° to rt; then HCl, MeOH

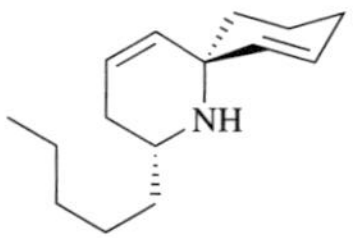

(60), 1 diastereomer

Ref. 59

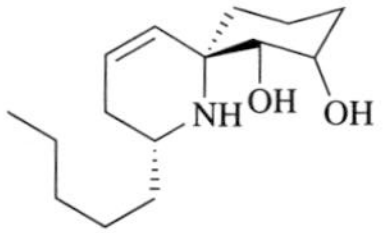

C_{15} NHCOCCl$_3$ geranyl-type substrate, OsO$_4$, TMEDA, CH$_2$Cl$_2$, –78° to rt; then HCl, MeOH

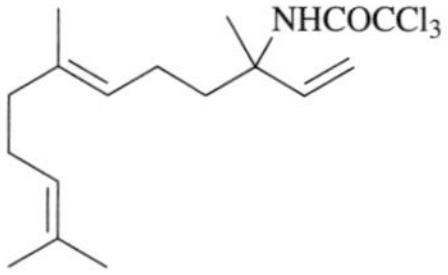

Products I and II (diol with NHCOCCl$_3$)

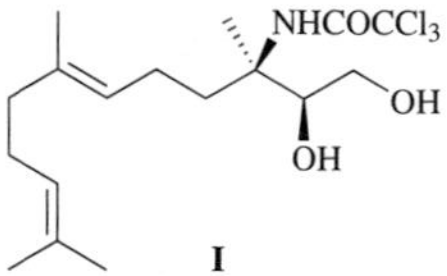

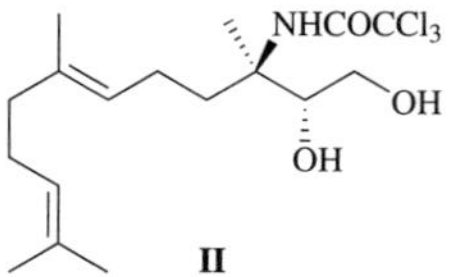

I + II (68), **I:II** = 13:1

Ref. 58

TABLE 6. DIRECTED DIHYDROXYLATION OF *N*-HOMOALLYLIC CYCLIC AMIDES

Substrate	Conditions	Product(s), Yield(s) (%), and dr (*syn:anti*)	Refs.
C$_5$  (cyclopentene with NHCOCCl$_3$)	1. OsO$_4$, TMEDA, CH$_2$Cl$_2$, –78° to rt; then HCl, MeOH 2. Ac$_2$O, py	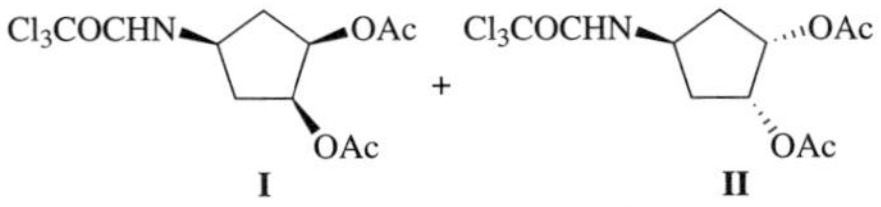 **I** + **II** (94), **I:II** = >25:1	22
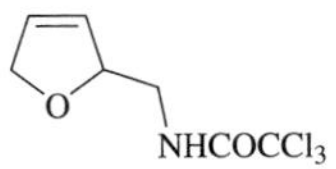 (dihydrofuran with CH$_2$NHCOCCl$_3$)	1. OsO$_4$, TMEDA, CH$_2$Cl$_2$, –78° to rt; then HCl, MeOH 2. Ac$_2$O, py	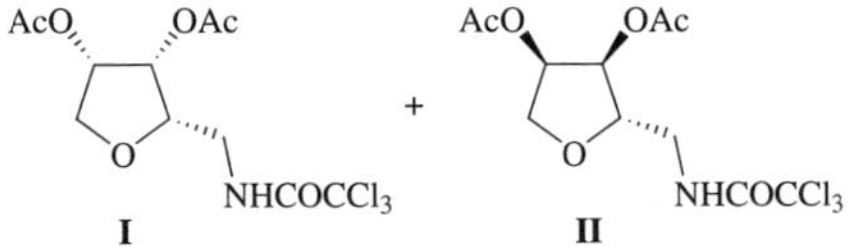 **I** + **II** (72), **I:II** = 2:1	22
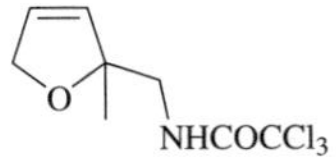 (methyl-substituted dihydrofuran with CH$_2$NHCOCCl$_3$)	1. OsO$_4$, TMEDA, CH$_2$Cl$_2$, –78° to rt; then HCl, MeOH 2. Ac$_2$O, py	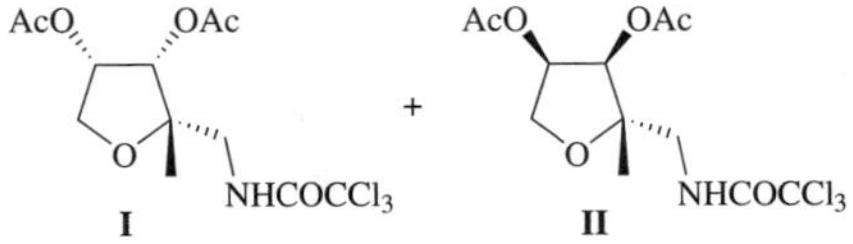 **I** + **II** (88), **I:II** = >25:1	22
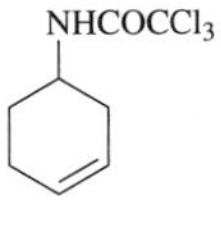 C$_6$ (cyclohexene with NHCOCCl$_3$)	1. OsO$_4$, TMEDA, CH$_2$Cl$_2$, –78° to rt; then HCl, MeOH 2. Ac$_2$O, py	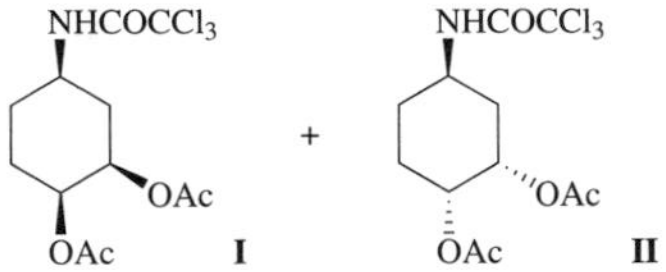 **I** + **II** (92), **I:II** = 1.2:1	22
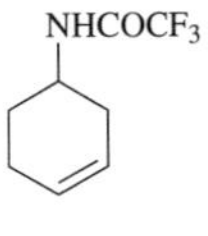 (cyclohexene with NHCOCF$_3$)	1. OsO$_4$, TMEDA, CH$_2$Cl$_2$, –78° to rt; then HCl, MeOH 2. Ac$_2$O, py	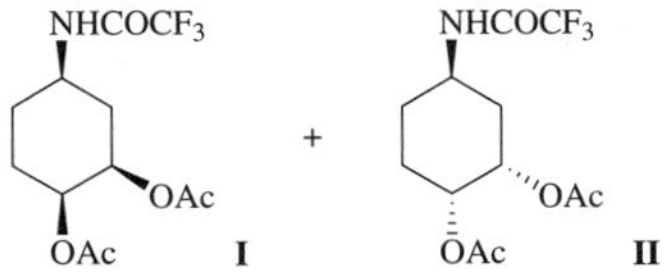 **I** + **II** (90), **I:II** = >20:1	22

TABLE 6. DIRECTED DIHYDROXYLATION OF *N*-HOMOALLYLIC CYCLIC AMIDES (*Continued*)

Substrate	Conditions	Product(s), Yield(s) (%), and dr (*syn:anti*)	Refs.
C_6 TBSO, AcHN, MeS (cyclopentene) 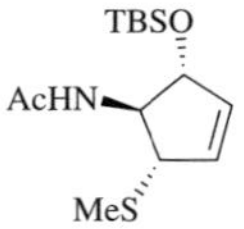	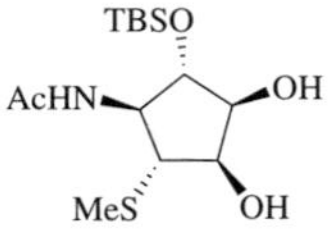 OsO$_4$•py, py, rt; then Na$_2$S$_2$O$_5$, THF/H$_2$O, 65°	TBSO, AcHN, MeS, OH, OH (81), 1 diastereomer	53
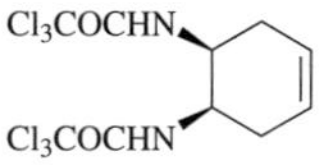 Cl$_3$COCHN, Cl$_3$COCHN (cyclohexene)	1. OsO$_4$, TMEDA, CH$_2$Cl$_2$, –78° to rt; then HCl, MeOH 2. Ac$_2$O, py 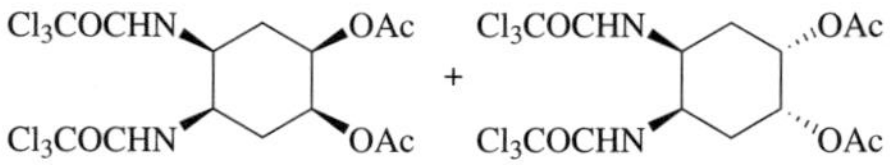	Cl$_3$COCHN—OAc / Cl$_3$COCHN—OAc (I) + Cl$_3$COCHN—OAc / Cl$_3$COCHN—OAc (II) **I + II** (100), **I:II** = 3:1	22
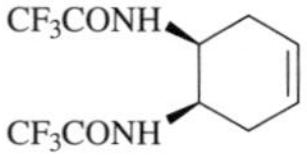 CF$_3$CONH, CF$_3$CONH (cyclohexene)	1. OsO$_4$, TMEDA, CH$_2$Cl$_2$, –78° to rt; then HCl, MeOH 2. Ac$_2$O, py 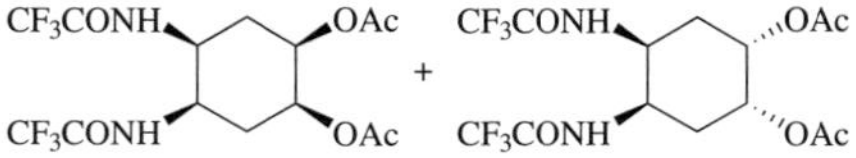	CF$_3$CONH—OAc / CF$_3$CONH—OAc (I) + CF$_3$CONH—OAc / CF$_3$CONH—OAc (II) **I + II** (95), **I:II** = >25:1	22
C_7 NHCOCF$_3$ (methylcyclohexene)	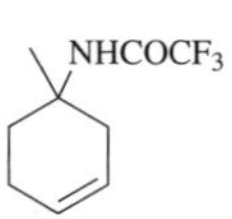 1. OsO$_4$, TMEDA, CH$_2$Cl$_2$, –78° to rt; then HCl, MeOH 2. Ac$_2$O, py	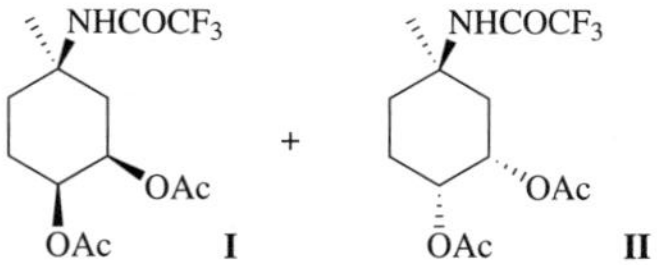 NHCOCF$_3$—OAc, OAc (I) + NHCOCF$_3$—OAc, OAc (II) **I + II** (67), **I:II** = 5:1	22

REFERENCES

[1] Donohoe, T. J. *Synlett* **2002**, 1223.
[2] Criegee, R. *Liebigs Ann.* **1936**, *522*, 75.
[3] Criegee, R. *Angew. Chem.* **1938**, *51*, 519.
[4] Vanrheenen, V.; Kelly, R. C.; Cha, D. Y. *Tetrahedron Lett.* **1976**, *17*, 1973.
[5] Kolb, H. C.; Vannieuwenhze, M. S.; Sharpless, K. B. *Chem. Rev.* **1994**, *94*, 2483.
[6] Cha, J. K.; Christ, W. J.; Kishi, Y. *Tetrahedron Lett.* **1983**, *24*, 3943.
[7] Christ, W. J.; Cha, J. K.; Kishi, Y. *Tetrahedron Lett.* **1983**, *24*, 3947.
[8] Cha, J. K.; Christ, W. J.; Kishi, Y. *Tetrahedron* **1984**, *40*, 2247.
[9] Cha, J. K.; Kim, N. S. *Chem. Rev.* **1995**, *95*, 1761.
[10] See ref. 7
[11] Jacobsen, E. N.; Marko, I.; Mungall, W. S.; Schroeder, G.; Sharpless, K. B. *J. Am. Chem. Soc.* **1988**, *110*, 1968.
[12] Woodward, R. B.; Brutcher, F. V. *J. Am. Chem. Soc.* **1958**, *80*, 209.
[13] Corey, E. J.; Sarshar, S.; Azimioara, M. D.; Newbold, R. C.; Noe, M. C. *J. Am. Chem. Soc.* **1996**, *118*, 7851.
[14] See ref. 13
[15] Donohoe, T. J.; Blades, K.; Moore, P. R.; Waring, M. J.; Winter, J. J. G.; Helliwell, M.; Newcombe, N. J.; Stemp, G. *J. Org. Chem.* **2002**, *67*, 7946.
[16] Andersson, P. G.; Sharpless, K. B. *J. Am. Chem. Soc.* **1993**, *115*, 7047.
[17] Brennan, N. K.; Guo, X.; Paquette, L. A. *J. Org. Chem.* **2005**, *70*, 732.
[18] Fürstner, A.; Wuchrer, M. *Chem.—Eur. J.* **2006**, *12*, 76.
[19] Blades, K.; Donohoe, T. J.; Winter, J. J. G.; Stemp, G. *Tetrahedron Lett.* **2000**, *41*, 4701.
[20] Donohoe, T. J.; Blades, K.; Helliwell, M.; Moore, P. R.; Winter, J. J. G.; Stemp, G. *J. Org. Chem.* **1999**, *64*, 2980.
[21] Donohoe, T. J.; Moore, P. R.; Waring, M. J. *Tetrahedron Lett.* **1997**, *38*, 5027.
[22] Donohoe, T. J.; Mitchell, L.; Waring, M. J.; Helliwell, M.; Bell, A.; Newcombe, N. J. *Org. Biomol. Chem.* **2003**, *1*, 2173.
[23] Berti, G. *Topics in Stereochemistry* **1973**, *7*, 93.
[24] Hoveyda, A.; Evans, D. A.; Fu, G. C. *Chem. Rev.* **1993**, *93*, 1307.
[25] Donohoe, T. J.; Mitchell, L.; Waring, M. J.; Helliwell, M.; Bell, A.; Newcombe, N. J. *Tetrahedron Lett.* **2001**, *42*, 8951.
[26] Ell, A. H.; Closson, A.; Adolfsson, H.; Backvall, J. E. *Adv. Synth. Catal.* **2003**, *345*, 1012.
[27] Xu, D.; Park, C. Y.; Sharpless, K. B. *Tetrahedron Lett.* **1994**, *35*, 2495.
[28] Johnson, C. R.; Barbachyn, M. R. *J. Am. Chem. Soc.* **1984**, *106*, 2459.
[29] Trost, B. M.; Kuo, G. H.; Benneche, T. *J. Am. Chem. Soc.* **1988**, *110*, 621.
[30] Poli, G. *Tetrahedron Lett.* **1989**, *30*, 7385.
[31] Kallatsa, O. A.; Koskinen, A. M. P. *Tetrahedron Lett.* **1997**, *38*, 8895.
[32] Plietker, B.; Niggemann, M. *Org. Lett.* **2003**, *5*, 3353.
[33] Vijaykumar, D.; Mao, W.; Kirschbaum, K. S.; Katzenellenbogen, J. A. *J. Org. Chem.* **2002**, *67*, 4904.
[34] Paquette, L. A.; Lo, H. Y. *J. Org. Chem.* **2003**, *68*, 2282.
[35] Stoianova, D. S.; Whitehead, A.; Hanson, P. R. *J. Org. Chem.* **2005**, *70*, 5880.
[36] Donohoe, T. J.; Johnson, P. D.; Pye, R. J.; Keenan, M. *Org. Lett.* **2005**, *7*, 1275.
[37] Donohoe, T. J.; Fisher, J. W.; Edwards, P. J. *Org. Lett.* **2004**, *6*, 465.
[38] Bataille, C.; Begin, G.; Guillam, A.; Lemiegre, L.; Lys, C.; Maddaluno, J.; Toupet, L. *J. Org. Chem.* **2002**, *67*, 8054.
[39] Hodgson, R.; Majid, T.; Nelson, A. *J. Chem. Soc., Perkin Trans. 1* **2002**, 1631.
[40] Donohoe, T. J.; Blades, K.; Helliwell, M.; Waring, M. J.; Newcombe, N. J. *Tetrahedron Lett.* **1998**, *39*, 8755.
[41] Haukaas, M. H.; O'Doherty, G. A. *Org. Lett.* **2001**, *3*, 3899.
[42] Harris, J. M.; Keranen, M. D.; Nguyen, H.; Young, V. G.; O'Doherty, G. A. *Carbohydr. Res.* **2000**, *328*, 17.
[43] Hodgson, R.; Majid, T.; Nelson, A. *J. Chem. Soc., Perkin Trans. 1* **2002**, 1444.

[44] Cook, M. J.; Fletcher, M. J. E.; Gray, D.; Lovell, P. J.; Gallagher, T. *Tetrahedron* **2004**, *60*, 5085.
[45] Kennedy, A.; Nelson, A.; Perry, A. *Chem. Commun.* **2005**, 1646.
[46] Kennedy, A.; Nelson, A.; Perry, A. *Beilstein J. Org. Chem.* **2005**, *1*.
[47] Harding, M.; Nelson, A. *Chem. Commun.* **2001**, 695.
[48] Harding, M.; Hodgson, R.; Majid, T.; McDowall, K. J.; Nelson, A. *Org. Biomol. Chem.* **2003**, *1*, 338.
[49] Anderson, E. A.; Alexanian, E. J.; Sorensen, E. J. *Angew. Chem., Int. Ed.* **2004**, *43*, 1998.
[50] Donohoe, T. J.; Newcombe, N. J.; Waring, M. J. *Tetrahedron Lett.* **1999**, *40*, 6881.
[51] Hodgson, D. M.; Bray, C. D.; Kindon, N. D. *Org. Lett.* **2005**, *7*, 2305.
[52] Kim, Y.; Fuchs, P. L. *Org. Lett.* **2007**, *9*, 2445.
[53] Cho, S. J.; Ling, R.; Kim, A.; Mariano, P. S. *J. Org. Chem.* **2000**, *65*, 1574.
[54] Takahata, H.; Banba, Y.; Ouchi, H.; Nemoto, H.; Kato, A.; Adachi, I. *J. Org. Chem.* **2003**, *68*, 3603.
[55] Landais, Y.; Mahieux, C.; Schenk, K.; Surange, S. S. *J. Org. Chem.* **2003**, *68*, 2779.
[56] Bennett, N. J.; Prodger, J. C.; Pattenden, G. *Tetrahedron* **2007**, *63*, 6216.
[57] Donohoe, T. J.; Blades, K.; Helliwell, M. *Chem. Commun.* **1999**, 1733.
[58] Donohoe, T. J.; Winter, J. J. G.; Helliwell, M.; Stemp, G. *Tetrahedron Lett.* **2001**, *42*, 971.
[59] Wybrow, R. A. J.; Edwards, A. S.; Stevenson, N. G.; Adams, H.; Johnstone, C.; Harrity, J. P. A. *Tetrahedron* **2004**, *60*, 8869.

CHAPTER 2

TRANSITION-METAL-CATALYZED α-ARYLATION OF ENOLATES

Damien Prim, Sylvain Marque, and Anne Gaucher

Université de Versailles-Saint-Quentin-en-Yvelines, Institut Lavoisier de Versailles UMR CNRS 8180, 45, Avenue des Etats-Unis, F-78035 Versailles, France

Jean-Marc Campagne

Ecole Nationale Supérieure de Chimie de Montpellier, 8, Rue de l'Ecole Normale, 34296, Montpellier Cédex, France

CONTENTS

	Page
Acknowledgments	50
Introduction	50
Mechanism and Stereochemistry	53
Scope and Limitations	59
Palladium-Catalyzed α-Arylation	59
α-Arylation of Ketones	59
α-Arylation of Aldehydes	64
α-Arylation of Amides	64
α-Arylation of Esters	67
α-Arylation of Amino Acids and Derivatives	68
α-Arylation of Nitriles	70
α-Arylation of Active Methylene Compounds	71
Vinylogy, Phenylogy. Site Selectivity in γ/ω-Arylation	73
α-Arylation via α-Stannylmethyl Carbonyl Compounds	74
α-Arylation of Silyl Enol Ethers and Ketene Acetals	75
Copper-Catalyzed α-Arylation	77
Nickel-Catalyzed α-Arylation	78
Applications to Synthesis	79
Comparison with other Methods	81
Palladium-Catalyzed α-Arylation via α-Halomethyl Carbonyl Compounds	81
Stoichiometric Transition-Metal- and Metal-Promoted α-Arylation	81
Copper-Mediated α-Arylation	81
Magnesium- and Titanium-Mediated α-Arylation	83

---

prim@chimie.uvsq.fr
Organic Reactions, Vol. 76, Edited by Scott E. Denmark et al.
© 2012 Organic Reactions, Inc. Published 2012 by John Wiley & Sons, Inc.

Bismuth-Mediated α-Arylation 83
Lead-Mediated α-Arylation 84
Miscellaneous α-Arylation Methods 85
EXPERIMENTAL CONDITIONS 87
EXPERIMENTAL PROCEDURES 87
   2-(2-Bromophenyl)cyclohexanone [Palladium-Catalyzed α-Arylation of a Ketone]   87
   1-Phenyl-2-arylpropan-1-one [α-Arylation of a Ketone Using an N-Heterocyclic
      Carbene-Based Catalytic System] 88
   1-Vinyl-3-*tert*-butyl-1*H*-isochromene [Palladium-Catalyzed α-Arylation of a Ketone]   88
   tert-Butyl 2-Phenyl-2,3-dihydro-1*H*-isoindole-1-carboxylate [Palladium-Catalyzed
      Intramolecular α-Arylation of an α-Amino Acid Ester] 89
   1-Benzyloxindole [Preparation of an Oxindole by Palladium-Catalyzed Intramolecular
      α-Arylation of an Amide] 90
   (1-Nitroethyl)benzene [Palladium-Catalyzed α-Arylation of Nitroethane] . . 90
   Diethyl 2-Phenylmalonate [Palladium-Catalyzed α-Arylation of a Malonate with an
      Aryl Bromide] 91
   Ethyl 2-(2-Methylphenyl)cyanoacetate [Palladium-Catalyzed α-Arylation of Ethyl
      Cyanoacetate] 91
   (R)-(+)-2-Methyl-2-(4-anisyl)-5-(N-methylanilinomethylene)cyclopentanone
      [Palladium-Catalyzed Asymmetric α-Arylation of a Ketone Enolate] . . . 92
   (−)-3-(3-Methoxyphenyl)-3-methyldihydrofuran-2-one [Nickel-Catalyzed Asymmetric
      α-Arylation of a Lactone] 93
   (S)-Ethyl 2-Methyl-3-oxo-2-[2-(2,2,2-trifluoroacetamido)phenyl]butanoate
      [Copper-Catalyzed Enantioselective α-Arylation of an Acetoacetate] . . . 94
TABULAR SURVEY 94
   Chart 1. Structures of Ligands Used in Tables 96
   Chart 2. Structures of Palladacycles and Palladium Complexes Used
      in Tables 98
   Chart 3. Structures of Polymer-Supported Catalysts Used in Tables . . . 99
   Table 1. Arylation of Aldehydes, Ketones, and Enol Ethers 100
   Table 2. Arylation of Esters, Amides, Lactones, Lactams, Nitriles, Ketene Acetals, and
      Preformed Enolates 168
   Table 3. Arylation of 1,3-Dicarbonyls and Cyanoacetates 241
REFERENCES 276

ACKNOWLEDGMENTS

The authors are greatly indebted to Professors Dale L. Boger, Stuart W. McCombie, Peter Wipf, Engelbert Ciganek, and Scott E. Denmark for helpful guidance and insightful comments.

INTRODUCTION

The last three decades have witnessed increasing efforts to develop highly efficient and selective tools for the catalyzed formation of carbon−carbon and carbon−heteroatom bonds. Among the latter, outstanding results have been obtained in the field of soft, non-organometallic nucleophiles.[1] The formation of carbon−oxygen, −nitrogen, −sulfur, −phosphorus, −boron, or −silicon bonds has become as widely used as the well-known Suzuki, Corriu−Kumada−Tamao,

Stille, Negishi, Hiyama, and Sonogashira cross-coupling reactions. One of the major challenges is the α-arylation of soft carbon nucleophiles such as stabilized carbon enolates and related functional groups. Although α-arylated carboxylic acids (and keto derivatives) are prevalent in natural products (for example, lucuminic acid,[2] welwistatin,[3] and chloropeptin and vancomycin[4]) and are important building blocks in a number of drugs (such as the anti-inflammatory agents Naproxen and Ibuprofen,[5] the anesthetic Scopolamine,[6] and *p*-malonylphenylalanine (Pmf,[7] a potent phosphotyrosine mimetic), catalytic α-arylation of stabilized carbon enolates had, until recently, been described only rarely (Fig. 1).

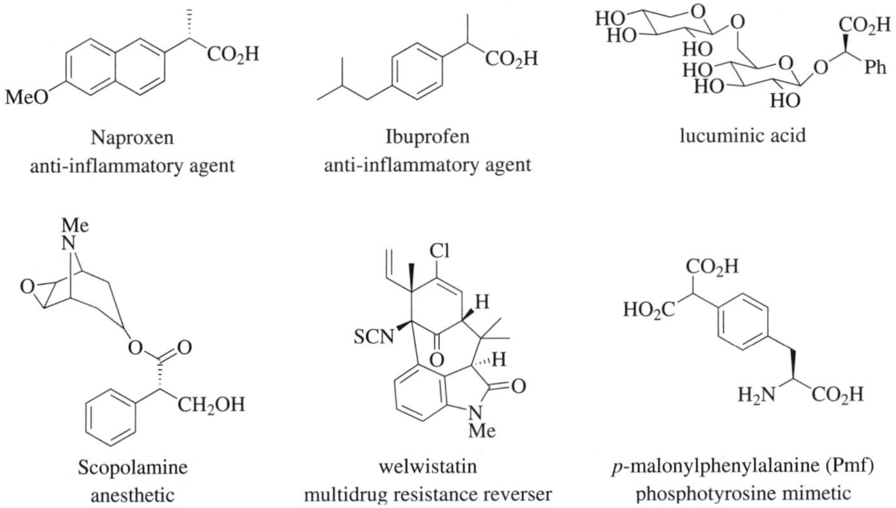

Figure 1. Natural products and drugs bearing α-arylated carboxylic acid derivatives.

The first examples of both inter- and intramolecular arylation of ketone enolates were reported in 1973 by Semmelhack using a nickel catalyst.[8] The reaction of the oxidative addition product of bromobenzene and $Ni(PPh_3)_4$ with the lithium enolate of acetophenone afforded the arylated ketone in modest yield (Eq. 1). An intramolecular version was also reported as the key step of the total synthesis of cephalotaxinine. The natural product was obtained in only 30% yield and required the use of a stoichiometric amount of $Ni(cod)_2$.[8]

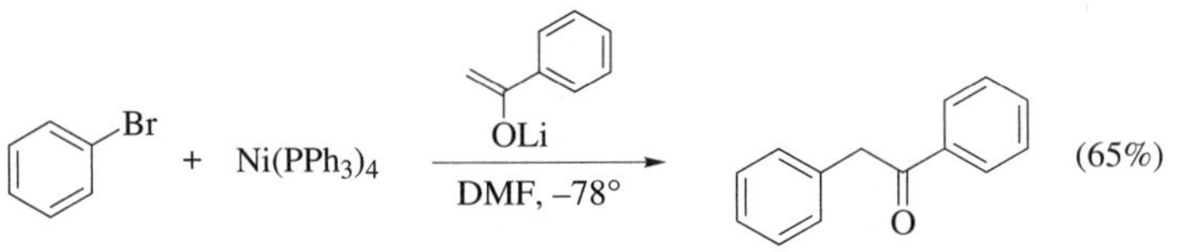

This pioneering work was followed by several examples of transition-metal-mediated α-arylations of ketones or their derivatives including the catalyzed arylation of vicinal bromotrimethylsilylalkenes[9] and acetonyltributyltin (Eq. 2),[10]

as well as the arylation of 2-trimethylsilyloxyallyl halides with stoichiometric reagents.[11]

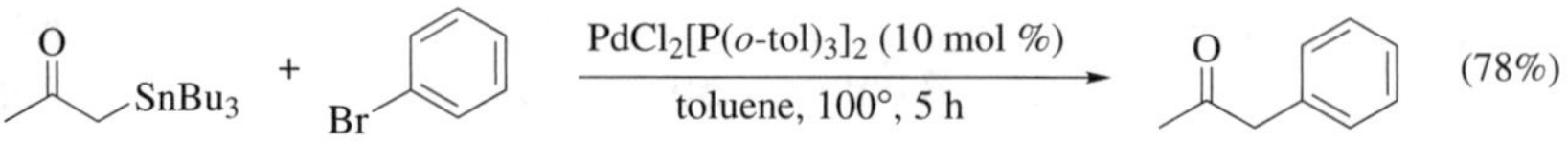

$$\underset{\text{SnBu}_3}{\overset{\text{O}}{\bigwedge}} + \underset{\text{Br}}{\bigcirc} \xrightarrow[\text{toluene, }100°, 5\text{ h}]{\text{PdCl}_2[\text{P}(o\text{-tol})_3]_2 \text{ (10 mol \%)}} \underset{}{\overset{\text{O}}{\bigwedge}}\bigcirc \quad (78\%)$$

(Eq. 2)

One variant of these reactions involves the site selective arylation of silyl enol ethers of methyl ketones with aryl halides under palladium catalysis in the presence of trialkyltin fluorides (Eq. 3).[12,13] The latter are assumed to generate tin enolates in situ via silyl/stannyl exchange followed by arylation with the aryl halide.

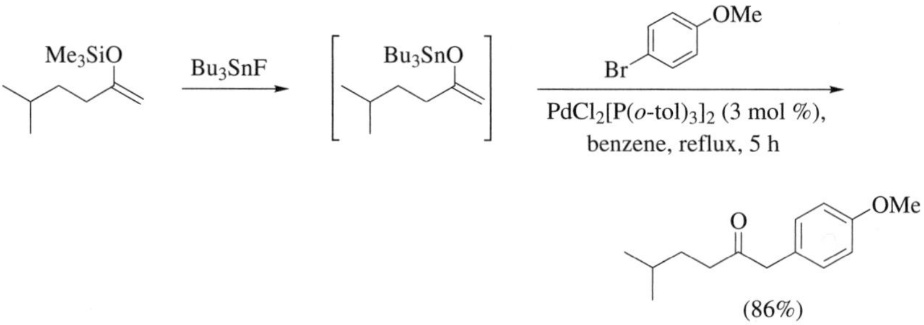

(Eq. 3)

Later, the use of malonate-type nucleophiles in the palladium-catalyzed α-arylation of ketones in inter- and intramolecular modes was reported (Eq. 4).[14,15]

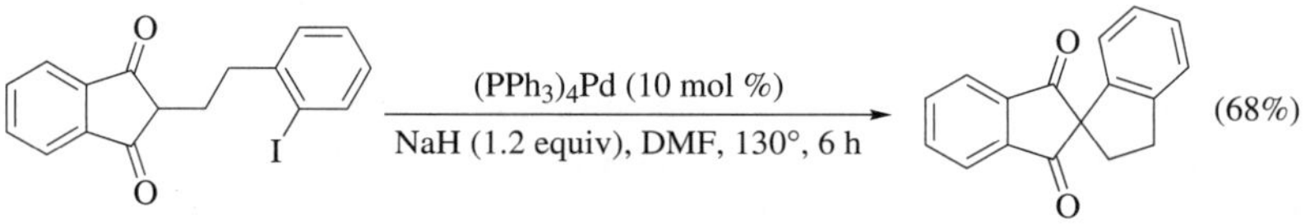

(Eq. 4)

In 1997, five seminal papers dealing with inter- and intramolecular palladium-catalyzed arylation of ketones appeared.[16–20] This method could then be extended to ketones and esters under milder conditions and with a wider scope (Eq. 5).

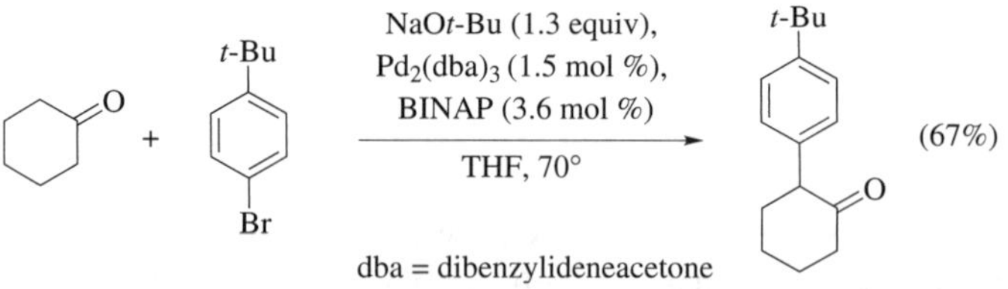

(Eq. 5)

dba = dibenzylideneacetone

The more recent developments of this method concern not only the use of a large number of related nucleophiles such as amides, lactones and lactams, malonates, cyanoacetates, aldehydes, α-amino esters, nitriles, sulfones, and nitroalkanes, but also activated benzylic and vinylogous γ-arylations. Current efforts are mainly devoted to multiple arylation sequences, intramolecular α-arylations, and enantioselective α-arylations. Although palladium-catalyzed reactions have received a great deal of attention, several other efficient, competitive, and enantioselective nickel-, copper-, or ruthenium-based catalytic systems have recently emerged.

In addition, the transition-metal-catalyzed introduction of aromatic groups at the position α to enolizable moieties has found applications in the total synthesis of natural products, demonstrating the utility of this general synthetic method. The aim of this chapter is to present an up-to-date overview of the transition-metal-catalyzed α-arylation of enolates and their derivatives. Because palladium is the transition metal predominantly employed, this chapter focuses on palladium-assisted synthetic transformations. However, the most relevant α-arylations involving other catalytic systems are also detailed. Several accounts have recently appeared covering some aspects of palladium-catalyzed α-arylations.[21–23] The literature up to the beginning of 2008 is covered in the Tabular Survey.

MECHANISM AND STEREOCHEMISTRY

Palladium is the most widely used transition metal in catalyzed α-arylations. Relevant mechanistic studies are almost exclusively carried out with this metal. The main features of reactions mediated by other transition metals are discussed in the corresponding sections.

The general mechanism of the palladium-catalyzed arylation of enolates is rather classical involving the well-known oxidative addition, transmetalation, and reductive elimination sequence (Scheme 1).[23] Although the general mechanism is well accepted, a number of questions about particular aspects of the catalytic cycle are still under discussion.

The first issue concerns the number of phosphines coordinated to the palladium atom. Initially, bidentate phosphines such as dppf (**1**) or BINAP (**2**) (Fig. 2) were used to form strongly chelated intermediates to prevent β-elimination.[16] However, during further synthetic and mechanistic studies, a monochelated intermediate involving the DtBPF ligand (**3**) was observed by ^{31}P NMR spectroscopy.[23] This observation led to the introduction of very efficient mono- and bidentate ligands **4–11** (Fig. 2). Such bulky, electron-rich monodentate ligands are involved in the formation of monoligated (12 electrons) [PdL] species that can undergo rapid oxidative additions with ArX, including aryl chlorides.[24,25] Ligandless reactions are also possible in some non-demanding reactions.[18]

The second issue concerns the nature of the palladium species undergoing reductive elimination in the catalytic cycle. In general, main group metal and early transition metals favor the *O*-bound enolate whereas late transition metals favor

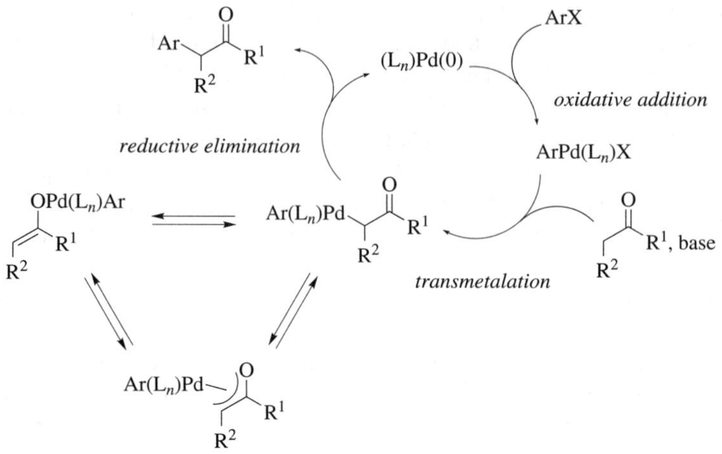

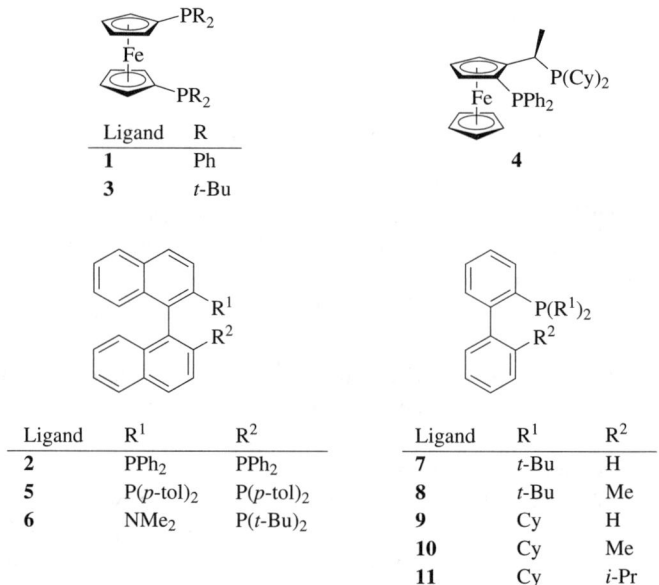

Scheme 1

Ligand	R
1	Ph
3	*t*-Bu

4

Ligand	R^1	R^2
2	PPh$_2$	PPh$_2$
5	P(*p*-tol)$_2$	P(*p*-tol)$_2$
6	NMe$_2$	P(*t*-Bu)$_2$

Ligand	R^1	R^2
7	*t*-Bu	H
8	*t*-Bu	Me
9	Cy	H
10	Cy	Me
11	Cy	*i*-Pr

Figure 2. Phosphine ligands for palladium-catalyzed α-arylations.

the *C*-bound form.[23] Indeed, with the exception of hindered α,α′-disubstituted ketones and complexes where the enolate oxygen is located *trans* with respect to the aryl group, the *C*-bound isomers are favored.[23] Competition experiments involving isolated palladium enolates showed that their relative stability is controlled by the number of substituents on the α-position of the keto group (Fig. 3).[23]

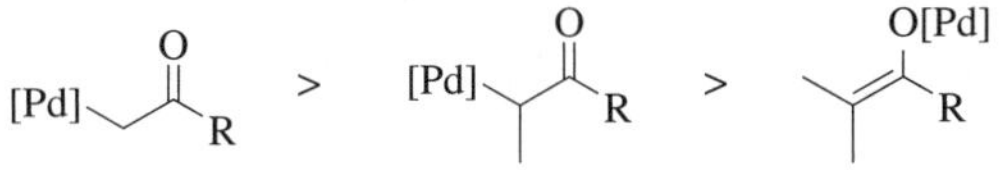

Figure 3. Relative stability of palladium enolates.

The reductive elimination step was investigated for various isolated *C*- and *O*-bound palladium enolates.[26] By comparison of the elimination rates, alternative mechanisms (such as isomerization to the enol or migratory insertion could be ruled out and the classical concerted reductive elimination was confirmed (Scheme 2).[23,27]

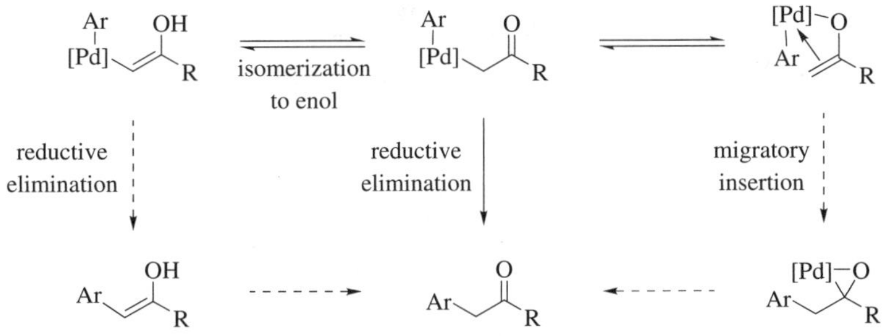

Scheme 2

The influence of electronic effects was also investigated. As illustrated in Fig. 4, acyl substituents (aryl, alkoxy, amino) on the palladium enolates have little influence on the reductive elimination rates.[28] In these cases, the observed differences in reactivity in the palladium-catalyzed arylation of ketones, esters, and amides must result from the rate of formation of the α-palladated species. On the other hand, striking differences are observed in the reductive eliminations of methyl, ketomethyl, cyanomethyl, and malonyl arylpalladium complexes.[23,29] Increasing the electron-withdrawing ability of the substituent lowers the reductive elimination rate. This effect is particularly impressive in malonate-type complexes where no reductive elimination is observed. In such reactions, this effect must be counteracted by ligands that are efficient in promoting reductive elimination, such as bulky, electron-rich monophosphines, as exemplified by the DPPBz ligand (**12**) (Fig. 4).[23,29]

Because enantiomerically enriched α-substituted carbonyl compounds are important structural features in drugs and natural products, the creation of tertiary and quaternary stereogenic centers α to carbonyl groups through the asymmetric arylation of enolates is of significant interest. Initial attempts employed BINAP-type ligands for this transformation.[30] Modest enantioselectivities are obtained using 10 to 20 mol % of palladium precatalyst, and 12 to 24 mol % of ligands **2** or **13** (Eq. 6).

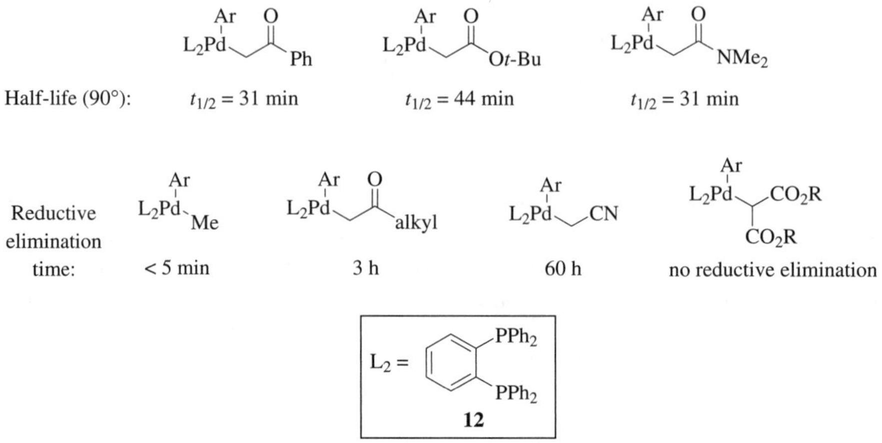

	Ar O	Ar O	Ar O
	L$_2$Pd—Ph	L$_2$Pd—O*t*-Bu	L$_2$Pd—NMe$_2$
Half-life (90°):	$t_{1/2}$ = 31 min	$t_{1/2}$ = 44 min	$t_{1/2}$ = 31 min

	Ar	Ar O	Ar	Ar
Reductive elimination time:	L$_2$Pd—Me	L$_2$Pd—alkyl	L$_2$Pd—CN	L$_2$Pd(CO$_2$R)(CO$_2$R)
	< 5 min	3 h	60 h	no reductive elimination

L$_2$ = 1,2-bis(diphenylphosphino)benzene (**12**)

Figure 4. Electronic effects on the rate of reductive elimination.

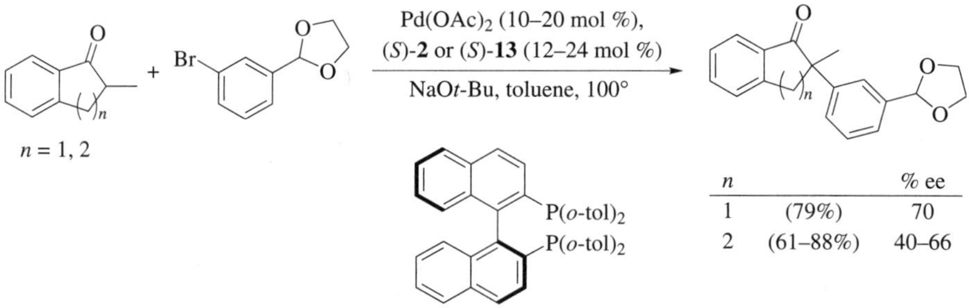

Pd(OAc)$_2$ (10–20 mol %), (*S*)-**2** or (*S*)-**13** (12–24 mol %), NaO*t*-Bu, toluene, 100°

n = 1, 2

(*S*)-**13**

n		% ee
1	(79%)	70
2	(61–88%)	40–66

(Eq. 6)

These transformations require the use of an excess of the aryl halide to ensure complete conversions because of the competitive homocoupling of the aryl bromide to afford biphenyl derivatives.[30a] In addition, these arylations are restricted to the formation of quaternary stereocenters. The acidity of the remaining proton in the tertiary arylation products leads to lower enantioselectivities under basic conditions.[30a] The arylation reactions proceed with high yields and high to excellent enantioselectivities in the case of blocked α-methylcyclopentanones (Eq. 7).[31] In such substrates, the use of a methylene group that can be easily removed can prevent arylation at the less substituted carbon of α-methylcyclopentanones. Toluene and sodium *tert*-butoxide is the most suitable combination of solvent and base in these transformations. Whereas arylations using *m*- and *p*-substituted aryl bromides generally proceed with high enantioselectivities (80–94% enantiomeric excess), the *o*-substituted derivatives react with low yields and enantioselectivities. Similar enantiomeric excesses (88–93%) are observed in couplings of ketones bearing methyl, *n*-propyl, and *n*-pentyl α-substituents, indicating that enantiomeric excesses are insensitive to the size of the α-alkyl moiety.[30a] Although BINAP is often used, reactions using palladium catalysts based on dialkyl monophosphines such as **14** can be employed under milder conditions (room temperature) and are more efficient in terms of substrate range and precatalyst loadings. As exemplified in Eq. 7, best results are obtained using monophosphine ligands. Although attempts to rationalize the sense of asymmetric induction are still inconclusive, it should be noted that reversed stereoselectivity is observed using diphosphine-type ligands such as BINAP.[30a] These observations suggest an asymmetric induction mechanism that is significantly different for monophosphine and diphosphine ligands, respectively, depending on the chelation mode in the enantiodetermining step.

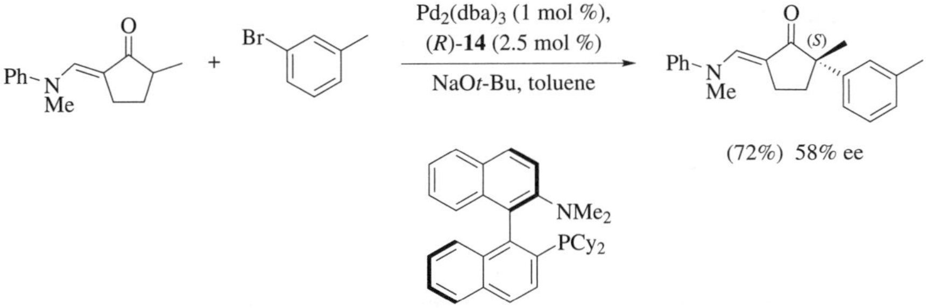

(Eq. 7)

An interesting alternative approach involves the use of chiral N-heterocyclic carbene (NHC) ligands in intramolecular arylation reactions for the asymmetric synthesis of oxindoles. Ligands **15** and **16**, derived from (−)-isopinocampheylamine and (+)-norbornylamine, respectively, display complementary induction under mild conditions, and afford products of opposite absolute configuration (Eq. 8).[32]

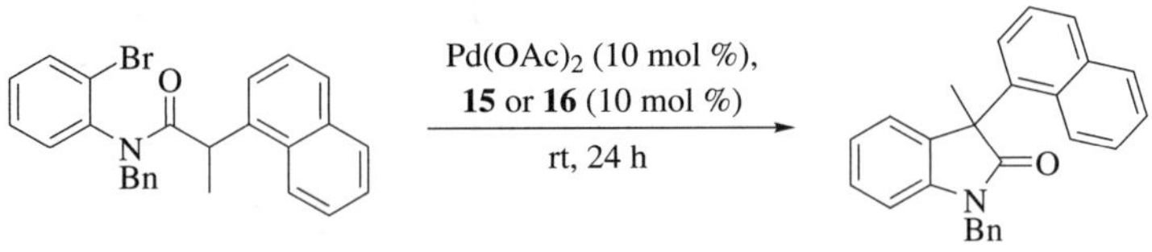

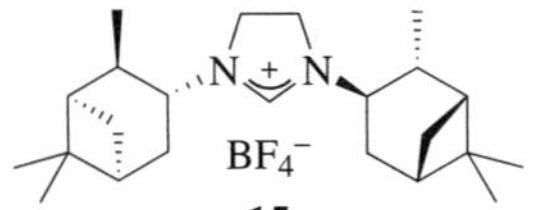

Ligand		% ee
15	(88%)	67
16	(80%)	71

(Eq. 8)

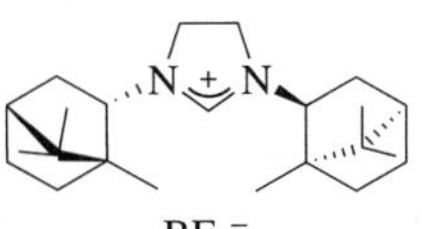

In addition to palladium-catalyzed asymmetric arylations, nickel- and copper-catalyzed processes have recently emerged as alternative approaches.[33] Substitution of palladium(0) with nickel(0) catalysts leads to very efficient, enantioselective α-arylation of γ-butyrolactones. The use of Ni(BINAP) catalytic systems provides the products in modest to good yields, but with high enantiomeric excesses (Eq. 9).[33] The increase in the enantioselectivity is attributed to the greater influence of the ligand in the stereodetermining step resulting from the smaller nickel-center. Interestingly, aryl chlorides can be used instead of bromides with comparable reaction rates.[34] Moreover, addition of substoichiometric amounts of zinc salts increases the reaction rate. Zinc halide species are assumed to act as Lewis acids to facilitate halide abstraction from the oxidative addition product, thus generating a more reactive cationic complex at an early stage of the transmetalation step (Eq. 10).[33]

$$\text{(Eq. 9)}$$

Ni(cod)$_2$ (5 mol %), (S)-BINAP **2** (8 mol %), ZnBr$_2$ (15 mol %), NaHMDS

(76%) 94% ee

$$(\text{BINAP})\text{Ni}(\text{Ar})(\text{Br}) \xrightarrow{\text{ZnBr}_2} [(\text{BINAP})\text{Ni}(\text{Ar})]^+ \text{ZnBr}_3^- \qquad (\text{Eq. 10})$$

The highly enantioselective arylation of ketones using ligand **17** and electron-poor aryl triflates in the presence of nickel-based catalysts, or electron-rich aryl triflates in the presence of palladium-based complexes, has been recently described (Eq. 11).[34]

M(0) (5–10 mol %), **17** (6–12 mol %), NaOt-Bu (2 equiv), toluene, 80–100°

M(0)		% ee
Ni(cod)	(84%)	95
Pd(dba)$_2$	(79%)	78

17

(Eq. 11)

Highly enantioselective nickel-catalyzed α-arylations of ketone enolates involving the Ni-P-Phos catalytic system (P-Phos = structure **18**) have been applied to the 2-methylindanone, -tetralone, and -benzosuberone series leading to the formation of quaternary centers with enantiomeric excesses up to 98% (Eq. 12).[35]

$$\text{(Eq. 12)}$$

n		% ee
1	(69–80%)	67–88
2	(59–94%)	70–98
3	(70–74%)	60–73

Copper-catalyzed arylations of acetoacetates in the presence of L-*trans*-4-hydroxyproline as the ligand proceed in high yield (29–87%) and enantiomeric excesses (60–93%) (Eq. 13).[36] It is worth noting that traces of water in the solvents are beneficial to the reaction rates, allowing the reactions to proceed smoothly at low temperature. Temperatures as low as −45° provide the carbon–carbon bond formation with enhanced enantiomeric composition.[36]

(79%) 93% ee

(Eq. 13)

SCOPE AND LIMITATIONS

Palladium-Catalyzed α-Arylation

α-Arylation of Ketones. This section focuses mainly on the direct, catalyzed arylation of ketone enolates; recent developments on the use of stoichiometric reagents[37–39] are not covered. In the period since their seminal 1997 papers, Hartwig, Buchwald, and Miura have tried to develop "universal conditions" for a wide range of substrates under mild reaction conditions.

On the basis of their mechanistic studies (see above), the Hartwig group has reported a simple and active catalytic system for the arylation of ketones: **3**/Pd(dba)$_2$ or (*t*-Bu)$_3$P/Pd(OAc)$_2$ (in a 1:1.25 Pd/L ratio) in THF (Eq. 14).[40] These conditions allow the reaction of a wide range of aryl bromides and chlorides (including electron-rich ones) to take place. One example of the coupling of an aryl tosylate with propiophenone is also described, but requires the use of ligand **19** at 70° (Eq. 15).[40] Under these conditions, monoarylated ketones can be obtained in good yields starting from cyclohexanone, 3-methyl-2-propanone, acetophenone, propiophenone, and isobutyrophenone. The limits of these conditions are found with methyl alkyl ketones, which form diarylated compounds.

(Eq. 14)

[Eq. 15 scheme: propiophenone + 4-methylphenyl tosylate with Pd(OAc)₂ (2 mol %), **19** (2.5 mol %), NaO*t*-Bu (2.2 equiv), THF, 70° → α-(4-tolyl)propiophenone (60%); ligand **19** is a ferrocene bearing P(*t*-Bu)₂ and PPh₂ substituents]

(Eq. 15)

Buchwald et al. have developed electron-rich monophosphine ligands **7–11** (Fig. 2) and **20–23** (Fig. 5) that possess a biphenyl skeleton. These ligands facilitate rapid oxidative additions with aryl halides, including aryl chlorides.[41]

Ligand	R^1	R^2	R^3
20	Ph	NMe$_2$	H
21	*t*-Bu	NMe$_2$	H
22	Cy	NMe$_2$	H
23	Cy	OMe	OMe

Figure 5. Electron-rich monophosphine ligands.

A wide range of substrates (electron-rich, electron-poor, and 2-substituted aryl bromides and chlorides with aromatic, aliphatic, and cyclic ketones) can be successfully combined under typical conditions: Xantphos (**24**), Pd(OAc)$_2$, NaO*t*-Bu, 70° in toluene (Eq. 16).[42] For reactions involving base-sensitive substrates, a milder base such as K$_3$PO$_4$ can be used in the presence of **24**/Pd$_2$(dba)$_3$.[43] Most of the non-demanding reactions (typically propiophenone and pinacolone with aryl bromides) can be carried out in the absence of a ligand. However, the use of bidentate ligands such as Xantphos (**24**) is essential in α-arylations of ketones bearing two enolizable methylene or methyl groups.[42] A major limitation of this method is the use of cyclopentanone, where self-aldolization is the major process (for an example where the cyclopentanone enolate is generated in situ through a 1,4-addition, see Ref. 44). The α-arylations of methyl or cyclic ketones with 2-halonitroarenes require the use of a substoichiometric amount (20%) of 4-methoxyphenol to be efficient.[45]

[Eq. 16 scheme: 3-methyl-2-butanone + 2-bromo-5-(dimethylamino)-toluene with Pd(OAc)$_2$ (0.1 mol %), **24** (0.2 mol %), NaO*t*-Bu (1.3 equiv), toluene, 85° → α-arylated ketone (70%); ligand **24** = Xantphos]

(Eq. 16)

Miura et al. have mainly focused their efforts on multiple arylations of phenyl ketone derivatives. Benzyl phenyl ketones undergo a triarylation sequence: the expected α-arylation but also two *o*-arylations (Eq. 17).[46] With alkyl aryl ketones, multiple arylation is also observed: after the initial α-arylation process, a second palladium-enolate is formed, which undergoes a β-elimination to form a double bond. This double bond then reacts further through a Heck reaction (Eq. 18), or a vinylogous γ-arylation (Eq. 19).[47]

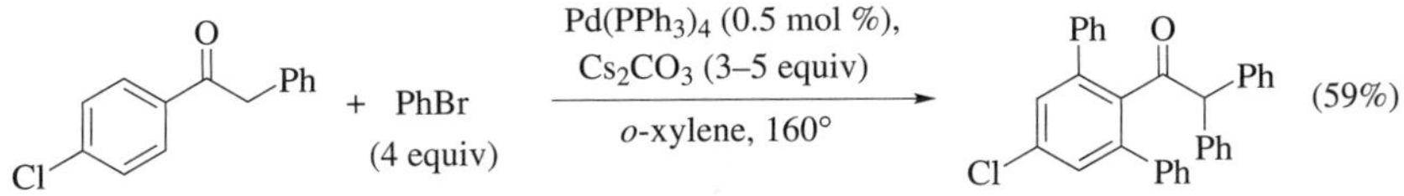

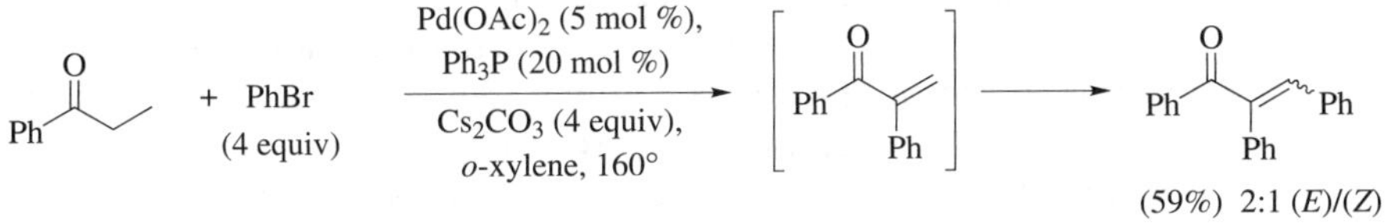

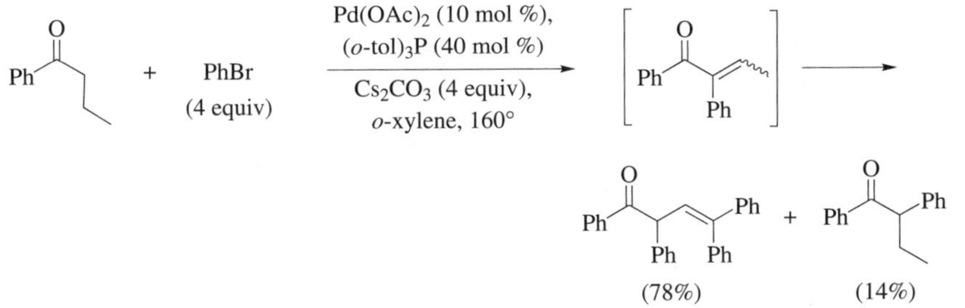

Robust catalytic systems capable of practical (industrial) applications, especially with aryl chlorides, are of great value. The use of the *n*-BuP(adamantyl)$_2$ electron-rich phosphine,[48] palladacycles possessing a tertiary phosphine, such as **25**,[49] or secondary phosphines **26–28**,[49] NHC ligand **29**,[50,51] or palladacycle **30**[52] provide efficient arylation of non- and deactivated aryl chlorides (Eqs. 20, 21).

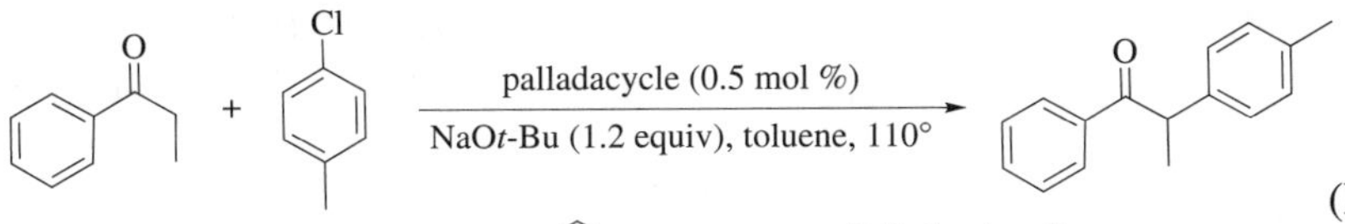

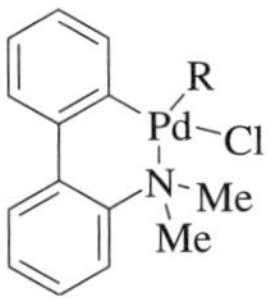

Palladacyle	R	
25	P(Cy)$_3$	(99%)
26	PH(Cy)$_2$	(69%)
27	PH(*t*-Bu)$_2$	(60%)
28	PH(2-norbornyl)$_2$	(100%)

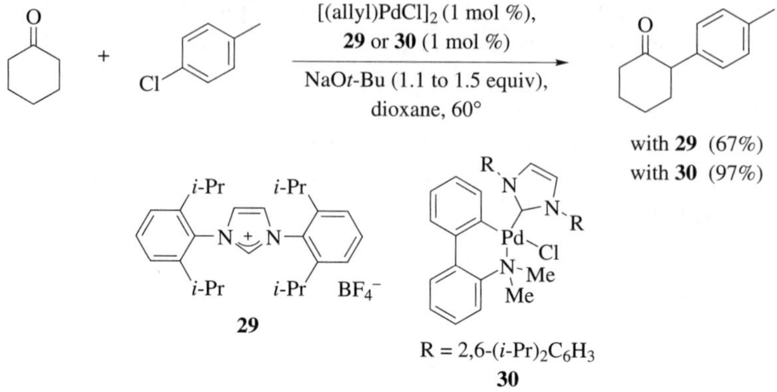

(Eq. 21)

with **29** (67%)
with **30** (97%)

R = 2,6-(*i*-Pr)$_2$C$_6$H$_3$

The selective mono-arylation of 1,2-diarylethanones has been studied, and byproducts (*o*-arylation, multiple arylation, and dehalogenation) can be minimized by a careful choice of reaction conditions (Eq. 22). Originally, homogeneous conditions (Pd(OAc)$_2$, Ph$_3$P, Cs$_2$CO$_3$) were described, but these reactions can also be performed with polystyrene-derived catalyts **31**–**33** (Eq. 23).[53–55] More recently, the same group has also described the use of phosphinite PCP-pincer complexes.[56] In addition, arylations on solid supports have been described using immobilized 4-bromobenzamide.[57]

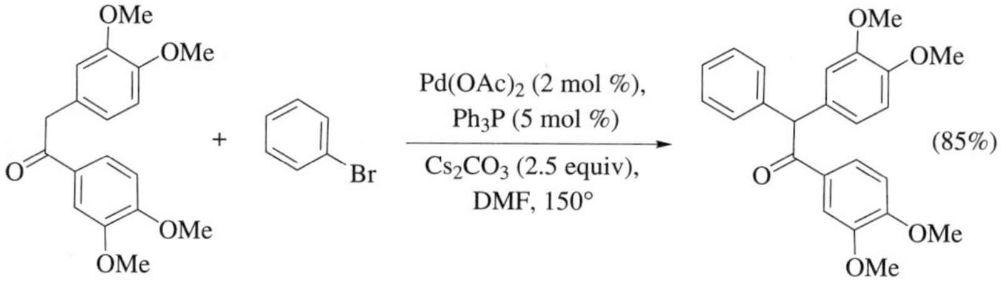

(Eq. 22)

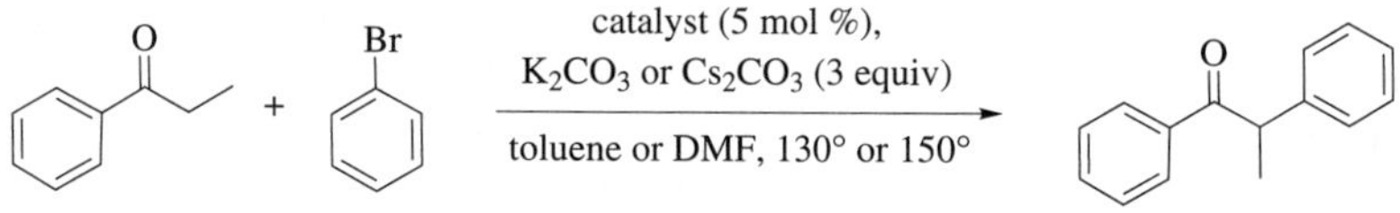

Catalyst	
31	(43%)
32	(16%)
33	(85%)

(Eq. 23)

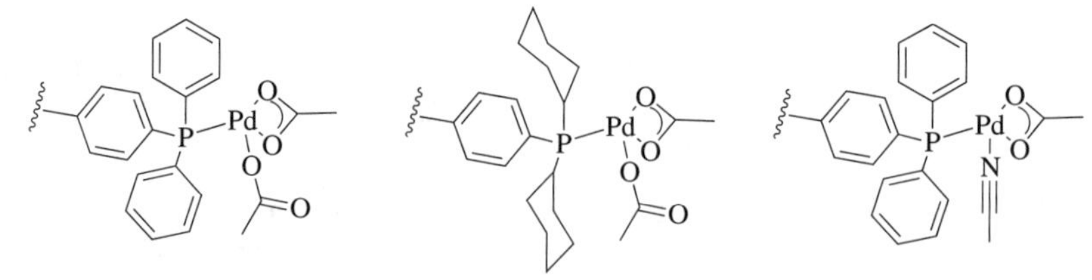

In some instances, either the ketone enolate or the aryl palladium species can be generated from alternative precursors. In Eq. 24, the ketone enolate is obtained in situ through a cyclopropanol ring-opening reaction.[58] Biphenylene can also be used as an aryl halide surrogate in the presence of *p*-cresol (Eq. 25).[59]

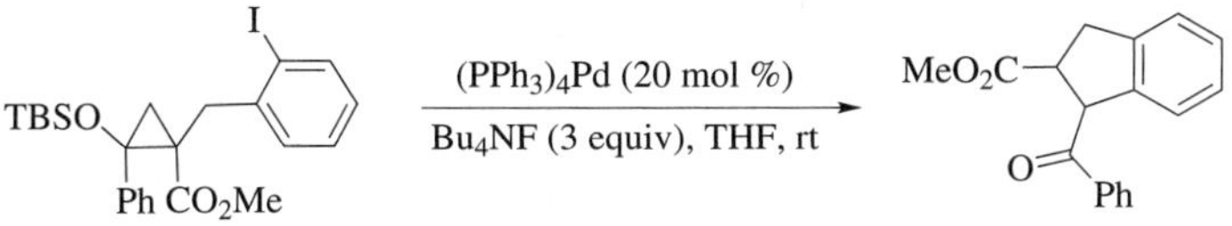

(Eq. 24)

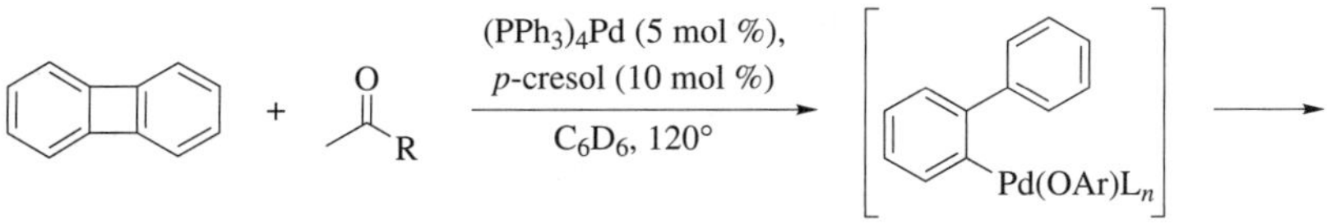

(Eq. 25)

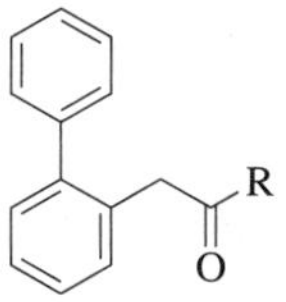

R	
Ph	(84%)
2-thienyl	(94%)
t-Bu	(88%)

Using 1,2-dibromobenzenes, tandem reactions can be envisioned through two successive palladium-catalyzed reactions. In the formation of benzofurans from aryl benzyl ketones, the mechanism involves first an α-arylation followed by a second deprotonation of the more acidic proton and finally the intramolecular arylation of the *O*-enolate intermediate **34** (Eq. 26).[60] An extension of this method to non-aromatic cyclic and acyclic ketones has been described using DPEphos (**35**).[61,62] A tandem alkylation of the *O*-enolate intermediate is reported in the synthesis of a small library of 1-vinyl-1*H*-isochromene derivatives as illustrated in Eq. 27.[63] Two carbon–carbon single bond formations are also possible as illustrated in Eq. 28.[60]

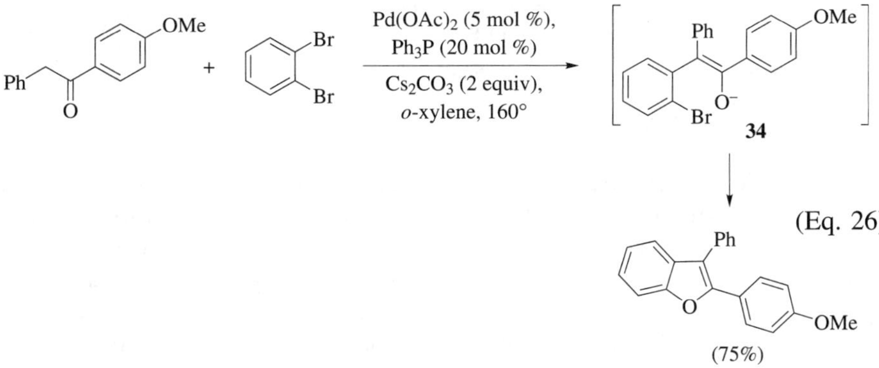

(Eq. 26)

(75%)

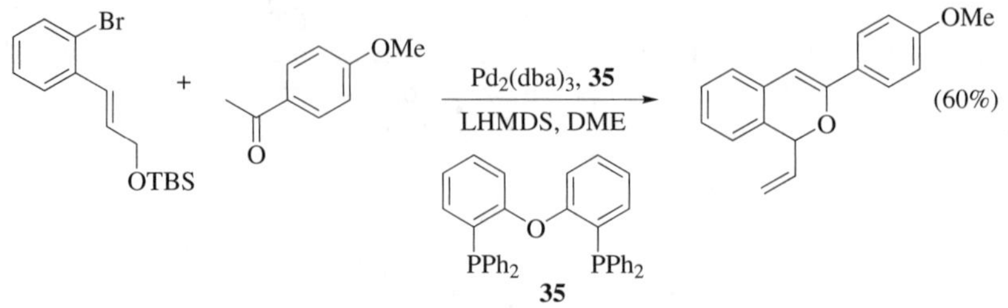

[Eq. 27 scheme]

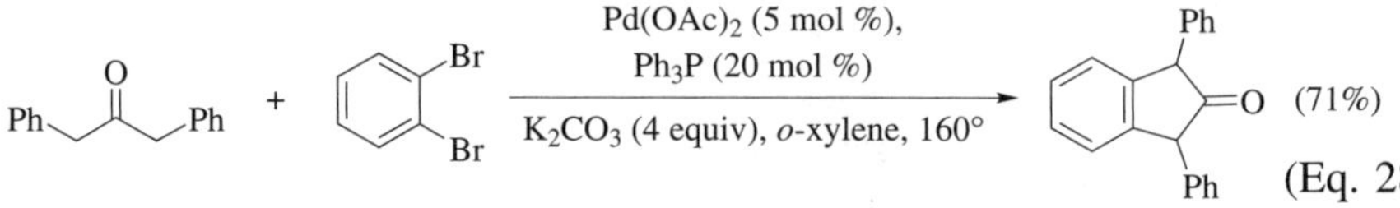

[Eq. 28 scheme]

α-Arylation of Aldehydes. Because of the propensity of aldehydes to undergo aldol self-condensations under basic conditions, arylation of aldehydes has received less attention. In the intramolecular arylation of aldehydes, a mixture of α-arylation and carbonyl-arylation is obtained (Eq. 29).[64]

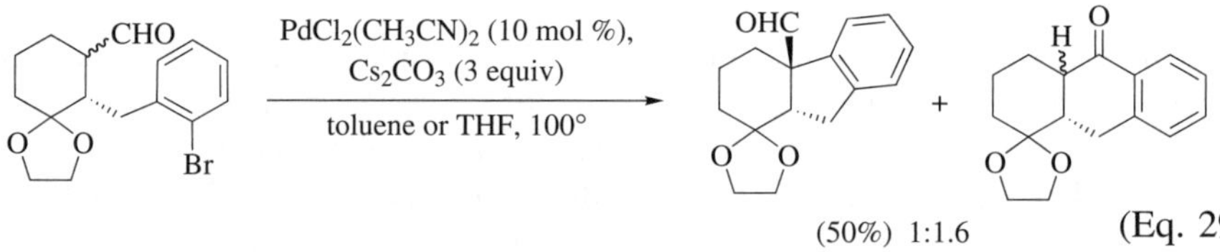

[Eq. 29 scheme]

In intermolecular, α-selective arylation of aldehydes (Eq. 30), the use of dioxane and $(t\text{-}Bu)_3P$ are essential to prevent aldolization and to promote the reaction (no reaction is observed in the presence of Cy_3P).[65]

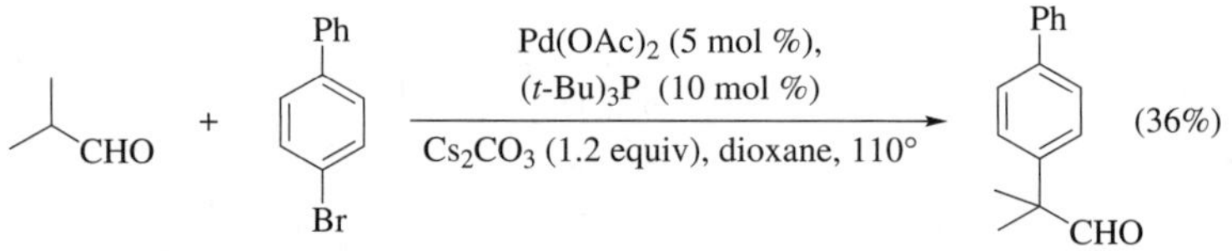

[Eq. 30 scheme]

α-Arylation of Amides. On the basis of the arylation of ketones and aldehydes, several groups have explored the extension of the method to carboxylic acid derivatives. In these studies and in contrast to ketones, the arylation of amides and esters is not plagued by site selectivity and such reactions would appear to be less challenging. Disappointingly, the first attempts to arylate amides gave rather low yields, even in intramolecular versions. This lack of reactivity can be attributed to the higher pK_a's of the amide moiety compared to that of ketones,[66] requiring the use of stronger bases. KHMDS is more efficient than LiTMP (lithium 2,2,6,6-tetramethylpiperidide) and the more classical NaOt-Bu. The α-arylation of N,N-dialkylacetamide has been reported in moderate to good yields.[67] Side-products such as diarylated amides and/or dehydrohalogenated arenes are observed along with the expected arylation products (Eq. 31).

$$\text{Naphthyl-Br} + \underset{\text{NMe}_2}{\overset{\text{O}}{\bigwedge}} \xrightarrow[\text{dioxane, 100°}]{\substack{\text{Pd(dba)}_2 \text{ (5 mol \%)}, \\ \textbf{2} \text{ (7.5 mol \%)} \\ \text{KHMDS (2.0 equiv),}}} \text{Naphthyl-CH}_2\text{C(O)NMe}_2 \text{ (70\%)} + \text{bis-naphthyl product (9\%)} \quad (\text{Eq. 31})$$

The scope of the reaction is limited by the use of strongly basic conditions that preclude the use of base-sensitive and/or electrophilic substituents and lead to catalyst deactivation or decomposition. The aforementioned problems have recently been overcome by moving to the less basic zinc enolates. The higher functional group tolerance of these nucleophiles allows an extended scope of such coupling processes. Indeed, ketone, nitrile, ester, and nitro derivatives can be used as substrates in one-pot procedures starting from N,N-dialkylamides or α-bromo N,N-dialkylamides.[68] Under these conditions (Pd(dba)$_2$, **36**), no side-products from diarylation of the enolate are observed (Eq. 32).

$$\underset{R}{\text{Ar-Br}} + \underset{\text{NMe}_2}{\overset{\text{O}}{\bigwedge}} \xrightarrow[\substack{\text{1. } s\text{-BuLi (1.2 equiv), THF, }-78°, 1\text{ h} \\ \text{2. ZnCl}_2 \text{ (2.4 equiv), rt, 10 min} \\ \text{3. Pd(dba)}_2 \text{ (1 mol \%), } \textbf{36} \text{ (1 mol \%), rt}}]{} \underset{R}{\text{Ar-CH}_2\text{C(O)NMe}_2} \quad (\text{Eq. 32})$$

36: 1,1′,2,2′,3,3′,4,4′-octaphenyl-5-(di-t-butylphosphino)ferrocene (Ph$_4$C$_5$-Fe-C$_5$Ph$_4$-P(t-Bu)$_2$)

R	
4-CO$_2$Me	(95%)
2-CN	(84%)
4-CN	(89%)
4-COPh	(92%)
4-NO$_2$	(90%)

Apart from the use of zinc enolates, the α-arylation of amides has been limited to intramolecular reactions and/or lactam substrates. Intramolecular reactions have been used in the synthesis of oxindoles and tetrahydroisoquinoline derivatives. In the oxindole series, a wide range of aryl substituents (with respect to both steric and electronic properties) are tolerated in these intramolecular reactions. Standard conditions involving BINAP and NaOt-Bu are used, but sterically hindered alkyl phosphines[67] or hindered imidazolium carbene precursors like **37**[32,69,70] generate more active catalytic systems, providing higher arylation rates (Eq. 33).

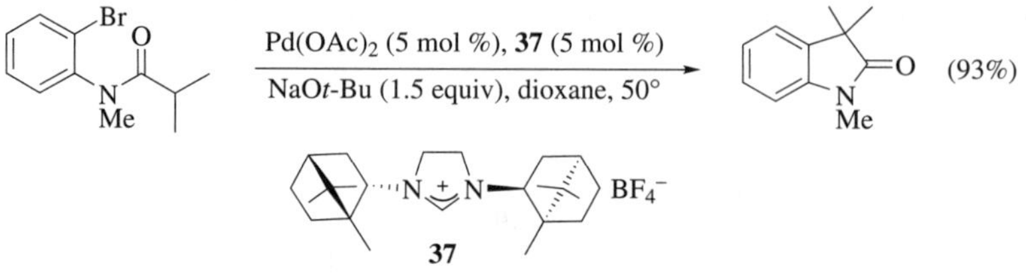

$$\text{(Eq. 33)}$$

In the tetrahydroisoquinoline series, more disparate yields have been observed. Poor yields (5%) of the expected cyclized product from a non-stabilized enolate using the Pd(dba)$_2$, BINAP, NaOt-Bu catalytic system are observed (Eq. 34).[67] In contrast, stabilization of the enolate allows yields ranging from 54 to 81% (Eq. 35).[71]

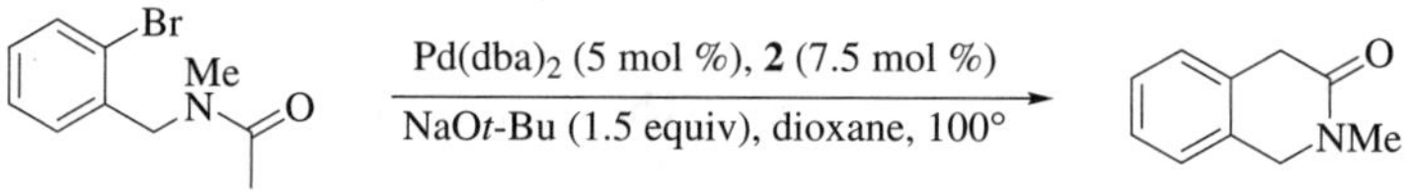

$$\text{(Eq. 34)}$$

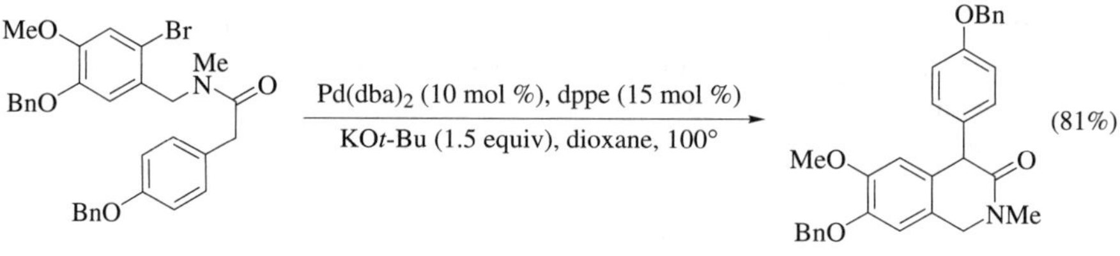

$$\text{(Eq. 35)}$$

Arylation of lactams is also possible. N-Methylpyrrolidinone (NMP) reacts with bromobenzene to give the expected arylation product in 49% yield, along with 9% of the diarylation product (Eq. 36).[67]

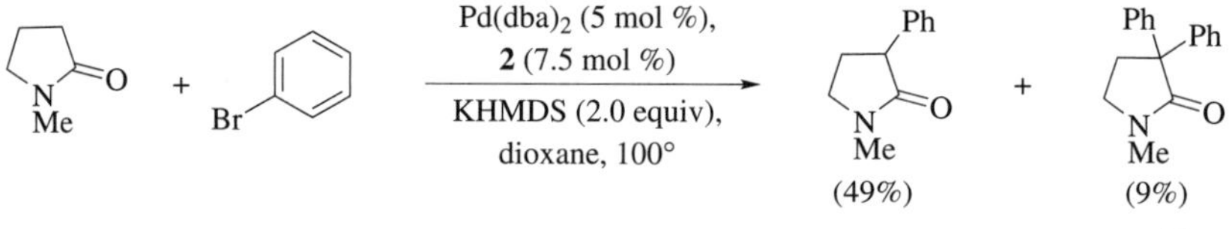

$$\text{(Eq. 36)}$$

Arylpiperidones can also be prepared by palladium-assisted α-arylation of the derived zinc enolates (Eq. 37).[72] No diarylation byproducts are observed. The α-arylation process seems not to be dependent on the nitrogen atom substitution. In contrast, steric hindrance around the aryl halide has a significant impact on yields: a modest 46% yield is observed with 2-methylbromobenzene whereas no arylation occurs with 2,6-dimethylbromobenzene.

$$\text{(piperidinone-NTs)} \xrightarrow[\substack{\text{1. LiHMDS (2.5 equiv), THF, }-20° \\ \text{2. ZnCl}_2 \text{ (2.2 equiv), THF, }-20° \\ \text{3. PhBr, Pd(dba)}_2 \text{ (5 mol \%),} \\ \textbf{21} \text{ (7.5 mol \%), THF, 65°}}]{} \text{(α-Ph-piperidinone-NTs)} \quad (78\%) \quad \text{(Eq. 37)}$$

α-Arylation of Esters. α-Arylation of methyl phenyl acetate, which involves a doubly-stabilized anion, using a $PdCl_2–Ph_3P$ catalytic system and Cs_2CO_3 as the base, affords the expected arylated ester in a modest 56% yield (Eq. 38).[73]

$$\text{Ph-CH}_2\text{-CO-OMe} \xrightarrow[\substack{\text{PhI, PdCl}_2 \text{ (5 mol \%),} \\ \text{Ph}_3\text{P (10–20 mol \%)} \\ \text{Cs}_2\text{CO}_3 \text{ (1.2 equiv), DMF, 100°}}]{} \text{Ph}_2\text{CH-CO-OMe} \quad (56\%) \quad \text{(Eq. 38)}$$

The use of hindered phosphines (Eq. 39), bidentate phosphines, or carbene-based ligands (Eq. 40) has been extended to the α-arylation of ester substrates.[74,75] *tert*-Butyl acetate or propionate reacts with a wide range of aryl bromides at room temperature in high yields. Although 2.2 to 2.5 equivalents of base are necessary to ensure complete conversion, diarylated byproducts are not observed.

$$\text{CH}_3\text{-CO-O}t\text{-Bu} + \text{2-Br-naphthalene} \xrightarrow[\substack{\text{Pd(OAc)}_2 \text{ (3 mol \%), } \textbf{22} \text{ (6 mol \%)} \\ \text{LHMDS (2.5 equiv), toluene, rt}}]{} \text{naphthyl-CH}_2\text{-CO-O}t\text{-Bu} \quad (84\%) \quad \text{(Eq. 39)}$$

$$\text{CH}_3\text{-CO-O}t\text{-Bu} + \text{3-CF}_3\text{-C}_6\text{H}_4\text{-Br} \xrightarrow[\substack{\text{Pd(dba)}_2 \text{ (0.2–1 mol \%),} \\ (t\text{-Bu})_3\text{P or } \textbf{29} \text{ (0.5–5 mol \%)} \\ \text{LHMDS or NaHMDS (2.2 equiv),} \\ \text{toluene, rt}}]{} \text{3-CF}_3\text{-C}_6\text{H}_4\text{-CH}_2\text{-CO-O}t\text{-Bu} \quad (88\%) \quad \text{(Eq. 40)}$$

Generation of the enolate prior to the coupling with a stronger hindered amide base, such as $LiNCy_2$, greatly improves the reaction and allows the formation of quaternary carbon atoms (Eq. 41).[23,74] Heterocycles such as furans, thiophenes, and pyridines are well-tolerated using this method. Moreover, high yields are generally obtained using low catalyst loadings and slight excesses of both ester and base.

$$\text{Et-CH(Me)-CO-OMe} \xrightarrow[\substack{\text{PhBr, Pd(dba)}_2 \text{ (1 mol \%),} \\ (t\text{-Bu})_3\text{P (1 mol \%)} \\ \text{LiNCy}_2 \text{ (1.3 equiv), toluene, rt}}]{} \text{Ph-C(Me)(Et)-CO-OMe} \quad (97\%) \quad \text{(Eq. 41)}$$

For aryl halides bearing base-sensitive or electrophilic substituents, the use of zinc enolates again improves the yield (Eq. 42).[76] In addition, these conditions also prove satisfactory for aryl halides bearing acidic and basic substituents, such as bromophenols, bromoanilines, and substituted pyridines.

[Zn]-enolate of O*t*-Bu ester + 4-bromo-nitrobenzene → Pd(dba)₂ (1 mol %), **36** (1 mol %), THF or dioxane, rt → α-(4-nitrophenyl) O*t*-Bu ester (96%) (Eq. 42)

The arylation of lactones under palladium catalysis appears in a single report; however, only modest yields are obtained using the Pd/BINAP, NaHMDS catalytic system (Eq. 43).[33]

3-methyl-γ-butyrolactone + 3-bromoanisole → Pd(dba)₂ (2.5 mol %), **2** (15 mol %), NaHMDS, toluene, 50° → 3-methyl-3-(3-methoxyphenyl)-γ-butyrolactone (31%) (Eq. 43)

More successful nickel-catalyzed α-arylations of esters have been developed, and this point will be discussed below.

α-Arylation of Amino Acids and Derivatives.

Ethyl N,N-dimethylglycinate (**38**) undergoes facile α-arylation even when using K_3PO_4 as the base (Eq. 44).[77]

Me₂N-CH₂-CO₂Et (**38**) + 4-bromo-*t*-butylbenzene → Pd(dba)₂ (2 mol %), (*t*-Bu)₃P (2 mol %), K₃PO₄ (2.3 equiv), 100° → Me₂N-CH(Ar)-CO₂Et, Ar = 4-*t*-Bu-C₆H₄ (84%) (Eq. 44)

Amino acids with more common nitrogen protecting groups have been evaluated subsequently. Because of their convenient preparation, imino esters derived from benzophenone and benzaldehyde are of particular value. Under the aforementioned conditions ((*t*-Bu)₃P, Pd(dba)₂, K₃PO₄), imino ester derivatives are arylated in high yields (Eq. 45).[77] Coordination of the substrate nitrogen atom is assumed to assist both the formation of the enolate and the arylation reaction.

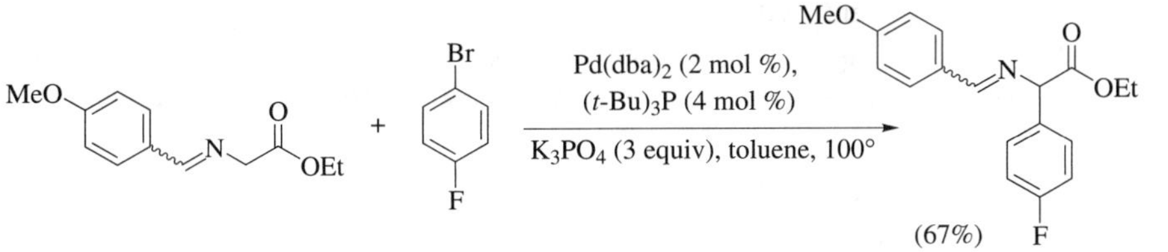

[Eq. 45 scheme]

(Eq. 45)

Intramolecular α-arylation of α-amino acid esters provides easy access to 5- and 6-membered dihydroisoindole and tetrahydroisoquinoline derivatives in good to excellent yields using a Pd$_2$(dba)$_3$/**20** or **22** and LiO*t*-Bu catalytic system (Eq. 46). In these reactions, LiO*t*-Bu is superior to other bases such as NaO*t*-Bu, NaHMDS, phosphates, or carbonates (which cause decomposition of the starting material or lead to poor conversion).[78] In addition, this method allows flexible access to fused tricyclic systems in fair to good yields. Although quaternary carbon centers are accessible via the aforementioned intramolecular route, the analogous intermolecular α-arylation of α-alkyl α-amino acid derivatives is still unreported.

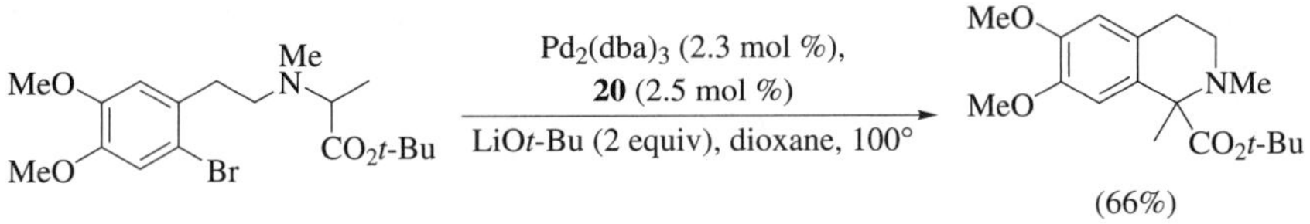

[Eq. 46 scheme]

(Eq. 46)

An elegant route to quaternary amino acids, involving a cyclodehydration–arylation–hydrolysis sequence, has been described (Scheme 3).[79] α-Alkyl α-amino acids are first transformed into azalactones. In the second step, the palladium-catalyzed α-arylation affords quaternary carbon centers in good yields

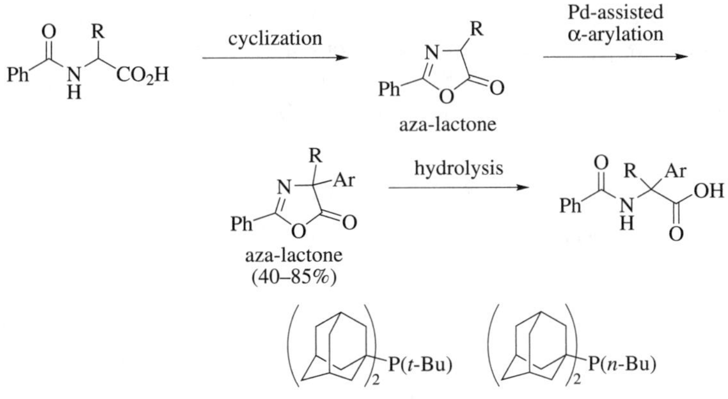

[Scheme 3 showing cyclization of α-amino acid to aza-lactone, Pd-assisted α-arylation to arylated aza-lactone (40–85%), then hydrolysis to quaternary amino acid; ligands **39** P(*t*-Bu) and **40** P(*n*-Bu) adamantyl phosphines]

Scheme 3

(40–85%). Subsequent hydrolysis gives rise to the quaternary amino acids. The highest yields and fastest rates occur with catalytic systems based on 5 mol % Pd(dba)$_2$ or Pd(OAc)$_2$, 10 mol % of ligand **39** or **40** and 3.3 equivalents of K$_2$CO$_3$ or K$_3$PO$_4$.

α-Arylation of Nitriles.   Aliphatic nitriles are less acidic than ketones[26,66] and thus require the use of stronger bases. For these substrates, the development of catalytic systems was first based on the general method established for ketones and amino acid derivatives. However, hindered alkyl phosphines are less effective than the more common BINAP ligand **2**. Thus, arylation of nitriles is possible using the Pd(OAc)$_2$/**2** catalytic system and affords high to quantitative yields of the expected arylated nitriles (Eq. 47).[26]

$$\text{ArCH(Me)CN} + \text{PhBr} \xrightarrow[\text{NaHMDS (1.3 equiv), toluene, 100°}]{\text{Pd(OAc)}_2 \text{ (1 mol \%), 2 (1 mol \%)}} \text{Ar(Me)C(CN)Ph} \quad (89\%) \qquad \text{(Eq. 47)}$$

A major limitation to this method is the arylation of linear nitriles. Only the diarylated product is obtained under the previously described conditions because of the increased acidity of the α-CH of the monoarylated product compared to the starting material (Eq. 48).[26]

$$\text{PrCN} + \text{PhBr} \xrightarrow[\text{NaHMDS (1.3 equiv), toluene, 100°}]{\text{Pd(OAc)}_2 \text{ (1 mol \%), 2 (1 mol \%)}} \text{Et(Ph)}_2\text{CCN} \quad (69\%) \qquad \text{(Eq. 48)}$$

One elegant solution that circumvents the formation of diarylation byproducts is the coupling of aryl halides with silylaceto- or propionitriles.[80] Under these conditions, monoarylation occurs selectively in the presence of Pd$_2$(dba)$_3$/(t-Bu)$_3$P or **24** and ZnF$_2$ (Eq. 49). Highly hindered silyl derivatives fail to react under these conditions and require the use of a stronger fluoride source such as KF. Indeed, α-arylation of α-trimethylsilylcyclohexanecarbonitrile occurs in good yields with various aryl bromides as exemplified by Eq. 50.

$$\text{Me}_3\text{Si-CH(Me)CN} + \text{4-EtO}_2\text{C-C}_6\text{H}_4\text{-Br} \xrightarrow[\text{ZnF}_2 \text{ (0.5 equiv), DMF, 90°}]{\text{Pd}_2\text{(dba)}_3 \text{ (2 mol \%), (t-Bu)}_3\text{P or 24 (2–4 mol \%)}} \text{4-EtO}_2\text{C-C}_6\text{H}_4\text{-CH(Me)CN} \quad (84\%) \qquad \text{(Eq. 49)}$$

$$\text{1-(SiMe}_3\text{)-1-(CN)-cyclohexane} + \text{PhBr} \xrightarrow[\text{DMF, 90°}]{\text{Pd}_2\text{(dba)}_3 \text{ (2 mol \%), 24 (2 mol \%), KF (1 equiv)}} \text{1-(CN)-1-Ph-cyclohexane} \quad (68\%) \qquad \text{(Eq. 50)}$$

α-**Arylation of Active Methylene Compounds.** Along with ketones and carboxylic acid derivatives, active methylene compounds such as malonates, β-cyanoesters, malononitriles, and β-diketones represent an important class of nucleophiles that can be arylated, thus considerably expanding the scope of the metal-assisted α-arylation reaction. An example of α-arylation of malononitriles is shown in Eq. 51. The mechanism is assumed to proceed as described in Scheme 1.[14,81]

$$\text{CH}_2(\text{CN})_2 + \text{4-MeC}_6\text{H}_4\text{Br} \xrightarrow[\text{NaH (1.5 equiv), THF, reflux}]{\text{PdCl}_2(\text{Ph}_3\text{P})_2 \ (3\text{–}10 \text{ mol \%})} \text{4-MeC}_6\text{H}_4\text{CH}(\text{CN})_2 \quad (69\%) \quad \text{(Eq. 51)}$$

Recent work has significantly expanded the range of nucleophiles and now allows malonates, acetoacetates, as well as β-cyanoesters to couple with aromatic halides in both an intra- and intermolecular fashion. Complete conversions are obtained using $(t\text{-Bu})_3\text{P}$, **3**, **36**, or **41** as ligands (with a palladium to ligand ratio of 1:1 or 1:2) in association with mild bases such as NaH or K_3PO_4 (Eq. 52).[43,82]

$$\text{EtO}_2\text{CCH}_2\text{CO}_2\text{Et} + \text{2-MeC}_6\text{H}_4\text{Br} \xrightarrow[\text{THF, 70°}]{\substack{\text{Pd(dba)}_2 \ (2 \text{ mol \%}), \\ \textbf{41} \ (4 \text{ mol \%}), \text{NaH (1 equiv)}}} \text{EtO}_2\text{CCH}(2\text{-MeC}_6\text{H}_4)\text{CO}_2\text{Et} \quad (83\%)$$

Adamantyl–P$(t\text{-Bu})_2$
41

(Eq. 52)

With a view to industrial applications where catalyst and purification costs are of economic importance, the use of heterogeneous conditions has also been explored. Although the results do not rival the most recent developments in homogeneous catalysis, $[\text{Pd}(\text{NH}_3)_4]$-zeolite in combination with NaOt-Bu, and PdCl_4^{-2} in combination with Ba(OH)$_2$, allow the arylation of diethyl malonates in yields ranging from 38–84% and 93–99%, respectively.[83,84] It is worth noting that no diarylated products are observed.

The formation of quaternary carbon centers can be achieved starting from fluoro malonates as shown in Eq. 53.[40] However, attempts to couple alkyl-substituted malonates failed. This lack of reactivity is mainly attributed to steric hindrance, which inhibits the formation of the carbon-bound palladium–malonate complex. This limitation is overcome through a sequential arylation–alkylation process (Eq. 54).[40]

$$\text{EtO}_2\text{CCHFCO}_2\text{Et} + \text{PhBr} \xrightarrow[\text{NaH (1 equiv), THF, 70°}]{\substack{\text{Pd(dba)}_2 \ (2 \text{ mol \%}), \\ (t\text{-Bu})_3\text{P} \ (4 \text{ mol \%})}} \text{EtO}_2\text{CCF(Ph)CO}_2\text{Et} \quad (88\%)$$

(Eq. 53)

$$\text{EtO}\underset{O}{\overset{O}{\diagup}}\text{OEt} + \text{Br-C}_6\text{H}_4\text{-CH}_3 \xrightarrow[\text{2. MeI}]{\substack{\text{1. Pd}_2(\text{dba})_3\ (1\ \text{mol \%}),\\ (t\text{-Bu})_3\text{P}\ (4\ \text{mol \%}),\\ \text{K}_3\text{PO}_4\ (4.5\ \text{equiv}),\ \text{toluene},\ 70°}} \text{EtO-CO-C(CH}_3\text{)(Ar)-CO-OEt}$$

(89%)

(Eq. 54)

α-Arylation of 1,3-diketones can also be carried out. 1,3-Cyclohexanedione and 1,3-cyclopentanedione undergo monoarylation using the aforementioned procedure (Eq. 55).[43]

$$\text{1,3-cyclopentanedione} + \text{Br-C}_6\text{H}_4\text{-CO}_2\text{Me} \xrightarrow[\text{K}_3\text{PO}_4\ (2.3\ \text{equiv}),\ \text{dioxane},\ 100°]{\text{Pd(OAc)}_2\ (1\ \text{mol \%}),\ \mathbf{8}\ (2\ \text{mol \%})} \text{arylated product (96\%)}$$

(Eq. 55)

Intramolecular arylation of 1,3-diketones has been described (Eq. 56). Although this strategy requires relatively high temperatures, precluding the use of thermally sensitive substrates, these results paved the way to more recent developments.[15,85]

$$\text{aryl iodide-tethered indanedione} \xrightarrow[\text{NaH, DMF, 135°}]{(\text{PPh}_3)_4\text{Pd}\ (5\ \text{mol \%})} \text{spirocyclic product} \quad (76\%)$$

(Eq. 56)

Palladium-catalyzed α-arylations of β-keto esters are rare. However, the key step of an elegant synthesis of the N-methylwelwitindolinone skeleton is based on an intramolecular palladium-catalyzed α-arylation of a β-keto ester (see below).[86] Moreover, arylacetic acid esters can be prepared through a one-pot two-step strategy involving a palladium-catalyzed α-arylation of acetoacetate followed by an in situ base-catalyzed deacylation of the arylacetoacetate intermediate **42** (Eq. 57).[87]

$$\text{MeCOCH}_2\text{CO}_2\text{Et} + \text{PhBr} \xrightarrow[\substack{\text{K}_3\text{PO}_4\ (2.5\ \text{equiv}),\\ \text{toluene},\ 90°}]{\substack{\text{Pd}_2(\text{dba})_3\ (1\ \text{mol \%}),\\ (t\text{-Bu})_3\text{P}\ (2\ \text{mol \%})}} \left[\text{MeCO-CH(Ph)-CO}_2\text{Et} \atop \mathbf{42} \right] \longrightarrow \text{PhCH}_2\text{CO}_2\text{Et} \quad (55\%)$$

(Eq. 57)

Although the arylation of cyanoacetates using $\text{Pd(PPh}_3)_2\text{Cl}_2/\text{KO}t\text{-Bu}$ as the catalytic system in 1,2-dimethoxyethane (monoglyme) proceed in low to good yields (8–88%),[88,89] recent advances allow selective access to monoarylated as well as symmetrical and unsymmetrical diarylated cyanoacetates. Optimization using high-throughput screening led to the development of a high-yielding arylation of cyanoacetates using $Pd(dba)_2/(t\text{-}Bu)_3P$ (Eq. 58).[90] However, reactions of aryl halides bearing electron-withdrawing groups generate varying amounts of diarylated side products. The selectivity of this method can be improved by switching from $(t\text{-}Bu)_3P$ to ligand **36**, without a noticeable decrease in the arylation yield. The reaction of 2 equivalents of aryl halide and ethyl cyanoacetate produces symmetrical diarylcyanoesters (Eq. 59) and the reaction of 1 equivalent of aryl halide with monoarylcyanoacetates cleanly affords the corresponding unsymmetrical diarylated cyanoacetates (Eq. 60).[90]

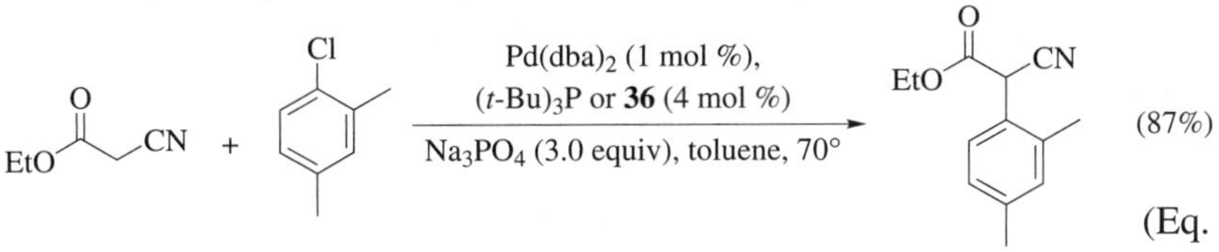

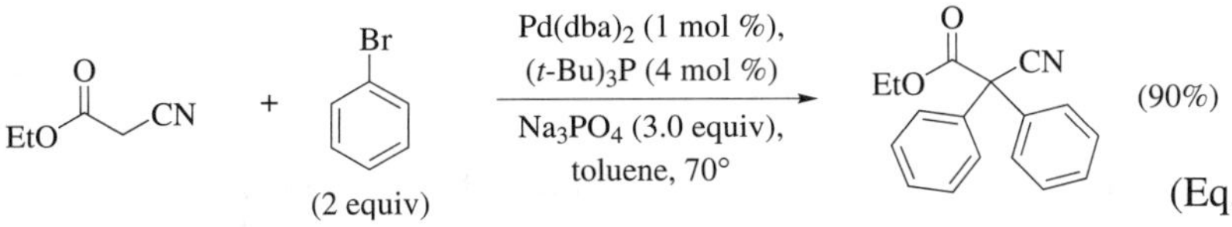

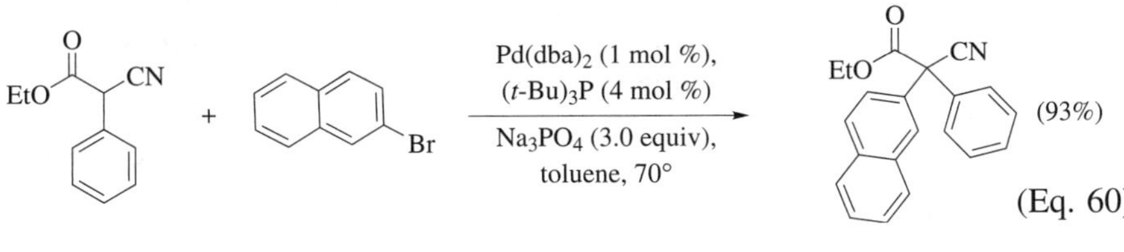

Vinylogy, Phenylogy. Site Selectivity in γ/ω-Arylation. In substrates incorporating vinyl and phenyl groups, arylation may occur at the γ- and ω-position according to vinylogy and phenylogy principles.[91] In α,β-unsaturated carbonyl substrates, where both α- and γ-positions are prone to arylation, only γ-arylation occurs even in substrates having hydrogen available for syn-β-elimination (Eq. 61).[92] Arylation at the β-position (Heck reaction) is not observed. Cyclic enones are also good substrates for γ-arylation (Eq. 62).[92] Mono- or diarylated compounds can be obtained selectively by changing the amount of aryl bromide.

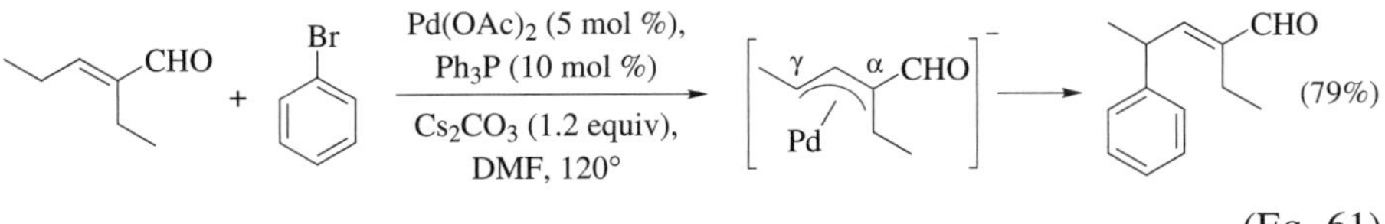

[Eq. 62: cyclopentenone with n-C5H11 and methyl + PhBr, Pd(OAc)2 (5 mol %), Ph3P (10 mol %), Cs2CO3 (1.2 equiv), DMF, 120° → benzylated product (58%)]

Alkylated nitroaromatic compounds undergo arylation in the side chains (Eq. 63).[93] Although good yields are obtained, the arylation process often affords mixtures of mono- and diarylated products. It is worth noting that selective diarylation may occur when 2 equivalents of the aryl bromide are used. However, this reaction is sensitive to steric factors and depends on the aromatic substitution pattern. Indeed, despite the use of 2 equivalents of the aryl bromide, selective monoarylation is obtained in the presence of neighboring substituents in either the aromatic halide or the nitro-substrate, or at the benzylic position.

[Eq. 63: nitrotoluene derivative + PhBr (1.2 equiv), Pd(OAc)2 (5 mol %), Ph3P (10 mol %), Cs2CO3 (1.2 equiv), DMF, 120° → monobenzylated nitroarene (70%)]

α-Arylation via α-Stannylmethyl Carbonyl Compounds. α-Stannylmethyl ketones are usually considered masked tin enolates. Although this strategy also requires the preparation of the stannylated coupling partner (in some cases such tin derivatives can be prepared in situ), α-aryl ketones are cleanly obtained in yields ranging from 51–91% in the presence of $PdCl_2[P(o\text{-tol})_3]_2$ (Eq. 64).[10]

[Eq. 64: CH3C(O)CH2SnBu3 + 4-Me2N-C6H4-Br, $PdCl_2[P(o\text{-tol})_3]_2$ (10 mol %), toluene, 100° → CH3C(O)CH2-C6H4-NMe2 (74%)]

Products arylated at the γ-position of α,β-unsaturated esters have been obtained from the corresponding tin derivatives (Eq. 65). The reaction takes place at the carbon directly bonded to tin. The allylic transposition product (α-arylation) is not detected.[94,95]

[Eq. 65: γ-stannyl α,β-unsaturated ester + 4-BrC6H4OMe, Pd(OAc)2 (5 mol %), Ph3P (20 mol %), toluene, reflux → γ-aryl product (30%)]

A similar strategy introduces an aromatic group on an acetate moiety.[96,97] Indeed, the preparation of the arylacetic fragment of elisabethin A involves arylation of an acetate ester through tin–zinc transmetallation and a subsequent Negishi-type coupling (Eq. 66).

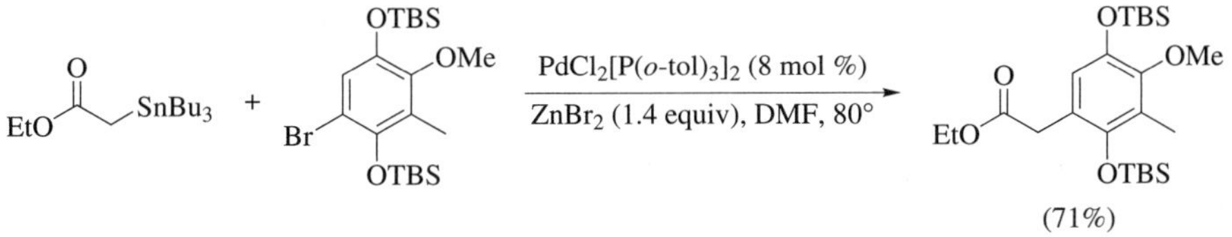

(Eq. 66)

α-Arylation of Silyl Enol Ethers and Ketene Acetals. The α-arylation of silyl enol ethers with aryl halides requires the use of a stoichiometric amount of Bu$_3$SnF as an additive. The reaction is assumed to proceed by in situ generation of tin enolate **43** via silicon–tin exchange followed by a palladium-catalyzed α-arylation (Eq. 67).[12,13]

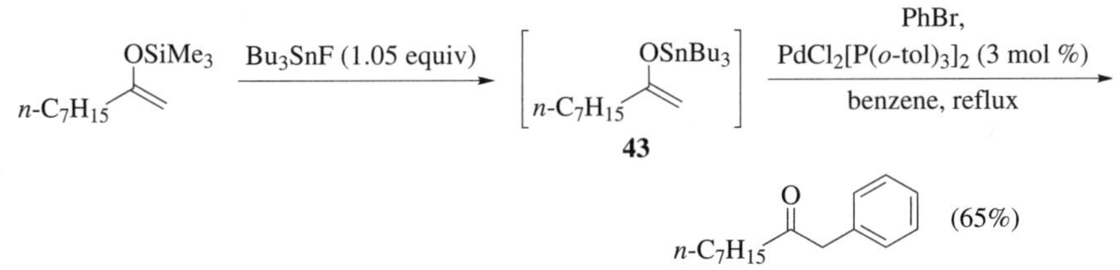

(Eq. 67)

Interestingly, good selectivity is observed in substrates bearing two silyl enol ether groups. Steric interactions between large alkyl groups and the approaching tin fluoride have been postulated to explain the observed selective arylation (Eq. 68).[12,13]

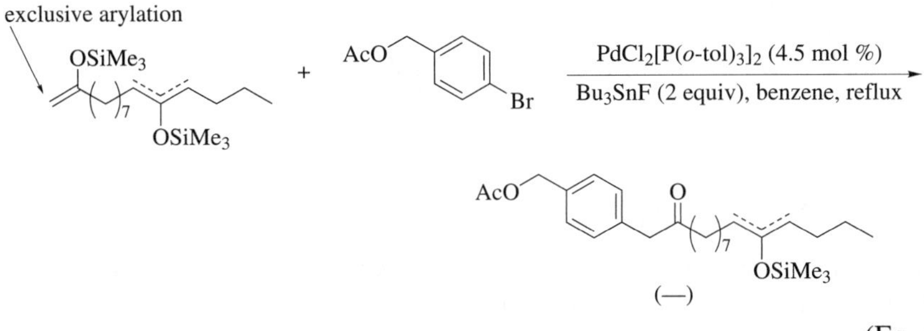

(Eq. 68)

α-Arylation of silyl ketene acetals also requires the use of additives, such as Bu$_3$SnF or CuF$_2$ (Eq. 69).[98]

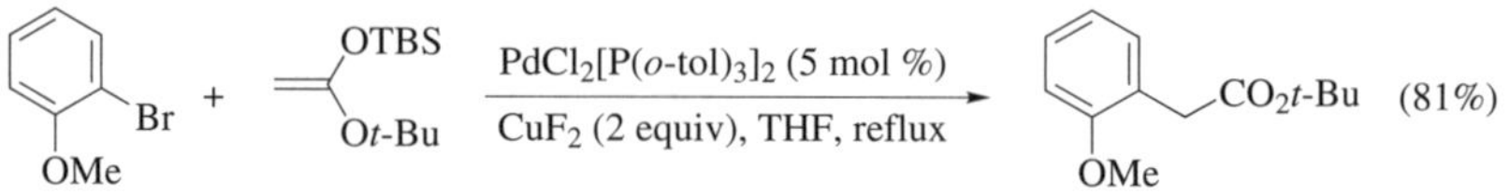

(Eq. 69)

More recently, a general α-arylation procedure for silyl ketene acetals that uses catalytic amounts of a Lewis acid promoter was reported. Thus, 0.25–0.50 equivalents of ZnF$_2$ or Zn(Ot-Bu)$_2$ allows complete reaction and arylation in high yields (Eq. 70). Since transmetalation products or accumulation of zinc enolates could not be detected, the exact role of the zinc-based additives is still unclear.[99]

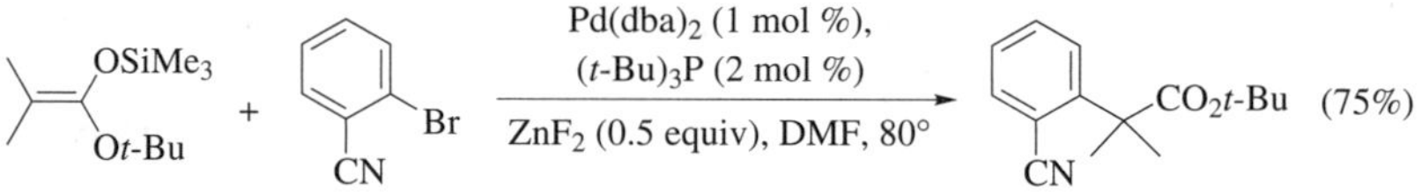

(Eq. 70)

This method has been extended to ketimines and silyl ketene acetals bearing chiral auxiliaries and thus to the diastereoselective formation of tertiary stereogenic centers (Eqs. 71 and 72).

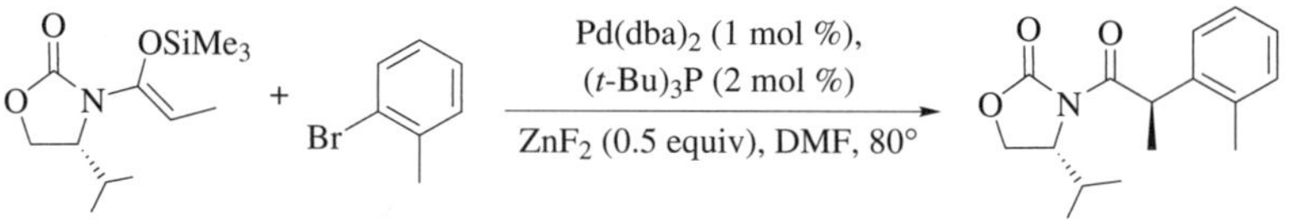

(78%) 92:8 dr

(Eq. 71)

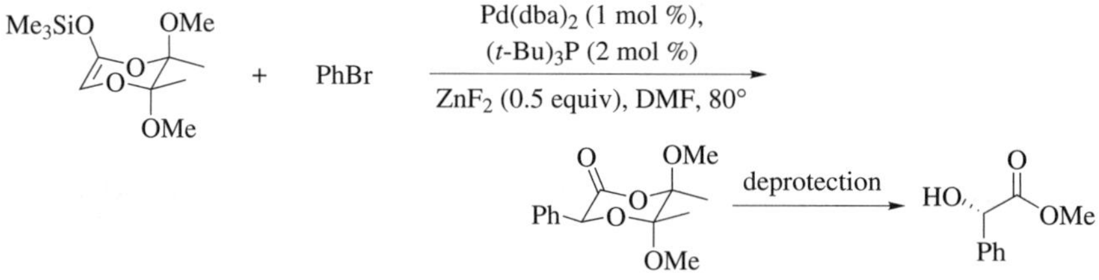

(67%) >50:1 dr   >99.5% ee

(Eq. 72)

The requirement for metal-based additives can be avoided by the use of aryl triflates. In the presence of palladium(II) and lithium acetate, α-arylated esters are often obtained in high yields (Eq. 73).[100]

$$\text{OSiMe}_3/\text{OMe} + \text{PhOTf} \xrightarrow[\text{THF, reflux}]{\substack{\mathbf{1}\ (8\ \text{mol \%}),\\ (\eta^3\text{-C}_4\text{H}_7\text{PdOAc})_2\ (2\ \text{mol \%})\\ \text{LiOAc (2 equiv)},}} \text{PhCH}_2\text{CO}_2\text{Me} \quad (53\%) \quad \text{(Eq. 73)}$$

Copper-Catalyzed α-Arylation

Although palladium complexes are the most studied catalytic systems for α-arylation of enolates, nickel and copper catalysts were initially used. Reaction of various halobenzoic acids in the presence of CuBr allows the formation of the expected arylation product (Eq. 74).[101]

$$\text{acac} + \text{2-BrC}_6\text{H}_4\text{CO}_2\text{H} \xrightarrow[\text{NaH (2 equiv)}]{\text{CuBr (6 mol \%)}} \text{product} \quad (74\%)$$

(Eq. 74)

The arylation is assumed to proceed through a copper-assisted nucleophilic displacement of the bromide anion. This strategy is limited to active methylene compounds and requires the presence of an *ortho* carboxylic acid group. This protocal has been extended to the arylation of malononitriles and related substrates,[102] and to the formation of benzofuran-2-ones.[103] The α-arylation of malononitrile and its derivatives proceeds in the presence of a catalytic amount of CuI as shown in Eq. 75.[102]

$$\text{NC-CH}_2\text{-CN} + \text{PhBr} \xrightarrow[\text{K}_2\text{CO}_3\ (8\ \text{equiv})]{\text{CuI (10 mol \%)}} \text{Ph-CH(CN)}_2 \quad (80\%) \quad \text{(Eq. 75)}$$

Catalytic amounts of CuBr promote the reaction between dimethyl malonate and 2-bromophenols to afford the corresponding benzofuran-2-ones in one step (Eq. 76). Access to benzofurans from the copper-catalyzed arylation of β-keto esters has also been described (Eq. 77).[104]

$$\text{MeO}_2\text{C-CH}_2\text{-CO}_2\text{Me} + \text{2-Br-4-Me-C}_6\text{H}_3\text{OH} \xrightarrow[\text{NaH (2 equiv)}]{\text{CuBr (20 mol \%)}} [\text{intermediate}] \longrightarrow \text{benzofuran-2-one} \quad (60\%)$$

(Eq. 76)

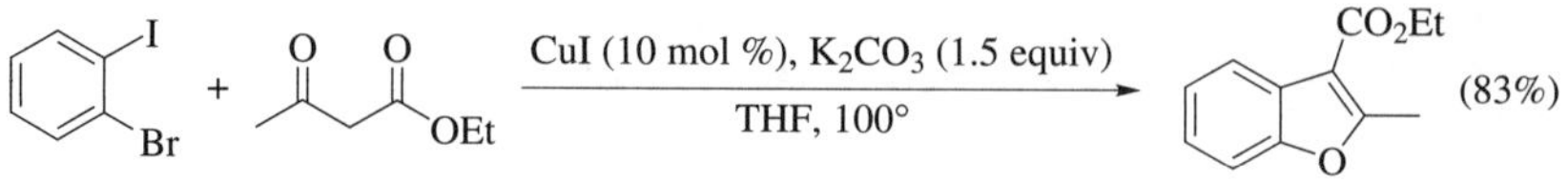

(Eq. 77)

The use of CuI in combination with 2-phenylphenol[105] or 2-nicotinic acid[106] as the ligand allows mild arylations of malonates in high to excellent yields at room temperature (Eq. 78).

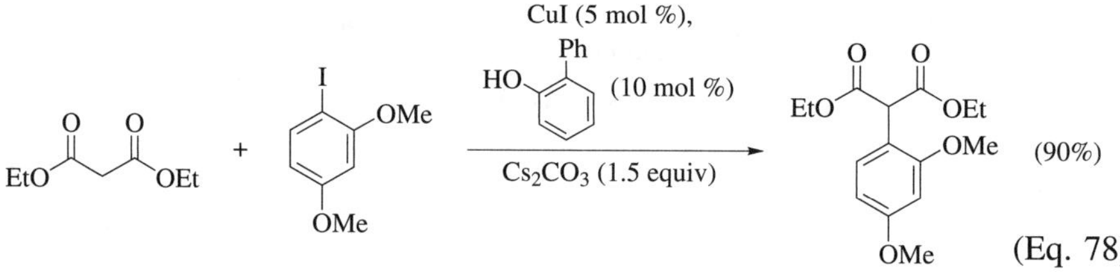

(Eq. 78)

The efficient copper-catalyzed enantioselective α-arylation of β-keto esters was presented earlier (Eq. 13).[36]

Nickel-Catalyzed α-Arylation

Until the recent development of a Ni–BINAP catalytic, enantioselective α-arylation of lactones (Eq. 9) and the *N*-heterocyclic carbene-based nickel complex **44** promoted α-arylation of acyclic ketones (Eq. 79),[107] examples of nickel-catalyzed arylations of enolates were rare. Two papers independently described the arylation of lithium (Eq. 80)[108] and zinc enolates (Eq. 81).[109]

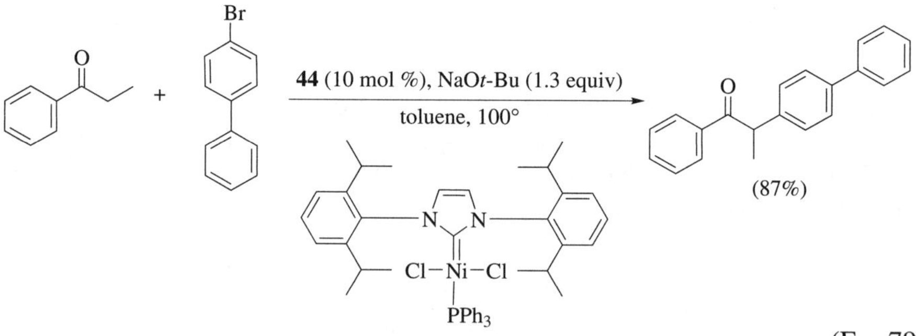

(Eq. 79)

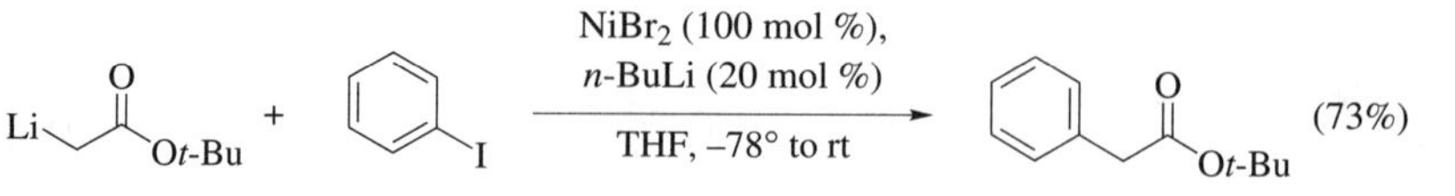

(Eq. 80)

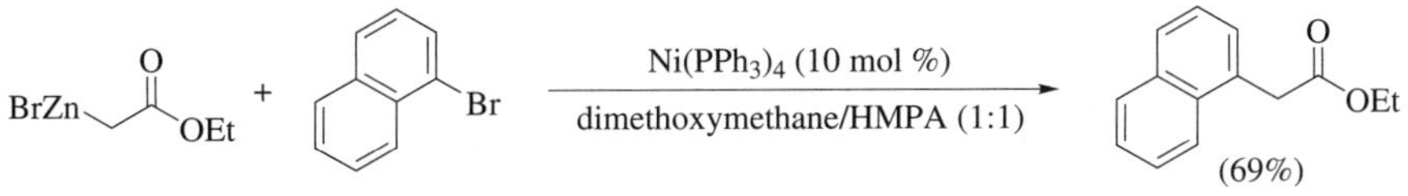

$$\text{BrZn} \diagdown \underset{\text{OEt}}{\overset{\text{O}}{\diagup}} + \text{(1-bromonaphthalene)} \xrightarrow[\text{dimethoxymethane/HMPA (1:1)}]{\text{Ni(PPh}_3)_4 \text{ (10 mol \%)}} \text{(naphthyl-CH}_2\text{-CO-OEt)}$$

(69%)

(Eq. 81)

Although the nature of the catalytic system (n-BuLi–NiBr$_2$) is still unclear, the arylation of preformed lithium ester enolates with aryl halides occurs in modest to high yields and is viable for both arylation and vinylation reactions (Eq. 80).[108] Preformed zinc enolates (Reformatsky-type reagents) can also be arylated in the presence of nickel(0) catalysts. However, the use of unstable and air sensitive Ni(PPh$_3$)$_4$, even at 10 mol % loading, affords good yields of the expected arylated products (Eq. 81).[109]

An efficient α-arylation of malononitrile has also been described using Ni(PPh$_3$)$_4$ (generated in situ from NiBr$_2$(PPh$_3$)$_2$, Ph$_3$P, and zinc) as the catalyst (Eq. 82).[110]

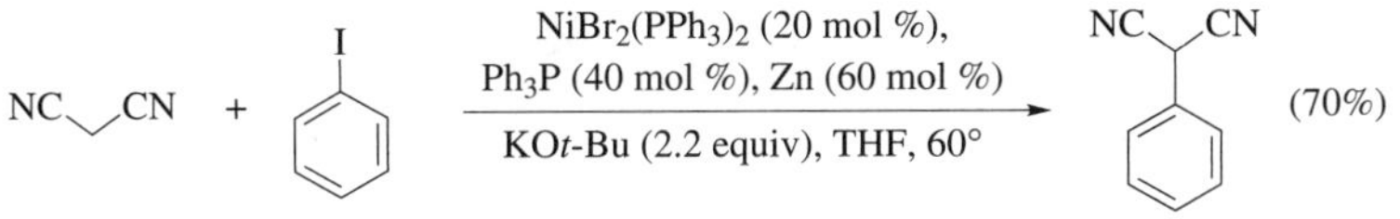

$$\text{NC} \diagdown \text{CN} + \text{PhI} \xrightarrow[\text{KO}t\text{-Bu (2.2 equiv), THF, 60°}]{\substack{\text{NiBr}_2(\text{PPh}_3)_2 \text{ (20 mol \%)},\\ \text{Ph}_3\text{P (40 mol \%), Zn (60 mol \%)}}} \text{Ph-CH(CN)}_2$$

(70%)   (Eq. 82)

APPLICATIONS TO SYNTHESIS

Transition-metal-catalyzed α-arylations of carbonyl compounds have found many useful synthetic applications,[15,111] particularly in the synthesis of natural products and biologically active substances. Bonjoch and Wills have described approaches to the syntheses of bridged azabicyclic and 1-vinyl-1H-isochromene compounds, respectively (Fig. 6).[63,112,113] As previously discussed, access to benzofurans (Eq. 26) and benzothiophenes is possible through a C-arylation, O-arylation sequence.[61,62,64]

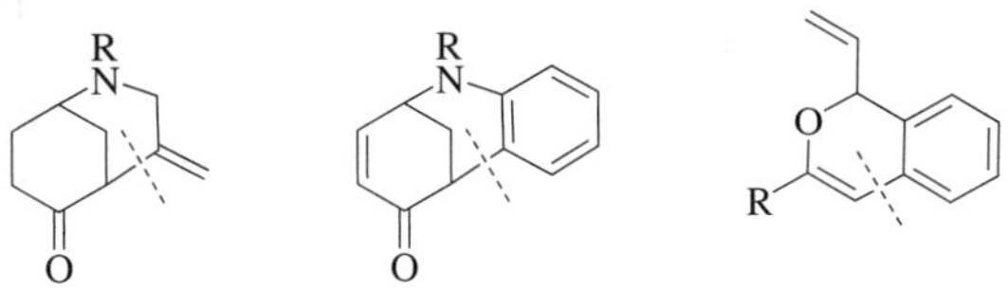

Figure 6. Palladium-assisted bond formation.

An efficient synthesis of trileptal, one of the most widely prescribed drugs for the treatment of epilepsy, involves a palladium-catalyzed α-arylation followed by a palladium-catalyzed N-arylation (Eq. 83).[114]

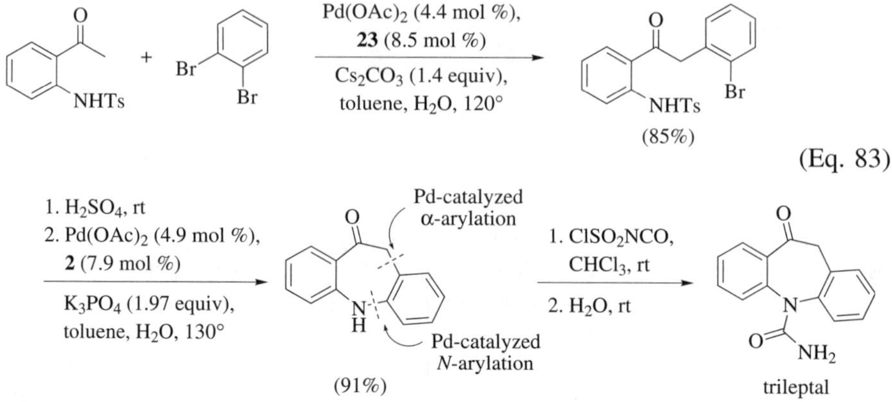

(Eq. 83)

The key step in an approach to the *N*-methylwelwitindoline skeleton involves an intramolecular palladium-catalyzed α-arylation of a cyclic β-keto ester derivative (Eq. 84).[86]

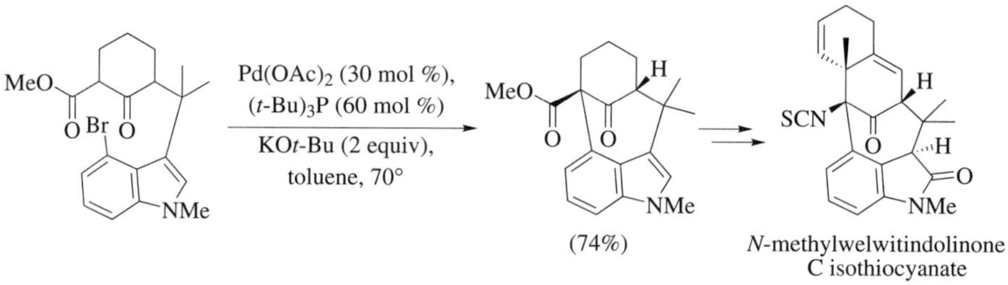

(Eq. 84)

Efficient syntheses of cherylline and latifine use the intramolecular α-arylation of amides (Eq. 85),[71] and a total synthesis of lennoxamine (Eq. 86) employs intramolecular α-arylation of an α-amido ketone.[115]

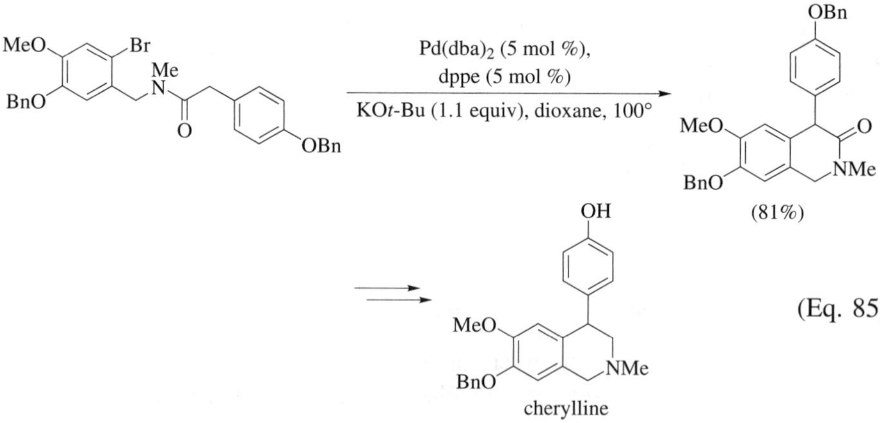

(Eq. 85)

(Eq. 86) reaction scheme showing synthesis of lennoxamine using Pd₂(dba)₃ (5 mol %), **2** (10 mol %), KOt-Bu (1.5 equiv), dioxane, 100°, (65%).

COMPARISON WITH OTHER METHODS

The arylation of enolates has been a challenging subject of research for more than fifty years. The advent of transition-metal-catalyzed carbon–carbon bond formation overcomes many of the drawbacks of existing methods. However, some catalytic, stoichiometric, and metal-free methods represent useful, complementary strategies that also allow the α-arylation of enolates. The following section focuses on: (1) palladium-catalyzed α-arylation from α-halomethyl ketones, (2) α-arylation involving the stoichiometric use of transition metals or metals, (3) miscellaneous methods including photo-induced α-arylation, iodonium salt α-arylation, and nucleophilic aromatic substitution using enolates.

Palladium-Catalyzed α-Arylation via α-Halomethyl Carbonyl Compounds

As shown in Eq. 87, oxidative addition of α-chloro ketones to a palladium(0) species affords C-bound "palladium enolates", some of which have been isolated and fully characterized.[116] Adducts of this type, such as that derived from ethyl bromoacetate, are able to couple with arylboronic acids under Suzuki-type reaction conditions (Pd(OAc)$_2$, P(1-naphthyl)$_3$, K$_3$PO$_4$) to give the corresponding arylated acetic acid ester derivatives (Eq. 88).[117] The main drawback of this strategy is the required preparation of α-halo carbonyl compounds prior to the coupling reaction.

PhCOCH₂Cl + (PPh₃)₄Pd → PhCOCH₂Pd(PPh₃)₂Cl (in toluene) (Eq. 87)

4-OHC-C₆H₄-B(OH)₂ + BrCH₂CO₂Et → 4-OHC-C₆H₄-CH₂CO₂Et, Pd(OAc)₂ (3 mol %), (1-Naphthyl)₃P (9 mol %), K₃PO₄ (5 equiv), THF, 20°, (74%) (Eq. 88)

Stoichiometric Transition-Metal- and Metal-Promoted α-Arylation

Copper-Mediated α-Arylation. Stoichiometric, copper-mediated α-arylation of ketones and other carbonyl derivatives has been known for almost a century.

Early reports deal with arylation of acetylacetone, ethyl malonate, and active methylene anions in the presence of copper-bronze or copper acetate.[118–120] More recently, reactions of α-halo tosylhydrazones with an excess of phenylcopper (Eq. 89)[121] or a bromination–phenylation sequence on enamines (Eq. 90)[122] smoothly afford the products, which are converted into the corresponding α-phenyl cycloalkanones.

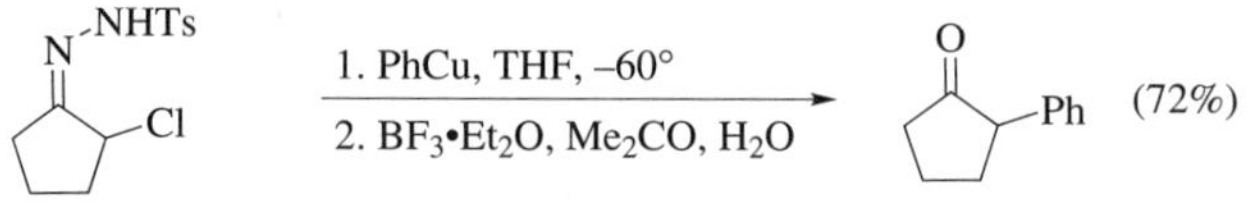

(Eq. 89)

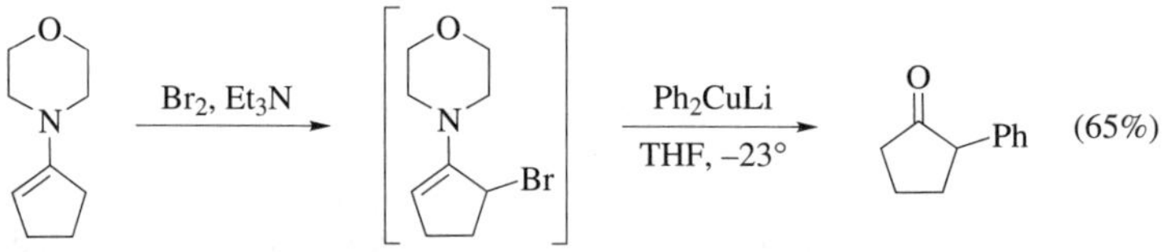

(Eq. 90)

α-Chloromethyl trimethylsilyl enol ethers also undergo arylation using lithium diarylcuprate. Depending on the reaction temperature, the arylated silyl enol ether **45** or the corresponding arylated ketone **46** may be obtained (Eq. 91).[11]

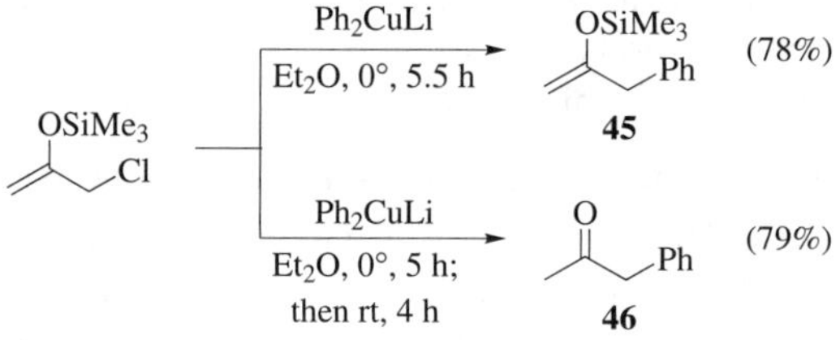

(Eq. 91)

Activation of symmetrical and unsymmetrical active methylene compounds with 1.2 to 1.3 equivalents of CuBr in refluxing dioxane allows the arylation of malonates (Eq. 92).[123,124]

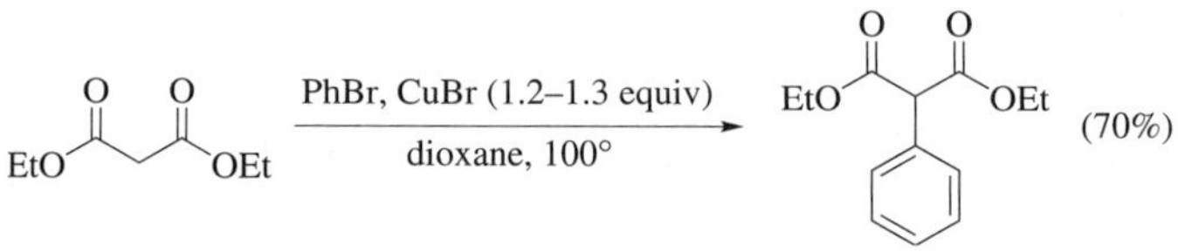

(Eq. 92)

CuI is used as a stoichiometric coupling agent in arylation reactions of unsymmetrical, phosphonyl-stabilized carbanions (Eq. 93). Two equivalents of both CuI and the active methylene reactant are required to obtain good to high yields.[125]

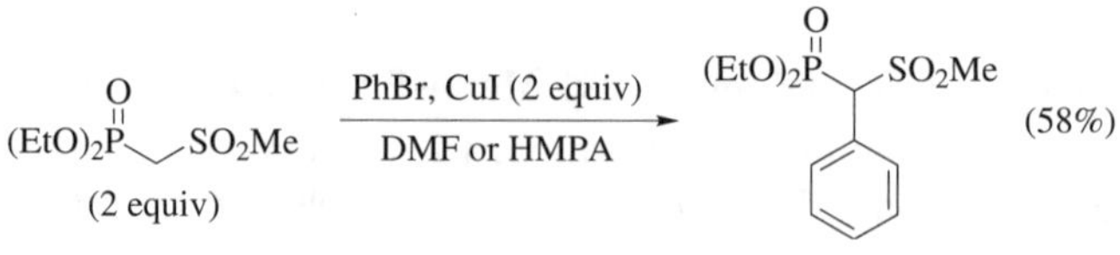

(Eq. 93)

Magnesium- and Titanium-Mediated α-Arylation. Although stoichiometric magnesium- or titanium-based arylations are rare, their use allows easy access to arylated ketones or diarylated esters in some cases. An example is the reaction of 2-chlorocyclohexanone with phenylmagnesium bromide (Eq. 94).[126]

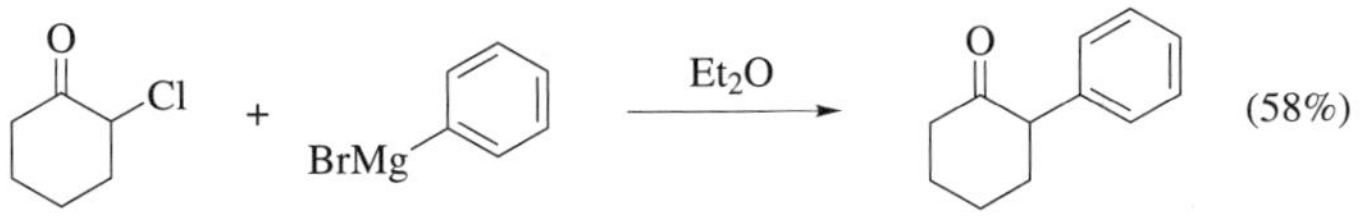

$$\text{2-chlorocyclohexanone} + \text{PhMgBr} \xrightarrow{\text{Et}_2\text{O}} \text{2-phenylcyclohexanone} \quad (58\%) \quad \text{(Eq. 94)}$$

The reaction of α-arylacetic esters with tertiary arylamines in the presence of stoichiometric amounts of $TiCl_4$ affords the corresponding α,α-diarylated ester (Eq. 95).[127] The proposed mechanism involves both the titanium ester enolate **47** and aryltitanium species **48**.

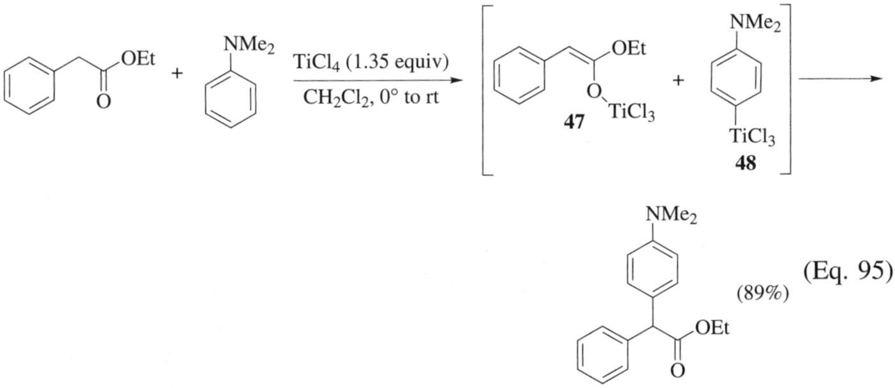

(Eq. 95) (89%)

Bismuth-Mediated α-Arylation. Ketones, active methylene derivatives, silyl enol ethers, and silyl ketene acetals are arylated under basic or neutral conditions using stoichiometric amounts of polyarylbismuth(V) compounds such as pentaphenylbismuth, tetraphenylbismuth chloride or triphenylbismuth dichloride as phenylating agents.[128,129] The postulated mechanism involves the O–Bi bonded intermediate **49** (Eq. 96). Internal ligand coupling between the aryl group and the enolate carbon accounts for the observed α-arylated products.

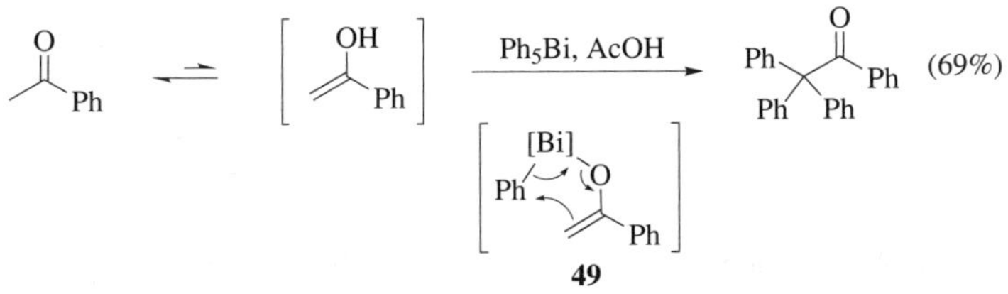

(Eq. 96) (69%)

α,β-Unsaturated carbonyl derivatives undergo site selective α-arylation with concomitant deconjugation of the olefinic moiety (Eq. 97).[37,130]

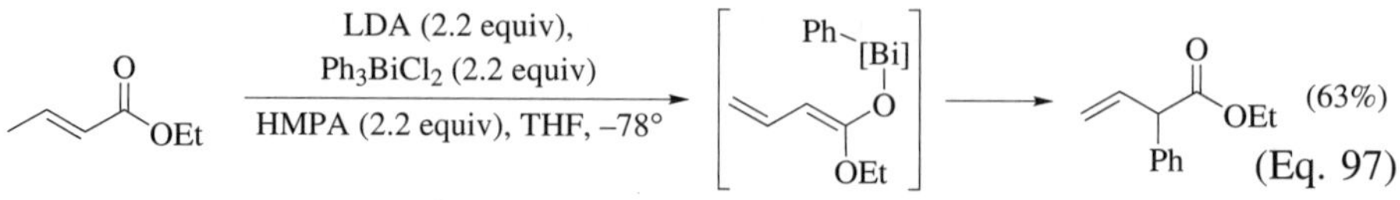

$$\text{(Eq. 97)}$$

Cyclic α,β-unsaturated ketones react with arylbismuth reagents in the presence of tributylphosphine and Hünig's base to afford α-arylation products (Eq. 98). The reaction likely proceeds by formation of phosphonium intermediate **50** followed by the generation of Bi-enolate **51**, the α-arylation reaction, and subsequent elimination of the phosphine.[38]

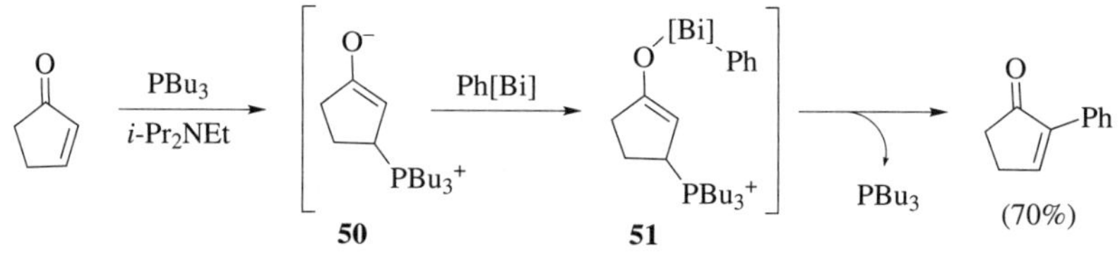

$$\text{(Eq. 98)}$$

Lead-Mediated α-Arylation. Aryllead triacetates effect the construction of α-aryl carbonyl compounds under mild conditions. As shown in Fig. 7, this method introduces aryl groups into various substrates, including β-dicarbonyls, active methylene compounds, lactones, dihydroindoles, benzofurans, and neoflavones.[131–135] In most cases, 1.1 equivalents of the aryllead derivative, obtained through Hg–Pb or Sn–Pb exchange, are required to ensure complete transformation. Reaction rates are significantly enhanced by the use of additional ligands such as 1,10-phenanthrolines.[136]

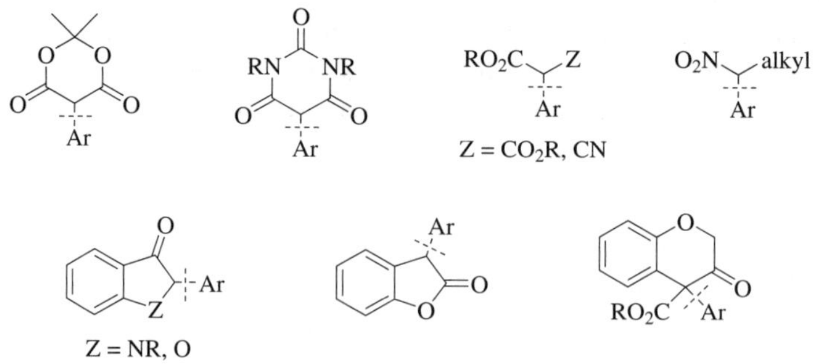

Figure 7. α-Aryl carbonyl compounds formed with aryllead triacetates.

Asymmetric coupling of aryllead reagents with β-keto esters is achieved by replacement of one or more of the labile acetate ligands with enantiomerically pure carboxylic acids such as **52** or the binaphthyl dicarboxylic acid **53** (Eq. 99).[133,137]

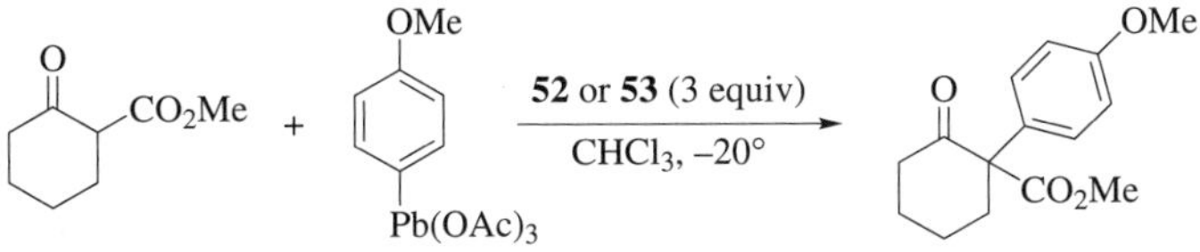

(Eq. 99)

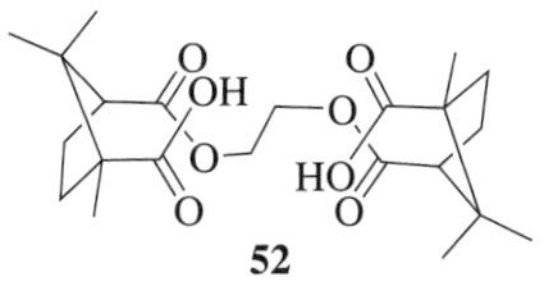

Miscellaneous α-Arylation Methods

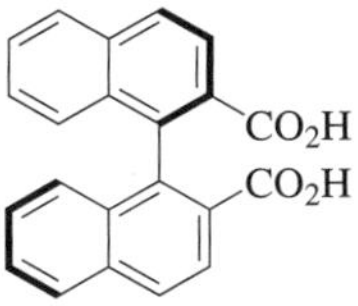

Introduction of aryl groups at the α-carbon of carbonyl compounds may also be accomplished through several non-metallic procedures. The following section briefly illustrates the most relevant ones, including nucleophilic or photo-induced aromatic substitution and α-arylation using iodonium salts.

Nucleophilic aromatic substitution of activated aryl halides by carbon nucleophiles proceeds through Meisenheimer-type intermediates and is followed by rearomatization. This process requires the presence of strong electron-withdrawing groups or complexation of the aromatic system with metallic fragments such as $Cr(CO)_3$. In this context, efficient arylation of ketone hydrazones with η^6-(chlorobenzene)$Cr(CO)_3$ followed by decomplexation and hydrolysis has been reported (Eq. 100).[138]

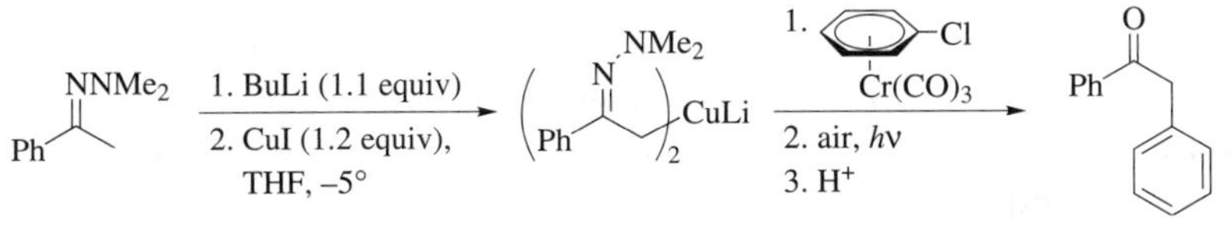

(Eq. 100)

In addition, the diastereoselective arylation of protected mandelic acids with fluoronitrobenzenes under basic conditions affords the expected arylated products (Eq. 101).[139]

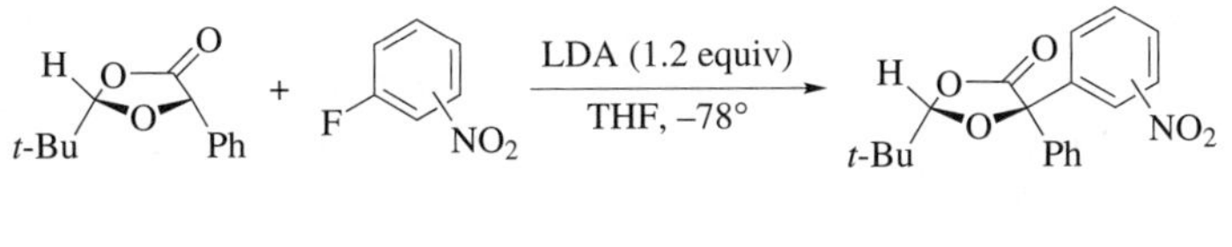

(Eq. 101)

Photo-stimulated formation of carbon–carbon bonds between aromatics and enolate equivalents has also received considerable attention. Photo-induced generation of phenyl cations and subsequent reaction with enamines, silyl enol ethers, or ketene silyl acetals afford the corresponding α-arylated carbonyl compounds (Eqs. 102 and 103).[96,140,141]

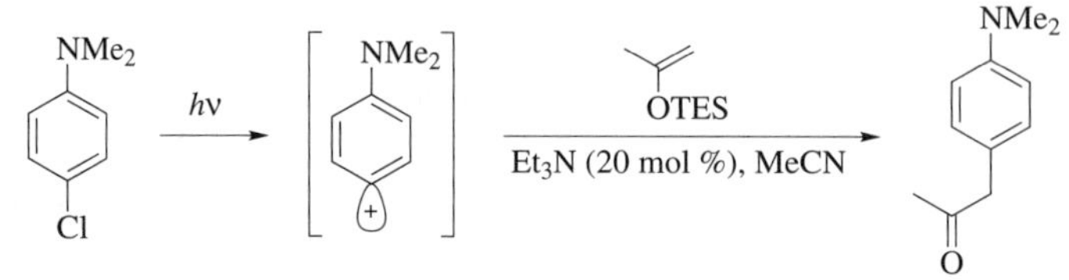

$$\text{(Eq. 102)} \quad (63\%)$$

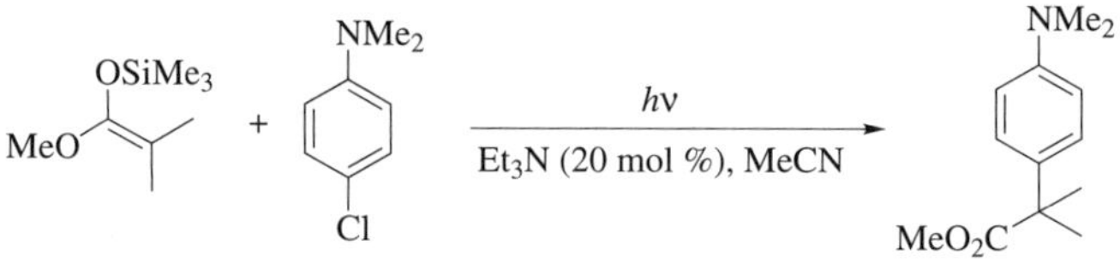

$$\text{(Eq. 103)} \quad (60\%)$$

An alternative strategy involves the use of diaryliodonium salts as the "phenyl cation" species. Such species exploit the excellent nucleofugality of the phenyliodonium group, which is about 10 times higher than that of triflate.[142] The treatment of silyl enol ethers with diaryliodonium fluorides affords the expected arylated ketones. Although the mechanism is still unclear, it might involve a trivalent iodine intermediate **54** as suggested in Eq. 104.[143,144]

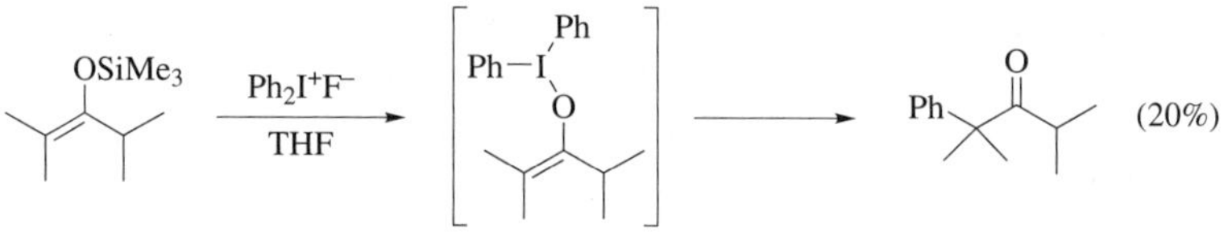

$$\text{(Eq. 104)} \quad (20\%)$$

A second alternative, involving deprotonation of cyclic ketones followed by generation of a copper enolate and subsequent coupling using diaryliodonium salts, is shown in Eq. 105.[145]

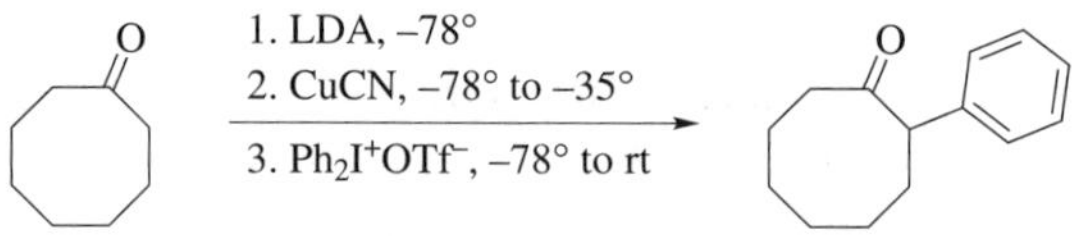

$$\text{(Eq. 105)} \quad (55\%)$$

A recent synthesis of (−)-epibatidine employs a chiral amide base to desymmetrize 4-substituted cyclohexanone derivatives, which then react with a diaryliodonium salt to give the arylated products in high yields, and modest dr and ee (Eq. 106).[146]

(Eq. 106)

EXPERIMENTAL CONDITIONS

An attractive aspect of the transition-metal-catalyzed α-arylation of enolates is its experimental simplicity. Although the use of a glove-box was initially recommended, most of the reactions can be carried out efficiently using normal Schlenk techniques in anhydrous solvents. As discussed above, a wide variety of different conditions (palladium source, ligand, base, solvent) have been developed, but it appears that $Pd_2(dba)_3$ [(or $Pd(OAc)_2$), $(t\text{-Bu})_3P$ in dioxane (or toluene)] in the presence of an excess of NaOt-Bu [(NaHMDS or LiNCy$_2$ in the arylation of esters and amides)] are often the first experimental conditions to be tried.

EXPERIMENTAL PROCEDURES

2-(2-Bromophenyl)cyclohexanone [Palladium-Catalyzed α-Arylation of a Ketone].[62] Cesium carbonate (4.570 g, 14.00 mmol) was added to a flask charged with $Pd_2(dba)_3$ (0.030 g, 0.033 mmol) and Xantphos (**24**) (0.040 g, 0.080 mmol) under nitrogen. The reagents were suspended in anhydrous dioxane (6.4 mL); 1-Bromo-2-iodobenzene (1.80 g, 6.37 mmol, 0.82 mL) and cyclohexanone (1.25 g, 12.74 mmol, 1.3 mL) were added under nitrogen, and the reaction mixture was heated at 80° for 24 h. After cooling, the reaction mixture was diluted with Et$_2$O (ca. 10 mL), filtered through Celite, and the solvents were removed in vacuo. The residue was purified by flash column chromatography (5–10% Et$_2$O/petroleum ether) to give 1.26 g (78% yield) of the title product: mp 57–58° (MeOH); IR (Nujol) 2920, 2855, 1709, 1566 (w), 1462, 1377, 1281, 1196, 1121, 1070, 1027, 977, 940, 769, 746, 722, 674 cm^{-1}; ^{1}H NMR (300 MHz, CDCl$_3$) δ 7.56 (td, J = 7.9, 1.5 Hz, 1H, Ar–H), 7.31 (td, J = 7.9, 1.1 Hz, 1H,

Ar–H), 7.21 (dd, J = 7.9, 1.9 Hz, 1H, Ar–H), 7.12 (ddd, J = 7.9, 7.2, 1.9 Hz, 1H, Ar–H), 4.11 (app. dd, J = 12.4, 5.3 Hz, 1H, Ar–CH), 2.89–2.51 (m, 2H, CH_2CO), 2.35–2.15 (m, 2H, ArCHCH$_2$), 2.10–1.71 (m, 4H, CH_2); ^{13}C NMR (75 MHz, CDCl$_3$) δ 208.3, 137.8, 132.1, 128.9, 127.8, 126.8, 124.6, 56.0, 41.8, 33.6, 27.1, 25.1; LRMS (CI$^+$, NH$_3$) m/z: [M + NH$_4$]$^+$ 270, [M + H:^{79}Br]$^+$ 253, [M − ^{79}Br]$^+$ 173, [M − ^{79}Br–CO]$^+$ 145, [M − ^{79}Br–CO–C$_2$H$_4$]$^+$ 115; HRMS (ES$^+$): [M + H]$^+$ calcd for C$_{12}$H$_{14}$BrO, 253.0223; found, 253.0225.

1-Phenyl-2-arylpropan-1-one [α-Arylation of a Ketone Using an N-Heterocyclic Carbene-Based Catalytic System].[51] In a drybox, 1.5 mmol of base (typically NaOt-Bu) was added to a screw-cap vial charged with 1 mol % of [N,N'-(2,6-diisopropyl phenyl)imidazol-2-ylidene]Pd(OAc)$_2$ complex. Dioxane (1.5 mL) was added and the vial sealed with a rubber septum. Outside the drybox, propiophenone (1.2 mmol) followed by the aryl halide (1 mmol) were injected into the vial with a syringe. The reaction mixture was shaken on a Lab-Line Orbit Shaker at 60° (J-Kem Scientific, Kem-Lab Controller) or stirred over a magnetic plate in an oil bath set at 60° for the indicated time. The reactions were monitored by gas chromatography. After reaching maximum conversion, the reaction mixture was allowed to cool to rt and it was then quenched with water. The water layer was extracted with methyl $tert$-butyl ether or Et$_2$O and dried over magnesium sulfate. The solvent was then evaporated in vacuo. When necessary the product was purified by flash chromatography on silica gel.

1-Vinyl-3-*tert*-butyl-1H-isochromene [Palladium-Catalyzed α-Arylation of a Ketone].[63] A solution of LiHMDS (1 M in THF, 3 equiv) was treated slowly with a solution of 3,3-dimethylbutan-2-one (2 equiv) at 5°. A solution of Pd$_2$(dba)$_3$ (5 mol %) and dppf (**1**) (10 mol %) in solvent was added at rt, followed by a solution of (E/Z)-*tert*-butyldimethyl[3-(2-bromophenyl)allyloxy]silane (1 equiv). The reaction mixture was heated at 100° overnight and quenched with 1 M HCl solution at rt. The mixture was twice extracted with CH$_2$Cl$_2$, the combined organic layers were dried, and the solvent was evaporated under reduced pressure yielding the crude product, which was purified by chromatography (details not reported) to give the title compound in 71% yield: IR (NaCl) 2954, 1638, 1085, 914, 750 cm^{-1}; ^{1}H NMR (300 MHz, CDCl$_3$) δ 7.30–6.90 (m, 4H), 6.15 (ddd, $J = 17.1, 10.4, 6.6$ Hz, 1H), 5.67 (s, 1H), 5.46 (d, $J = 6.6$ Hz, 1H), 5.16 (dt, $J = 17.1, 1.3$ Hz, 1H), 5.23 (dt, $J = 10.4, 1.3$ Hz, 1H), 1.16 (s, 9H); ^{13}C NMR (75 MHz, CDCl$_3$) δ 164.2, 136.7, 132.0, 129.4, 128.5, 126.2, 124.9, 123.8, 118.2, 97.7, 79.3, 35.5, 28.3; MS (EI) m/z: M$^+$ 214, 187, 129; HRMS (EI) m/z: calcd for C$_{15}$H$_{18}$O, 214.135765; found, 214.136173.

tert-Butyl 2-Phenyl-2,3-dihydro-1H-isoindole-1-carboxylate [Palladium-Catalyzed Intramolecular α-Arylation of an α-Amino Acid Ester].[78] An oven-dried resealable Schlenk tube containing a magnetic stir bar was evacuated and purged with argon while cooling to ambient temperature. The Schlenk tube was charged with LiOt-Bu (2.0 equiv), Pd$_2$(dba)$_3$ (2.25 mol %), and 2-(dicyclohexylphosphino)-2'-(N,N-dimethylamino)biphenyl (5 mol %). The reaction vessel was evacuated, backfilled with argon, and dioxane (2 mL) was added via syringe under argon. The mixture was stirred for 5 min at rt, followed by addition of a solution of dodecane (0.045 mL, 0.2 mmol) and the substrate (0.188 g, 0.5 mmol) in dioxane (0.5 mL) via syringe. The solution was 0.2 M in dioxane as solvent. The Schlenk tube was sealed and placed in a preheated oil bath at 85° for 1 h. After complete conversion, as judged by either GC or TLC analysis, the reaction mixture was cooled to ambient temperature. The crude mixture was filtered through a silica gel plug, and the plug then washed thoroughly with ether. The filtrate was concentrated and further purified by flash column chromatography (hexanes/ether 15:1) to give 0.115 g (78% yield) of the title product: mp 72°; IR (CH$_2$Cl$_2$) 3043, 2979, 2933, 2875, 2840, 1740, 1603, 1505, 1468, 1368, 1146, 1094, 1036, 1003, 955, 839, 750, 690 cm^{-1}; ^{1}H NMR (400 MHz, C$_6$D$_6$) δ 7.47–7.41 (m, 1H), 7.32–7.26 (m, 2H), 7.10–7.03 (m, 2H), 6.97–6.91 (m, 1H), 6.86–6.81 (m, 1H), 6.70–6.55 (m, 2H), 5.36–5.33 (d, $J = 3.5$ Hz, 1H), 4.59–4.53 (dd, $J = 3.5, 12.9$ Hz, 1H), 4.36–4.31 (d, $J = 12.9$ Hz, 1H), 1.19 (s, 9H); ^{13}C NMR (100 MHz, C$_6$D$_6$) δ 171.2, 146.9, 139.1, 138.0, 130.0, 128.7, 127.8, 123.4, 123.0, 117.7, 112.8, 81.5, 68.3, 57.8, 54.3, 28.1. Anal. Calcd for C$_{19}$H$_{21}$NO$_2$: C, 77.26; H, 7.17. Found: C, 77.12; H, 7.17.

1-Benzyloxindole [Preparation of an Oxindole by Palladium-Catalyzed Intramolecular α-Arylation of an Amide].[67] In a drybox, Pd(dba)$_2$ (57.5 mg, 0.10 mmol), BINAP (**2**) (93.4 mg, 0.15 mmol), and potassium *tert*-butoxide (288 mg, 3.00 mmol) were combined in a round-bottom flask. Dioxane (18 mL) was added, and the flask was sealed with a septum. After removing the flask from the drybox, 2-bromo-*N*-benzylacetanilide (609 mg, 2.00 mmol) was added. The flask was placed in an oil bath at 100° for 3 h. The reaction mixture was poured into 50 mL of saturated NH$_4$Cl solution and extracted (3 × 30 mL) with ether. The combined ether extracts were washed with brine (50 mL), dried over MgSO$_4$, and filtered. The solvent was removed under vacuum, and the resulting crude product was purified by flash chromatography on silica gel (hexanes/EtOAc 85:15); recrystallization from hexanes gave 0.297 g (66% yield) of the title product as pale yellow needles: mp 66.5–67°; FTIR (neat, NaCl plate) 1717 cm^{-1}; ^{1}H NMR (500 MHz, CDCl$_3$) δ 7.34–7.33 (m, 3H), 7.30–7.26 (m, 2H), 7.19 (t, *J* = 7.8 Hz, 1H), 7.03 (t, *J* = 7.3 Hz, 1H), 6.75 (d, *J* = 7.5 Hz, 1H), 4.94 (s, 2H), 3.65 (s, 2H); ^{13}C NMR (125 MHz, CDCl$_3$) δ 175.3, 144.4, 135.9, 128.8, 127.9, 127.7, 127.4, 124.5, 124.5, 122.5, 109.2, 43.8, 35.8. Anal. Calcd for C$_{15}$H$_{13}$NO: C, 80.68; H, 5.87; N, 6.27. Found: C, 80.59; H, 5.92; N, 6.16.

(1-Nitroethyl)benzene [Palladium-Catalyzed α-Arylation of Nitroethane].[147] A dried, resealable Schlenk tube containing a magnetic stir bar was charged with Pd$_2$(dba)$_3$ (27.4 mg, 0.03 mmol, 3 mol %), 2-(di-*tert*-butylphosphino)-2′-methylbiphenyl (**8**) (19.2 mg, 0.06 mmol, 6 mol %), and 1.2 equiv of Cs$_2$CO$_3$. The reaction vessel was capped with a rubber septum, evacuated, and backfilled three times with argon, and 1.0 mmol of bromobenzene, dimethoxyethane (5 mL), and 1.0–2.0 equiv of nitroethane were added sequentially via syringe under argon. The mixture was stirred vigorously for 1 min at rt, the Schlenk tube was sealed and placed in a preheated oil bath at 50° for 8–30 h. After completion of the reaction, as judged by either GC or TLC analysis, the reaction mixture was allowed to cool to ambient temperature. The crude mixture was quenched with a solution of saturated aqueous NH_4Cl (2 × 2 mL, two times), the aqueous phase was extracted with ether (2 mL), and the combined organic phases were washed with brine. The solvent was removed, and the remaining oil was purified by flash silica gel column chromatography (hexanes/Et_2O 20:1) to provide 0.136 g (90% yield) of the title product: ^{1}H NMR (300 MHz, $CDCl_3$) δ 7.48–7.40 (m, 5H), 5.62 (q, J = 6.9 Hz, 1H), 1.89 (d, J = 6.9 Hz, 3H); ^{13}C NMR (75 MHz, $CDCl_3$) δ 135.7, 129.9, 129.1, 127.5, 86.3, 19.5.

Diethyl 2-Phenylmalonate [Palladium-Catalyzed α-Arylation of a Malonate with an Aryl Bromide].[53,82] To a screw-capped vial in a dry box containing diethyl malonate (176 mg, 1.1 mmol) was added tetrahydrofuran (1.0 mL) followed by NaH (1.1 mmol). Upon completion of hydrogen evolution (ca. 2 min), bromobenzene (157 mg, 1.00 mmol), (t-Bu)$_3$P (0.040 mmol), $Pd(dba)_2$ (0.020 mmol), and THF (2.0 mL) were added. The vial was sealed with a cap containing a PTFE septum and removed from the drybox. The homogeneous reaction mixture was stirred at 70° and monitored by GC. After complete conversion of the aryl halide, the crude reaction mixture was filtered through a plug of Celite and concentrated in vacuo. The residue was purified by column chromatography on silica gel (hexanes/CH_2Cl_2 2:1) to give 0.204 g (87% yield) of the title compound: ^{1}H NMR ($CDCl_3$) δ 7.43–7.34 (m, 5H), 4.62 (s, 1H), 4.27–4.18 (m, 4H), 1.27 (t, J = 7.2 Hz, 6H); ^{13}C NMR ($CDCl_3$) δ 168.17, 132.81, 129.27, 128.59, 128.20, 61.80, 57.96, 14.01.

Ethyl 2-(2-Methylphenyl)cyanoacetate [Palladium-Catalyzed α-Arylation of Ethyl Cyanoacetate].[82] Into a screw-capped vial containing ethyl cyanoacetate (125 mg, 1.10 mmol) and 2-bromotoluene (172 mg, 1.00 mmol), were added (*t*-Bu)₃P (40 μL, 1.0 M in toluene, 0.040 mmol), Pd(dba)₂ (3.7 mg, 0.010 mmol), and Na₃PO₄ (3.0 mmol), followed by toluene (3.0 mL). The vial was sealed with a cap containing a PTFE septum and removed from the drybox. The heterogeneous reaction mixture was stirred at 70° and monitored by GC. After complete conversion of the aryl bromide, the reaction mixture was filtered through a plug of Celite and concentrated in vacuo. The residue was purified by chromatography on silica gel (hexanes/CH_2Cl_2, 3:1) to give the title product (88% yield): ¹H NMR ($CDCl_3$) δ 7.47–7.45 (m, 1 H), 7.32–7.22 (m, 3H), 4.89 (s, 1H), 4.31–4.19 (m, 2H), 2.40 (s, 3H), 1.28 (t, $J = 7.2$ Hz, 3H); ¹³C NMR ($CDCl_3$) δ 165.03, 136.23, 131.24, 129.34, 128.93, 128.62, 127.05, 115.90, 63.25, 41.06, 19.42, 13.92.

(*R*)-(+)-2-Methyl-2-(4-anisyl)-5-(*N*-methylanilinomethylene)cyclopentanone [Palladium-Catalyzed Asymmetric α-Arylation of a Ketone Enolate].[41]

An oven-dried Schlenk tube equipped with a rubber septum was evacuated and backfilled with argon. The tube was charged with tris(dibenzylideneacetone)dipalladium (0.005 mmol), (*S*)-BINAP (0.125 mmol), and (*E*)-2-methyl-5-((methyl(phenyl)amino)methylene)cyclopentanone (0.50 mmol). The tube was evacuated and backfilled three times with argon. Toluene (2 mL) was added and the mixture was stirred for 15 min at rt. 4-Bromoanisole (1.00 mmol) and sodium *tert*-butoxide (96 mg, 1.00 mmol) were added to the tube. The tube was capped with a septum, purged with argon, and additional toluene (1 mL) was added through the septum. The mixture was stirred at rt until the starting ketone had been completely consumed as judged by GC analysis. The reaction mixture was quenched with saturated aqueous ammonium chloride (10 mL) and diluted with ether (20 mL). The aqueous layer was extracted with ether (20 mL) and the combined organic layers were washed with brine (20 mL), dried over anhydrous magnesium sulfate, filtered, and concentrated in vacuo. The crude material was purified by silica gel chromatography (10–20% EtOAc/hexanes) to give the title product (74% yield, 57% ee): $[\alpha]^{20}$+1.53 (*c* 10.0, $CHCl_3$); ¹H NMR (300 MHz, $CDCl_3$) δ 7.62 (t, $J = 1.5$ Hz, 1H), 7.33–7.29 (m, 4H), 7.13–7.08 (m, 3H), 6.85–6.80 (m, 2H), 3.75 (s, 3H), 3.45 (s, 3H), 2.56–2.32 (m, 3H), 1.89–1.79 (m, 1H), 1.41 (s, 3H); ¹³C NMR (75 MHz, $CDCl_3$) δ 207.5, 157.8, 146.1, 142.4, 136.1, 129.1, 127.4, 124.6, 121.1, 113.6, 108.0, 55.3, 51.9, 40.1,

36.6, 25.4, 24.9; MS (EI) m/z: M$^+$ 321. Anal. Calcd for $C_{21}H_{23}NO$: C, 82.58; H, 7.59. Found: C, 82.53; H, 7.68.

[Scheme: α-methyl-γ-butyrolactone + 3-chloroanisole, Ni(cod)$_2$ (5 mol %), (S)-2 (8.5 mol %), ZnBr$_2$ (15 mol %), NaHMDS (2.3 equiv) → 3-(3-methoxyphenyl)-3-methyldihydrofuran-2-one (86%) 96% ee; (S)-2 = (S)-BINAP with PPh$_2$, PPh$_2$ groups]

(−)-3-(3-Methoxyphenyl)-3-methyldihydrofuran-2-one [Nickel-Catalyzed Asymmetric α-Arylation of a Lactone].[33] An oven-dried, resealable Schlenk tube containing a magnetic stir bar was allowed to cool to rt and was then charged with (S)-BINAP (2) (13.2 mg, 21.3 μmol). The tube was sealed, evacuated, and backfilled with argon. From a freshly prepared, yellow, homogeneous stock solution of Ni(cod)$_2$ (0.05 M, toluene), 250 μL (12.5 μmol) was added by syringe while purging with argon. The tube was sealed and heated to 60° for 5 min during which time the solution turned dark red. The reaction vessel was removed from the oil bath, and sequentially α-methyl-γ-butyrolactone (47.0 μL, 0.5 mmol), dodecane (50 μL, internal standard) and NaHMDS (105.4 mg, 0.575 mmol) were added under argon. From a stock solution, ZnBr$_2$ (0.51 M in THF, 250 μL 37.5 μmol) was added by syringe while purging with argon; the mixture was then stirred for 5 min at rt. 3-Chloroanisole (0.25 mmol) was added by syringe followed by toluene (500 μL) while purging with argon. The tube was sealed and heated at 60° for 20 h. After complete conversion had been accomplished, as judged by GC analysis, the reaction mixture was allowed to cool to rt and was then filtered through a pad of silica gel, eluting with EtOAc. The eluate was concentrated under reduced pressure, followed by chromatography of the residue on silica gel (1.5 × 30 cm, EtOAc/hexanes), to give 42.6 mg of the title compound (86% yield, 96% ee) as a colorless oil: HPLC (OD-col., 10% i-PrOH/hexanes, 0.7 mL/min) t_r (major) = 13.7 min, t_r (minor) = 17.0 min; $[\alpha]^{20}$ −7.3 (c 2.5, CH$_2$Cl$_2$); ^{1}H NMR (400 MHz, CDCl$_3$) δ 18.30 (m, 1H), 6.99 (m, 2H), 6.84 (m, 1H), 4.34 (ddd, J = 9.0, 7.8, 3.9 Hz, 1H), 4.16 (ddd, J = 8.9, 8.9, 6.5 Hz, 1H), 3.82 (s, 3H), 2.69 (ddd, J = 12.9, 6.5, 3.9 Hz, 1H), 2.41 (ddd, J = 12.9, 8.7, 7.8 Hz, 1H), 1.62 (s, 3H); ^{13}C NMR (100 MHz, CDCl$_3$) δ 180.3, 160.3, 143.0, 130.3, 118.5, 112.8, 112.6, 66.7, 65.5, 55.7, 47.9, 38.5, 25.9. Anal. Calcd. for $C_{12}H_{14}O_2$: C, 69.88; H, 6.84. Found: C, 69.58; H, 6.91.

(S)-Ethyl 2-Methyl-3-oxo-2-[2-(2,2,2-trifluoroacetamido)phenyl]butanoate [Copper-Catalyzed Enantioselective α-Arylation of an Acetoacetate].[36] A Schlenk tube was charged with 2-(2,2,2-trifluoroacetamido)phenyliodide (0.5 mmol), CuI (19 mg, 0.2 mmol), trans-4-hydroxy-L-proline (26 mg, 0.4 mmol), and NaOH (80 mg, 2 mmol), evacuated and backfilled with argon. After injection of H_2O (5 µL) and DMF (1 mL), the tube was immersed in a cooling bath and ethyl 2-methylacetoacetate (0.75 mmol) was injected. The reaction mixture was stirred at −45° until the conversion was complete as monitored by TLC. The mixture was partitioned between EtOAc and saturated NH_4Cl; the organic layer was washed with brine, dried over Na_2SO_4, and concentrated in vacuo. The residue was purified by column chromatography on silica gel (petroleum ether/EtOAc 50:1 to 20:1) to provide the title product (80% yield, 71% ee): HPLC (Chiralpak AD, 95:5 hexanes/i-PrOH, 0.7 mL/min) t_r (major) = 10.2 min, t_r (minor) = 9.7 min; $[\alpha]^{20}$ +191 (c 1.0, $CHCl_3$); ^1H NMR (400 MHz, $CDCl_3$) δ 10.27 (br s, 1H), 7.71–7.34 (m, 4H), 4.32–4.20 (m, 2H), 1.95 (s, 3H), 1.88 (s, 3H), 1.31 (t, J = 6.8 Hz, 3H); ^{13}C NMR (100 MHz, $CDCl_3$) δ 206.2, 173.0, 155.8, 134.2, 131.8, 129.6, 127.5, 127.5, 127.0, 114.5, 63.9, 63.1, 24.5, 20.3, 13.9; ESI-MS m/z: $[M + Na]^+$ 354; HRMS: $[M + Na]^+$ calcd for $C_{15}H_{16}NO_4F_3Na$, 354.0924; found, 354.0920.

TABULAR SURVEY

The tabular survey includes all examples found in the literature through mid-2008. The literature survey was conducted by computer search of Beilstein and SciFinder and by direct inspection of the literature.

Table 1 compiles the arylation of ketones, aldehydes, and enol ethers. Table 2 lists the arylation of esters, amides, lactones, lactams, nitriles, ketene acetals, and preformed enolates. Table 3 compiles the arylation of 1,3-dicarbonyl compounds and cyanoacetates; other types of active methylene compounds can be found in Tables 1 and 2.

In all the tables, the examples are listed according to the nucleophiles involved, in order of increasing total number of carbon atoms. Protecting groups are included in the carbon count. Intramolecular couplings are included in the tables and are given after intermolecular reactions for the same carbon count. Aryl compounds are ordered according to the class of the structure in the following order: phenyl- > naphthyl- > heteroaromatic compounds. Within the same aryl class, the entries are ordered according to the nature of the leaving group in order of, Cl > Br > I > OTf. Entries of the same aryl halide are further ordered by increasing number of substituents and then by the position of those substituents (2 > 3 > 4). Finally, if all other functionalization of the aryl compound is the same, the entries are arranged by the substituent attached to the aryl ring, heteroatom-substituted (by increasing atomic number) > carbon-substituted. A sub-table contains examples where a comparison between halides was possible, following the same ordering rules as previously stated. The entry containing the sub-table is placed within the aryl compound ordering based on the first entry in the subtable.

Unreported yields are indicated using "(—)". There are a number of entries where ee or de values are reported but for which the authors did not report the absolute configuration of the created center. In several entries substituents ("R") or halide ("X") functions were not reported. All these entries are faithful to the original literature data.

The following abbreviations are used in the tables:

Ad	adamantyl
All	allyl
biPh	biphenyl
p-cresol	4-methylphenol
Cy	cyclohexyl
dba	dibenzylideneacetone
diglyme	diethylene glycol dimethyl ether
dioxane	1,4-dioxane
dppb	1,4-bis(diphenylphosphino)butane
dppe	1,2-bis(diphenylphosphino)ethane
dppp	1,3-bis(diphenylphosphino)propane
eq	equivalents
ether	diethyl ether
LiHMDS	lithium bis(trimethylsilyl)amide
LiTMP	lithium 2,2,6,6-tetramethylpiperidide
monoglyme	ethylene glycol dimethyl ether
MS	molecular sieves
Nap	naphthyl
TBDPS	*tert*-butyldiphenylsilyl
tol	methylphenyl
xyl	dimethylphenyl

CHART 1. STRUCTURES OF LIGANDS USED IN TABLES

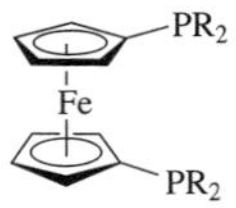

R	Ligand
Ph	L1
o-tol	L2
t-Bu	L3
Cy	L4

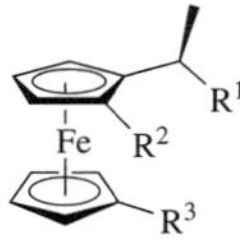

R¹	R²	R³	Ligand
OMe	PPh$_2$	H	L5
NMe$_2$	P(t-Bu)$_2$	H	L6
PPh$_2$	PCy$_2$	H	L7
PCy$_2$	PPh$_2$	H	L8
PCy$_2$	PCy$_2$	H	L9
P(t-Bu)$_2$	P(o-tol)$_2$	H	L10
NMe$_2$	PPh$_2$	PPh$_2$	L11

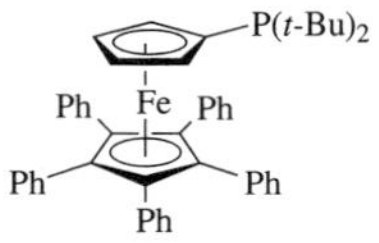

L12

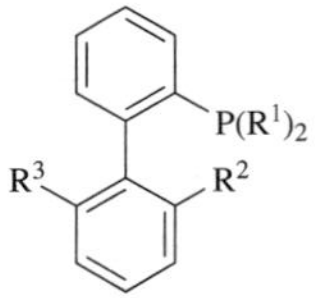

R¹	R²	R³	Ligand
Ph	NMe$_2$	H	L13
t-Bu	NMe$_2$	H	L14
t-Bu	H	H	L15
t-Bu	Me	H	L16
Cy	NMe$_2$	H	L17
Cy	H	H	L18
Cy	Me	H	L19
Cy	OMe	OMe	L20
Cy	i-Pr	i-Pr	L21

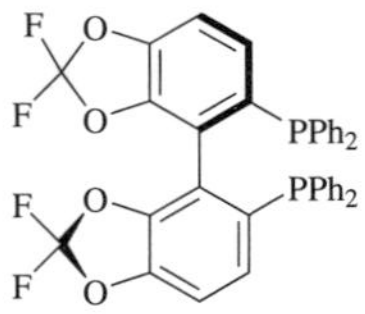

L22

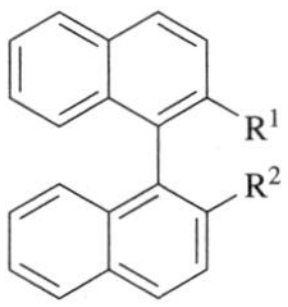

R¹	R²	Ligand
PPh$_2$	PPh$_2$	L23
P(o-tol)$_2$	P(o-tol)$_2$	L24
P(p-tol)$_2$	P(p-tol)$_2$	L25
NMe$_2$	P(t-Bu)$_2$	L26
NMe$_2$	PPh(t-Bu)	L27
NMe$_2$	PCy$_2$	L28
OH	PPh$_2$	L29
OMe	PPh$_2$	L30
—OP(NEt$_2$)O—		L31

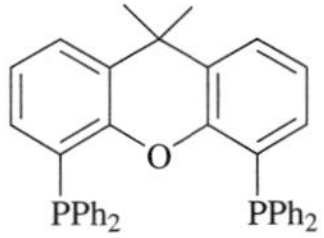

L32

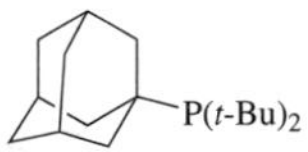

L33

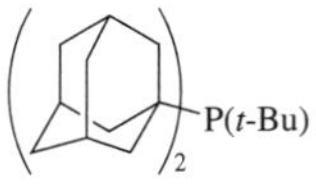

L34

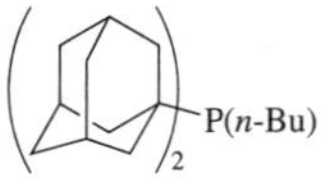

L35

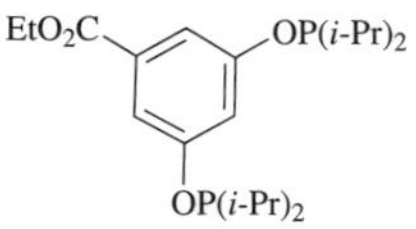

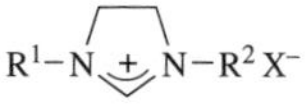

L36

L37

L38

L39

L40

L41

L42

L43

L44

L45 R = PPh_2

R^1	R^2	X	Ligand
Bu	Bu	PF_6	**L46**
2,6-(i-Pr)$_2C_6H_3$	2,6-(i-Pr)$_2C_6H_3$	BF_4	**L47**

L48

L49

L50

L51

L52

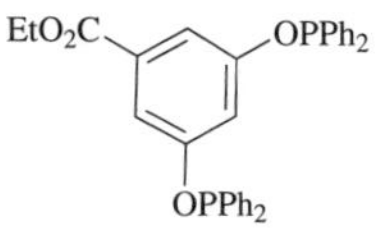

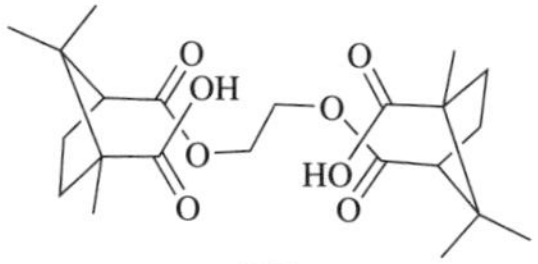

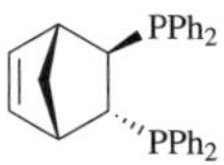

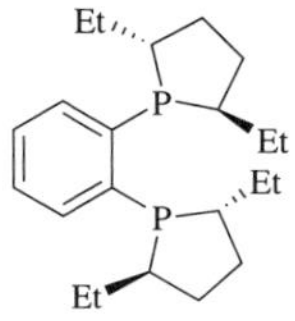

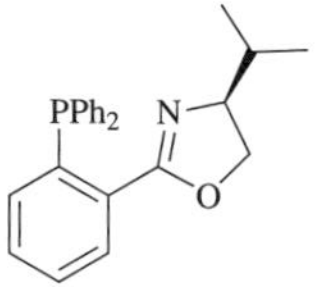

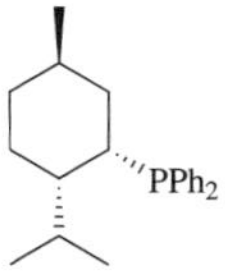

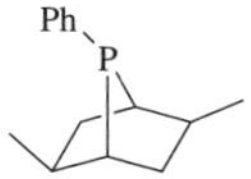

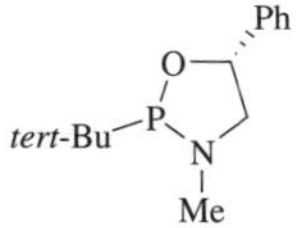

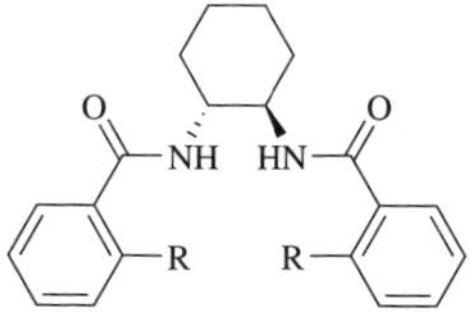

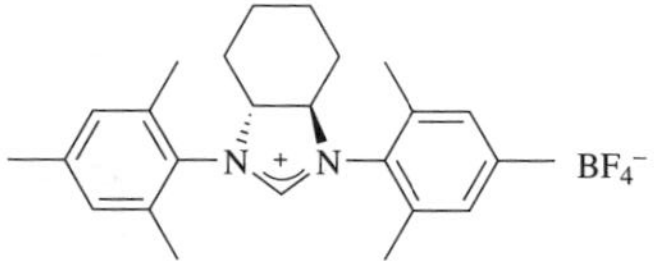

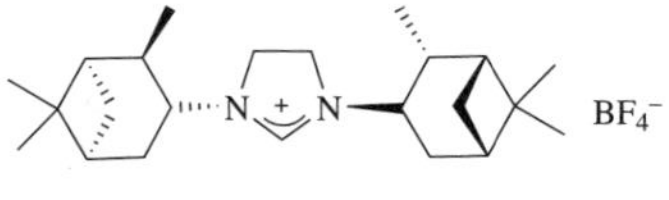

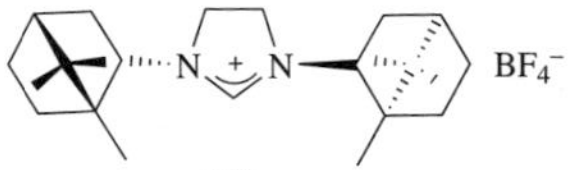

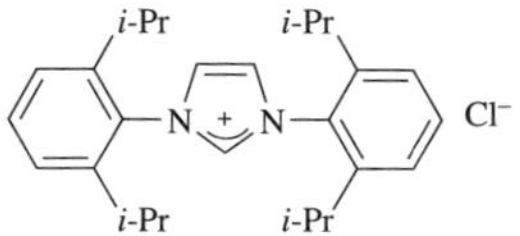

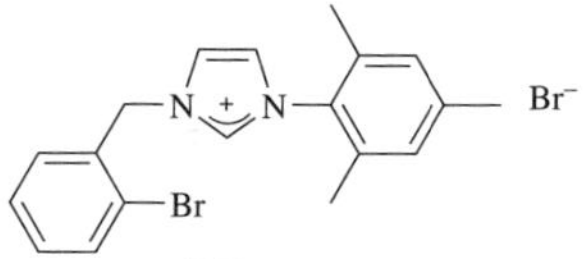

CHART 2. STRUCTURES OF PALLADACYCLES AND PALLADIUM COMPLEXES USED IN TABLES

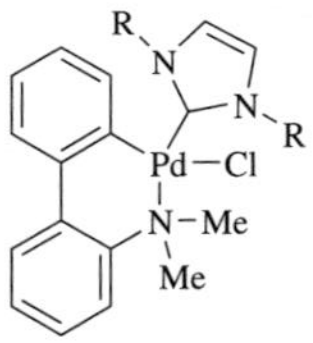

Pd cat **1**

R = 2,6-(i-Pr)$_2$C$_6$H$_3$

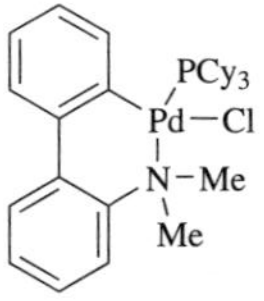

Pd cat **2**

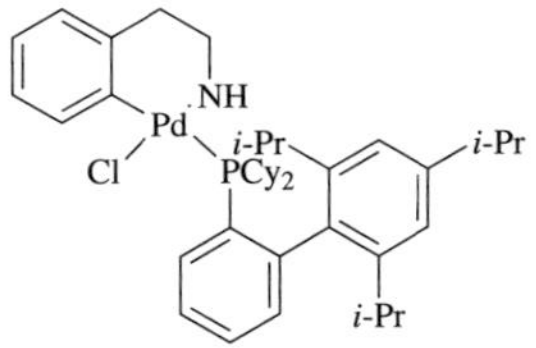

Pd cat **3**

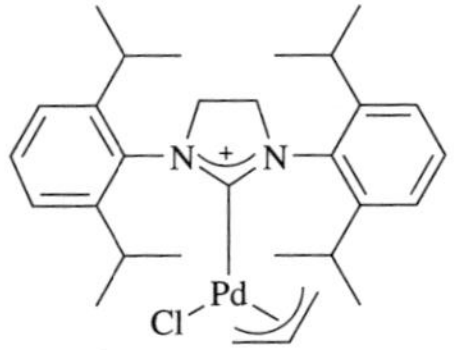

Pd cat **4**

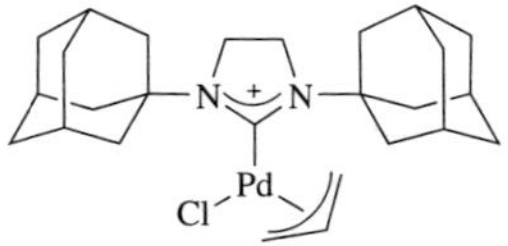

Pd cat **5**

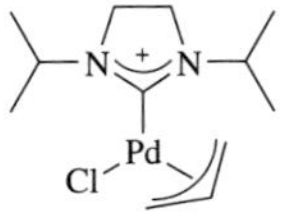

Pd cat **6**

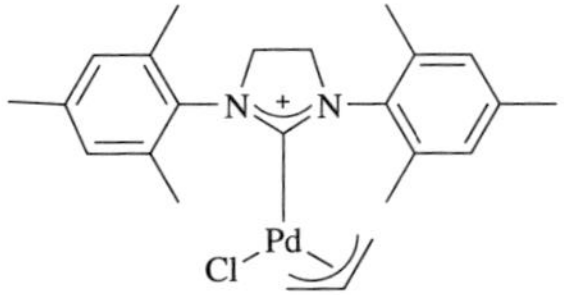

Pd cat **7**

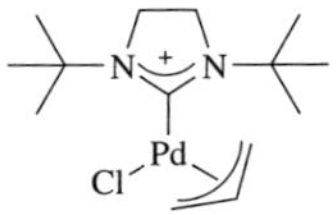

Pd cat **8**

CHART 3. STRUCTURES OF POLYMER-SUPPORTED CATALYSTS USED IN TABLES

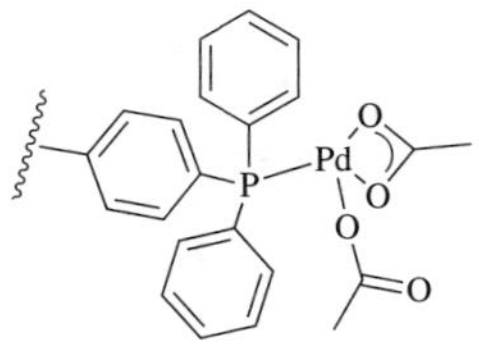

Pd cat **9**

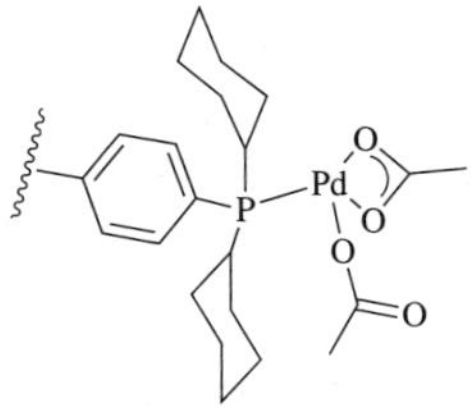

Pd cat **10**

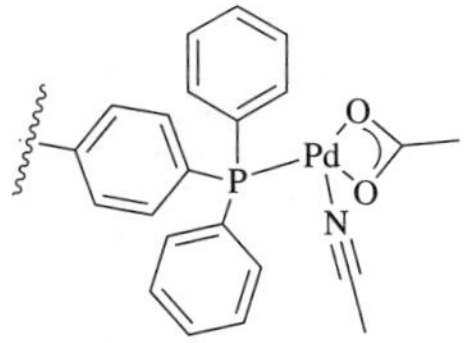

Pd cat **11**

TABLE 1. ARYLATION OF ALDEHYDES, KETONES, AND ENOL ETHERS

Substrate	Aryl Compound	Conditions	Product(s) and Yield(s) (%)	Refs.
 C₃ acetone	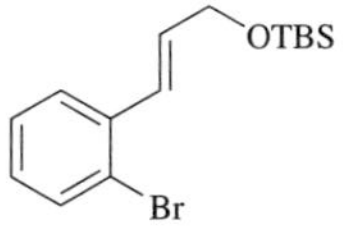 2-bromocinnamyl OTBS ether	Pd₂(dba)₃ (5 mol %), P(t-Bu)₃ (10 mol %), LiHMDS, dioxane, 90°, 3 h	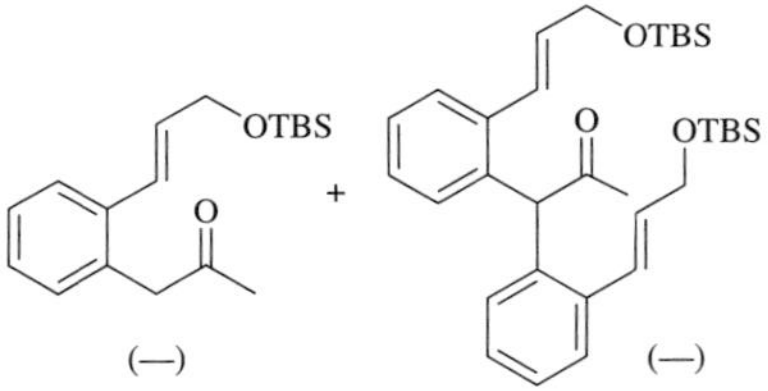 cinnamyl OTBS ketone (—) + bis-arylated OTBS ketone (—)	63
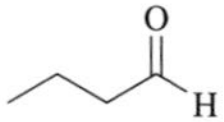 C₄ butyraldehyde	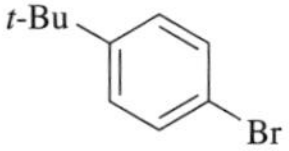 4-t-Bu-bromobenzene	Pd(OAc)₂ (2 mol %), L23 (3 mol %), Cs₂CO₃ (1.2 eq), dioxane, 24 h	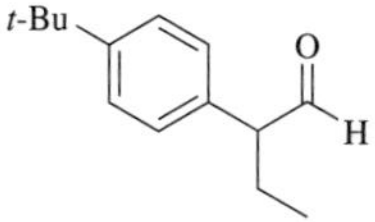 2-(4-t-Bu-phenyl)butanal (55)	148
butyraldehyde	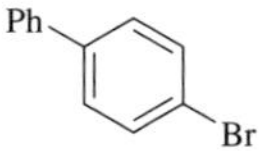 4-Ph-bromobenzene	Pd(OAc)₂ (0.05 mol %), P(t-Bu)₃ (10 mol %), Cs₂CO₃ (1.2 eq), dioxane, 4 h	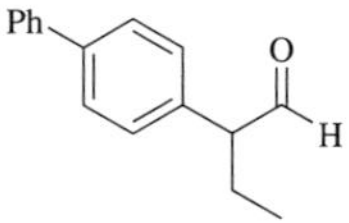 2-(4-Ph-phenyl)butanal (54)	65
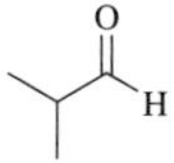 isobutyraldehyde	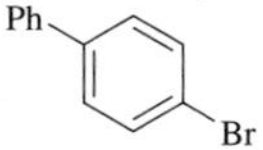 4-Ph-bromobenzene	Pd(OAc)₂ (0.05 mol %), P(t-Bu)₃ (10 mol %), Cs₂CO₃ (1.2 eq), dioxane, 4 h	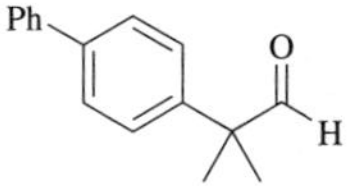 2-methyl-2-(4-Ph-phenyl)propanal (43)	65

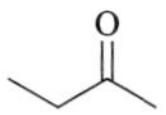

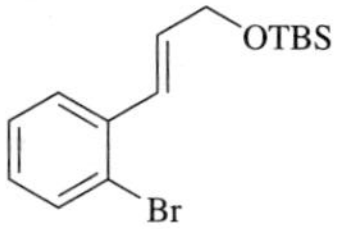

![ketone structure] + ![aryl bromide with OTBS]	Pd₂(dba)₃ (5 mol %), P(t-Bu)₃ (10 mol %), LiHMDS, dioxane, 90°, 3 h	products (—), (—), (—) + 63

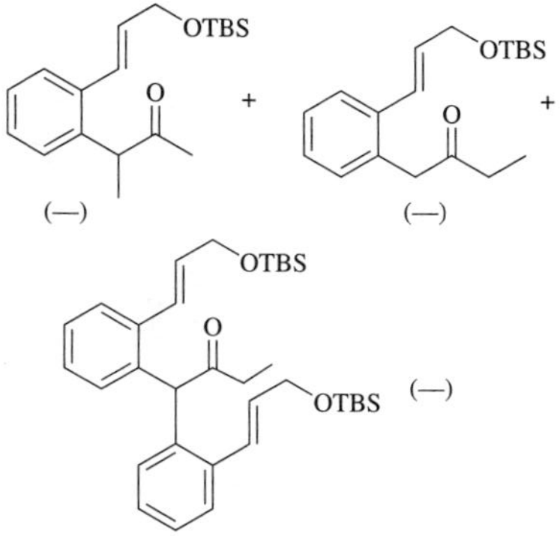

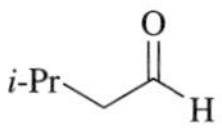

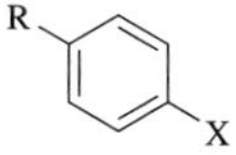

C₅

i-Pr-CH₂-CHO + 4-R-C₆H₄-X → α-aryl aldehyde

Catalyst, ligand, Cs₂CO₃ (1.2 eq), dioxane, 80–100°   148

X	R	Catalyst	Ligand	
Cl	MeO	Pd cat **3**	none	(51)
Cl	MeS	Pd cat **3**	none	(50)
Cl	t-Bu	Pd cat **3**	none	(61)
Br	MeO	[Pd(allyl)Cl]₂	**L32**	(64)
Br	MeS	Pd(OAc)₂	**L32**	(58)
Br	t-Bu	Pd(OAc)₂	**L32**	(65)

3-pentanone + 1,2-dichlorobenzene → α-aryl ketone (2-chlorophenyl)

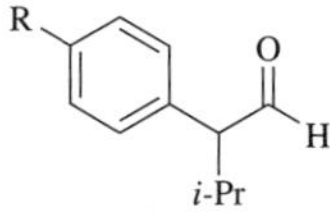

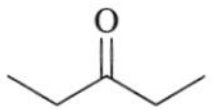

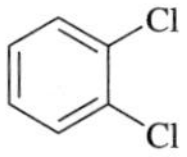

Pd(OAc)₂ (1 mol %), **L35** (2 mol %), K₃PO₄ (2.2 eq), dioxane, 100°, 20 h   48

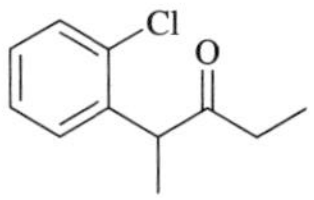

(58)

TABLE 1. ARYLATION OF ALDEHYDES, KETONES, AND ENOL ETHERS (*Continued*)

Substrate	Aryl Compound	Conditions	Product(s) and Yield(s) (%)	Refs.
C₅				
(3-pentanone)	3-ClC₆H₄OH	Pd(OAc)₂ (0.1 mol %), **L19** (0.2 mol %), NaO*t*-Bu (2 eq), THF, 70°, 24 h	3-hydroxyphenyl pentan-3-one (88)	43
	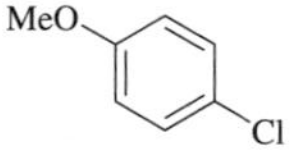 4-Cl-C₆H₄-OMe	Pd(OAc)₂ (0.1 mol %), **L19** (0.2 mol %), NaO*t*-Bu (1.3 eq), toluene, 70°, 24 h	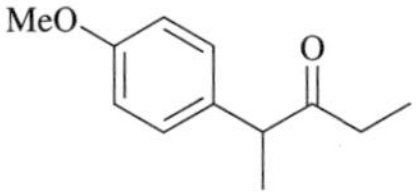 4-methoxyphenyl pentan-3-one (74)	43
	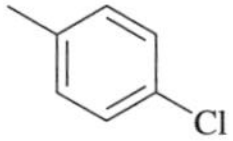 4-Cl-C₆H₄-Me	Pd(OAc)₂ (1 mol %), **L35** (2 mol %), K₃PO₄ (2.2 eq), dioxane, 100°, 20 h	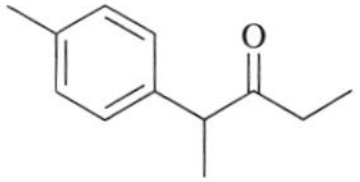 4-tolyl pentan-3-one (55)	48
	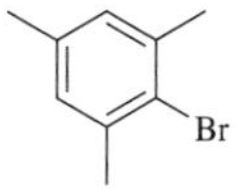 2-Br-mesitylene	Pd cat **4** (1 mol %), NaO*t*-Bu (1.05 eq), THF, 60°, 0.5 h	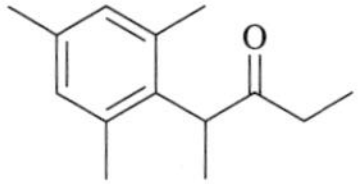 2-mesityl pentan-3-one (68)	50
(3-methyl-2-butanone)	PhBr	1. CH₂=C=O 2. Bu₃SnOMe (1 eq), PdCl₂[P(*o*-tol)₃]₂ (0.7 mol %), toluene, 100°, 5 h	3-methyl-1-phenylbutan-2-one (60)	149
	2-Br-C₆H₄-NO₂	Pd₂(dba)₃, **L17**, K₃PO₄ (2.2 eq), PhOH or PhOK (0.2 eq), toluene, 50°, 14 h	3-methyl-1-(2-nitrophenyl)butan-2-one (—)	45

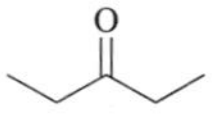

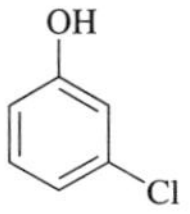

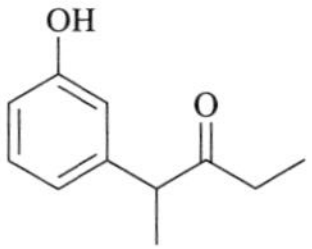

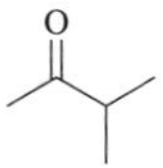

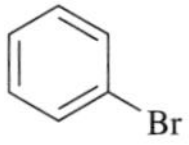

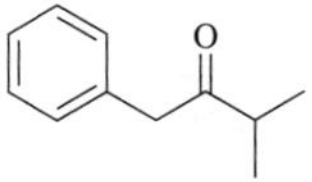

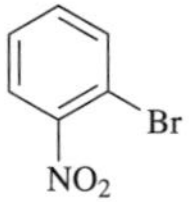

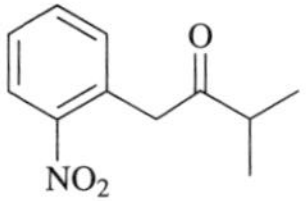

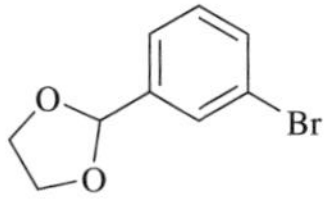 3-bromo-dioxolane	Pd₂(dba)₃ (1.5 mol %), **L25** (3.6 mol %), NaO-t-Bu (1.3 eq), THF, 70° 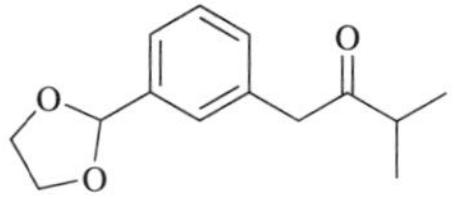	isobutyl ketone product (76)	16
2-(OTBS-allyl)bromobenzene	 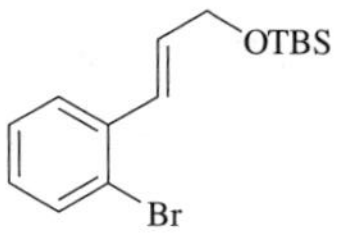 Pd₂(dba)₃ (5 mol %), P(t-Bu)₃ (10 mol %), LiHMDS, dioxane, 90°, 3 h 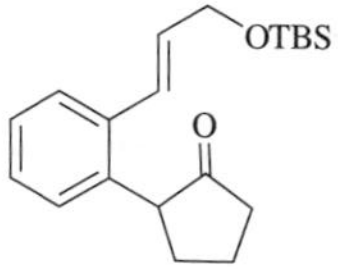	OTBS-allyl cyclopentanone product (—)	63
PhBr	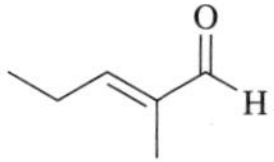 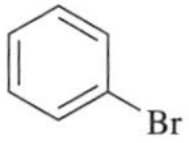 Pd(OAc)₂ (5 mol %), P(t-Bu)₃ (20 mol %), Cs₂CO₃ (1.2 eq), DMF, 60°, 7 h 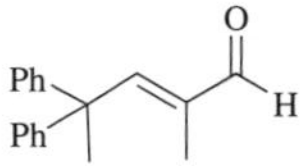	Ph,Ph-dimethyl enal (—)	47
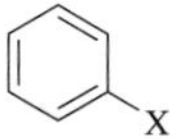 PhX	Pd(OAc)₂ (5 mol %), PPh₃ (0.1 eq), base (y eq), DMF	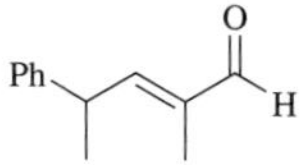 Ph-methyl enal (—)	92

X	x	Base	y	Temp (°)	Time (h)
Br	1.0	Cs₂CO₃	1.2	120	1 (84)
Br	1.0	Cs₂CO₃	1.2	60	4 (80)
Br	2.0	Cs₂CO₃	2.0	120	1 (96)
Br	2.0	K₂CO₃	2.0	120	4 (69)
Br	2.0	Na₂CO₃	2.0	120	5 (12)
I	1.0	Cs₂CO₃	1.2	60	4 (34)

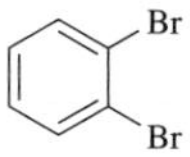 1,2-dibromobenzene	Pd(OAc)₂ (5 mol %), P(t-Bu)₃ (10 mol %), Cs₂CO₃ (2 eq), DMF 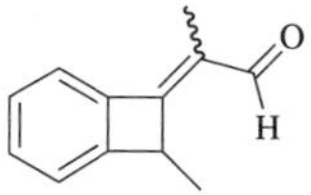	benzocyclobutane aldehyde (45)	60

C₆: cyclopentanone, 2-methyl-2-pentenal, x eq

103

TABLE 1. ARYLATION OF ALDEHYDES, KETONES, AND ENOL ETHERS (*Continued*)

Substrate	Aryl Compound	Conditions	Product(s) and Yield(s) (%)	Refs.
C₆ hexanal (O=CH-pentyl)	R^2-C₆H₃(R^1)-X	Catalyst, ligand, Cs₂CO₃ (1.2 eq), dioxane, 80–100°	2-aryl hexanal products with R^1, R^2 substituents	148

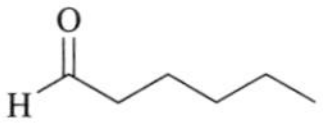

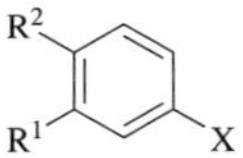

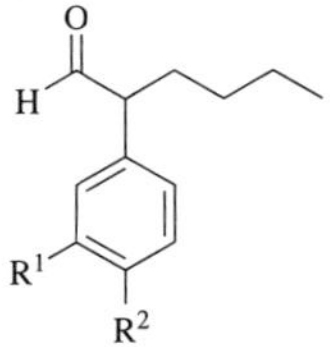

X	R^1	R^2	Catalyst	Ligand	Yield
Cl	H	OCH₂CO₂Et	Pd cat **3**	none	(45)
Cl	Me	NMe₂	Pd cat **3**	none	(62)
Br	Me	NMe₂	Pd(OAc)₂	**L32**	(65)
Br	Me	OCH₂CO₂Et	[Pd(allyl)Cl]₂	**L32**	(84)

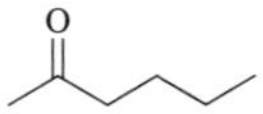

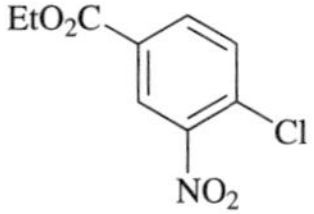

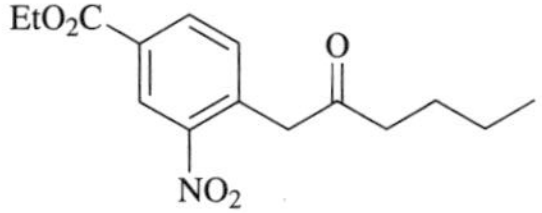

Substrate	Aryl Compound	Conditions	Product(s) and Yield(s) (%)	Refs.
2-hexanone	4-Cl-3-NO₂-C₆H₃-CO₂Et (Cl with EtO₂C and NO₂)	Pd₂(dba)₃, **L17**, K₃PO₄ (2.5 eq), PhOH (0.2 eq), toluene, 50°, 24 h	EtO₂C-C₆H₃(NO₂)-CH₂C(O)-C₄H₉ (65)	45
2-hexanone	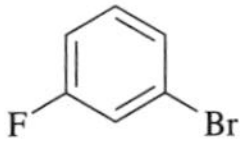 3-F-C₆H₄-Br	Pd₂(dba)₃ (1 mol %), **L32** (1.2 mol %), NaO*t*-Bu (1.3 eq), THF, 70°, 12 h	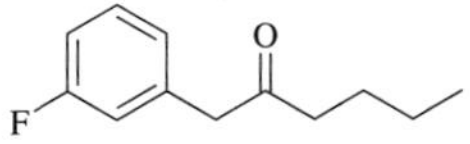 F-C₆H₄-CH₂C(O)-C₄H₉ (74)	43
2-hexanone	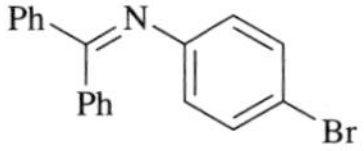 4-(PhC(=NPh))-C₆H₄-Br	Pd₂(dba)₃ (1.5 mol %), **L23** (3.6 mol %), NaO*t*-Bu (1.3 eq), THF, 70°	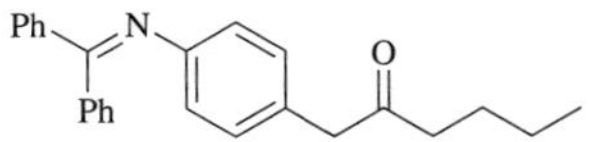 Ph-N=C(Ph)-C₆H₄-CH₂C(O)-C₄H₉ (64)	16
2-hexanone	2-Br-3-NO₂-4-MeO-C₆H₃ (MeO, Br, NO₂)	Pd₂(dba)₃, **L17**, K₃PO₄ (2.5 eq), PhOH (0.2 eq), toluene, 50°, 24 h	MeO-C₆H₃(NO₂)-CH₂C(O)-C₄H₉ (67)	45

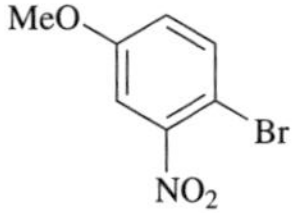

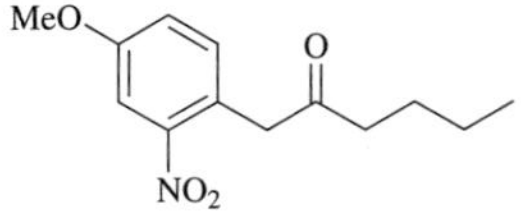

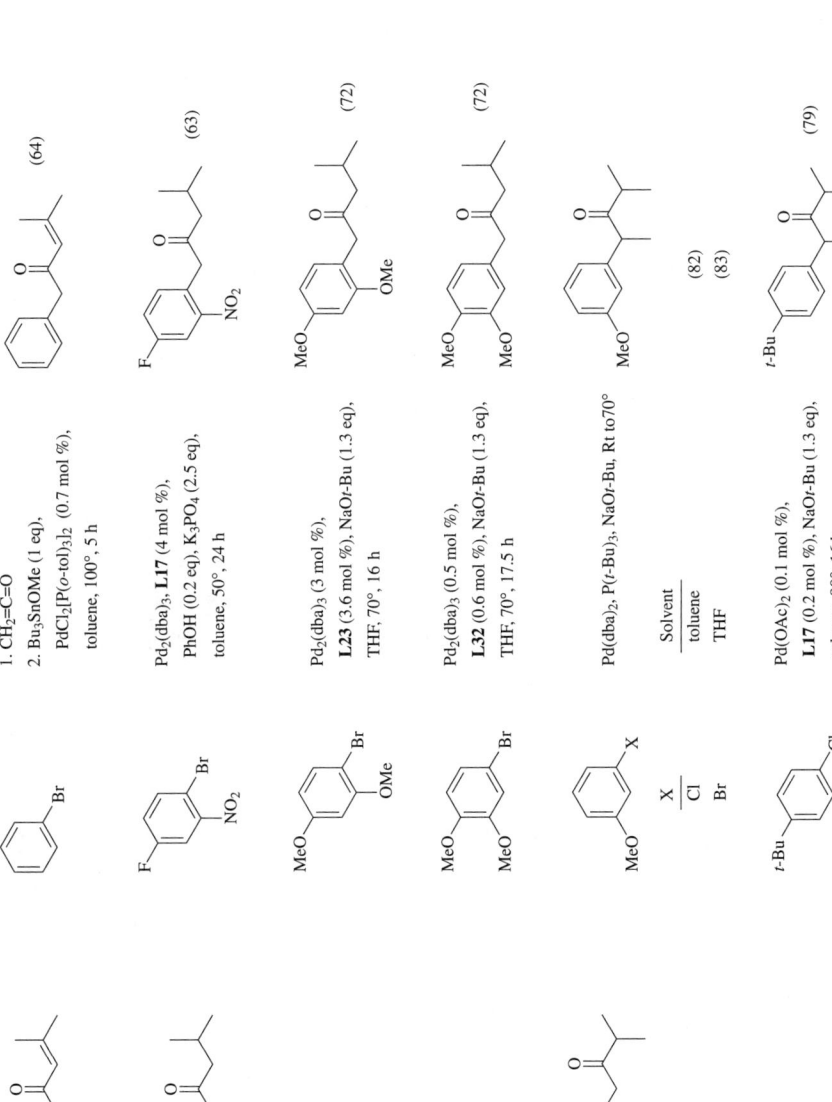

TABLE 1. ARYLATION OF ALDEHYDES, KETONES, AND ENOL ETHERS (*Continued*)

Substrate	Aryl Compound	Conditions	Product(s) and Yield(s) (%)	Refs.
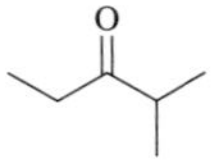 C₆ (3-methyl-2-pentanone structure)	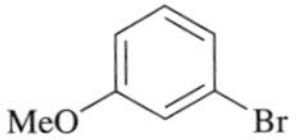 3-BrC₆H₄OMe	Pd(dba)₂ (2 mol %), **L3** (2.5 mol %), 70°, 12 h	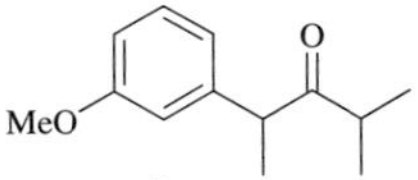 (92) [product with MeO-aryl]	40
	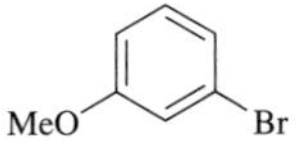 3-BrC₆H₄OMe	Pd(OAc)₂ (1 mol %), P(*t*-Bu)₃ (2.5 mol %), 50°, 12 h	**I** (83)	40
	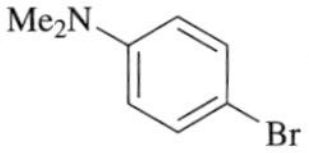 4-BrC₆H₄NMe₂	Pd(OAc)₂ (0.1 mol %), **L17** (0.2 mol %), NaO*t*-Bu (1.3 eq), toluene, 85°, 24 h	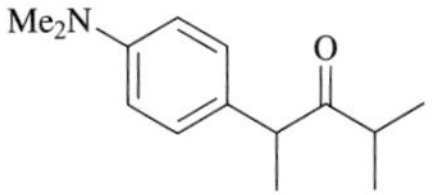 (70) [product with Me₂N-aryl]	43
	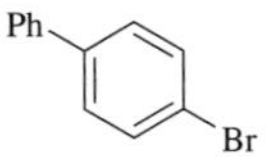 4-BrC₆H₄Ph	Pd₂(dba)₃ (1.5 mol %), **L25** (3.6 mol %), NaO*t*-Bu (1.3 eq), THF, 70°	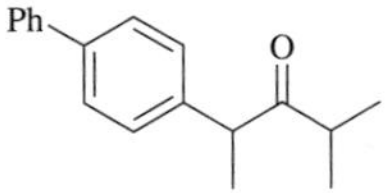 (93) [product with Ph-aryl]	16
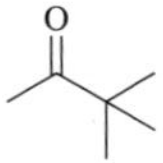 (pinacolone / *t*-Bu methyl ketone)	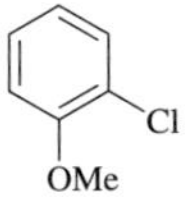 2-Cl-C₆H₄OMe	Pd cat **4** (1 mol %), NaO*t*-Bu (1.05 eq), THF, 60°, 1 h	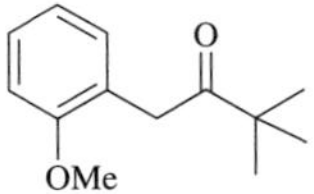 (91) [product with OMe-aryl]	50
	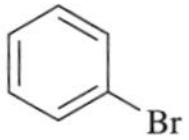 PhBr	Pd(dba)₂ (7.5 mol %), **L2** (9 mol %), KHMDS (2.2 eq), THF, reflux, 0.75 h	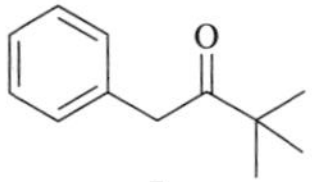 (51)	17
	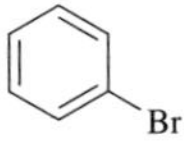 PhBr	1. CH₂=C=O 2. Bu₃SnOMe (1 eq), PdCl₂[P(*o*-tol)₃]₂ (0.7 mol %), toluene, 100°, 5 h	**I** (86)	149

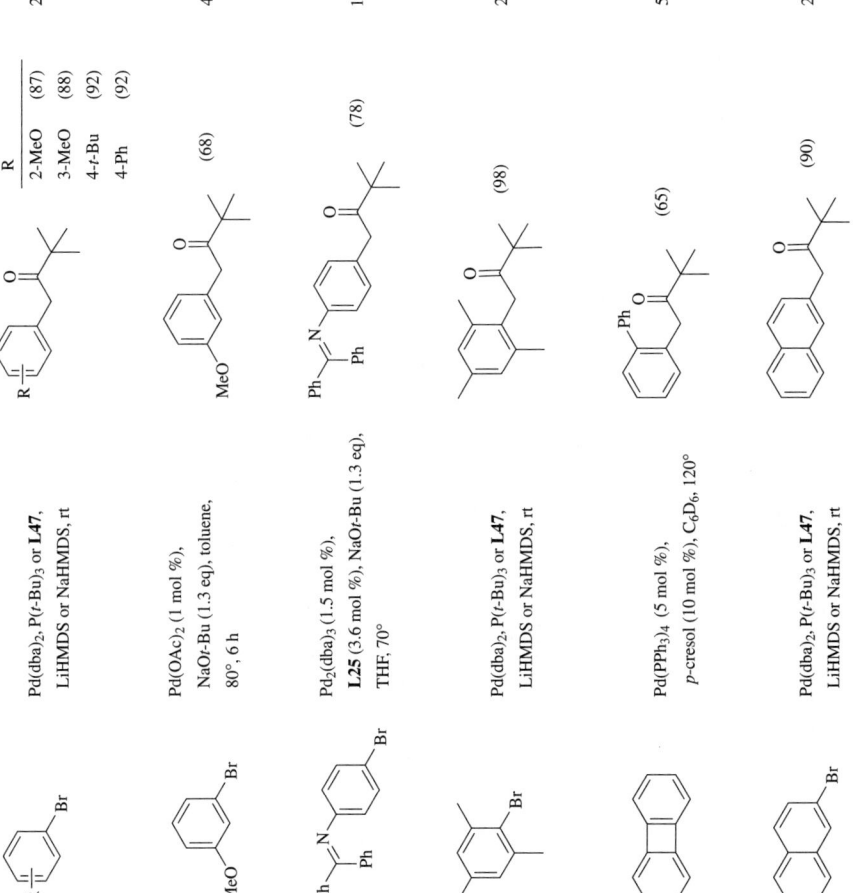

TABLE 1. ARYLATION OF ALDEHYDES, KETONES, AND ENOL ETHERS (*Continued*)

Substrate	Aryl Compound	Conditions	Product(s) and Yield(s) (%)	Refs.
C₆				
(2-methylcyclopent-2-enone)	4-R-C₆H₄-Br	1. CuCl (1 mol %), (*S*)-**L25** (1 mol %), NaO*t*-Bu (1 mol %), Ph₂SiH₂ (0.51 eq), THF/pentane (1:1), −78° 2. Pd(OAc)₂ (5 mol %), **L15** (10 mol %), CsF (1.1 eq), THF, rt, 18 h	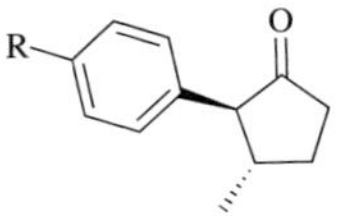 (trans-2-aryl-3-methylcyclopentanone) R / H (72) / dr^a 96.5:3.5 ; MeO (71) 97:3 ; EtO₂C (54) 97.5:2.5 ; *t*-Bu (75) 96.5:3.5	44
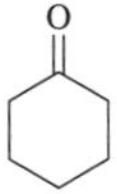 cyclohexanone	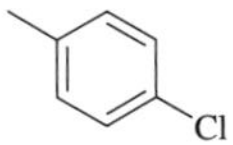 4-methyl-C₆H₄-Cl	Pd cat **2** (1 mol %), NaO*t*-Bu, dioxane, 70°, 2 h	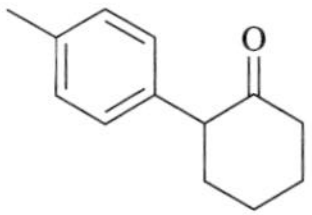 **I** (97)	52
cyclohexanone	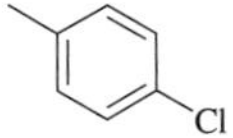 4-methyl-C₆H₄-Cl	Pd(OAc)₂ (1 mol %), **L35** (2 mol %), K₃PO₄, dioxane, 100°, 20 h	**I** (59)	48
cyclohexanone	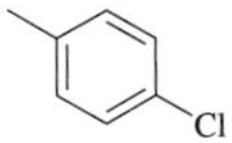 4-methyl-C₆H₄-Cl	Pd(OAc)₂ (1 mol %), **L51** (1 mol %), NaO*t*-Bu (1.5 eq), dioxane, 60°, 2 h	**I** (67)	51
cyclohexanone	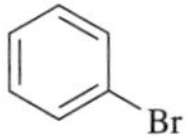 C₆H₅-Br	Pd(dba)₂ (2 mol %), **L3** (2.5 mol %), NaO*t*-Bu, THF, 70°, 12 h	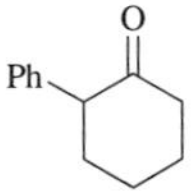 (2-phenylcyclohexanone) (70)	40
cyclohexanone	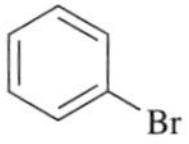 C₆H₅-Br	Pd(OAc)₂ (1 mol %), P(*t*-Bu)₃ (1.25 mol %), NaO*t*-Bu, THF, 50°, 3 h	**I** (73)	40
cyclohexanone	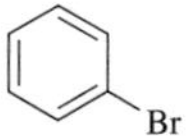 C₆H₅-Br	Pd(dba)₂, **L3**, NaO*t*-Bu, rt to 70°	**I** (70)	23

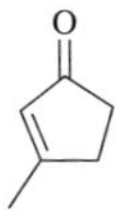

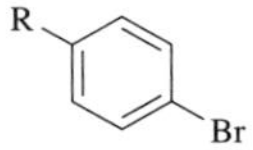

Substrate	Conditions	Product(s) and Yield(s) (%)	Refs.
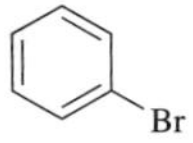 PhBr	PdCl$_2$(cod), ligand, Cs$_2$CO$_3$, DMF, 153°, 1 h	I — Ligand / L36 (88) / L37 (83)	56
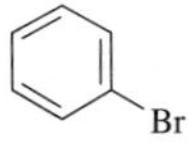 PhBr	1. CH$_2$=C=O 2. Bu$_3$SnOMe (1 eq), PdCl$_2$[P(o-tol)$_3$]$_2$ (0.7 mol %), toluene, 100°, 5 h	I (54)	149
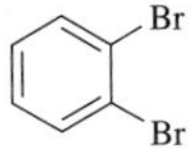 1,2-dibromobenzene	Pd$_2$(dba)$_3$ (0.5 mol %), L24 (1 mol %), Cs$_2$CO$_3$, toluene, 80°, 24 h	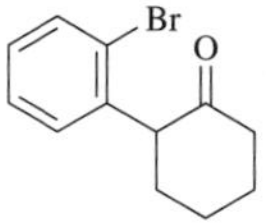 2-(2-bromophenyl)cyclohexanone (—)	62
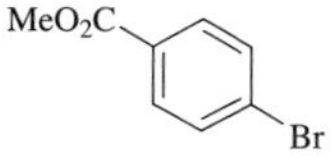 4-BrC$_6$H$_4$CO$_2$Me	Pd$_2$(dba)$_3$ (1 mol %), ligand (1.1 mol %), K$_3$PO$_4$, toluene, 80°, 15 h	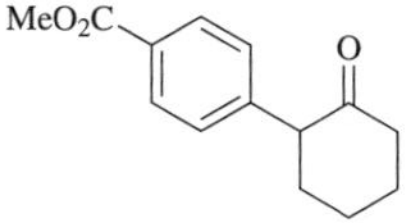 2-(4-(methoxycarbonyl)phenyl)cyclohexanone — Ligand / L19 (70) / L23 (74)	43
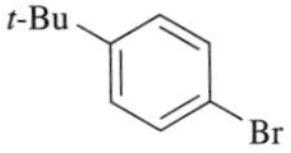 4-t-Bu-C$_6$H$_4$Br	Pd$_2$(dba)$_3$ (1.5 mol %), L25 (3.6 mol %), NaOt-Bu (1.3 eq), THF, 70°	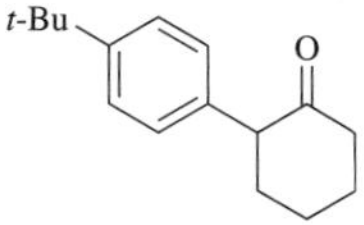 2-(4-t-butylphenyl)cyclohexanone (83)	16
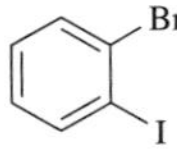 2-Br-1-I-C$_6$H$_4$	Pd(dba)$_2$ (0.5 mol %), L32 (1.2 mol %), Cs$_2$CO$_3$, toluene, 80°	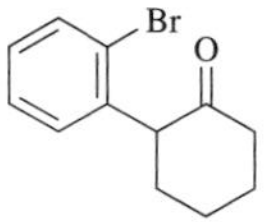 2-(2-bromophenyl)cyclohexanone (78)	61

TABLE 1. ARYLATION OF ALDEHYDES, KETONES, AND ENOL ETHERS (*Continued*)

Substrate	Aryl Compound	Conditions	Product(s) and Yield(s) (%)	Refs.
C₆ cyclohexanone	2-bromoiodobenzene	Pd₂(dba)₃ (0.5 mol %), ligand (1 mol %), Cs₂CO₃, toluene, 80°, 24 h	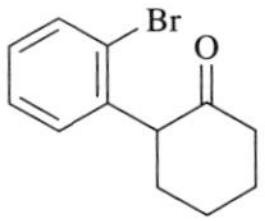 2-(2-bromophenyl)cyclohexanone Ligand **L1** (—) **L3** (25) **L9** (—) **L15** (—) **L16** (—) **L17** (9) **L23** (78) **L24** (—)	62
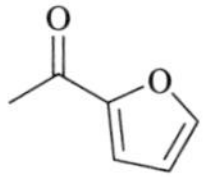 2-acetylfuran	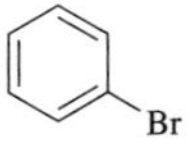 bromobenzene	Pd(dba)₂ (7.5 mol %), **L2** (9 mol %), KHMDS (2.2 eq), THF, reflux, 0.75 h	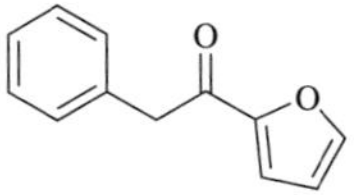 1-(furan-2-yl)-2-phenylethanone (57)	17
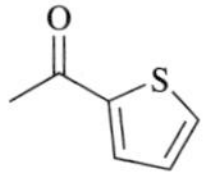 2-acetylthiophene	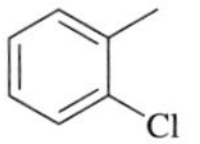 2-chlorotoluene	Pd(OAc)₂ (1 mol %), **L51** (1 mol %), NaO*t*-Bu (1.5 eq), dioxane, 60°, 12 h	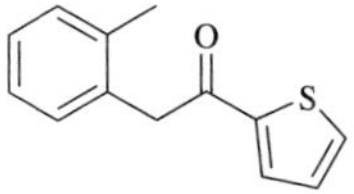 1-(thiophen-2-yl)-2-(o-tolyl)ethanone (13)	51
2-acetylthiophene	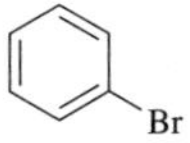 bromobenzene	Pd(dba)₂ (7.5 mol %), **L2** (9 mol %), KHMDS (2.2 eq), THF, reflux, 0.75 h	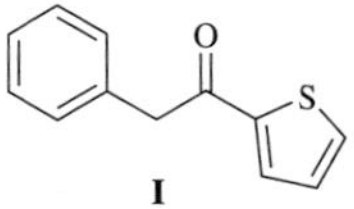 **I** 1-(thiophen-2-yl)-2-phenylethanone (68)	17
	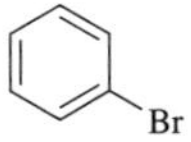 bromobenzene	Pd(dba)₂, **L1** or **L3**, KHMDS, 70°	**I** (68)	23
	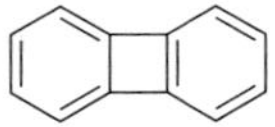 biphenylene	Pd(PPh₃)₄ (5 mol %), *p*-cresol (10 mol %), C₆D₆, 120°	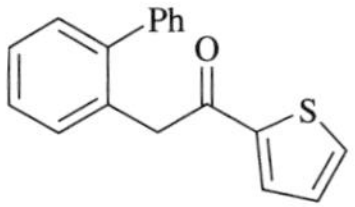 1-(thiophen-2-yl)-2-(2-biphenyl)ethanone (75)	59

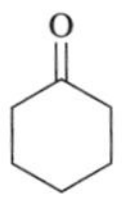

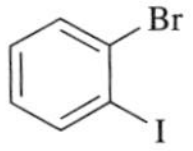

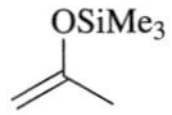

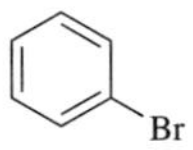

C₇

Substrate	Conditions	Product	(Yield)	Ref
PhCH=C(OSiMe₃)Me	Pd(dba)₂ (3 mol %), P(t-Bu)₃ (5.4 mol %), ZnF₂ (1.4 eq), MnF₂ (1.4 eq), DMF, 70° with PhBr	PhCH₂C(O)Me	(68)	150

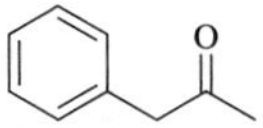

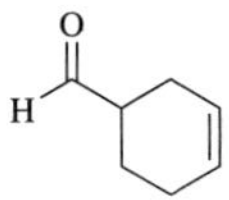

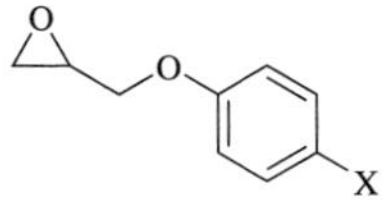

Cyclohex-3-enecarbaldehyde + 4-X-C₆H₄-OCH₂-epoxide:

Catalyst, ligand, Cs₂CO₃ (1.2 eq), dioxane, 80–100°

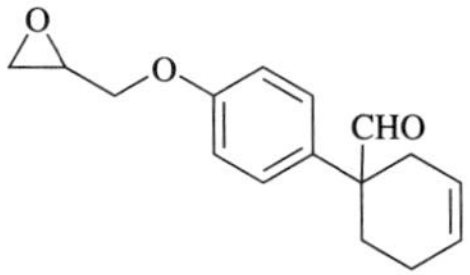

X	Catalyst	Ligand	Yield
Cl	Pd cat **3**	none	(51)
Br	Pd(OAc)₂	**L22**	(54)

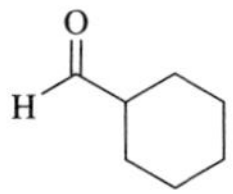

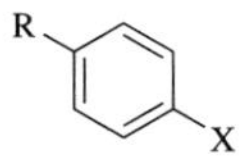

Ref: 148

Cyclohexanecarbaldehyde + 4-R-C₆H₄-X:

Catalyst, ligand, Cs₂CO₃ (1.2 eq), dioxane, 80–100°

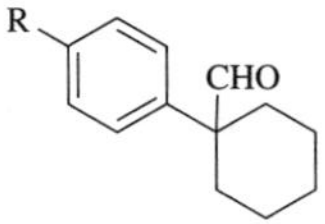

X	R	Catalyst	Ligand	Yield
Cl	Me₂N	Pd cat **3**	none	(54)
Cl	MeO	Pd cat **3**	none	(60)
Cl	MeS	Pd cat **3**	none	(63)
Br	Me₂N	Pd(OAc)₂	**L20**	(68)
Br	MeO	Pd(OAc)₂	**L20**	(65)
Br	MeS	Pd(OAc)₂	**L20**	(59)

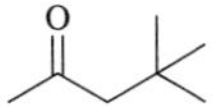

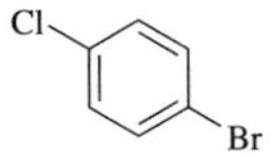

Ref: 148

Pinacolone + 4-Cl-C₆H₄-Br:
Pd₂(dba)₃ (1.5 mol %), **L25** (3.6 mol %), NaO-t-Bu (1.3 eq), THF, 70° 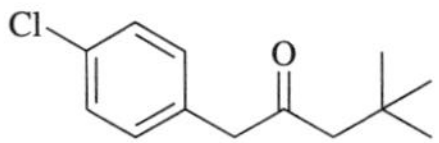
→ 4-ClC₆H₄CH₂C(O)C(CH₃)₃ (88)

Ref: 16

TABLE 1. ARYLATION OF ALDEHYDES, KETONES, AND ENOL ETHERS (*Continued*)

Substrate	Aryl Compound	Conditions	Product(s) and Yield(s) (%)	Refs.
C₇				
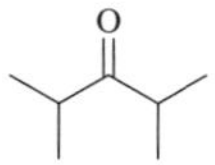 (2,4-dimethyl-3-pentanone)	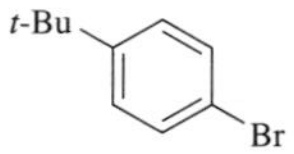 4-*t*-Bu-C₆H₄-Br	Pd(OAc)₂ (0.5 mol %), **L19** (0.2 mol %), NaO*t*-Bu (1.3 eq), toluene, 85°, 24 h	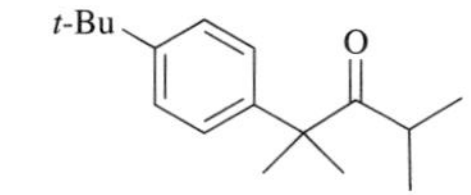 (61)	43
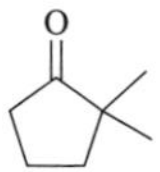 (2,2-dimethylcyclopentanone)	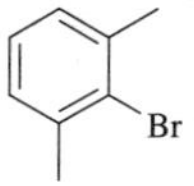 2,6-dimethyl-C₆H₃-Br	Pd(OAc)₂ (0.5 mol %), **L19** (0.2 mol %), NaO*t*-Bu (1.3 eq), toluene, 70°, 23 h	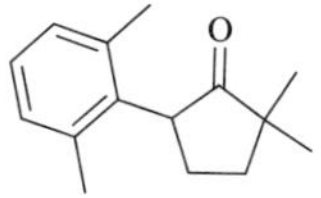 (61)	43
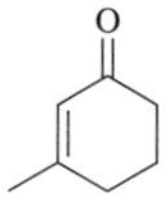 (3-methyl-2-cyclohexenone)	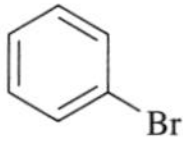 C₆H₅-Br	Pd(OAc)₂ (0.05 eq), PPh₃ (0.1 eq), Cs₂CO₃ (1.2–4 eq), DMF, 60°, 2 h	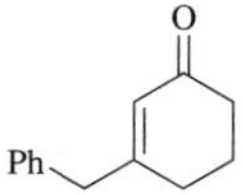 (50) (Ph-CH₂ product)	92
(cyclohexanone)	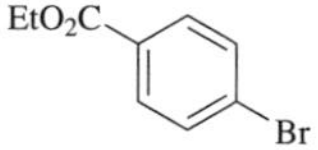 4-EtO₂C-C₆H₄-Br	1. CuCl (1 mol %), (*S*)-**L25** (1 mol %), NaO*t*-Bu (0.01 eq), Ph₂SiH₂ (0.51 eq), THF/pentane (1:1), −78° 2. Pd(OAc)₂ (5 mol %), **L15** (10 mol %), CsF (1.1 eq), THF, rt, 18 h	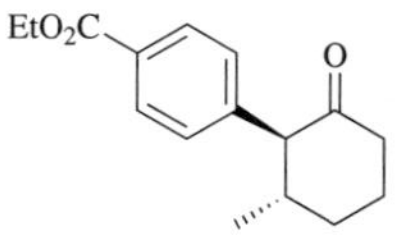 (42), 96:4 dr[b]	44
(cyclohexanone)	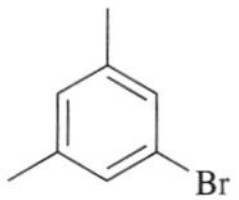 3,5-dimethyl-C₆H₃-Br	1. CuCl (1 mol %), (*S*)-**L25** (1 mol %), NaO*t*-Bu (0.01 eq), Ph₂SiH₂ (0.51 eq), THF/pentane (1:1), −78° 2. Pd(OAc)₂ (5 mol %), **L15** (10 mol %), CsF (1.1 eq), THF, rt, 18 h	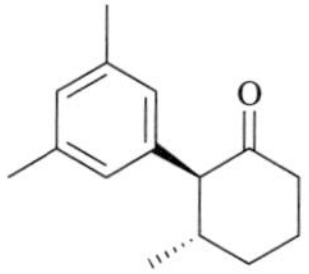 (52), 97:3 dr[b]	44

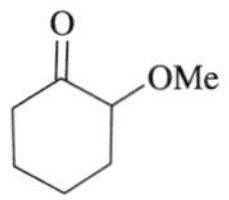

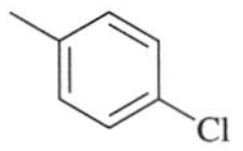

Substrate	Conditions	Product	Yield (%)	Ref

2-methoxycyclohexanone + 4-chlorotoluene, Pd(OAc)₂ (1 mol %), **L51** (1 mol %), NaO*t*-Bu (1.5 eq), dioxane, 60°, 12 h → 2-(4-methylphenyl)-6-methoxycyclohexanone (—), 51

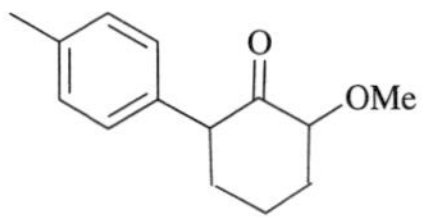

cycloheptanone + 4-bromo-R-benzene (R = NC, *t*-Bu), Pd₂(dba)₃, ligand, THF, 80°

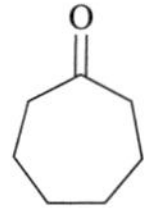

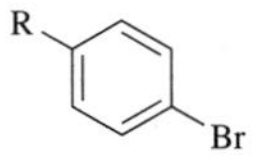

Ligand	Time (h)
L32	16
L19	17

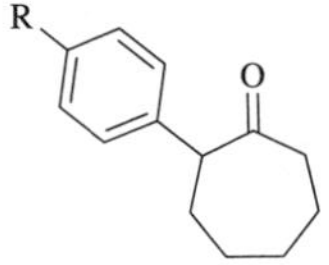

(72), (81); 43

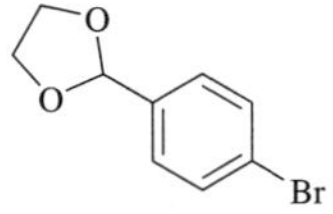

cycloheptanone + 2-(4-bromophenyl)-1,3-dioxolane, Pd(OAc)₂, ligand, NaO*t*-Bu (1.3 eq), toluene

Ligand	Temp (°)	Time (h)
L18	60	18.5
L19	45	20

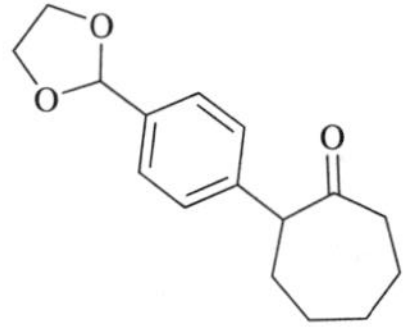

(66), (80); 43

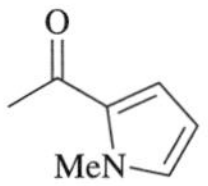

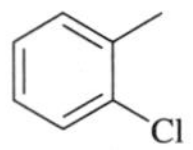

2-methylacetophenone-type + 2-chlorotoluene, Pd(OAc)₂ (1 mol %), **L51** (1 mol %), NaO*t*-Bu (1.5 eq), dioxane, 60°, 6 h → (46); 51

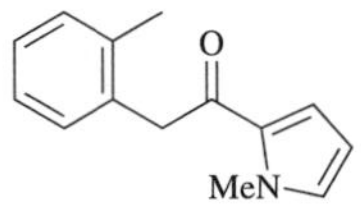

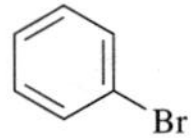

1-(1-methyl-1H-pyrrol-2-yl)ethanone + bromobenzene, Pd(dba)₂, **L1** or **L3**, KHMDS, 70° → (79); 23

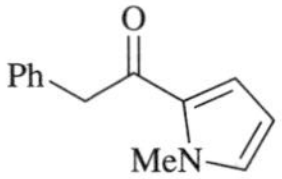

1-(1-methyl-1H-pyrrol-2-yl)ethanone + bromobenzene, Pd(dba)₂ (7.5 mol %), **L3** (9 mol %), KHMDS (2.2 eq), THF, reflux, 0.75 h → **I** (79); 17

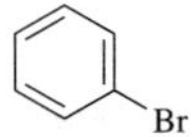

113

TABLE 1. ARYLATION OF ALDEHYDES, KETONES, AND ENOL ETHERS (*Continued*)

Substrate	Aryl Compound	Conditions	Product(s) and Yield(s) (%)	Refs.
C$_8$				
(2-butenal with Bu)	PhBr	Pd(OAc)$_2$, PPh$_3$, Cs$_2$CO$_3$	(—) + (—)	47
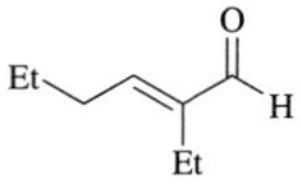 (2-ethyl pentenal)	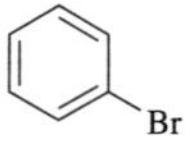 PhBr	Pd(OAc)$_2$ (5 mol %), P(t-Bu)$_3$ (0.2 eq), Cs$_2$CO$_3$ (1.2 eq), DMF, 7 h, 60°	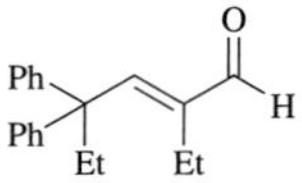 (—)	47
	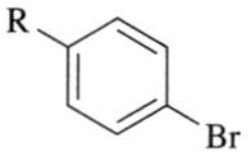 4-R-C$_6$H$_4$-Br	Pd(OAc)$_2$ (0.05 eq), PPh$_3$ (0.1 eq), Cs$_2$CO$_3$ (1.2-4 eq), DMF	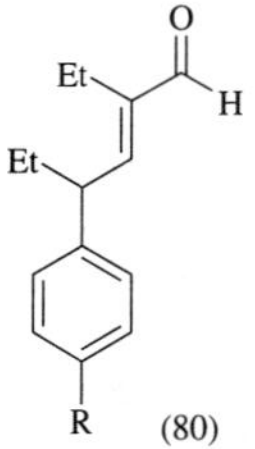	92

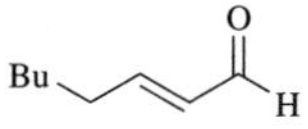

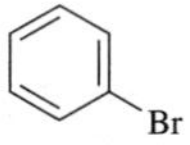

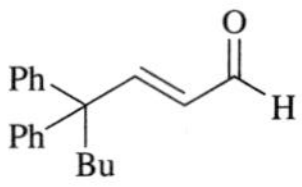

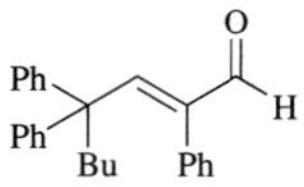

R	Temp (°)	Time (h)	Yield
H	120	1	(80)
MeO	60	2	(60)
Cl	60	4	(64)
Me	60	5	(69)

| | 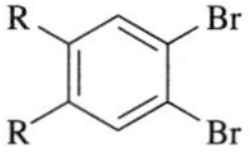 1,2-diBr-4,5-R$_2$-C$_6$H$_2$ | Pd(OAc)$_2$ (5 mol %), ligand, Cs$_2$CO$_3$ (2 eq), DMF, 80°, 4 h | 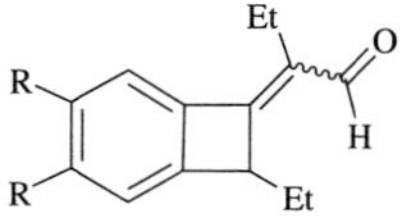 | 60 |

R	Ligand	Yield
H	PPh$_3$	(12)
H	P(o-tol)$_3$	(—)
H	P(t-Bu)$_3$	(77)
Me	P(t-Bu)$_3$	(50)

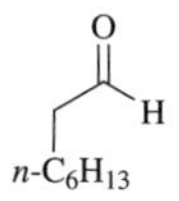

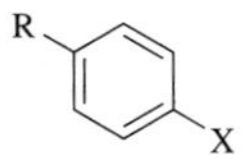

148

Catalyst, ligand, Cs_2CO_3 (1.2 eq), dioxane, 80–100°

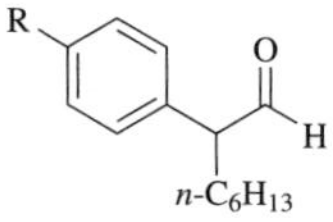

X	R	Catalyst	Ligand
Cl	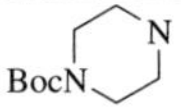 BocN-N	Pd cat **3**	none	(50)
Cl	MeO	Pd cat **3**	none	(65)
Br	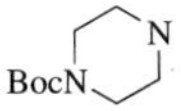 BocN-N	[Pd(cinnamyl)Cl]$_2$	**L32**	(56)
Br	MeO	[Pd(allyl)Cl]$_2$	**L32**	(70)

65

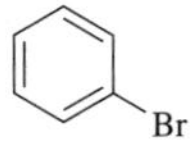

$Pd(OAc)_2$ (0.05 mol %), ligand (10 mol %), base (1.2 eq), 110°

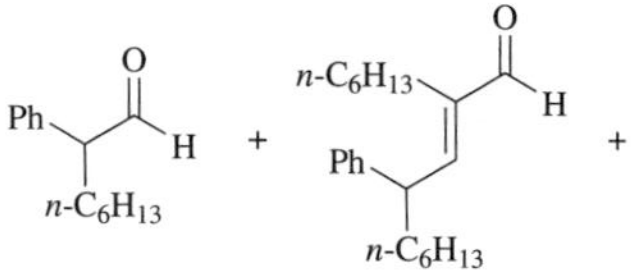

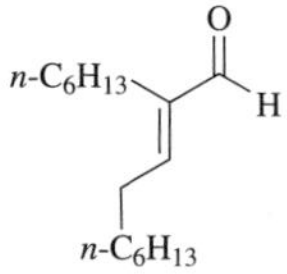

x	Ligand	Base	Solvent	Time (h)	I	II	III
2.0	P(t-Bu)$_3$	Cs$_2$CO$_3$	DMF	2	(49)	(20)	(31)
1.0	P(t-Bu)$_3$	Cs$_2$CO$_3$	dioxane	2	(53)	(1)	(2)
2.0	P(t-Bu)$_3$	Cs$_2$CO$_3$	dioxane	2	(77)	(1)	(5)
2.0	P(t-Bu)$_3$	K$_2$CO$_3$	dioxane	24	(70)	(2)	(12)
2.0	PPh$_3$	Cs$_2$CO$_3$	DMF	2	(0)	(45)	(22)
2.0	PPh$_3$	Cs$_2$CO$_3$	dioxane	2	(0)	(6)	(15)
2.0	PPh$_3$	Cs$_2$CO$_3$	dioxane	23	(0)	(32)	(11)
2.0	PCy$_3$	Cs$_2$CO$_3$	dioxane	2	(0)	(4)	(18)
2.0	PCy$_3$	Cs$_2$CO$_3$	dioxane	21	(0)	(46)	(10)

TABLE 1. ARYLATION OF ALDEHYDES, KETONES, AND ENOL ETHERS (*Continued*)

Substrate	Aryl Compound	Conditions	Product(s) and Yield(s) (%)	Refs.
C₈				
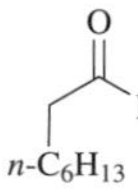 n-C₆H₁₃CH₂CHO	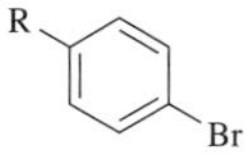 R-C₆H₄-Br	Pd(OAc)₂ (0.05 mol %), P(t-Bu)₃ (10 mol %), Cs₂CO₃ (1.2 eq), dioxane, 4 h	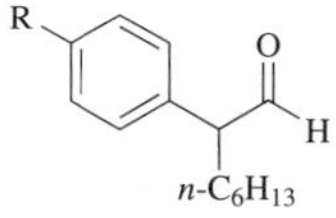 R-C₆H₄-CH(n-C₆H₁₃)CHO R: OMe (61), Me (67), Ph (59)	65
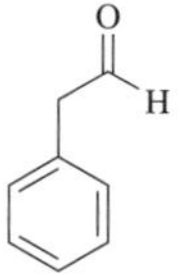 PhCH₂CHO	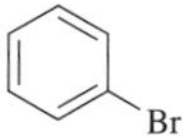 PhBr	Pd(OAc)₂ (0.05 mol %), P(t-Bu)₃ (10 mol %), Cs₂CO₃ (1.2 eq), dioxane, 2 h	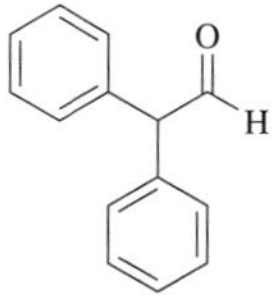 Ph₂CHCHO (56)	65
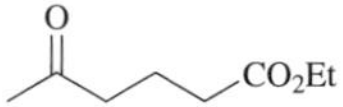 CH₃CO(CH₂)₃CO₂Et	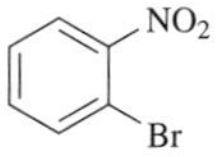 2-NO₂-C₆H₄-Br	Pd₂(dba)₃, **L17** (4 mol %), K₃PO₄ (2.5 eq), PhOH (0.2 eq), toluene, 50°, 23 h	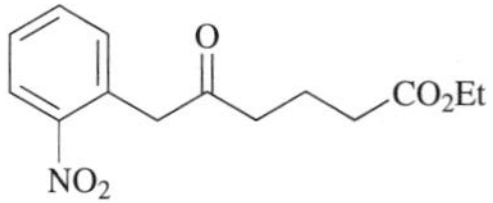 2-NO₂-C₆H₄-CH₂CO(CH₂)₃CO₂Et (65)	45
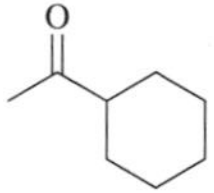 cyclohexyl methyl ketone	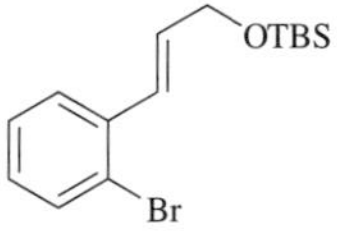 2-(CH=CHCH₂OTBS)-C₆H₄-Br	Pd₂(dba)₃ (5 mol %), P(t-Bu)₃ (10 mol %), LiHMDS, dioxane, 90°, 3 h	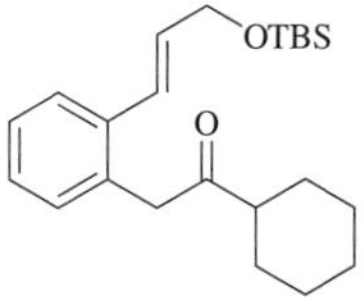 2-(CH=CHCH₂OTBS)-C₆H₄-CH₂C(O)-cyclohexyl (—)	63
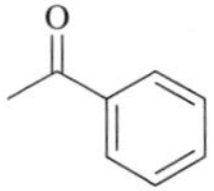 PhC(O)CH₃ (acetophenone)	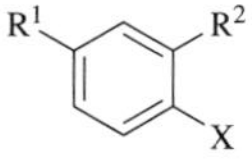 R¹,R²-substituted C₆H₃-X	Pd cat **4** (1 mol %), NaOt-Bu (1.05 eq), THF, 0.5 h	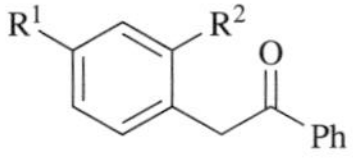 R¹,R²-C₆H₃-CH₂C(O)Ph X / R¹ / R² : Cl / H / H (93); Cl / H / Me (90); Cl / MeO / H (78); Cl / Me / H (88); OTf / H / MeO (80)	50

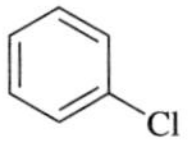

[Chlorobenzene] + [PhCH2C(O)Ph] →

Pd(dba)₂, ligand, base, 20 h

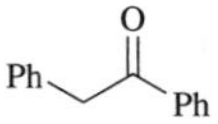

I 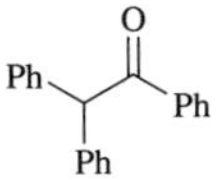 + II

Ligand	Base	Solvent	Temp (°)	I	II
L35	NaOt-Bu	toluene	120	(70)	(—)
L35	Na₂CO₃	dioxane	100	(0)	(0)
L35	K₂CO₃	dioxane	100	(40)	(25)
L35	K₃PO₄	dioxane	100	(16)	(51)
L35	Cs₂CO₃	dioxane	100	(22)	(62)
L35	CaO	dioxane	100	(0)	(0)
L35	K₃PO₄	dioxane	100	(16)	(51)
L35	K₃PO₄	dioxane	100	(59)	(31)
P(t-Bu)₂(n-Bu)	K₃PO₄	dioxane	100	(9)	(20)
P(t-Bu)₃	K₃PO₄	dioxane	100	(0)	(19)
PCy₃	K₃PO₄	dioxane	100	(33)	(32)
PCy₂(n-Bu)	K₃PO₄	dioxane	100	(17)	(3)
PCy₂Ph	K₃PO₄	dioxane	100	(31)	(31)
PCy₂(biPh)	K₃PO₄	dioxane	100	(17)	(19)

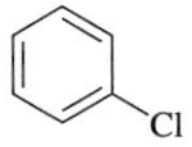

48

Pd(dba)₂ (0.1 mol %), **L35**, NaOt-Bu, toluene, 120°, 20 h   **I** (70)

48

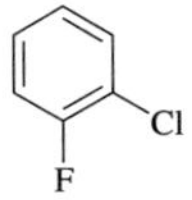

Pd(OAc)₂, **L35**, K₃PO₄, dioxane, 100°, 20 h

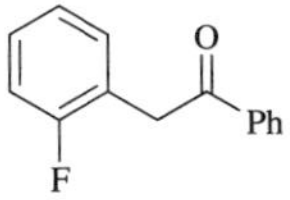

(57) 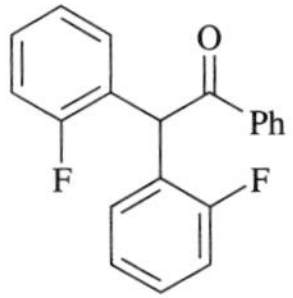 + (23)

48

TABLE 1. ARYLATION OF ALDEHYDES, KETONES, AND ENOL ETHERS (Continued)

Substrate	Aryl Compound	Conditions	Product(s) and Yield(s) (%)	Refs.
C_8 acetophenone	4-MeO-C₆H₄-Cl	Pd(OAc)₂, **L35**, K₃PO₄, dioxane, 100°, 20 h	4-MeO-C₆H₄-CH(C(O)Ph)-C₆H₄-OMe (57) + 4-MeO-C₆H₄-CH₂-C(O)Ph (25)	48
	2-NO₂, R¹, R² substituted aryl halide (X = Cl, Br)	Pd₂(dba)₃ (1 mol %), **L17** (4 mol %), K₃PO₄ (2.5 eq), p-methoxyphenol (20%), toluene, 35–50°, 22 h	R¹(NO₂)-C₆H₂(R²)-CH₂-C(O)Ph	45

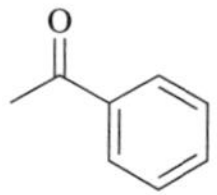

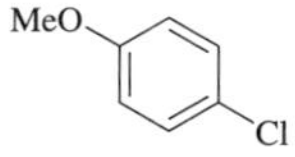

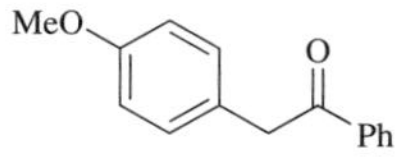

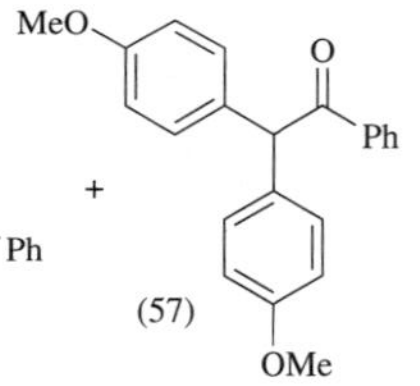

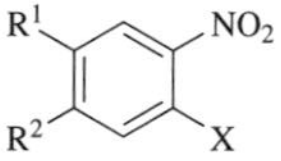

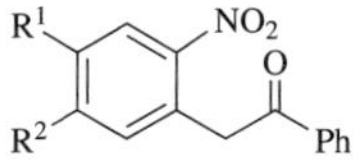

X	R¹	R²	Yield
Cl	EtO₂C	H	(61)
Br	Me	H	(90)
Br	MeO	MeO	(65)

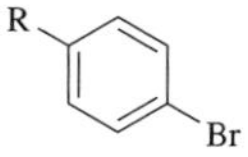

Substrate				
R-C₆H₄-Br	Catalyst, ligand, base, toluene, 130°		R-C₆H₄-CH₂-C(O)Ph	56

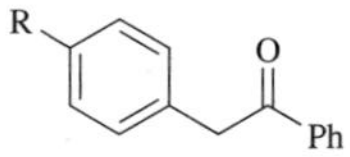

Catalyst	Ligand	Base	R	Yield
PdCl₂(cod)	L36	K₃PO₄	H	(>99)
Pd(OCOCF₃)₂	L37	K₃PO₄	H	(94)
PdCl₂(cod)	L36	Cs₂CO₃	H	(>99)
Pd(OCOCF₃)₂	L37	Cs₂CO₃	H	(88)
PdCl₂(cod)	L36	K₃PO₄	MeO	(>99)
Pd(OCOCF₃)₂	L37	K₃PO₄	MeO	(93)
PdCl₂(cod)	L36	K₃PO₄	F	(>99)
Pd(OCOCF₃)₂	L37	K₃PO₄	F	(94)

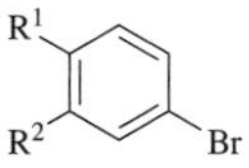

R¹	R²		
H	MeO	(96)	
MeO	H	(85)	

Pd(OAc)$_2$ (1 mol %), PPh$_3$ (8 mol %), Cs$_2$CO$_3$ (2.5 eq), DMF, 150°, 0.5–1 h

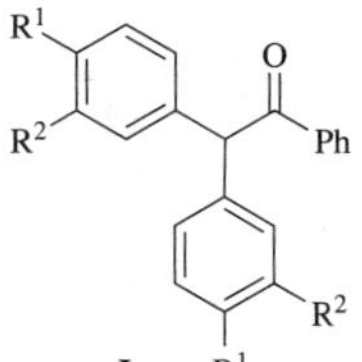

53

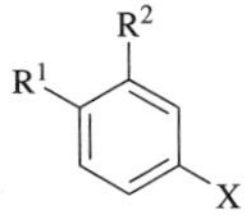

See table

54, 55

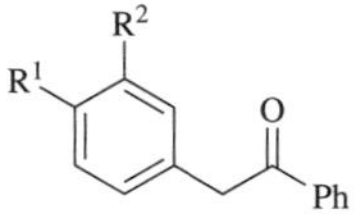

X	R¹	R²	Catalyst	Ligand	Base	Solvent	Temp (°)	I	II
Br	H	H	Pd/C (5 mol %)	none	Na$_2$CO$_3$	DMF	150	(2)^c	(3)^c
Br	H	H	PdCl$_2$ (3 mol %)	PPh$_3$	K$_2$CO$_3$	DMF	130	(11)^c	(—)^c
Br	H	H	Pd(OAc)$_2$	PPh$_3$	Cs$_2$CO$_3$	DMF	153	(91)	(—)
Br	H	H	Pd(OCOCF$_3$)$_2$ (5 mol %)	PPh$_3$	Cs$_2$CO$_3$	DMF	150	(82)^c	(4)^c
Br	H	H	Pd(PPh$_3$)$_4$ (1 mol %)	none	Cs$_2$CO$_3$	xylene	153	(71)^c	(2)^c
Br	H	H	Pd cat **9** (5 mol %)	none	K$_2$CO$_3$	toluene	130	(8)^c	(43)^c
Br	H	H	Pd cat **10** (1 mol %)	none	K$_2$CO$_3$	toluene	130	(45)^c	(16)^c
Br	H	H	Pd cat **10** (2 mol %)	none	K$_2$CO$_3$	xylene	153	(15)^c	(44)^c
Br	H	H	Pd cat **10** (2 mol %)	none	Cs$_2$CO$_3$	DMF	153	(—)^c	(3)^c
Br	H	H	Pd cat **11**	none	Cs$_2$CO$_3$	DMF	153	(89)	(—)
Br	H	MeO	Pd cat **11** (5 mol %)	none	Cs$_2$CO$_3$	DMF	153	(85)^c	(2)^c
Br	H	MeO	Pd(OAc)$_2$	P(t-Bu)$_3$	NaOt-Bu	THF	80	(5)^c	(70)^c
Br	H	MeO	Pd(OAc)$_2$	PPh$_3$	Cs$_2$CO$_3$	DMF	153	(71)	(—)

TABLE 1. ARYLATION OF ALDEHYDES, KETONES, AND ENOL ETHERS (*Continued*)

Substrate	Aryl Compound	Conditions	Product(s) and Yield(s) (%)	Refs.
C$_8$ acetophenone	R^2–C$_6$H$_3$(R^1)–X	See table (*cont. from previous page*)	I (PhC(O)CH(Ar)(Ar')) + II (PhC(O)CH$_2$Ar)	54, 55

X	R^1	R^2	Catalyst	Ligand	Base	Solvent	Temp (°)	I	II
Br	H	MeO	Pd(OCOCF$_3$)$_2$ (5 mol %)	PPh$_3$	Cs$_2$CO$_3$	DMF	150	(54)^c	(11)^c
Br	H	MeO	Pd(PPh$_3$)$_4$ (1 mol %)	none	Cs$_2$CO$_3$	xylene	150	(25)^c	(—)^c
Br	H	MeO	Pd cat **11**	none	Cs$_2$CO$_3$	DMF	153	(79)	(—)
Br	F	H	Pd(OAc)$_2$	PPh$_3$	Cs$_2$CO$_3$	DMF	153	(63)	(—)
Br	F	H	Pd cat **11**	none	Cs$_2$CO$_3$	DMF	153	(73)	(—)
Br	MeO	MeO	Pd(OAc)$_2$	PPh$_3$	Cs$_2$CO$_3$	DMF	153	(61)	(—)
Br	MeO	MeO	Pd cat **11**	none	Cs$_2$CO$_3$	DMF	153	(80)	(—)
I	H	H	Pd cat **11** (7 mol %)	PPh$_3$	Cs$_2$CO$_3$	DMF	100	(49)^c	(—)

Substrate	Aryl Compound	Conditions	Product(s) and Yield(s) (%)	Refs.
	R^2–C$_6$H$_3$(R^1)–Br	See table	I + II	55

R^1	R^2	Catalyst	Ligand	Base	Solvent	Temp (°)	I	II
H	H	Pd(PPh$_3$)$_4$ (0.5 mol %)	none	Cs$_2$CO$_3$	DMF	150	(18)^c	(1)^c
H	H	Pd(OAc)$_2$ (5 mol %)	PEt$_3$	Cs$_2$CO$_3$	DMF	150	(27)^c	(—)^c
H	H	Pd(OAc)$_2$ (5 mol %)	P(o-tol)$_3$	Cs$_2$CO$_3$	DMF	153	(8)^c	(22)^c
H	H	Pd(OAc)$_2$ (5 mol %)	P(o-tol)$_3$	NaO-t-Bu	THF	80	(80)	(—)

H	H	Pd cat **9** (1 mol %)	none	K_2CO_3	toluene	130	(4)^c (13)^c
H	H	Pd cat **10** (5 mol %)	none	NaOt-Bu	toluene	85	(—) (87)
H	H	Pd cat **10** (1 mol %)	none	K_2CO_3	xylene	130	(32)^c (17)^c
H	MeO	Pd cat **9** (5 mol %)	none	K_2CO_3	toluene	130	(6)^c (32)^c
H	MeO	Pd cat **10**	none	NaOt-Bu	THF	85	(—) (92)
F	H	Pd(OAc)$_2$	PPh$_3$	Cs$_2$CO$_3$	DMF	153	(63) (—)
F	H	Pd cat **10**	none	NaOt-Bu	THF	85	(—) (80)
MeO	MeO	Pd cat **10**	none	NaOt-Bu	THF	85	(—) (91)

TABLE 1. ARYLATION OF ALDEHYDES, KETONES, AND ENOL ETHERS (*Continued*)

Substrate	Aryl Compound	Conditions	Product(s) and Yield(s) (%)	Refs.
C_8 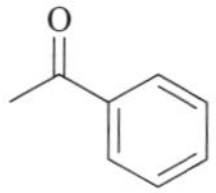 acetophenone	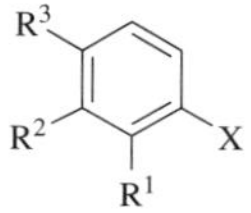 see table	Pd(dba)$_2$ (7.5 mol %), ligand, KHMDS (2.2 eq), THF, reflux, 0.75 h	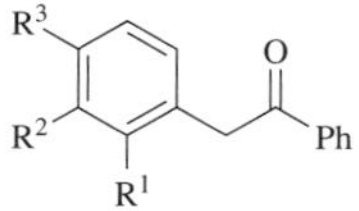 product with R^1, R^2, R^3, COCH$_2$Ph	17
	X R^1 R^2 R^3 Ligand			
	Br H H H **L2**		(84)	
	Br H H H **L1**		(76)	
	Br Me H H **L2**		(94)	
	Br H NC H **L2**		(73)	
	Br H H t-Bu **L2**		(85)	
	Br H H MeO **L2**		(69)	
	I H H H **L2**		(79)	
	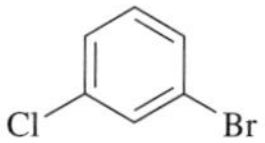 3-Cl-C$_6$H$_4$Br	Pd(PPh$_3$)$_4$ (0.5 mol %), Cs$_2$CO$_3$ (3–5 eq), *o*-xylene, 23 h	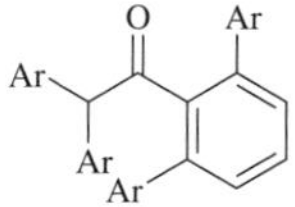 Ar-substituted product, Ar = *m*-ClC$_6$H$_4$ (61)	46
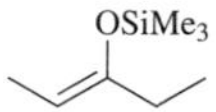 enol silane (OSiMe$_3$)	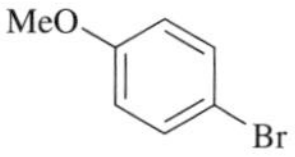 4-MeO-C$_6$H$_4$Br	Pd$_2$(dba)$_3$ (2.5 mol %), P(*t*-Bu)$_3$ (6 mol %), Bu$_3$SnF (2 eq), benzene, reflux, 26 h	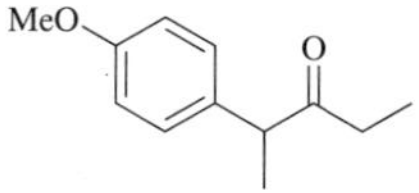 MeO-aryl ethyl ketone (70)	151
C_9 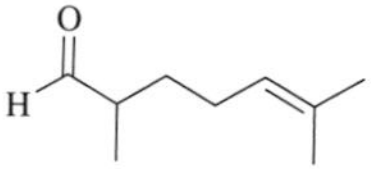 citronellal	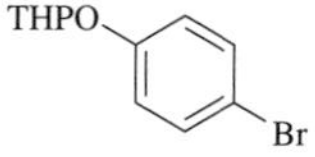 4-THPO-C$_6$H$_4$Br	Pd(OAc)$_2$ (2 mol %), **L20** (3 mol %), Cs$_2$CO$_3$ (1.2 eq), dioxane, 100°	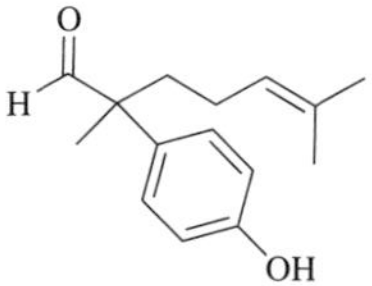 4-HO-C$_6$H$_4$ substituted aldehyde (—)	148

Substrate	Aryl halide	Conditions	Product	Yield (%)
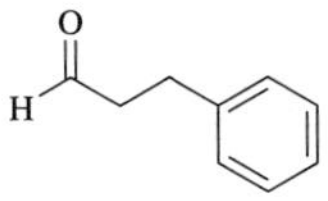 PhCH2CH2CHO	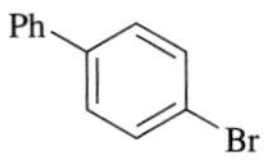 4-Ph-C6H4-Br	Pd(OAc)2 (0.05 mol %), P(t-Bu)3 (10 mol %), Cs2CO3 (1.2 eq), dioxane, 3 h	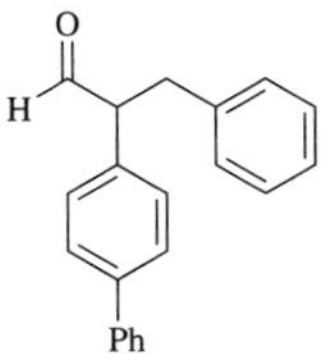 PhCH2CH(4-Ph-C6H4)CHO (45)	65
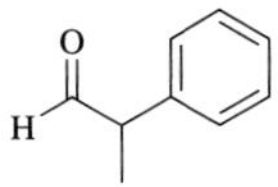 PhCH(CH3)CHO	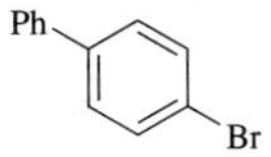 4-Ph-C6H4-Br	Pd(OAc)2 (0.05 mol %), P(t-Bu)3 (10 mol %), Cs2CO3 (1.2 eq), dioxane, 3 h	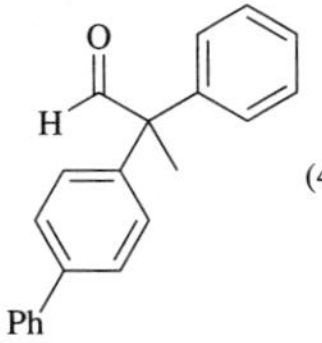 Ph-C(CH3)(4-Ph-C6H4)CHO (43)	65
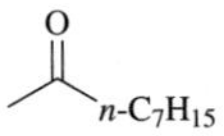 CH3C(O)-n-C7H15	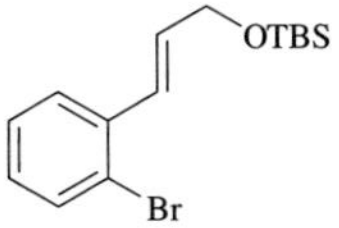 2-(TBSO-CH2CH=CH)-C6H4-Br	Pd2(dba)3 (5 mol %), P(t-Bu)3 (10 mol %), LiHMDS, dioxane, 90°, 3 h	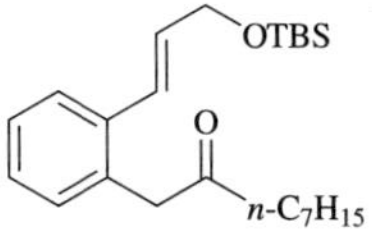 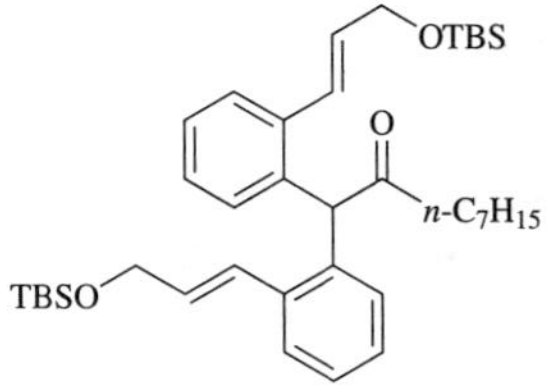 mono + bis arylation products	63 (—)
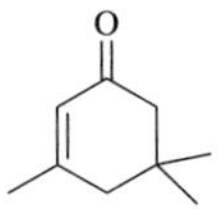 3,5,5-trimethylcyclohex-2-enone	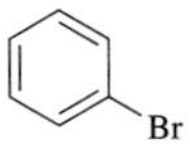 PhBr	Pd(OAc)2, PPh3 or P(t-Bu)3, Cs2CO3	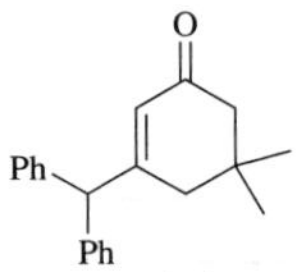 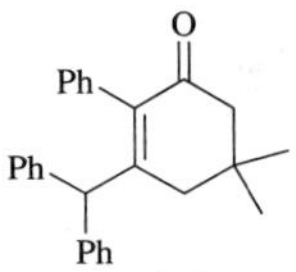 α-benzhydryl and α-benzyl dimethylcyclohexenones	47 (—)

TABLE 1. ARYLATION OF ALDEHYDES, KETONES, AND ENOL ETHERS (*Continued*)

Substrate	Aryl Compound	Conditions	Product(s) and Yield(s) (%)	Refs.
C$_9$ [4,4-dimethylcyclohex-2-enone]	[PhBr]	Pd(OAc)$_2$ (0.05 eq), PPh$_3$ (10 mol %), Cs$_2$CO$_3$, DMF	Temp (°) 60 / 80, Time (h) 6 / 5, Yields (56)/(56) [benzylated cyclohexenone product]	92
	[dibromo-dialkyl benzene]	Pd(OAc)$_2$, PPh$_3$ (5 mol %), Cs$_2$CO$_3$ (4 eq), DMF, 80°, 5–6 h	[fluorenone product], R = H (94), Me (80)	60
[4′-methylacetophenone]	[aryl bromide with R^1, R^2]	See table	**I** + **II**	55

R^1	R^2	Catalyst	Ligand	Base	Solvent	Temp (°)	I	II
H	H	Pd(OAc)$_2$	PPh$_3$	Cs$_2$CO$_3$	DMF	153	(—)	(91)
H	H	Pd cat **10**	none	NaO*t*-Bu	THF	85	(86)	(—)
H	H	Pd cat **11**	none	Cs$_2$CO$_3$	DMF	153	(—)	(93)
H	MeO	Pd(OAc)$_2$	PPh$_3$	Cs$_2$CO$_3$	DMF	153	(—)	(71)
H	MeO	Pd cat **10**	none	NaO*t*-Bu	THF	85	(86)	(—)
H	MeO	Pd cat **11**	none	Cs$_2$CO$_3$	DMF	153	(—)	(75)
F	H	Pd(OAc)$_2$	PPh$_3$	Cs$_2$CO$_3$	DMF	153	(—)	(63)
F	H	Pd cat **10**	none	NaO*t*-Bu	THF	85	(74)	(—)
F	H	Pd cat **11**	none	Cs$_2$CO$_3$	DMF	153	(—)	(80)
MeO	MeO	Pd(OAc)$_2$	PPh$_3$	Cs$_2$CO$_3$	DMF	153	(—)	(61)
MeO	MeO	Pd cat **10**	none	NaO*t*-Bu	THF	85	(82)	(—)
MeO	MeO	Pd cat **11**	none	Cs$_2$CO$_3$	DMF	153	(—)	(92)

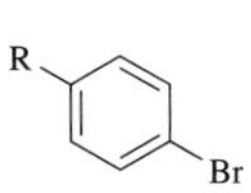

Catalyst, ligand, K$_3$PO$_4$, toluene, 130°

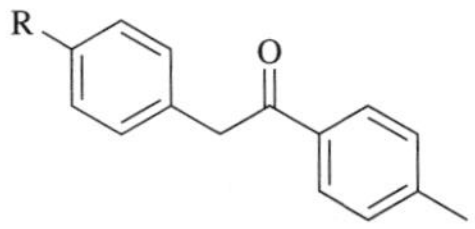

R	Catalyst	Ligand	
H	PdCl$_2$(cod)	L36	(>99)
H	Pd(OCOCF$_3$)$_2$	L37	(90)
MeO	PdCl$_2$(cod)	L36	(>99)
MeO	Pd(OCOCF$_3$)$_2$	L37	(90)
F	PdCl$_2$(cod)	L36	(>99)
F	Pd(OCOCF$_3$)$_2$	L37	(91)

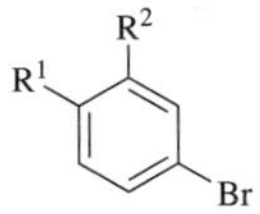

56

Catalyst, ligand, Cs$_2$CO$_3$, DMF, 153°

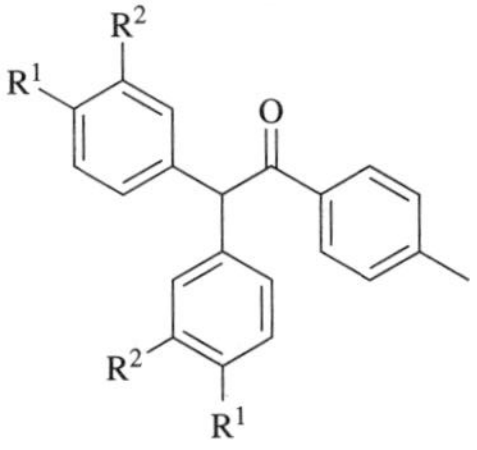

R^1	R^2	Catalyst	Ligand	
H	H	Pd(OAc)$_2$	PPh$_3$	(87)
H	H	Pd cat 11	none	(93)
H	MeO	Pd(OAc)$_2$	PPh$_3$	(69)
H	MeO	Pd cat 11	none	(75)
F	H	Pd(OAc)$_2$	PPh$_3$	(68)
F	H	Pd cat 11	none	(80)
MeO	MeO	Pd(OAc)$_2$	PPh$_3$	(60)
MeO	MeO	Pd cat 11	none	(92)

54

TABLE 1. ARYLATION OF ALDEHYDES, KETONES, AND ENOL ETHERS (*Continued*)

Substrate	Aryl Compound	Conditions	Product(s) and Yield(s) (%)	Refs.
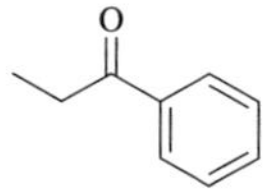 C₉	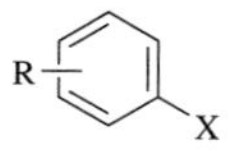	Pd(OAc)₂ (1 mol %), **L51** (1 mol %), base, 60°	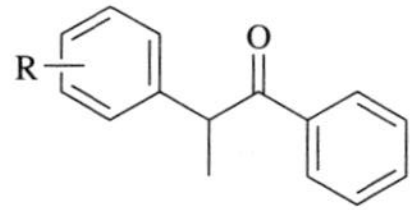	51

X	R	Base	Solvent	Time (h)	
Cl	H	NaO*t*-Bu	dioxane	6	(71)
Cl	2-MeO	NaO*t*-Bu	dioxane	1	(94)
Cl	2-Me	NaO*t*-Bu	dioxane	2	(90)
Cl	2-CF₃	NaO*t*-Bu	dioxane	12	(71)
Cl	3-MeO	NaO*t*-Bu	dioxane	1	(81)
Cl	4-MeO	NaO*t*-Bu	dioxane	3	(90)
Cl	4-Me	NaO*t*-Bu	toluene	1	(77)
Cl	4-Me	NaO*t*-Bu	DME	2	(85)
Cl	4-Me	NaO*t*-Bu	dioxane	1	(87)
Cl	4-Me	NaO*t*-Bu	THF	1	(84)
Cl	4-Me	NaO*t*-Bu	MTBE	1	(82)
Cl	4-Me	NaO*t*-Bu	dioxane	3	(96)
Cl	4-Me	KO*t*-Bu	dioxane	4	(85)
Cl	4-Me	KOMe	dioxane	5	(60)
Cl	4-Me	NaH	dioxane	2	(85)
Cl	4-Me	KH	dioxane	12	(78)
Cl	4-Me	Cs₂CO₃	dioxane	3	(10)
Cl	4-Me	K₃PO₄	dioxane	3	(27)
Cl	4-CF₃	NaO*t*-Bu	dioxane	12	(69)
Br	4-Me	NaO*t*-Bu	dioxane	1	(88)
I	4-Me	NaO*t*-Bu	dioxane	1	(83)

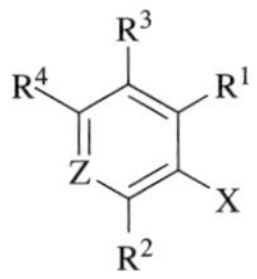

Catalyst (1 mol %), NaOt-Bu, THF

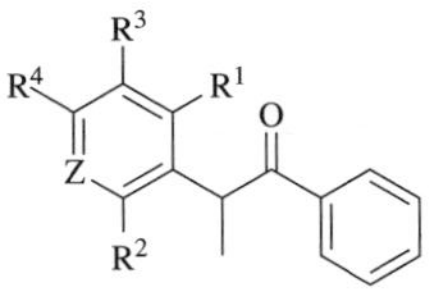

X	R¹	R²	R³	R⁴	Z	Catalyst	Temp (°)	
Cl	H	H	H	H	CH	Pd cat **4**	70	(100)
Cl	H	H	H	H	CH	Pd cat **4**	50	(91)
Cl	H	H	H	H	CH	Pd cat **5**	70	(95)
Cl	H	H	H	H	CH	Pd cat **6**	70	(95)
Cl	H	H	H	H	CH	Pd cat **7**	70	(93)
Cl	H	H	H	H	CH	Pd cat **8**	70	(99)
Cl	H	MeO	H	H	CH	Pd cat **4**	60	(80)
Cl	H	Me	H	H	CH	Pd cat **4**	60	(87)
Cl	H	H	MeO	H	CH	Pd cat **4**	70	(88)
Cl	H	H	H	CF₃	CH	Pd cat **4**	70	(81)
Cl	H	H	H	PhC(O)	CH	Pd cat **4**	70	(71)
Br	Me	Me	H	Me	CH	Pd cat **4**	60	(72)
OTf	H	H	H	H	CH	none	60	(91)
OTf	H	H	H	Me	CH	none	60	(93)
Cl	H	H	H	H	N	Pd cat **4**	50	(60)

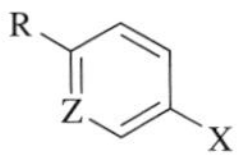

Pd cat **1** (1 mol %), NaOt-Bu, dioxane, 70°, 2 h

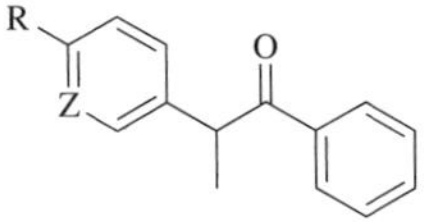

X	Z	R	
Cl	CH	Me	(89)
OTf	CH	Me	(78)
Cl	N	H	(90)

TABLE 1. ARYLATION OF ALDEHYDES, KETONES, AND ENOL ETHERS (*Continued*)

Substrate	Aryl Compound	Conditions	Product(s) and Yield(s) (%)	Refs.
C₉				
(propiophenone)	4-chlorotoluene	Pd(dba)₂ (x mol %), **L35**, base, 20 h x / Base / Solvent / Temp (°) 0.01 / NaO*t*-Bu / toluene / 80 0.01 / NaO*t*-Bu / toluene / 120 0.05 / NaO*t*-Bu / toluene / 80 0.1 / NaO*t*-Bu / toluene / 80 1 / K₃PO₄ / dioxane / 100	2-(p-tolyl)-1-phenylpropan-1-one (40) (41) (97) (100) (90)	48
	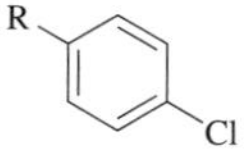 4-R-chlorobenzene (R = CN, CO₂Me)	Pd(OAc)₂ (0.5 mol %), **L19** (0.2 mol %), NaO*t*-Bu (1.3 eq), 80°, toluene Time (h) 18 16	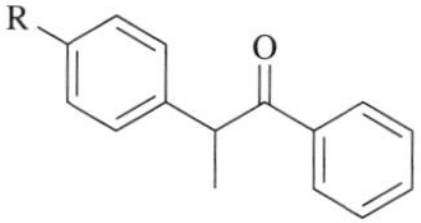 2-(4-R-phenyl)-1-phenylpropan-1-one (78) (76)	43
	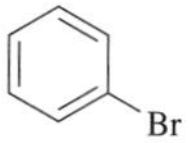 bromobenzene	Pd(dba)₂, **L1** or **L3**, KHMDS, 70°	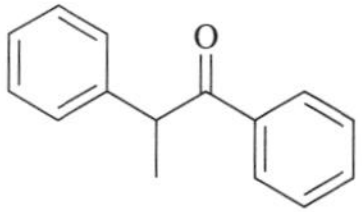 1,2-diphenyl-2-phenylpropan-1-one (71) **I**	23
	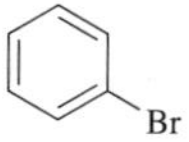 bromobenzene	Pd(dba)₂ (7.5 mol %), ligand, KHMDS (2.2 eq), THF, reflux, 0.75 h	**I** Ligand **L2** (71) **L1** (47)	17

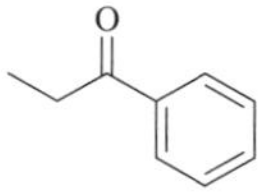

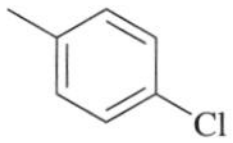

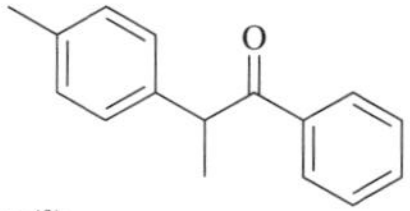

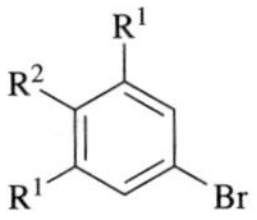

[43]

Catalyst (x mol %),
NaOt-Bu (1.3 eq), toluene

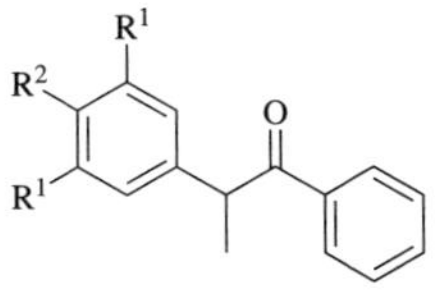

R¹	R²
H	H
H	MeO
Me	H
Me	H

Catalyst	x	Temp (°)	Time (h)	
Pd(OAc)₂	0.001	120	24	(74)
Pd(OAc)₂	1.0	80	3	(84)
Pd(OAc)₂	1.0	80	2.3	(76)
Pd₂(dba)₃	1.0	80	14	(93)

[152]

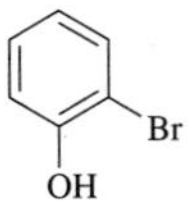

Pd(OAc)₂ (5 mol %),
ligand (10 mol %),
NaOt-Bu (2.3 eq), 80°

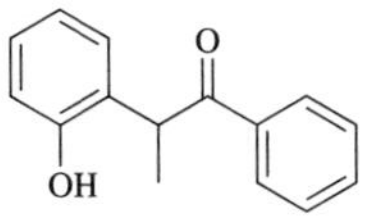

Ligand	Solvent	
L9	THF	(85)
L9	DME	(78)
L9	dioxane	(84)
L9	DMF	(54)
L9	DMA	(37)
L9	t-BuOH	(45)
L9	DCE	(0)
L9	toluene	(64)
L9	toluene	(97)
L14	toluene	(42)
L15	toluene	(30)
L16	toluene	(0)
L18	toluene	(0)
L19	toluene	(62)

TABLE 1. ARYLATION OF ALDEHYDES, KETONES, AND ENOL ETHERS (*Continued*)

Substrate	Aryl Compound	Conditions	Product(s) and Yield(s) (%)	Refs.
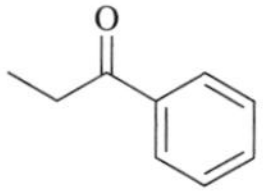 C₉ phenyl propyl ketone	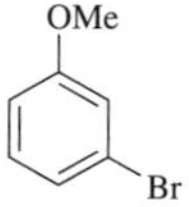 3-bromoanisole	Pd₂(dba)₃ (1.5 mol %), **L25** (3.6 mol %), NaO*t*-Bu (1.3 eq), THF, 70°	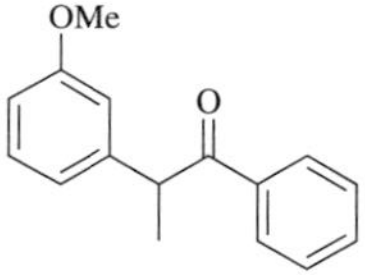 α-(3-methoxyphenyl)propiophenone (91)	16
	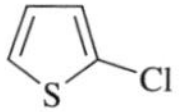 2-chlorothiophene	Pd(OAc)₂ (1 mol %), **L51** (1 mol %), NaO*t*-Bu, dioxane, 60°, 12 h	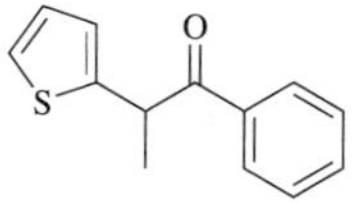 α-(2-thienyl)propiophenone (42)	51
	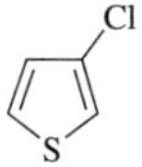 3-chlorothiophene	Pd(OAc)₂ (1 mol %), **L51** (1 mol %), NaO*t*-Bu, dioxane, 60°, 24 h	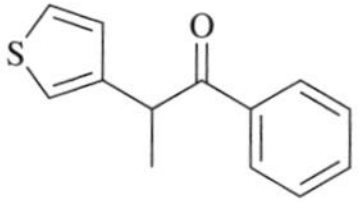 α-(3-thienyl)propiophenone (36)	51
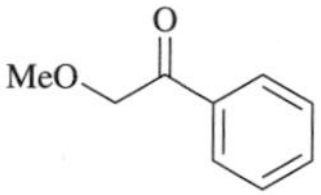 2-methoxyacetophenone	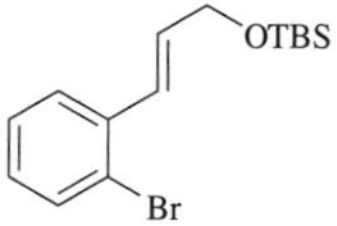 2-bromo-(OTBS-allyl)benzene	Pd₂(dba)₃ (5 mol %), LiHMDS, P(*t*-Bu)₃ (10 mol %), dioxane, 90°, 3 h	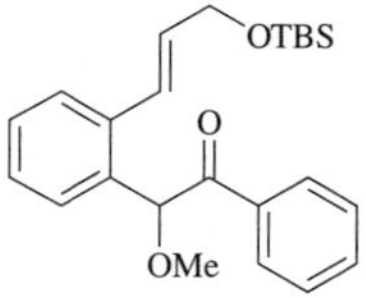 (—)	63
	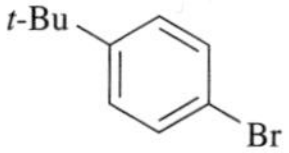 4-*t*-Bu-bromobenzene	Pd(OAc)₂ (0.5 mol %), **L19** (0.2 mol %), NaO*t*-Bu (1.3 eq), toluene, 70°, 17 h	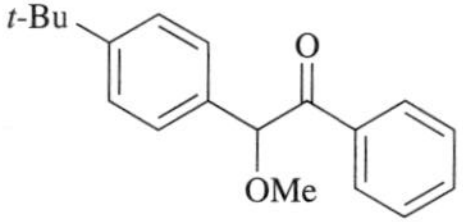 (83)	43
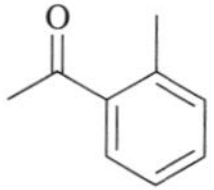 2-methylacetophenone	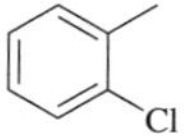 2-chlorotoluene	Pd(OAc)₂ (1 mol %), **L51** (1 mol %), NaO*t*-Bu (1.5 eq), dioxane, 60°, 24 h	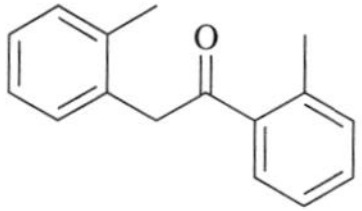 (75)	51

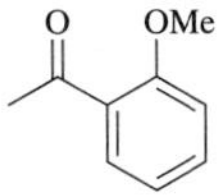

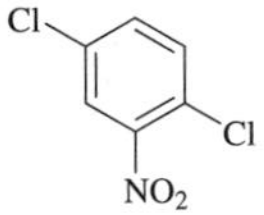

Ketone	Aryl halide	Conditions	Product	Yield (%)
2'-methoxyacetophenone	2,5-dichloronitrobenzene	Pd₂(dba)₃ (1 mol %), **L17** (4 mol %), K₃PO₄ (2.5 eq), *p*-methoxyphenol (20 mol %), toluene, 30–80°	aryl ketone with OMe and Cl, NO₂ substituents	45
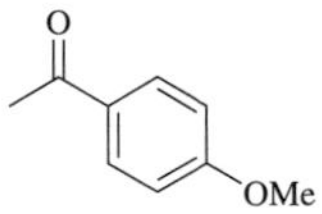 4'-methoxyacetophenone	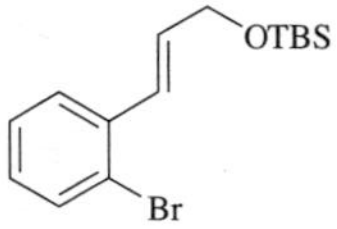 (E)-(3-(2-bromophenyl)allyloxy)(tert-butyl)dimethylsilane	Pd₂(dba)₃ (5 mol %), P(*t*-Bu)₃ (10 mol %), LiHMDS, dioxane, 90°, 3 h 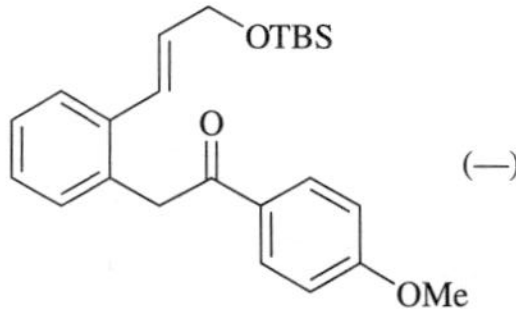	α-arylated ketone with OTBS allyl and 4-OMe aryl	63
4'-fluoroacetophenone	4-bromofluorobenzene	Catalyst, Cs₂CO₃, DMF, 153° Catalyst: Pd(OAc)₂ / Ligand: PPh₃ (52); Pd cat **11** / none (70)	α-arylated ketone with two 4-F aryl groups	54
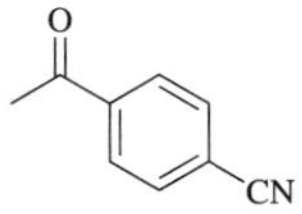 4'-cyanoacetophenone	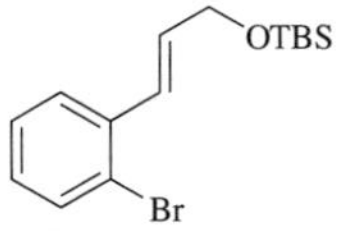 (E)-(3-(2-bromophenyl)allyloxy)(tert-butyl)dimethylsilane	Pd₂(dba)₃ (5 mol %), P(*t*-Bu)₃ (10 mol %), LiHMDS, dioxane, 90°, 3 h 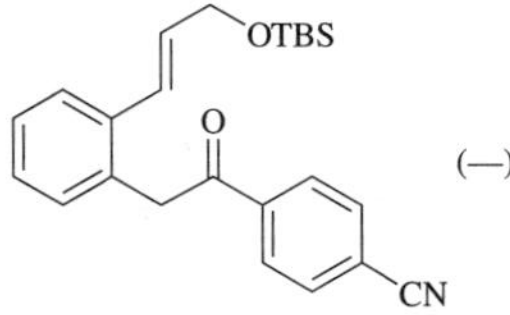	α-arylated ketone with OTBS allyl and 4-CN aryl	63
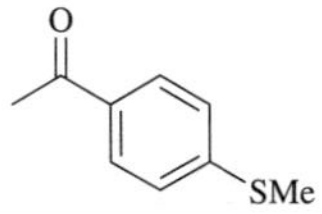 4'-(methylthio)acetophenone	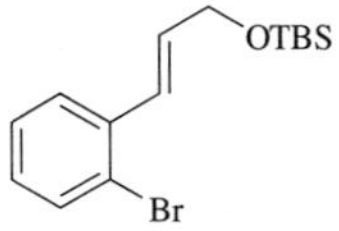 (E)-(3-(2-bromophenyl)allyloxy)(tert-butyl)dimethylsilane	Pd₂(dba)₃ (5 mol %), P(*t*-Bu)₃ (10 mol %), LiHMDS (3 eq), dioxane, 90°, 3 h 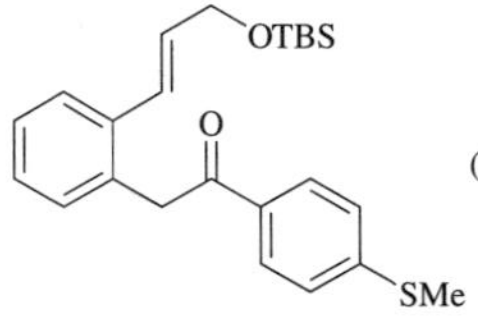	α-arylated ketone with OTBS allyl and 4-SMe aryl	63

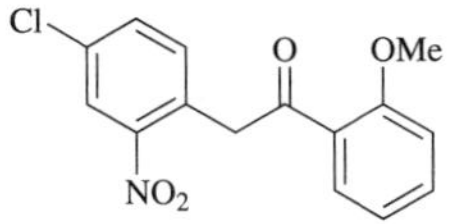

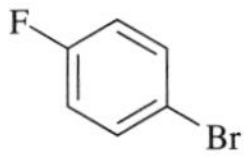

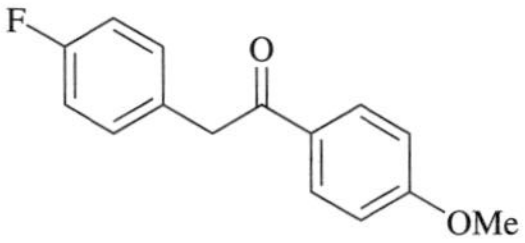

TABLE 1. ARYLATION OF ALDEHYDES, KETONES, AND ENOL ETHERS (*Continued*)

Substrate	Aryl Compound	Conditions	Product(s) and Yield(s) (%)	Refs.
C9				
(2,3-dihydro-1H-inden-1-one)	4-chlorotoluene	Pd(OAc)$_2$, **L35**, base, dioxane, 100°, 20 h	2-(p-tolyl)-2,3-dihydro-1H-inden-1-one; Base: K$_3$PO$_4$ (42), Cs$_2$CO$_3$ (26)	48
	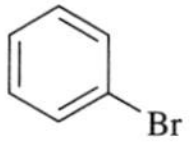 bromobenzene	Catalyst, ligand, K$_3$PO$_4$, xylene, 153°, 22 h Catalyst / Ligand: PdCl$_2$(cod) / **L36**; Pd(OCOCF$_3$)$_2$ / **L37**	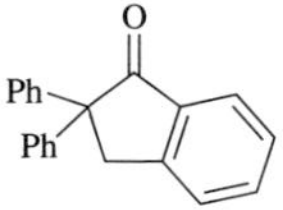 2,2-diphenyl-2,3-dihydro-1H-inden-1-one (89) (89)	56
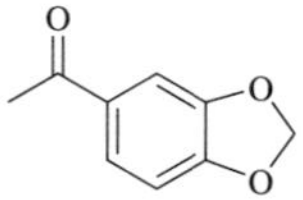 (1-(benzo[d][1,3]dioxol-5-yl)ethanone TMS enol ether)	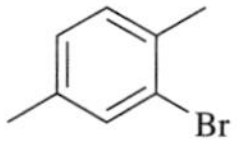 2-bromo-p-xylene	Pd$_2$(dba)$_3$ (5 mol %), **L23** (6 mol %), NaO-t-Bu (1.3 eq), THF, 70°	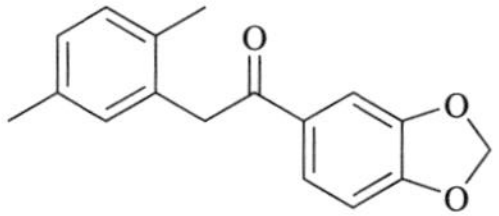 1-(benzo[d][1,3]dioxol-5-yl)-2-(2,5-dimethylphenyl)ethanone (84)	16
(Me$_3$SiO, 4-methylpent-3-en-2-one TMS enol ether)	bromobenzene	PdCl$_2$[P(o-tol)$_3$]$_2$ (3 mol %), Bu$_3$SnF, benzene	4-methyl-1-phenylpent-3-en-2-one (56)	13
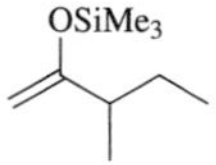 (OSiMe$_3$, 3-methylpentan-2-one TMS enol ether)	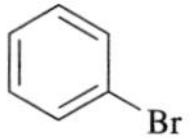 bromobenzene	PdCl$_2$[P(o-tol)$_3$]$_2$ (3 mol %), Bu$_3$SnF, benzene	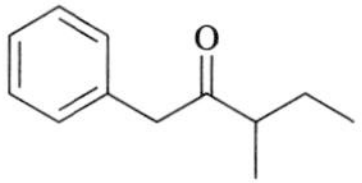 3-methyl-1-phenylpentan-2-one (47)	13
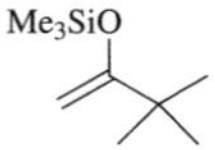 (Me$_3$SiO, pinacolone TMS enol ether)	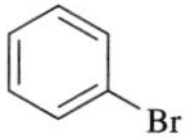 bromobenzene	PdCl$_2$[P(o-tol)$_3$]$_2$ (3 mol %), Bu$_3$SnF, benzene	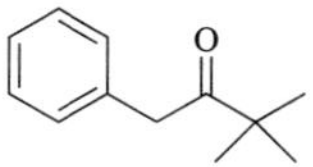 3,3-dimethyl-1-phenylbutan-2-one (29)	13

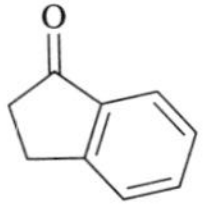

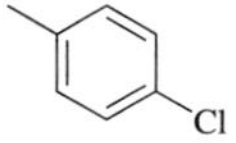

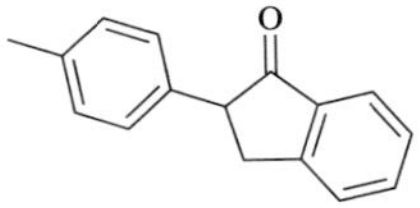

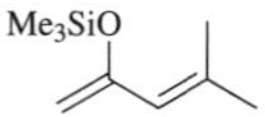

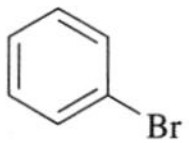

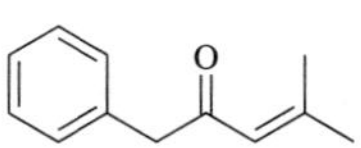

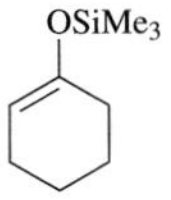

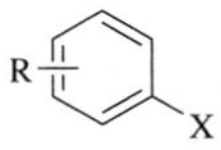

		Pd$_2$(dba)$_3$ (2.5 mol %), P(t-Bu)$_3$ (6 mol %), Bu$_3$SnF, benzene		151

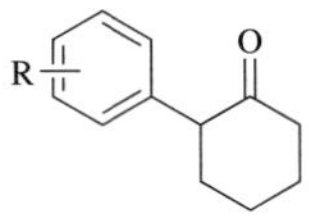

OSiMe$_3$ (cyclohexenyl)

R—(aryl)—X

Product: 2-aryl cyclohexanone (R-substituted)

X	R	Time (h)		
Cl	2-Me	3.5	(25)	
Cl	2-Me	20	(55)	
Cl	4-O$_2$N	5	(61)	
Cl	4-O$_2$N	20	(64)	
Br	2-Me	21	(70)	
Br	4-O$_2$N	8	(71)	
Br	4-Me$_2$N	9	(54)	
Br	4-MeO	9	(78)	
Br	4-MeO	19	(64)	
Br	4-MeO$_2$C	20	(62)	
I	H	21	(75)	
I	2-MeO	21	(71)	
I	3-MeO	20	(53)	
I	4-O$_2$N	8	(70)	
I	4-MeO	19	(82)	

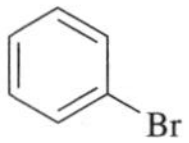

	PdCl$_2$[P(o-tol)$_3$]$_2$ (3 mol %), Bu$_3$SnF, benzene		13

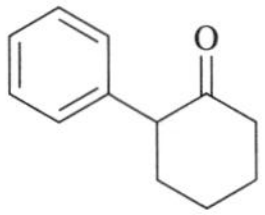

PhBr → 2-phenylcyclohexanone (<15)

TABLE 1. ARYLATION OF ALDEHYDES, KETONES, AND ENOL ETHERS (*Continued*)

Substrate	Aryl Compound	Conditions	Product(s) and Yield(s) (%)	Refs.

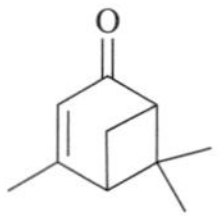

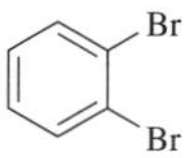

C_{10}

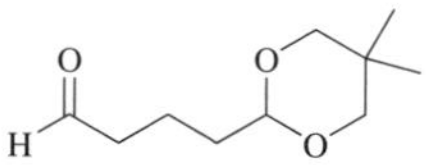

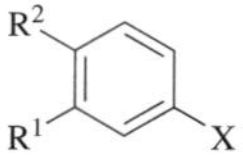

Substrate: 2-(3-oxopropyl)-5,5-dimethyl-1,3-dioxane derivative

Aryl Compound: R^2-/R^1-substituted aryl-X

Conditions: Catalyst, ligand, Cs_2CO_3 (1.2 eq), dioxane, 80–100°

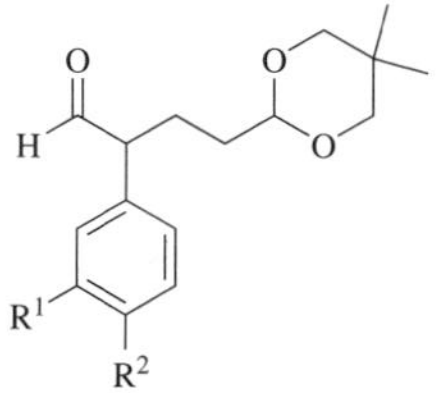

X	R^1	R^2	Catalyst	Ligand	
Cl	H	$MeOCH_2CH_2O$	Pd cat **3**	none	(60)
Cl	Me	MeO	Pd cat **3**	none	(58)
Br	H	$MeOCH_2CH_2O$	[Pd(cinnamyl)Cl]$_2$	**L32**	(66)
Br	Me	MeO	[Pd(cinnamyl)Cl]$_2$	**L32**	(74)

Ref: 148

Aryl Compound: 1,2-dibromobenzene

Conditions: Pd(OAc)$_2$, PPh$_3$ (5 mol %), Cs$_2$CO$_3$ (4 eq), DMF, 80°, 21 h

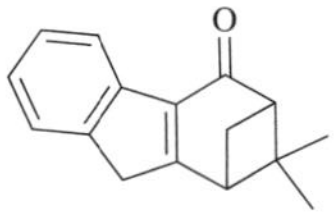

Product: tricyclic ketone (62)

Ref: 60

Aryl Compound: 4-chlorotoluene

Conditions: Pd(OAc)$_2$ (1 mol %), **L51** (1 mol %), NaO*t*-Bu, dioxane, 60°, 12 h

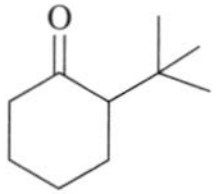

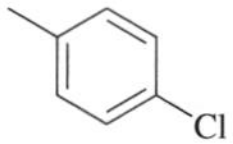

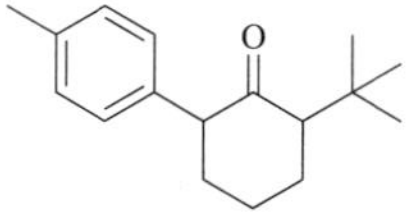

Product: 2-*tert*-butyl-6-(*p*-tolyl)cyclohexanone (20)

Ref: 51

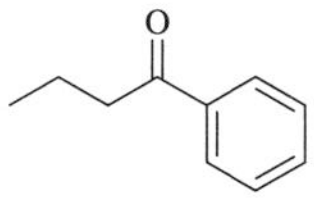

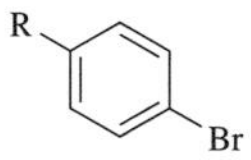

Pd(OAc)$_2$ (x mol %), phosphine, Cs$_2$CO$_3$, xylene, 140°

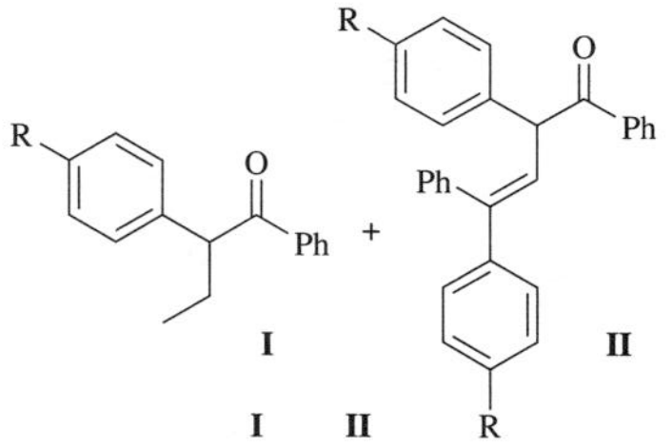

R	x	Phosphine	Time (h)	I	II
H	0.1	PPh$_3$	2	(20)	(23)
H	0.1	P(o-tol)$_3$	2	(14)	(78)
H	0.1	P(t-Bu)$_3$	2	(78)	(19)
H	0.075	P(o-tol)$_3$	4.5	(12)	(81)
F	0.075	P(o-tol)$_3$	2	(—)	(64)
Me	0.075	P(o-tol)$_3$	2	(—)	(50)

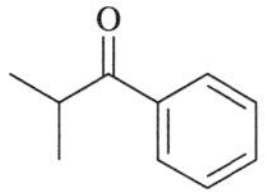

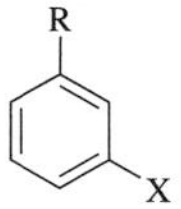

47

Pd(dba)$_2$, ligand, NaOt-Bu, rt to 70°

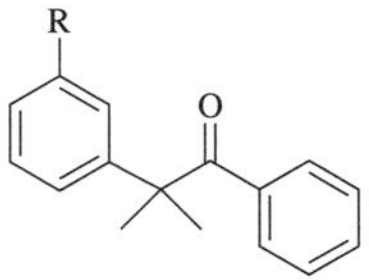

X	R	Ligand	
Cl	MeO	L3	(82)
Cl	MeO	P(t-Bu)$_3$	(92)
Br	H	L3	(78)
Br	H	P(t-Bu)$_3$	(87)

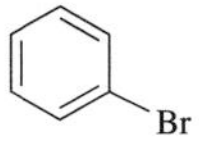

23

Pd(dba)$_2$ (2 mol %), L2 (15 mol %), KHMDS (1.2 eq), THF, reflux, 5 h

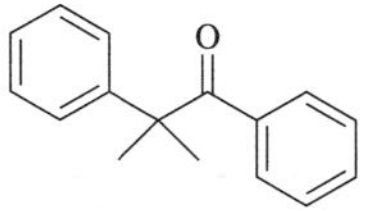

(55)

40

TABLE 1. ARYLATION OF ALDEHYDES, KETONES, AND ENOL ETHERS (*Continued*)

Substrate	Aryl Compound	Conditions	Product(s) and Yield(s) (%)	Refs.
C$_{10}$ *isobutyrophenone*	PhBr	Catalyst (2 mol %), ligand (2.5 mol %), NaO-*t*-Bu, THF, 50° Catalyst / Ligand / Time (h) Pd(dba)$_2$ / **L3** / 6 Pd(OAc)$_2$ / P(*t*-Bu)$_3$ / 12	PhC(O)C(CH$_3$)$_2$Ph (87) (92)	40
	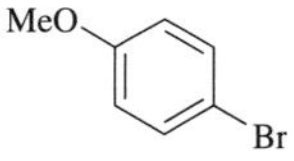 4-MeO-C$_6$H$_4$-Br	Pd(OAc)$_2$ (0.5 mol %), **L15** (0.2 mol %), NaO-*t*-Bu (1.3 eq), toluene, 80°, 1 h	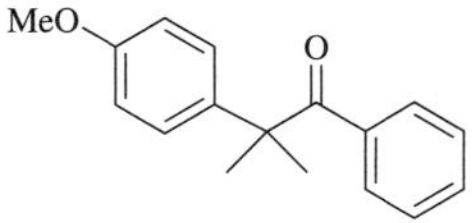 4-MeO-C$_6$H$_4$-C(CH$_3$)$_2$C(O)Ph (72)	43
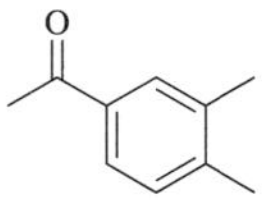 3,4-dimethylacetophenone	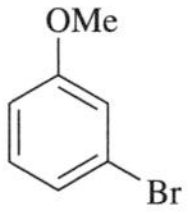 3-MeO-C$_6$H$_4$-Br	Pd(OAc)$_2$ (1 mol %), PPh$_3$ (8 mol %), Cs$_2$CO$_3$ (2.5 eq), DMF, 150°, 0.5–1 h	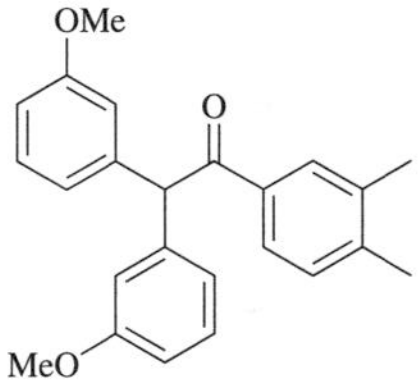 (3,4-Me$_2$C$_6$H$_3$)C(O)CH(3-MeOC$_6$H$_4$)$_2$ (74)	53
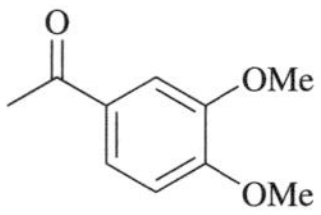 3,4-dimethoxyacetophenone	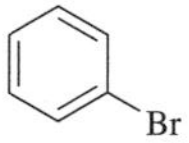 PhBr	Pd(OAc)$_2$ (1 mol %), PPh$_3$ (8 mol %), Cs$_2$CO$_3$ (2.5 eq), DMF, 150°, 0.5–1 h	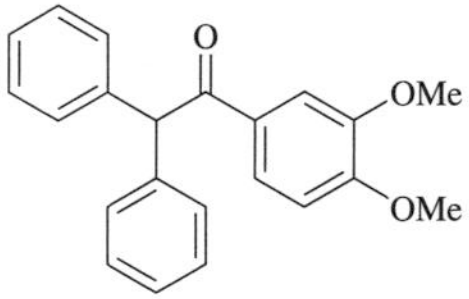 (3,4-(MeO)$_2$C$_6$H$_3$)C(O)CHPh$_2$ (85)	53

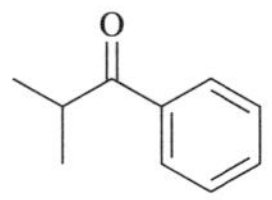

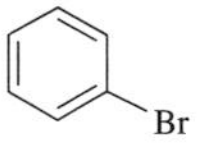

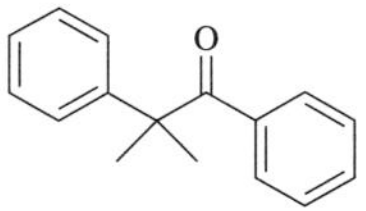

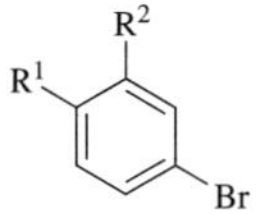

See table

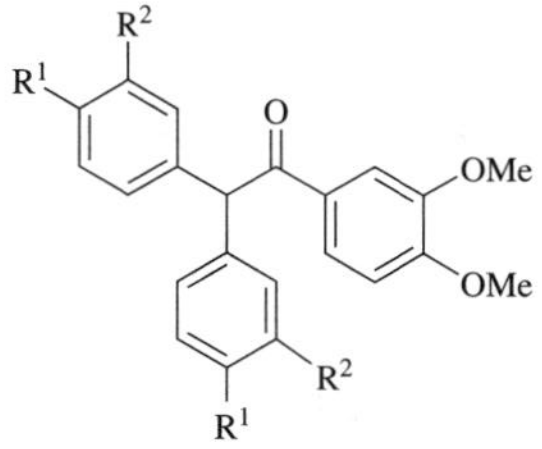

R[1]	R[2]	Catalyst	Ligand	Base	Solvent	Temp (°)	
H	H	Pd(OAc)$_2$	PPh$_3$	Cs$_2$CO$_3$	DMF	153	(62)
H	H	Pd cat **10**	none	NaOt-Bu	THF	85	(90)
H	H	Pd cat **11**	none	Cs$_2$CO$_3$	DMF	153	(57)
H	MeO	Pd cat **10**	none	NaOt-Bu	THF	85	(82)
H	MeO	Pd cat **11**	none	Cs$_2$CO$_3$	DMF	153	(47)
H	MeO	Pd(OAc)$_2$	PPh$_3$	Cs$_2$CO$_3$	DMF	153	(64)
O$_2$N	H	Pd cat **11**	none	Cs$_2$CO$_3$	DMF	153	(70)
O$_2$N	H	Pd(OAc)$_2$	PPh$_3$	Cs$_2$CO$_3$	DMF	153	(72)
F	H	Pd cat **10**	none	NaOt-Bu	THF	85	(35)
F	H	Pd cat **11**	none	Cs$_2$CO$_3$	DMF	153	(20)
F	H	Pd(OAc)$_2$	PPh$_3$	Cs$_2$CO$_3$	DMF	153	(52)
MeO	MeO	Pd cat **10**	none	NaOt-Bu	THF	85	(90)
MeO	MeO	Pd cat **11**	none	Cs$_2$CO$_3$	DMF	153	(88)
MeO	MeO	Pd(OAc)$_2$	PPh$_3$	Cs$_2$CO$_3$	DMF	153	(74)

TABLE 1. ARYLATION OF ALDEHYDES, KETONES, AND ENOL ETHERS (Continued)

Substrate	Aryl Compound	Conditions	Product(s) and Yield(s) (%)	Refs.
C_{10}				
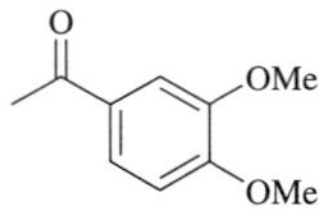 (3,4-dimethoxyacetophenone)	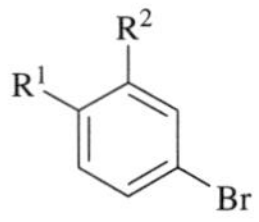 (4-bromo, R¹/R² substituted arene)	Catalyst, ligand, Cs_2CO_3, DMF, 153°	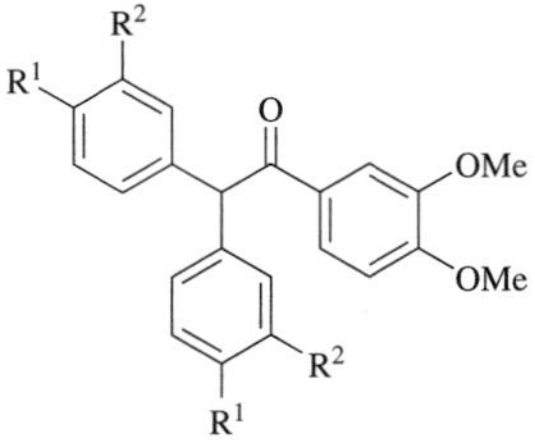 (triarylated ketone product with OMe groups)	54

R¹	R²	Catalyst	Ligand	
H	H	$Pd(OAc)_2$	PPh_3	(62)
H	H	Pd cat **11**	none	(90)
H	MeO	$Pd(OAc)_2$	PPh_3	(57)
H	MeO	Pd cat **11**	none	(82)
O_2N	H	$Pd(OAc)_2$	PPh_3	(35)
O_2N	H	Pd cat **11**	none	(20)
MeO	MeO	$Pd(OAc)_2$	PPh_3	(47)
MeO	MeO	Pd cat **11**	none	(64)

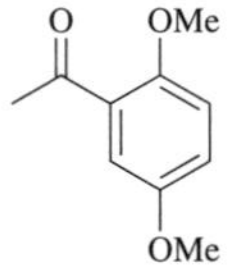 (2',5'-dimethoxyacetophenone)	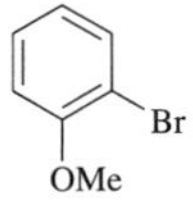 (2-bromoanisole)	$Pd(OAc)_2$ (1 mol %), **L19** (2 mol %), NaO-t-Bu, toluene, 80°	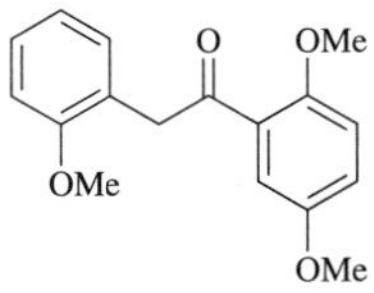 (coupled product, 2-OMe benzyl) (91)	43
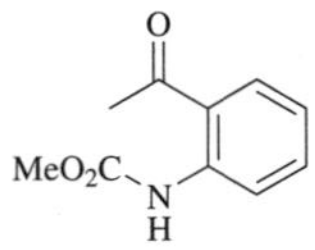 (N-(2-acetylphenyl)methyl carbamate)	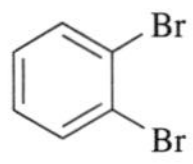 (1,2-dibromobenzene)	$Pd(OAc)_2$ (4.9 mol %), **L32** (2 eq), Cs_2CO_3, toluene, 130°, 20 h	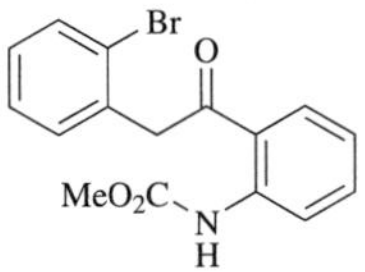 (2-bromo benzyl coupled product) (17)	114

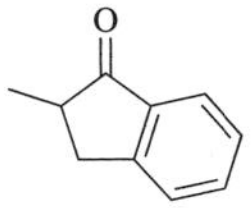

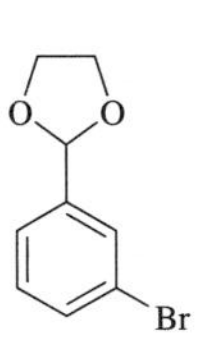

Substrate	Reagent	Conditions	Product	Refs.
2-methyl-1-indanone	3-bromo-phenyl-1,3-dioxolane	Pd(0), **L23**, NaO*t*-Bu, toluene, 100°	2-(3-(1,3-dioxolan-2-yl)phenyl)-2-methyl-1-indanone (79), 70% ee	30
2-methyl-1-indanone	aryl triflate (R¹, R², R³)	Catalyst, **L22**, NaO*t*-Bu (2 eq), toluene	2-aryl-2-methyl-1-indanone; see table below	34
4,6-dimethyl-3(2H)-benzofuranone	4-bromoaryl (R)	Pd(OAc)₂, **L19**, NaO*t*-Bu, toluene or THF, 60–80°	2-aryl-4,6-dimethyl-3(2H)-benzofuranone; R = H (15), MeO (22), Me (51)	153
4-methyl-6-methoxy-3(2H)-benzofuranone	4-bromotoluene	Pd(OAc)₂, **L19**, NaO*t*-Bu, toluene or THF, 60–80°	4-methyl-6-methoxy-2-(4-methylphenyl)-3(2H)-benzofuranone (39)	153

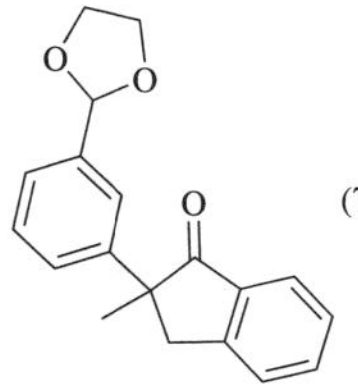

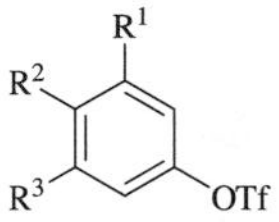

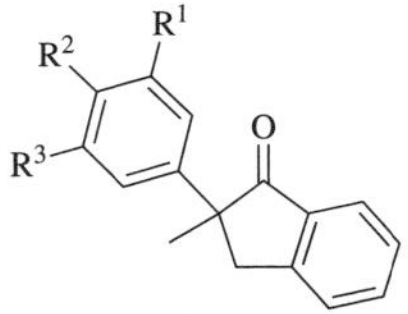

Table for **L22** reactions:

R¹	R²	R³	Catalyst	Temp (°)	% ee
H	H	H	Pd(dba)₂	60	(77) 70
H	H	CF₃	Ni(cod)₂	80	(70) 86
H	Me	H	Pd(dba)₂	60	(79) 78
H	NC	H	Ni(cod)₂	80	(84) 95
H	CF₃	H	Ni(cod)₂	80	(69) 96
MeO	MeO	H	Pd(dba)₂	60	(78) 82
t-Bu	H	*t*-Bu	Pd(dba)₂	60	(84) 89

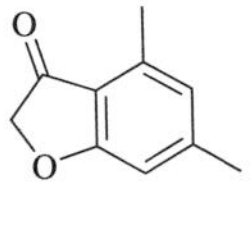

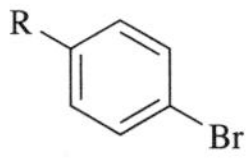

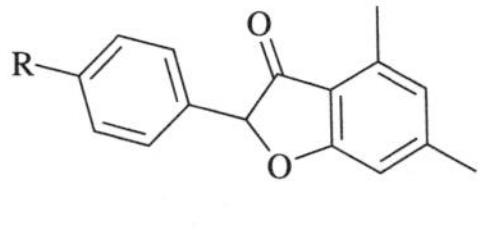

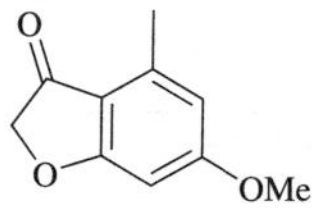

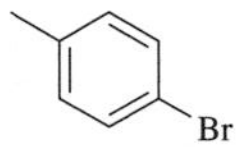

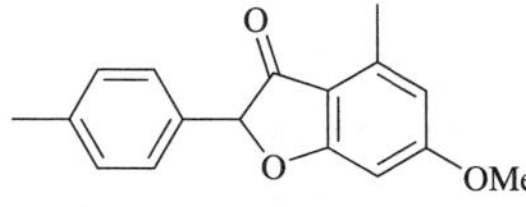

TABLE 1. ARYLATION OF ALDEHYDES, KETONES, AND ENOL ETHERS (*Continued*)

Substrate	Aryl Compound	Conditions	Product(s) and Yield(s) (%)	Refs.
C_{10}				
(4,6-dimethoxy-3(2H)-benzofuranone)	1,4-R-C6H4-Br; R = OMe, Me	Pd(OAc)2, **L19**, NaO*t*-Bu, toluene or THF, 60–80°	2-aryl-4,6-dimethoxy-benzofuran-3(2H)-one; R = H (21), MeO (35), Me (50)	153
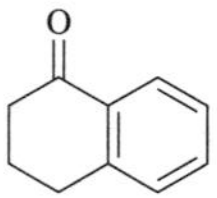 α-tetralone	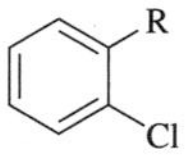 2-R-C6H4-Cl; R = OMe, Me	Pd cat **4** (1 mol %), NaO*t*-Bu (1.05 eq), THF; Temp 70° / 6 h; 60° / 1 h	2-(2-R-phenyl)tetralone; OMe (35), Me (50)	50
	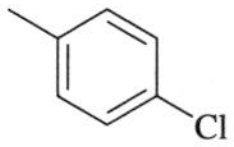 4-Me-C6H4-Cl	Pd(OAc)2 (1 mol %), **L51** (1 mol %), NaO*t*-Bu, dioxane, 60°, 6 h	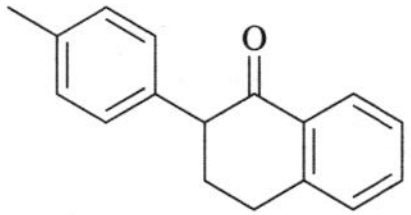 2-(4-methylphenyl)tetralone (90)	51
	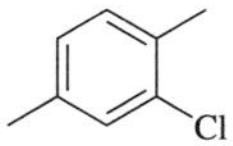 2,5-Me2-C6H3-Cl	Pd(OAc)2, ligand, NaO*t*-Bu, toluene, 80°; **L17** 22 h, **L19** 5 h	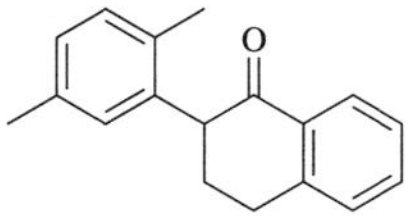 2-(2,5-dimethylphenyl)tetralone (76), (93)	43

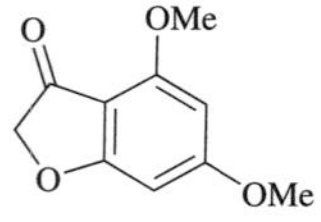

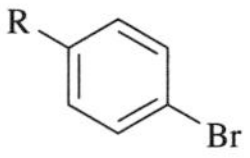

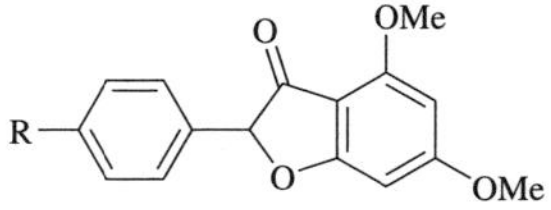

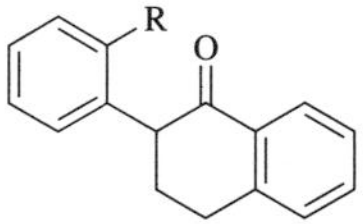

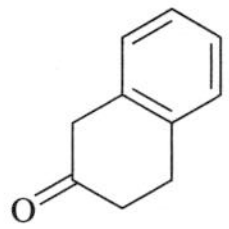

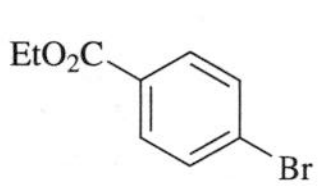

Substrate	ArX	Conditions	Product	Ref.
2-tetralone	4-BrC6H4-CO2Et	Pd(OAc)2, L19, base	1-(4-CO2Et-phenyl)-2-tetralone	43

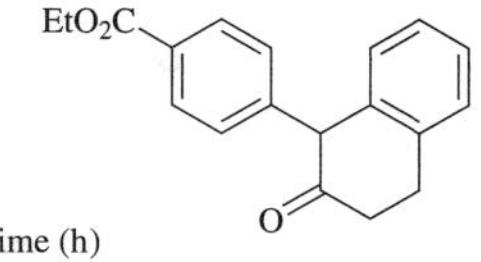

Base	Solvent	Temp (°)	Time (h)	(Yield)
NaO-t-Bu	THF	80	22	(85)
K3PO4	toluene	100	23	(91)

Substrate: 2-(trimethylsilyloxy)-5-methyl-1-hexene (OSiMe3)

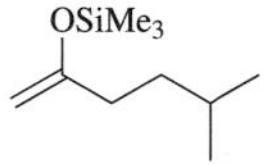

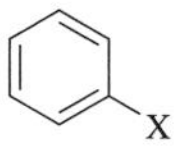

PdCl2[P(o-tol)3]2 (3 mol %), Bu3SnF, benzene

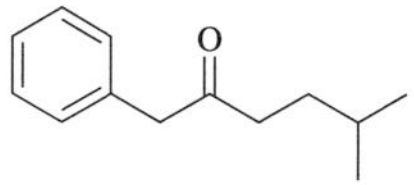

Product: PhCH2C(O)CH2CH2CH(CH3)2

X	Time (h)	(Yield)
Br	3	(85)
Br	4	(65)
I	3.5	(22)

Ref. 13

Substrate: 1-(trimethylsilyloxy)-3-methylcyclohexene

ArX: 3-nitrochlorobenzene

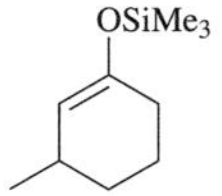

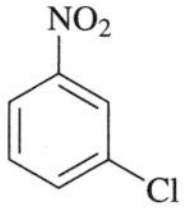

Pd(dba)2 (3 mol %), P(Bu-t)3 (5.4 mol %), ZnF2 (1.4 eq), MnF2 (1.4 eq), DMF, 90°

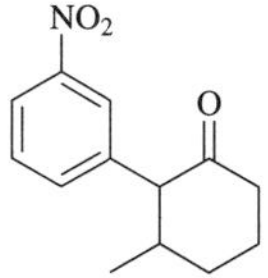

Product: 2-(3-nitrophenyl)-3-methylcyclohexanone (77)

Ref. 150

Substrate: 1-(trimethylsilyloxy)-3-methylcyclohexene

ArX: 4-R-chlorobenzene

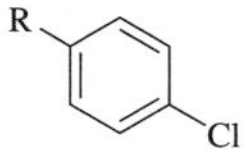

Pd(OAc)2 (3 mol %), P(t-Bu)3 (5.4 mol %), Bu3SnF (1.4 eq), CsF (1.4 eq), toluene

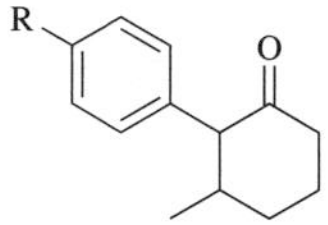

Product: 2-(4-R-phenyl)-3-methylcyclohexanone

R	Temp (°)	(Yield)
O2N	85	(84)
Me(O)C	90	(70)
MeO2C	90	(80)

Ref. 150

TABLE 1. ARYLATION OF ALDEHYDES, KETONES, AND ENOL ETHERS (*Continued*)

Substrate	Aryl Compound	Conditions	Product(s) and Yield(s) (%)	Refs.
C_{10} 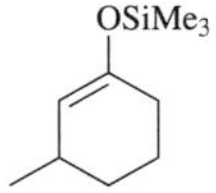 3-methyl-1-(trimethylsilyloxy)cyclohexene	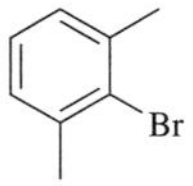 2-bromo-1,3-dimethylbenzene	Pd(OAc)$_2$ (3 mol %), P(t-Bu)$_3$ (5.4 mol %), Bu$_3$SnF (1.4 eq), CsF (1.4 eq), toluene	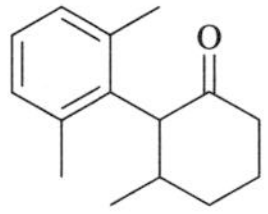 2-(2,6-dimethylphenyl)-3-methylcyclohexanone (85)	150
	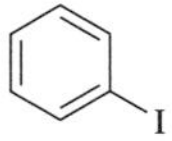 iodobenzene	Pd$_2$(dba)$_3$ (2.5 mol %), P(t-Bu)$_3$ (6 mol %), Bu$_3$SnF, benzene, 21 h	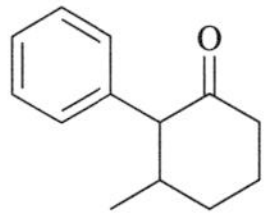 3-methyl-2-phenylcyclohexanone (85)	151
	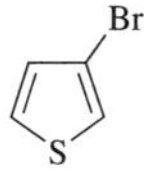 3-bromothiophene	Pd(OAc)$_2$ (3 mol %), P(t-Bu)$_3$ (5.4 mol %), Bu$_3$SnF (1.4 eq), CsF (1.4 eq), toluene	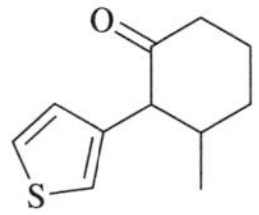 3-methyl-2-(thiophen-3-yl)cyclohexanone (78)	150
 norbornene (Me$_3$SiO-substituted)	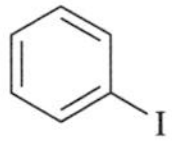 iodobenzene	Pd$_2$(dba)$_3$ (2.5 mol %), P(t-Bu)$_3$ (6 mol %), Bu$_3$SnF, benzene, 21 h	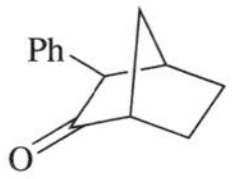 phenyl norbornanone (58)	151
C_{11} 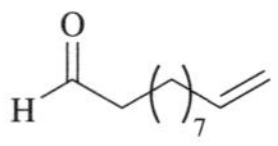 aldehyde (long chain with terminal alkene)	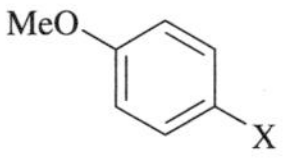 4-halo-anisole (X = Cl, Br)	Catalyst, ligand, Cs$_2$CO$_3$ (1.2 eq.), dioxane, 80–100° Catalyst / Ligand: Pd cat **3** / none; [Pd(allyl)Cl]$_2$ / **L32**	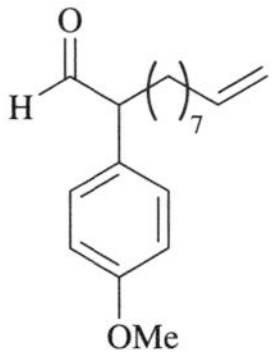 2-(4-methoxyphenyl) aldehyde with terminal alkene (58) (70)	148

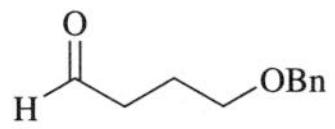

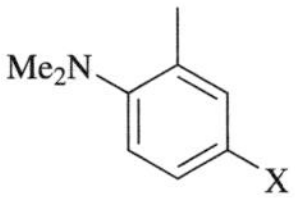

Catalyst	Ligand	
Pd cat **3**	none	(57)
Pd(OAc)₂	**L32**	(73)

Catalyst, ligand, Cs₂CO₃ (1.2 eq), dioxane, 80–100°

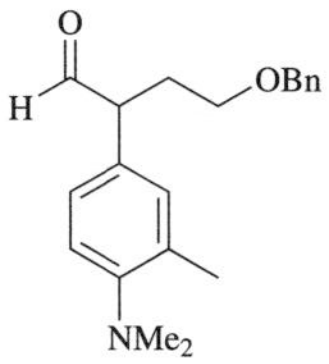

148

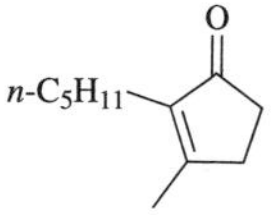

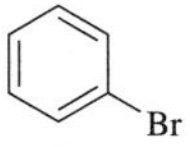

Pd(OAc)₂ (5 mol %), PPh₃ (10 mol %), Cs₂CO₃, DMF, 60°, 21 h

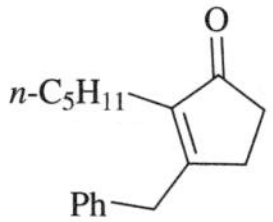

(58)

92

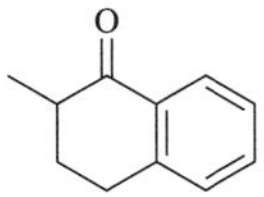

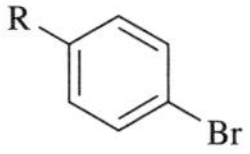

Pd(0), (S)-**L23**, NaO*t*-Bu, toluene, 100°

R	% ee
H	(66) 73
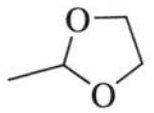 2-methyl-1,3-dioxolane	(74) 84
CN	(73) 88
t-Bu	(40) 61

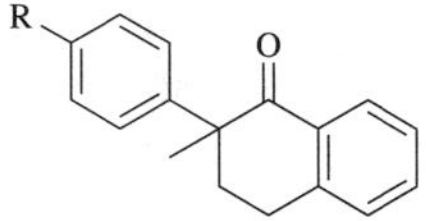

30

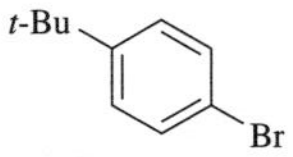

Pd(0), (S)-**L23**, NaO*t*-Bu, toluene, 100°

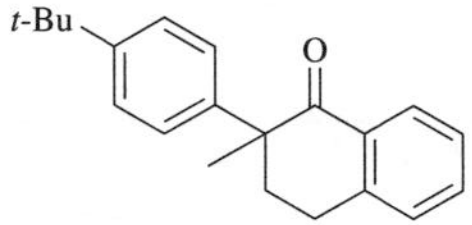

(56), 77% ee

30

TABLE 1. ARYLATION OF ALDEHYDES, KETONES, AND ENOL ETHERS (*Continued*)

Substrate	Aryl Compound	Conditions	Product(s) and Yield(s) (%)	Refs.
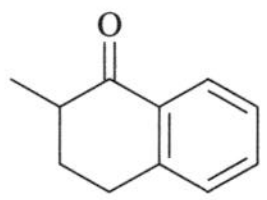 C11 (2-methyl-tetralone)	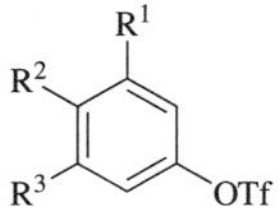 R1, R2, R3 substituted aryl OTf	Catalyst, **L22**, NaO-*t*-Bu (2 eq), toluene	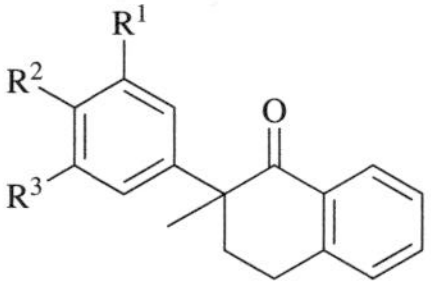 2-aryl-2-methyl-tetralone product	34

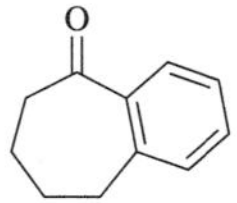

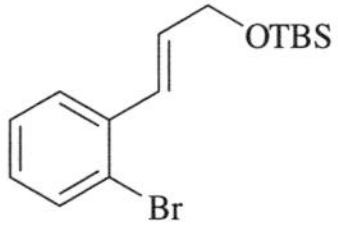

R1	R2	R3	Catalyst	Temp (°)		% ee
H	H	H	Pd(dba)₂	60	(81)	90
H	Me	H	Pd(dba)₂	60	(79)	92
H	NC	H	Ni(cod)₂	100	(55)	97
H	CF₃	H	Ni(cod)₂	100	(40)	98
H	*t*-Bu	H	Ni(cod)₂	60	(85)	92
MeO	MeO	H	Pd(dba)₂	60	(79)	91
t-Bu	H	*t*-Bu	Pd(dba)₂	60	(83)	95

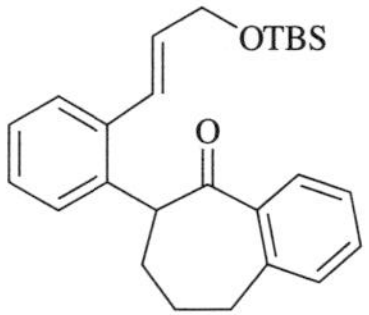

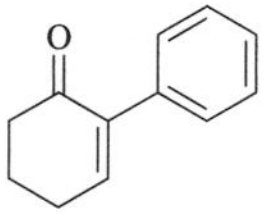

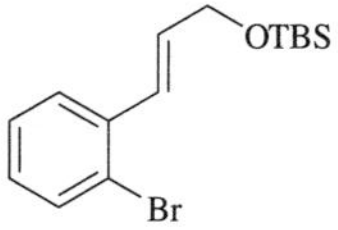

benzosuberone	2-Br-styryl-CH₂OTBS	Pd₂(dba)₃ (5 mol %), P(*t*-Bu)₃ (10 mol %), LiHMDS, dioxane, 90°, 3 h	α-aryl benzosuberone with OTBS allyl (—)	63
1-phenyl-1-(trimethylsilyloxy)ethylene	PhBr	PdCl₂[P(*o*-tol)₃]₂ (3 mol %), Bu₃SnF, benzene	deoxybenzoin (35)	13
C12 2-phenylcyclohex-2-enone	2-Br-styryl-CH₂OTBS	Pd₂(dba)₃ (5 mol %), P(*t*-Bu)₃ (10 mol %), LiHMDS, dioxane, 90°, 3 h	α-aryl product with OTBS allyl (—)	63

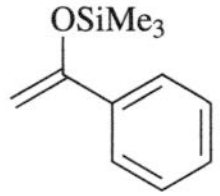

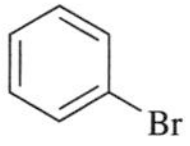

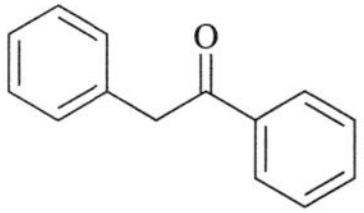

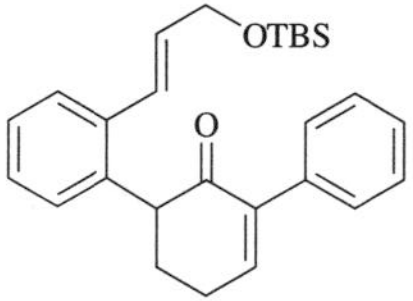

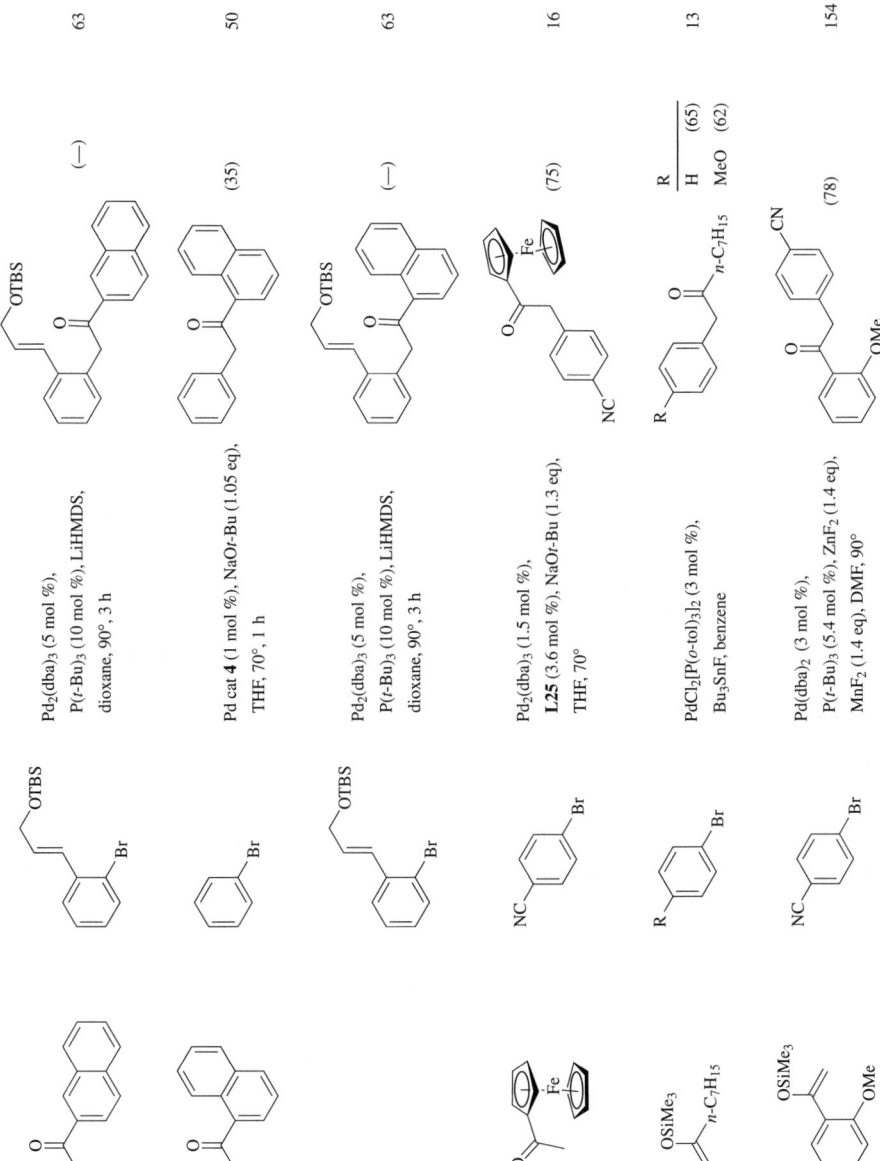

TABLE 1. ARYLATION OF ALDEHYDES, KETONES, AND ENOL ETHERS (Continued)

Substrate	Aryl Compound	Conditions	Product(s) and Yield(s) (%)	Refs.
C$_{12}$ 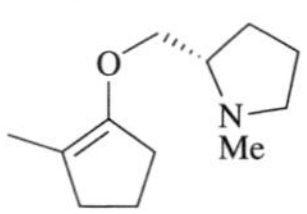 (pyrrolidine-N-Me with CH$_2$O-cyclopentenyl-Me)	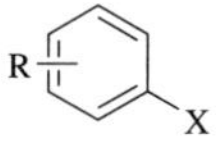 R-C$_6$H$_4$-X	1. Pd(OAc)$_2$, DMF, H$_2$O 2. H$_3$O$^+$	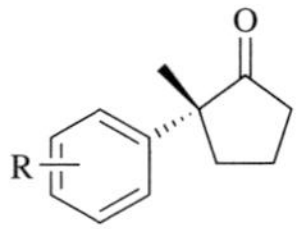 2-methyl-2-aryl cyclopentanone	154

R	X	Temp (°)	Time (h)		% ee
4-PhCO	Br	80	24	(47)	97
H	I	70	68	(61)	94
2-MeO	I	70	24	(67)	98
2-MeO	I	80	30	(50)	94
3-MeO	I	70	18	(68)	93
4-MeO	I	70	18	(54)	93
4-PhCO	I	100	24	(78)	94

| 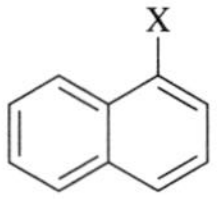 | 1-naphthyl-X | 1. Pd(OAc)$_2$, DMF, H$_2$O, 48 h
2. H$_3$O$^+$ | 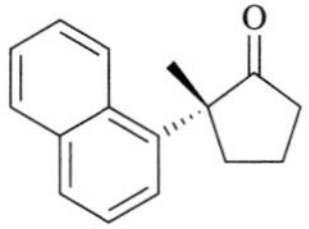 2-methyl-2-(1-naphthyl)cyclopentanone | 154 |

X	Temp (°)		% ee
Br	100	(49)	91
I	80	(45)	90

| 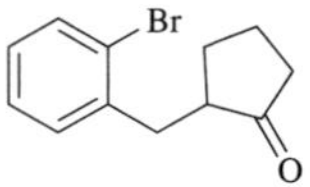 2-(2-bromobenzyl)cyclopentanone | | PdCl$_2$(Ph$_3$P)$_2$ (10 mol %),
CsCO$_3$ (3 eq), THF, 100°, 16 h | 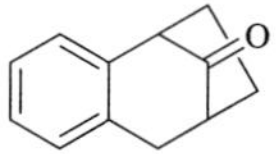 bicyclic ketone (26) | 20 |

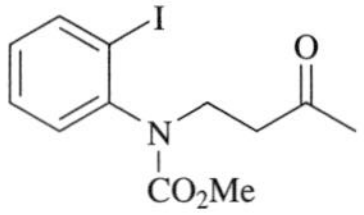

substrate	conditions	product (yield), refs

C13 substrate (2-iodophenyl N-methylcarbamate with ketone side chain):

Reagents: Et₃N, toluene, 110°, 24 h

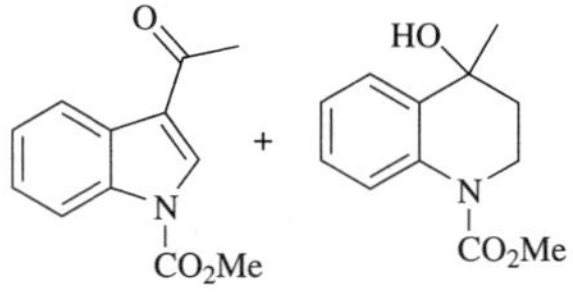

Product (44): 4-hydroxy-4-methyl-1-(CO₂Me)-1,2,3,4-tetrahydroquinoline + 3-acetyl-1-(CO₂Me)-indole (29)

Refs: 155, 113

Substrate: 3-(2-phenylethyl)cyclopent-2-enone + 4-R-bromobenzene

Conditions:
1. CuCl (1 mol %), (S)-**L25** (1 mol %), NaO*t*-Bu (1 mol %), Ph₂SiH₂ (0.51 eq), 0°
2. Pd(OAc)₂ (5 mol %), **L15** (10 mol %), CsF (1.1 eq), THF, rt, 18 h

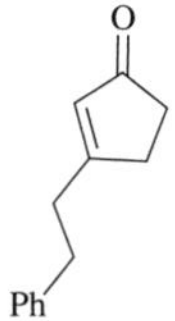

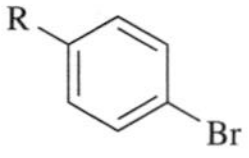

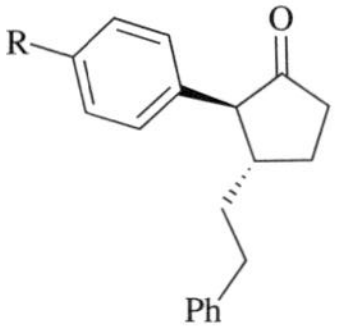

R	Solvent		% ee, dr
H	THF	(75)	95, 98:2
MeO	toluene	(55)	95, 99:1
t-Bu	THF	(80)	76, —
t-Bu	THF/pentane	(75)	95, —
t-Bu	toluene	(—)	95, —

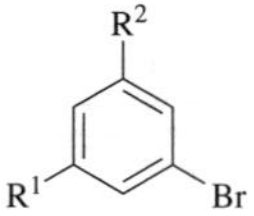

Ref: 44

Substrate: 3,5-disubstituted bromobenzene (R¹, R²)

Conditions:
1. CuCl (1 mol %), (S)-**L25** (1 mol %), P(*t*-Bu)₃ (1 mol %), Ph₂SiH₂ (0.51 eq), THF/pentane, 0°
2. Pd(OAc)₂ (5 mol %), **L15** (10 mol %), CsF (1.1 eq), THF, rt, 18 h

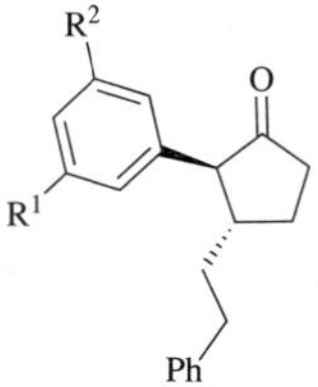

R¹	R²		% ee, dr
EtO₂C	H	(44)	95, 98.5:1.5
Me	Me	(58)	95, 99:1

Ref: 44

TABLE 1. ARYLATION OF ALDEHYDES, KETONES, AND ENOL ETHERS (*Continued*)

Substrate	Aryl Compound	Conditions	Product(s) and Yield(s) (%)	Refs.
C₁₃ (2-methyl-benzylidenecyclopentanone with Ph)	1,2-dibromobenzene with R	Catalyst, (S)-**L23**, base R | Catalyst | Base ---|---|--- 3-Me | Pd(OAc)₂ | NaO*t*-Bu 3-(dioxolane) | Pd₂(dba)₃ | NaHMDS 4-*t*-Bu | Pd(OAc)₂ | NaO*t*-Bu	(2-bromoaryl-methyl-benzylidenecyclopentanone) | % ee ---|--- (86) | 95 (80) | 94 (75) | 98	30
	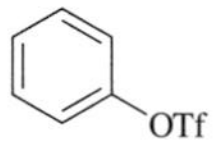 phenyl triflate	Pd(OAc)₂, **L22**, NaO*t*-Bu (2 eq), toluene, 100°	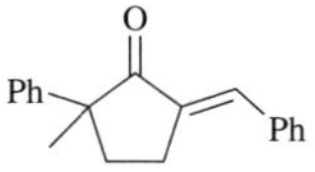 (2,2-dimethyl-5-benzylidene-cyclopentanone with Ph) (70), 95% ee	34
(trimethylsilyl enol ether of 4-phenyl-2-butanone)	4-substituted aryl halide (X, R)	Pd(OAc)₂ (3 mol %), P(*t*-Bu)₃, additive, toluene, 85° X | R | Additive ---|---|--- Cl | CN | Bu₃SnF (1.4 eq), CsF (1.4 eq) Cl | CO₂Me | Bu₃SnF (1.4 eq), CsF (1.4 eq) Br | NO₂ | Bu₃SnF (1.4 eq), CsF (1.4 eq) Br | NMe₂ | Bu₃SnF (1.4 eq), CsF (1.4 eq) Br | OH | Bu₃SnF (1.4 eq), CsF (1.4 eq) Br | CF₃ | Bu₃SnF (1.4 eq), CsF (1.4 eq) Br | C(O)Me | Bu₃SnF (1.4 eq), CsF (1.4 eq) Br | *t*-Bu | none Br | *t*-Bu | Me₄NF (1.2 eq)	(α-aryl ketone product) (80) (89) (84) (96) (55) (91) (97) (0) (0)	150

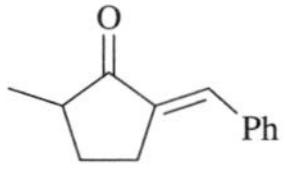

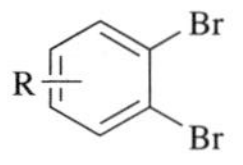

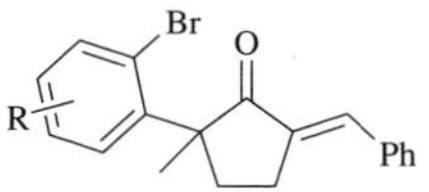

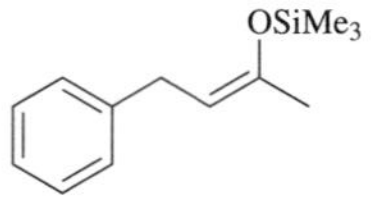

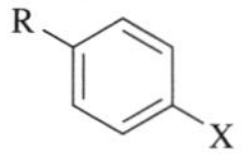

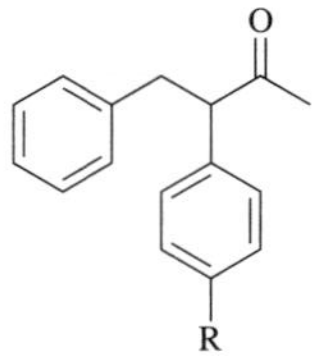

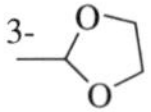

Br	t-Bu	CsF (1.2 eq)	(18)
Br	t-Bu	CsF (1.4 eq)	(81)
Br	t-Bu	ZnF$_2$ (1.2 eq)	(38)
Br	t-Bu	Bu$_3$SnF (1.2 eq)	(34)
Br	t-Bu	Bu$_3$SnF (1.4 eq), CsF (1.4 eq)	(67)
Br	t-Bu	Bu$_3$SnF (1.2 eq), CsF (1.2 eq)	(65)

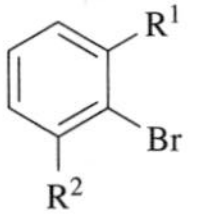

Pd(OAc)$_2$ (3 mol %), P(t-Bu)$_3$, Bu$_3$SnF (1.4 eq), CsF (1.4 eq), toluene, 85°

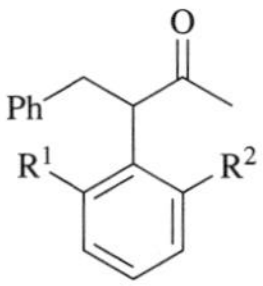

R^1	R^2	
H	OMe	(97)
Me	Me	(93)

150

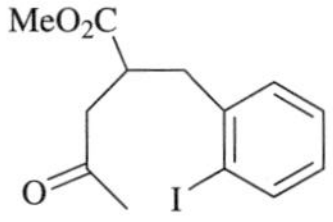

Pd(PPh$_3$)$_4$ (20 mol %), KOt-Bu

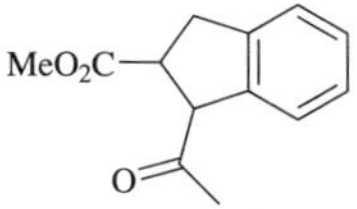

(6)   58

C$_{14}$

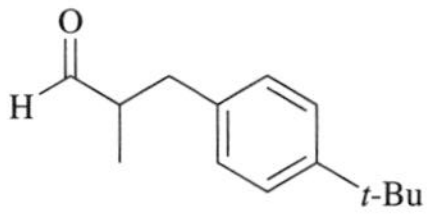

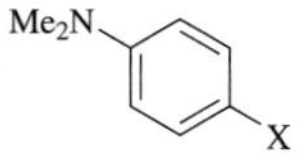

Catalyst, ligand, Cs$_2$CO$_3$ (1.2 eq), dioxane, 80–100°

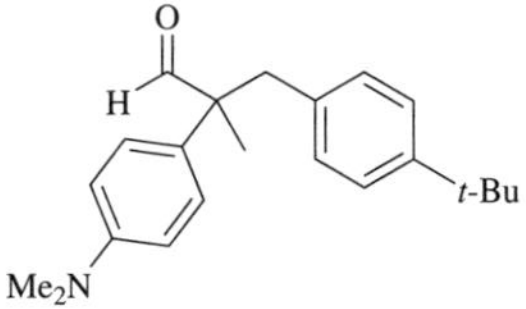

Catalyst	Ligand	
Pd cat **3**	none	
Pd(OAc)$_2$	**L20**	

X	
Cl	(55)
Br	(68)

148

TABLE 1. ARYLATION OF ALDEHYDES, KETONES, AND ENOL ETHERS (*Continued*)

Substrate	Aryl Compound	Conditions	Product(s) and Yield(s) (%)	Refs.
C_{14}				
(2-methyl-6-benzylidenecyclohexanone)	R²–C₆H₃(R¹)–OTf	Pd(dba)₂, **L22**, NaO-*t*-Bu (2 eq), toluene, 60°	(benzylidene cyclohexanone product) %ee: (80) 78; (70) 77	34
	R¹ R² H H MeO MeO			
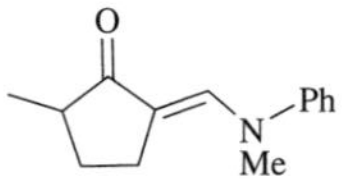 (2-methyl-cyclopentanone N-phenyl methylidene imine)	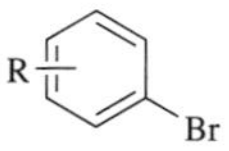 R–C₆H₄–Br	Catalyst (*x* mol %), (*S*)-**L23**, NaO-*t*-Bu, toluene, 100°	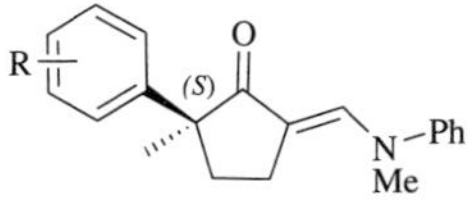 (S-configured product with N–Ph, Me)	41

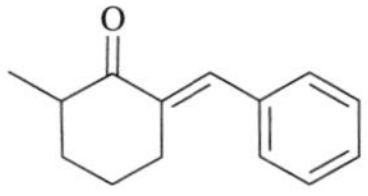

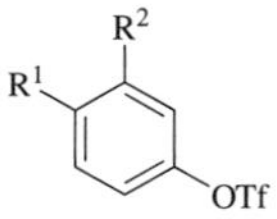

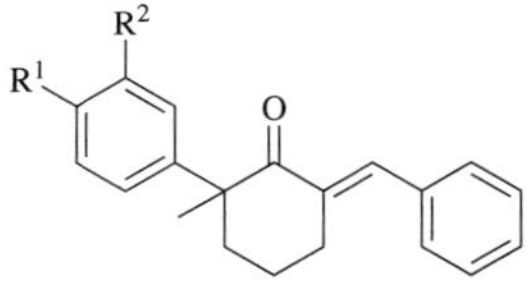

R	Catalyst	*x*		%ee
2-Me	Pd₂(dba)₃	2.5	(45)	8
2-Me	Pd(OAc)₂	5	(52)	10
3-MeO	Pd₂(dba)₃	5	(87)	85
3-Me	Pd₂(dba)₃	2.5	(70)	80
3-(1,3-dioxolan-2-yl) 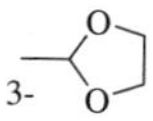	Pd₂(dba)₃	5	(96)	86
4-MeO	Pd₂(dba)₃	5	(74)	57
4-Me	Pd₂(dba)₃	5	(65)	63
4-CF₃	Pd₂(dba)₃	5	(93)	53
4-*t*-Bu	Pd₂(dba)₃	5	(65)	88
4-*t*-Bu	Pd(OAc)₂	10	(70)	89

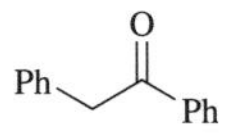

			31
(substrate: 3-bromotoluene)	Pd₂(dba)₃, ligand, NaO-t-Bu, toluene	(product: 2-(N-methyl-N-phenyl-aminomethylene)-3-methyl-3-(m-tolyl)cyclopentanone, (S))	

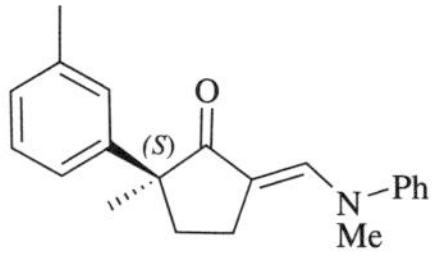

Ligand	% ee
(R,R_p)-**L27**	(57) 5
(R,S_p)-**L27**	(83) 10
(R)-**L28**	(72) 58

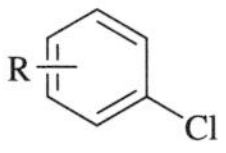

			48
(substrate: R-C₆H₄-Cl)	Pd(dba)₂ (10 mol %), **L35** (10 mol %), base (2.2 eq), toluene, 20 h	(product: R-C₆H₄-CH(Ph)-C(O)-Ph)	

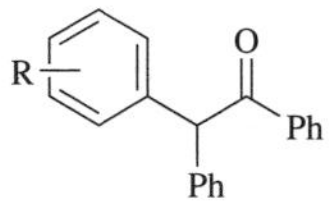

(starting ketone: PhCH₂C(O)Ph)

R	Base	
2-F	K₃PO₄	(72)
4-MeO	K₃PO₄	(76)
4-Me	K₃PO₄	(90)
4-Me	Cs₂CO₃	(99)

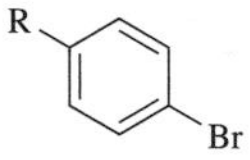

			56
(substrate: 4-R-C₆H₄-Br)	Catalyst, ligand, Cs₂CO₃, toluene, 130°	(product: 4-R-C₆H₄-CH(Ph)-C(O)-Ph)	

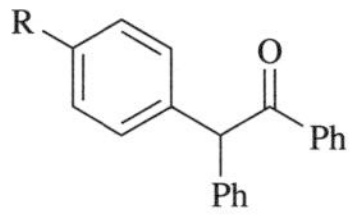

R	Catalyst	Ligand	
H	PdCl₂(cod)	**L36**	(78)
H	Pd(OCOCF₃)₂	**L37**	(84)
MeO	PdCl₂(cod)	**L36**	(95)
MeO	Pd(OCOCF₃)₂	**L37**	(96)
Me	PdCl₂(cod)	**L36**	(88)
Me	Pd(OCOCF₃)₂	**L37**	(96)

TABLE 1. ARYLATION OF ALDEHYDES, KETONES, AND ENOL ETHERS (*Continued*)

Substrate	Aryl Compound	Conditions	Product(s) and Yield(s) (%)	Refs.
C_{14} Ph-CO-CH2-Ph	R—C6H4—Br	Catalyst (0.5 mol %), ligand, base	**I** R-C6H4-CH(Ph)-CO-Ph + **II** (2-aryl biphenyl ketone with Ph) + **III** (diaryl substituted)	46

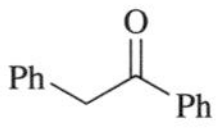

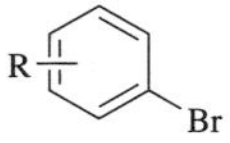

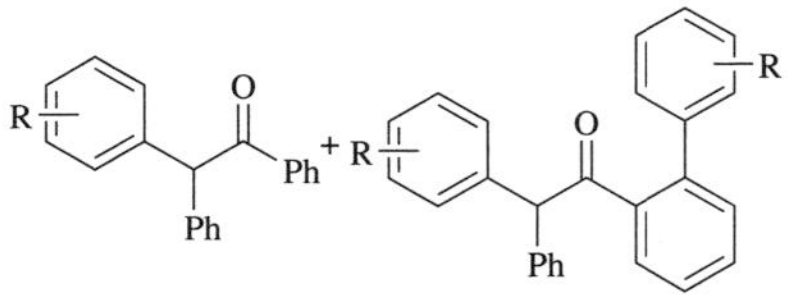

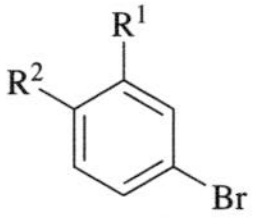

R	Catalyst	Ligand	Base	Solvent	Time (h)	I	II	III
H	Pd(OAc)$_2$	PPh$_3$	Cs$_2$CO$_3$	xylene	5	(47)	(35)	(9)
H	Pd(OAc)$_2$	PPh$_3$	Cs$_2$CO$_3$	xylene	22	(1)	(5)	(54)
H	Pd(OAc)$_2$	PPh$_3$	Cs$_2$CO$_3$	DMF	2	(79)	(5)	(0)
H	Pd(OAc)$_2$	PPh$_3$	Cs$_2$CO$_3$	DMF	22	(3)	(1)	(0)
H	Pd(PPh$_3$)$_4$	none	Cs$_2$CO$_3$	xylene	20	(6)	(7)	(61)
H	Pd(OAc)$_2$	P(*o*-tol)$_3$	Cs$_2$CO$_3$	xylene	22	(91)	(7)	(0)
H	Pd(PPh$_3$)$_4$	none	K$_2$CO$_3$	xylene	24	(96)	(3)	(0)
3-Cl	Pd(PPh$_3$)$_4$	none	Cs$_2$CO$_3$	xylene	20	(—)	(—)	(54)
3-CF$_3$	Pd(PPh$_3$)$_4$	none	Cs$_2$CO$_3$	xylene	44	(—)	(—)	(41)
4-Cl	Pd(PPh$_3$)$_4$	none	Cs$_2$CO$_3$	xylene	20	(—)	(—)	(25)

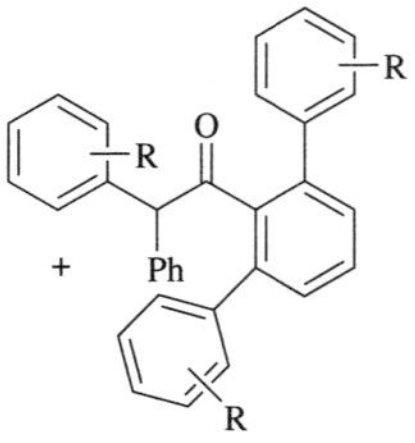

R^1,R^2-C6H3-Br, Pd(OAc)$_2$ (1 mol %), PPh$_3$ (8 mol %), Cs$_2$CO$_3$ (2.5 eq), DMF, 150°, 0.5–1 h

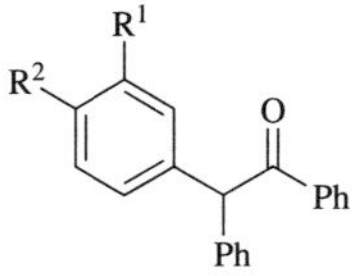

R^1-C6H3(R^2)-CH(Ph)-CO-Ph

R^1	R^2	
H	H	(80)
OMe	H	(73)
—OCH$_2$O—		(70)

53

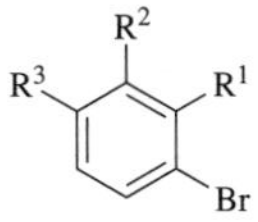

Catalyst, ligand, Cs₂CO₃, DMF

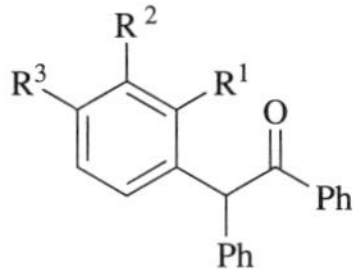

R¹	R²	R³	Catalyst	Ligand	Temp (°)		
H	H	H	Pd(OAc)₂	PPh₃	150	(80)	55
H	H	H	Pd cat **10**	none	153	(57)	
H	OMe	H	Pd(OAc)₂	PPh₃	150	(73)	
H	OMe	H	Pd cat **10**	none	153	(54)	
H	H	NO₂	Pd(OAc)₂	PPh₃	150	(54)	
H	H	OMe	Pd(OAc)₂	PPh₃	150	(71)	
H	H	OMe	Pd cat **10**	none	153	(60)	
H	—OCH₂O—		Pd(OAc)₂	PPh₃	150	(70)	
H	—OCH₂O—		Pd cat **10**	none	153	(45)	
H	OMe	OMe	Pd cat **10**	none	153	(37)	
F	OMe	OMe	Pd(OAc)₂	PPh₃	150	(57)	

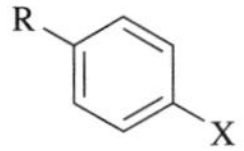

PdCl₂, Cs₂CO₃, DMF, 2 h

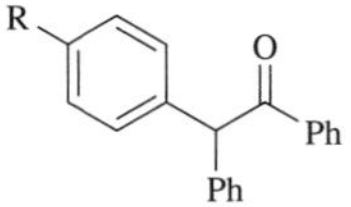

X	R	Temp (°)		
Br	H	130	(96)	73
I	H	100	(90)	
I	MeO	100	(93)	
I	Cl	100	(82)	

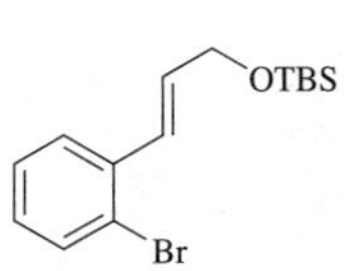

Pd₂(dba)₃ (5 mol %), P(t-Bu)₃ (10 mol %), LiHMDS, dioxane, 90°, 3 h

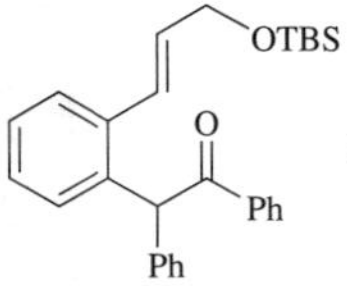

(—) 63

TABLE 1. ARYLATION OF ALDEHYDES, KETONES, AND ENOL ETHERS (*Continued*)

Substrate	Aryl Compound	Conditions	Product(s) and Yield(s) (%)	Refs.
C$_{14}$ 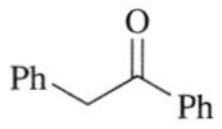 PhCOCH$_2$Ph	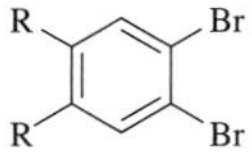 1,2,4,5-tetrasubstituted: R, Br, R, Br	Pd(OAc)$_2$ (0.5 mol %), PPh$_3$ (20 mol %), Cs$_2$CO$_3$ (2 eq), xylene, 160°	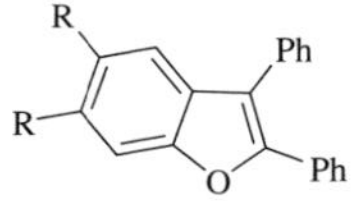 2,3-diaryl benzofuran with R substituents; R=F: Time 2 h (40); R=Me: Time 4 h (69)	60
	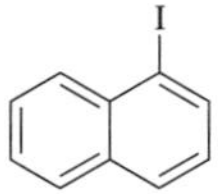 1-iodonaphthalene	PdCl$_2$, PPh$_3$, K$_2$CO$_3$, DMF, 100°, 2 h	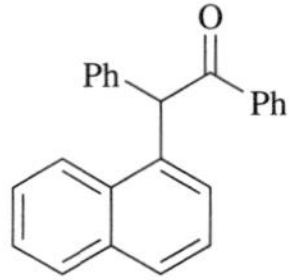 Ph$_2$CH-CO-(1-naphthyl) (75)	73
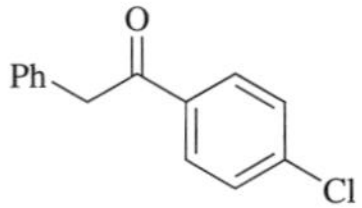	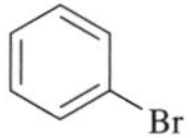 bromobenzene	Pd(PPh$_3$)$_4$ (0.5 mol %), Cs$_2$CO$_3$, xylene, 160°	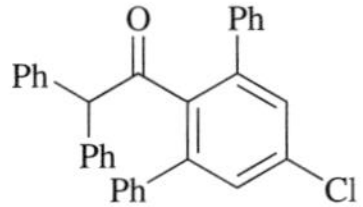 4-Cl-C$_6$H$_3$(Ph)(CHPh)C=O; Time 6 h (59), 20 h (40)	47
4-ClC$_6$H$_4$COCH$_2$Ph	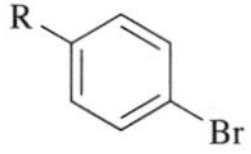 4-R-C$_6$H$_4$Br	Pd(PPh$_3$)$_4$ (0.5 mol %), CsCO$_3$ (3–5 eq), xylene	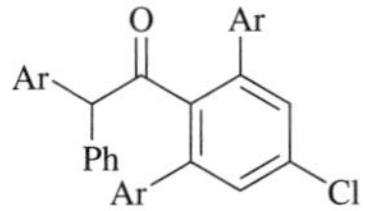 Ar, Ph, Ar diarylated product; Ar = 4-RC$_6$H$_4$; R=H: Time 6 (59); R=Ph: Time 21 (30)	46
	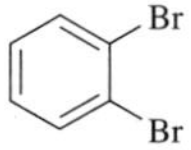 1,2-dibromobenzene	Pd(OAc)$_2$ (0.5 mol %), PPh$_3$ (0.2 eq), Cs$_2$CO$_3$ (2 eq), xylene, 160°, 2 h	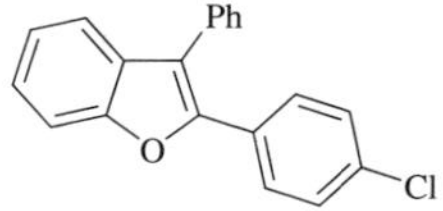 2-(4-chlorophenyl)-3-phenylbenzofuran (49)	60

Substrate	Aryl halide	Conditions	Product(s) (Yield %)	Ref.
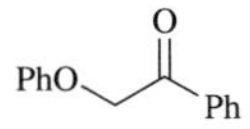 PhOCH2C(O)Ph	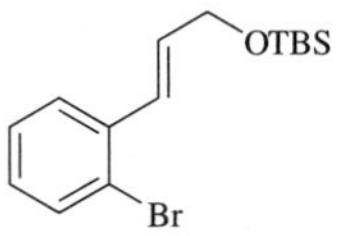 2-Br-C6H4-CH=CH-CH2-OTBS	Pd2(dba)3 (5 mol %), P(t-Bu)3 (10 mol %), LiHMDS, dioxane, 90°, 3 h	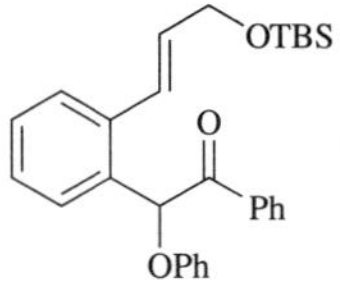 [product with OTBS, Ph, OPh substituents] (—)	63
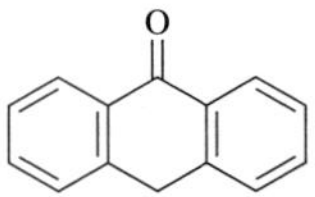 anthracen-9(10H)-one	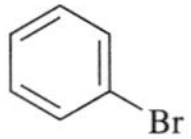 PhBr	Pd(OAc)2 (2.5 mol %), PPh3 (10 mol %), Cs2CO3 (1 eq), xylene, 160°, 5 h	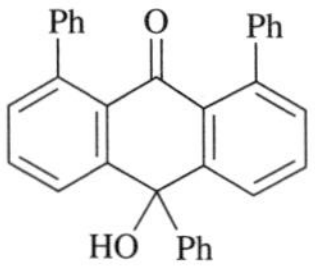 [1,8-diphenyl-10-hydroxy-10-phenyl-anthracenone] (47)	47
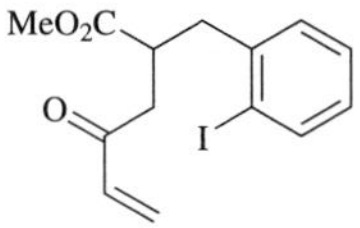 MeO2C-CH(CH2-C6H4-I)-CH2-C(O)-CH=CH2	—	Pd(OAc)2, PPh3, additive, Et3N, MeCN, 82°	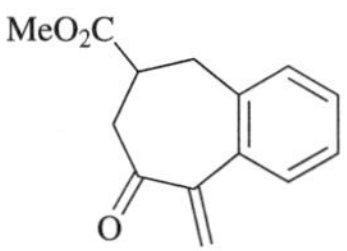 [benzocycloheptanone product]	58

Additive	Time (h)	
none	4	(22)
Bu4NCl	3	(23)
none	0.8	(24)

C15

Substrate	Aryl halide	Conditions	Product(s) (Yield %)	Ref.
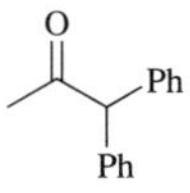 PhCH(Ph)C(O)CH3	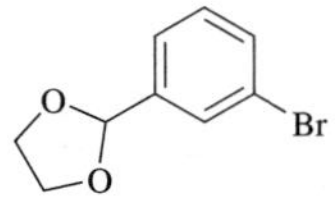 3-Br-C6H4-(1,3-dioxolan-2-yl)	Pd2(dba)3 (1.5 mol %), L25 (3.6 mol %), NaOt-Bu (1.3 eq), THF, 70°	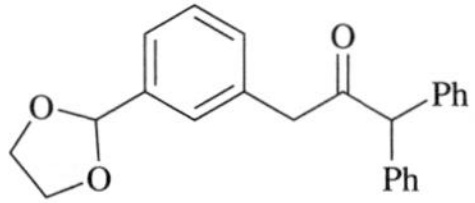 [aryl ketone with dioxolane, CHPh2] (72)	16

TABLE 1. ARYLATION OF ALDEHYDES, KETONES, AND ENOL ETHERS (*Continued*)

Substrate	Aryl Compound	Conditions	Product(s) and Yield(s) (%)	Refs.
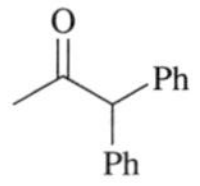 C₁₅ PhCH(Ph)C(O)CH₃	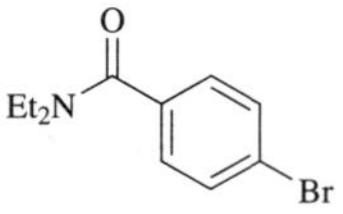 4-Br-C₆H₄-C(O)NEt₂	Pd₂(dba)₃ (1.5 mol %), **L25** (3.6 mol %), NaO*t*-Bu (1.3 eq), THF, 70°	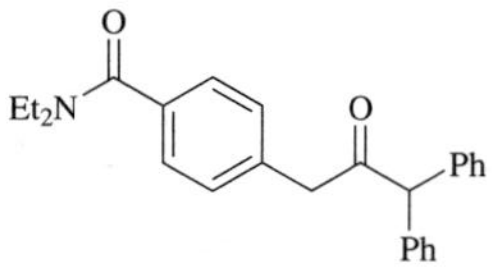 Et₂N-C(O)-C₆H₄-CH₂-C(O)-CH(Ph)(Ph) (69)	16
4-MeC₆H₄-C(O)-CH₂-Ph	1,2-dibromobenzene	Pd(OAc)₂ (0.5 mol %), PPh₃ (20 mol %), Cs₂CO₃ (2 eq), xylene, 160°, 2 h	3-Ph-2-(4-tolyl)benzofuran (74)	60
4-MeOC₆H₄-C(O)-CH₂-Ph	1,2-dibromobenzene	Pd(OAc)₂ (0.5 mol %), PPh₃ (20 mol %), Cs₂CO₃ (2 eq), xylene, 160°, 2 h	3-Ph-2-(4-methoxyphenyl)benzofuran (75)	60
PhCH₂-C(O)-CH₂-Ph	1,2-dibromobenzene	Pd(OAc)₂ (0.5 mol %), PPh₃ (20 mol %), K₂CO₃ (2 eq), xylene, 160°, 4 h	1,3-diphenylindan-2-one (64)	60
PhCH₂-C(O)-CH₂-Ph	PhI	PdCl₂, LiCl (4 mol %), Cs₂CO₃	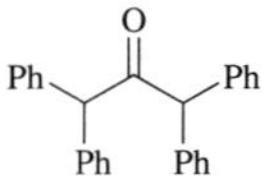 Ph-CH(Ph)-C(O)-CH(Ph)-Ph **I** (48)	18
PhCH₂-C(O)-CH₂-Ph	PhI	PdCl₂, Cs₂CO₃, DMF, 100°, 2 h	**I** (59)	73

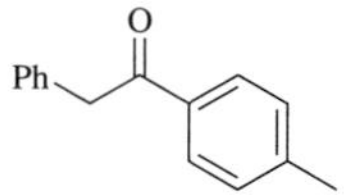

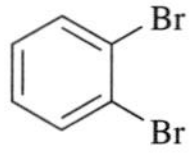

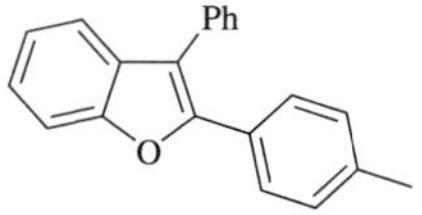

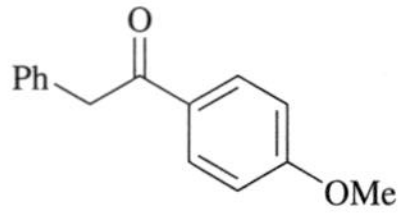

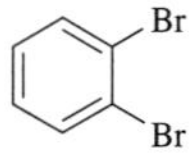

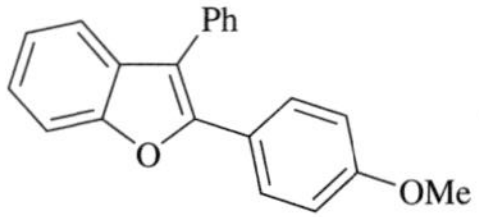

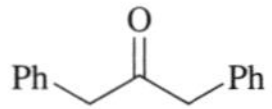

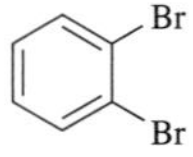

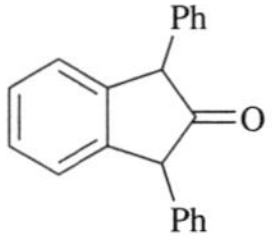

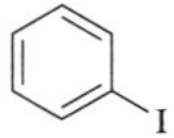

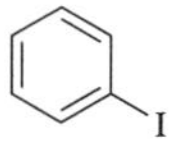

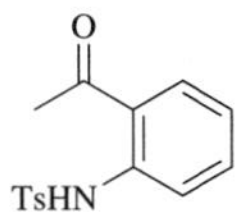

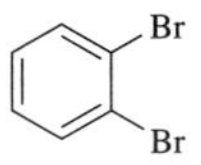

Catalyst, ligand, base

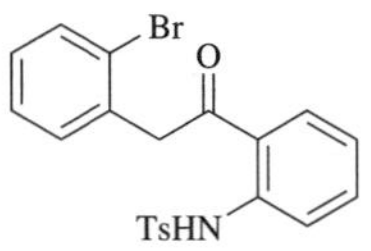

Catalyst	Ligand	Base	Solvent	Temp (°)	Time (h)	
Pd(dba)₂	P(t-Bu)₃	K₃PO₄	toluene	80	24	(0)
Pd(dba)₂	L1	NaOt-Bu	THF	80	8	(0)
Pd(OAc)₂	PPh₃	Cs₂CO₃	DMF	80	22	(0)
Pd₂(dba)₃	L23	NaOt-Bu	THF	80	22	(0)
Pd(OAc)₂	L32	K₃PO₄	toluene	130	5	(76)
Pd(OAc)₂	L32	K₃PO₄	toluene	130	29	(72)
Pd(OAc)₂	L32	Cs₂CO₃	toluene	120	48	(86)
Pd(OAc)₂	L32	Cs₂CO₃	THF	130	5	(61)
Pd(OAc)₂	L51	NaOt-Bu	dioxane	130	5	(0)

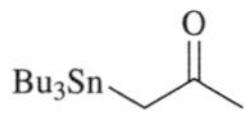

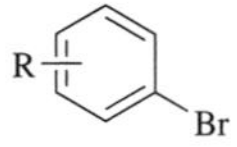

114

Catalyst, toluene, 100°, 5 h

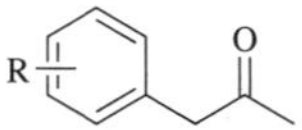

R	Catalyst	
H	Pd(PPh₃)₄	(22)
H	PdCl₂(PPh₃)₂	(15)
H	PdCl₂[P(o-tol)₃]₂	(78)
H	PdCl₂[P(p-tol)₃]₂	(16)
H	PdCl₂[P(2,4,6-Me₃C₆H₂)₃]₂	(0)
2-Cl	PdCl₂[P(o-tol)₃]₂	(80)
2-Me	PdCl₂[P(o-tol)₃]₂	(91)
3-Me	PdCl₂[P(o-tol)₃]₂	(88)
4-Me₂N	PdCl₂[P(o-tol)₃]₂	(71)
4-MeO	PdCl₂[P(o-tol)₃]₂	(51)

149

TABLE 1. ARYLATION OF ALDEHYDES, KETONES, AND ENOL ETHERS (*Continued*)

Substrate	Aryl Compound	Conditions	Product(s) and Yield(s) (%)	Refs.
C_{15}				
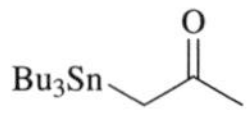 Bu$_3$Sn-CH$_2$-C(O)-CH$_3$	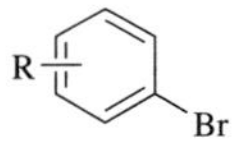 R-C$_6$H$_4$-Br (R varies)	Catalyst, toluene, 100°, 5 h	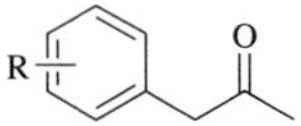 R-C$_6$H$_4$-CH$_2$-C(O)-CH$_3$	149
		Catalyst: PdCl$_2$[P(o-tol)$_3$]$_2$ (all entries)	R = 4-Cl (73); 4-Me (80); 4-Ac (64); 2,4,6-Me$_3$ (94)	
MeO$_2$C–(chain with aryl iodide, methacryloyl)		Pd(PPh$_3$)$_4$ (10–12 mol %), base	indanone-type product with MeO$_2$C and methacryloyl	58
			Base: Et$_3$N (6); K$_2$CO$_3$ (23)	
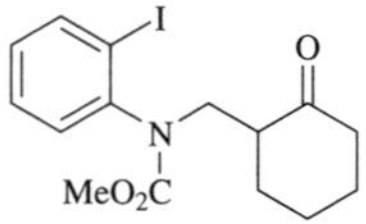 2-iodoaniline derivative with cyclohexanone and N-CO$_2$Me		PdCl$_2$(PPh$_3$)$_2$ (0.2 eq), Et$_3$N	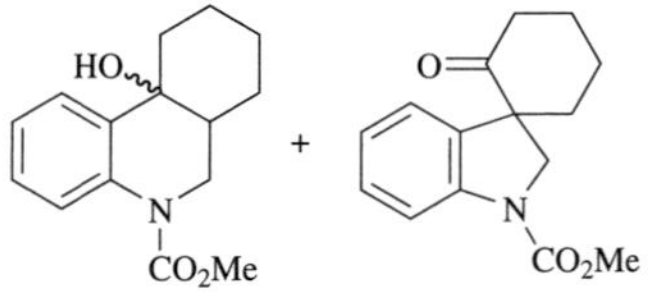 I (hydroxy tetrahydroquinoline) + II (spiro indoline)	155, 113
		Solvent / Time (h): THF / 65; toluene / 24	I : II = (32):(24) THF; (45):(31) toluene	
C_{16}				
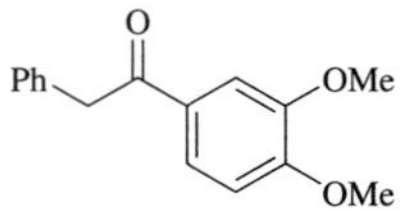 3,4-(MeO)$_2$C$_6$H$_3$-C(O)-CH$_2$-Ph	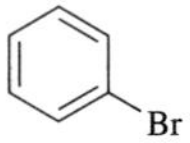 PhBr	Pd(OAc)$_2$ (1 mol %), PPh$_3$ (8 mol %), Cs$_2$CO$_3$ (2.5 eq), DMF, 150°, 0.5–1 h	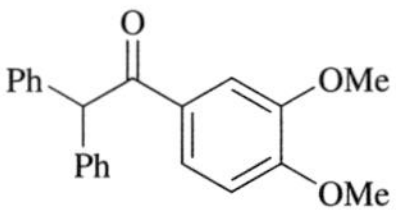 3,4-(MeO)$_2$C$_6$H$_3$-C(O)-CHPh$_2$ (74) **I**	53, 55

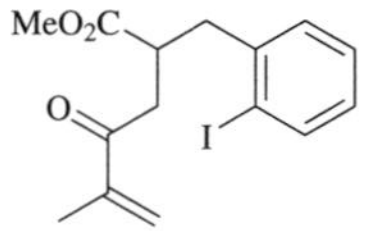

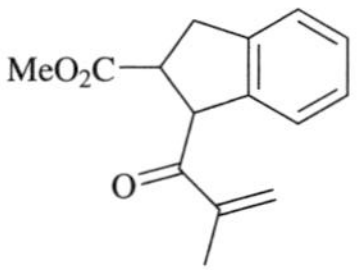

158

Substrate (ArBr)	Alkene/Reagent	Conditions	Product(s)	Refs.
PhBr	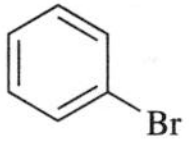 PhCH=CHCH₂Ph (cinnamyl-type, Ph-CH=CH-CH(Ph)-)	Pd cat **11**, Cs₂CO₃, DMF, 153°, 0–8 h	**I** (51)	55
PhBr	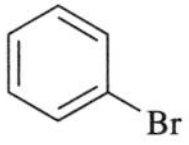 same	Catalyst, ligand, Cs₂CO₃, toluene, 130°, 75 min	Catalyst / Ligand / **I**: PdCl₂(cod) / **L36** / (92); Pd(OCOCF₃)₂ / **L37** / (95)	56
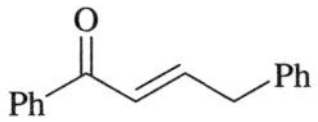 PhBr	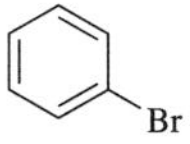 same	Pd(OAc)₂ (1.7 mol %), PPh₃ (6.8 mol %), Cs₂CO₃ (1 eq), xylene, 160°, 2 h	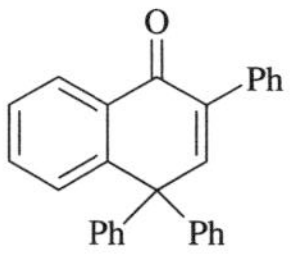 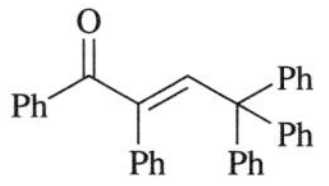 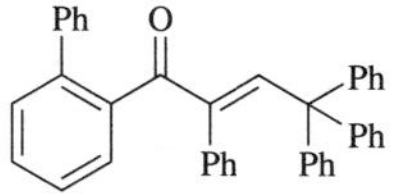 **I** (8) + naphthalenone product + **II** (12)	47
PhBr	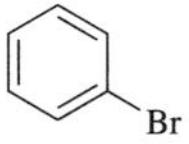 same	Pd(OAc)₂ (1 mol %), P(t-Bu)₃ (4 mol %), Cs₂CO₃ (1 eq), xylene, 160°, 2 h	**II** (85) + **III** (34)	47
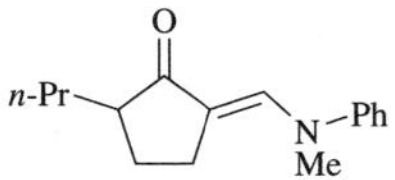 3-BrC₆H₄Me	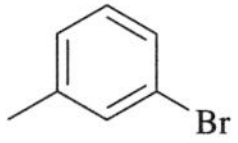 2-(PhN(Me)=CH)cyclopentanone (n-Pr substituent)	Pd₂(dba)₃ (1 mol %), (S)-**L23** (2.5 mol %), NaOt-Bu (2 eq), toluene	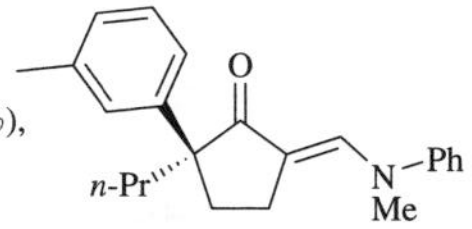 (74), 91% ee	41
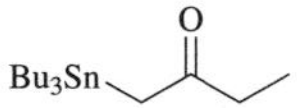 PhBr	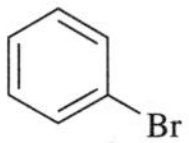 Bu₃SnCH₂C(O)Et	PdCl₂[P(o-tol)₃]₂ (10 mol %), toluene, 100°, 5 h	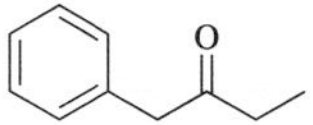 PhCH₂C(O)Et (67)	149

159

TABLE 1. ARYLATION OF ALDEHYDES, KETONES, AND ENOL ETHERS (*Continued*)

Substrate	Aryl Compound	Conditions	Product(s) and Yield(s) (%)	Refs.
C_{16}				
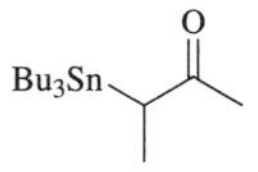 Bu$_3$Sn-CH(CH$_3$)-C(O)-CH$_3$	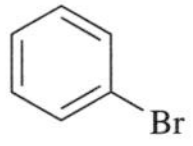 Ph–Br	PdCl$_2$[P(*o*-tol)$_3$]$_2$ (10 mol %), toluene, 100°, 5 h	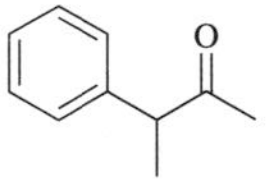 Ph-CH(CH$_3$)-C(O)-CH$_3$ (60)	149
(cyclohexanone-spiro-dioxolane with CHO and CH$_2$-Ar-Br substituent)		PdCl$_2$(PPh$_3$)$_2$ (10 mol %), Cs$_2$CO$_3$ (3 eq)	**I** + **II** + **III** Solvent / Temp / Time (h): toluene / reflux / 5; THF / 100° / 24 I / II / III: (76) / (11) / —; (29) / (9) / (17)	64
(same cyclohexanone-spiro-dioxolane substrate)		PdCl$_2$(PPh$_3$)$_2$ (10 mol %), Cs$_2$CO$_3$ (3 eq)	**I** + **II** + **III** Solvent / Temp / Time (h): toluene / reflux / 6; THF / 100° / 22 I / II / III: (32) / (17) / (13); (31) / (10) / (19)	64
C_{17}				
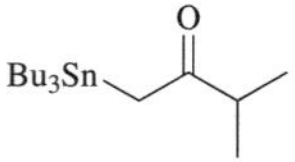 Bu$_3$Sn-CH$_2$-CH(CH$_3$)-C(O)-CH$_3$ (isobutyl ketone stannane)	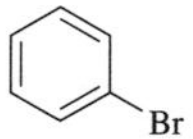 Ph–Br	PdCl$_2$[P(*o*-tol)$_3$]$_2$ (10 mol %), toluene, 100°, 5 h	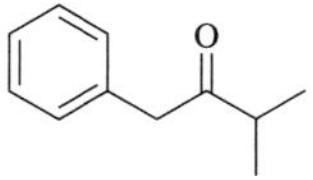 PhCH$_2$-C(O)-CH(CH$_3$)$_2$ (87)	149

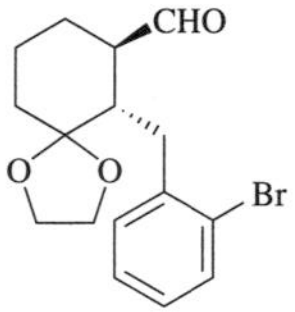

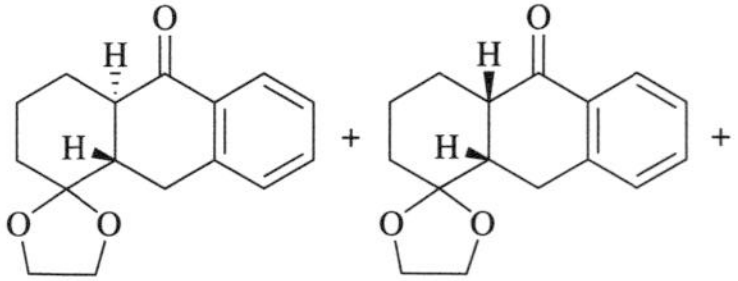

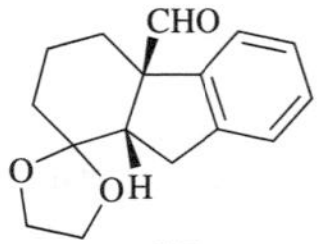

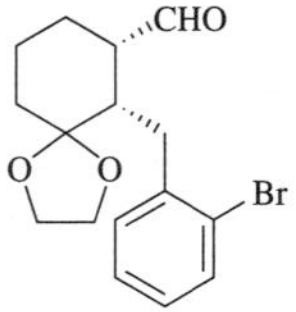

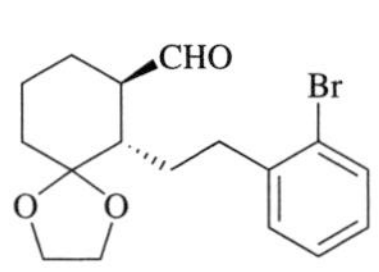

I	II	III
(49)	(4)	(22)
(52)	(9)	(—)

PdCl$_2$(PPh$_3$)$_2$ (10 mol %), Cs$_2$CO$_3$ (3 eq), 3 h

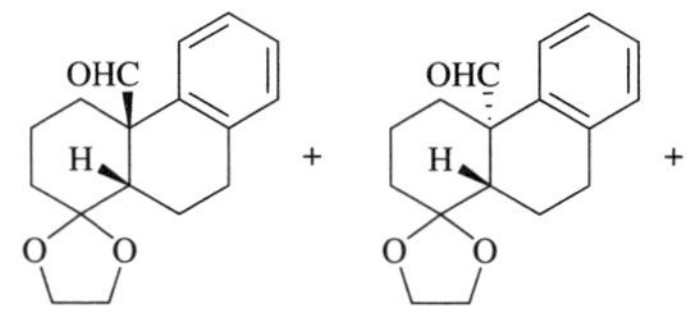

Solvent	Temp
toluene	reflux
THF	100°

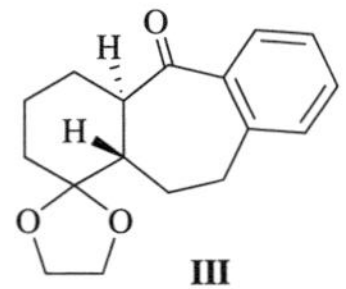

I	II	IV
(52)	(22)	(6)
(48)	(6)	(—)

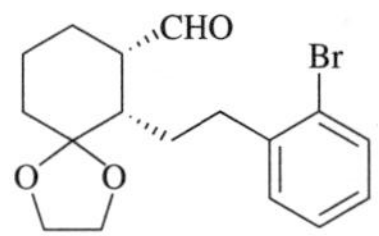

PdCl$_2$(PPh$_3$)$_2$ (10 mol %), Cs$_2$CO$_3$ (3 eq), 3 h

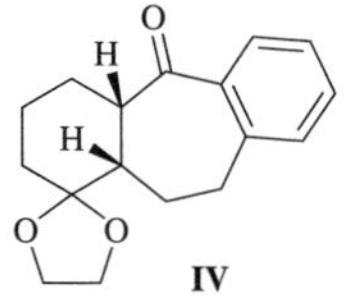

Solvent	Temp
toluene	reflux
THF	100°

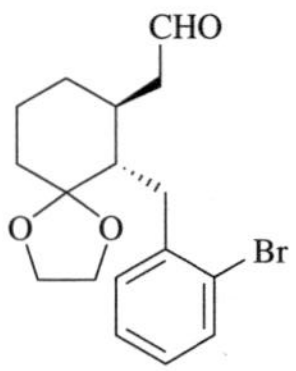

PdCl$_2$(PPh$_3$)$_2$ (10 mol %), Cs$_2$CO$_3$ (3 eq), toluene, reflux, 1.5 h

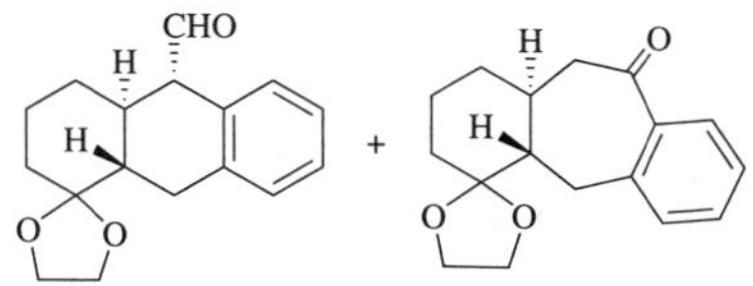

64

64

64

TABLE 1. ARYLATION OF ALDEHYDES, KETONES, AND ENOL ETHERS (*Continued*)

Substrate	Conditions	Product(s) and Yield(s) (%)	Refs.
C17	PdCl₂(PPh₃)₂ (10 mol %), Cs₂CO₃ (3 eq), THF, 100°, 14 h	(23) + (30)	64
C18	Pd₂(dba)₃ (1 mol %), (S)-**L23** (2.5 mol %), NaOt-Bu (2 eq), toluene	(75), 93% ee	41
	Catalyst, ligand, Cs₂CO₃, toluene, 130°		56

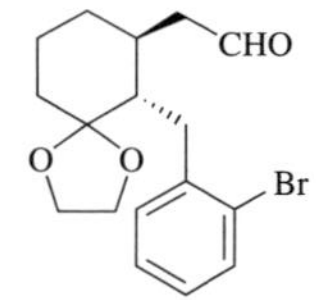

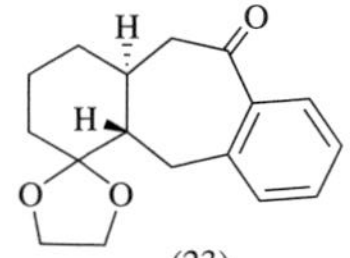

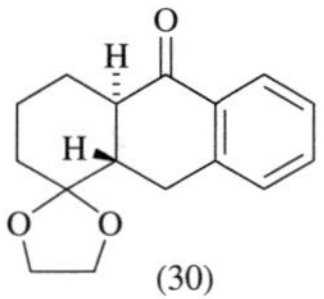

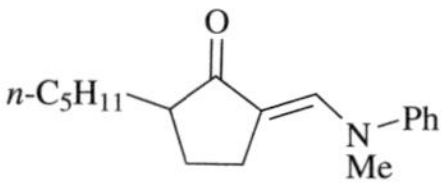

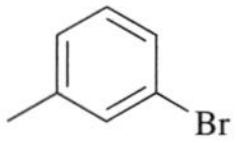

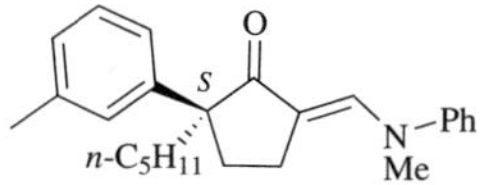

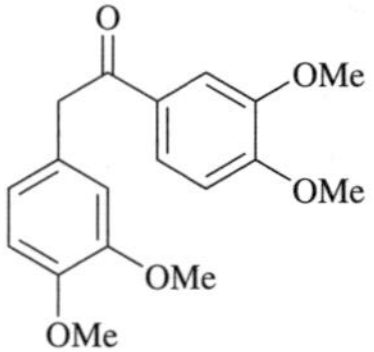

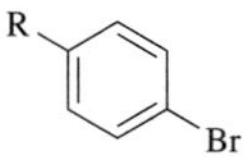

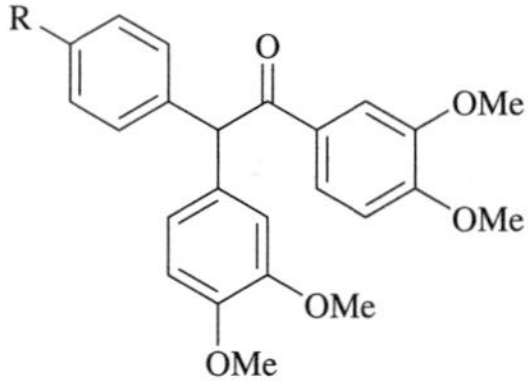

Catalyst	Ligand	R	
PdCl₂(cod)	**L36**	H	(93)
Pd(OCOCF₃)₂	**L37**	H	(92)
PdCl₂(cod)	**L36**	MeO	(95)
Pd(OCOCF₃)₂	**L37**	MeO	(91)

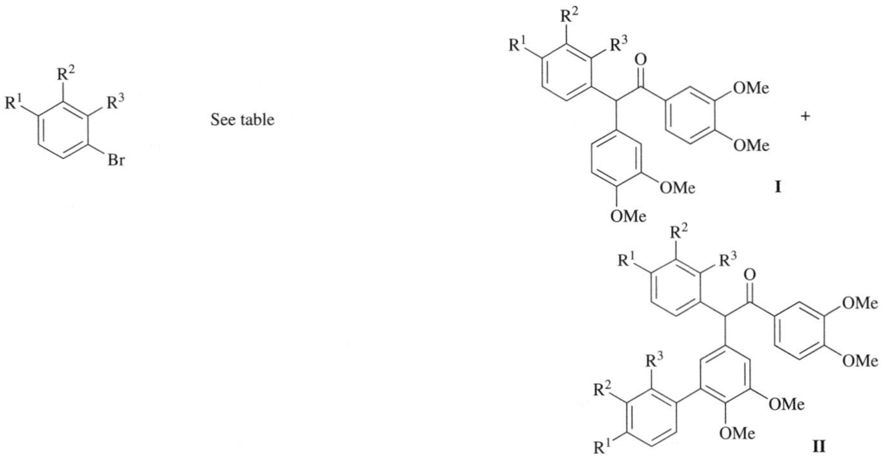

R^1	R^2	R^3	Catalyst	Ligand	Base	Solvent	Temp (°)	I	II
H	H	H	PdCl$_2$	none	Cs$_2$CO$_3$	DMF	100	(20)	(—)
H	H	H	PdCl$_2$	PPh$_3$	K$_2$CO$_3$	DMF	170	(24)	(—)
H	H	H	Pd(OAc)$_2$	PPh$_3$	K$_2$CO$_3$	xylene	170	(55)	(29)
H	H	H	Pd(OAc)$_2$	PPh$_3$	K$_2$CO$_3$	DMF	170	(83)	(—)
H	H	H	Pd(OAc)$_2$	PPh$_3$	Cs$_2$CO$_3$	xylene	150	(85)	(—)
H	H	H	Pd(OAc)$_2$	PPh$_3$	Cs$_2$CO$_3$	DMF	150	(85)	(—)
H	H	H	Pd(OAc)$_2$	PPh$_3$	Cs$_2$CO$_3$	DMF	170	(52)	(35)
H	H	O$_2$N	Pd(OAc)$_2$	PPh$_3$	Cs$_2$CO$_3$	DMF	150	(44)	(—)
H	H	MeO	Pd(OAc)$_2$	PPh$_3$	Cs$_2$CO$_3$	DMF	150	(46)	(—)
H	MeO	H	Pd(OAc)$_2$	PPh$_3$	Cs$_2$CO$_3$	DMF	150	(51)	(—)
H	MeO	MeO	Pd(OAc)$_2$	PPh$_3$	Cs$_2$CO$_3$	DMF	150	(47)	(—)
—OCH$_2$O—	H	Pd(OAc)$_2$	PPh$_3$	Cs$_2$CO$_3$	DMF	150	(55)	(—)
MeO	MeO	MeO	Pd(OAc)$_2$	PPh$_3$	Cs$_2$CO$_3$	DMF	150	(12)	(—)

TABLE 1. ARYLATION OF ALDEHYDES, KETONES, AND ENOL ETHERS (*Continued*)

Substrate	Aryl Compound	Conditions	Product(s) and Yield(s) (%)	Refs.
C₁₈ (aryl ketone with OMe groups)	(bromoarene with R¹, R², R³)	See table	(products I, II, III shown)	55

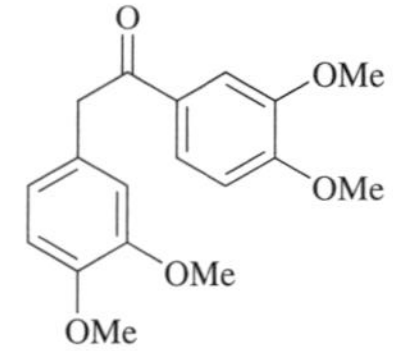

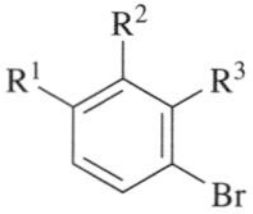

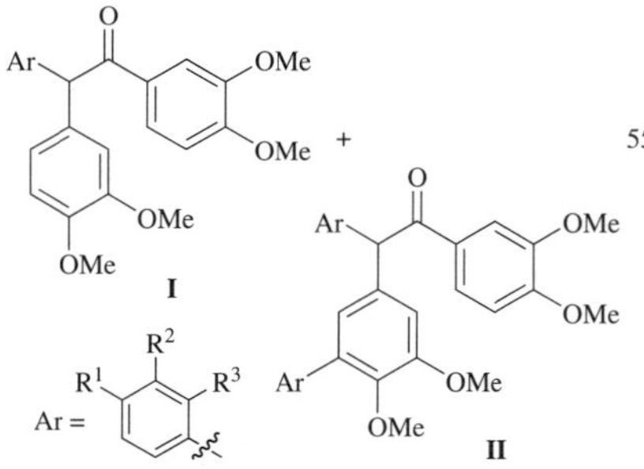

R¹	R²	R³	Catalyst	Ligand	Base	Solvent	Temp (°)	I	II	III
H	H	H	Pd(OAc)₂	PPh₃	K₂CO₃	xylene	150	(86)	(—)	(—)
H	H	H	Pd(OAc)₂	PPh₃	K₂CO₃	xylene	170	(58)	(31)	(6)
H	H	H	Pd(OAc)₂	PPh₃	Cs₂CO₃	xylene	170	(23)	(32)	(25)
H	H	H	Pd(OAc)₂	PPh₃	Cs₂CO₃	DMF	150	(89)	(—)	(—)
H	H	H	Pd(OAc)₂	PPh₃	Cs₂CO₃	DMF	170	(56)	(38)	(—)
H	H	H	Pd(OAc)₂	PEt₃	Cs₂CO₃	DMF	150	(32)	(29)	(—)
H	H	O₂N	Pd cat 11	none	Cs₂CO₃	DMF	150	(53)	(—)	(—)
H	H	MeO	Pd(OAc)₂	PPh₃	Cs₂CO₃	DMF	150	(51)	(—)	(—)
H	MeO	H	Pd cat 11	none	Cs₂CO₃	DMF	150	(40)	(—)	(—)
H	MeO	H	Pd(OAc)₂	PPh₃	Cs₂CO₃	DMF	153	(47)	(—)	(—)
H	MeO	MeO	Pd(OAc)₂	PPh₃	Cs₂CO₃	DMF	150	(44)	(—)	(—)
MeO	MeO	H	Pd cat 11	none	Cs₂CO₃	DMF	150	(53)	(—)	(—)
—OCH₂O—		H	Pd(OAc)₂	PPh₃	Cs₂CO₃	DMF	153	(55)	(—)	(—)
—OCH₂O—		H	Pd cat 11	none	Cs₂CO₃	DMF	150	(38)	(—)	(—)
MeO	MeO	MeO	Pd(OAc)₂	PPh₃	Cs₂CO₃	DMF	153	(12)	(—)	(—)

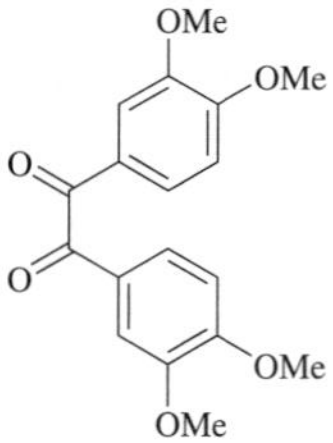

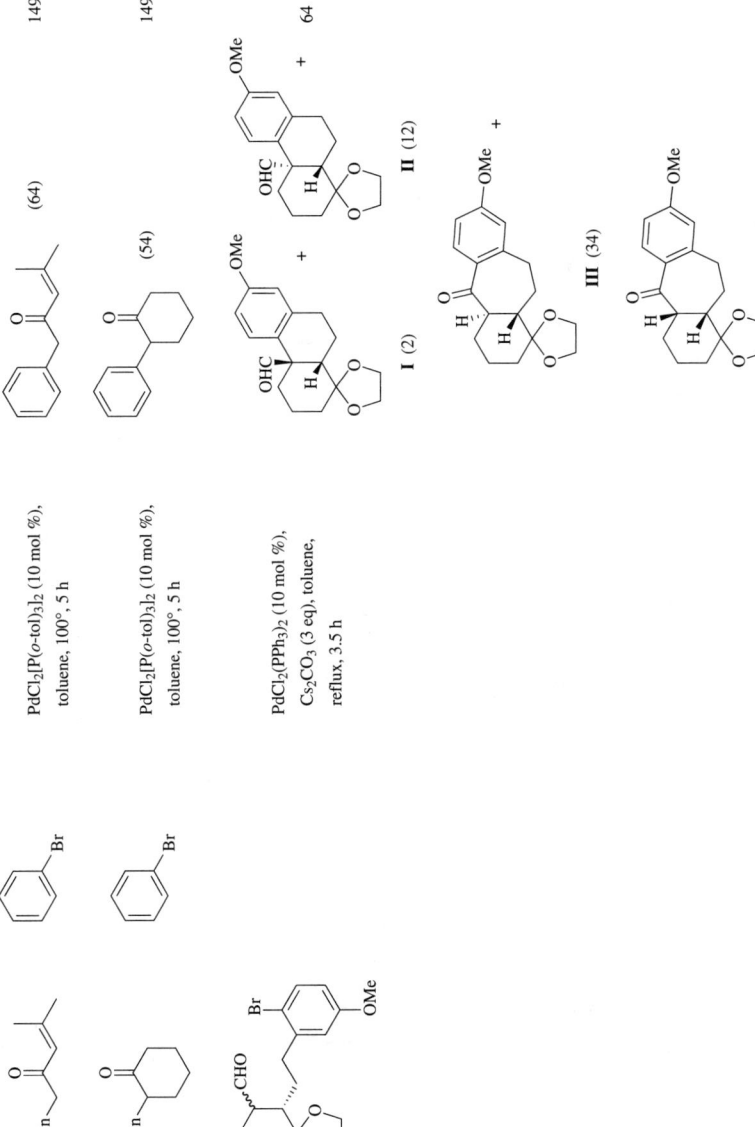

TABLE 1. ARYLATION OF ALDEHYDES, KETONES, AND ENOL ETHERS (*Continued*)

Substrate	Conditions	Product(s) and Yield(s) (%)	Refs.
C₁₈			
(2-bromobenzyl cyclohexane spiro-dioxolane with CHO)	PdCl₂(PPh₃)₂ (10 mol %), Cs₂CO₃ (3 eq), THF, 100°, 14 h	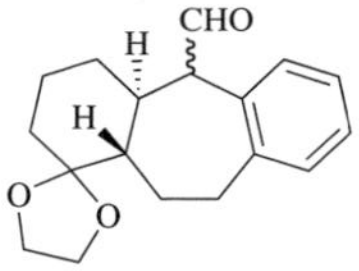 (40) tricyclic CHO product	64
(MeO₂C, iodoaryl, t-Bu enone)	Pd(PPh₃)₄ (20 mol %), Bu₄NF (3 eq), THF, rt, 15 h	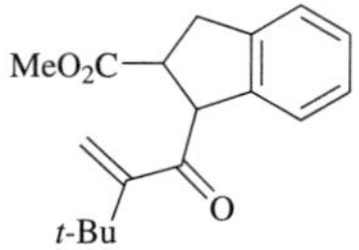 (68) indane product with MeO₂C and t-Bu enone	58
(iodoaryl-N(Bn)-CH₂-CH₂-CH₂-C(O)Me)	PdCl₂(PPh₃)₂ (0.2 eq), Cs₂CO₃ (3 eq), THF, 100–110°, 48 h	HO-tetrahydroquinoline acetyl N-Bn (9) + acetyl tetrahydroquinoline N-Bn (20)	155, 113
C₂₀			
(Ph₂CH-C(O)-Ph) + 2-bromonaphthalene	Pd(PPh₃)₄ (0.5 mol %), Cs₂CO₃ (3–5 eq), xylene, 23 h	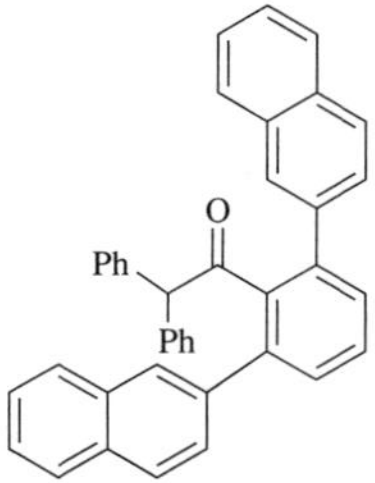 (43) bis-naphthyl aryl diphenyl ketone	46

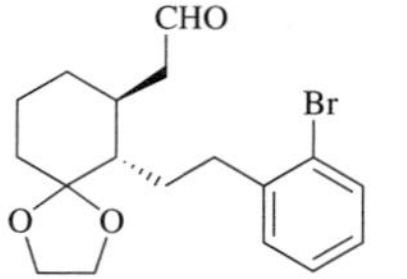

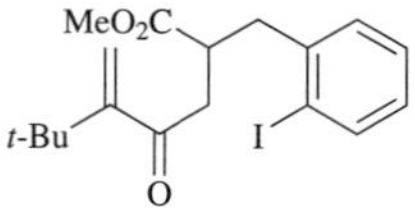

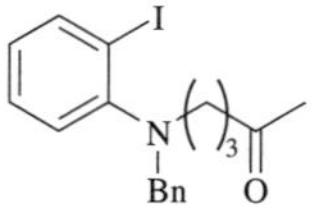

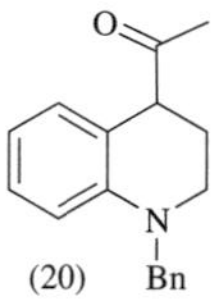

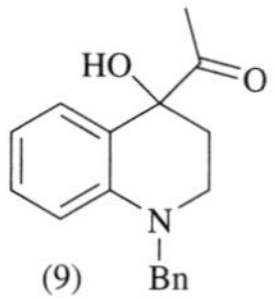

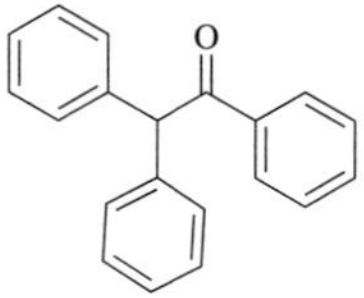

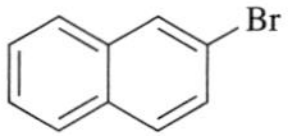

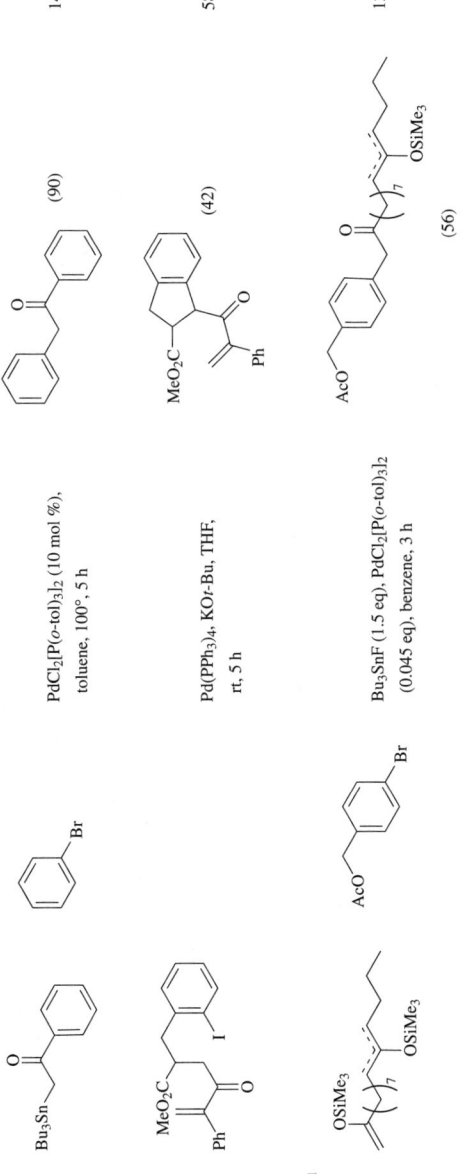

[a] The intermediate dimeric silyl enol ether was formed with 97% ee.
[b] The intermediate dimeric silyl enol ether was formed with 84% ee.
[c] These yields were determined by GC-MS.

TABLE 2. ARYLATION OF ESTERS, AMIDES, LACTONES, LACTAMS, NITRILES, KETENE ACETALS, AND PREFORMED ENOLATES

Substrate	Aryl Compound	Conditions	Product(s) and Yield(s) (%)	Refs.
C₂				
CH_3CN	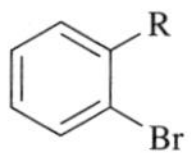 2-Br, R-aryl	$Pd(OAc)_2$ (5 mol %), **L23** (5 mol %), NaHMDS (1.3 eq), toluene, 100°, 16 h	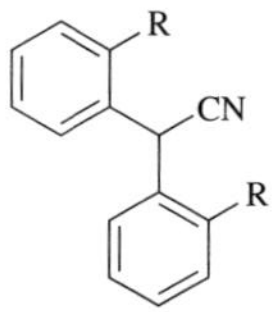 R: H (62); Me (60)	26
C₄				
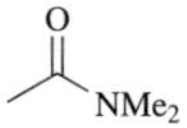 $CH_3C(O)NMe_2$	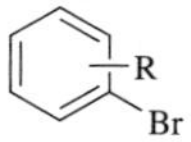 2-Br, R-aryl	1. s-BuLi (1.2 eq), THF, −78°, 1 h; 2. $ZnCl_2$ (2.4 eq), rt, 10 min; 3. $Pd(dba)_2$ (1 mol %), **L12** (1 mol %), rt, 24 h	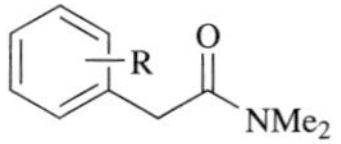 R: 2-MeO (91); 4-NC (89); 4-MeO₂C (95)	68
$CH_3C(O)NMe_2$	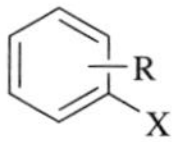 R-aryl-X	$Pd(dba)_2$ (5 mol %), **L23** (7.5 mol %), KHMDS (2 eq), dioxane, 100°	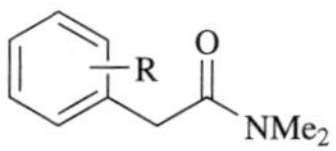 **I** 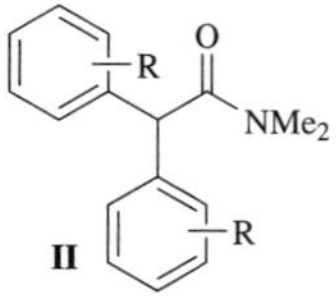 + **II** (see below)	67

X	R	Time (h)	**I**	**II**
Br	2-Me	1.5	(72)	(4)
Br	4-MeO	4	(48)	(18)
Br	4-Me	1.5	(72)	(10)
Br	4-Me	1.5[a]	(24)	(74)
Br	4-Ph	2	(66)	(13)
I	4-Me	1	(70)	(4)

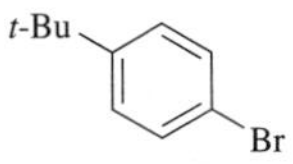

Substrate	Conditions	Product(s)	Yield(%)
4-t-Bu-C6H4-Br	Pd(OAc)2 (7.5 mol %), ligand (9 mol %), base (1.2 eq), 85°, 2 h	t-Bu-C6H4-CH(C(O)NMe2)-C6H4-t-Bu (**I**) + t-Bu-C6H4-Ph (**II**)	67

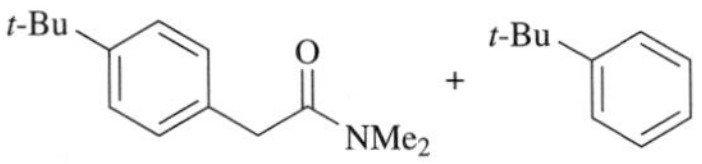

Ligand	Base	Solvent	I	II
L1	KHMDS	THF	(22)	(22)
L1	KHMDS	dioxane	(48)	(2)
L1	LTMP	THF	(5)	(4)
L1	LTMP	dioxane	(38)	(5)
L23	KHMDS	THF	(32)	(8)

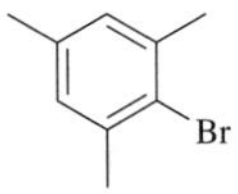

Substrate	Conditions	Product	Yield(%)
2-Br-mesitylene	1. s-BuLi (1.2 eq), THF, −78°, 1 h; 2. ZnCl2 (2.4 eq), rt, 10 min; 3. Pd(dba)2 (1 mol %), L12 (1 mol %), rt, 24 h	mesityl-CH2-C(O)NMe2 (96)	68
2-bromonaphthalene	Pd(dba)2 (5 mol %), L23 (7.5 mol %), KHMDS (2 eq), dioxane, 100°, 2 h	2-Naph-CH(C(O)NMe2)-2-Naph + 2-Naph-CH2-C(O)NMe2 (70)	67
		(9)	
PhBr	Pd2(dba)3·CHCl3 (1 mol %), P(t-Bu)3 (2 mol %), LiHMDS (2.2 eq), toluene, 70°, 6 h	Ph2C(Et)CN (69)	26

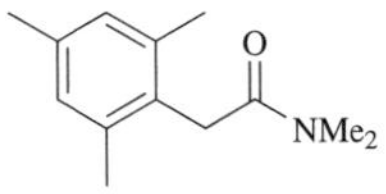

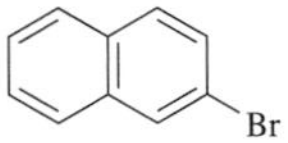

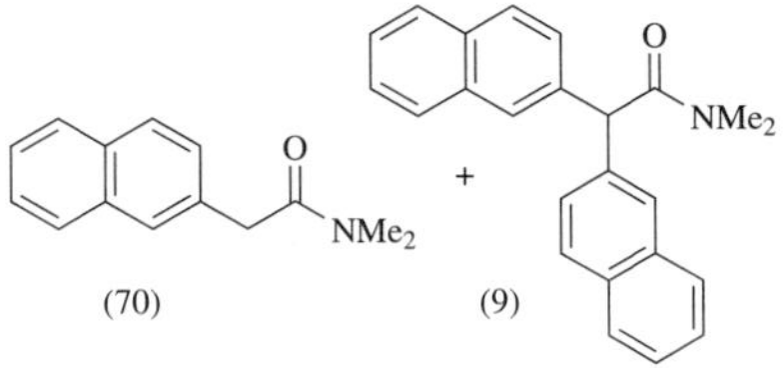

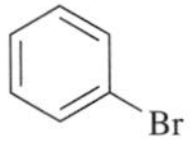

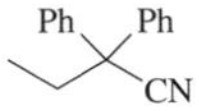

169

TABLE 2. ARYLATION OF ESTERS, AMIDES, LACTONES, LACTAMS, NITRILES, KETENE ACETALS, AND PREFORMED ENOLATES (*Continued*)

Substrate	Aryl Compound	Conditions	Product(s) and Yield(s) (%)	Refs.
C_4				
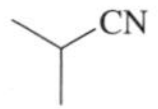 (i-Pr-CN)	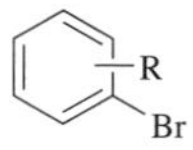 Ar-R-Br	Pd(OAc)$_2$ (x mol %), **L23** (x mol %), NaHMDS (1.3 eq), toluene, 100°	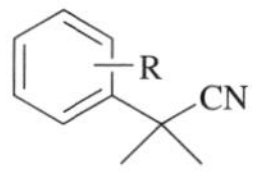 (Ar-R-C(Me)$_2$CN)	26
		R x Time (h)		
		2-Me 1 6	(70)	
		4-MeO 1 8	(83)	
		4-NC 0.5 1	(99)	
		4-*t*-Bu 1 2	(87)	
BrZn-CH$_2$CO$_2$Et	Ph-X	Catalyst (10 mol %)	PhCH$_2$CO$_2$Et	109
		Catalyst Solvent Temp Time (h)		
	X = Cl	Pd(PPh$_3$)$_4$ CH$_2$(OMe)$_2$/HMPA (1:1) reflux 3	(0)	
	X = Cl	Ni(PPh$_3$)$_4$ CH$_2$(OMe)$_2$/HMPA (1:1) reflux 3	(65)	
	X = Br	Pd(PPh$_3$)$_4$ benzene/HMPA (1:1) reflux 7	(15)	
	X = Br	Ni(PPh$_3$)$_4$ CH$_2$(OMe)$_2$/HMPA (1:1) reflux 3	(67)	
	X = I	Pd(PPh$_3$)$_4$ CH$_2$(OMe)$_2$/HMPA (1:1) reflux 6	(47)	
	X = I	Ni(PPh$_3$)$_4$ CH$_2$(OMe)$_2$/HMPA (1:1) rt 3	(55)	
BrZn-CH$_2$CO$_2$Et	HO$_2$C-C$_6$H$_4$-I	Pd(PPh$_3$)$_4$ (10 mol %), CH$_2$(OMe)$_2$/HMPA (1:1), reflux, 3 h	HO$_2$C-C$_6$H$_4$-CH$_2$CO$_2$Et (85)	109

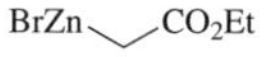

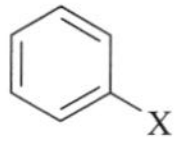

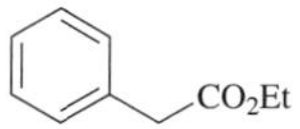

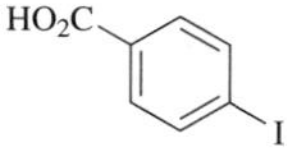

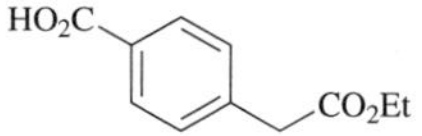

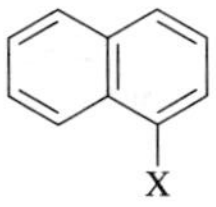

C5

1-BrNaphthalene / 1-I-Naphthalene + CH₂(OMe)₂/HMPA (1:1), Catalyst (10 mol %), reflux, 3 h → 1-naphthyl-CH(CO₂Et)(–)

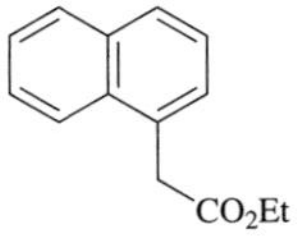

Catalyst	X	
Pd(PPh₃)₄	Br	(0)
Ni(PPh₃)₄	Br	(69)
Pd(PPh₃)₄	I	(47)

109

PhBr + methyl methoxyacetate → PhCH(OMe)CO₂Et (<5)

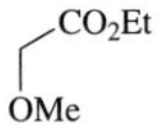

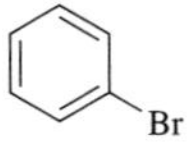

Pd(dba)₂ (5 mol %), P(t-Bu)₃ (10 mol %), LiNCy₂, toluene, rt or Pd(dba)₂, carbene,[b] NaHMDS, toluene, rt

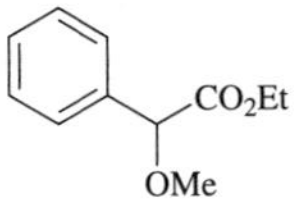

99

PhCl + methyl isobutyrate → PhC(Me)₂CO₂Me

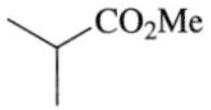

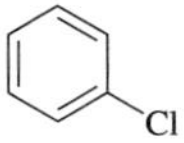

1. Base (1.3 eq), toluene, rt, 10 min
2. [PdP(t-Bu)₃Br]₂ (0.1 mol %), 4 h

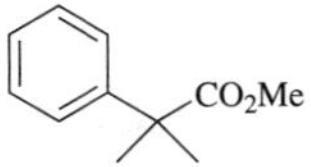

Base	Temp	
NaHMDS	rt	(0)
NaHMDS	60°	trace
NaHMDS	100°	(90)
LiNCy₂	rt	(0)
LiNCy₂	100°	trace

156

TABLE 2. ARYLATION OF ESTERS, AMIDES, LACTONES, LACTAMS, NITRILES, KETENE ACETALS, AND PREFORMED ENOLATES (*Continued*)

Substrate	Aryl Compound	Conditions	Product(s) and Yield(s) (%)	Refs.
C_5 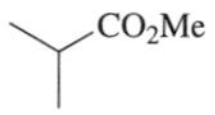 ![CO2Me isobutyrate]	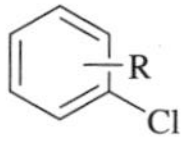 ![ArCl with R]	1. NaHMDS (1.2 eq), toluene, rt, 10 min 2. Catalyst (x mol %), ligand (y mol %), 100°, 4 h	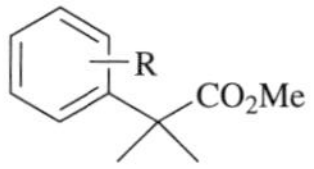 ![product R-C6H4-C(Me)2-CO2Me]	156

R	Catalyst	x	Ligand	y	
H	Pd(dba)$_2$	0.2	P(t-Bu)$_3$	0.2	(89)
H	[PdP(t-Bu)$_3$Br]$_2$	0.1	none	—	(92)
3-MeO	Pd(dba)$_2$	0.4	P(t-Bu)$_3$	0.4	(77)
3-MeO	[PdP(t-Bu)$_3$Br]$_2$	0.2	none	—	(74)
3-F	Pd(dba)$_2$	0.4	P(t-Bu)$_3$	0.4	(77)
3-F	[PdP(t-Bu)$_3$Br]$_2$	0.2	none	—	(81)
3-CF$_3$	Pd(dba)$_2$	0.4	P(t-Bu)$_3$	0.4	(81)
3-CF$_3$	[PdP(t-Bu)$_3$Br]$_2$	0.2	none	—	(75)
4-MeO	Pd(dba)$_2$	0.4	P(t-Bu)$_3$	0.4	(67)
4-MeO	[PdP(t-Bu)$_3$Br]$_2$	0.2	none	—	(71)
4-F	Pd(dba)$_2$	0.4	P(t-Bu)$_3$	0.4	(90)
4-F	[PdP(t-Bu)$_3$Br]$_2$	0.2	none	—	(87)
4-CF$_3$	Pd(dba)$_2$	0.4	P(t-Bu)$_3$	0.4	(67)
4-CF$_3$	[PdP(t-Bu)$_3$Br]$_2$	0.2	none	—	(71)
4-Me	Pd(dba)$_2$	0.4	P(t-Bu)$_3$	0.4	(79)
4-Me	[PdP(t-Bu)$_3$Br]$_2$	0.2	none	—	(84)

| | 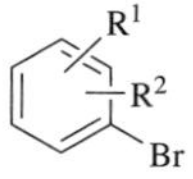 ![ArBr with R1,R2] | Pd(dba)$_2$ (x mol %),
P(t-Bu)$_3$ (0.1–5.0 mol %),
LiNCy$_2$ (1.3 eq), toluene, rt | 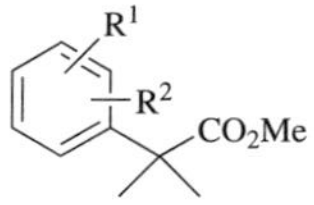 ![product with R1,R2] | 75 |

R^1	R^2	x	Time (h)	
H	H	0.1	18	(87)
3-MeO	H	0.1	24	(97)
3-F	H	0.1	16	(93)
3-Cl	H	0.1	12	(71)
3-Me	H	0.1	24	(97)
3-CF$_3$	H	0.1	24	(89)
4-Me$_2$N	H	0.05	18	(88)
4-MeO	H	0.1	24	(90)
4-PhO	H	0.1	12	(95)
4-F	H	0.1	16	(90)
4-Cl	H	0.1	14	(72)
4-Me	H	0.1	24	(90)
4-CF$_3$	H	0.1	24	(86)
4-*t*-Bu	H	0.1	12	(91)
4-Ph	H	0.05	18	(96)
4-PhCO	H	3.0	16	(73)
3-F	4-Ph	0.2	14	(91)

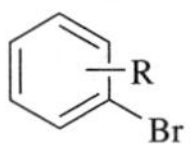

1. LiNCy$_2$ (1.3 eq), toluene, rt, 10 min
2. [PdP(*t*-Bu)$_3$Br]$_2$ (x mol %), rt, 4 h

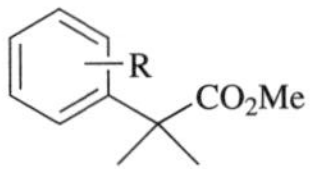

R	x	
3-MeO	0.5	(88)
3-F	0.5	(72)
4-Me$_2$N	0.05	(88)
4-MeO	0.5	(85)
4-F	0.5	(85)
4-Cl	0.5	(89)
4-CF$_3$	0.5	(60)
4-*t*-Bu	0.05	(72)

TABLE 2. ARYLATION OF ESTERS, AMIDES, LACTONES, LACTAMS, NITRILES, KETENE ACETALS, AND PREFORMED ENOLATES (*Continued*)

Substrate	Aryl Compound	Conditions	Product(s) and Yield(s) (%)	Refs.
C₅ ⟨CO₂Me (isobutyrate)⟩	3-(1,3-dioxolan-2-yl)phenyl Br	Pd(dba)₂ (2.0 mol %), P(*t*-Bu)₃ (2.0 mol %), LiNCy₂ (1.3 eq), toluene, rt, 16 h	aryl–C(CH₃)₂CO₂Me with 1,3-dioxolane (77)	75
	4-Cl–C₆H₄–Br	1. NaHMDS (1.2 eq), toluene, rt, 10 min 2. Catalyst (*x* mol %), ligand (*y* mol %), 100°, 4 h Catalyst | *x* | Ligand | *y* Pd(dba)₂ | 0.4 | P(*t*-Bu)₃ | 0.4 (69) [PdP(*t*-Bu)₃Br]₂ | 0.2 | none | — (82)	4-Cl–C₆H₄–C(CH₃)₂CO₂Me	156
	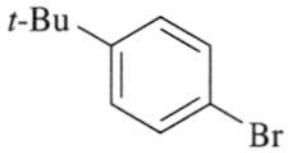 4-*t*-Bu–C₆H₄–Br	Pd(dba)₂ (0.5 mol %), P(*t*-Bu)₃ (0.5 mol %), LiHMDS (2.3 eq), rt, 12 h	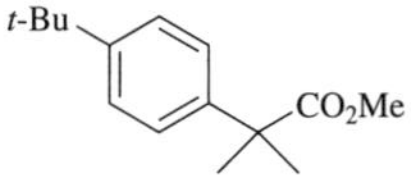 4-*t*-Bu–C₆H₄–C(CH₃)₂CO₂Me (92)	77
	4-*t*-Bu–C₆H₄–Br	1. LiNCy₂ (1.3 eq), toluene, 0°, 20 min 2. [PdP(*t*-Bu)₃Br]₂ (0.05 mol %), rt, 4 h	**I** (86)	157
	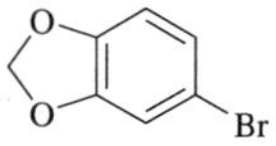 3,4-methylenedioxyphenyl Br	Pd(dba)₂ (0.1 mol %), P(*t*-Bu)₃ (0.1 mol %), LiNCy₂ (1.3 eq), toluene, rt, 20 h	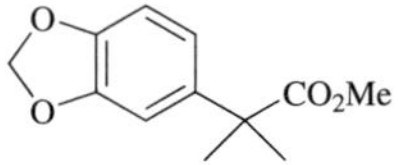 3,4-(methylenedioxy)C₆H₃–C(CH₃)₂CO₂Me (95)	75

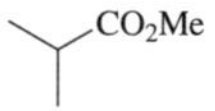

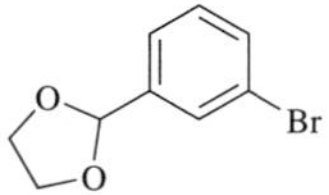

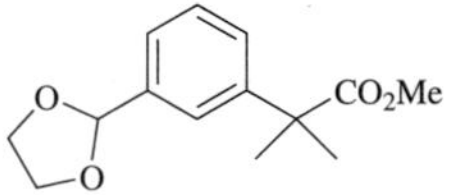

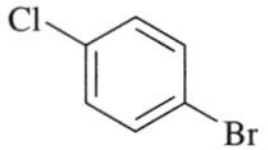

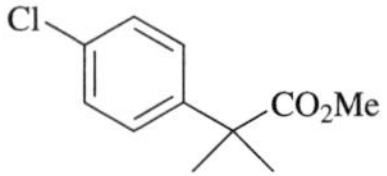

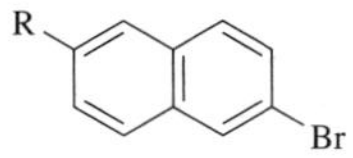

Substrate	Conditions	Product(s) and Yield(s) (%)	Refs.
R-naphthyl-Br	Pd(dba)₂ (x mol %), P(t-Bu)₃ (0.1–0.2 mol %), LiNCy₂ (1.3 eq), toluene, rt	R-naphthyl-C(Me)₂CO₂Me R / x / Time (h) H / 0.1 / 8 (94) MeO / 0.2 / 24 (93)	75
2-chloropyridine	1. NaHMDS (1.2 eq), toluene, rt, 10 min 2. Catalyst (x mol %), ligand (y mol %), 100°, 4 h	2-pyridyl-C(Me)₂CO₂Me **I** Catalyst / x / Ligand / y Pd(dba)₂ / 0.4 / P(t-Bu)₃ / 0.4 (66) [PdP(t-Bu)₃BrI]₂ / 0.2 / none / — (71)	156
2-bromopyridine	1. LiNCy₂ (1.3 eq), toluene, rt, 10 min 2. [PdP(t-Bu)₃BrI]₂ (0.5 mol %), rt, 4 h	**I** (71)	156
X-pyrimidine	Pd(dba)₂ (5.0 mol %), P(t-Bu)₃ (5.0 mol %), LiNCy₂ (1.3 eq), toluene, rt	pyrimidyl-C(Me)₂CO₂Me X / Time (h) 2-Br / 7 (94) 3-Br / 12 (80) 4-Br / 16 (51)	75
3-bromoquinoline	Pd(dba)₂ (5.0 mol %), P(t-Bu)₃ (5.0 mol %), LiNCy₂ (1.3 eq), toluene, rt, 12 h	3-quinolyl-C(Me)₂CO₂Me (70)	75

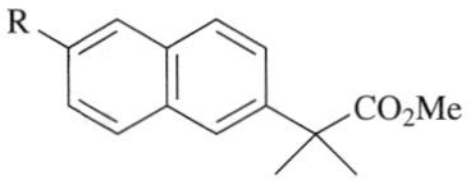

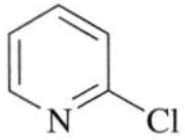

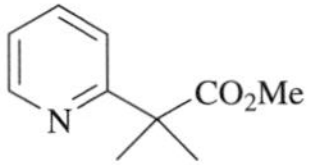

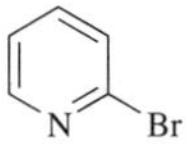

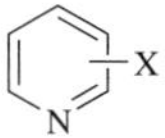

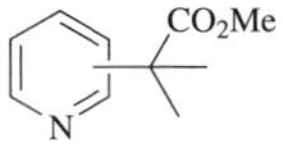

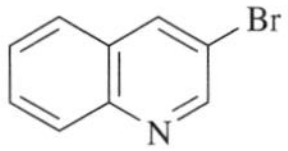

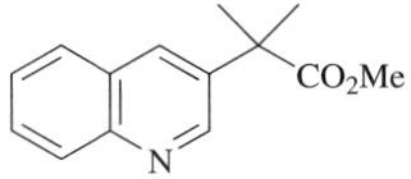

175

TABLE 2. ARYLATION OF ESTERS, AMIDES, LACTONES, LACTAMS, NITRILES, KETENE ACETALS, AND PREFORMED ENOLATES (*Continued*)

Substrate	Aryl Compound	Conditions	Product(s) and Yield(s) (%)	Refs.
C$_5$ 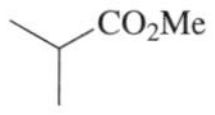 iBu-CO$_2$Me	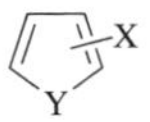 Y—X (heteroaryl)	Pd(dba)$_2$ (5.0 mol %), P(t-Bu)$_3$ (5.0 mol %), LiNCy$_2$ (1.3 eq), toluene, rt	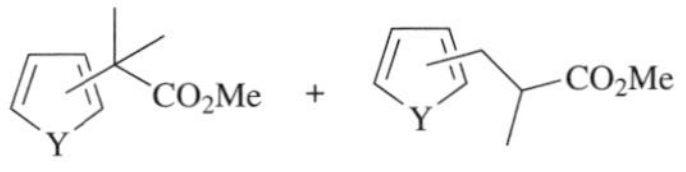 I + II (I / II)	75
	Y / X / —	Time (h)	(72) (—) (71) (—) (48) (43) (71) (—)	
	O / 2-Br O / 3-Br S / 2-Br S / 3-Br	7 9 9 9		
	3-bromothiophene	1. LiNCy$_2$ (1.3 eq), toluene, rt, 10 min 2. [PdP(t-Bu)$_3$Br]$_2$ (0.5 mol %), rt, 4 h	thienyl-C(Me)$_2$CO$_2$Me (75)	156
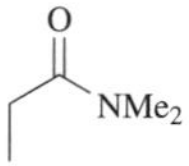 n-Bu–C(O)NMe$_2$	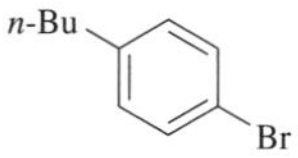 n-Bu–C$_6$H$_4$–Br	Pd(dba)$_2$ (5 mol %), P(t-Bu)$_3$ (7.5 mol %), KHMDS (2 eq), dioxane, 100°, 3 h	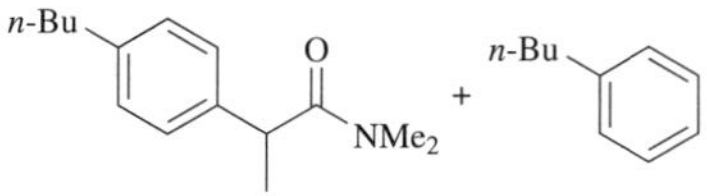 n-Bu-C$_6$H$_4$-CH(Me)C(O)NMe$_2$ (16) + n-Bu-C$_6$H$_4$-Ph (78)	67
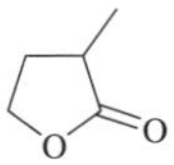 3-methyl-γ-butyrolactone	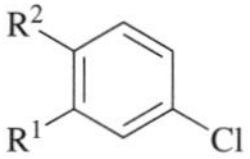 R^2/R^1-C$_6$H$_3$-Cl	Ni(cod)$_2$ (5 mol %), (*S*)-**L23** (8.5 mol %), ZnBr$_2$ (15 mol %), NaHMDS (2.3 eq), toluene/THF (3:1), 60°, 17–20 h	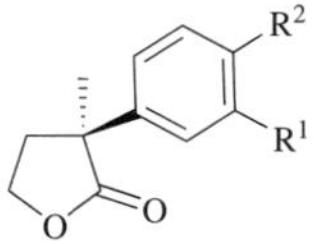 aryl lactone	33
	R^1 / R^2 H / H Me$_2$N / H MeO / H H / TBSO H / MeO		% ee (86) >97 (81) 97 (86) 96 (67) 95 (76) 94	

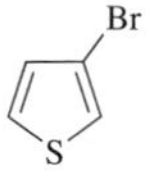

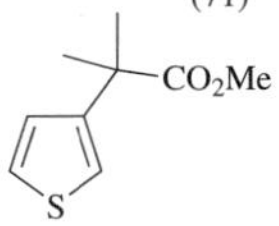

176

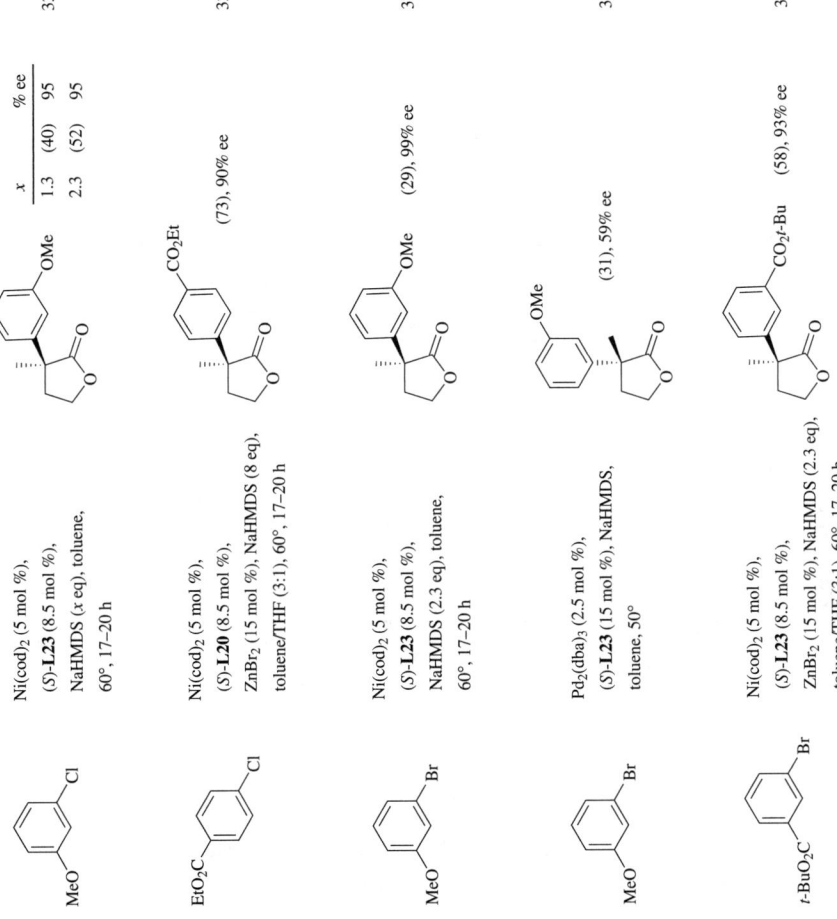

TABLE 2. ARYLATION OF ESTERS, AMIDES, LACTONES, LACTAMS, NITRILES, KETENE ACETALS, AND PREFORMED ENOLATES (*Continued*)

Substrate	Aryl Compound	Conditions	Product(s) and Yield(s) (%)	Refs.
C₅				
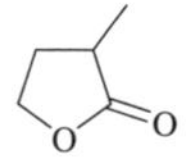 (3-methyl-γ-butyrolactone)	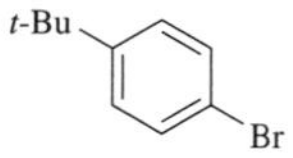 4-*t*-Bu-C₆H₄-Br	Ni(cod)₂ (5 mol %), (*S*)-**L23** (8.5 mol %), ZnBr₂ (15 mol %), NaHMDS (2.3 eq), toluene/THF (3:1), 60°, 17–20 h	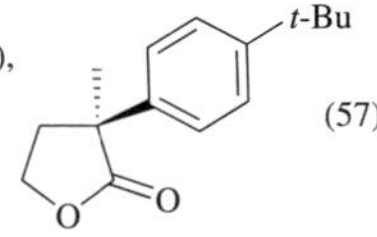 (*t*-Bu-aryl lactone) (57), 94% ee	33
	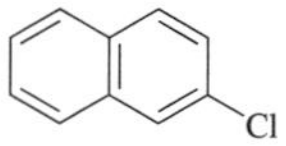 2-Cl-naphthalene	Ni(cod)₂ (5 mol %), (*S*)-**L23** (8.5 mol %), ZnBr₂ (15 mol %), NaHMDS (2.3 eq), toluene/THF (3:1), 60°, 17–20 h	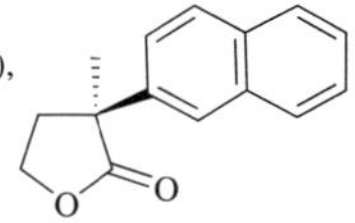 **I** (95), 94% ee	33
	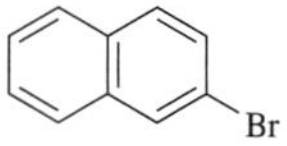 2-Br-naphthalene	Ni(cod)₂ (5 mol %), (*S*)-**L23** (8.5 mol %), NaHMDS (2.3 eq), toluene, 60°, 17–20 h	**I** (33), 98% ee	33
	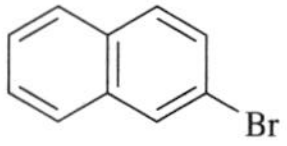 2-Br-naphthalene	Pd₂(dba)₃ (2.5 mol %), (*S*)-**L23** (15 mol %), NaHMDS, toluene, 50°	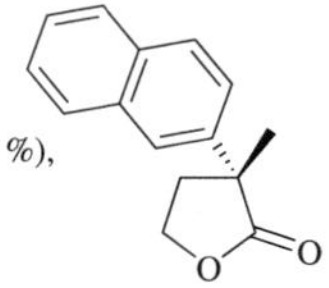 (naphthyl-methyl lactone) (62), 60% ee	33
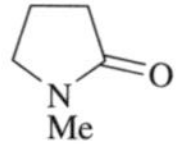 (*N*-methylpyrrolidinone)	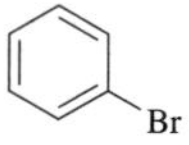 Ph-Br	Pd(dba)₂ (5 mol %), **L23** (7.5 mol %), KHMDS (2 eq), dioxane, 100°, 3 h	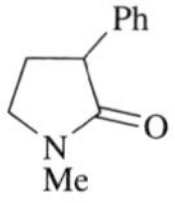 (3-Ph-N-Me-pyrrolidinone) (49) 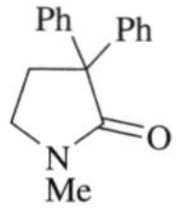 + (3,3-diPh-N-Me-pyrrolidinone) (9)	67

Me₃Si⁓CN +

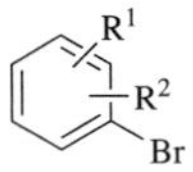

[Ar-Br structure with R¹, R² substituents]

Pd₂(dba)₃ (2 mol %),
ligand (x mol %), ZnF₂ (0.5 eq),
DMF, 90°

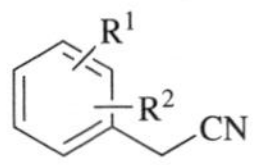

→ [ArCH₂CN product with R¹, R²]   80

R¹	R²	Ligand	x	Time (h)	
2-Me	H	P(t-Bu)₃	4	24	(83)
2-i-Pr	H	P(t-Bu)₃	4	24	(69)
2-Cy	H	P(t-Bu)₃	4	24	(71)
4-O₂N	H	L23	2	8	(68)
4-Me₂N	H	PPh(t-Bu)₂	4	18	(83)
4-MeO	H	L23	2	18	(64)
4-F	H	L23	2	18	(79)
4-NC	H	L23	2	8	(81)
4-EtO₂C	H	L23	2	8	(84)
4-CF₃	H	L23	2	18	(78)
4-MeCO	H	L23	2	8	(78)
4-t-Bu	H	L23	2	18	(87)
4-PhCO	H	L23	2	8	(92)
2-Me	4-Me	P(t-Bu)₃	4	24	(78)
2-Me	6-Me	P(t-Bu)₃	4	24	(84)
3-Me	5-Me	L23	2	18	(74)

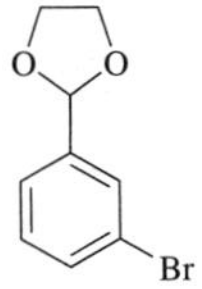

[3-(1,3-dioxolan-2-yl)phenyl bromide] → [3-(1,3-dioxolan-2-yl)phenyl acetonitrile] (78)

Pd₂(dba)₃ (2 mol %), L23 (2 mol %),
ZnF₂ (0.5 eq), DMF, 90°, 18 h   80

[6-bromo-2-methoxynaphthalene] → [6-methoxynaphthalen-2-yl acetonitrile] (87)

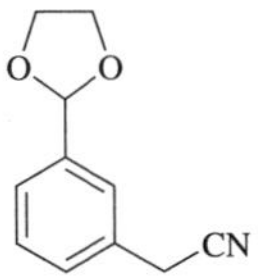

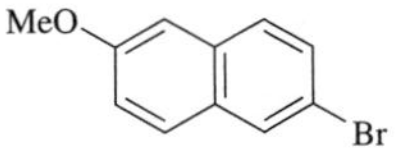

Pd₂(dba)₃ (2 mol %), L23 (2 mol %),
ZnF₂ (0.5 eq), DMF, 90°, 18 h   80

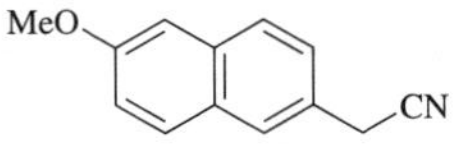

TABLE 2. ARYLATION OF ESTERS, AMIDES, LACTONES, LACTAMS, NITRILES, KETENE ACETALS, AND PREFORMED ENOLATES (*Continued*)

Substrate	Aryl Compound	Conditions	Product(s) and Yield(s) (%)	Refs.
C₆				
MeCO₂t-Bu	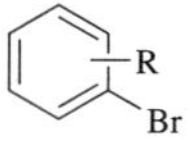 R–C₆H₄–Br	Pd(dba)₂ (x mol %), **L47** (x mol %), LiHMDS (2.3 eq), rt, 12 h	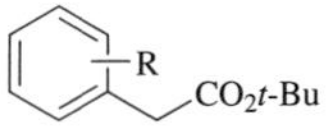 R–C₆H₄–CH₂CO₂t-Bu	77

R	x	
H	1	(87)
2-MeO	2	(87)
3-MeO	0.5	(88)
4-MeO	1	(85)
4-t-Bu	0.5	(92)
4-Ph	0.5	(92)

Substrate	Aryl Compound	Conditions	Product(s) and Yield(s) (%)	Refs.
	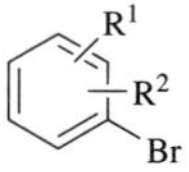 R¹,R²–C₆H₃–Br	Pd(dba)₂ (x mol %), P(t-Bu)₃ (0.2–1.0 mol %), LiNCy₂ (1.3 eq), toluene, rt	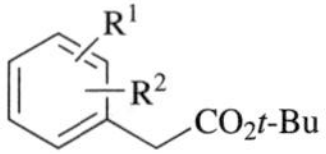 R¹,R²–C₆H₃–CH₂CO₂t-Bu	75

R¹	R²	x	Time (h)	
H	H	0.2	24	(92)
2-F	H	1.0	15	(88)
2-CF₃	H	1.0	15	(82)
3-MeO	H	0.5	24	(94)
3-F	H	1.0	15	(95)
3-CF₃	H	0.5	24	(85)
4-MeO	H	0.5	24	(95)
4-F	H	1.0	15	(95)
4-CF₃	H	1.0	10	(94)
4-t-Bu	H	0.5	8	(87)
4-Ph	H	0.5	15	(96)
3-F	4-Ph	1.0	15	(93)

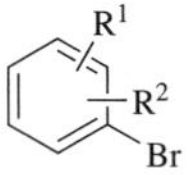

Pd(dba)$_2$ (x mol %), **L47** (0.5–2 mol %), LiHMDS (2.3 eq), toluene, rt

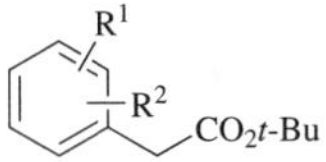

Ar-Br with R^1, R^2 → product with R^1, R^2, CH$_2$CO$_2$t-Bu

R^1	R^2	x	Time (h)	
H	H	1.0	12	(87)
2-MeO	H	2.0	12	(87)
3-MeO	H	0.5	12	(88)
3-F	H	1.0	12	(91)
3-CF$_3$	H	1.0	12	(88)
4-MeO	H	1.0	12	(85)
4-F	H	1.0	12	(90)
4-t-Bu	H	1.0	12	(92)
4-Ph	H	0.5	12	(93)
3-F	4-Ph	0.5	15	(80)
		1.0	12	

75

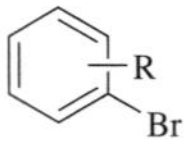

Pd(dba)$_2$ (3 mol %), **L17** (6.3 mol %), LiHMDS (2.5 eq), toluene

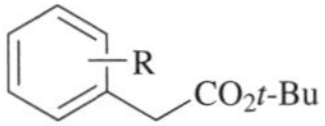

Ar-Br with R → product with R, CH$_2$CO$_2$t-Bu

R	Temp	Time (h)	
2-Me	80°	0.2	(78)
4-F	rt	6	(71)
4-Me	rt	4	(81)

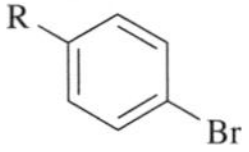

74

1. LiNCy$_2$ (1.3 eq), toluene, rt, 10 min
2. [PdP(t-Bu)$_3$Br]$_2$ (x mol %), rt, 4 h

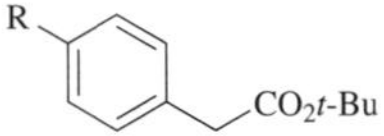

4-R-C$_6$H$_4$-Br → 4-R-C$_6$H$_4$-CH$_2$CO$_2$t-Bu

R	x	
MeO	0.2	(86)
F	0.4	(82)
CF$_3$	0.4	(73)
t-Bu	0.2	(83)

157

TABLE 2. ARYLATION OF ESTERS, AMIDES, LACTONES, LACTAMS, NITRILES, KETENE ACETALS, AND PREFORMED ENOLATES (*Continued*)

Substrate	Aryl Compound	Conditions	Product(s) and Yield(s) (%)	Refs.
C₆				
MeCO₂*t*-Bu	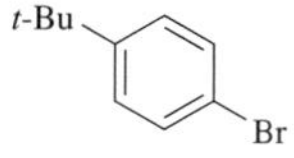 *t*-Bu–C₆H₄–Br	1. LiNCy₂ (1.3 eq), toluene, 0°, 20 min 2. [PdP(*t*-Bu)₃Br]₂ (0.04 mol %), rt, 4 h	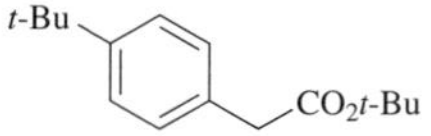 *t*-Bu–C₆H₄–CH₂CO₂*t*-Bu (87)	157
	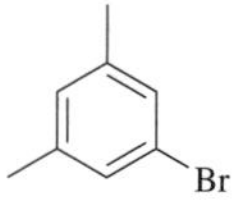 3,5-dimethylphenyl bromide	Pd(dba)₂ (0.2 mol %), P(*t*-Bu)₃ (0.2 mol %), LiNCy₂ (1.3 eq), toluene, rt, 10 h	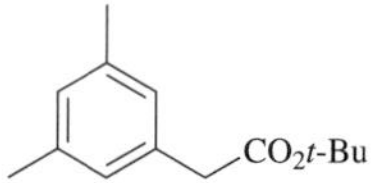 3,5-Me₂C₆H₃–CH₂CO₂*t*-Bu (97)	75
	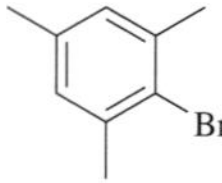 2,4,6-trimethylphenyl bromide	Pd(dba)₂ (0.5 mol %), **L47** (0.5 mol %), LiHMDS (2.3 eq), rt, 12 h	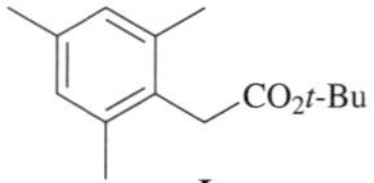 mesityl–CH₂CO₂*t*-Bu (98) **I**	77
	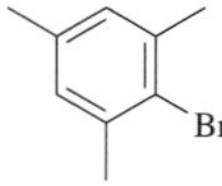 2,4,6-trimethylphenyl bromide	Pd(dba)₂ (0.5 mol %), P(*t*-Bu)₃ (0.5 mol %), LiHMDS (2.3 eq), toluene, rt	**I** (98)	75
	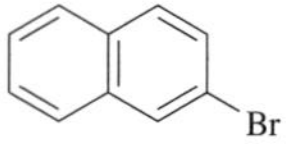 2-bromonaphthalene	Pd(dba)₂ (0.5 mol %), **L47** (0.5 mol %), LiHMDS (2.3 eq), rt, 12 h	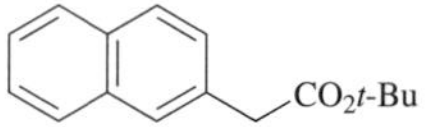 2-Naphthyl–CH₂CO₂*t*-Bu (90)	77
	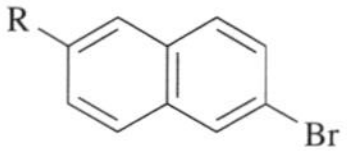 6-R-2-bromonaphthalene	Pd(dba)₂ (3 mol %), **L17** (6.3 mol %), LiHMDS (2.5 eq), toluene, rt	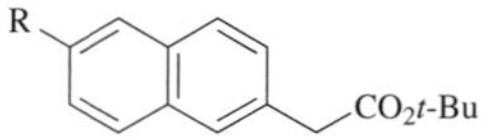 6-R-2-Naphthyl–CH₂CO₂*t*-Bu **I** R / Time (h) H / 1 (84) MeO / 4 (90)	74

Substrate	Conditions	Product	Ref
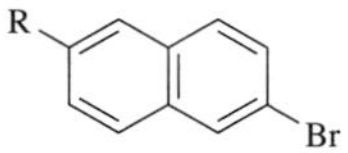 R–[naphthalene]–Br	Pd(dba)$_2$ (x mol %), **L47** (0.2–0.5 mol %), LiNCy$_2$ (1.3 eq), toluene, rt, 15 h	**I** R x H 0.2 (93) MeO 0.5 (96)	75
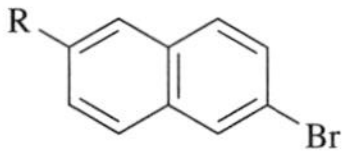 R–[naphthalene]–Br	Pd(dba)$_2$ (x mol %), **L47** (0.5–1.0 mol %), LiHMDS (2.3 eq), toluene, rt, 12 h	**I** R x H 0.5 (90) MeO 1.0 (60)	75
t-Bu–[C$_6$H$_4$]–Br	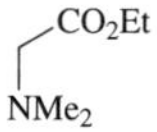 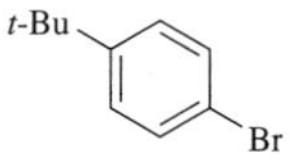 Pd(dba)$_2$ (2 mol %), P(t-Bu)$_3$ (2 mol %), K$_3$PO$_4$ (2.3 eq), 100°, 12 h	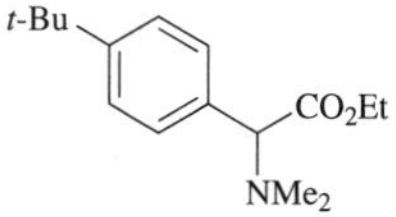 t-Bu–[C$_6$H$_4$]–CH(CO$_2$Et)(NMe$_2$) (84)	77

| R^1–[C$_6$H$_3$(R^2)]–Br | 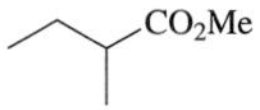 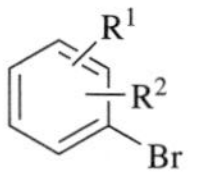 Pd(dba)$_2$ (x mol %), P(t-Bu)$_3$ (0.1–0.2 mol %), LiNCy$_2$ (1.3 eq), toluene, rt | 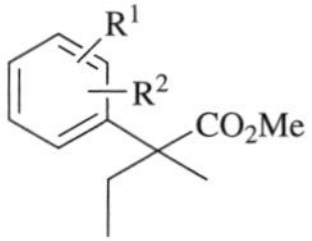 R^1–[C$_6$H$_3$(R^2)]–C(CH$_3$)(Et)(CO$_2$Me) | 75 |

R^1	R^2	x	Time (h)	
3-MeO	H	0.1	16	(96)
3-F	H	0.2	12	(91)
3-Me	H	0.1	18	(87)
3-CF$_3$	H	0.1	10	(76)
4-MeO	H	0.2	10	(99)
4-F	H	0.2	12	(90)
4-Me	H	0.1	14	(81)
4-CF$_3$	H	0.1	10	(83)
4-Ph	H	0.1	10	(78)
3-F	4-Ph	0.1	10	(83)

CH(CO$_2$Et)(NMe$_2$)

CH(CH$_3$)(Et)(CO$_2$Me)

TABLE 2. ARYLATION OF ESTERS, AMIDES, LACTONES, LACTAMS, NITRILES, KETENE ACETALS, AND PREFORMED ENOLATES (*Continued*)

Substrate	Aryl Compound	Conditions	Product(s) and Yield(s) (%)	Refs.
C_6				
MeO₂C-CH(CH₃)-CO₂Me	R-naphthyl-Br	Pd(dba)₂ (x mol %), P(t-Bu)₃ (0.1–0.2 mol %), LiNCy₂ (1.3 eq), toluene, rt R / x / Time (h) H / 0.1 / 12 (88) MeO / 0.2 / 10 (99)	R-naphthyl-C(CH₃)₂CO₂Me	75
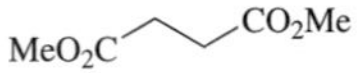 MeO₂C-(CH₂)₂-CO₂Me	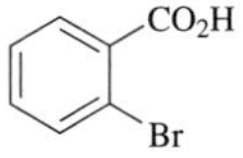 2-Br-C₆H₄-CO₂H	CuBr (cat), NaH (2 eq), 0.5 h	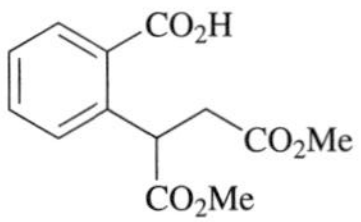 2-(CO₂H)C₆H₄-CH(CO₂Me)CH₂CO₂Me (60)	101
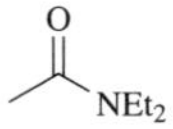 CH₃C(O)NEt₂	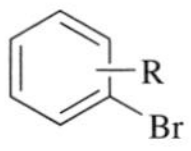 R-C₆H₄-Br	1. s-BuLi (1.2 eq), THF, −78°, 1 h 2. ZnCl₂ (2.4 eq), rt, 10 min 3. Pd(dba)₂ (x mol %), **L12** (x mol %), rt, 24 h	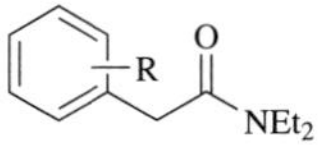 R-C₆H₄-CH₂C(O)NEt₂ R / x 2-F / 3 (92) 2-NC / 2 (84) 4-Me₂N / 3 (90)[c] 4-O₂N / 1 (90) 4-MeO / 1 (93) 4-MeS / 1 (91) 4-Cl / 1 (94) 4-NC / 1 (98) 4-CF₃ / 1 (90) 4-MeO₂C / 1 (97) 4-MeCO / 2 (92) 4-t-Bu / 1 (93) 4-PhCO / 1 (92)	68

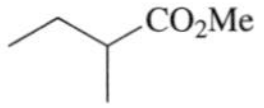

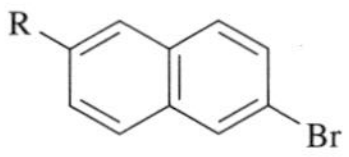

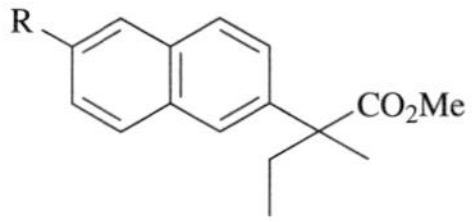

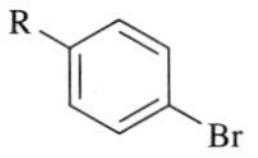

Substrate	Conditions	Product	Yield (%)	Refs.
4-R-C6H4-Br	1. s-BuLi (1.2 eq), THF, −78°, 1 h 2. ZnCl2 (2.4 eq), rt, 10 min 3. [PdP(t-Bu)3Br]2 (0.5 mol %), KH (1.05 eq), rt, 24 h	4-R-C6H4-CH2C(O)NEt2	R = H2N (80) R = HO (91)	68
4-t-Bu-C6H4-Br	1. Base, rt, 1 min 2. ZnCl2, rt, 5 min 3. Pd(dba)2 (1 mol %), **L12** (1 mol %), THF, rt, 4 h	4-t-Bu-C6H4-CH2C(O)NEt2	Base LDA (2 eq): (20) LiNCy2 (1.2 eq): (100)	68
4-t-Bu-C6H4-Br	1. s-BuLi (1.2 eq), THF, −78°, 1 h 2. ZnCl2 (2.4 eq), rt, 10 min 3. Pd(dba)2 (1–5 mol %), ligand (x mol %), 24 h	**I** = 4-t-Bu-C6H4-CH2C(O)NEt2	see table below	68
2-Br-mesitylene (2,6-Me2,4-Me)	1. s-BuLi (1.2 eq), THF, −78°, 1 h 2. ZnCl2 (2.4 eq), rt, 10 min 3. Pd(dba)2 (1 mol %), **L12** (1 mol %), rt, 24 h	2,4,6-Me3-C6H2-CH2C(O)NEt2	(96)	68

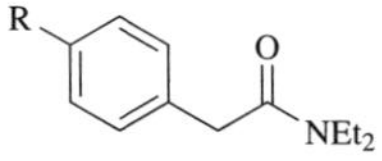

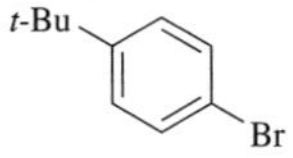

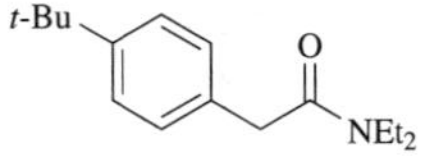

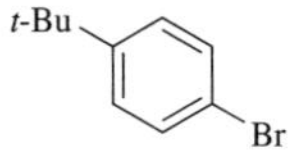

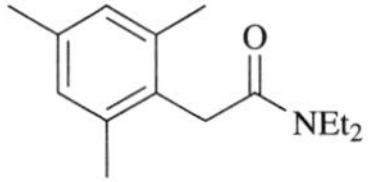

Ligand screen for **I**:

Ligand	x	Temp	(%)
P(t-Bu)3	1	rt	(25)
P(t-Bu)3	2	rt	(34)
PCy3	1	rt	(0)
PCy3	2	rt	(0)
L12	1	rt	(100)
L12	2	rt	(100)
L17	1	rt	(25)
L17	1	70°	(57)
L17	7.5	rt	(40)
L17	7.5	70°	(76)
L23	1	rt	(0)
L23	2	70°	(60)

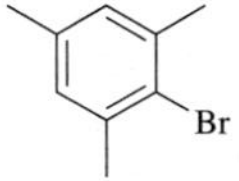

TABLE 2. ARYLATION OF ESTERS, AMIDES, LACTONES, LACTAMS, NITRILES, KETENE ACETALS, AND PREFORMED ENOLATES (*Continued*)

Substrate	Aryl Compound	Conditions	Product(s) and Yield(s) (%)	Refs.
C_6				
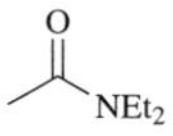 ![O=C(NEt2)CH3]	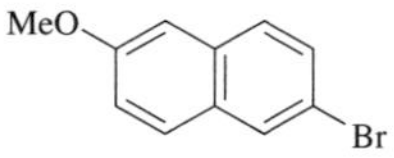 MeO–naphthyl–Br	1. *s*-BuLi (1.2 eq), THF, −78°, 1 h 2. $ZnCl_2$ (2.4 eq), rt, 10 min 3. $Pd(dba)_2$ (1 mol %), **L12** (1 mol %), rt, 24 h	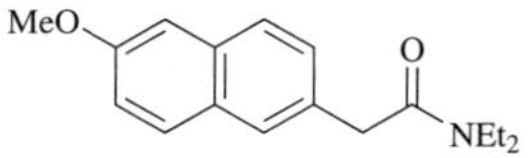 MeO–naphthyl–$CH_2C(O)NEt_2$ (95)	68
	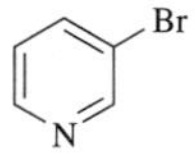 3-Br-pyridine	1. *s*-BuLi (1.2 eq), THF, −78°, 1 h 2. $ZnCl_2$ (2.4 eq), rt, 10 min 3. $Pd(dba)_2$ (2 mol %), **L12** (2 mol %), 70°, 24 h	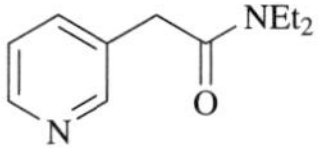 pyridyl–$CH_2C(O)NEt_2$ (91)	68
	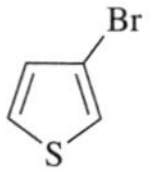 3-Br-thiophene	1. *s*-BuLi (1.2 eq), THF, −78°, 1 h 2. $ZnCl_2$ (2.4 eq), rt, 10 min 3. $Pd(dba)_2$ (2 mol %), **L12** (2 mol %), 70°, 24 h	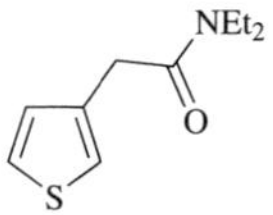 thienyl–$CH_2C(O)NEt_2$ (89)	68
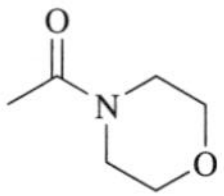 ![O=C(N-morpholine)CH3]	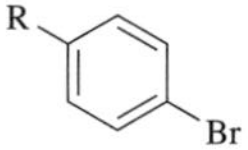 R–C6H4–Br	1. *s*-BuLi (1.2 eq), THF, −78°, 1 h 2. $ZnCl_2$ (2.4 eq), rt, 10 min 3. $Pd(dba)_2$ (x mol %), **L12** (x mol %), rt, 12 h	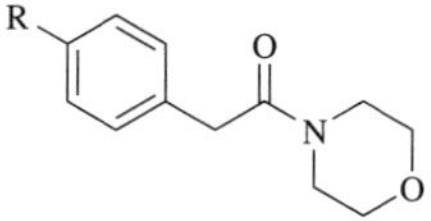 R–C6H4–$CH_2C(O)$-morpholine	68

R	x	
O_2N	2	(89)
MeO	2	(89)
MeS	1	(89)
CF_3S	1	(83)
NC	2	(89)
CF_3	3	(88)
MeO_2C	2	(89)
PhCO	2	(92)

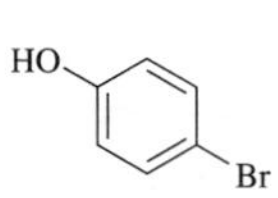

Aryl Bromide	α-Bromo Carbonyl	Conditions	Product (Yield %)	Refs.
4-HO-C6H4-Br	BrCH2C(O)N(morpholinyl)	1. s-BuLi (1.2 eq), THF, −78°, 1 h 2. ZnCl2 (2.4 eq), rt, 10 min 3. [PdP(t-Bu)3Br]2 (1 mol %), KH (1.05 eq), 70°, 12 h	4-HO-C6H4-CH2-C(O)-morpholine (92)	68
2-Br-mesitylene	BrCH2C(O)N(morpholinyl)	1. s-BuLi (1.2 eq), THF, −78°, 1 h 2. ZnCl2 (2.4 eq), rt, 10 min 3. Pd(dba)2 (2 mol %), L12 (2 mol %), rt, 12 h	mesityl-CH2-C(O)-morpholine (90)	68
R-C6H4-Br	BrCH2C(O)NEt2	1. Zn (1.5 eq), THF, rt, 0.5 h 2. Pd(dba)2 (1 mol %), L12 (1 mol %), THF, rt, 6 h	R-C6H4-CH2-C(O)NEt2 R: 2-F (91), 4-O2N (81), 4-NC (87), 4-MeO2C (89), 4-t-Bu (90)	68
2-Br-mesitylene	BrCH2C(O)NEt2	1. Zn (1.5 eq), THF, rt, 0.5 h 2. Pd(dba)2 (1 mol %), L12 (1 mol %), THF, rt, 6 h	mesityl-CH2-C(O)NEt2 (91)	68
R-C6H4-Br	BrC(Me)2C(O)NMe2	1. Zn (1.5 eq), THF, rt, 0.5 h 2. [PdP(t-Bu)3Br]2 (2.5 mol %), THF/toluene (2:8), 12 h	R-C6H4-C(Me)2-C(O)NMe2 R / Temp: 3-F / rt (91); 4-Me2N / 70° (93); 4-MeO / rt (94); 4-Cl / 70° (91); 4-CF3 / 70° (86); 4-t-Bu / rt (94); 4-PhCO / rt (88)	68

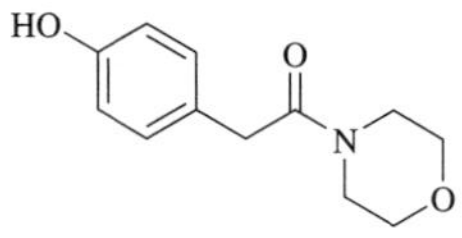

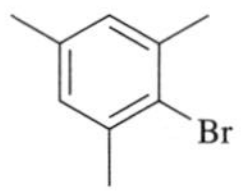

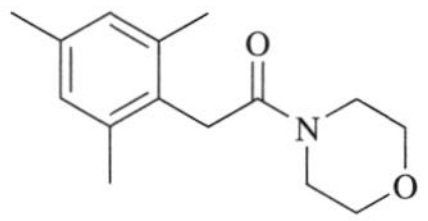

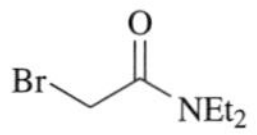

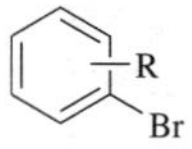

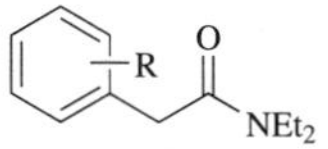

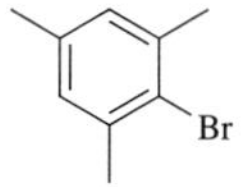

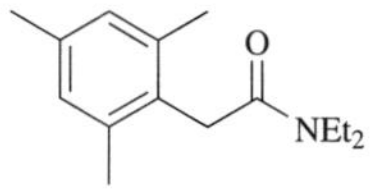

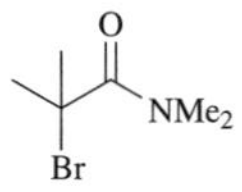

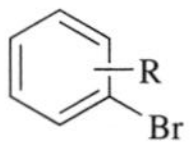

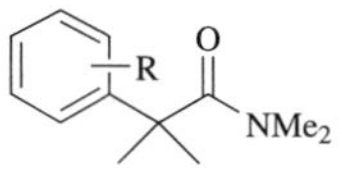

TABLE 2. ARYLATION OF ESTERS, AMIDES, LACTONES, LACTAMS, NITRILES, KETENE ACETALS, AND PREFORMED ENOLATES (*Continued*)

Substrate	Aryl Compound	Conditions	Product(s) and Yield(s) (%)	Refs.
C$_6$				
(2-bromoamide with NMe$_2$)	2-bromopyridine	1. Zn (1.5 eq), THF, rt, 0.5 h 2. [PdP(*t*-Bu)$_3$Br]$_2$ (2.5 mol %), THF/toluene (2:8), 70°, 12 h	(pyridyl-C(Me)$_2$-C(O)NMe$_2$) (94)	68
	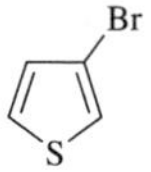 3-bromothiophene	1. Zn (1.5 eq), THF, rt, 0.5 h 2. [PdP(*t*-Bu)$_3$Br]$_2$ (2.5 mol %), THF/toluene (2:8), rt, 12 h	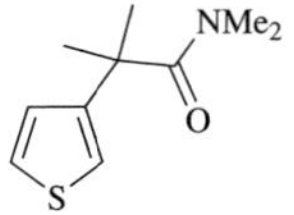 (thienyl-C(Me)$_2$-C(O)NMe$_2$) (82)	68
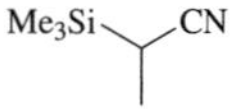 Me$_3$Si-CH(Me)-CN	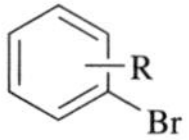 Ar–Br (R-substituted)	Pd$_2$(dba)$_3$ (2 mol %), ligand, ZnF$_2$ (0.5 eq), DMF, 90°	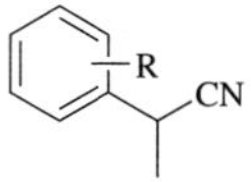 Ar-CH(Me)-CN	80

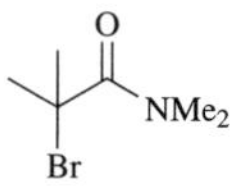

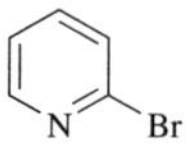

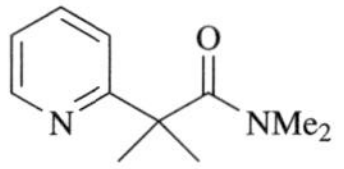

	Ligand	Time (h)	
2-Me	P(*t*-Bu)$_3$ (4 mol %)	24	(71)
2-*i*-Pr	P(*t*-Bu)$_3$ (4 mol %)	24	(60)
2-Cy	P(*t*-Bu)$_3$ (4 mol %)	24	(64)
4-MeO	L23 (2 mol %)	18	(71)
4-NC	L23 (2 mol %)	8	(82)
4-EtO$_2$C	L23 (2 mol %)	8	(84)
4-*t*-Bu	L23 (2 mol %)	18	(77)
4-PhCO	L23 (2 mol %)	8	(87)

| | 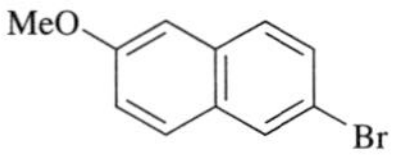 6-bromo-2-methoxynaphthalene | Pd$_2$(dba)$_3$ (2 mol %), L23 (2 mol %), ZnF$_2$ (0.5 eq), DMF, 90°, 18 h | 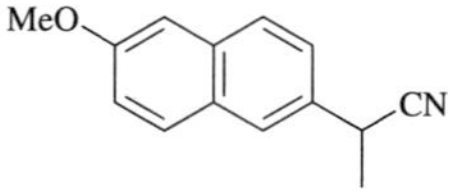 6-methoxy-2-(1-cyanoethyl)naphthalene (98) | 80 |

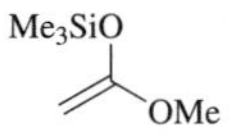

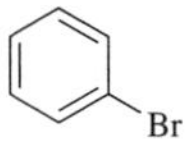

Substrate	Nucleophile	Conditions	Product(s) and Yield(s) (%)	Refs.
PhBr	Me₃SiO-C(=CH₂)-OMe	[η³-C₄H₇Pd(OAc)]₂ (5 mol %), **L1** (20 mol %), TlOAc (1 eq)	Ph-CH₂-CO₂Me (56)	100
4-R-C₆H₄-OTf		[η³-C₄H₇Pd(OAc)]₂ (5 mol %), **L1** (20 mol %), LiOAc (2 eq), THF, reflux, 6 h	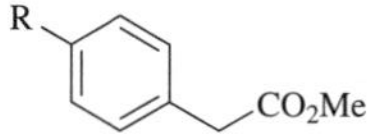 4-R-C₆H₄-CH₂-CO₂Me; R: H (53), MeO (50)	100
6-MeO-naphthyl-2-OTf		[η³-C₄H₇Pd(OAc)]₂ (5 mol %), **L1** (20 mol %), LiOAc (2 eq), THF, reflux, 6 h	6-MeO-naphthyl-2-CH₂-CO₂Me (40)	100
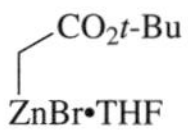 4-R-C₆H₄-Cl	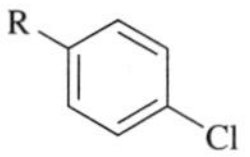 t-BuO₂C-CH₂-ZnBr·THF	Catalyst (x mol %), ligand (1 mol %), THF, 70°, 12 h	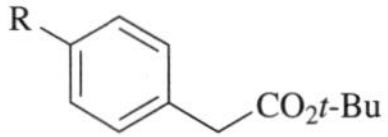 4-R-C₆H₄-CH₂-CO₂t-Bu	156

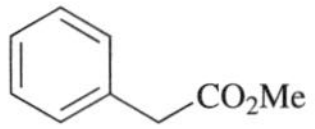

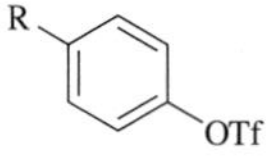

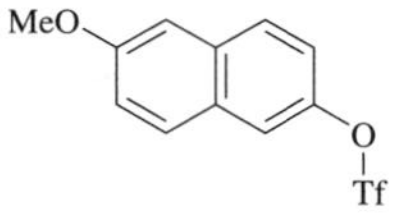

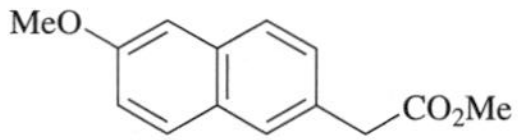

R	Catalyst	x	Ligand	Yield
H	[PdP(t-Bu)₃Br]₂	0.5	—	trace
H	Pd(dba)₂	1	**L12**	(95)
O₂N	Pd(dba)₂	1	**L12**	(83)
NC	Pd(dba)₂	1	**L12**	(86)
MeO₂C	Pd(dba)₂	1	**L12**	(85)

Substrate	Conditions	Product	Yield	Refs.
R-C₆H₄-Br	Pd(dba)₂ (1 mol %), **L12** (1 mol %), THF, rt, 4 h	R-C₆H₄-CH₂-CO₂t-Bu		76
2-O₂N			(87)	
2-NC			(91)	
4-O₂N			(96)	
4-EtCO			(94)	
4-PhCO			(72)[d]	

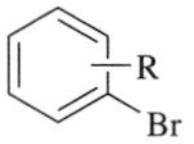

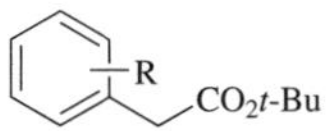

TABLE 2. ARYLATION OF ESTERS, AMIDES, LACTONES, LACTAMS, NITRILES, KETENE ACETALS, AND PREFORMED ENOLATES (*Continued*)

Substrate	Aryl Compound	Conditions	Product(s) and Yield(s) (%)	Refs.
C_6 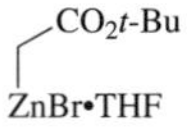 CO$_2$*t*-Bu ZnBr·THF	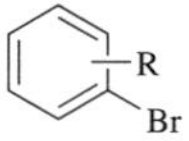 Ar-Br (R)	Pd(dba)$_2$ (1–2 mol %), **L12**, THF, rt	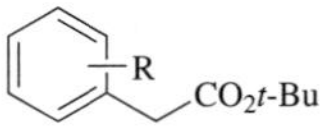 Ar-CH$_2$CO$_2$*t*-Bu R: 2-O$_2$N (87); 2-HO (83); 2-NC (91); 4-H$_2$N (80); 4-O$_2$N (96); 4-HO (91); 4-PhCO (75)	22
	Ar-Br (R)	[PdP(*t*-Bu)$_3$Br]$_2$ (0.5 mol %), KH (1.05 eq), THF, rt R / Time (h): 2-HO / 4; 4-H$_2$N / 24; 4-HO / 4	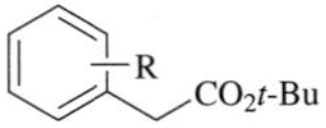 Ar-CH$_2$CO$_2$*t*-Bu (83), (85), (91)	76
	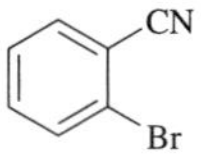 2-CN-C$_6$H$_4$-Br	[PdP(*t*-Bu)$_3$Br]$_2$ (0.5 mol %), THF, rt, 4 h	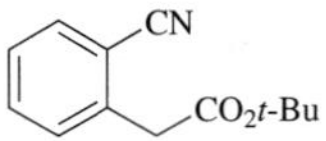 2-CN-C$_6$H$_4$-CH$_2$CO$_2$*t*-Bu (81)	76
	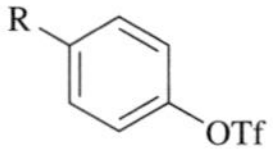 4-R-C$_6$H$_4$-OTf	PdCl$_2$(PPh$_3$)$_2$ (20 mol %), DIBALH, THF/HMPA (2:1), overnight	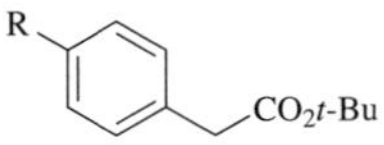 4-R-C$_6$H$_4$-CH$_2$CO$_2$*t*-Bu R: MeS (50); Cl (55)	22
	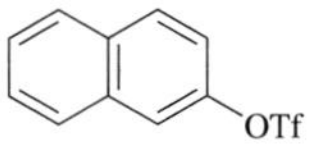 2-naphthyl-OTf	PdCl$_2$(PPh$_3$)$_2$ (20 mol %), DIBALH, THF/HMPA (2:1), overnight	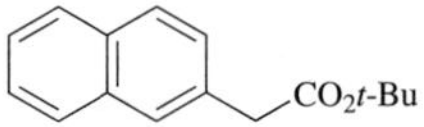 2-naphthyl-CH$_2$CO$_2$*t*-Bu (60)	158

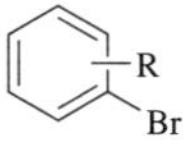

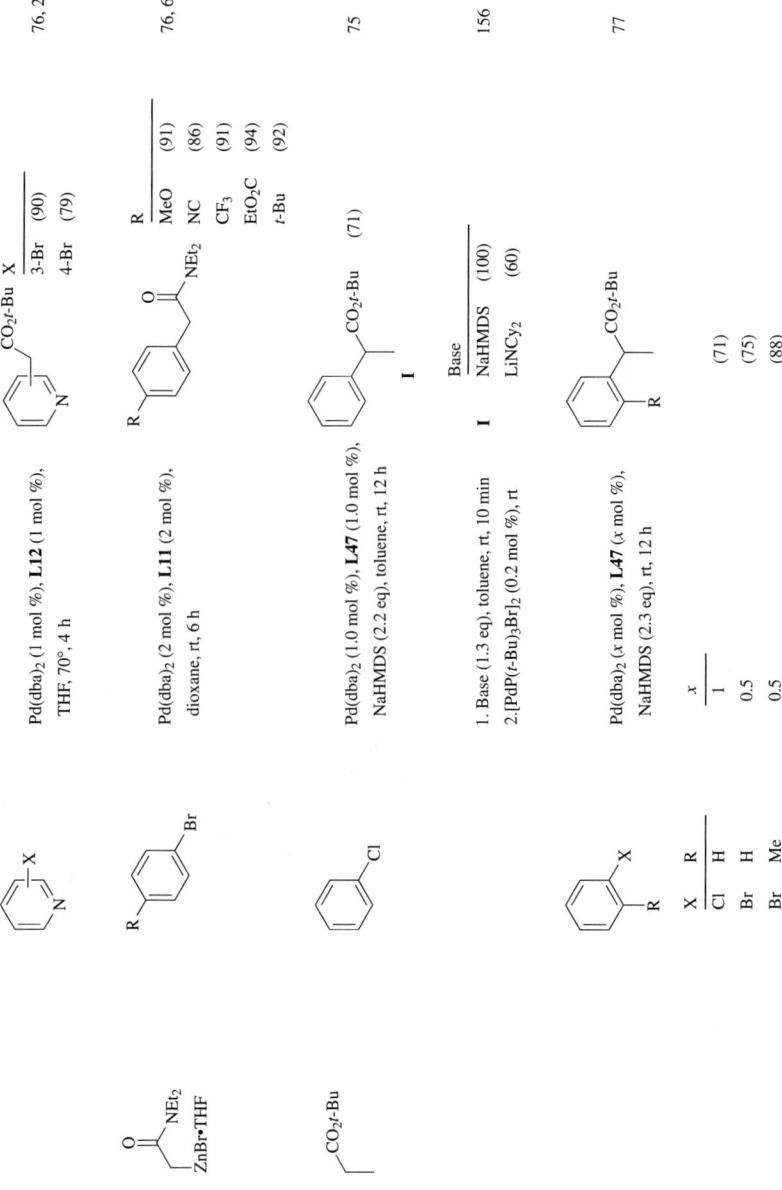

TABLE 2. ARYLATION OF ESTERS, AMIDES, LACTONES, LACTAMS, NITRILES, KETENE ACETALS, AND PREFORMED ENOLATES (*Continued*)

Substrate	Aryl Compound	Conditions	Product(s) and Yield(s) (%)	Refs.
C_7 CO$_2$*t*-Bu	R—⟨⟩—Cl	1. NaHMDS (1.2 eq), toluene, rt, 10 min 2. Catalyst (x mol %), ligand (y mol %), 4 h	R—⟨⟩—CH(CH$_3$)CO$_2$*t*-Bu	156

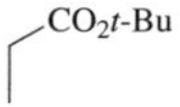

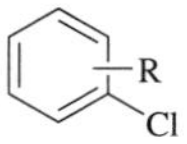

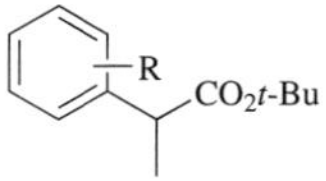

R	Catalyst	x	Ligand	y	Temp	
H	Pd(dba)$_2$	0.4	P(*t*-Bu)$_3$	0.4	rt	(91)
H	[PdP(*t*-Bu)$_3$Br]$_2$	0.2	none	—	rt	(94)
2-MeO	Pd(dba)$_2$	1	P(*t*-Bu)$_3$	1	rt	(62)
2-MeO	[PdP(*t*-Bu)$_3$Br]$_2$	0.4	none	—	rt	(77)
3-MeO	Pd(dba)$_2$	1	P(*t*-Bu)$_3$	1	rt	(41)
3-MeO	[PdP(*t*-Bu)$_3$Br]$_2$	0.4	none	—	rt	(75)
4-MeO	Pd(dba)$_2$	1	P(*t*-Bu)$_3$	1	rt	(69)
4-MeO	[PdP(*t*-Bu)$_3$Br]$_2$	0.4	none	—	rt	(88)
4-F	[PdP(*t*-Bu)$_3$Br]$_2$	0.15	none	—	60°	(73)
4-Br	Pd(dba)$_2$	0.3	P(*t*-Bu)$_3$	0.3	60°	(84)
4-Br	[PdP(*t*-Bu)$_3$Br]$_2$	0.15	none	—	60°	(81)
2,6-Me$_2$	Pd(dba)$_2$	0.4	P(*t*-Bu)$_3$	0.4	60°	(42)
2,6-Me$_2$	[PdP(*t*-Bu)$_3$Br]$_2$	0.2	none	—	60°	(81)

| | 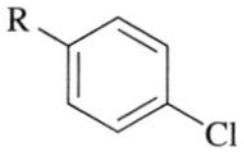 R—⟨⟩—Cl | Pd$_2$(dba)$_3$ (1.5 mol %), **L14** (6.3 mol %), LiHMDS (2.5 eq), toluene, 80° | 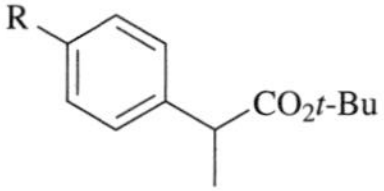 R—⟨⟩—CH(CH$_3$)CO$_2$*t*-Bu | 74 |

R	Time (h)	
MeO	5	(56)
PhO	3	(54)

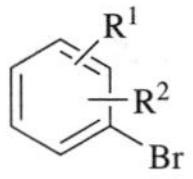

Pd(dba)$_2$ (x mol %),
L47 (0.5–5 mol %),
NaHMDS (2.2 eq), toluene, rt, 12 h

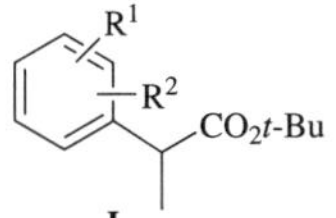

Ar–Br (with R^1, R^2) → **I** (Ar-CH(Me)-CO$_2$t-Bu)

R^1	R^2	x	
H	H	0.5	(75)
2-Me	H	1.0	(88)
4-MeO	H	1.0	(82)
4-CF$_3$	H	5.0	(74)
3-F	4-Ph	2.0	(66)

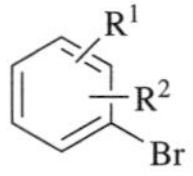

75

Pd(OAc)$_2$ (3.0 mol %),
L17 (6.3 mol %),
LiHMDS (2.5 eq), toluene, 80°

Ar–Br → **I**

R^1	R^2	Time (h)	
2-Me	H	2	(82)
2-H$_2$C=CH	H	2	(77)
2-i-Pr	H	2	(88)
3-CF$_3$	H	1	(81)
4-PhO	H	2	(71)
4-Ph	H	0.5	(86)
2-Me	6-Me	2	(68)
3-F	4-Ph	0.3	(86)

74

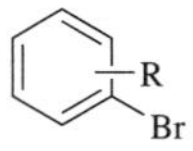

1. LiNCy$_2$ (1.3 eq), toluene, rt, 10 min
2. [PdP(t-Bu)$_3$Br]$_2$ (x mol %), rt, 4 h

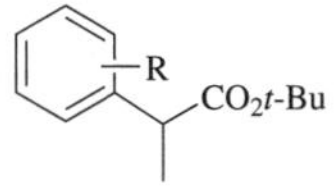

Ar–Br → Ar-CH(Me)-CO$_2$t-Bu

R	x	
2-MeO	0.25	(87)
3-MeO	0.25	(84)
3-F	0.2	(90)
4-MeO	0.25	(87)
4-F	0.2	(88)
4-Cl	0.2	(83)

157

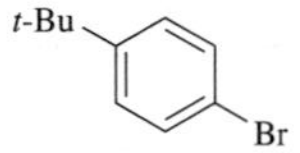

1. LiNCy$_2$ (1.3 eq), toluene, 0°, 20 min
2. [PdP(t-Bu)$_3$Br]$_2$ (0.2 mol %), rt, 4 h

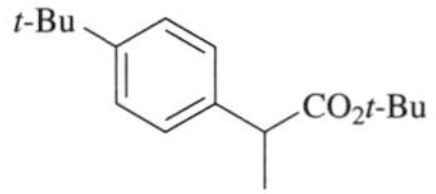

t-Bu-C$_6$H$_4$-Br → t-Bu-C$_6$H$_4$-CH(Me)-CO$_2$t-Bu (79)

157

TABLE 2. ARYLATION OF ESTERS, AMIDES, LACTONES, LACTAMS, NITRILES, KETENE ACETALS, AND PREFORMED ENOLATES (*Continued*)

Substrate	Aryl Compound	Conditions	Product(s) and Yield(s) (%)	Refs.
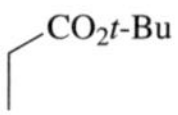 C₇ CO_2t-Bu	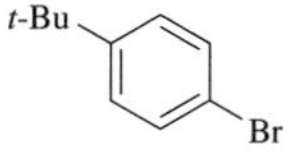 t-Bu—⟨Ar⟩—Br	1. Base (1.3 eq), toluene, rt, 10 min 2. Catalyst (x mol %), ligand (y mol %), rt, 4 h	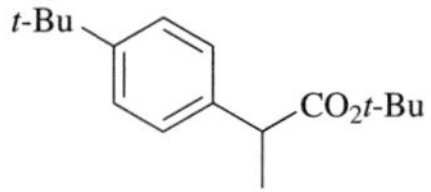 t-Bu—⟨Ar⟩—CH(CH₃)—CO_2t-Bu	157
		Base / Catalyst / x / Ligand / y		
		NaHMDS / [PdP(t-Bu)₃Br]₂ / 0.25 / none / — (100)		
		LiHMDS / [PdP(t-Bu)₃Br]₂ / 0.25 / none / — (20)		
		LiNCy₂ / [PdP(t-Bu)₃Br]₂ / 0.25 / none / — (100)		
		LiNCy₂ / [PdP(t-Bu)₃Br]₂ / 0.05 / none / — (80)		
		LiNCy₂ / Pd(dba)₂ / 0.1 / P(t-Bu)₃ / 0.1 (75)		
		LiNCy₂ / Pd(dba)₂ / 0.25 / P(t-Bu)₃ / 0.25 (91)		
		LiNCy₂ / Pd(dba)₂ / 0.25 / [HP(t-Bu)₃]BF₄ / 0.25 (60)		
	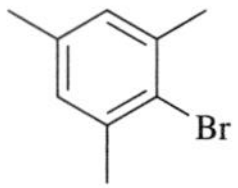 ⟨mesityl⟩—Br	Pd(dba)₂ (1 mol %), P(t-Bu)₃ (1 mol %), NaHMDS (2.3 eq), rt, 12 h	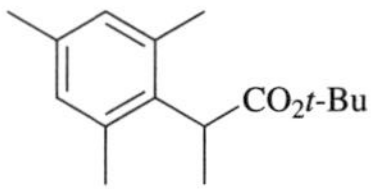 ⟨mesityl⟩—CH(CH₃)—CO_2t-Bu **I** (74)	77
		Pd(dba)₂ (1 mol %), **L47** (1 mol %), NaHMDS (2.2 eq), toluene, rt, 12 h	**I** (74)	75
		1. LiNCy₂ (1.3 eq), toluene, rt, 10 min 2. [PdP(t-Bu)₃Br]₂ (0.05 mol %), rt, 4 h	**I** (72)	157
	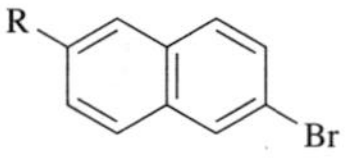 R—⟨naphthyl⟩—Br	Pd(OAc)₂ (3 mol %), **L17** (6.3 mol %), LiHMDS (2.5 eq), toluene	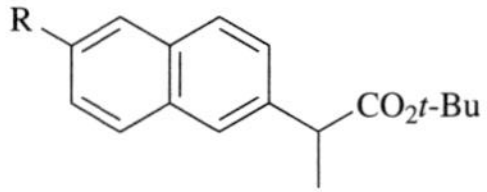 R—⟨naphthyl⟩—CH(CH₃)—CO_2t-Bu	74
		R / Temp / Time (h)		
		H / rt / 17 (84)		
		H / 80° / 0.25 (92)		
		MeO / rt / 15 (79)		
		MeO / 80° / 0.5 (74)		

ArBr	Nucleophile	Conditions	Product (%)	Refs.
6-bromo-2-methoxynaphthalene	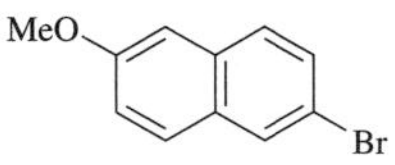 i-Pr-CH2-CO2t-Bu	Pd(dba)$_2$ (1 mol %), **L47** (1 mol %), NaHMDS (2.2 eq), toluene, rt, 12 h	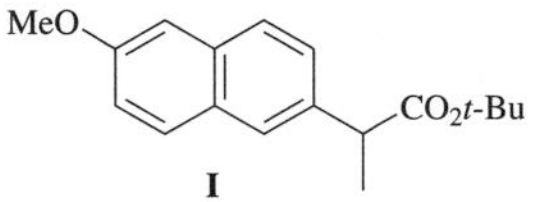 2-(6-methoxynaphthalen-2-yl)-CH(Me)-CO$_2$t-Bu **I** (83)	75
"	"	Pd(dba)$_2$ (1 mol %), **L47** (1 mol %), NaHMDS (2.3 eq), rt, 12 h	**I** (83)	77
"	"	1. LiNCy$_2$ (1.3 eq), toluene, rt, 10 min 2. [PdP(t-Bu)$_3$Br]$_2$ (0.2 mol %), rt, 4 h	**I** (75)	157
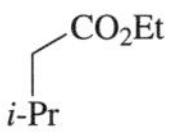 4-t-Bu-C$_6$H$_4$Br	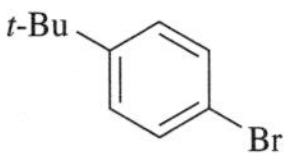 sec-Bu-CO$_2$Et (with i-Pr branch)	Pd(dba)$_2$ (5 mol %), **L47** (5 mol %), NaHMDS (2.3 eq), rt, 12 h	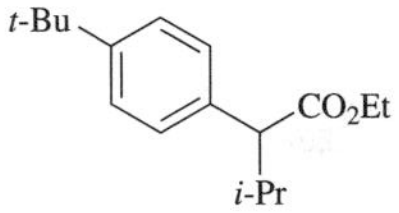 4-t-Bu-C$_6$H$_4$-CH(i-Pr)-CO$_2$Et (95)	77
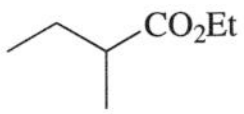 4-Me-C$_6$H$_4$Br	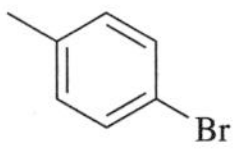 sec-Bu-CO$_2$Et	Pd(OAc)$_2$ (3.0 mol %), **L26** (6.3 mol %), LiHMDS (2.5 eq), toluene, 80°, 0.5 h	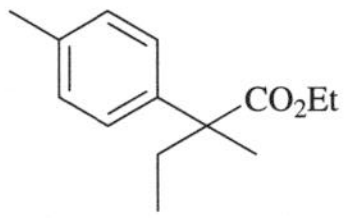 4-Me-C$_6$H$_4$-C(Me)(Et)-CO$_2$Et (48)	74
2-bromonaphthalene	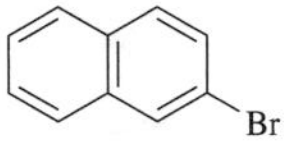 sec-Bu-CO$_2$Et	Pd(OAc)$_2$ (3.0 mol %), **L26** (6.3 mol %), LiHMDS (2.5 eq), toluene, 40°, 17 h	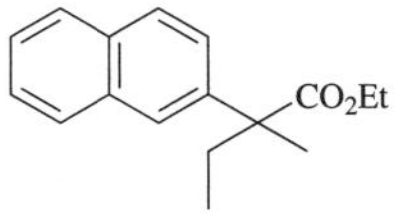 2-naphthyl-C(Me)(Et)-CO$_2$Et (54)	74
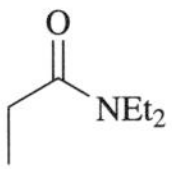 R-C$_6$H$_4$Br	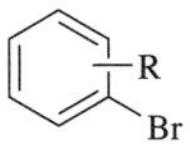 CH$_3$CH$_2$C(O)NEt$_2$	1. s-BuLi (1.2 eq), THF, −78°, 1 h 2. ZnCl$_2$ (2.4 eq), rt, 10 min 3. Pd(dba)$_2$ (x mol %), **L12** (x mol %), rt, 24 h	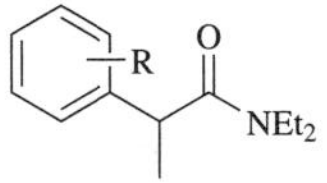 R-C$_6$H$_4$-CH(Me)-C(O)NEt$_2$	68

R	x	(%)
2-MeO	3	(88)
2-F	3	(86)
2-Me	3	(98)
4-O$_2$N	3	(87)
4-MeO	1	(94)
4-MeS	1	(91)
4-Cl	1	(87)
4-NC	3	(91)
4-CF$_3$	1	(90)

TABLE 2. ARYLATION OF ESTERS, AMIDES, LACTONES, LACTAMS, NITRILES, KETENE ACETALS, AND PREFORMED ENOLATES (*Continued*)

Substrate	Aryl Compound	Conditions	Product(s) and Yield(s) (%)	Refs.
C₇ O=C(NEt₂)CH₂CH₃ 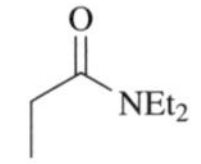	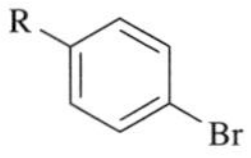 R–C₆H₄–Br	1. *s*-BuLi (1.2 eq), THF, −78°, 1 h 2. ZnCl₂ (2.4 eq), rt, 10 min 3. [PdP(*t*-Bu)₃Br]₂ (1 mol %), KH (1.05 eq), rt, 24 h	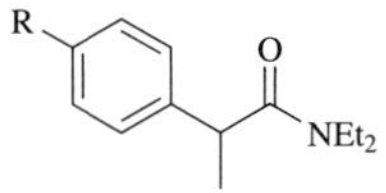 R-C₆H₄-CH(CH₃)C(O)NEt₂ R: H₂N (90), HO (95)	68
	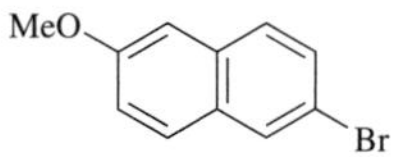 MeO-naphthyl-Br	1. *s*-BuLi (1.2 eq), THF, −78°, 1 h 2. ZnCl₂ (2.4 eq), rt, 10 min 3. Pd(dba)₂ (2 mol %), **L12** (2 mol %), rt, 24 h	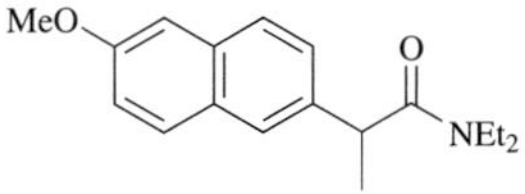 MeO-naphthyl-CH(CH₃)C(O)NEt₂ (92)	68
	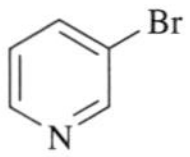 3-Br-pyridine	1. *s*-BuLi (1.2 eq), THF, −78°, 1 h 2. ZnCl₂ (2.4 eq), rt, 10 min 3. Pd(dba)₂ (3 mol %), **L12** (3 mol %), 70°, 24 h	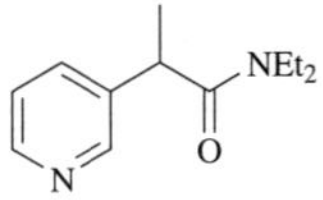 pyridin-3-yl-CH(CH₃)C(O)NEt₂ (86)	68
	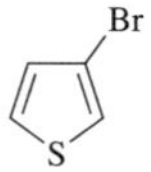 3-Br-thiophene	1. *s*-BuLi (1.2 eq), THF, −78°, 1 h 2. ZnCl₂ (2.4 eq), rt, 10 min 3. Pd(dba)₂ (1 mol %), **L12** (1 mol %), 70°, 24 h	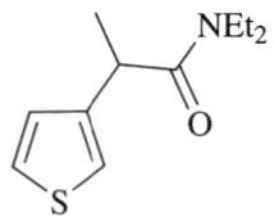 thien-3-yl-CH(CH₃)C(O)NEt₂ (92)	68
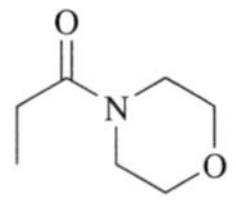 O=C(morpholine)CH₂CH₃	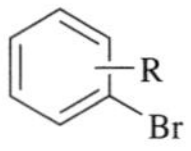 R–C₆H₄–Br	1. *s*-BuLi (1.2 eq), THF, −78°, 1 h 2. ZnCl₂ (2.4 eq), rt, 10 min 3. Pd(dba)₂ (4 mol %), **L12** (4 mol %), rt, 24 h	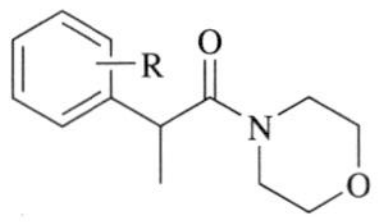 R-C₆H₄-CH(CH₃)C(O)(morpholine) R: 2-F (69) 2-Me (79) 4-MeO (91) 4-Cl (89) 4-CF₃ (82) 4-EtO₂C (85) 4-PhCO (65)	68

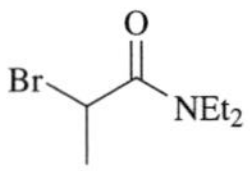

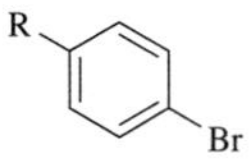

Substrate	Electrophile	Conditions	Product(s) and Yield(s) (%)	Refs.
Br-CH(CH3)-C(O)NEt2	4-R-C6H4-Br (R = Me2N, MeO2C)	1. Zn (1.5 eq), THF, rt, 0.5 h 2. Pd(dba)2 (1 mol %), L12 (1 mol %), THF, 6 h	R-C6H4-CH(CH3)-C(O)NEt2 Temp / 70° / rt (89) / (87)	68
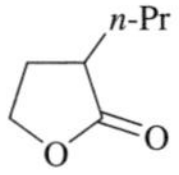 n-Pr-substituted γ-butyrolactone	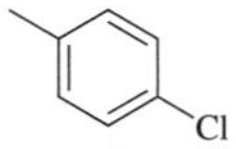 4-Cl-C6H4-Me	Ni(cod)2 (5 mol %), (S)-L23 (8.5 mol %), ZnBr2 (15 mol %), NaHMDS (2.3 eq), toluene/THF (3:1), 60°, 17–20 h	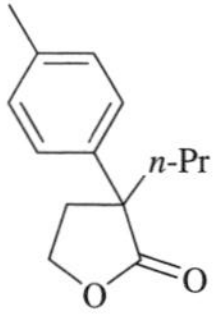 n-Pr, p-tolyl γ-butyrolactone (84), 98% ee	33
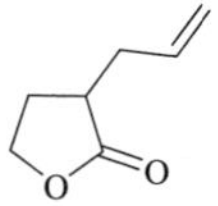 allyl-substituted γ-butyrolactone	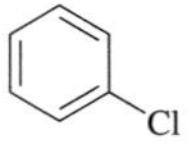 C6H5-Cl	Ni(cod)2 (5 mol %), (S)-L23 (8.5 mol %), ZnBr2 (15 mol %), NaHMDS (2.3 eq), toluene/THF (3:1), 60°, 17–20 h	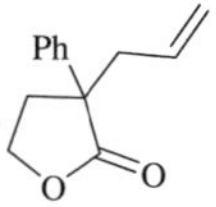 Ph, allyl γ-butyrolactone (25), 83% ee	33
allyl-substituted γ-butyrolactone	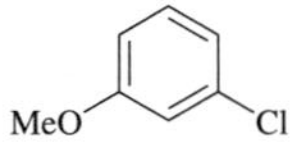 3-MeO-C6H4-Cl	Ni(cod)2 (5 mol %), (S)-L23 (8.5 mol %), ZnBr2 (15 mol %), NaHMDS (2.3 eq), toluene/THF (3:1), 60°, 17–20 h	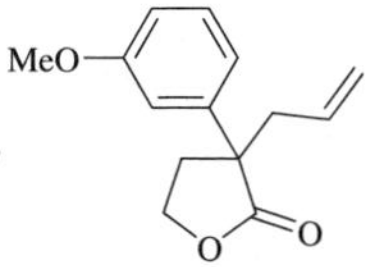 3-MeO-C6H4, allyl γ-butyrolactone (56), 95% ee	33
allyl-substituted γ-butyrolactone	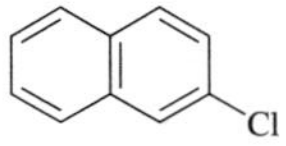 2-chloronaphthalene	Ni(cod)2 (5 mol %), (S)-L23 (8.5 mol %), ZnBr2 (15 mol %), NaHMDS (2.3 eq), toluene/THF (3:1), 60°, 17–20 h	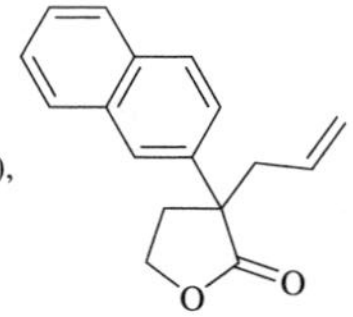 2-naphthyl, allyl γ-butyrolactone (80), 89% ee	33

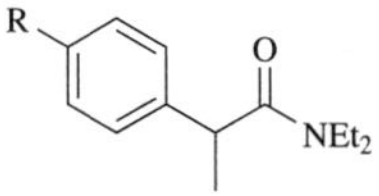

TABLE 2. ARYLATION OF ESTERS, AMIDES, LACTONES, LACTAMS, NITRILES, KETENE ACETALS, AND PREFORMED ENOLATES (*Continued*)

Substrate	Aryl Compound	Conditions	Product(s) and Yield(s) (%)	Refs.
C₇ NC-cyclohexyl	R¹,R²-Br-aryl; R¹ / R²: 2-Me / H; 4-Me₂N / H; 4-MeO / H; 4-NC / H; 4-EtO₂C / H; 4-t-Bu / H; 4-PhCO / H; 2-Me / 5-t-Bu	Pd(OAc)₂ (2 mol %), P(t-Bu)₃ (4 mol %), ZnCl₂ (1.2 eq), THF, rt	R¹,R²-aryl-C(CN)(cyclohexyl) (87), (82), (72), (71), (63), (85), (75), (91)	80
	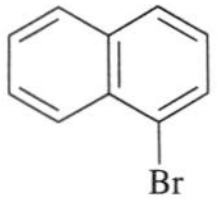 1-Br-naphthyl	Pd(OAc)₂ (2 mol %), P(t-Bu)₃ (4 mol %), ZnCl₂ (1.2 eq), THF, rt	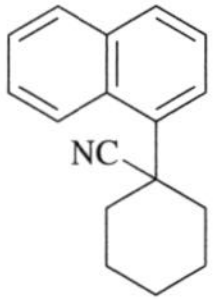 1-naphthyl-C(CN)(cyclohexyl) (91)	80
	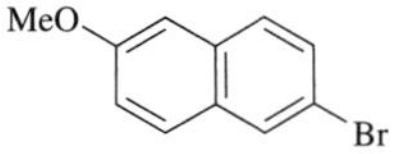 6-MeO-2-Br-naphthyl	Pd(OAc)₂ (2 mol %), P(t-Bu)₃ (4 mol %), ZnCl₂ (1.2 eq), THF, rt	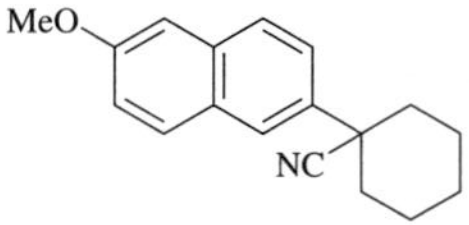 6-MeO-2-naphthyl-C(CN)(cyclohexyl) (81)	80
Me₃SiO-C(OMe)=CHMe	R-C₆H₄-Br	[η³-C₄H₇-Pd(OAc)]₂ (5 mol %), L1 (20 mol %), TlOAc (1 eq)	R-C₆H₄-CH(Me)CO₂Me; R = H (70), O₂N (40), MeO (70)	100

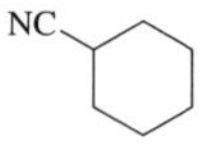

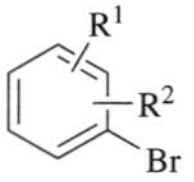

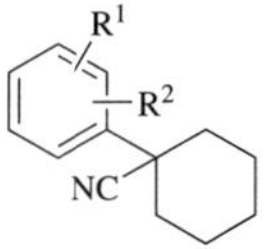

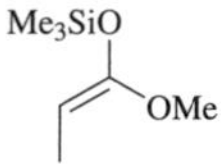

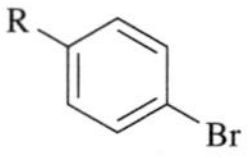

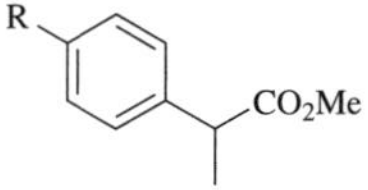

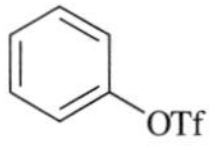

[η³-C₄H₇Pd(OAc)]₂ (x mol %), ligand (y mol %), LiOAc (2 eq), THF, reflux

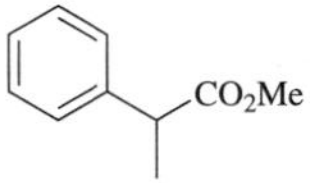

PhOTf → PhCH(Me)CO₂Me (100)

x	Ligand	y	Time (h)	
2	PPh₃	6	6	(15)
2	dppe	4	6	trace
2	dppp	4	12	(10)
2	dppb	4	6	(24)
2	L1	2	6	(13)
2	L1	4	6	(73)
1	L1	4	6	(72)
0.5	L1	4	6	(70)

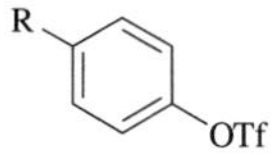

[η³-C₄H₇Pd(OAc)]₂ (x mol %), L1 (y mol %), LiOAc (2 eq), reflux, 6 h

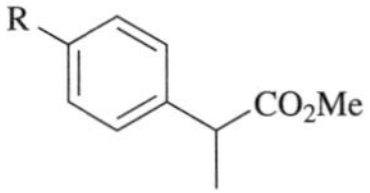

4-R-C₆H₄-OTf → 4-R-C₆H₄-CH(Me)CO₂Me (100)

x	y	Solvent	R = O₂N	R = MeO
2	8	DME	(25)	
5	20	THF		(30)

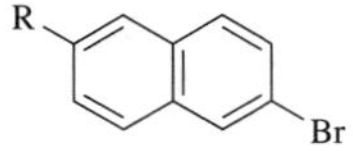

[η³-C₄H₇Pd(OAc)]₂ (5 mol %), L1 (20 mol %), TlOAc (1 eq)

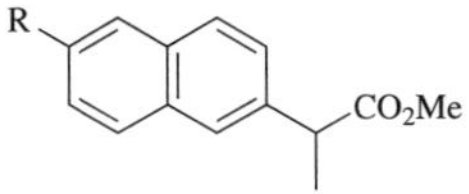

6-R-naphthalen-2-yl-Br → 6-R-naphthalen-2-yl-CH(Me)CO₂Me (100)

R	
H	(80)
MeO	(80)

TABLE 2. ARYLATION OF ESTERS, AMIDES, LACTONES, LACTAMS, NITRILES, KETENE ACETALS, AND PREFORMED ENOLATES (*Continued*)

Substrate	Aryl Compound	Conditions	Product(s) and Yield(s) (%)	Refs.
C_7				
Me$_3$SiO–C(OMe)=CH– (ketene acetal)	R–naphthyl–OTf	[η^3-C$_4$H$_7$Pd(OAc)]$_2$ (x mol %), **L1** (y mol %), LiOAc (2 eq), reflux, 6 h	R–naphthyl–CH(CH$_3$)CO$_2$Me	100
		<table><tr><td>R</td><td>x</td><td>y</td><td>Solvent</td></tr><tr><td>H</td><td>2</td><td>8</td><td>THF</td></tr><tr><td>H</td><td>2</td><td>8</td><td>DME</td></tr><tr><td>MeO</td><td>5</td><td>20</td><td>THF</td></tr></table>	trace (70) (40)	
THF•BrZn–CH(CH$_3$)CO$_2$t-Bu	R–C$_6$H$_4$–Br	Pd(dba)$_2$ (1–2 mol %), **L12**, THF	R–C$_6$H$_4$–CH(CH$_3$)CO$_2$t-Bu **I**	22
		<table><tr><td>R</td><td>Temp</td></tr><tr><td>2-HO</td><td>rt</td></tr><tr><td>2-NC</td><td>rt</td></tr><tr><td>4-H$_2$N</td><td>rt</td></tr><tr><td>4-MeO$_2$C</td><td>rt</td></tr><tr><td>4-PhCO</td><td>70°</td></tr></table>	(95) (85) (67) (87) (89)	
	R–C$_6$H$_4$–Br	Pd(dba)$_2$ (1 mol %), **L12** (1 mol %), THF, 4 h	**I** <table><tr><td>R</td><td colspan="2">Temp</td></tr><tr><td>2-NC</td><td>rt</td><td>(85)^e</td></tr><tr><td>4-PhCO</td><td>70°</td><td>(89)</td></tr><tr><td>4-MeO$_2$C</td><td>rt</td><td>(87)</td></tr></table>	76
	R–C$_6$H$_4$–Br	[PdP(t-Bu)$_3$Br]$_2$ (0.5 mol %), THF, rt	**I** <table><tr><td>R</td><td colspan="2">Time (h)</td></tr><tr><td>4-H$_2$N</td><td>12</td><td>(70)</td></tr><tr><td>4-MeO$_2$C</td><td>4</td><td>(81)</td></tr></table>	76

Substrate	Conditions	Product	Yield (%)
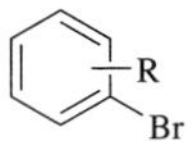 Ar-Br (R)	[PdP(t-Bu)₃Br]₂ (0.5 mol %), KH (1.05 eq), THF, rt	I; R: 4-H₂N (24 h, 66); 4-HO (4 h, 95)	76
2-naphthyl-OTf	PdCl₂, **L1** (15 mol %), THF, rt, overnight	2-naphthyl-CH(Me)CO₂t-Bu (18)	159
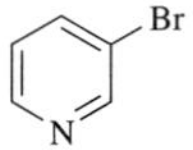 3-Br-pyridine	Pd(dba)₂ (1–2 mol %), **L12**, THF, rt	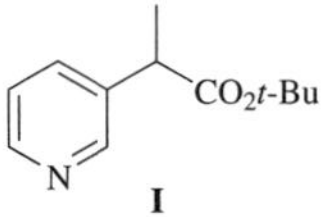 3-pyridyl-CH(Me)CO₂t-Bu (91)	22
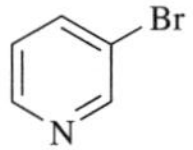 3-Br-pyridine	Pd(dba)₂ (1 mol %), **L12** (1 mol %), THF, 70°, 4 h	I (91)	76
Ar-Br (R)	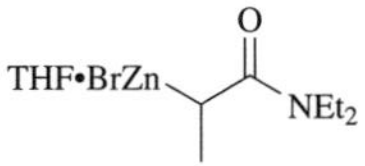 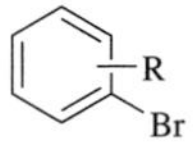 Pd(dba)₂ (2 mol %), **L12** (2 mol %), dioxane, rt, 6 h	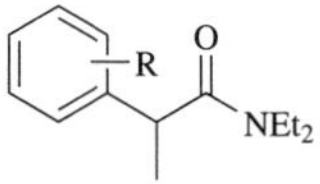 Ar-CH(Me)C(O)NEt₂; R: 3-MeO (88); 4-O₂N (97); 4-EtO₂C (95); 4-CF₃ (88); 4-t-Bu (88)	68, 76
2-Br-naphthalene	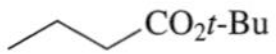 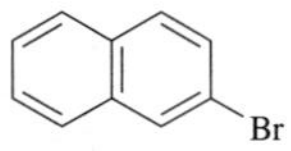 Pd(OAc)₂ (3.0 mol %), **L17** (6.3 mol %), LiHMDS (2.5 eq), toluene, 80°, 2 h	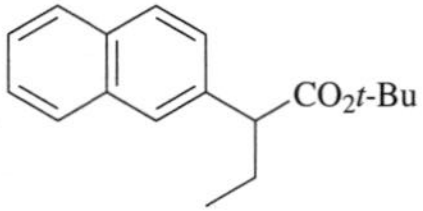 2-naphthyl-CH(Et)CO₂t-Bu (81)	74

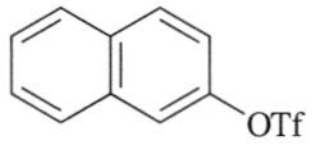

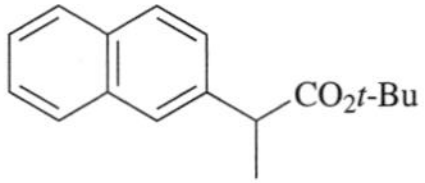

C₈

THF·BrZn–CH(Me)C(O)NEt₂

CH₃CH₂CH₂–CO₂t-Bu

TABLE 2. ARYLATION OF ESTERS, AMIDES, LACTONES, LACTAMS, NITRILES, KETENE ACETALS, AND PREFORMED ENOLATES (*Continued*)

Substrate	Aryl Compound	Conditions	Product(s) and Yield(s) (%)	Refs.
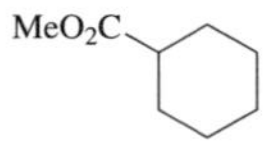 C8 MeO2C-cyclohexyl	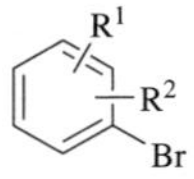 R1,R2-substituted aryl bromide	Pd(dba)2 (x mol %), P(t-Bu)3 (0.1–0.5 mol %), LiNCy2 (1.3 eq), toluene, rt	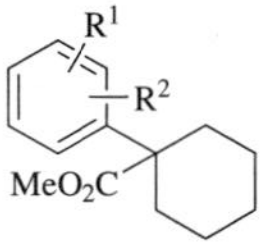 MeO2C-cyclohexyl-aryl(R1,R2)	75

R1	R2	x	Time (h)	Yield
3-MeO	H	0.1	20	(94)
3-F	H	0.1	24	(94)
3-Me	H	0.2	24	(92)
3-CF3	H	0.5	8	(90)
4-MeO	H	0.2	12	(94)
4-F	H	0.1	24	(82)
4-Me	H	0.2	24	(95)
4-CF3	H	0.5	10	(78)
4-Ph	H	0.5	18	(86)
3-F	4-Ph	0.5	18	(90)

Substrate	Aryl Compound	Conditions	Product	Refs.
MeO2C-cyclohexyl	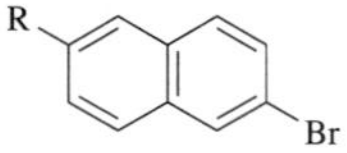 R-naphthyl-Br (2,6)	Pd(dba)2 (0.1 mol %), P(t-Bu)3 (0.1 mol %), LiNCy2 (1.3 eq), toluene, rt	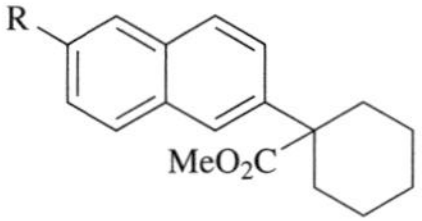 R-naphthyl-C(cyclohexyl)-CO2Me	75

R	Time (h)	Yield
H	18	(91)
MeO	10	(97)

Substrate	Aryl Compound	Conditions	Product	Refs.
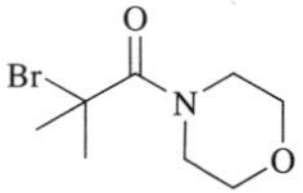 Br-C(Me)2-C(O)-morpholine	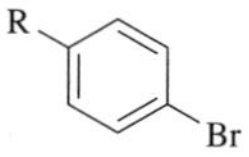 R-C6H4-Br	1. Zn (1.5 eq), THF, rt, 0.5 h; 2. [PdP(t-Bu)3Br]2 (2.5 mol %), THF/toluene (2:8), 12 h	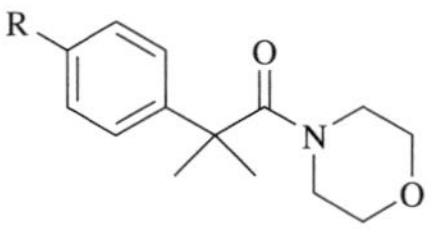 R-C6H4-C(Me)2-C(O)-morpholine	68

R	Temp	Yield
Me2N	70°	(88)
MeO	rt	(97)
Cl	rt	(91)
CF3	rt	(85)
t-Bu	rt	(93)

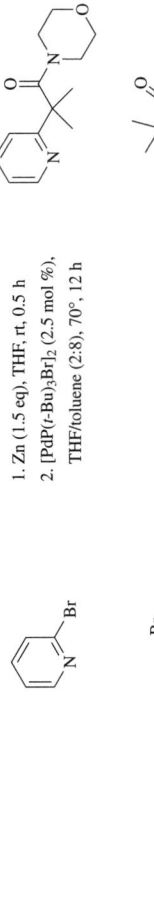

TABLE 2. ARYLATION OF ESTERS, AMIDES, LACTONES, LACTAMS, NITRILES, KETENE ACETALS, AND PREFORMED ENOLATES (*Continued*)

Substrate	Aryl Compound	Conditions	Product(s) and Yield(s) (%)	Refs.
C_8				
 [norbornyl-CN]	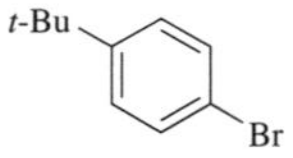 [4-t-Bu-C6H4-Br]	Pd(OAc)$_2$ (2 mol %), **L23** (2 mol %), NaHMDS (1.3 eq), toluene, 100°, 4 h	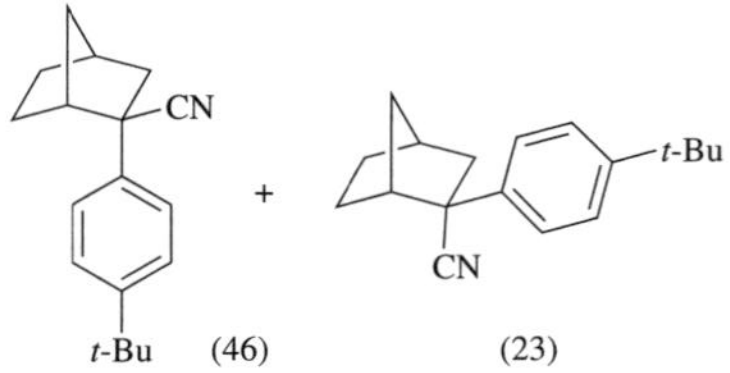 [norbornyl(CN)(4-t-Bu-C6H4)] (46) + [norbornyl(CN)-t-Bu isomer] (23)	26
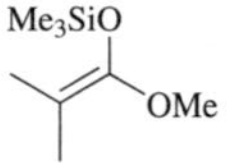 Me$_3$SiO-C(OMe)=CMe$_2$	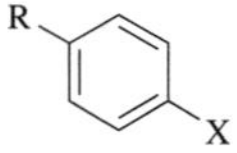 [R-C6H4-X]	Pd(dba)$_2$ (1 mol %), P(t-Bu)$_3$ (2 mol %), ZnF$_2$ (0.5 eq), DMF, 80°, 12 h	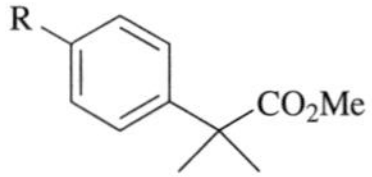 [R-C6H4-C(Me)$_2$-CO$_2$Me]	
	X / R			
	Cl / O$_2$N		(95)^e	99
	Cl / MeO$_2$C		(96)^e	99
	Br / H		(91)	99, 76
	Br / O$_2$N		(98)	99, 76
	Br / MeO		(88)	99, 76
	Br / CF$_3$		(95)	76
	Br / MeO$_2$C		(94)	99, 76, 22
	Br / MeCO		(78)	99, 76, 22
	Br / PhCO		(99)	99, 76, 22
	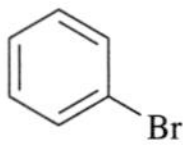 [Ph-Br]	Pd(dba)$_2$ (5 mol %), P(t-Bu)$_3$ (10 mol %), additive (x eq), 80°, 12 h	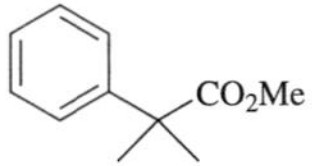 [Ph-C(Me)$_2$-CO$_2$Me]	99

t-Bu–C₆H₄–Br + Me₃SiO–C(OEt)=CH–OMe

Pd(dba)₂ (5 mol %), P(t-Bu)₃ (10 mol %), additive, dioxane, 100°, 12 h

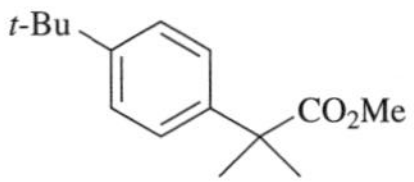

Additive	x	Solvent	
none	—	DMF	(—)
LiCl	1	DMF	(—)
ZnF₂	0.5	DMF/DME (1:1)	(95)
ZnF₂	0.5	DMF	(100)
ZnF₂	0.25	DMF	(93)
ZnCl₂	0.5	DME	(0)
ZnCl₂	0.5	DMF/DME (1:1)	(—)
ZnCl₂	0.25	DMF/DME (1:1)	(66)
ZnBr₂	0.25	DMF/DME (1:1)	(38)

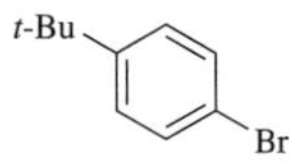

Product: t-Bu–C₆H₄–C(Me)₂–CO₂Me   99

Ph–OTf + Me₃SiO–C(OEt)=CH–OMe

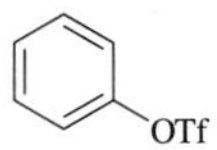

[η³-C₄H₇-Pd(OAc)]₂ (5 mol %), **L1** (20 mol %), LiOAc (2 eq), THF, reflux, 6 h

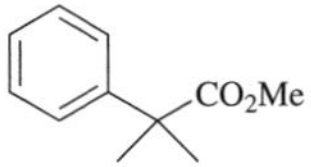

Product: Ph–C(Me)₂–CO₂Me

Additive	
none	(0)
ZnCl₂	
CuCl	

100

R–C₆H₄–Br + Me₃SiO–C(OEt)=CH–OMe

Pd(dba)₂ (5 mol %), P(t-Bu)₃ (10 mol %), ZnF₂ (0.5 eq), DMF; 80°, 12 h

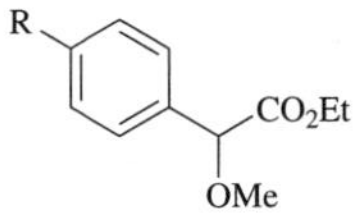

Product: R–C₆H₄–CH(OMe)–CO₂Et

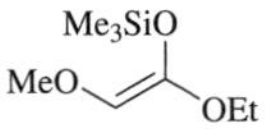

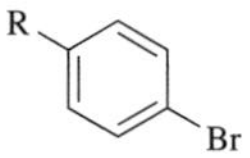

R	
MeO₂C	(63)
t-Bu	(54)

99

C₉: cyclohexyl-CH₂-CO₂Me

t-Bu–C₆H₄–Br + cyclohexyl-CH₂-CO₂Me

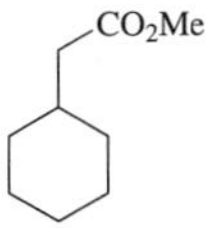

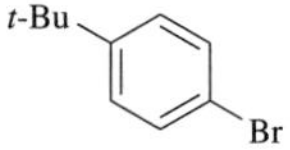

Pd(dba)₂ (5 mol %), **L47** (5 mol %), NaHMDS (2.3 eq), rt, 12 h

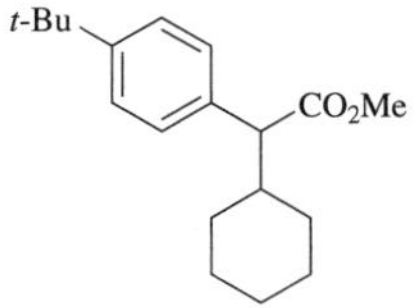

Product: t-Bu–C₆H₄–CH(C₆H₁₁)–CO₂Me   (98)

77

TABLE 2. ARYLATION OF ESTERS, AMIDES, LACTONES, LACTAMS, NITRILES, KETENE ACETALS, AND PREFORMED ENOLATES (*Continued*)

Substrate	Aryl Compound	Conditions	Product(s) and Yield(s) (%)	Refs.
C₉				
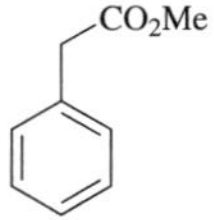 PhCH₂CO₂Me	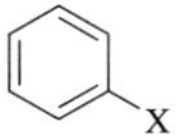 Ph–X	PdCl₂ (5 mol %), ligand (10–20 mol %), Cs₂CO₃ (1.2 eq), additive Ligand / Additive / Temp (°) / Time (h) PPh₃ / 4 Å MS / 130 / 4 none / none / 100 / 50 none / 4 Å MS / 100 / 2	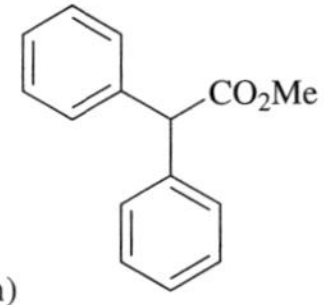 Ph₂CHCO₂Me (9) (7) (60)	73
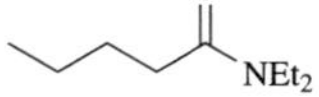 Pentanamide NEt₂ (valeryl diethylamide)	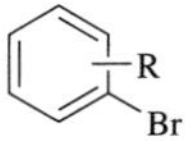 Ar–Br (R-substituted)	1. *s*-BuLi (1.2 eq), THF, −78°, 1 h 2. ZnCl₂ (2.4 eq), rt, 10 min 3. Pd(dba)₂ (3 mol %), **L12** (3 mol %), rt, 24 h	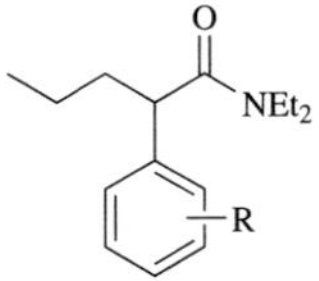 α-aryl-pentanamide NEt₂ R 2-MeO (78) 4-MeO (87) 4-NC (97) 4-MeO₂C (88) 4-*t*-Bu (85)	68
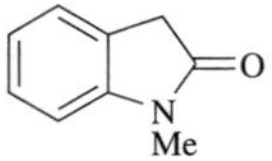 1-methyl-2-oxindole	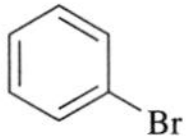 PhBr	Pd(OAc)₂ (2 mol %), PCy₃ (3 mol %), NaO-*t*-Bu, dioxane, 70°, 3 h	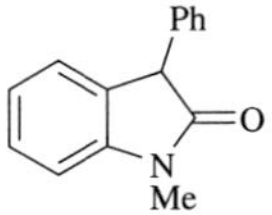 3-Ph-1-methyl-2-oxindole **I** (0)	160
1-methyl-2-oxindole	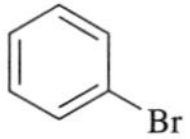 PhBr	Pd(dba)₂ (2 mol %), ligand (3 mol %), base (*x* eq), THF/toluene Ligand / Base / *x* / Temp / Time (h) PCy₃ / KHMDS / 2.0 / 70° / 3 P(*t*-Bu)₃•HBF₄ / KHMDS / 2.0 / 70° / 3 **L1** / KHMDS / 2.0 / 70° / 3	**I** (0) (<2) (0)	160

L20	KHMDS	2.0	70°	3	(20)
L21	NaH	1.1	70°	3	(44)
L21	NaHMDS	1.1	70°	3	(81)
L21	KHMDS	1.1	rt	3	(0)
L21	KHMDS	1.1	70°	0.5	(95)
L21	KHMDS	1.1	70°	1	(90)
L21	KHMDS	1.1	70°	3	(90)
L21	KHMDS	2.0	70°	3	(91)
L23	KHMDS	2.0	70°	3	(0)

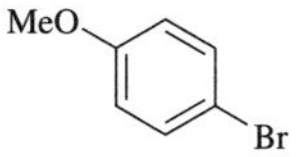

Pd(dba)$_2$ (2 mol %), **L21** (3 mol %), KHMDS, THF/toluene, 70°   160

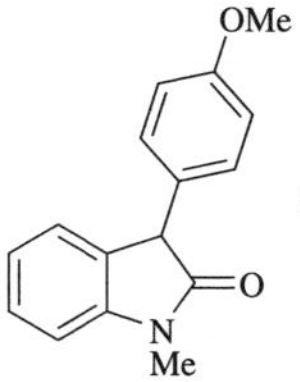

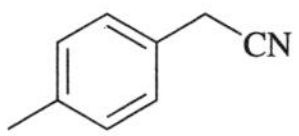

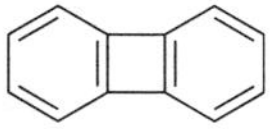

(—)

Pd(PPh$_3$)$_4$ (5 mol %),
p-cresol (10 mol %), C$_6$D$_6$, 120°, 1 d   59

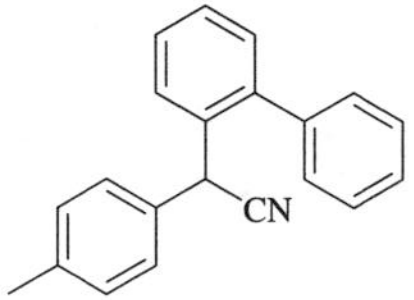

(85)

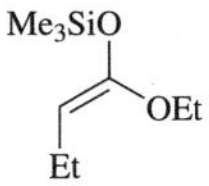

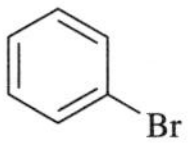

[η^3-C$_4$H$_7$Pd(OAc)]$_2$ (5 mol %),
ligand (x mol %), TlOAc (1 eq)   100

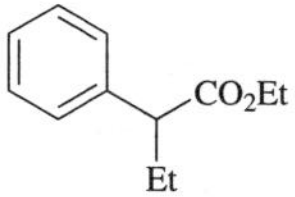

Ligand	x	
PPh$_3$	15	(10)
dppe	20	trace
dppb	20	(8)
L1	20	(80)

TABLE 2. ARYLATION OF ESTERS, AMIDES, LACTONES, LACTAMS, NITRILES, KETENE ACETALS, AND PREFORMED ENOLATES (*Continued*)

Substrate	Aryl Compound	Conditions	Product(s) and Yield(s) (%)	Refs.
C₉ Me₃SiO / OEt / Et	4-R-C₆H₄-OTf	[η³-C₃H₅Pd(OAc)]₂ (2 mol %), **L1** (10 mol %), LiOAc (2 eq), solvent, reflux, 6 h	4-R-C₆H₄-CH(Et)CO₂Et R | Solvent H | THF (80) O₂N | DME (30)	100
2-Br-C₆H₄-N(Me)C(O)Me	R-C₆H₄-Cl	Pd(OAc)₂ (10 mol %), PCy₃ (10 mol %), NaO*t*-Bu (1.5 eq), dioxane, 70°, 12 h	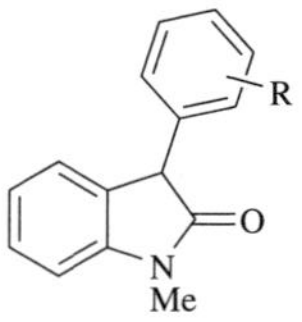 3-aryl-1-methylindolin-2-one R | H (61) 2-MeO (60) 2-Me (56) 4-Me (55)	32
	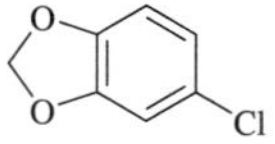 5-Cl-benzo[d][1,3]dioxole	Pd(OAc)₂ (10 mol %), PCy₃ (10 mol %), NaO*t*-Bu (1.5 eq), dioxane, 70°, 12 h	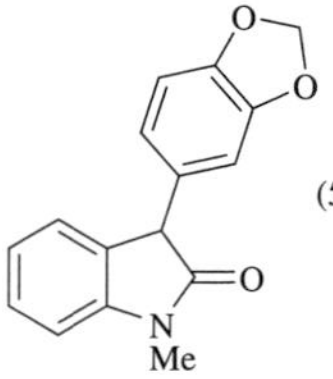 3-(benzo[d][1,3]dioxol-5-yl)-1-methylindolin-2-one (56)	32
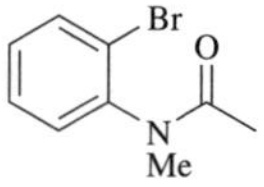 2-Br-C₆H₄-N(Me)C(O)Me		Pd(dba)₂ (5 mol %), **L23** (7.5 mol %), NaO*t*-Bu (1.5 eq), 2 h	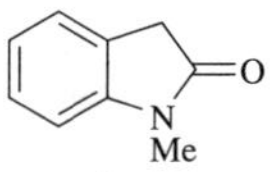 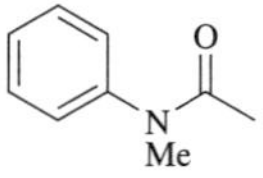 1-methylindolin-2-one (60) + PhN(Me)C(O)Me (1) **I**	67
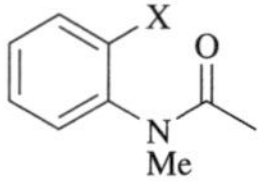 2-X-C₆H₄-N(Me)C(O)Me		Pd(OAc)₂ (5 mol %), ligand (5 mol %), NaO*t*-Bu (1.5 eq), dioxane	**I**	32

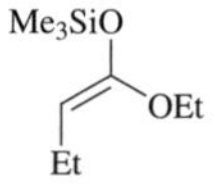

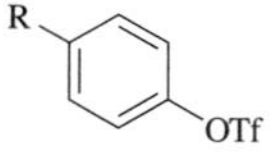

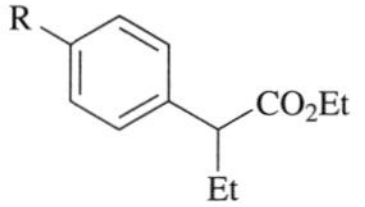

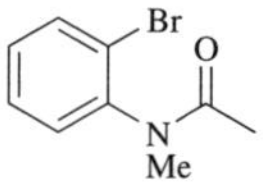

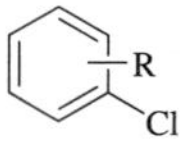

C_{10}

PhCH2-CO2Et + Ar-X →

X	Ligand	Temp (°)	Time (h)	Product	Yield
Cl	PCy3	70	5		(68)
Br	P(t-Bu)3	50	10		(9)
Br	PCy3	50	3		(62)
Br	PCy3	50	10		(62)
Br	L15	50	10		(33)
Br	L18	50	10		(29)
Br	L47	50	10		(39)
Br	L51	50	10		(10)

Product: 4-R-C6H4-CH(Ph)-CO2Et

Pd2(dba)3 (1.5 mol %), L14 (6.3 mol %), LiHMDS (2.5 eq), toluene, 80°

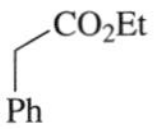

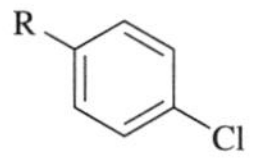

R	Time (h)	Yield
MeO	1	(87)
Me	2	(82)

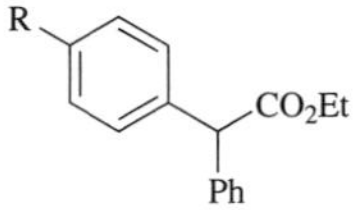

74

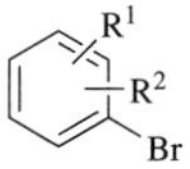

Pd(OAc)2 (3.0 mol %), L17 (6.3 mol %), LiHMDS (2.5 eq), toluene, 80°

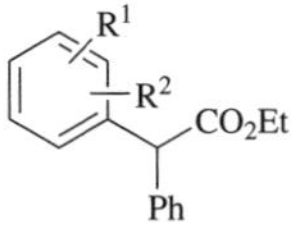

Product: Ar-CH(Ph)-CO2Et (with R^1, R^2 substituents on aryl, from ArBr)

R^1	R^2	Time (h)	Yield
3-Cl	H	1	(79)
4-Me2N	H	3	(49)
4-Me	H	0.5	(85)
4-Et2NCO	H	1	(88)
4-t-BuO2C	H	0.5	(75)
2-Me	6-Me	2	(71)

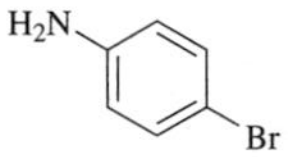

74

Pd2(dba)3 (1.5 mol %), L14 (6.3 mol %), LiHMDS (2.5 eq), toluene, 80°, 0.2 h

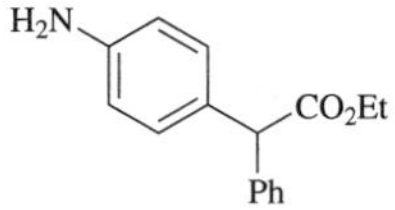

From 4-H2N-C6H4-Br: 4-H2N-C6H4-CH(Ph)-CO2Et (90)

74

TABLE 2. ARYLATION OF ESTERS, AMIDES, LACTONES, LACTAMS, NITRILES, KETENE ACETALS, AND PREFORMED ENOLATES (*Continued*)

Substrate	Aryl Compound	Conditions	Product(s) and Yield(s) (%)	Refs.
C_{10} Ph-CH$_2$-CO$_2$Et	2-Br-naphthyl	Pd(OAc)$_2$ (3 mol %), **L17** (6.3 mol %), LiHMDS (2.5 eq), toluene, 80°, 3 h	naphthyl-CH(Ph)-CO$_2$Et (88)	74
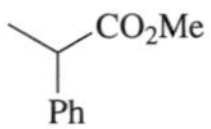 Ph-CH(Me)-CO$_2$Me	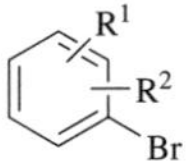 R^1,R^2-C$_6$H$_3$-Br	Pd(dba)$_2$ (x mol %), P(t-Bu)$_3$ (0.5–1.0 mol %), LiNCy$_2$ (1.3 eq), toluene, rt	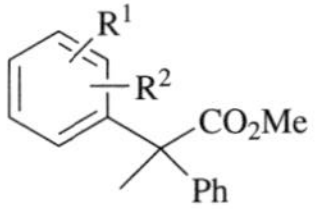 C$_6$H$_3$(R^1,R^2)-C(Me)(Ph)-CO$_2$Me	75

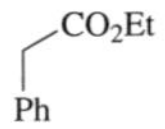

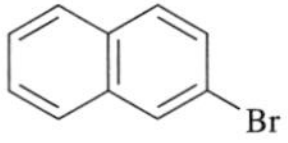

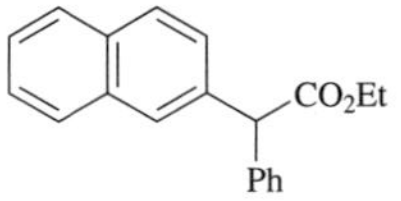

R^1	R^2	x	Time (h)	
3-F	H	1.0	15	(82)
3-Me	H	0.5	8	(89)
4-F	H	1.0	15	(84)
4-Me	H	0.5	8	(87)
4-Ph	H	0.5	15	(93)
3-F	4-Ph	1.0	15	(97)

Substrate	Aryl Compound	Conditions	Product(s) and Yield(s) (%)	Refs.
Ph-CH(Me)-CO$_2$Me	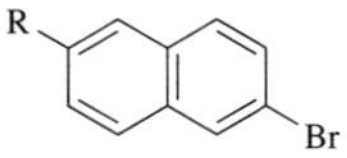 6-R-2-Br-naphthyl	Pd(dba)$_2$ (x mol %), P(t-Bu)$_3$ (0.5–1.0 mol %), LiNCy$_2$ (1.3 eq), toluene, rt, 15 h	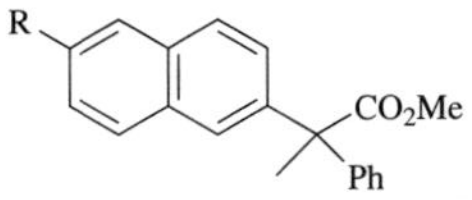 6-R-naphthyl-C(Me)(Ph)-CO$_2$Me	75

R	x	
H	0.5	(78)
MeO	1.0	(82)

Substrate	Aryl Compound	Conditions	Product(s) and Yield(s) (%)	Refs.
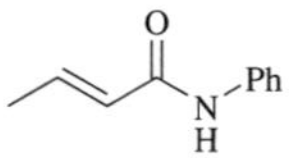 crotonamide (N-Ph)	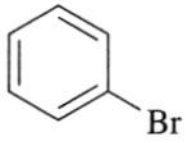 PhBr	Pd(OAc)$_2$, P(t-Bu)$_3$, Cs$_2$CO$_3$, *o*-xylene	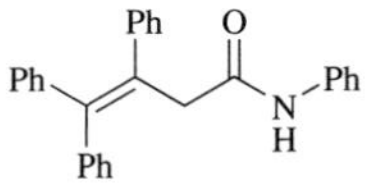 Ph$_2$C=C(Ph)-CH$_2$-C(O)NHPh (—)	47

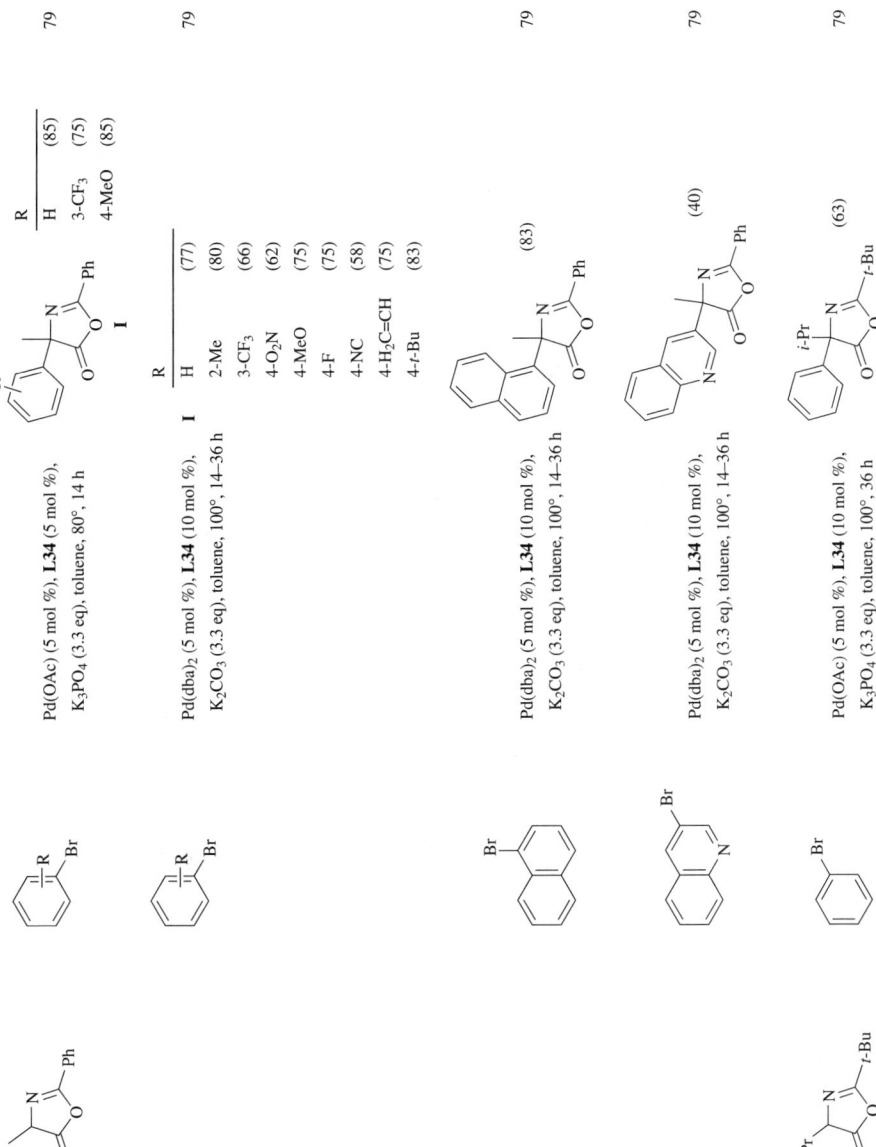

TABLE 2. ARYLATION OF ESTERS, AMIDES, LACTONES, LACTAMS, NITRILES,
KETENE ACETALS, AND PREFORMED ENOLATES (*Continued*)

Substrate	Aryl Compound	Conditions	Product(s) and Yield(s) (%)	Refs.
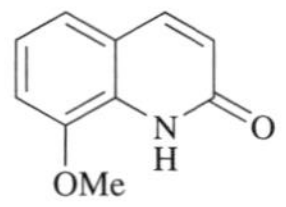 C₁₀ 2(1H)-quinolinone with 8-OMe	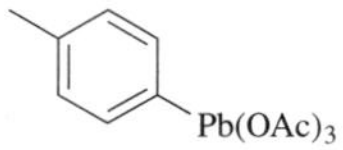 4-MeC₆H₄-Pb(OAc)₃	Cu(OAc)₂ (0.1 eq)	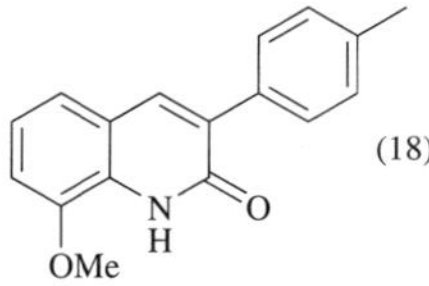 3-(4-MeC₆H₄)-8-OMe-2(1H)-quinolinone (18)	133
Ph-CH(CN)-Et	PhBr	Pd₂(dba)₃·CHCl₃ (1 mol %), P(t-Bu)₃ (2 mol %), LiHMDS (1.3 eq), toluene, 70°, 3 h	Ph₂C(CN)Et (89)	26
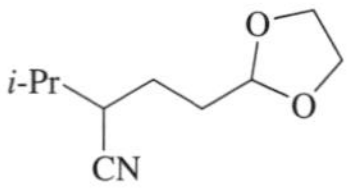 i-Pr-CH(CN)-CH₂CH₂-(1,3-dioxolan-2-yl)	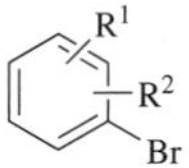 R¹,R²-C₆H₃-Br	Pd(OAc)₂ (2 mol %), ligand (4 mol %), ZnCl₂ (1.2 eq), THF, rt	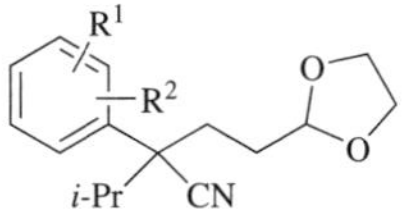 aryl-C(i-Pr)(CN)-CH₂CH₂-(1,3-dioxolan-2-yl)	80
	R¹ R² Ligand			
	4-t-Bu H P(t-Bu)₃		(76)	
	4-t-Bu H L33		(80)	
	4-PhCO H P(t-Bu)₃		(86)	
	4-PhCO H L33		(77)	
	3-MeO 4-MeO P(t-Bu)₃		(82)	
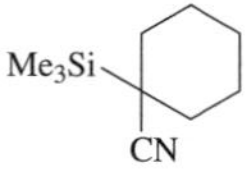 Me₃Si-C(CN)(cyclohexyl)	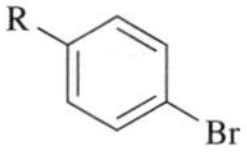 R-C₆H₄-Br	Pd₂(dba)₃ (2 mol %), L23 (2 mol %), KF (1 eq), DMF, 90°, 24 h	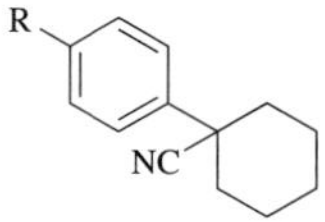 R-C₆H₄-C(CN)(cyclohexyl)	80
	R			
	H		(68)	
	MeO		(77)	
	EtO₂C		(84)	
	t-Bu		(62)	

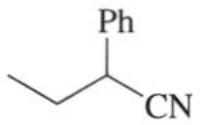

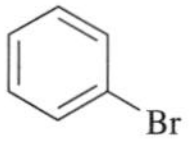

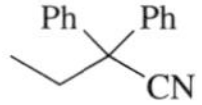

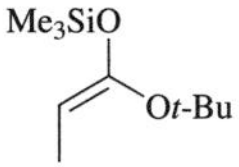

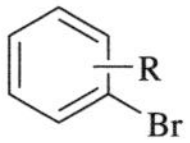

Me₃SiO—C(=CHMe)—Ot-Bu	Ar-Br (R substituted)	Pd(dba)₂ (1 mol %), P(t-Bu)₃ (2 mol %), ZnF₂ (0.5 eq), DMF, 80°, 12 h

Product: Ar–CH(Me)–CO₂t-Bu

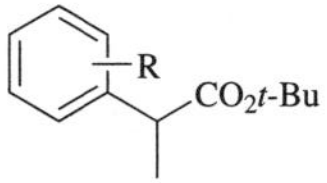

R	
2-NC	(75)
3-NC	(67)
4-O₂N	(76)
4-MeO₂C	(80)
4-MeCO	(68)

76, 99

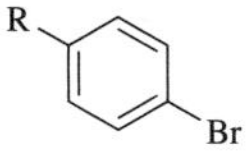

4-R-C₆H₄-Br, Pd(dba)₂ (1 mol %), P(t-Bu)₃ (2 mol %), ZnF₂ (0.5 eq), DMF, 80°, 12 h

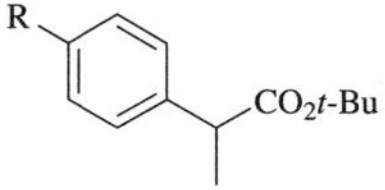

Product: 4-R-C₆H₄–CH(Me)–CO₂t-Bu

R	
MeO₂C	(80)
MeOC	(68)

22

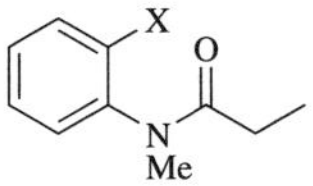

2-X-C₆H₄-N(Me)-C(=O)-Et, Pd(OAc)₂ (5 mol %), ligand (5 mol %), NaOt-Bu (1.5 eq), dioxane

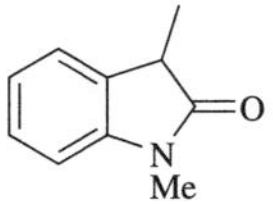

Product **I**: 1,3-dimethyl-2-oxindole

X	Ligand	Temp (°)	Time (h)	
Cl	PCy₃	70	24	(8)
Br	P(t-Bu)₃	50	10	(3)
Br	PCy₃	50	3	(10)
Br	PCy₃	50	10	(6)
Br	**L15**	50	10	(10)
Br	**L18**	50	10	(14)
Br	**L47**	50	10	(29)
Br	**L51**	50	10	(11)

32

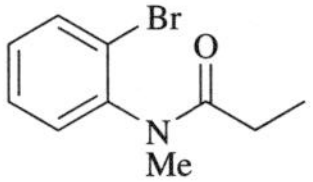

2-Br-C₆H₄-N(Me)-C(=O)-Et, Pd(dba)₂ (5 mol %), **L23** (7.5 mol %), NaOt-Bu (1.5 eq), 15 h

Product: **I** (52) + PhN(Me)C(=O)Et (8)

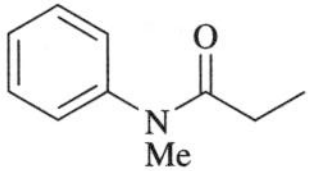

67

TABLE 2. ARYLATION OF ESTERS, AMIDES, LACTONES, LACTAMS, NITRILES, KETENE ACETALS, AND PREFORMED ENOLATES (*Continued*)

Substrate	Aryl Compound	Conditions	Product(s) and Yield(s) (%)	Refs.
C_{10} 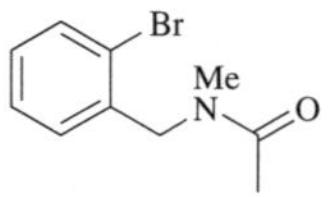 2-Br-C$_6$H$_4$-CH$_2$-N(Me)-C(O)-Me		Pd(dba)$_2$ (5 mol %), **L23** (7.5 mol %), NaO-t-Bu (1.5 eq), dioxane, 100°, 0.5 h	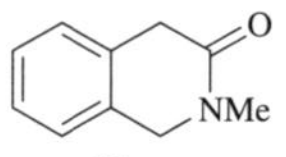 isoquinolinone product + N-benzyl-N-methylacetamide (5) 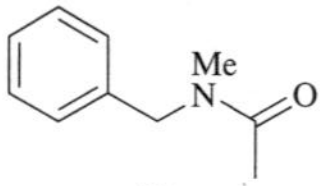 (7)	67
C_{11} BnO$_2$C-CH$_2$-CHMe$_2$	R^1,R^2-C$_6$H$_3$-Br R^1 R^2 3-Me H 3-CF$_3$ H 4-MeO H 4-F H 4-Me H 3-F 4-Ph	Pd(dba)$_2$ (1.0 mol %), P(t-Bu)$_3$ (1.0 mol %), LiNCy$_2$ (1.8 eq), toluene, rt, 12 h	R^1,R^2-C$_6$H$_3$-C(Me)$_2$-CO$_2$Bn (93) (80) (83) (82) (90) (90)	75
	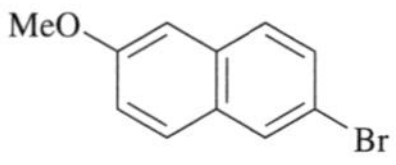 6-MeO-2-Br-naphthalene	Pd(dba)$_2$ (1.0 mol %), P(t-Bu)$_3$ (1.0 mol %), LiNCy$_2$ (1.8 eq), toluene, rt, 12 h	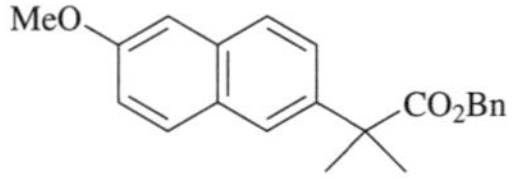 6-MeO-naphthyl-C(Me)$_2$-CO$_2$Bn (88)	75
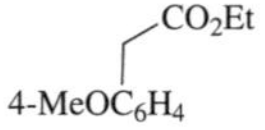 4-MeOC$_6$H$_4$-CH$_2$-CO$_2$Et	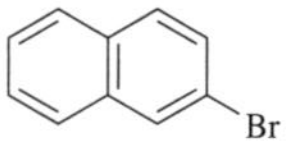 2-Br-naphthalene	Pd(OAc)$_2$ (3 mol %), **L17** (6.3 mol %), LiHMDS (2.5 eq), toluene, 80°, 0.5 h	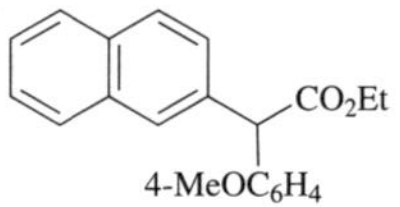 naphthyl-CH(CO$_2$Et)(4-MeOC$_6$H$_4$) (83)	74
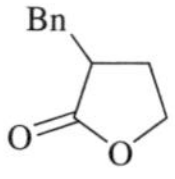 3-Bn-γ-butyrolactone	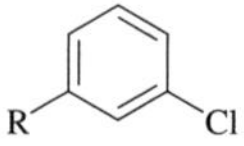 3-R-C$_6$H$_4$-Cl	Ni(cod)$_2$ (10 mol %), (*S*)-**L23** (17 mol %), ZnBr$_2$ (15 mol %), NaHMDS (1.15 eq), toluene/THF (3:1), 80°, 17–20 h	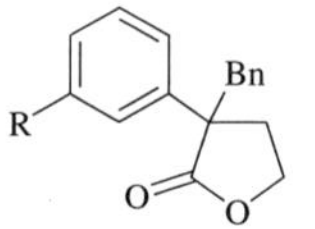 3-Bn-3-(3-R-C$_6$H$_4$)-γ-butyrolactone R % ee Me$_2$N (58) 96 MeO (63) 94	33

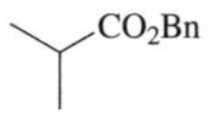

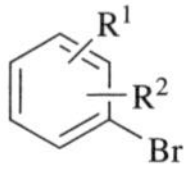

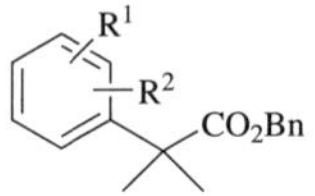

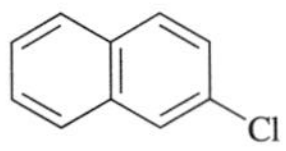

![2-chloronaphthalene]	Ni(cod)₂ (5 mol %), (S)-**L23** (8.5 mol %), ZnBr₂ (15 mol %), NaHMDS (1.15 eq), toluene/THF (3:1), 60°, 17–20 h	3-benzyl-3-(naphthalen-2-yl)dihydrofuran-2(3H)-one (91), 96% ee	33
4-*t*-Bu-C₆H₄-Br	Pd(OAc) (5 mol %), **L35** (10 mol %), K₃PO₄ (3.3 eq), toluene, 80°, 36 h	oxazoline product (82)	79

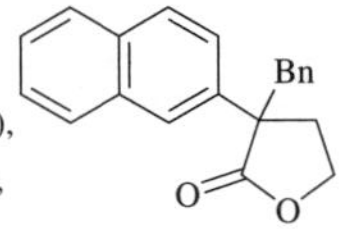

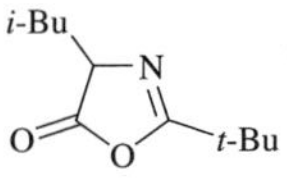

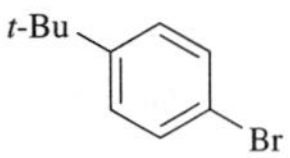

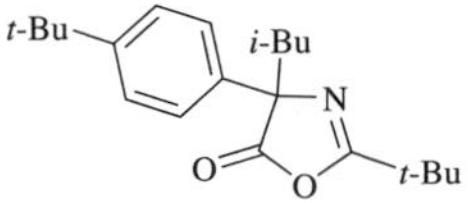

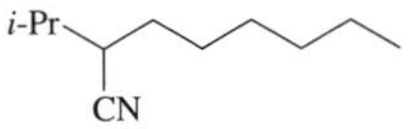

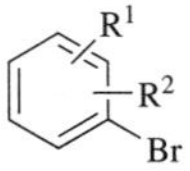

Pd(OAc)₂ (2 mol %),
ligand (4 mol %),
ZnCl₂ (1.2 eq), THF, rt

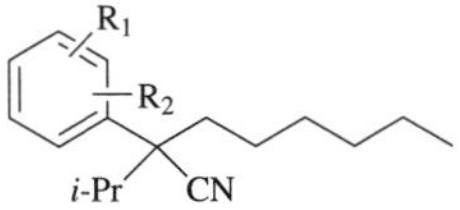

Substrate: aryl bromide with R¹, R²

R¹	R²	Ligand	Yield
4-Me₂N	H	P(*t*-Bu)₃	(69)
4-MeO	H	P(*t*-Bu)₃	(72)
4-EtO₂C	H	P(*t*-Bu)₃	(47)
4-EtO₂C	H	**L33**	(55)
4-*t*-Bu	H	P(*t*-Bu)₃	(85)
4-PhCO	H	P(*t*-Bu)₃	(78)
3-MeO	4-MeO	P(*t*-Bu)₃	(68)
3-MeO	4-MeO	**L33**	(72)

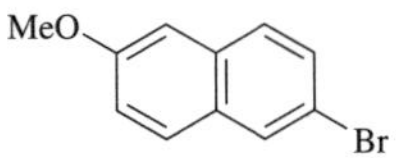

80

6-bromo-2-methoxynaphthalene	Pd(OAc)₂ (2 mol %), **L33** (4 mol %), ZnCl₂ (1.2 eq), THF, rt	arylated nitrile product (81)	80

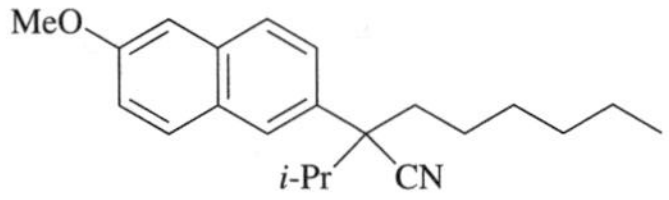

TABLE 2. ARYLATION OF ESTERS, AMIDES, LACTONES, LACTAMS, NITRILES,
KETENE ACETALS, AND PREFORMED ENOLATES (*Continued*)

Substrate	Aryl Compound	Conditions	Product(s) and Yield(s) (%)	Refs.
C_{11} 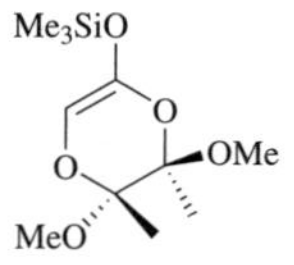 Me$_3$SiO—[dioxine with OMe, Me, Me, MeO]	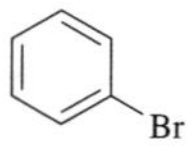 Ph—Br	Pd(dba)$_2$ (5 mol %), P(t-Bu)$_3$ (10 mol %), ZnF$_2$, DMF, 80° or Zn(Ot-Bu)$_2$, DMF, rt	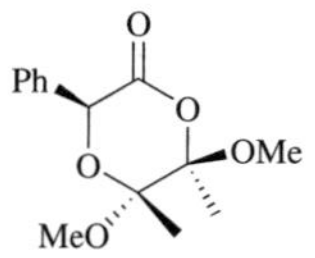 Ph—[dioxanone with OMe, Me, Me, MeO] (67–73)	99
	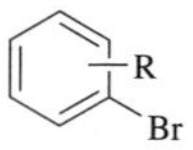 R—C$_6$H$_4$—Br	Pd(dba)$_2$ (5 mol %), P(t-Bu)$_3$ (10 mol %), zinc additive (x eq), DMF, 12 h	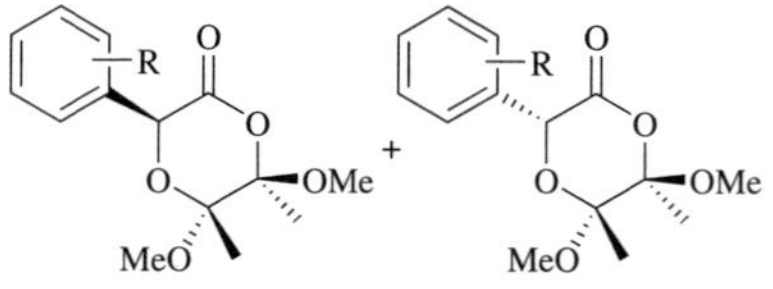 I + II (see below)	99

R	Zinc Additive	x	Temp	I + II	I:II
H	ZnF$_2$	0.5	80°	(67)	>50:1
H	Zn(Ot-Bu)$_2$	0.25	rt	(73)	>50:1
2-Cl	ZnF$_2$	1	80°	(78)	20:1
3-O$_2$N	ZnF$_2$	1	80°	(64)	>26:1
3-O$_2$N	Zn(Ot-Bu)$_2$	0.5	rt	(89)	>50:1
3-O$_2$N	Zn(Ot-Bu)$_2$	0.25	rt	(78)	>50:1
3-MeCO	ZnF$_2$	1	80°	(76)	25:1
4-MeO	ZnF$_2$	0.5	80°	(57)	>50:1
4-MeO$_2$C	Zn(Ot-Bu)$_2$	0.5	rt	(95)	>50:1
4-MeO$_2$C	Zn(Ot-Bu)$_2$	0.25	rt	(80)	>50:1

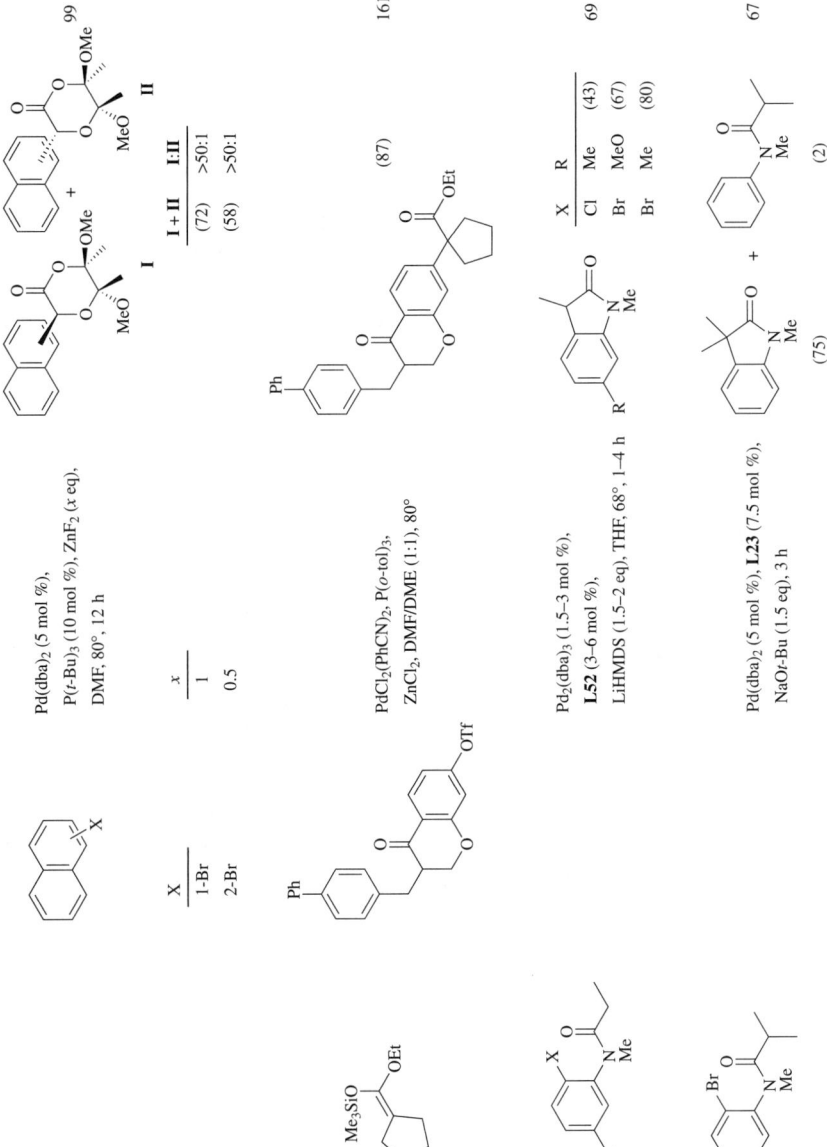

TABLE 2. ARYLATION OF ESTERS, AMIDES, LACTONES, LACTAMS, NITRILES, KETENE ACETALS, AND PREFORMED ENOLATES (*Continued*)

Substrate	Aryl Compound	Conditions	Product(s) and Yield(s) (%)	Refs.
C$_{11}$ 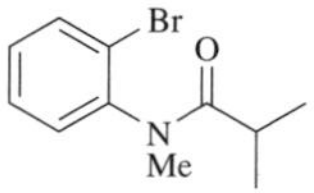 [structure: Br-C$_6$H$_4$-N(Me)-C(O)-CH(Me)$_2$]		Catalyst (5 mol %), ligand (5 mol %), NaO-t-Bu (1.5 eq), dioxane	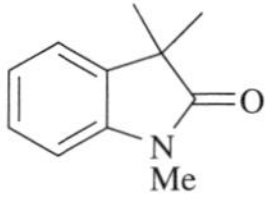 [structure: 3,3-dimethyl-1-methyl-oxindole with N-Me, =O]	32
		Catalyst / Ligand / Temp (°) / Time (h)		
		Pd(dba)$_2$ / P(t-Bu)$_3$ / 100 / 5	(62)	
		Pd(dba)$_2$ / PCy$_3$ / 50 / 3	(62)	
		Pd(dba)$_2$ / PCy$_3$ / 100 / 1	(82)	
		Pd(dba)$_2$ / **L3** / 100 / 5	(41)	
		Pd(dba)$_2$ / **L4** / 100 / 4	(42)	
		Pd(dba)$_2$ / **L23** / 100 / 3	(53)	
		Pd(OAc)$_2$ / P(t-Bu)$_3$ / 50 / 3	(21)	
		Pd(OAc)$_2$ / PCy$_3$ / 50 / 3	(99)	
		Pd(OAc)$_2$ / **L15** / 50 / 3	(67)	
		Pd(OAc)$_2$ / **L18** / 50 / 3	(11)	
		Pd(OAc)$_2$ / **L47** / 50 / 3	(99)	
		Pd(OAc)$_2$ / **L51** / 50 / 3	(99)	
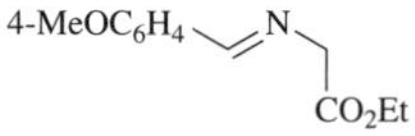 C$_{12}$ [4-MeOC$_6$H$_4$-N=CH-CO$_2$Et]	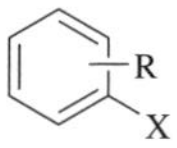 [X-C$_6$H$_4$-R]	Pd(dba)$_2$ (2 mol %), P(t-Bu)$_3$ (4 mol %), K$_3$PO$_4$ (3 eq), toluene, 20 h	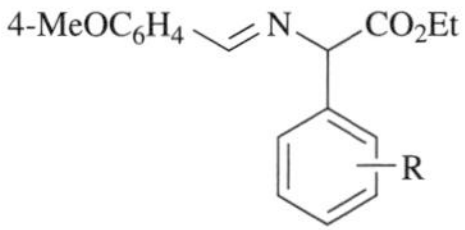 4-MeOC$_6$H$_4$-N=CH-C(Ar(R))-CO$_2$Et	77
	X / R / Temp (°)			
	Cl / H / 120		(67)	
	Cl / 4-CF$_3$ / 120		(67)	
	Br / H / 100		(75)	
	Br / 2-Me / 100		(72)	
	Br / 4-MeO / 100		(71)	
	Br / 4-F / 100		(67)	
	Br / 4-CF$_3$ / 100		(73)	
	Br / 4-Ph / 100		(80)	

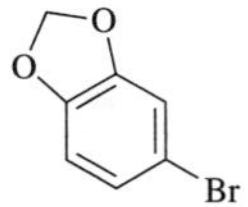 ![bromo-benzodioxole]	Pd(dba)₂ (2 mol %), P(t-Bu)₃ (4 mol %), K₃PO₄ (3 eq), toluene, 100°, 20 h	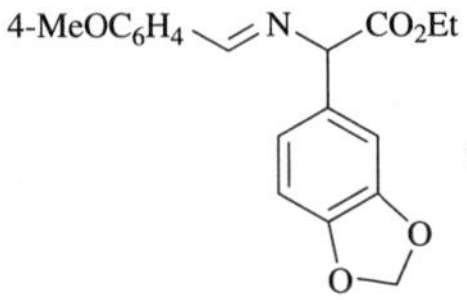 4-MeOC₆H₄—N=CH—CH(CO₂Et)—(benzodioxolyl) (71)	77
naphthyl-X	Pd(dba)₂ (2 mol %), P(t-Bu)₃ (4 mol %), K₃PO₄ (3 eq), toluene, 100°, 20 h	4-MeOC₆H₄—N=CH—CH(CO₂Et)—(naphthyl)	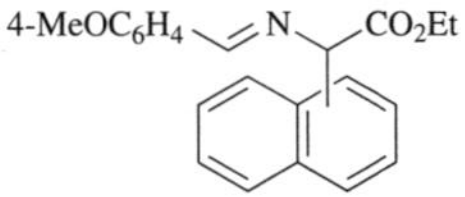 X 1-Br (74) 2-Br (74)	77
PhBr	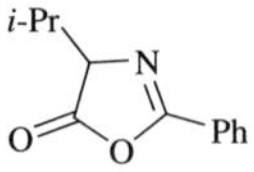 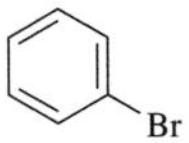 Pd(OAc) (5 mol %), **L3** (10 mol %), K₃PO₄ (3.3 eq), toluene, 80°, 14 h	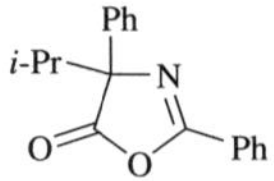 oxazolone-Ph product (29)	79

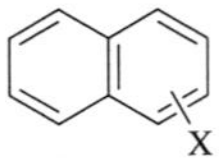

aryl bromide with R¹, R²	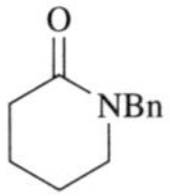 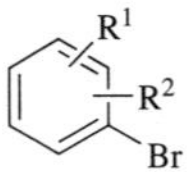 1. LiHMDS (2.0 eq), ZnCl₂ (2.2 eq), −20°; 2. Pd(dba)₂ (5 mol %), **L17** (7.5 mol %), THF, 65°	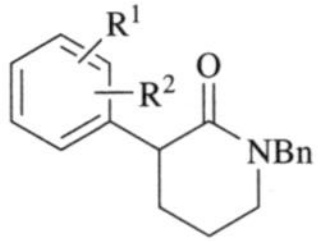 3-aryl-N-Bn-piperidin-2-one	72

R¹	R²	
H	H	(98)
2-Me	H	(46)
3-MeO	H	(50)
4-MeO	H	(77)
4-Me	H	(52)
2-MeO	4-MeO	(84)
2-Me	4-Me	(0)

TABLE 2. ARYLATION OF ESTERS, AMIDES, LACTONES, LACTAMS, NITRILES, KETENE ACETALS, AND PREFORMED ENOLATES (*Continued*)

Substrate	Aryl Compound	Conditions	Product(s) and Yield(s) (%)	Refs.
C_{12}				
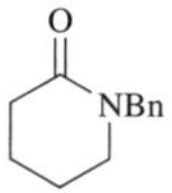 6-membered lactam with NBn	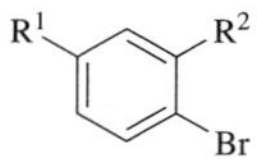 R^1, R^2-Br aryl	1. LiHMDS (2.5 eq), ZnCl$_2$ (2.2 eq), –20°; 2. Pd(OAc)$_2$ (3 mol %), **L17** (6.3 mol %), toluene, 80°	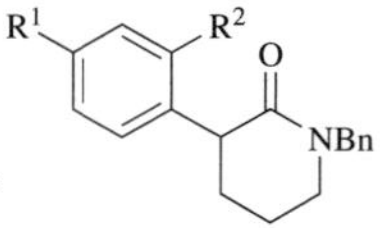 product with NBn lactam, aryl with R^1, R^2: R^1=H, R^2=H (86); R^1=Me, R^2=H (35); R^1=MeO, R^2=MeO (16)	72
6-membered lactam with NTs	PhBr	1. LiHMDS (x eq), ZnX$_2$ (y eq), –20°; 2. Pd(OAc)$_2$ (5 mol %), **L17** (7.5 mol %), THF, rt	product with NTs (yields below): x=2.0, ZnX$_2$=none, y=—: (22) x=0.95, ZnCl$_2$, y=1.1: (48) x=2.0, ZnCl$_2$, y=2.2: (48) x=2.0, ZnBr$_2$, y=2.2: (0)	72
	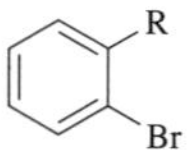 R-aryl-Br (ortho)	1. LiHMDS (2.0 eq), ZnX$_2$ (2.2 eq), –20°; 2. Pd(dba)$_2$ (5 mol %), **L17** (7.5 mol %), THF, 65°	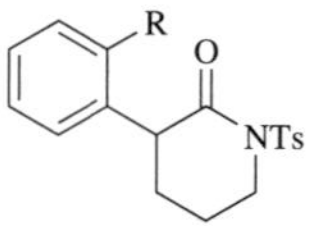 product with NTs, R-aryl: R=H (92); R=Me (41)	72
TBSO, Ot-Bu ketene acetal	R-aryl-Br	PdCl$_2$[P(o-tol)$_3$]$_2$ (2–5 mol %), Bu$_3$SnF (2 eq), benzene, reflux	R-aryl-CH$_2$CO$_2$-t-Bu **I** R=2-MeO, Time=20 h (42); R=4-MeO, 22 h (51); R=4-MeCO, 6 h (80); R=4-MeO$_2$C, 22 h (82)	98

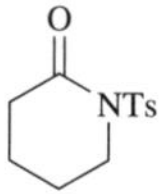

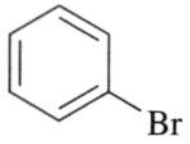

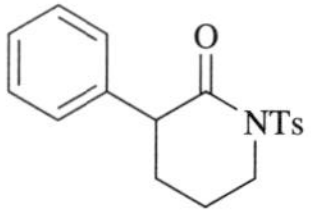

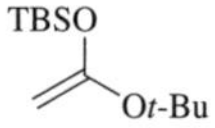

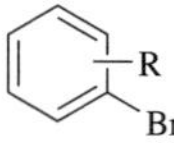

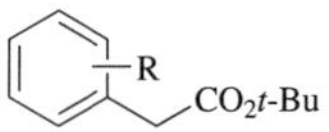

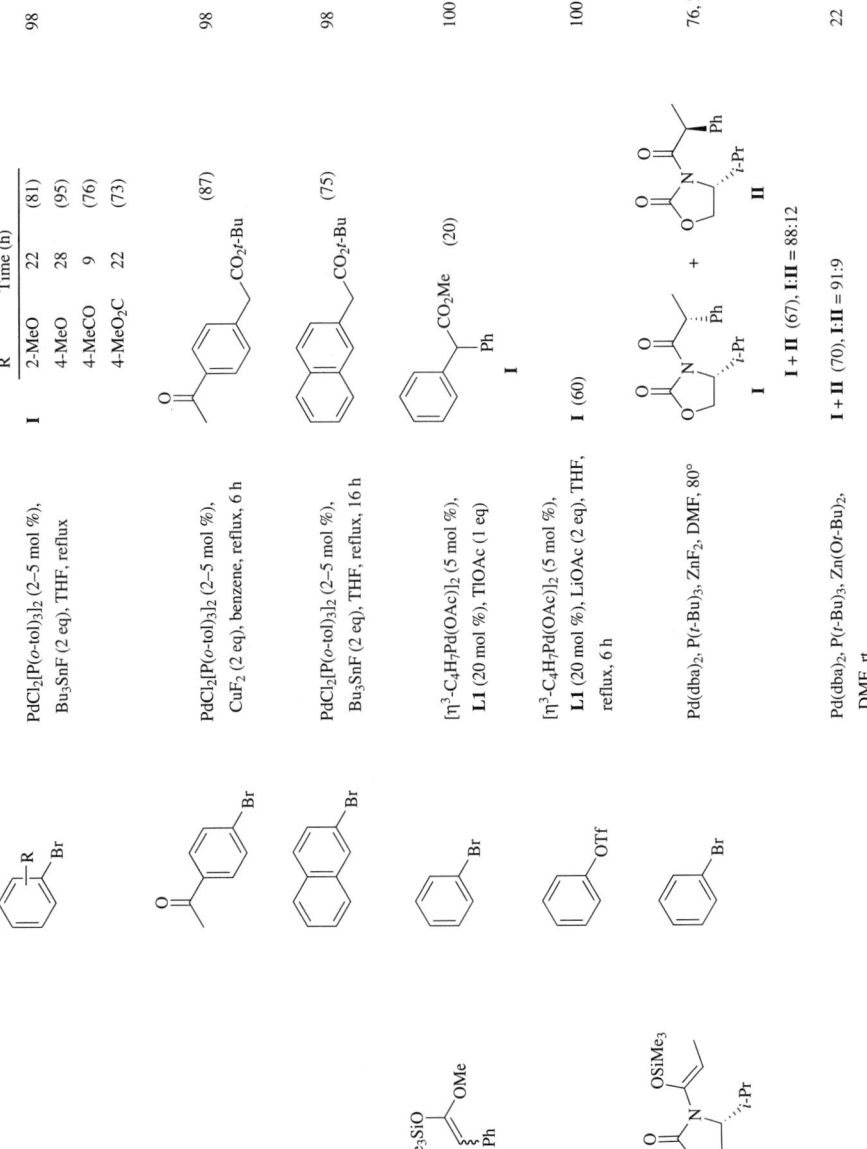

TABLE 2. ARYLATION OF ESTERS, AMIDES, LACTONES, LACTAMS, NITRILES, KETENE ACETALS, AND PREFORMED ENOLATES (*Continued*)

Substrate	Aryl Compound	Conditions	Product(s) and Yield(s) (%)	Refs.

C_{12}

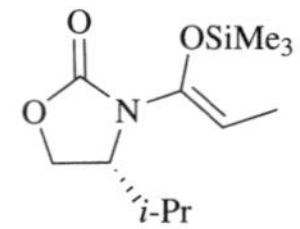

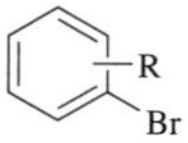

Substrate: oxazolidinone with OSiMe$_3$ and i-Pr group

Aryl compound: 2-bromo-R-pyridine/arene

Conditions: Pd(dba)$_2$ (5 mol %), P(t-Bu)$_3$ (10 mol %), ZnF$_2$ (0.5 eq), DMF, 80°, 12 h

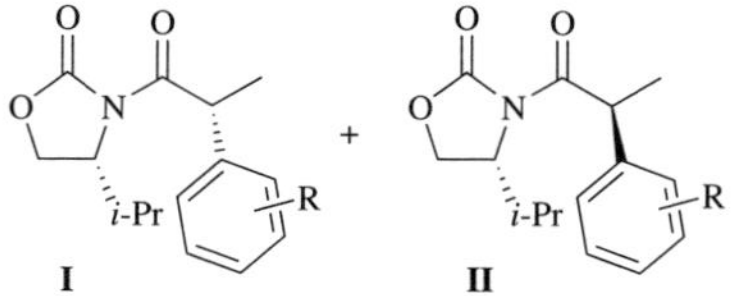

Products I + II (diastereomers)

R	I+II	I:II
H	(67)	87:13
H^f	(70)	91:9
2-Me	(78)	92:8
3-MeCO	(75)	84:16
4-NC	(65)	77:23
4-t-Bu	(57)	83:17

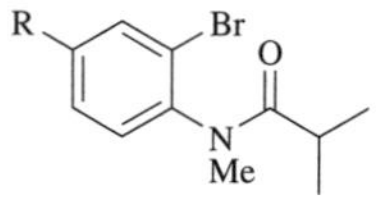

Refs. 99

Substrate: R-Br, C(=O)-N(Me)-i-Pr amide

Conditions: Pd(dba)$_2$ (5 mol %), **L23** (7.5 mol %), NaOt-Bu (1.5 eq)

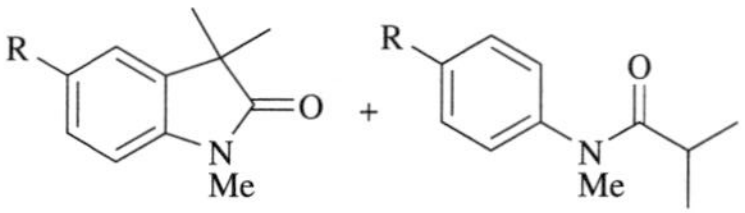

Products: 3,3-dimethyl oxindole I + N-methyl isobutyramide II

R	Time (h)	I	II
MeO	12	(80)	(2)
NC^g	4	(83)	(1)

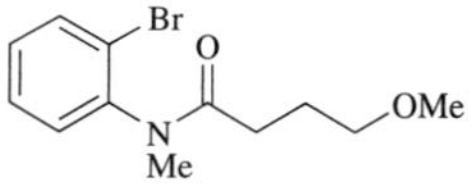

Refs. 67

Substrate: 2-bromo-N-methyl anilide with OMe chain

Conditions: Pd$_2$(dba)$_3$ (1.5–3 mol %), **L52** (3–6 mol %), LiHMDS (1.5–2 eq), THF, 68°, 1–4 h

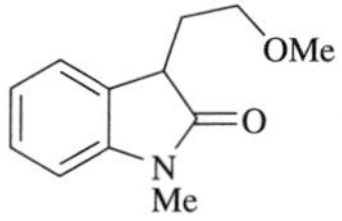

Product: oxindole with CH$_2$CH$_2$OMe side chain (82)

Refs. 69

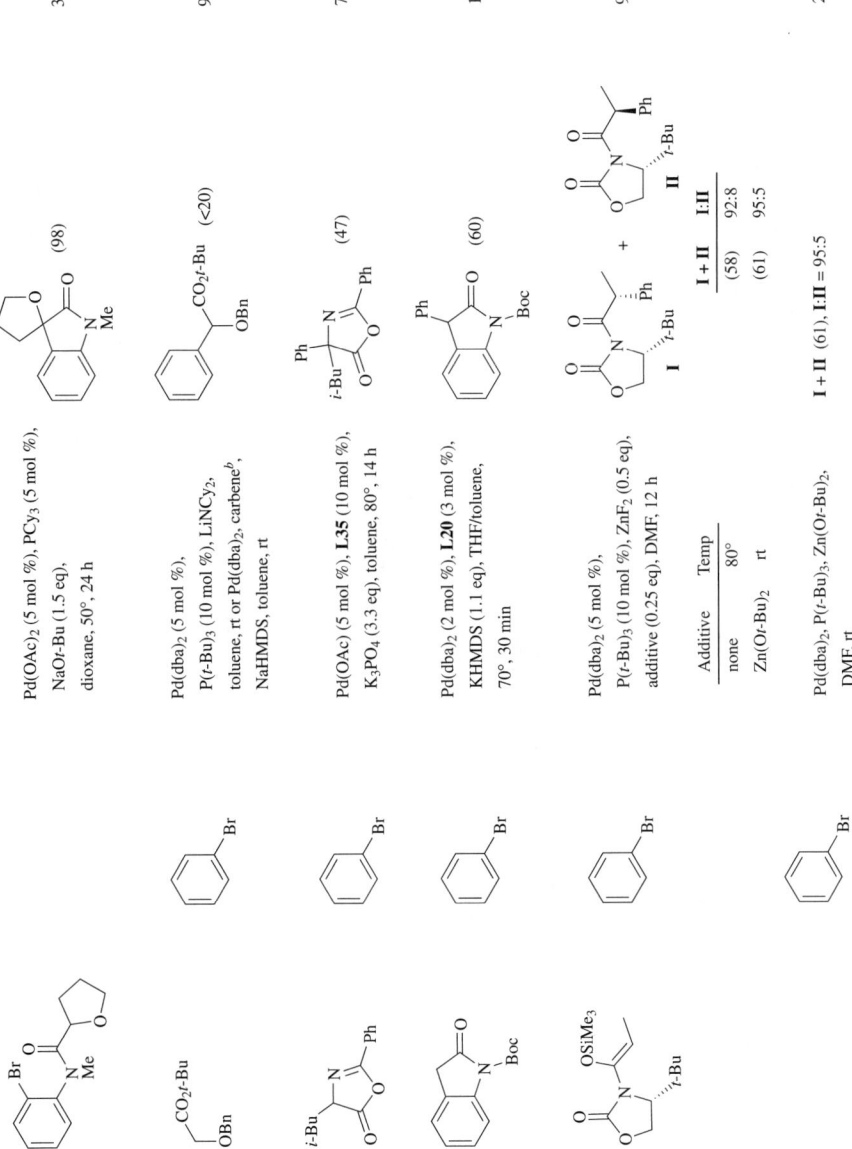

TABLE 2. ARYLATION OF ESTERS, AMIDES, LACTONES, LACTAMS, NITRILES, KETENE ACETALS, AND PREFORMED ENOLATES (*Continued*)

Substrate	Aryl Compound	Conditions	Product(s) and Yield(s) (%)	Refs.
C_{13}				
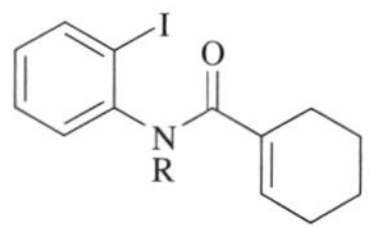 (2-iodoaryl N-R cyclohexenyl amide)		Pd(OAc)$_2$, PPh$_3$, AgNO$_3$, Et$_3$N, rt	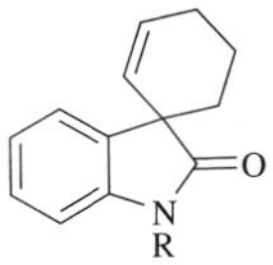 (spiroindolinone) (68)^j	111
C_{14}				
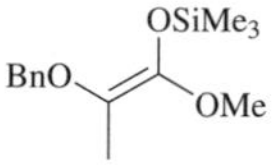 (BnO, OSiMe$_3$, OMe ketene acetal)	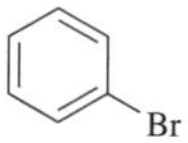 PhBr	Pd(dba)$_2$ (5 mol %), P(t-Bu)$_3$ (10 mol %), ZnF$_2$ (0.5 eq), DMF, 80°, 12 h	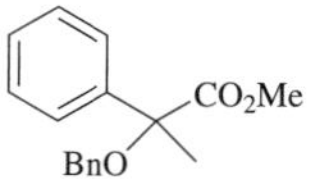 (PhC(Me)(OBn)CO$_2$Me) (75)	99
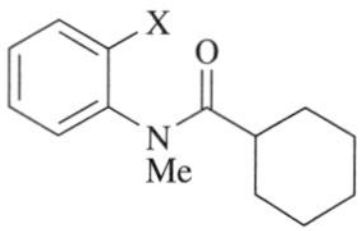 (2-haloaryl N-Me cyclohexanecarboxamide)		Pd(OAc)$_2$ (5 mol %), PCy$_3$ (5 mol %), NaO-t-Bu (1.5 eq), dioxane	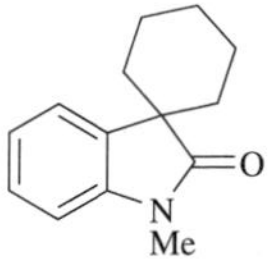 (spiroindolinone I) X Temp (°) Time (h) Cl 70 4 (90) Br 50 1 (93)	32
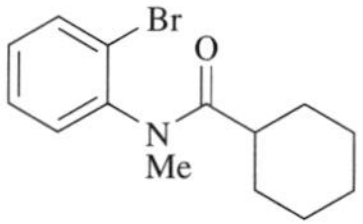 (2-bromoaryl N-Me cyclohexanecarboxamide)		Pd(dba)$_2$ (5 mol %), L23 (7.5 mol %), NaO-t-Bu (1.5 eq), 4 h	I (82) 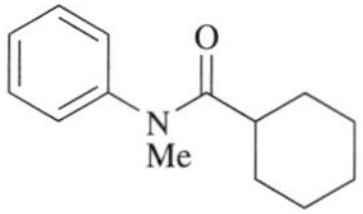 + (N-methyl cyclohexanecarboxanilide) (1)	67
C_{15}				
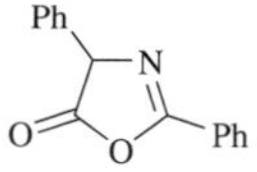 (4-phenyl-2-phenyloxazol-5(4H)-one)	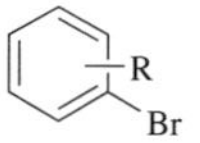 (ArBr, R-substituted)	Pd(OAc)$_2$ (5 mol %), L12 (5 mol %), K$_3$PO$_4$ (3.3 eq), toluene, 80°, 14 h	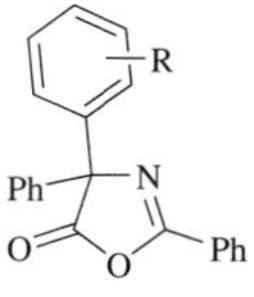 (4,4-diaryl oxazolone I) R H (94) 3-CF$_3$ (92) 4-MeO (75)	79

224

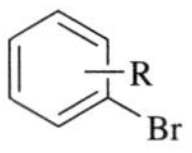

	R		
	H	(55)	
Pd(dba)$_2$ (5 mol %), **L35** (10 mol %), K$_2$CO$_3$ (3.3 eq), toluene, 100°, 14 h	3-CF$_3$	(55)	79
	4-MeO	(30)	

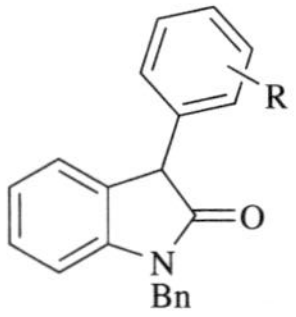

X	R		
Cl	H	(70)	
Cl	2-Me	(60)	
Cl	4-MeO	(70)	
Cl	4-CF$_3$	(66)	
Br	H	(91)	
Br	2-Me	(74)	
Br	3-MeO	(85)	
Br	3-Cl	(73)	160
Br	3-Me	(77)	
Br	4-MeO	(85)	
Br	4-Cl	(70)	
Br	4-CF$_3$	(80)	
Br	4-*t*-Bu	(79)	
OTf	H	(85)	
OTf	4-Cl	(70)	

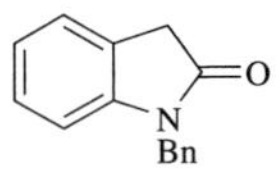

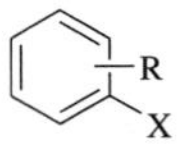

Pd(dba)$_2$ (2 mol %), **L21** (3 mol %), KHMDS (1.1 eq), THF/toluene, 70°, 30 min

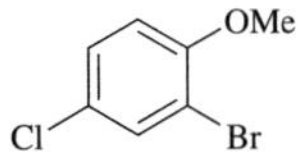

Pd(dba)$_2$ (2 mol %), **L21** (3 mol %), KHMDS (1.1 eq), THF/toluene, 70°, 30 min

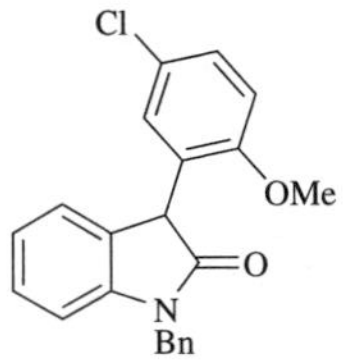

(77)   160

TABLE 2. ARYLATION OF ESTERS, AMIDES, LACTONES, LACTAMS, NITRILES, KETENE ACETALS, AND PREFORMED ENOLATES (*Continued*)

Substrate	Aryl Compound	Conditions	Product(s) and Yield(s) (%)	Refs.
C_{15}				
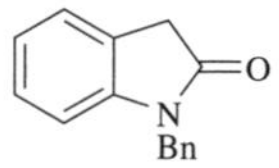 *N-benzyl oxindole*	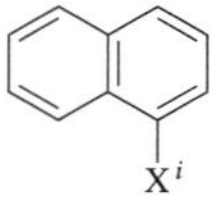 1-naphthyl-X[i]	Pd(dba)$_2$ (2 mol %), **L21** (3 mol %), KHMDS (1.1 eq), THF/toluene, 70°, 30 min	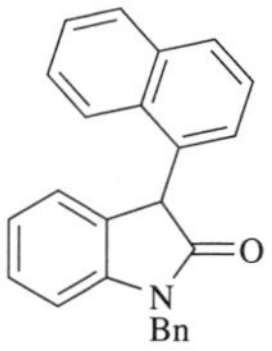 3-(1-naphthyl)-N-benzyl oxindole (80)	160
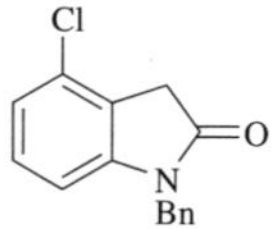 *N-benzyl-4-chloro oxindole*	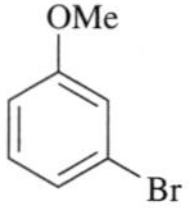 3-bromoanisole	Pd(dba)$_2$ (2 mol %), **L21** (3 mol %), KHMDS (1.1 eq), THF/toluene, 70°, 2.5 h	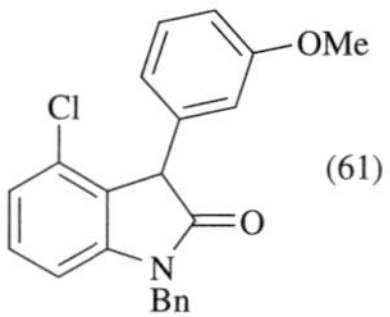 3-(3-methoxyphenyl)-4-chloro-N-benzyl oxindole (61)	160
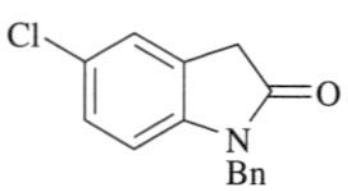 *N-benzyl-5-chloro oxindole*	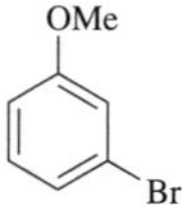 3-bromoanisole	Pd(dba)$_2$ (2 mol %), **L21** (3 mol %), KHMDS (1.1 eq), THF/toluene, 70°, 2.5 h	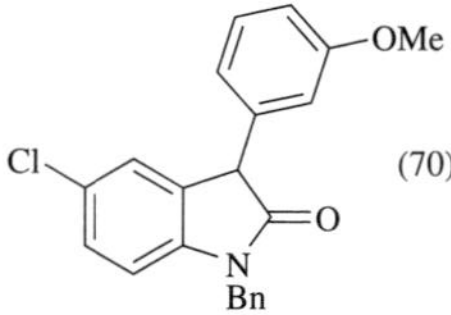 3-(3-methoxyphenyl)-5-chloro-N-benzyl oxindole (70)	160
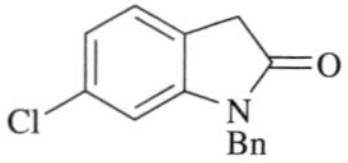 *N-benzyl-6-chloro oxindole*	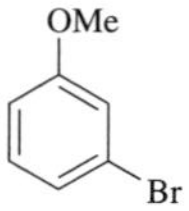 3-bromoanisole	Pd(dba)$_2$ (2 mol %), **L21** (3 mol %), KHMDS (1.1 eq), THF/toluene, 70°, 1 h	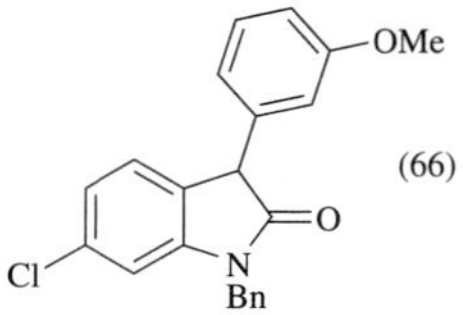 3-(3-methoxyphenyl)-6-chloro-N-benzyl oxindole (66)	160

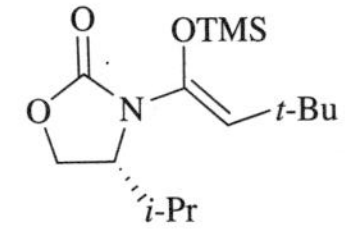

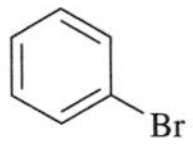

Pd(dba)₂ (5 mol %),
P(t-Bu)₃ (10 mol %), ZnF₂ (0.5 eq),
DMF, 80°, 12 h

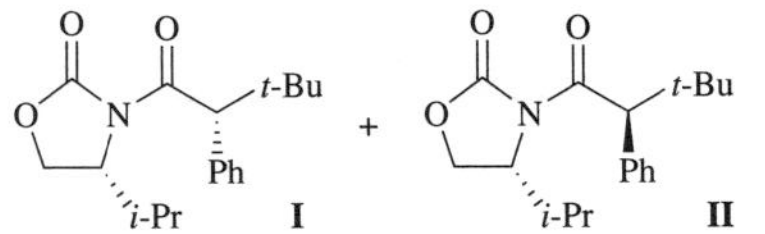

I + II (35), **I:II** = 89:11

99

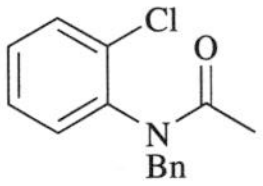

Pd(OAc)₂ (5 mol %), PCy₃ (5 mol %),
NaO-t-Bu (1.5 eq), dioxane, 70°, 5 h

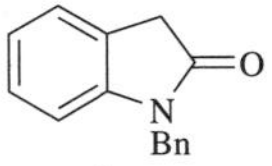

(64)

32

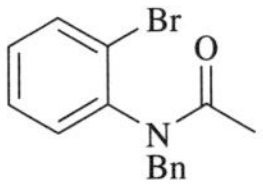

Pd(dba)₂ (5 mol %),
ligand (7.5 mol %),
base (1.5 eq), THF

I + II

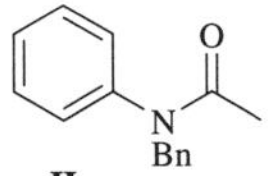

Ligand	Base	Temp (°)	Time (h)	I	II
L1	KHMDS	75	19	trace	(14)
L1	NaO-t-Bu	75	19	(56)	(13)
L1	NaO-t-Bu	100	3	(53)	(9)
L23	NaO-t-Bu	75	23	(65)	(2)
L23	NaO-t-Bu	100	3	(57)	(2)

67

Pd(dba)₂ (5 mol %),
ligand (7.5 mol %),
NaO-t-Bu (1.5 eq), dioxane

I + II

Ligand	Temp (°)	Time (h)	I	II
P(o-tol)₃	100	4	(6)	(—)
L1	75	23	(10)	(1)
L1	100	3	(43)	(10)
L23	75	23	(50)	(1)
L23	100	3	(64)	(1)

67

227

TABLE 2. ARYLATION OF ESTERS, AMIDES, LACTONES, LACTAMS, NITRILES, KETENE ACETALS, AND PREFORMED ENOLATES (*Continued*)

Substrate	Aryl Compound	Conditions	Product(s) and Yield(s) (%)	Refs.
C_{15}	(2-X-phenyl)-N-methyl cyclohexylacetamide (X = Cl, Cl, Br, Br, Br, Br, Br, Br, Br, Br)	Pd(OAc)$_2$ (5 mol %), ligand (5 mol %), NaO-t-Bu (1.5 eq), dioxane Ligand / Temp (°) / Time (h) PCy$_3$ / 70 / 24 **L31** / 70 / 24 P(t-Bu)$_3$ / 50 / 10 PCy$_3$ / 50 / 3 PCy$_3$ / 50 / 10 **L15** / 50 / 10 **L18** / 50 / 10 **L47** / 50 / 10 **L47** / 50 / 20 **L51** / 50 / 10	3-cyclohexyl-1-methyl-2-oxindole (35) (64) (34) (43) (45) (36) (31) (77) (75) (56)	32
(2-X-phenyl)-N-methyl phenylacetamide X = Cl Br		Pd(OAc)$_2$ (5 mol %), PCy$_3$ (5 mol %), NaO-t-Bu (1.5 eq), dioxane, 3 h Temp (°) 70 50	3-phenyl-1-methyl-2-oxindole **I** (89) (92)	32
	I	Pd(OAc)$_2$ (5 mol %), PCy$_3$ (5 mol %), NaO-t-Bu (1.5 eq), dioxane	X / Temp (°) / Time (h) Cl / 70 / 24 (69) Br / 50 / 1.5 (82)	32

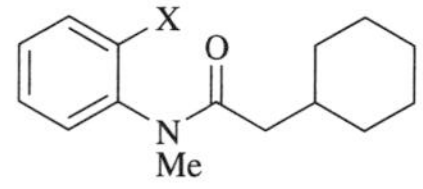

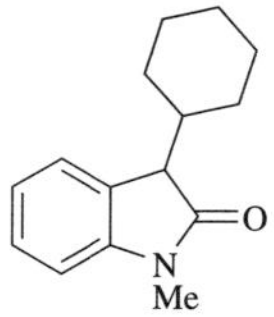

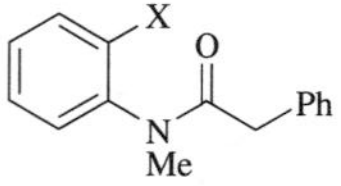

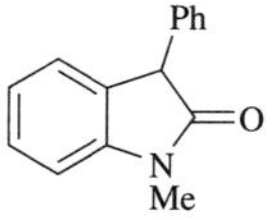

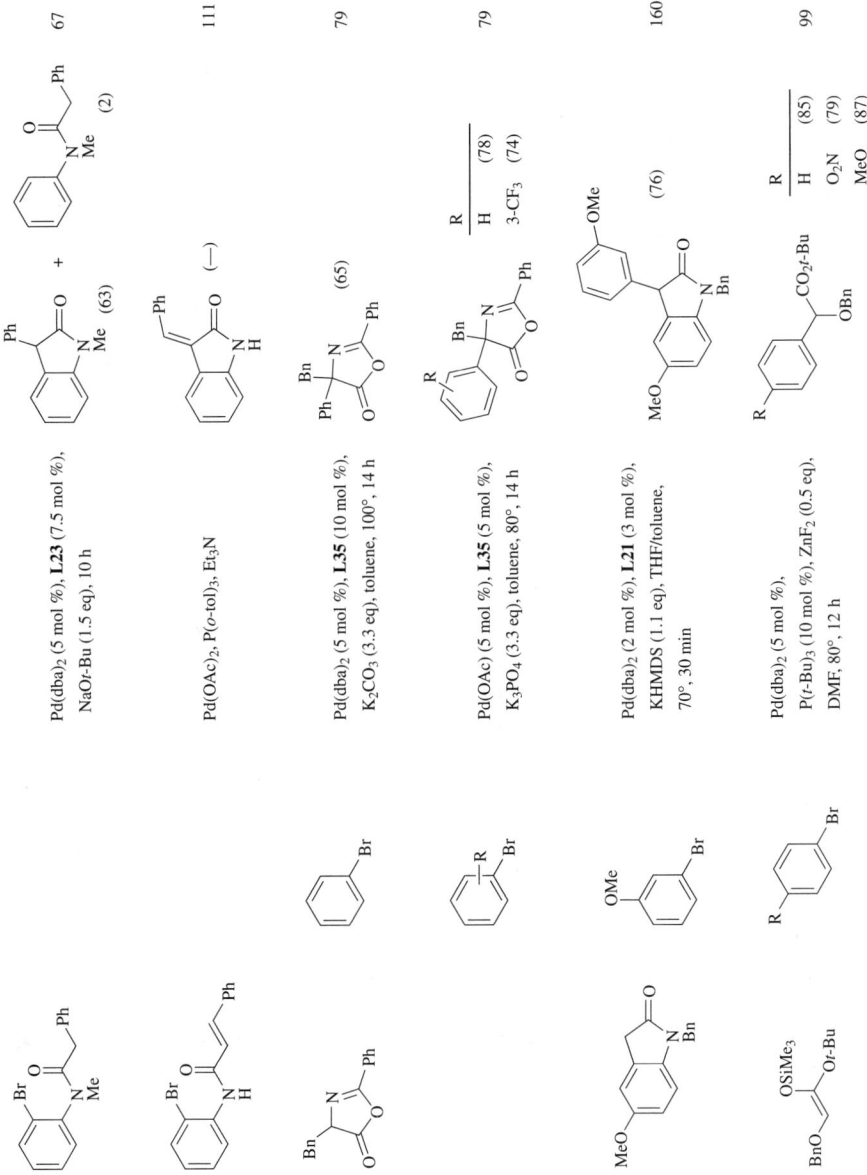

TABLE 2. ARYLATION OF ESTERS, AMIDES, LACTONES, LACTAMS, NITRILES, KETENE ACETALS, AND PREFORMED ENOLATES (*Continued*)

Substrate	Aryl Compound	Conditions	Product(s) and Yield(s) (%)	Refs.
C$_{16}$ 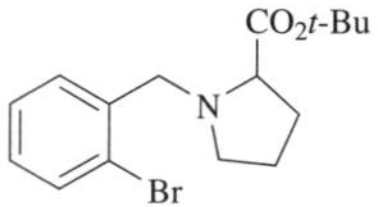		Pd$_2$(dba)$_3$ (x mol %), ligand (y mol %), LiOt-Bu (2 eq), dioxane, 85°, 20 h $\quad \begin{array}{ccc} x & \text{Ligand} & y \\ 3.8 & \text{PCy}_3 & 8 \\ 2.3 & \textbf{L17} & 2.5 \end{array}$	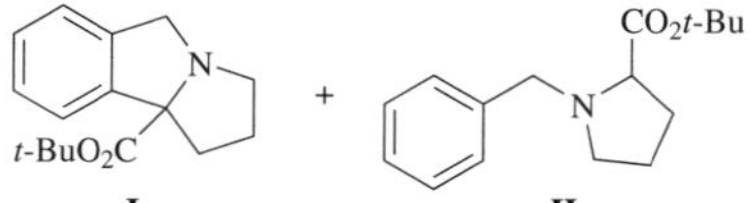 **I** **II** (55) (8) (62) (—)	78
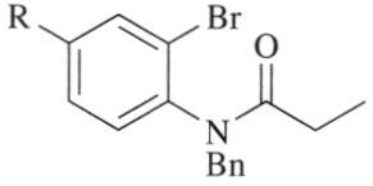		Pd$_2$(dba)$_3$ (1.5–3 mol %), **L52** (3–6 mol %), LiHMDS (1.5–2 eq), THF, 68°, 1–4 h	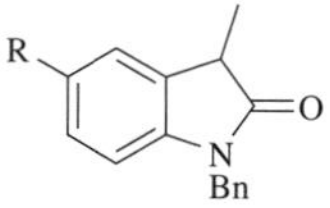 $\begin{array}{cc} R & \\ H & (90) \\ F & (40) \end{array}$	69
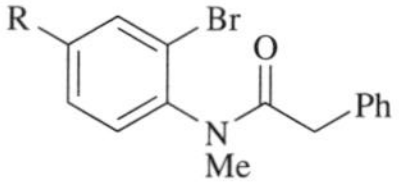		Pd(OAc)$_2$ (5 mol %), PCy$_3$ (5 mol %), NaOt-Bu (1.5 eq), dioxane, 50°	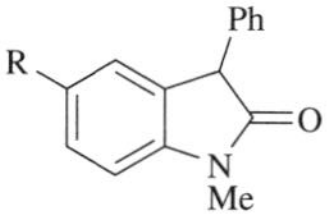 $\begin{array}{ccc} R & \text{Time (h)} & \\ \text{MeO} & 4 & (83) \\ \text{NC} & 9 & (66) \end{array}$	32
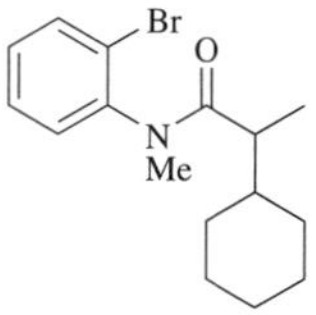		Pd(OAc)$_2$ (5 mol %), PCy$_3$ (5 mol %), NaOt-Bu (1.5 eq), dioxane, 50°, 14 h	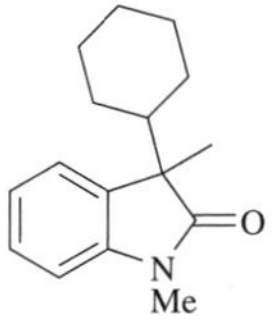 (64)	32
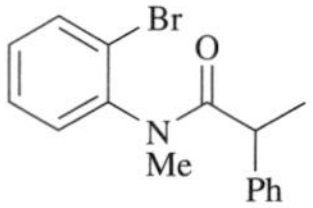		Pd(dba)$_2$ (x mol %), ligand, rt, 24 h	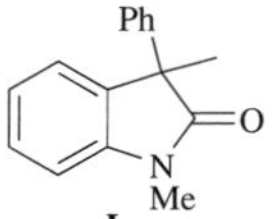 **I** $\begin{array}{cccc} x & \text{Ligand} & & \text{\% ee} \\ 10 & \textbf{L49} & (94) & 34 \\ 5 & \textbf{L50} & (74) & 57 \end{array}$	32

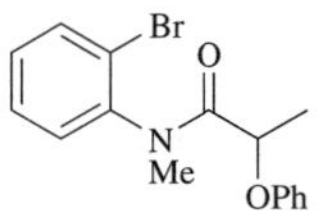

C17

	Conditions	Product	Yield (%)
(Br-aryl amide with OPh)	Pd(OAc)$_2$ (5 mol %), PCy$_3$ (5 mol %), NaO-t-Bu (1.5 eq), dioxane, 50°, 14 h	I (99)	32
	Pd(OAc)$_2$ (5 mol %), PCy$_3$ (5 mol %), NaO-t-Bu (1.5 eq), dioxane	(0)	32

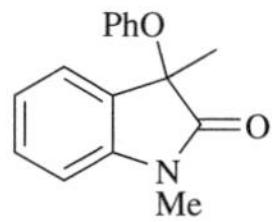

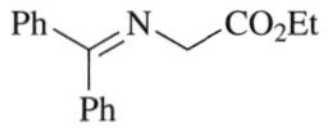

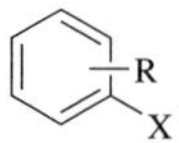

Pd(dba)$_2$ (2 mol %), P(t-Bu)$_3$ (4 mol %), K$_3$PO$_4$ (3 eq), toluene, 20 h — 77

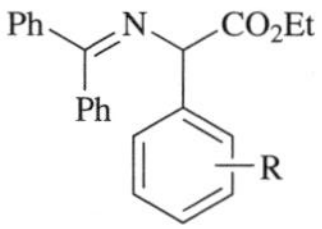

X	R	Temp (°)	Yield (%)
Cl	H	120	(82)
Cl	2-Me	120	(81)
Cl	4-MeO	120	(83)
Cl	4-F	120	(83)
Cl	4-NC	120	(85)
Cl	4-CF$_3$	120	(84)
Br	H	100	(88)
Br	2-MeO	100	(89)
Br	2-Me	100	(84)
Br	4-MeO	100	(85)
Br	4-PhO	100	(90)
Br	4-F	100	(86)
Br	4-NC	100	(89)
Br	4-MeO$_2$C	100	(89)
Br	4-CF$_3$	100	(86)
Br	4-Ph	100	(92)

TABLE 2. ARYLATION OF ESTERS, AMIDES, LACTONES, LACTAMS, NITRILES, KETENE ACETALS, AND PREFORMED ENOLATES (*Continued*)

Substrate	Aryl Compound	Conditions	Product(s) and Yield(s) (%)	Refs.
C$_{17}$				
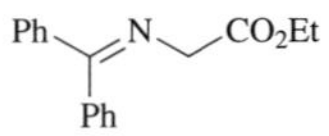 Ph-N=C(Ph)-CH$_2$-CO$_2$Et	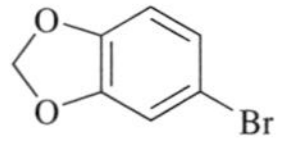 5-Br-benzo[d][1,3]dioxole	Pd(dba)$_2$ (2 mol %), P(t-Bu)$_3$ (4 mol %), K$_3$PO$_4$ (3 eq), toluene, 20 h	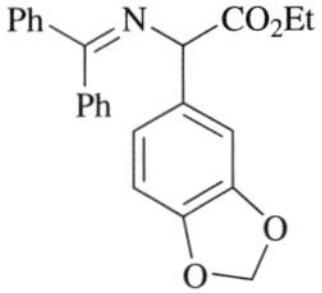 Ph-N=C(Ph)-CH(CO$_2$Et)-(benzo[d][1,3]dioxol-5-yl) (87)	77
	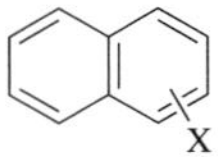 1- or 2-X-naphthalene	Pd(dba)$_2$ (2 mol %), P(t-Bu)$_3$ (4 mol %), K$_3$PO$_4$ (3 eq), toluene, 100°, 20 h	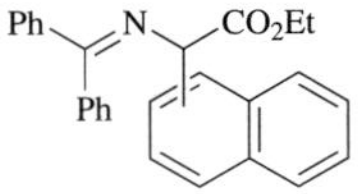 Ph-N=C(Ph)-CH(CO$_2$Et)-(naphthyl) X: 1-Br (87); 2-Br (89)	77
	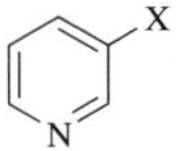 3-X-pyridine	Pd(dba)$_2$ (2 mol %), P(t-Bu)$_3$ (4 mol %), K$_3$PO$_4$ (3 eq), toluene, 20 h	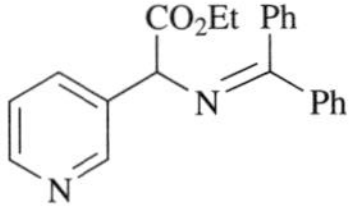 CO$_2$Et, Ph-N=C(Ph)-CH-(pyridin-3-yl) X Temp(°) Cl 120 (80) Br 100 (85)	77
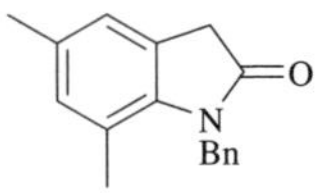 5,7-dimethyl-1-benzyl-indolin-2-one	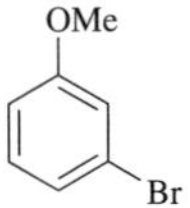 3-bromoanisole (OMe, Br)	Pd(dba)$_2$ (2 mol %), **L21** (3 mol %), KHMDS (1.1 eq), THF/toluene, 70°, 30 min	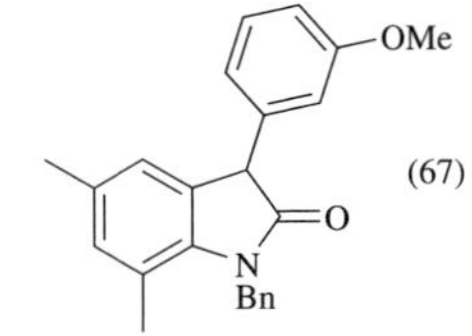 3-(3-methoxyphenyl)-5,7-dimethyl-1-benzyl-indolin-2-one (67)	160
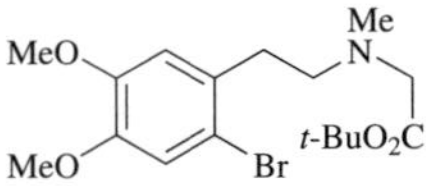 Me-N-CH$_2$CH$_2$-(aryl)-CH$_2$-CO$_2$t-Bu (aryl = 4-Br-2,5-(MeO)$_2$-C$_6$H$_2$)		Pd$_2$(dba)$_3$ (2.3 mol %), **L13** (2.5 mol %), LiO-t-Bu (2 eq), dioxane, 85°, 3 h	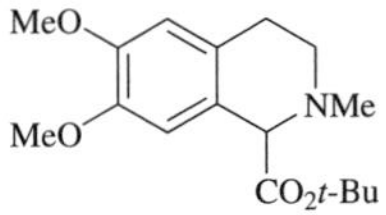 tetrahydroisoquinoline-NMe, CO$_2$t-Bu, (MeO)$_2$ (75)	78

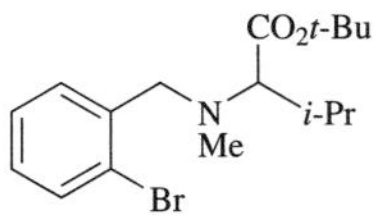

Conditions	Product	Yield (%)

Pd$_2$(dba)$_3$ (3.8 mol %), **L13** (8 mol %), LiOt-Bu (2 eq), dioxane, 110°, 48 h — (51) + (8), 78

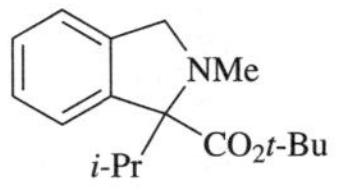

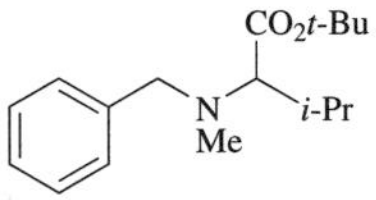

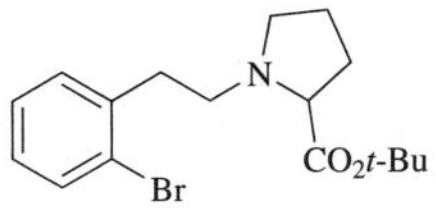

Pd$_2$(dba)$_3$ (2.3 mol %), **L17** (2.5 mol %), LiOt-Bu (2 eq), dioxane, 90°, 24 h — (74), 78

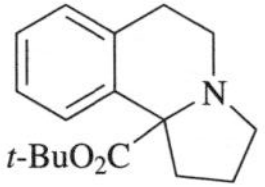

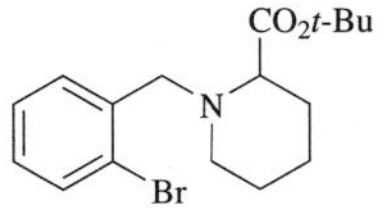

Pd$_2$(dba)$_3$ (2.3 mol %), ligand (2.5 mol %), LiOt-Bu (2 eq), dioxane, 85°, 20 h

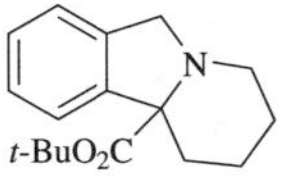

Ligand	
PCy$_3$	(39)
L17	(84)

78

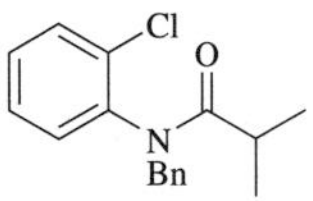

Pd(OAc)$_2$ (5 mol %), PCy$_3$ (5 mol %), NaOt-Bu (1.5 eq), dioxane, 70°, 3 h — (93), 32

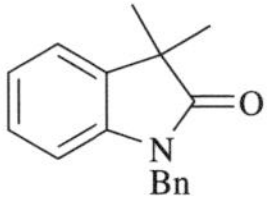

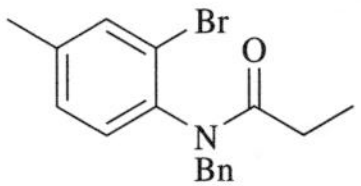

Pd$_2$(dba)$_3$ (1.5–3 mol %), **L52** (3–6 mol %), LiHMDS (1.5–2 eq), THF, 68°, 1–4 h — (83), 69

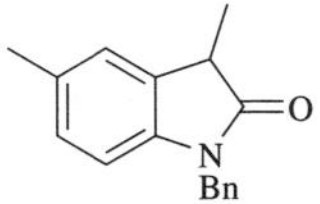

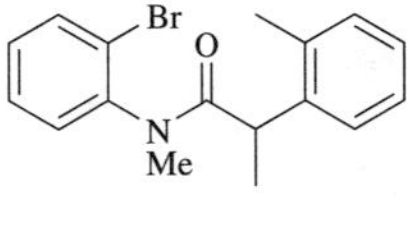

Pd(dba)$_2$ (x mol %), ligand, rt, 24 h

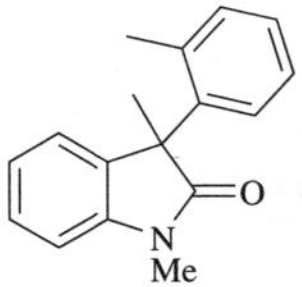

x	Ligand		% ee
10	**L49**	(97)	33
5	**L50**	(95)	42

32

TABLE 2. ARYLATION OF ESTERS, AMIDES, LACTONES, LACTAMS, NITRILES, KETENE ACETALS, AND PREFORMED ENOLATES (*Continued*)

Substrate	Aryl Compound	Conditions	Product(s) and Yield(s) (%)	Refs.
C_{18}				
(isobutyl ester with Me-N-ethylaryl, Br, di-MeO)		Pd$_2$(dba)$_3$ (2.3 mol %), L13 (2.5 mol %), LiOt-Bu (2 eq), dioxane, 100°, 24 h	(tetrahydroisoquinoline with NMe, CO$_2t$-Bu, di-MeO) (66)	78
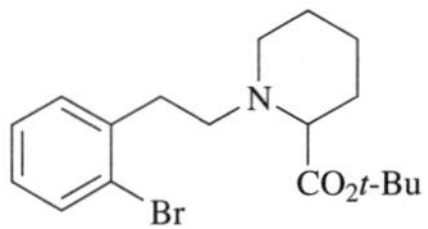 (piperidine-CO$_2t$-Bu with N-ethyl-2-bromoaryl)		Pd$_2$(dba)$_3$ (2.3 mol %), ligand (2.5 mol %), LiOt-Bu (2 eq), dioxane, 90°, 24 h	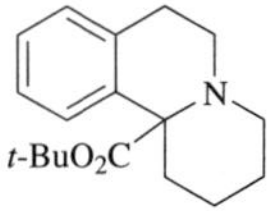 (fused tricyclic, t-BuO$_2$C) Ligand L13 (81) L17 (66)	78
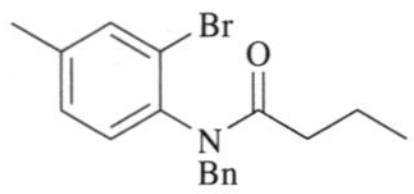 (2-bromo-4-methyl-N-Bn butyramide)		Pd$_2$(dba)$_3$ (1.5–3 mol %), L52 (3–6 mol %), LiHMDS (1.5–2 eq), THF, 68°, 1–4 h	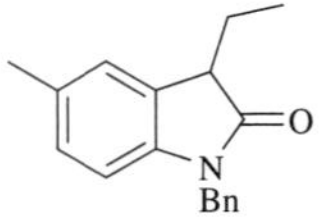 (3-ethyl-5-methyl-N-Bn oxindole) (62)	69
(N-Ph-N-(2-bromobenzyl) glycinate t-Bu ester)		Pd$_2$(dba)$_3$ (2.3 mol %), L17 (2.5 mol %), LiOt-Bu (2 eq), dioxane, 85°, 1 h	(isoindoline NPh, CO$_2t$-Bu) (79)	78
C_{19}				
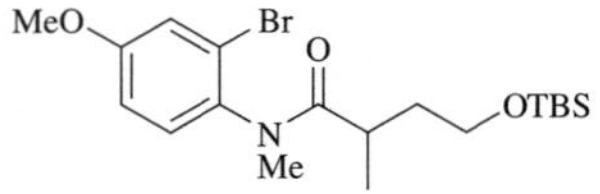 (2-bromo-4-MeO-N-Me anilide with α-Me, OTBS chain)		See table	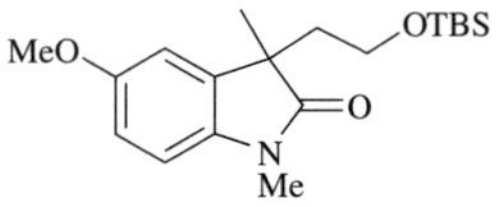 (5-MeO-N-Me-3-Me-3-(CH$_2$CH$_2$OTBS) oxindole)	70

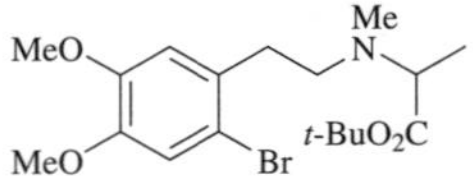

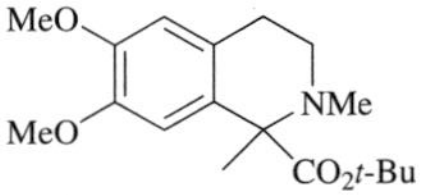

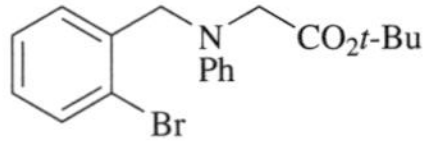

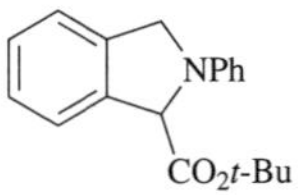

Catalyst	Ligand	Base	Solvent	Temp (°)	
Pd(OAc)$_2$	(R)-L23	LiHMDS	THF	68	(60), 11% ee
Pd$_2$(dba)$_3$	P(t-Bu)$_3$	LiHMDS	THF	68	(<1)
Pd$_2$(dba)$_3$	L1	LiHMDS	THF	68	(<1)
Pd$_2$(dba)$_3$	L23	LiHMDS	THF	68	(50)
Pd$_2$(dba)$_3$	L23	KHMDS	toluene	70	(<5)

C_{20}	Substrate	Conditions	Product(s)	Yield

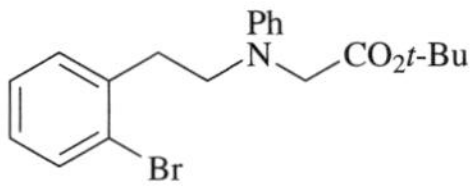

Substrate 1: 2-bromophenethyl-N(Ph)-CH2-CO2t-Bu

Pd2(dba)3 (2.3 mol %), **L13** (2.5 mol %), LiOt-Bu (2 eq), dioxane, 85°, 8 h

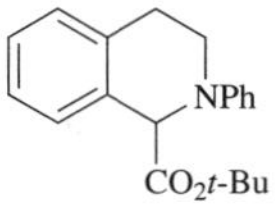

Product: tetrahydroisoquinoline with NPh and CO2t-Bu (79)

78

Substrate 2: 2-bromobenzyl-N(Ph)-CH(CH3)-CO2t-Bu

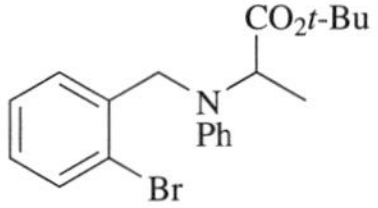

Pd2(dba)3 (2.3 mol %), **L17** (2.5 mol %), LiOt-Bu (2 eq), dioxane, 85°, 24 h

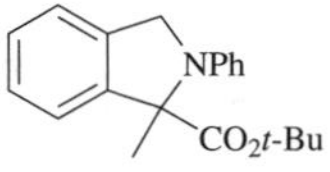

Product **I**: indoline with NPh, Me, CO2t-Bu (91)

78

Pd2(dba)3 (2.3 mol %), PCy3 (2.5 mol %), LiOt-Bu (2 eq), dioxane, 85°, 24 h

I (89)

78

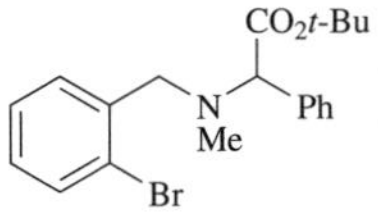

Substrate 3: 2-bromobenzyl-N(Me)-CH(Ph)-CO2t-Bu

Pd2(dba)3 (2.3 mol %), ligand (2.5 mol %), LiOt-Bu (2 eq), dioxane

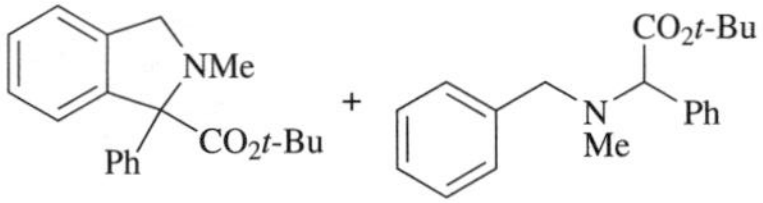

Products: **I** (indoline Ph, CO2t-Bu, NMe) + **II** (open-chain N(Me)(CH2Ph)-CH(Ph)-CO2t-Bu)

Ligand	Temp (°)	Time (h)	**I**	**II**
none	90	17	(35)	(5)
L13	50	20	(65)	(—)
L13	70	3.5	(95)	(—)
L13	90	2	(99)	(—)

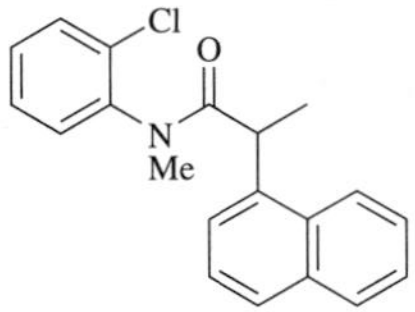

78

Substrate: N-methyl-2-chloro-N-(2-(naphthalen-1-yl)propanoyl)aniline

Pd(dba)2 (10 mol %), ligand

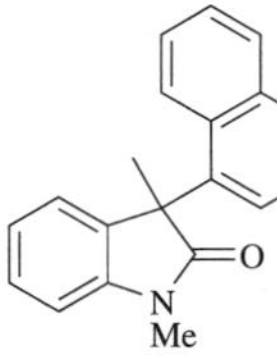

Product: 3-methyl-3-(naphthalen-1-yl)-1-methyl-2-indolinone

Ligand	Temp (°)	Time (h)		% ee
L49	50	26	(91)	69
L50	0	48	(5)	58

32

TABLE 2. ARYLATION OF ESTERS, AMIDES, LACTONES, LACTAMS, NITRILES, KETENE ACETALS, AND PREFORMED ENOLATES (*Continued*)

Substrate	Aryl Compound	Conditions	Product(s) and Yield(s) (%)	Refs.
C_{20}				
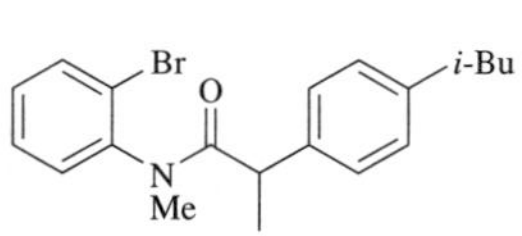 (structure: Br-aryl amide with i-Bu)		Pd(OAc)$_2$ (5 mol %), PCy$_3$ (5 mol %), NaO-t-Bu (1.5 eq), dioxane, 50°, 14 h	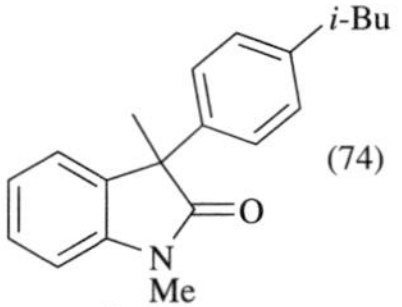 **I** (74)	32
		Pd(dba)$_2$ (x mol %), ligand, rt	**I** see table below	32
			x Ligand Time (h) % ee	
			10 **L49** 24 (73) 40	
			5 **L50** 20 (87) 50	
(structure: Br-aryl amide with naphthyl)		Catalyst (5 mol %), **L49** (5 mol %), base (1.5 eq)	**I** see table below	32

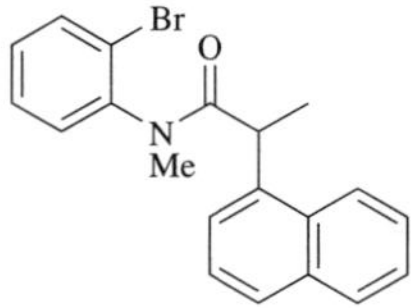

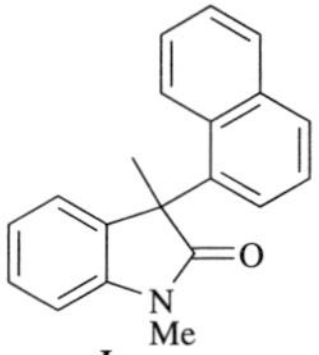

Catalyst	Base	Solvent	Temp	Time (h)		% ee
Pd(OAc)$_2$	NaO-t-Bu	toluene	50°	14	(50)	34
Pd(OAc)$_2$	NaO-t-Bu	m-xylene	50°	14	(60)	28
Pd(OAc)$_2$	NaO-t-Bu	THF	50°	14	(55)	42
Pd(OAc)$_2$	NaO-t-Bu	DME	50°	14	(64)	55
Pd(OAc)$_2$	NaO-t-Bu	diglyme	50°	14	(61)	51
Pd(dba)$_2$	LiO-t-Bu	DME	50°	36	(20)	50
Pd(dba)$_2$	NaO-t-Bu	DME	rt	14	(93)	67
Pd(dba)$_2$	NaO-t-Bu	DME	50°	16	(89)	59
Pd(dba)$_2$	KO-t-Bu	DME	50°	36	(5)	11
Pd(dba)$_2$	NaHMDS	DME	50°	14	(56)	48
Pd(dba)$_2$	KHMDS	DME	50°	16	(57)	6

Pd(dba)$_2$ (x mol %), ligand **I**

x	Ligand	Temp	Time (h)		% ee	
10	**L49**	50°	8	(93)	59	32
1	**L50**	rt	26	(91)	69	
10	**L50**	0°	40	(27)	70	32

Pd(OAc)$_2$ (5 mol %), ligand (5 mol %), NaO*t*-Bu (1.5 eq), dioxane **I**

Ligand	Temp (°)	Time (h)		% ee
PCy$_3$	50	12	(89)	—
L5	70	24	(57)	38 (+)
L6	70	24	(67)	41 (+)
L7	100	24	(70)	61 (+)
L8	100	6	(55)	31 (−)
L9	100	24	(56)	61 (+)
L10	100	14	(24)	38 (+)
L11	100	12	(60)	15 (−)
(*R*)-**L23**	100	3	(49)	46 (−)
(*R*)-**L24**	100	3	(45)	18 (−)
(*R*)-**L29**	100	24	(49)	7 (+)
(*R*)-**L30**	50	3	(63)	0
(*R*)-**L31**	100	24	(50)	2 (−)
L39	70	24	(72)	6 (−)
L40	100	12	(72)	2 (−)
L41	100	24	(66)	16 (+)
L42	70	9	(75)	53 (+)
L43	50	24	(82)	20 (−)
L44	100	36	(37)	4 (+)
L45	100	36	(18)	5 (−)
L48	70	20	(35)	4
L49	50	12	(56)	34

TABLE 2. ARYLATION OF ESTERS, AMIDES, LACTONES, LACTAMS, NITRILES, KETENE ACETALS, AND PREFORMED ENOLATES (*Continued*)

Substrate	Conditions	Product(s) and Yield(s) (%)	Refs.
C$_{20}$ 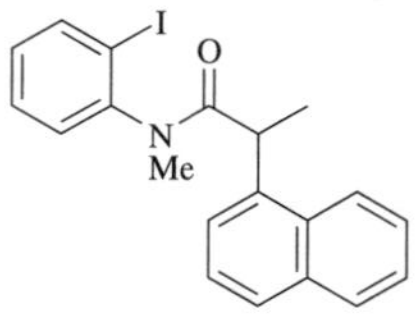	Pd(dba)$_2$ (10 mol %), ligand	(75) 0 (30) 40 — Ligand L49 Temp 50° Time 6 h; L50 Temp 0° Time 36 h; % ee values 0 and 40	32
C$_{21}$ 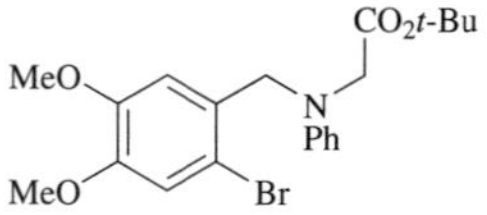	Pd$_2$(dba)$_3$ (2.3 mol %), L17 (2.5 mol %), LiOt-Bu (2 eq), dioxane, 85°, 24 h	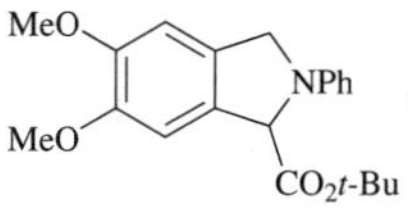 (89)	78
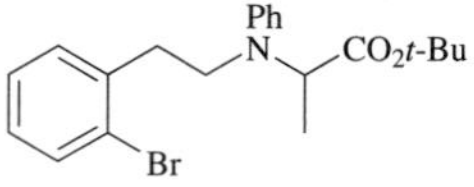	Pd$_2$(dba)$_3$ (2.3 mol %), ligand (2.5 mol %), LiOt-Bu (2 eq), dioxane, 85°, 24 h	I + II; Ligand PPh$_3$: I (12), II (7); L13: I (62), II (—); L18: I (27), II (—)	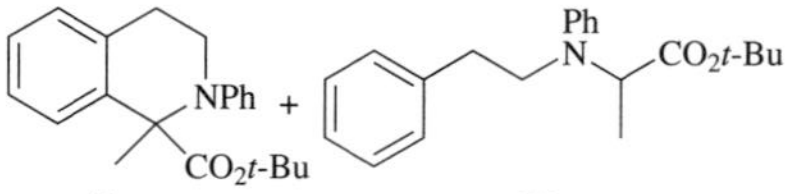 78
C$_{22}$ 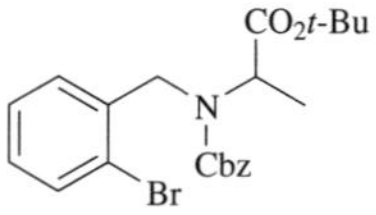	Pd$_2$(dba)$_3$ (2.3 mol %), L17 (2.5 mol %), LiOt-Bu (2 eq), dioxane, 90°, 24 h	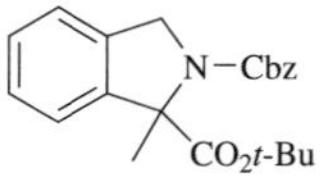 (67)	78

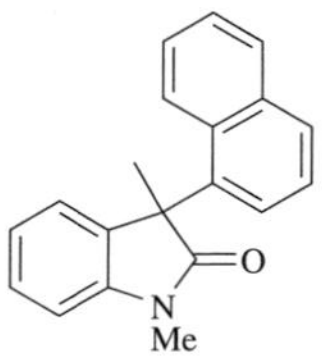

C$_{25}$

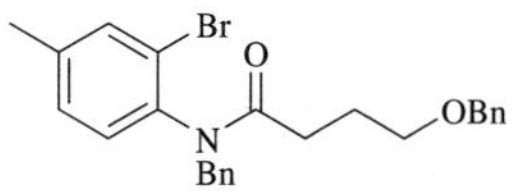

Substrate: 2-bromo-N-benzyl-N-(aryl)-4-(benzyloxy)butanamide with 4-methyl group

Pd$_2$(dba)$_3$ (1.5–3 mol %),
L52 (3–6 mol %), LiHMDS (1.5–2 eq),
THF, 68°, 1–4 h

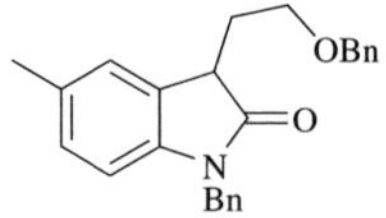

Product: 3-(2-(benzyloxy)ethyl)-1-benzyl-5-methylindolin-2-one (47)

70

Substrate: MeO-aryl bromide with N-Me, CH$_2$-C(O)-4-MeOC$_6$H$_4$

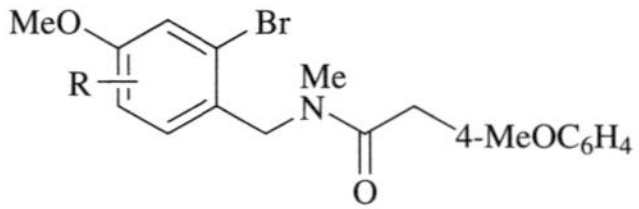

Pd(dba)$_2$, dppe, KOt-Bu, dioxane, 100°

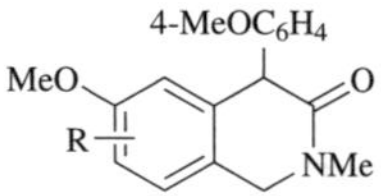

Product: 4-(4-MeOC$_6$H$_4$)-N-Me-MeO-isoquinolinone

R	
3-OBn	(54)
5-OBn	(81)

71

C$_{26}$

Substrate: 2-bromo-N-benzyl-2-(naphthalen-1-yl)propanamide

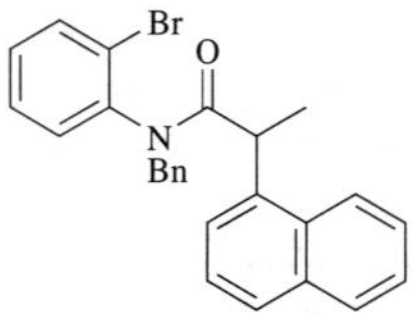

Pd(OAc)$_2$ (5 mol %),
PCy$_3$ (5 mol %), NaOt-Bu (1.5 eq),
dioxane, 50°, 12 h

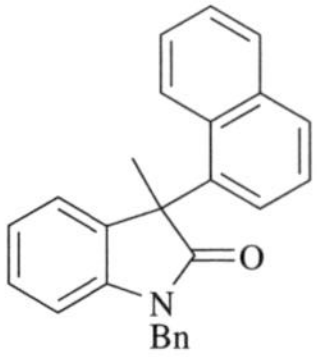

Product: 1-benzyl-3-methyl-3-(naphthalen-1-yl)indolin-2-one **I** (97)

32

Pd(dba)$_2$ (x mol %), ligand

I

x	Ligand	Temp	Time (h)		% ee
10	**L49**	rt	24	(88)	67
2	**L50**	rt	24	(80)	71
10	**L50**	10°	40	(75)	76

32

TABLE 2. ARYLATION OF ESTERS, AMIDES, LACTONES, LACTAMS, NITRILES, KETENE ACETALS, AND PREFORMED ENOLATES (*Continued*)

Substrate	Aryl Compound	Conditions	Product(s) and Yield(s) (%)	Refs.
C_{27} [structure with CO_2t-Bu, Ph, N-Cbz, Br]		Pd$_2$(dba)$_3$ (2.3 mol %), L13 (2.5 mol %), LiOt-Bu (2 eq), dioxane, 90°, 2 h	[structure with N–Cbz, CO_2t-Bu, Ph] (86)	78

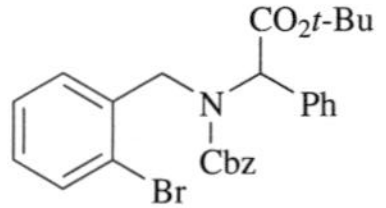

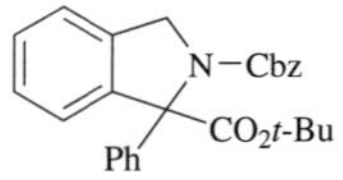

[a] The amount of amide used was 0.5 eq.
[b] The carbene was not identified.
[c] The reaction was performed at 70°.
[d] Pd(dba)$_2$ (2 mol %) and L12 (2 mol %) were used.
[e] The reaction was performed with 5 mol % of catalyst.
[f] The reaction was performed at room temperature with Zn(Ot-Bu)$_2$ (0.25 eq).
[g] The reaction was performed with 10 mol % of Pd(dba)$_2$.
[h] The R group was not defined.
[i] The X group was not defined.

TABLE 3. ARYLATION OF 1,3-DICARBONYLS AND CYANOACETATES

Substrate	Aryl Compound	Conditions	Product(s) and Yield(s) (%)	Refs.
C₄ CO₂Me–CH₂–CN	4-iodotoluene	PdCl₂(PPh₃)₂ (4 mol %), KO*t*-Bu (2.2 eq), monoglyme, 70°, 6 h	Ar-CH(CO₂Me)(CN) (83)	88
	2-iodonaphthalene	PdCl₂(PPh₃)₂ (4 mol %), KO*t*-Bu (2.2 eq), monoglyme, 70°, 6 h	Naph-CH(CO₂Me)(CN) (88)	88
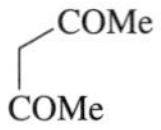 C₅ COMe–CH₂–COMe	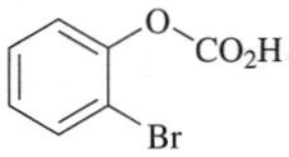 2-bromophenyl acetate (O-CO₂H on ring linked via O)	CuBr (cat), NaH (2 eq), 16 h	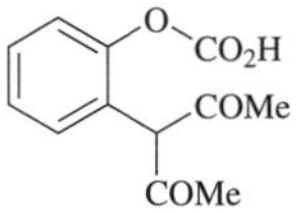 o-(OCO₂H)C₆H₄-CH(COMe)₂ (0)	101
	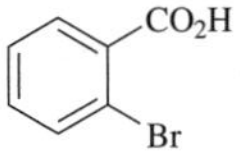 2-bromobenzoic acid	Cu (6 mol %), NaOEt, EtOH	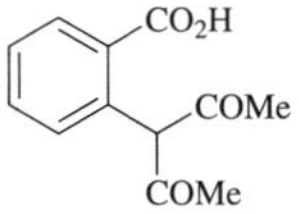 o-(CO₂H)C₆H₄-CH(COMe)₂ (34)	101
	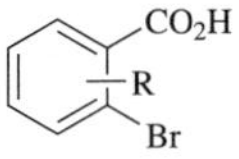 2-bromo-R-benzoic acid	CuBr (cat), NaH (2 eq), neat	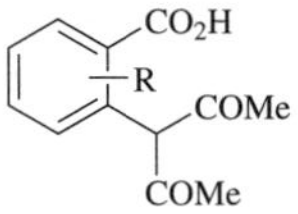 o-(CO₂H)(R)C₆H₃-CH(COMe)₂ R / Time (h) / Yield H — 1 (91) 3-O₂N — 5 (0) 3-HO₂C — 38 (40) 4-O₂N — 0.5 (91) 4-Me — 0.8 (82) 5-H₂N — 6 (73) 5-HO — 6.5 (77) 4,5-(MeO)₂ — 3 (59) 3,4,5-(MeO)₃ — 7 (84)	101

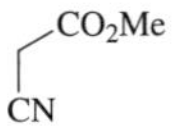

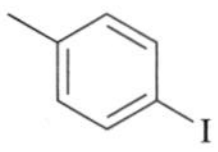

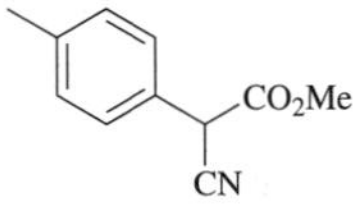

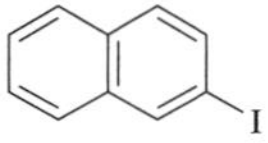

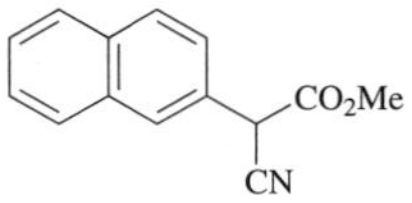

TABLE 3. ARYLATION OF 1,3-DICARBONYLS AND CYANOACETATES (*Continued*)

Substrate	Aryl Compound	Conditions	Product(s) and Yield(s) (%)	Refs.
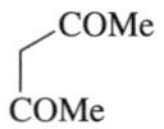 C₅ COMe / COMe	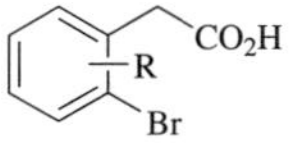 2-Br-C₆H₄-CH₂-CO₂H (R)	CuBr (cat), NaH (2 eq)	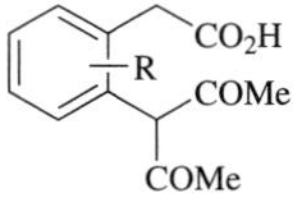 ArCH(CO₂H)-CH(COMe)₂ R: H, 18 (72); 4,5-(MeO)₂, 24 (25); 4,6-Br₂, 32 (—)	101
	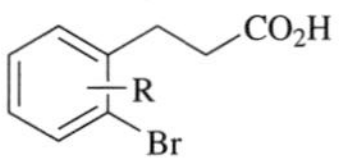 2-Br-C₆H₄-(CH₂)₃-CO₂H (R)	CuBr (cat), NaH (3 eq)	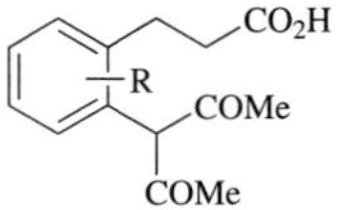 (0) R: H, 24; 4,5-(MeO)₂, 24; 4,6-Br₂, 30	101
	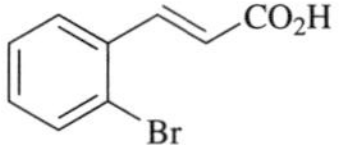 2-Br-C₆H₄-CH=CH-CO₂H	CuBr (cat), NaH (2 eq), 16 h	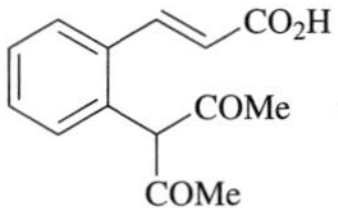 (0)	101
	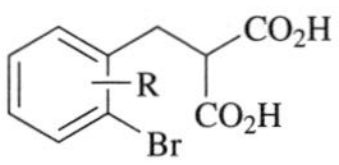 2-Br-C₆H₄-CH₂-CH(CO₂H)-CH₂-CO₂H (R)	CuBr (cat), NaH (4 eq)	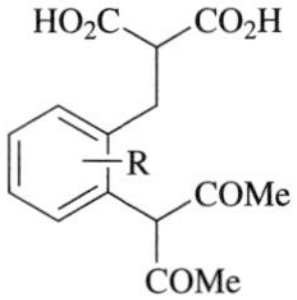 (0) R: H, 32; 4,6-Br₂, 30	101
	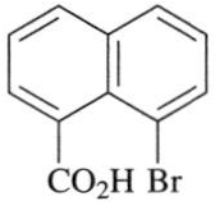 1,8-(CO₂H)₂-8-Br-naphthalene	CuBr (cat), NaH, 10 min	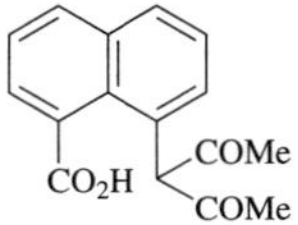 (78)	101
	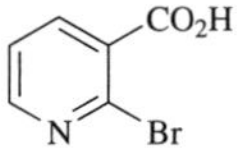 3-CO₂H-2-Br-pyridine	CuBr (cat), NaH, 5 min	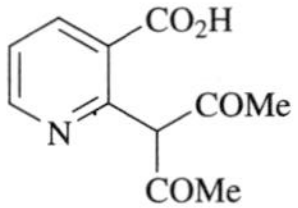 (—)	101

| | | 43 |

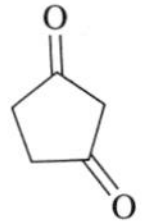

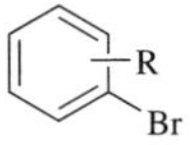

Pd(OAc)$_2$ (1 mol %),
L16 (2.2 mol %),
K$_3$PO$_4$ (2.3 eq)

Solvent	Temp (°)	Time (h)	
dioxane	100	19	(96)
THF	80	23	(96)

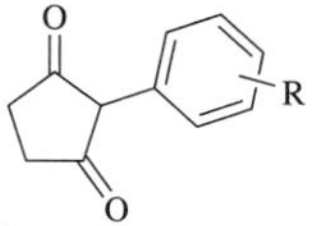

R
3-MeO$_2$C
4-t-Bu

| | | 162 |

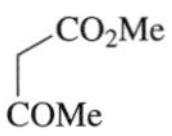

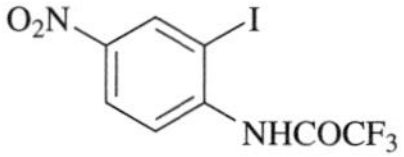

CuI (10 mol %),
L-proline (20 mol %),
Cs$_2$CO$_3$ (4 eq), DMSO,
50°, 12 h

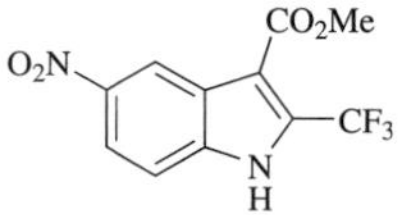

(78)

| | | 103 |

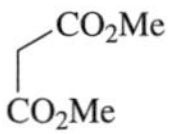

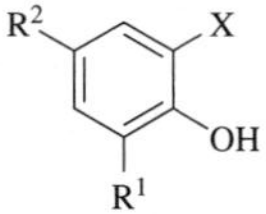

Catalyst (20 mol %),
NaH or NaOMe (2 eq),
80°, 4–6 d

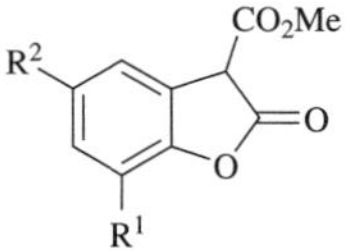

Catalyst	
CuBr	(79)
CuI	(27)
CuBr	(85)
CuBr•SMe$_2$	(36)
CuBr	(60)
CuBr	(48)
CuBr	(33)
CuI	(8)

X	R^1	R^2
Br	H	H
Br	H	H
Br	Br	H
Br	Br	H
Br	H	Me
Br	Br	F
I	H	H
I	H	H

| | | 163 |

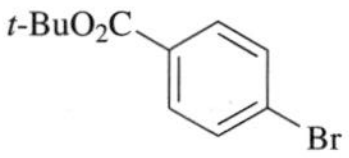

Pd$_2$(dba)$_3$, P(t-Bu)$_3$,
K$_3$PO$_4$, toluene

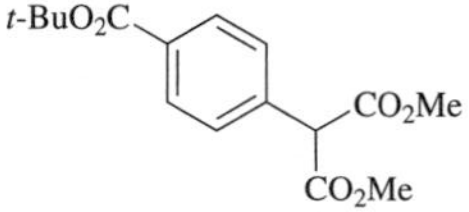

(—)

TABLE 3. ARYLATION OF 1,3-DICARBONYLS AND CYANOACETATES (Continued)

Substrate	Aryl Compound	Conditions	Product(s) and Yield(s) (%)	Refs.
C_5				
CO$_2$Me, CO$_2$Me	Ph–I	CuI (5 mol %), 2-picolinic acid (10 mol %), Cs$_2$CO$_3$ (3 eq), dioxane, 70°, 25 h	Ph-CH(CO$_2$Me)$_2$ (82)	106
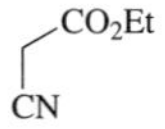 CO$_2$Et, CN	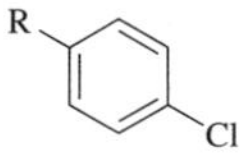 4-Cl-C$_6$H$_4$-R	[Pd(allyl)Cl]$_2$ (1 mol %), P(t-Bu)$_3$ (4 mol %), Na$_3$PO$_4$ (3 eq), toluene, 100°, 12 h	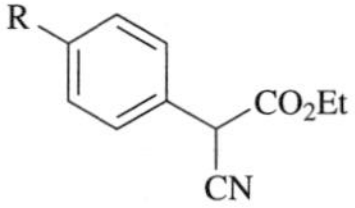 Ar-CH(CO$_2$Et)(CN) R: H (86), MeO (90), F (82)	82
	2-X-C$_6$H$_4$-R	[Pd(allyl)Cl]$_2$ (1 mol %), ligand (4 mol %), Na$_3$PO$_4$ (3 eq), toluene, 100° X R Cl H Cl 4-MeO Br 2-MeO Br 4-MeO Br 4-MeO	Ar-CH(CO$_2$Et)(CN) Ligand Time (h) P(t-Bu)$_3$ 12 (86) P(t-Bu)$_3$ 12 (90) P(t-Bu)$_3$ 8 (83) P(t-Bu)$_3$ 7 (89) **L33** 7 (85)	90
	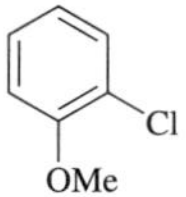 2-Cl-C$_6$H$_4$-OMe	[Pd(allyl)Cl]$_2$ (1 mol %), **L12** (4 mol %), Na$_3$PO$_4$ (3 eq), toluene, 100°, 12 h	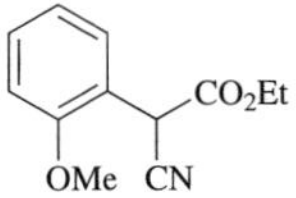 2-MeO-C$_6$H$_4$-CH(CO$_2$Et)(CN) (81)	82
	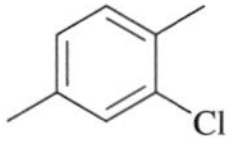 2-Cl-4-Me-C$_6$H$_3$-Me (2,5-dimethyl-Cl)	[Pd(allyl)Cl]$_2$ (1 mol %), P(t-Bu)$_3$ (4 mol %), Na$_3$PO$_4$ (3 eq), toluene, 100°, 12 h	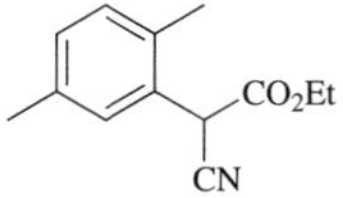 2,5-Me$_2$-C$_6$H$_3$-CH(CO$_2$Et)(CN) (87)	82, 90

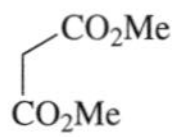

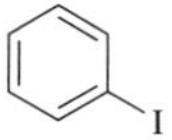

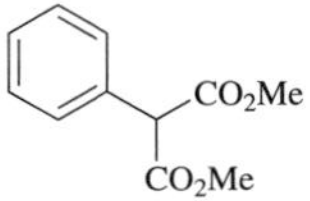

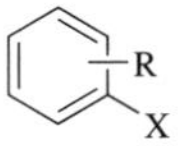

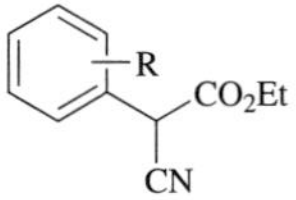

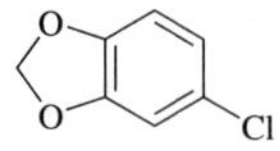

Aryl Halide	Product	Conditions	Ref.
(benzo[d][1,3]dioxol-5-yl chloride)	Ar-CH(CO₂Et)(CN) (86)	[Pd(allyl)Cl]₂ (1 mol %), **L12** (4 mol %), Na₃PO₄ (3 eq), toluene, 100°, 12 h	82
PhBr	Ph-CH(CO₂Et)(CN) (89)	[Pd(allyl)Cl]₂ (0.05 mol %), P(t-Bu)₃ (4 mol %), Na₃PO₄ (3 eq), toluene, 100°, 7 h	90
PhBr	Ph₂C(CO₂Et)(CN) (90)	Pd(dba)₂ (4 mol %), P(t-Bu)₃ (8 mol %), Na₃PO₄ (3 eq), toluene, 70°, 12 h	90
4-R-C₆H₄Br (R = H, H, F)	4-R-C₆H₄-CH(CO₂Et)(CN) R: H (87), H (87), F (91)	Pd(dba)₂ (2 mol %), ligand (4 mol %), Na₃PO₄ (3 eq), toluene, 4 h Ligand / Temp: P(t-Bu)₃ / 70° **L33** / rt P(t-Bu)₃ / 70°	90
R-C₆H₄Br	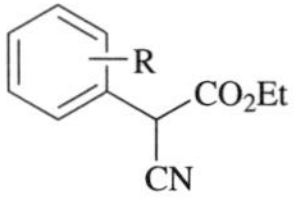 R-C₆H₄-CH(CO₂Et)(CN) R: H (90), 2-MeO (—), 2-Me (90), 4-F (—), 4-Cl (—)	Pd(OAc)₂ (1 mol %), **L1** (4 mol %), KOt-Bu (2.5 eq), dioxane, 70°, 1–4 h	164

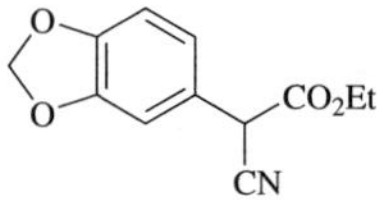

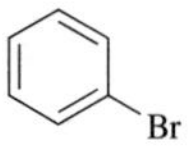

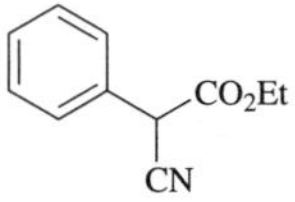

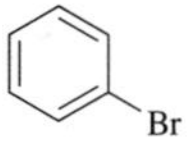

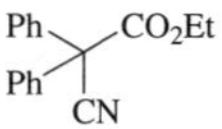

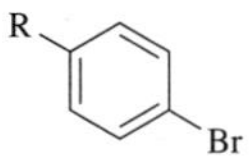

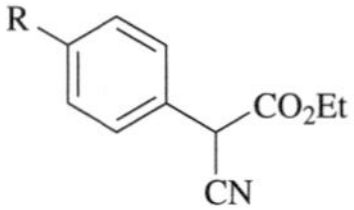

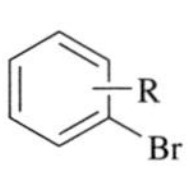

TABLE 3. ARYLATION OF 1,3-DICARBONYLS AND CYANOACETATES (*Continued*)

Substrate	Aryl Compound	Conditions	Product(s) and Yield(s) (%)	Refs.
C5 CO_2Et / CN	R¹, R², X aryl halide (X = Br or I; R¹ = H, 2-MeO, 2-F, 2-Me, 4-MeO, 4-F, 4-Cl, 4-Me, 3-Cl; R² = H or 4-Cl)	$PdCl_2(PPh_3)_2$ (4 mol %), KOt-Bu (2.2 eq), monoglyme, 70° X / R¹ / R² / Time (h) Br / H / H / 6 I / H / H / 5 I / 2-MeO / H / 24 I / 2-F / H / 6 I / 2-Me / H / 8 I / 4-MeO / H / 6 I / 4-F / H / 10 I / 4-Cl / H / 9 I / 4-Me / H / 5 I / 3-Cl / 4-Cl / 10	Ar-CH(CO$_2$Et)(CN) products: (8) (73) (44) (36) (45) (71) (46) (45) (78) (42)	88
	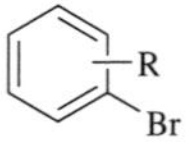 R-C$_6$H$_4$-Br	$Pd(dba)_2$ (2 mol %), $P(t$-Bu$)_3$ (4 mol %), Na_3PO_4 (3 eq), toluene, 70°, 6 h	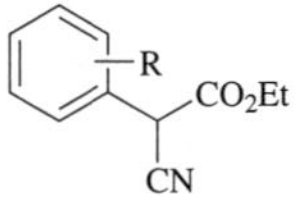 R / yield 2-Me (85) 3-CF$_3$ (87) 4-PhO (90)	82
	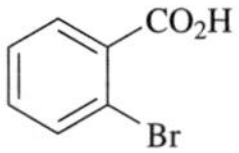 2-Br-C$_6$H$_4$-CO$_2$H	CuBr (cat), NaH	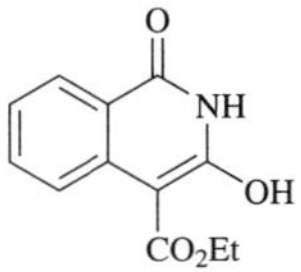 isoquinolinone product (36)	101
	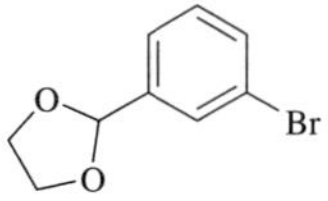 3-Br-C$_6$H$_4$-(1,3-dioxolan-2-yl)	$Pd(dba)_2$ (2 mol %), $P(t$-Bu$)_3$ (4 mol %), Na_3PO_4 (3 eq), toluene, 70°, 4 h	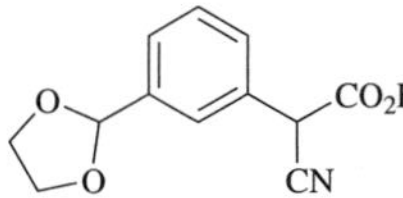 arylated product (81)	90

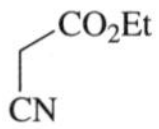

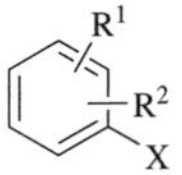

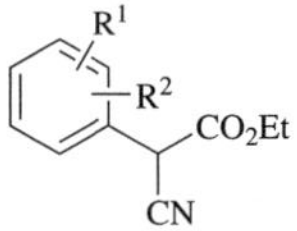

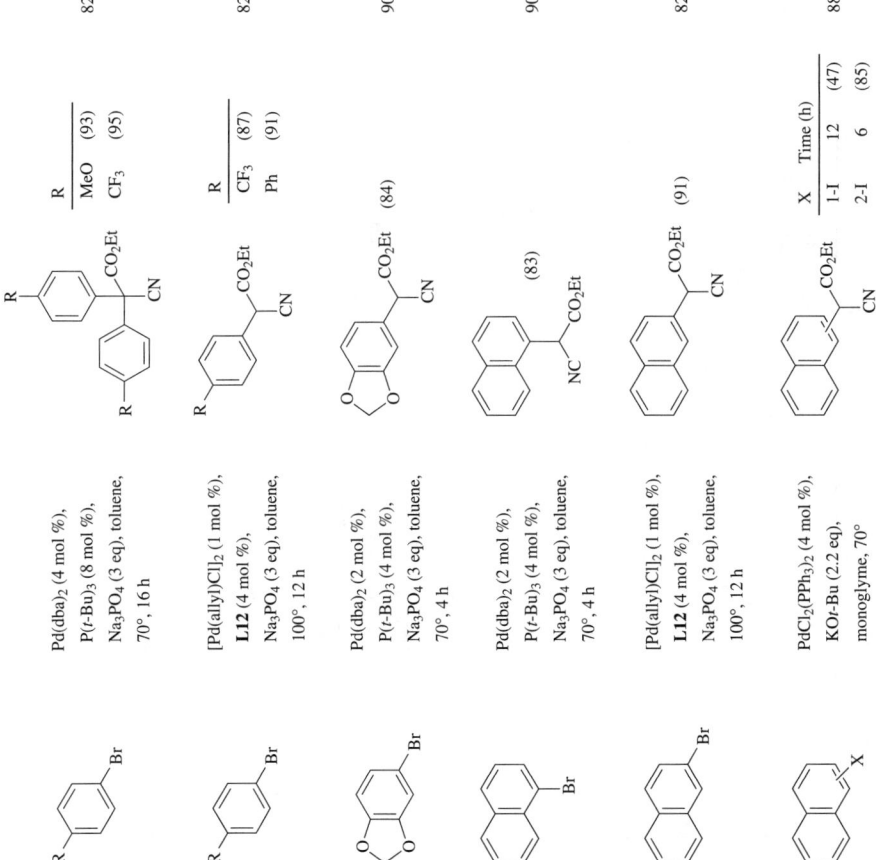

TABLE 3. ARYLATION OF 1,3-DICARBONYLS AND CYANOACETATES (*Continued*)

Substrate	Aryl Compound	Conditions	Product(s) and Yield(s) (%)	Refs.
C$_5$ 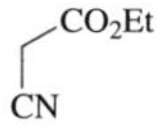 $\quad$ NC-CH(CO$_2$Et)	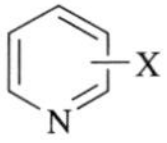 pyridine-X	Pd(OAc)$_2$ (1 mol %), L1 (4 mol %), KO*t*-Bu (2.5 eq), dioxane, 70°, 1–4 h	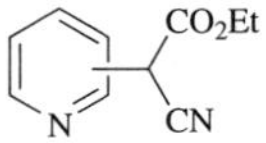 X: 2-Br (84), 3-Br (85)	164
	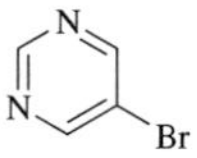 5-bromopyrazine	Pd(OAc)$_2$ (1 mol %), L1 (4 mol %), KO*t*-Bu (2.5 eq), dioxane, 70°, 1–4 h	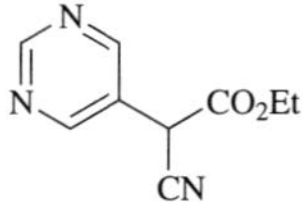 pyrazinyl-CH(CO$_2$Et)(CN) (85)	164
C$_6$ 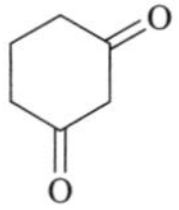 $\quad$ 1,3-cyclohexanedione	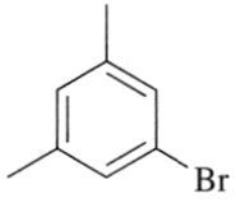 1-bromo-3,5-dimethylbenzene	Pd(OAc)$_2$ (1 mol %), L16 (2.2 mol %), K$_3$PO$_4$ (2.3 eq), THF, 80°, 15 h	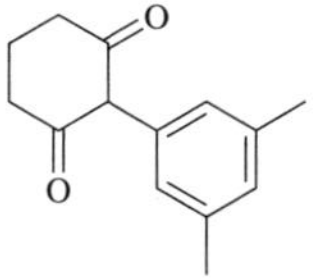 2-(3,5-dimethylphenyl)-1,3-cyclohexanedione (84)	43
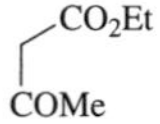 ethyl acetoacetate–type (CO$_2$Et, COMe)	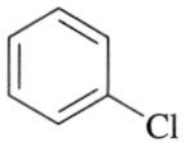 chlorobenzene	Pd(OAc)$_2$ (2 mol %), L16 (4 mol %), K$_3$PO$_4$ (2.8 eq), toluene, 90°, 16 h	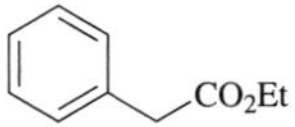 PhCH(CO$_2$Et) (93)	87
CO$_2$Et / COMe	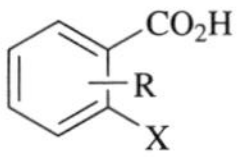 ArX (see table)	See table	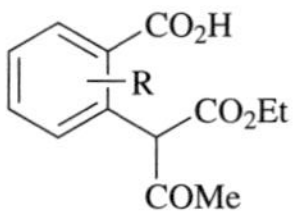 product (see table)	101

X	R	Catalyst	Base	Solvent	Time (h)	(Yield)
Cl	H	CuBr	NaH (2 eq)	none	0.8	(91)
Br	H	Cu (6 mol %)	NaOEt	EtOH	—	(56)
Br	H	CuBr	NaH (2 eq)	none	0.3	(96)
Br	3-O$_2$N	CuBr	NaH (2 eq)	none	2.5	(51)
Br	4-O$_2$N	CuBr	NaH (2 eq)	none	0.7	(98)
Br	4-Me	CuBr	NaH (2 eq)	none	1	(99)

Br	5-MeO	CuBr	NaH (2 eq)	none	2	(98)
Br	4,5-(MeO)$_2$	CuBr	NaH (2 eq)	none	2.5	(90)
Br	3,4,5-(MeO)$_3$	CuBr	NaH (2 eq)	none	7	(99)

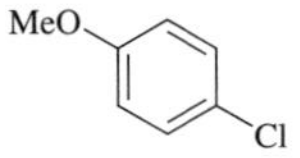

4-MeO-C$_6$H$_4$-Cl + Pd(OAc)$_2$ (0.5 mol %), L16 (1 mol %), K$_3$PO$_4$ (2.8 eq), toluene, 90°, 40 h → 4-MeO-C$_6$H$_4$-CH$_2$-CO$_2$Et (75)   87

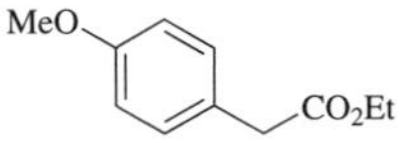

4-PhC(O)-C$_6$H$_4$-Cl + Pd(OAc)$_2$ (1 mol %), P(t-Bu)$_3$•HBF$_4$ (2 mol %), K$_3$PO$_4$ (2.8 eq), toluene, 90°, 16 h → 4-PhC(O)-C$_6$H$_4$-CH$_2$-CO$_2$Et (30) + PhC(O)Ph (40)   87

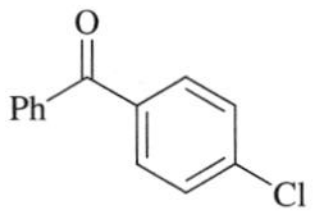

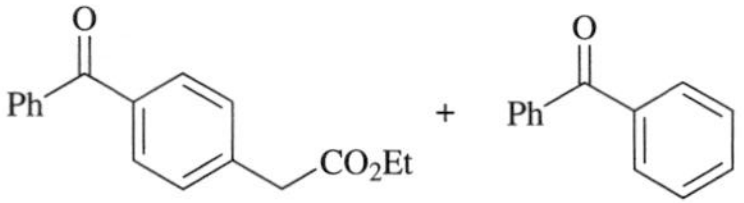

Ph-Br + Catalyst, ligand, K$_3$PO$_4$ (2.8 eq), toluene, 90°, 16 h → Ph-CH$_2$-CO$_2$Et   87

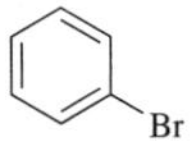

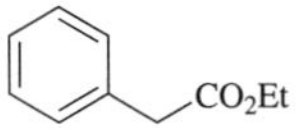

Catalyst	Ligand	
Pd(dba)$_2$ (1 mol %)	P(t-Bu)$_3$ (2 mol %)	(45)
Pd(OAc)$_2$ (1 mol %)	P(t-Bu)$_3$ (2 mol %)	(48)
Pd(OAc)$_2$ (5 mol %)	PPh$_3$ (20 mol %)	(0)
Pd(dba)$_2$ (1 mol %)	PCy$_3$ (2 mol %)	(0)
Pd(OAc)$_2$ (1 mol %)	L15 (2 mol %)	(56)
Pd(OAc)$_2$ (1 mol %)	L16 (2 mol %)	(89)
Pd(OAc)$_2$ (2 mol %)	L15 (4 mol %)	(68)
Pd(OAc)$_2$ (2 mol %)	L16 (4 mol %)	(93)

TABLE 3. ARYLATION OF 1,3-DICARBONYLS AND CYANOACETATES (*Continued*)

Substrate	Aryl Compound	Conditions	Product(s) and Yield(s) (%)	Refs.
C_6 — CH(CO$_2$Et)(COMe)	Aryl-X, X = Br/I, R = H, 4-MeO, 4-NC, 4-MeCO, 2-Me, 4-MeO, 4-EtO$_2$C, 4-MeCO	CuI (20 mol %), K$_2$CO$_3$ (7.5 eq), DMSO, 80°, 20 h	**I** Ar-CH(CO$_2$Et)(COMe) + **II** Ar-CH(CO$_2$Et)(COMe) — monoester products X R I II Br H (2) (—) Br 4-MeO (0) (0) Br 4-NC (6) (13) Br 4-MeCO (48) (22) I 2-Me (19) (41) I 4-MeO (93) (7) I 4-EtO$_2$C (53) (41) I 4-MeCO (63) (35)	165
	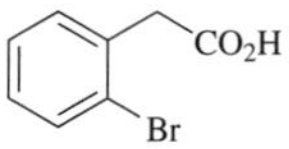 2-Br-C$_6$H$_4$-CO$_2$H	CuBr (cat), NaH (2 eq), 16 h	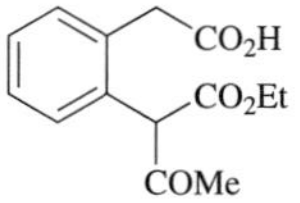 cyclized product with CO$_2$H, CO$_2$Et, COMe (—)	101
	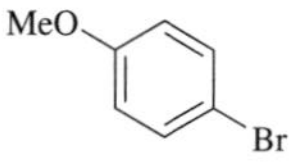 4-MeO-C$_6$H$_4$-Br	Pd(OAc)$_2$ (1 mol %), P(t-Bu)$_3$·HBF$_4$ (2 mol %), K$_3$PO$_4$ (2.8 eq), toluene, 90°, 16 h	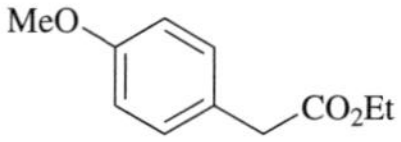 4-MeO-C$_6$H$_4$-CH$_2$-CO$_2$Et (40) + 4-MeO-C$_6$H$_4$-CH(CO$_2$Et)(—) (35)	87
	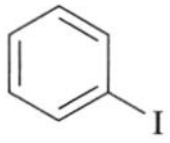 Ph-I	Copper catalyst (x mol %), K$_2$CO$_3$ (y eq), DMSO, 80°	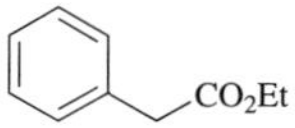 **I** Ph-CH$_2$-CO$_2$Et + **II** Ph-CH(CO$_2$Et)(COMe) Catalyst x y Time(h) I II CuBr 20 4 20 (56) (—) CuI 5 5 24 (85) (11) CuI 10 7.5 20 (85) (9)	165

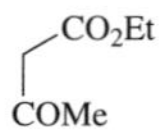

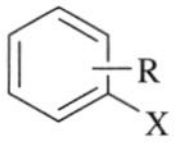

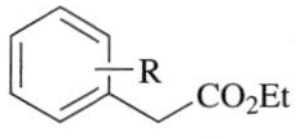

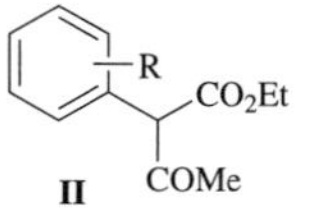

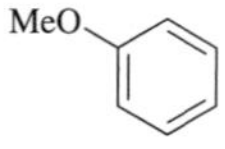

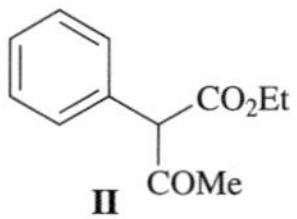

250

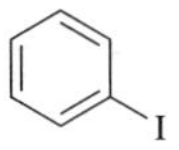

	CuI (20 mol %), ligand (40 mol %), K$_2$CO$_3$ (4 eq), 80°, 20 h		I + II	I	II
Ligand		Solvent			
none		DMSO		(54)	(18)
none		dioxane		(12)	(—)
none		NMP		(42)	(—)
none		DMF		(43)	(—)
ethylenediamine		DMSO		(29)	(—)
ethylenediamine		dioxane		(48)	(—)
ethylenediamine		toluene		(38)	(—)
N,N'-dimethylethylenediamine		dioxane		(26)	(—)
N,N'-dimethylethylenediamine		toluene		(23)	(—)
1,8-diaminonaphthalene		DMSO		(47)	(22)
1,1'-bipyridine		DMSO		(54)	(20)
1,10-phenanthroline		DMSO		(41)	(17)
2-phenylphenol		DMSO		(41)	(—)
2-phenylphenol		dioxane		(5)	(—)
proline		DMSO		(45)	(8)
phenylalanine		DMSO		(48)	(6)
phenylalanine		dioxane		(42)	(8)
tryptophan		DMSO		(37)	(11)
ornithine		DMSO		(17)	(7)
2,4-diaminobutyric acid		dioxane		(18)	(6)

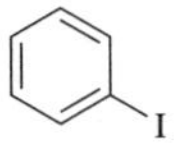

165

	CuI (5 mol %), 2-picolinic acid (10 mol %), Cs$_2$CO$_3$ (3 eq), dioxane, 70°, 25 h	II (78)

106

TABLE 3. ARYLATION OF 1,3-DICARBONYLS AND CYANOACETATES (*Continued*)

Substrate	Aryl Compound	Conditions	Product(s) and Yield(s) (%)	Refs.
C_6 — CO$_2$Et / COMe	Ph–I	CuI (20 mol %), ligand (40 mol %), base (x eq)	**I** Ph-CH(CO$_2$Et)(COMe) + **II** Ph-CH(CO$_2$Et)(COMe) (α-aryl isomer)	165

Ligand	Base	x	Solvent	Temp (°)	Time (h)	I	II
none	K$_3$PO$_4$	1	DMSO	80	20	(20)	(26)
none	Cs$_2$CO$_3$	4	DMSO	80	20	(7)	(17)
none	K$_2$CO$_3$	1	DMSO	80	7	(16)	(34)
none	K$_2$CO$_3$	2	DMSO	80	20	(53)	(16)
none	K$_2$CO$_3$	4	DMSO	40	40	(38)	(38)
none	K$_2$CO$_3$	4	DMSO	80	20	(54)	(18)
none	K$_2$CO$_3$	7.5	DMSO	40	20	(17)	(65)
none	K$_2$CO$_3$	7.5	DMSO	40	40	(37)	(56)
none	K$_2$CO$_3$	7.5	DMSO	80	20	(86)	(4)
none	K$_2$CO$_3$	7.5	dioxane	80	44	(58)	(7)
ethylenediamine	K$_2$CO$_3$	7.5	dioxane	80	40	(72)	(3)
ethylenediamine	K$_3$PO$_4$	5	toluene	80	44	(15)	(0)

Aryl compound: 2,6-disubstituted aniline with NHCOCF$_3$, I, R^2, R^1

Conditions: CuI (10 mol %), L-proline (20 mol %), Cs$_2$CO$_3$ (4 eq), DMSO

Product: indole with CO$_2$Et at 3, CF$_3$ at 2, R^1, R^2

R^1	R^2	Temp (°)	Time (h)	Yield
H	H	80	15	(27)
O$_2$N	H	50	12	(85)
I	MeO$_2$C	40	12	(78)

162

Aryl compound: 2-bromo-iodobenzene

Conditions: CuI (5 mol %), ligand (10 mol %), K$_2$CO$_3$ (1.5 eq)

Product: 2-bromophenyl-CH$_2$-CO$_2$Et

104

Ligand	Solvent	Temp (°)	
none	dioxane	100	(57)
none	DMSO	100	(19)
none	toluene	100	(25)
none	THF	100	(83)
none	THF	70	(28)
L-proline	THF	70	(9)
L-proline	DMSO	40	(0)

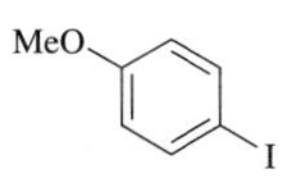

MeO–C6H4–I

CuI (5 mol %), K2CO3 (1.5 eq), THF, 100°, 24 h

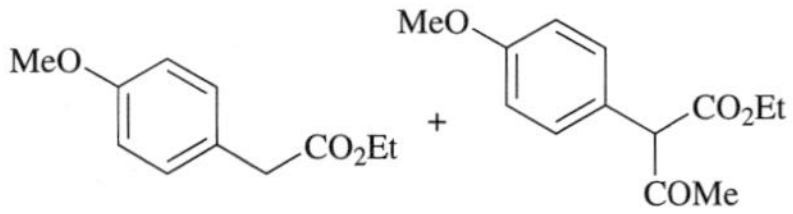

MeO–C6H4–CH2CO2Et (45)
+
MeO–C6H4–CH(CO2Et)(COMe) (29)   104

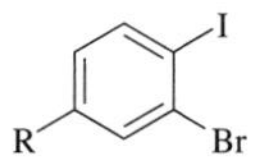

CuI (5 mol %), K2CO3 (1.5 eq), 100°

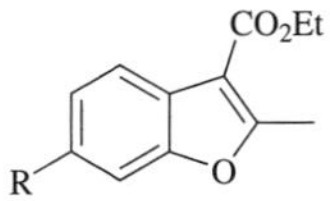

2-bromo-4-R-iodobenzene → 2-methyl-6-R-benzofuran-3-carboxylate (CO2Et)

R	Time (h)	
Cl	24	(75)
Me	24	(88)
CF3	26	(82)
HOCH2	26	(32)
TBSOCH2	26	(75)
MeCO	30	(78)
MeO2C	26	(79)

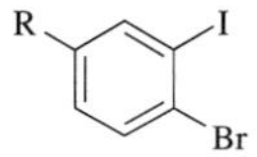

104

CuI (5 mol %), K2CO3 (1.5 eq), 100°

2-bromo-4-R-iodobenzene → 2-methyl-5-R-benzofuran-3-carboxylate (CO2Et)

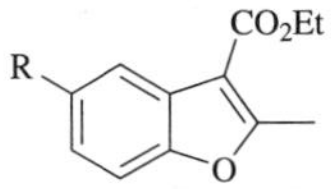

R	Time (h)	
O2N	30	(52)
MeO	26	(78)
F	24	(80)

104

TABLE 3. ARYLATION OF 1,3-DICARBONYLS AND CYANOACETATES (Continued)

Substrate	Aryl Compound	Conditions	Product(s) and Yield(s) (%)	Refs.
C_6				
CO$_2$Et, COMe	(methylenedioxybenzene with Br, I)	CuI (5 mol %), K$_2$CO$_3$ (1.5 eq), THF, 100°, 26 h	(benzofuran product with CO$_2$Et) (70)	104
	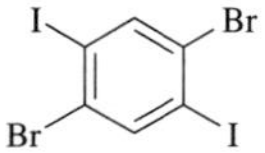 (benzene with I, Br, Br, I)	CuI (20 mol %), K$_2$CO$_3$ (1.5 eq), THF, 100°, 24 h	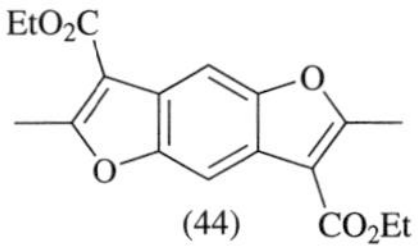 (dibenzofuran-type product) (44) + (bromobenzofuran product) (28)	104
	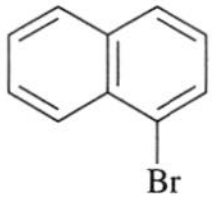 1-bromonaphthalene	Pd(OAc)$_2$ (1 mol %), P(t-Bu)$_3$·HBF$_4$ (2 mol %), K$_3$PO$_4$ (2.8 eq), toluene, 90°, 16 h	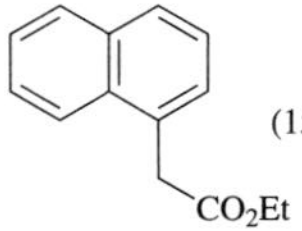 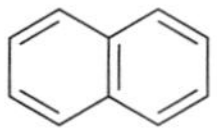 (naphthyl-CH$_2$CO$_2$Et) (58) + naphthalene (15)	87
	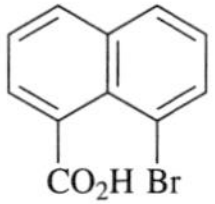 8-bromonaphthalene-1-carboxylic acid	CuBr (cat), NaH, 5 min	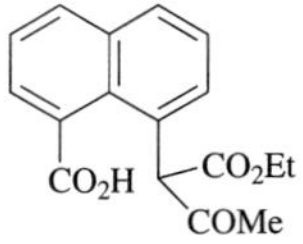 (naphthyl product with CO$_2$H, CO$_2$Et, COMe) (81)	101
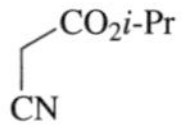	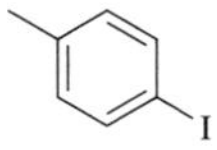 4-iodotoluene	PdCl$_2$(PPh$_3$)$_2$ (4 mol %), KOt-Bu (2.2 eq), monoglyme, 70°, 6 h	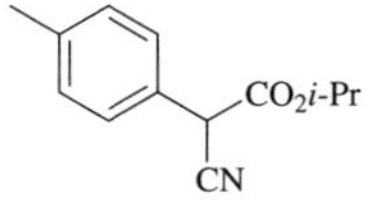 (tolyl product with CO$_2$i-Pr, CN) (74)	88
CO$_2$i-Pr, CN				
C_7				
(2-cyclopentanone with CO$_2$Me)	PhB(OH)$_2$	(BzO$_2$)$_2$Pb, **L38**, Hg(OAc)$_2$ (cat), pyridine (3 eq)	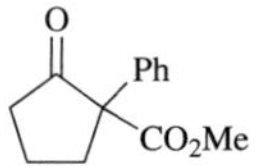 (cyclopentanone with Ph, CO$_2$Me) (69), 10% ee	133

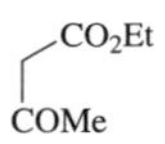

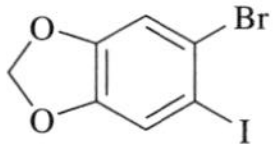

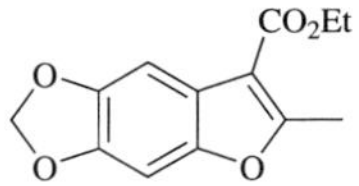

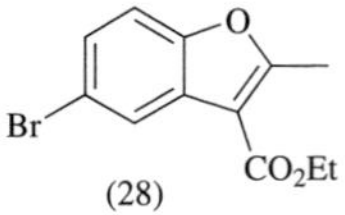

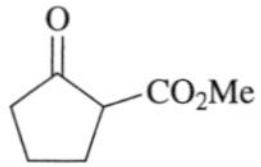

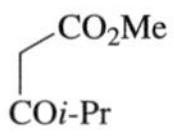

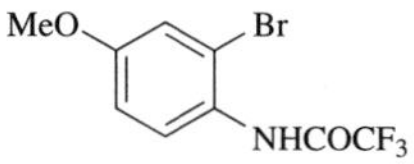

Substrate	Reagent	Conditions	Product(s) and Yield(s) (%)	Refs.
CH(CO₂Me)(COi-Pr)	MeO—C₆H₃(Br)(NHCOCF₃)	CuI (20 mol %), L-proline (40 mol %), Cs₂CO₃ (4 eq), DMSO, 80°, 18 h	5-MeO-2-CF₃-3-CO₂Me-indole (55)	162
	R²—C₆H₃(I)(NHCOCF₃) with R¹	CuI (10 mol %), L-proline (20 mol %), Cs₂CO₃ (4 eq), DMSO	indole product with R¹, R², 2-CF₃, 3-CO₂Me	162

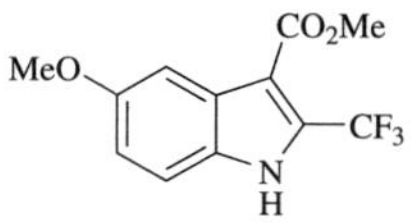

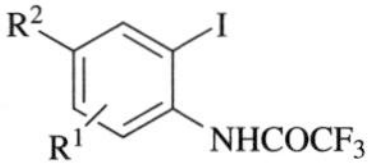

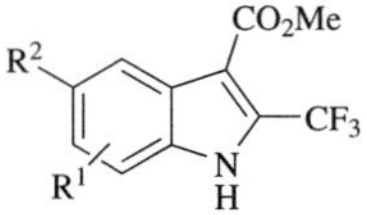

R^1, R^2, Temp (°), Time (h), (Yield):

R^1	R^2	Temp (°)	Time (h)	Yield
H	H	80	15	(80)
5-MeO	H	75	15	(70)
5-MeCO	H	70	15	(65)
H	O₂N	50	12	(84)
H	Me	70	18	(63)
H	MeCO	60	12	(93)
6-Cl	I	60	12	(77)
5-Me	I	75	15	(79)

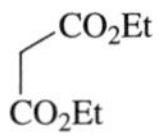

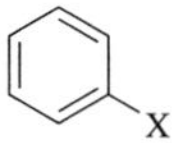

Substrate	Reagent	Conditions	Product(s) and Yield(s) (%)	Refs.
CH₂(CO₂Et)₂	C₆H₅X	Na₂PdCl₄ (2 mol %), Ba(OH)₂ (2 eq), DMA	PhCH(CO₂Et)₂	84

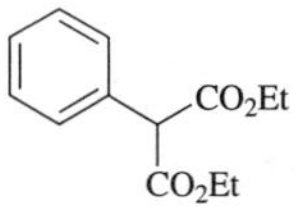

X	
Cl	(93)
Br	(99)
I	(100)

Substrate	Reagent	Conditions	Product(s) and Yield(s) (%)	Refs.
CH₂(CO₂Et)₂	4-R-C₆H₄Cl	Pd(dba)₂ (2 mol %), L33 (4 mol %), K₃PO₄ (3 eq), toluene, 100°, 16–21 h	4-R-C₆H₄CH(CO₂Et)₂	82

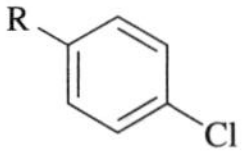

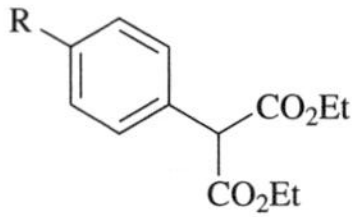

R	
H	(85)
MeO	(90)
CF₃	(86)

255

TABLE 3. ARYLATION OF 1,3-DICARBONYLS AND CYANOACETATES (*Continued*)

Substrate	Aryl Compound	Conditions	Product(s) and Yield(s) (%)	Refs.
C₇ EtO₂C–CH₂–CO₂Et	R²–C₆H₃(Cl)–R¹	Pd(dba)₂ (2 mol %), **L12** (4 mol %), K₃PO₄ (3 eq), toluene, 100°, 16–21 h	Ar–CH(CO₂Et)₂ with R¹/R² = H/H (81); MeO/H (87); H/MeO (86); H/CF₃ (89)	82
	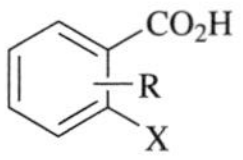 HO₂C–C₆H₃(X)–R	CuBr (cat), NaH (2 eq), neat	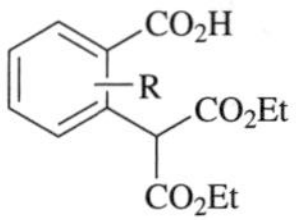 R–C₆H₃(CO₂H)–CH(CO₂Et)₂	101

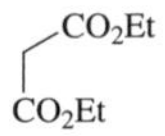

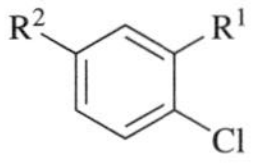

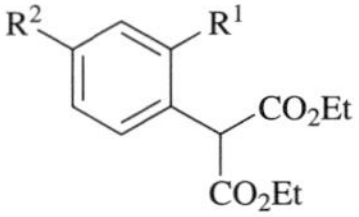

X	R	Time (h)	(%)
Cl	H	20	(80)
Br	H	1.7	(92)
Br	3-O₂N	2	(70)
Br	3-HO₂C	25	(85)
Br	4-O₂N	0.8	(88)
Br	4-Me	2	(90)
Br	5-H₂N	6	(70)
Br	5-HO	12	(27)
Br	4,5-(MeO)₂	4	(92)
Br	4,6-Br₂	4.5	(83)
Br	3,4,5-(MeO)₃	3	(84)

| | 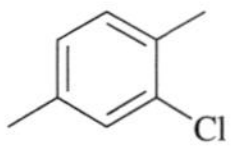 2,5-dimethyl-chlorobenzene | Pd(dba)₂ (2 mol %), **L12** (4 mol %), K₃PO₄ (3 eq), toluene, 100°, 16–21 h | 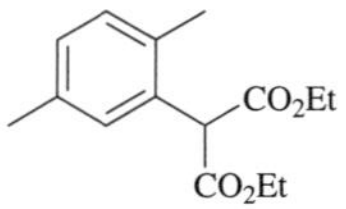 2,5-Me₂C₆H₃–CH(CO₂Et)₂ (91) | 82 |

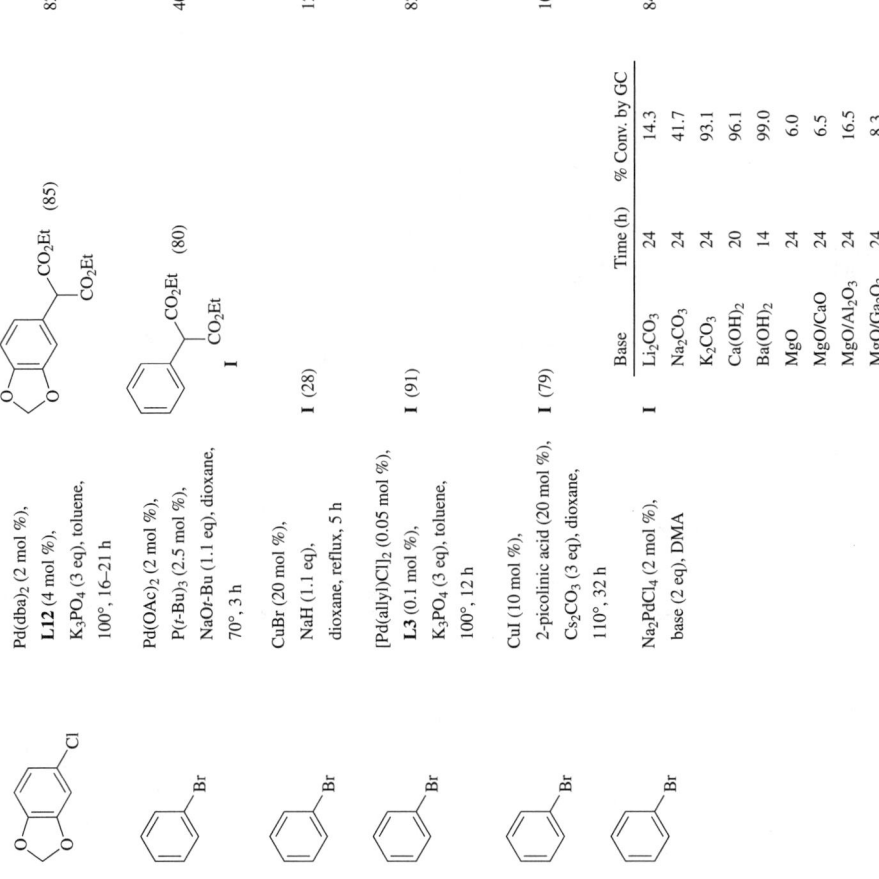

TABLE 3. ARYLATION OF 1,3-DICARBONYLS AND CYANOACETATES (Continued)

Substrate	Aryl Compound	Conditions	Product(s) and Yield(s) (%)	Refs.
C_7 CO$_2$Et, CO$_2$Et	R—C$_6$H$_4$—Br	Pd(dba)$_2$ (2 mol %), P(t-Bu)$_3$ (4 mol %), K$_3$PO$_4$ (3 eq), toluene, 70°	R—C$_6$H$_4$—CH(CO$_2$Et)$_2$ R / Time (h) / (%): H / 5 / (88); MeCO / 14 / (89); PhCO / 14 / (90)	82
	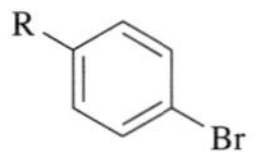 R—C$_6$H$_4$—Br	Catalyst (1 mol %), NaOt-Bu (2 eq), THF, 110°, 20 h	**I** = R—C$_6$H$_4$—CH(CO$_2$Et)$_2$; see below	83
			R / Catalyst / (Yield): H / Pd(OAc)$_2$–NaY zeolite / (41); H / Pd(OAc)$_2$, PPh$_3$ (4 mol %) / (50); O$_2$N / Pd(OAc)$_2$–NaY zeolite / (84); O$_2$N / Pd(OAc)$_2$, PPh$_3$ (4 mol %) / (78); MeO / Pd(OAc)$_2$–NaY zeolite / (45); MeO / Pd(OAc)$_2$, PPh$_3$ (4 mol %) / (48); F / Pd(OAc)$_2$–NaY zeolite / (50); F / Pd(OAc)$_2$, PPh$_3$ (4 mol %) / (60); Me / Pd(OAc)$_2$–NaY zeolite / (38); Me / Pd(OAc)$_2$, PPh$_3$ (4 mol %) / (56); MeCO / Pd(OAc)$_2$–NaY zeolite / (62); MeCO / Pd(OAc)$_2$, PPh$_3$ (4 mol %) / (64)	
	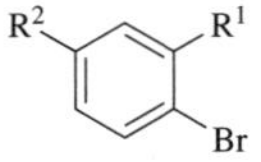 R^2—C$_6$H$_3$(R^1)—Br	Pd(dba)$_2$ (2 mol %), P(t-Bu)$_3$ (4 mol %), NaH (1.1 eq), THF, 70°	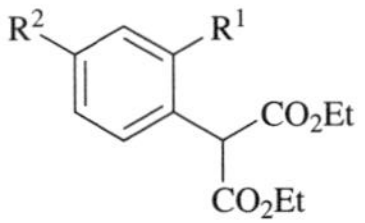 R^2—C$_6$H$_3$(R^1)—CH(CO$_2$Et)$_2$	82

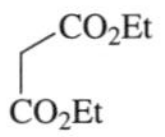

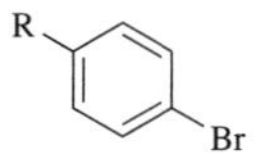

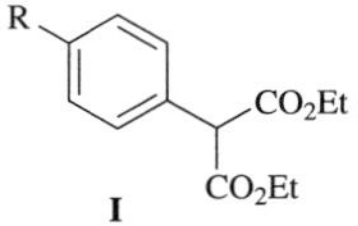

R^1	R^2	Time (h)	(%)
H	H	1	(89)
MeO	H	8	(89)
Me	H	8	(84)
H	Me$_2$N	6	(87)
H	MeO	3	(89)
H	PhO	8	(85)

H	F		3	(82)	
H	CF₃		3	(91)	
H	MeO₂C		3	(91)	
H	Ph		4	(91)	

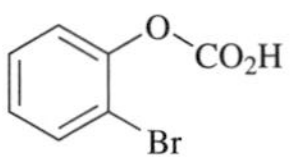

[Substrate: 2-bromophenol with O-CH₂-CO₂H] · CuBr (cat), NaH (2 eq), 16 h → product with O-CH(CO₂Et)₂ (—) · 101

[Substrate: 2-bromobenzoic acid derivative with CH₂CO₂H] · Cu(OAc)₂ (4 mol %), NaOEt, EtOH → product (34) · 101

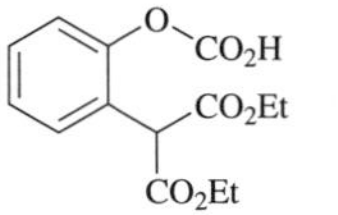

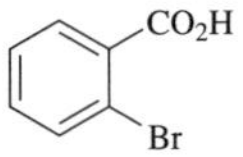

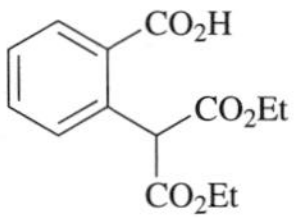

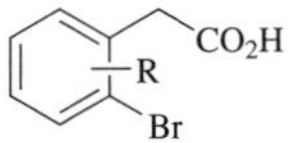

[Substrate: 2-bromoaryl-CH₂CO₂H with R] · CuBr (cat), NaH (2 eq) → product

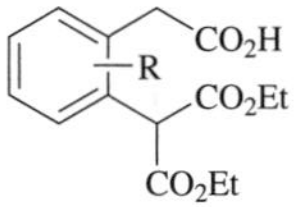

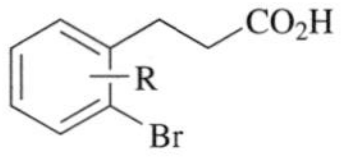

R	Time (h)	
H	24	(41)
4,5-(MeO)₂	7	(95)
4,6-Br₂	32	(—)

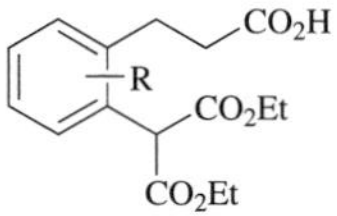

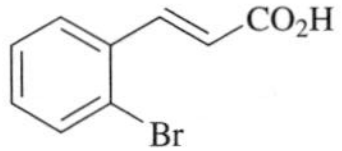

101

[Substrate: 2-bromoaryl-CH₂CH₂CO₂H with R] · CuBr (cat), NaH (3 eq), 30 h → product

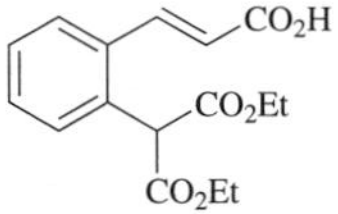

R		
H		(—)
4,6-Br₂		(—)

101

[Substrate: 2-bromocinnamic acid] · CuBr (cat), NaH (2 eq), 30 h → product (—) · 101

TABLE 3. ARYLATION OF 1,3-DICARBONYLS AND CYANOACETATES (*Continued*)

Substrate	Aryl Compound	Conditions	Product(s) and Yield(s) (%)	Refs.
C_7				
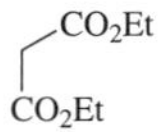 CO$_2$Et, CO$_2$Et	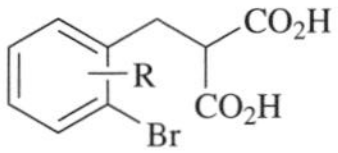 (aryl bromide with CH(CO$_2$H)$_2$ group, R substituent)	CuBr (cat), NaH (4 eq)	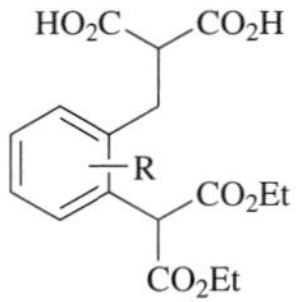 HO$_2$C—CH(CO$_2$H)—Ar—R—CH(CO$_2$Et)$_2$ R / Time (h) / Yield H / 32 / (0) 4,6-Br$_2$ / 30 / (0)	101
	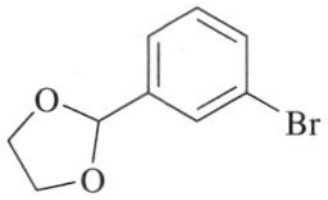 3-(1,3-dioxolan-2-yl)bromobenzene	Pd(dba)$_2$ (2 mol %), P(*t*-Bu)$_3$ (4 mol %), K$_3$PO$_4$ (3 eq), toluene, 70°, 6 h	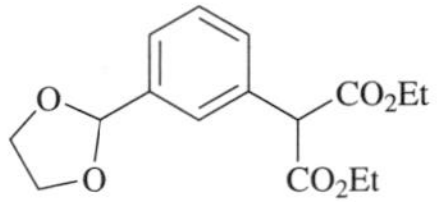 3-(1,3-dioxolan-2-yl)-C$_6$H$_4$-CH(CO$_2$Et)$_2$ (87)	82
	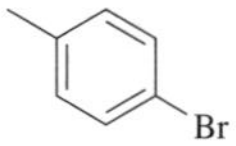 4-bromotoluene	Catalyst (1 mol %), ligand (4 mol %), NaO*t*-Bu (2 eq), THF, 110°, 20 h	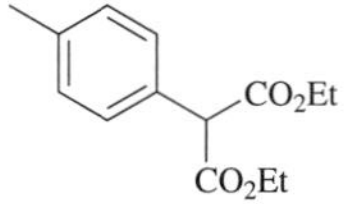 4-MeC$_6$H$_4$-CH(CO$_2$Et)$_2$ Catalyst / Ligand / Yield Pd(0)–NaY zeolite / none / (21) Pd(II)–NaY zeolite / none / (32) Pd(NH$_3$)$_4$–NaY zeolite / none / (38) Pd(OAc)$_2$–NaY zeolite / none / (29) Pd(OAc)$_2$ / PPh$_3$ / (56)	83
	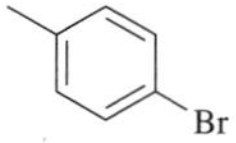 4-bromotoluene	Catalyst (1 mol %), base (2 eq), 110°, 20 h	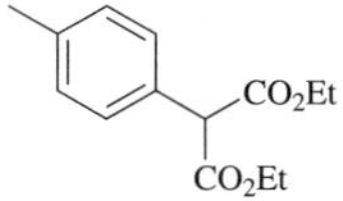 4-MeC$_6$H$_4$-CH(CO$_2$Et)$_2$	83

Catalyst	Base	Solvent	Yield
Pd(OAc)$_2$–NaY zeolite	NaOt-Bu	THF	(41)
Pd(OAc)$_2$–NaY zeolite	NaOt-Bu	DMF	(45)
Pd(OAc)$_2$–NaY zeolite	KOt-Bu	THF	(39)
Pd(OAc)$_2$–NaY zeolite	KOt-Bu	DMF	(46)
Pd(OAc)$_2$–NaY zeolite	K$_2$CO$_3$	THF	(6)
Pd(OAc)$_2$–NaY zeolite	K$_2$CO$_3$	DMF	(24)
Pd(OAc)$_2$ (1 mol %), PPh$_3$ (4 mol %)	NaOt-Bu	THF	(65)
Pd(OAc)$_2$ (1 mol %), PPh$_3$ (4 mol %)	NaOt-Bu	DMF	(67)
Pd(OAc)$_2$ (1 mol %), PPh$_3$ (4 mol %)	KOt-Bu	THF	(40)
Pd(OAc)$_2$ (1 mol %), PPh$_3$ (4 mol %)	KOt-Bu	DMF	(60)
Pd(OAc)$_2$ (1 mol %), PPh$_3$ (4 mol %)	K$_2$CO$_3$	THF	(8)
Pd(OAc)$_2$ (1 mol %), PPh$_3$ (4 mol %)	K$_2$CO$_3$	DMF	(32)

Conditions	Substrate (ArBr)	Product	Yield
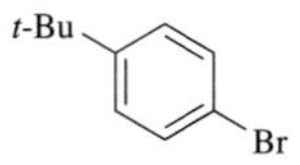 Pd(OAc)$_2$ (1 mol %), L16 (2.2 mol %), K$_3$PO$_4$ (2.3 eq), THF, 70°, 10 h	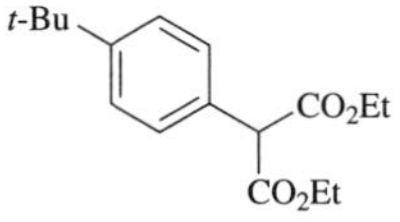 4-t-Bu-C$_6$H$_4$Br	t-Bu-C$_6$H$_4$-CH(CO$_2$Et)$_2$ (92)	43
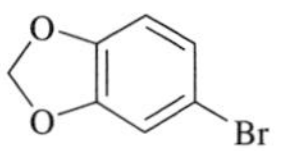 Pd(dba)$_2$ (2 mol %), P(t-Bu)$_3$ (4 mol %), NaH (1.1 eq), THF, 70°, 6 h	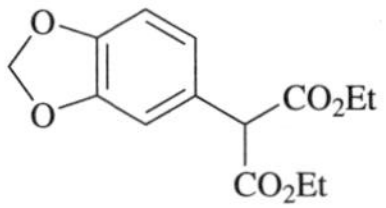 6-bromo-1,3-benzodioxole	benzodioxolyl-CH(CO$_2$Et)$_2$ (83)	82
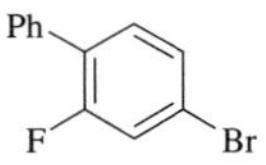 Pd(dba)$_2$ (2 mol %), P(t-Bu)$_3$ (4 mol %), K$_3$PO$_4$ (3 eq), toluene, 70°, 12 h	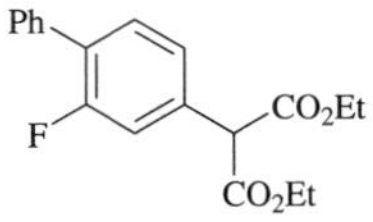 4-Ph-2-F-C$_6$H$_3$Br	(4-Ph-2-F-C$_6$H$_3$)-CH(CO$_2$Et)$_2$ (89)	82

TABLE 3. ARYLATION OF 1,3-DICARBONYLS AND CYANOACETATES (*Continued*)

Substrate	Aryl Compound	Conditions	Product(s) and Yield(s) (%)	Refs.
C_7 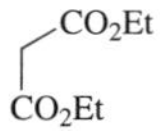 EtO$_2$C–CO$_2$Et	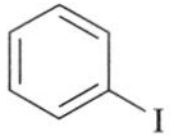 PhI	CuI (5 mol %), ligand (x mol %), Cs$_2$CO$_3$ (3 eq), dioxane, 20 h	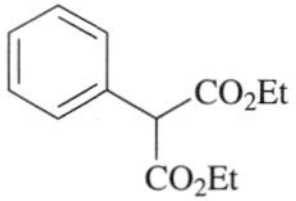 Ph-CH(CO$_2$Et)$_2$	106
			Ligand / x / Temp	
			phenol / 10 / 70° (2)	
			2-dioxolanephenol / 10 / 70° (15)	
			2-*t*-butylphenol / 10 / 70° (9)	
			2-phenylphenol / 10 / 70° (83)	
			2-phenylphenol / 20 / rt (48)	
			2-phenylphenol / 20 / 70° (90)	
			2,4-dimethylphenol / 10 / 70° (4)	
			2,6-di-*t*-butylphenol / 10 / 70° (<1)	
			benzoic acid / 10 / 70° (79)	
			thiophene-2-carboxylic acid / 10 / 70° (60)	
			indole-2-carboxylic acid / 10 / 70° (66)	
			2-picolinic acid / 10 / rt (92)	
			2-picolinic acid / 10 / 70° (99)	
			2-picolinic acid / 10 / 70° (93)[a]	
			2-picolinic acid / 10 / 70° (90)[b]	
	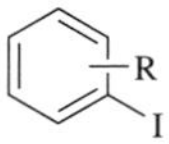 R-C$_6$H$_4$-I	CuI (5 mol %), 2-phenylphenol (10 mol %), Cs$_2$CO$_3$ (1.5 eq), THF, 70°	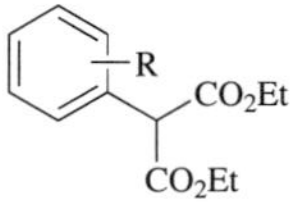 R-C$_6$H$_4$-CH(CO$_2$Et)$_2$ **I**	105
		R / Time (h)		
		H / 24	(91)	
		3-O$_2$N / 24	(84)	
		3-F / 24	(87)	
		3-NC / 24	(61)	

3-CF3	24	(89)
3-EtO2C	24	(86)
4-H2N	29	(70)
4-Cl	24	(94)
4-MeCO	24	(86)

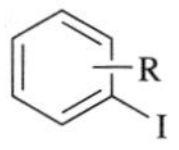

CuI (5 mol %), 2-picolinic acid (10 mol %), Cs$_2$CO$_3$ (3 eq), dioxane   **I**   106

R	Temp	Time (h)	
2-MeO	rt	20	(92)
2-Me	70°	20	(84)[c]
3-MeO	rt	20	(79)
3-EtO2C	rt	20	(96)
4-MeO	rt	20	(80)
4-F	rt	20	(73)
4-NC	rt	20	(82)
2,4-(MeO)2	70°	25	(88)
3,5-Me2	rt	28	(88)
3,4,5-(MeO)3	rt	28	(81)

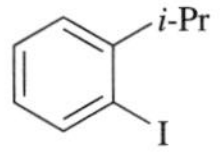

CuI (10 mol %), 2-phenylphenol (15 mol %), Cs$_2$CO$_3$ (1.5 eq), THF, 70°, 31 h   (84)   105

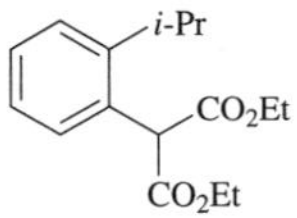

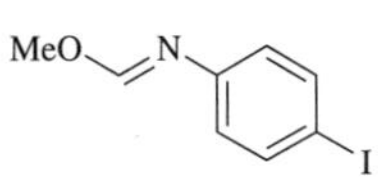

CuI (10 mol %), 2-phenylphenol (15 mol %), Cs$_2$CO$_3$ (2.5 eq), THF, 70°, 29 h   (75)   105

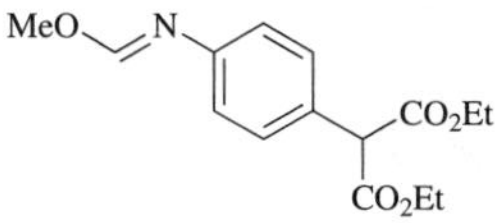

TABLE 3. ARYLATION OF 1,3-DICARBONYLS AND CYANOACETATES (*Continued*)

Substrate	Aryl Compound	Conditions	Product(s) and Yield(s) (%)	Refs.
C_7 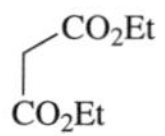 diethyl malonate (CO$_2$Et, CO$_2$Et)	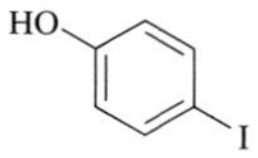 4-iodophenol	CuI (10 mol %), 2-phenylphenol (15 mol %), Cs$_2$CO$_3$ (3.5 eq), THF, 70°, 29 h	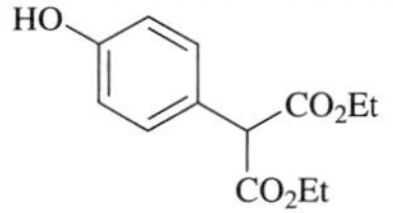 HO–C$_6$H$_4$–CH(CO$_2$Et)$_2$ (80)	105
	R^1, R^2 substituted aryl iodide; 2-MeO/4-MeO (Time 30 h), 3-Me/5-Me (Time 27 h)	CuI (5 mol %), 2-phenylphenol (10 mol %), Cs$_2$CO$_3$ (1.5 eq), THF, 70°	Ar–CH(CO$_2$Et)$_2$ (90), (95)	105
	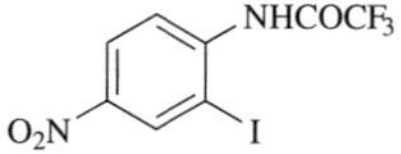 2-iodo-4-nitro-NHCOCF$_3$ aniline	CuI (10 mol %), L-proline (20 mol %), Cs$_2$CO$_3$ (4 eq), DMSO, 50°, 12 h	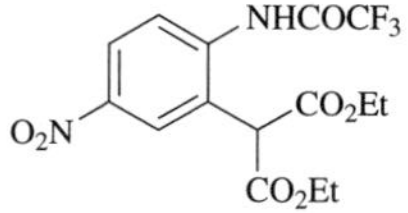 2-NHCOCF$_3$-5-NO$_2$-C$_6$H$_3$–CH(CO$_2$Et)$_2$ (64)	162
	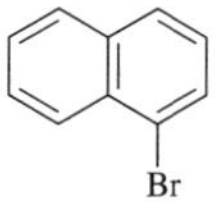 1-bromonaphthalene	Pd(dba)$_2$ (2 mol %), P(t-Bu)$_3$ (4 mol %), NaH (1.1 eq), THF, 70°, 6 h	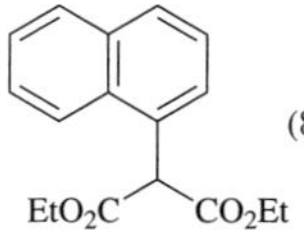 1-Naph–CH(CO$_2$Et)$_2$ (83)	82
	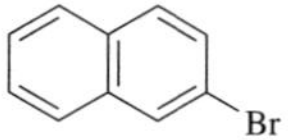 2-bromonaphthalene	Pd(dba)$_2$ (2 mol %), P(t-Bu)$_3$ (4 mol %), NaH (1.1 eq), THF, 70°, 6 h	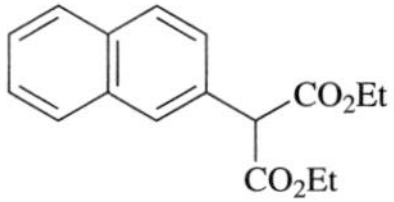 2-Naph–CH(CO$_2$Et)$_2$ (89)	82
	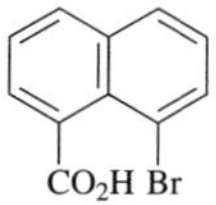 8-bromo-1-naphthoic acid	CuBr (cat), NaH, 50 min	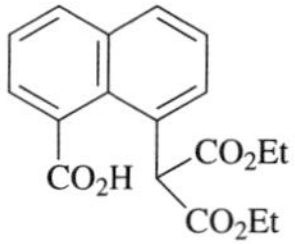 8-(CO$_2$H)-1-Naph–CH(CO$_2$Et)$_2$ (84)	101

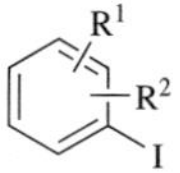

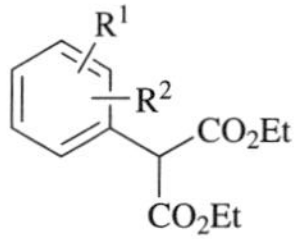

Substrate	Conditions	Product (Yield %)	Ref.
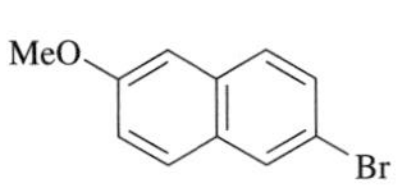 6-bromo-2-methoxynaphthalene	Pd(dba)$_2$ (2 mol %), P(t-Bu)$_3$ (4 mol %), K$_3$PO$_4$ (3 eq), toluene, 70°, 12 h	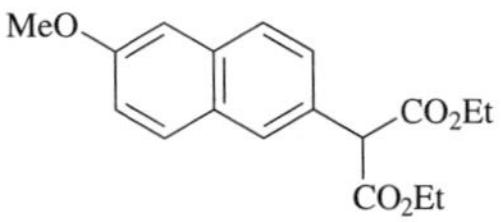 MeO-naphthyl-CH(CO$_2$Et)$_2$ (92)	82
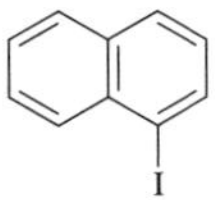 1-iodonaphthalene	CuI (10 mol %), 2-picolinic acid (10 mol %), Cs$_2$CO$_3$ (3 eq), dioxane, rt, 20 h	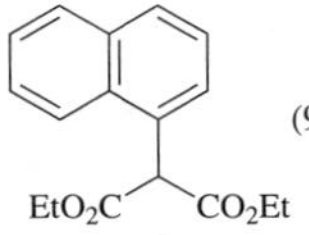 naphthyl-CH(CO$_2$Et)(CO$_2$Et) **I** (95)	106
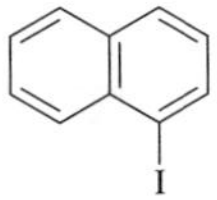 1-iodonaphthalene	CuI (5 mol %), 2-phenylphenol (10 mol %), Cs$_2$CO$_3$ (1.5 eq), THF, 70°, 30 h	**I** (98)	105
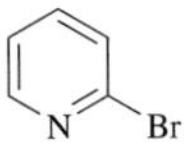 2-bromopyridine	CuI (5 mol %), 2-picolinic acid (20 mol %), Cs$_2$CO$_3$ (3 eq), dioxane, 110°, 32 h	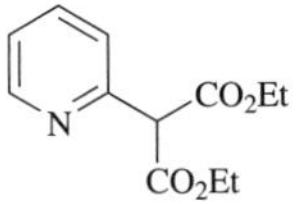 2-pyridyl-CH(CO$_2$Et)$_2$ (88)	106
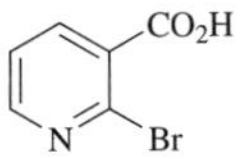 2-bromo-3-carboxypyridine	CuBr (cat), NaH, 30 min	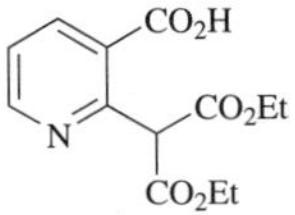 3-CO$_2$H-2-pyridyl-CH(CO$_2$Et)$_2$ (77)	101
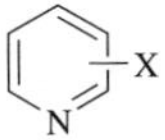 2-X-pyridine	CuI (5 mol %), 2-picolinic acid (10 mol %), Cs$_2$CO$_3$ (3 eq), dioxane, rt, 20 h	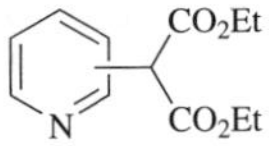 2-pyridyl-CH(CO$_2$Et)$_2$ X: 2-I (90), 3-I (77)	106

TABLE 3. ARYLATION OF 1,3-DICARBONYLS AND CYANOACETATES (*Continued*)

Substrate	Aryl Compound	Conditions	Product(s) and Yield(s) (%)	Refs.
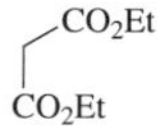 C7 EtO2C–CH2–CO2Et	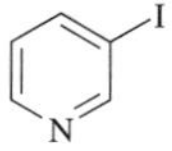 3-iodopyridine	CuI (5 mol %), 2-phenylphenol (10 mol %), Cs2CO3 (1.5 eq), THF, 70°, 26.5 h	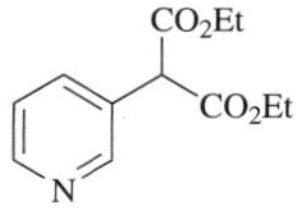 pyridyl-CH(CO2Et)2 (73)	105
	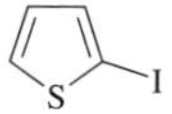 2-iodothiophene	CuI (5 mol %), 2-picolinic acid (10 mol %), Cs2CO3 (3 eq), dioxane, rt, 20 h	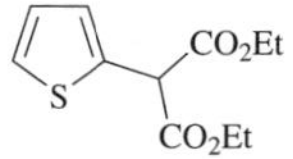 thienyl-CH(CO2Et)2 (68)	106
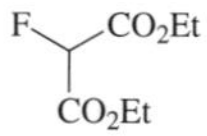 F–CH(CO2Et)2	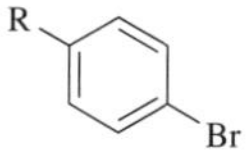 4-R-C6H4-Br	Pd(dba)2 (2 mol %), P(t-Bu)3 (4 mol %), NaH (1.1 eq), THF, 70°, 6 h	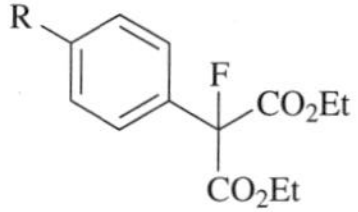 Ar-CF(CO2Et)2 R: H (84), MeO (88), MeO2C (89), t-Bu (86)	82
	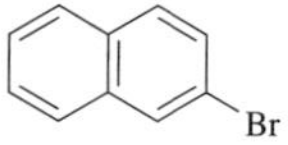 2-bromonaphthalene	Pd(dba)2 (2 mol %), P(t-Bu)3 (4 mol %), NaH (1.1 eq), THF, 70°, 6 h	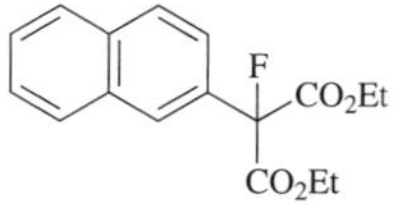 naphthyl-CF(CO2Et)2 (79)	82
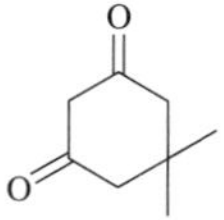 C8 5,5-dimethylcyclohexane-1,3-dione	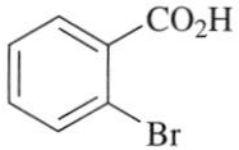 2-bromobenzoic acid	CuBr (cat), NaH	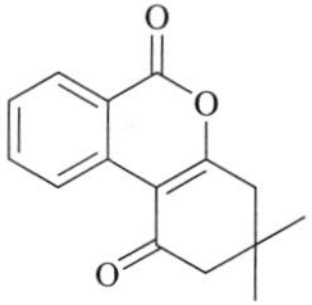 3,3-dimethyl-benzo-fused chromanone (63)	166
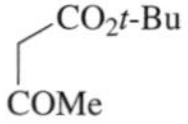 t-BuO2C-CH2-COMe	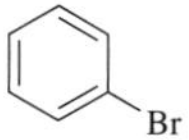 bromobenzene	Pd(dba)2 (1 mol %), P(t-Bu)3 (2 mol %), K3PO4 (2.8 eq), toluene, 90°, 16 h	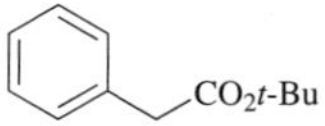 PhCH2CO2t-Bu (55)	87

| Substrate | Reagent | Conditions | Product(s) (%) | Ref. |

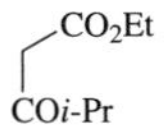

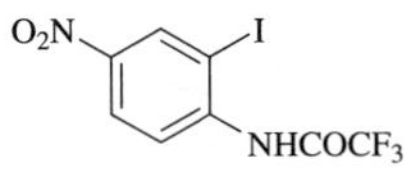

Table content (rendered as structured list due to complex layout):

- Substrate: 2-iodo-4-nitro-N-trifluoroacetyl aniline + ethyl 4-methyl-3-oxopentanoate (CO₂Et, COi-Pr)
  - Conditions: CuI (10 mol %), L-proline (20 mol %), Cs₂CO₃ (4 eq), DMSO, 50°, 12 h
  - Product: 5-nitro-2-CF₃-3-(CO₂t-Bu)-indole (81)
  - Ref: 162

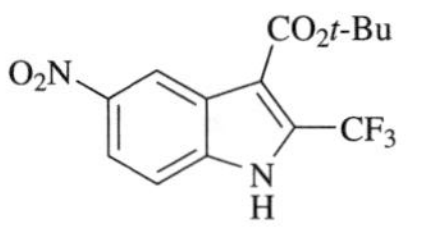

- Substrate: methyl 3-bromo-4-(trifluoroacetamido)benzoate + β-ketoester
  - Conditions: CuI (20 mol %), L-proline (40 mol %), Cs₂CO₃ (4 eq), DMSO, 70°, 15 h
  - Product: 5-(MeO₂C)-2-CF₃-3-(CO₂Et)-indole (72)
  - Ref: 162

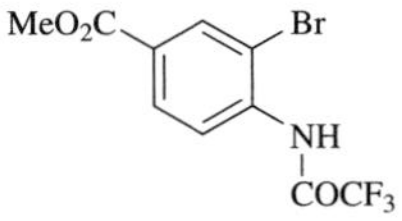

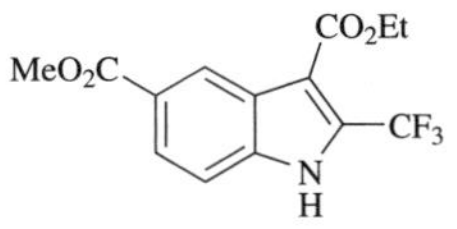

- Substrate: 1-(3-bromo-4-iodophenyl)ethanone + ethyl 4-methyl-3-oxopentanoate
  - Conditions: CuI (5 mol %), K₂CO₃ (1.5 eq), THF, 100°, 26 h
  - Product: 6-acetyl-2-(i-Pr)-3-(CO₂Et)-benzofuran (48)
  - Ref: 104

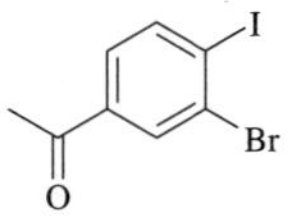

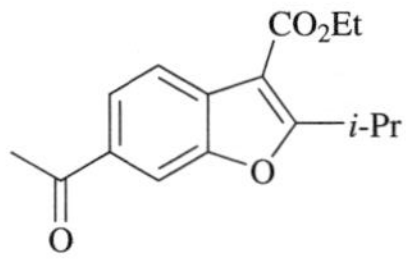

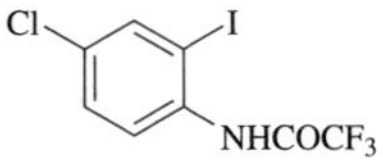

- Substrate: 4-chloro-2-iodo-N-trifluoroacetyl aniline + β-ketoester
  - Conditions: CuI (10 mol %), L-proline (20 mol %), Cs₂CO₃ (4 eq), DMSO, 70°, 15 h
  - Product: 5-chloro-2-CF₃-3-(CO₂Et)-indole (81)
  - Ref: 162

- Substrate: 2-bromobenzoic acid + ethyl 2-oxocyclopentanecarboxylate
  - Conditions: CuBr (cat), NaH, 80°, 6 h

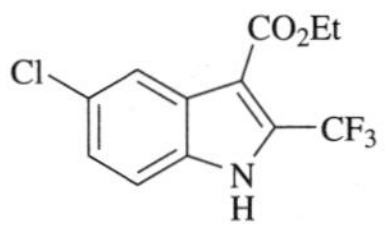

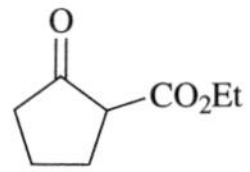

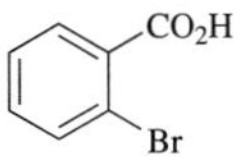

Solvent	I	II
none[d]	(38)	(37)
benzene	(87)	(0)

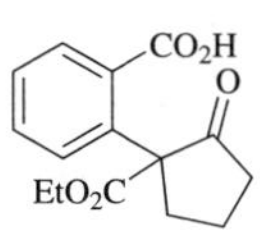

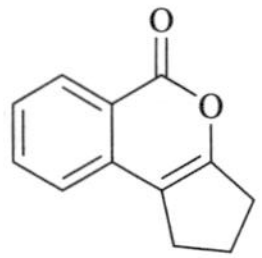

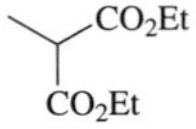

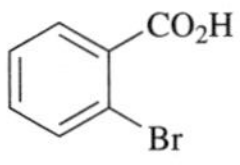

  Products I (α-arylated ketoester with CO₂H) and II (tricyclic lactone)
  - Ref: 166

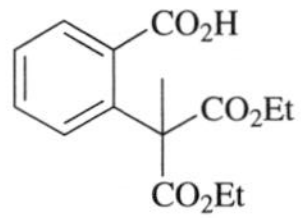

- Substrate: 2-bromobenzoic acid + diethyl methylmalonate (CH(CH₃)(CO₂Et)₂)
  - Conditions: CuBr (cat), NaH
  - Product: 2-[1,1-bis(ethoxycarbonyl)ethyl]benzoic acid (83)
  - Ref: 166

TABLE 3. ARYLATION OF 1,3-DICARBONYLS AND CYANOACETATES (*Continued*)

Substrate	Aryl Compound	Conditions	Product(s) and Yield(s) (%)	Refs.
C₉				
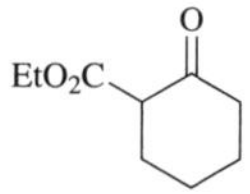 (EtO₂C-cyclohexanone)	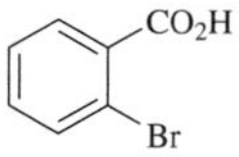 2-bromobenzoic acid (CO₂H, Br)	CuBr (cat), NaH, 80°, 6 h	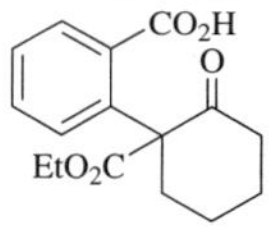 2-(1-(ethoxycarbonyl)-2-oxocyclohexyl)benzoic acid — Solvent: none^d (72); benzene (58)	166
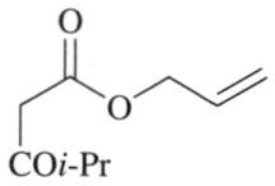 (allyl isopropyl malonate-like: O=C-CH₂-C(O)O-allyl, CO-*i*-Pr)	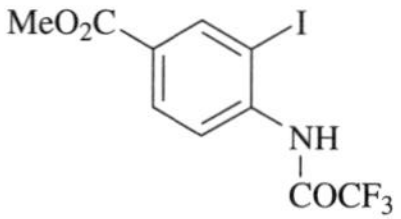 MeO₂C-C₆H₃(I)(NHCOCF₃)	CuI (10 mol %), L-proline (20 mol %), Cs₂CO₃ (4 eq), DMSO, 70°, 15 h	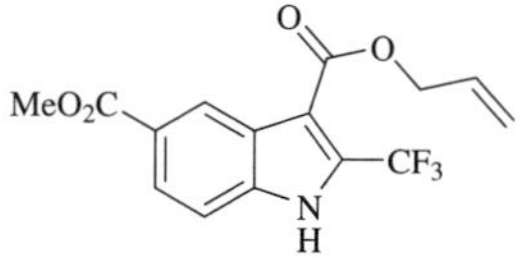 allyl 5-(methoxycarbonyl)-2-(trifluoromethyl)-1H-indole-3-carboxylate (72)	162
(EtO₂C-CH₂-C(O)-CH₂CH₂CH=CH₂)	R-C₆H₃(I)(Br)	CuI (5 mol %), K₂CO₃ (1.5 eq), THF, 100°	2-(but-3-enyl)-3-(ethoxycarbonyl)benzofuran derivative	104
			R Time (h)	
			H 24 (75)	
			MeO 26 (74)	
C₁₀				
(COPh, COMe)	2-bromobenzoic acid	Cu (6 mol %), NaOEt, EtOH	2-(1-benzoyl-2-oxopropyl)benzoic acid (78)	101
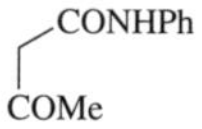 (CONHPh, COMe)	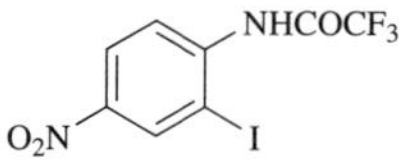 3-iodo-4-(NHCOCF₃)-nitrobenzene	CuI (10 mol %), L-proline (20 mol %), Cs₂CO₃ (4 eq), DMSO, 50°, 12 h	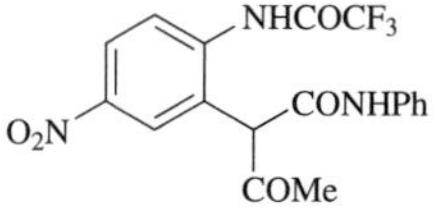 (78)	162
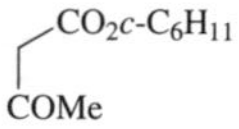 (CO₂-*c*-C₆H₁₁, COMe)	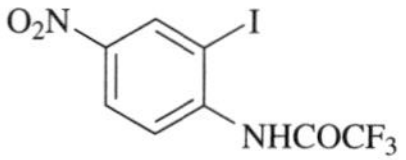 3-iodo-4-(NHCOCF₃)-nitrobenzene	CuI (10 mol %), L-proline (20 mol %), Cs₂CO₃ (4 eq), DMSO, 50°, 12 h	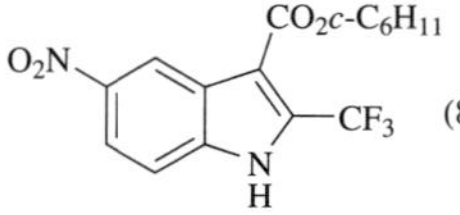 cyclohexyl 5-nitro-2-(trifluoromethyl)-1H-indole-3-carboxylate (80)	162

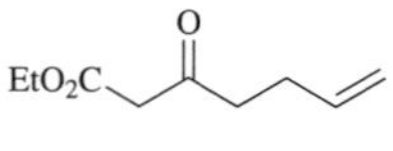

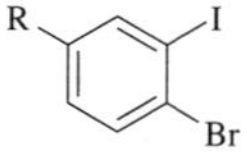

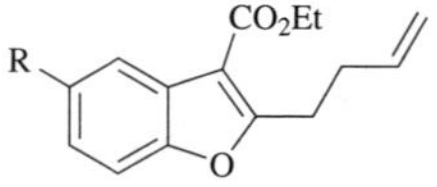

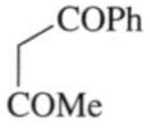

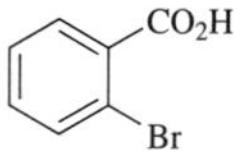

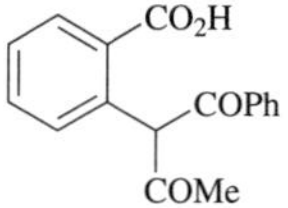

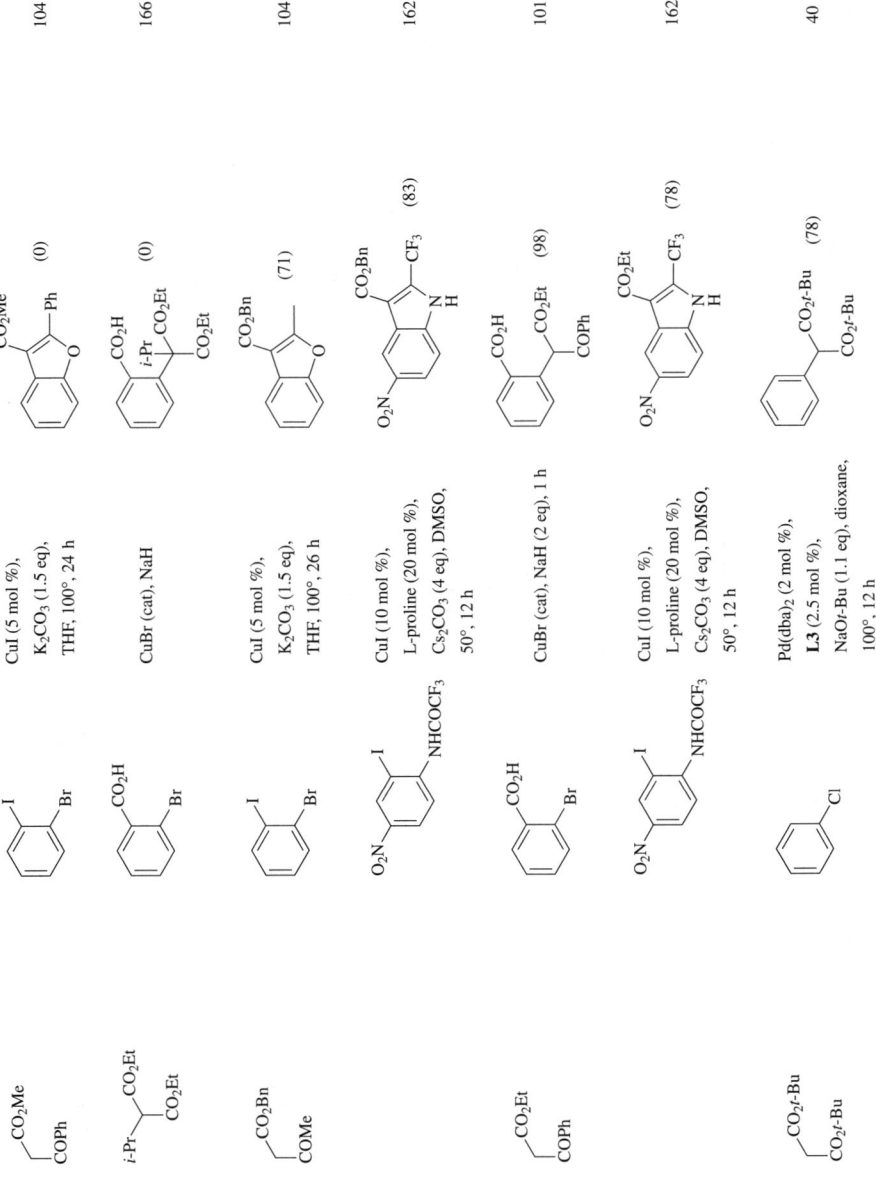

TABLE 3. ARYLATION OF 1,3-DICARBONYLS AND CYANOACETATES (*Continued*)

Substrate	Aryl Compound	Conditions	Product(s) and Yield(s) (%)	Refs.
C_{11} 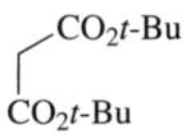 CO_2t-Bu CO_2t-Bu	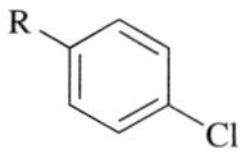 4-Cl-C$_6$H$_4$-R	[Pd(allyl)Cl]$_2$ (1 mol %), P(t-Bu)$_3$ (4 mol %), NaOt-Bu (1.1 eq), dioxane, 100°, 12 h	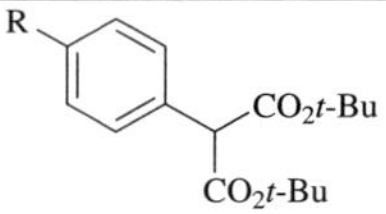 Ar-CH(CO$_2t$-Bu)$_2$ R H (88) MeO (84) CF$_3$ (86)	82
	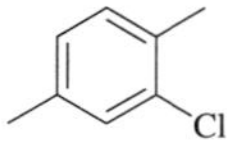 2-Me-4-Cl-C$_6$H$_3$(Me)	[Pd(allyl)Cl]$_2$ (1 mol %), P(t-Bu)$_3$ (4 mol %), NaOt-Bu (1.1 eq), dioxane, 100°, 12 h	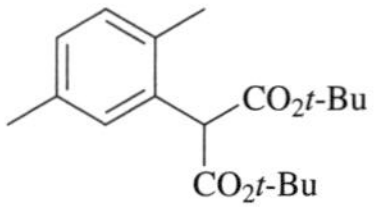 2,5-dimethylphenyl-CH(CO$_2t$-Bu)$_2$ (90)	82
	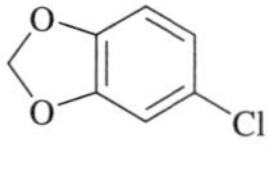 5-Cl-benzodioxole	[Pd(allyl)Cl]$_2$ (1 mol %), P(t-Bu)$_3$ (4 mol %), NaOt-Bu (1.1 eq), dioxane, 100°, 12 h	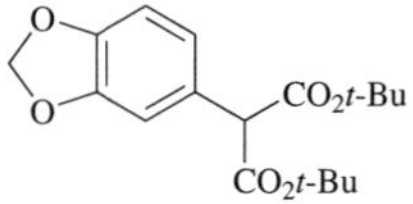 benzodioxol-5-yl-CH(CO$_2t$-Bu)$_2$ (87)	82
	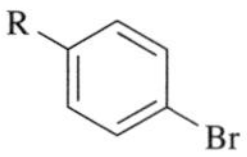 4-Br-C$_6$H$_4$-R	[Pd(allyl)Cl]$_2$ (1 mol %), P(t-Bu)$_3$ (4 mol %), NaOt-Bu (1.1 eq), dioxane, 45°, 8 h	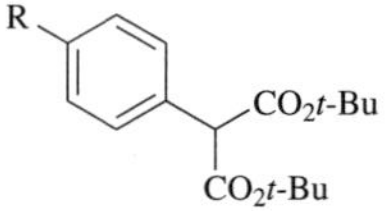 Ar-CH(CO$_2t$-Bu)$_2$ R H (91) CF$_3$ (90)	82
	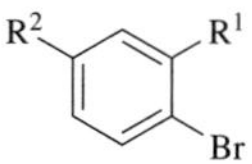 R^1,R^2-C$_6$H$_3$-Br	Pd(dba)$_2$ (2 mol %), P(t-Bu)$_3$ (4 mol %), NaH (1.1 eq), THF, 70°	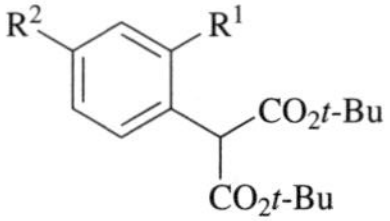 R^1,R^2-C$_6$H$_3$-CH(CO$_2t$-Bu)$_2$	82

R^1	R^2	Time (h)	Yield
H	H	6	(89)
MeO	H	12	(86)
Me	H	12	(85)
H	Me$_2$N	6	(90)
H	MeO	6	(90)
H	CF$_3$	6	(88)

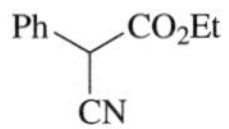

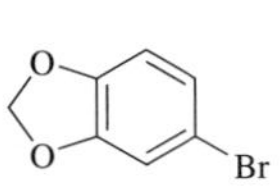

Nucleophile: Ph–CH(CO$_2$Et)(CN)

Aryl halide	Conditions	Product	Ref
5-bromo-1,3-benzodioxole	Pd(dba)$_2$ (2 mol %), P(t-Bu)$_3$ (4 mol %), NaH (1.1 eq), THF, 70°, 6 h	Ar–CH(CO$_2$$t$-Bu)$_2$ (87)	82
4-R-C$_6$H$_4$–Cl	Pd(dba)$_2$ (2 mol %), P(t-Bu)$_3$ (4 mol %), Na$_3$PO$_4$ (3 eq), toluene, 100°, 16 h	Ar–C(Ph)(CO$_2$Et)(CN) **I** R = H (89); MeO (91)	82
4-R-C$_6$H$_4$–Br	Pd(dba)$_2$ (2 mol %), P(t-Bu)$_3$ (4 mol %), Na$_3$PO$_4$ (3 eq), toluene, 70°, 16 h	**I** R = MeO (93); F (89); CF$_3$ (91); MeO$_2$C (92); MeCO (92); Ph (90); PhCO (93)	82
4-(1,3-dioxolan-2-yl)-C$_6$H$_4$–Br	Pd(dba)$_2$ (2 mol %), P(t-Bu)$_3$ (4 mol %), Na$_3$PO$_4$ (3 eq), toluene, 70°, 16 h	Ar–C(Ph)(CO$_2$Et)(CN) (94)	82
2-bromonaphthalene	Pd(dba)$_2$ (2 mol %), P(t-Bu)$_3$ (4 mol %), Na$_3$PO$_4$	Naphthyl–C(Ph)(CO$_2$Et)(CN) (93)	90
2-bromonaphthalene	Pd(dba)$_2$ (2 mol %), P(t-Bu)$_3$ (4 mol %), Na$_3$PO$_4$ (3 eq), toluene, 70°, 12 h	**I** (93)	90

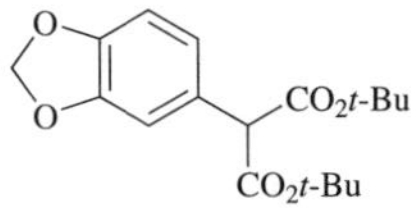

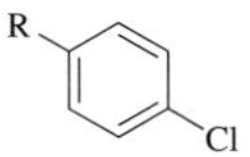

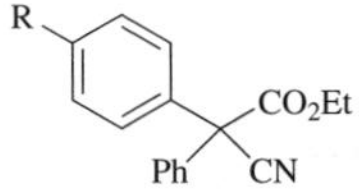

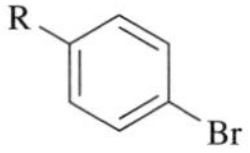

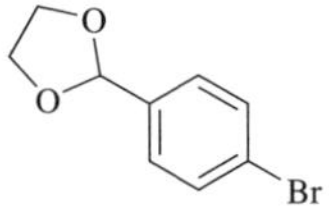

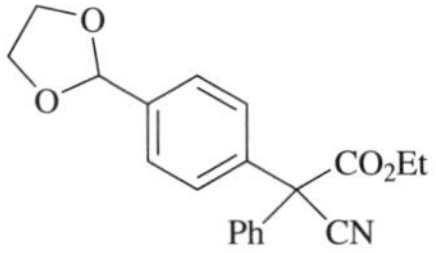

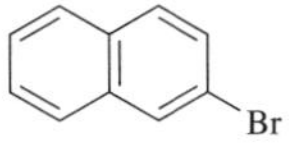

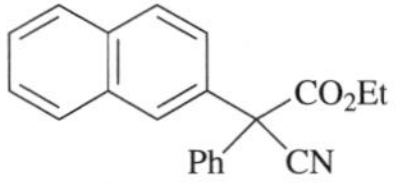

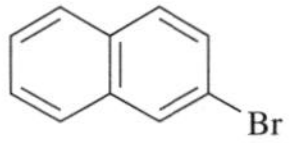

TABLE 3. ARYLATION OF 1,3-DICARBONYLS AND CYANOACETATES (*Continued*)

Substrate	Aryl Compound	Conditions	Product(s) and Yield(s) (%)	Refs.
C₁₁		Pd(PPh₃)₄, NaH, DMF, 130–140°	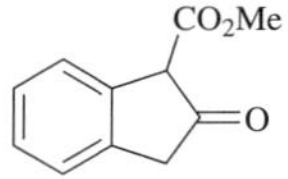 (0)	85
		Pd(PPh₃)₄, NaH, DMF, 130–140°	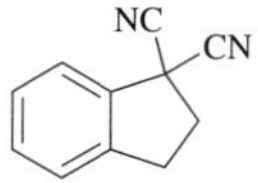 (54)	85
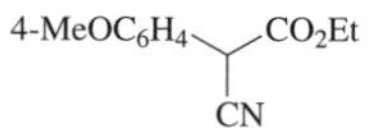 C₁₂	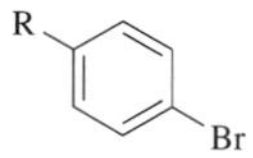 4-Br-C₆H₄-R	Pd(dba)₂ (2 mol %), P(t-Bu)₃ (4 mol %), Na₃PO₄ (3 eq), toluene, 70°	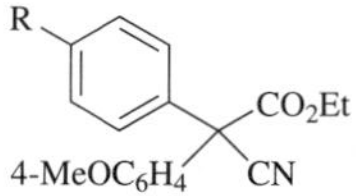 R = MeO, Time 20 h (94); R = CF₃, Time 16 h (90)	82
	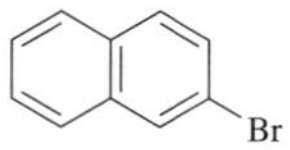 2-bromonaphthalene	Pd(dba)₂ (2 mol %), P(t-Bu)₃ (4 mol %), Na₃PO₄ (3 eq), toluene, 70°, 20 h	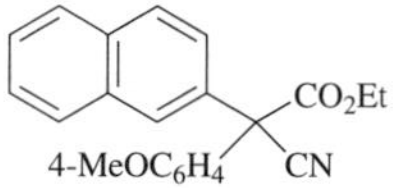 (91)	82
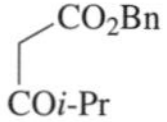 C₁₃	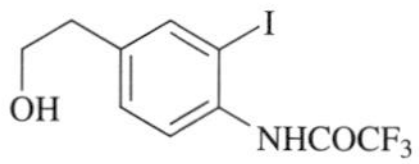 2-iodo-4-(2-hydroxyethyl)-NHCOCF₃ benzene	CuI (10 mol %), L-proline (20 mol %), Cs₂CO₃ (4 eq), DMSO, 80°, 15 h	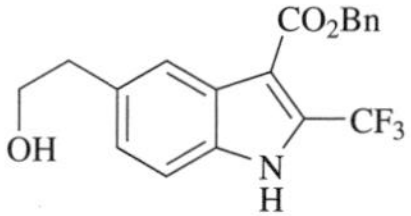 (67)	162
	2-bromo-1-iodo-4-chlorobenzene	CuI (5 mol %), K₂CO₃ (1.5 eq), THF, 100°, 26 h	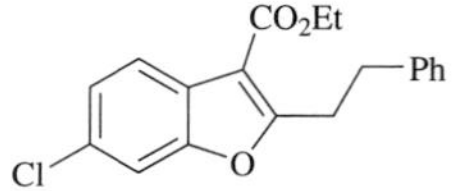 (78)	104

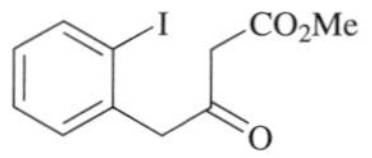

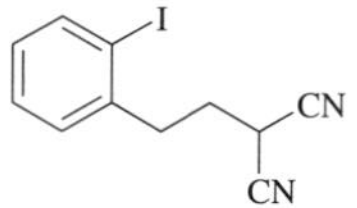

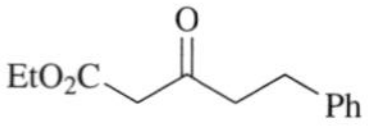

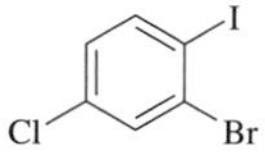

Substrate	Reagents	Product (Yield %)	Ref.
Ph-CH(CO₂Et)₂ + 2-bromobenzoic acid	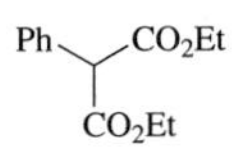 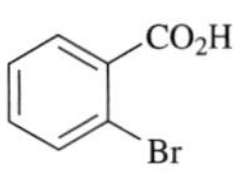 CuBr (cat), NaH	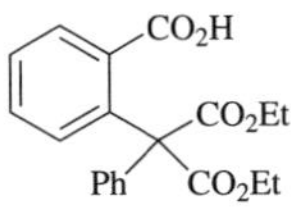 2-(Ph)(CO₂Et)(CO₂Et)C-C₆H₄-CO₂H (45)	166
Aryl iodide with CH(CO₂Me)₂ tether	Pd(PPh₃)₄, NaH, DMF, 130–140°	indane-1,1-bis(CO₂Me) (53)	85
Aryl iodide N-Me amide with CH(CO₂Et)Me	Pd(PPh₃)₄, NaH, DMF, 130–140°	3-methyl-3-(CO₂Et)-N-Me-oxindole (31)	85
Aryl iodide with CH(CN)(CO₂Et) tether	Pd(PPh₃)₄, NaH, DMF, 130–140°	1-CN-1-(CO₂Et)-indane (38)	85
Aryl iodide benzyl sulfone pent-4-en-2-yl ester	Pd(PPh₃)₄, NaH, DMF, 130–140°	pent-4-en-2-yl ester benzosulfolane (37)	85
Aryl iodide with CH(CN)(CO₂Et) (longer tether)	Pd(PPh₃)₄, NaH, DMF, 130–140°	1-CN-1-(CO₂Et)-tetralin (49)	85
Aryl halide X with CH(CO₂Et)₂ tether	Pd(PPh₃)₄, NaH, DMF, 130–140°	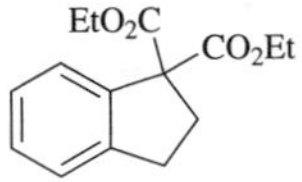 1,1-bis(CO₂Et)-indane; X=Br (48), X=I (62)	85

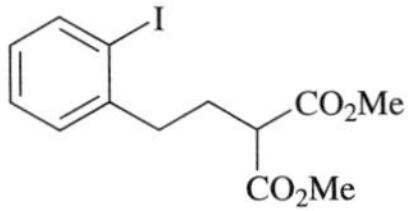

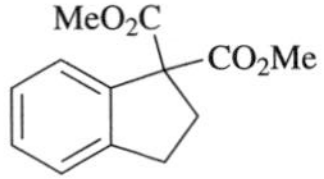

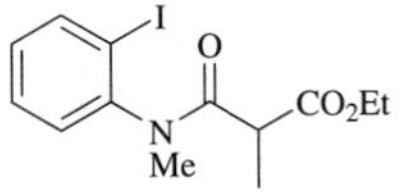

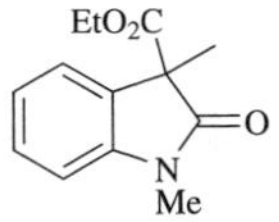

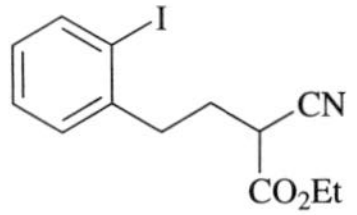

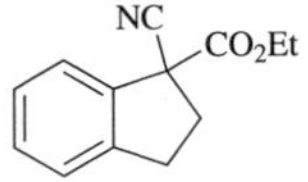

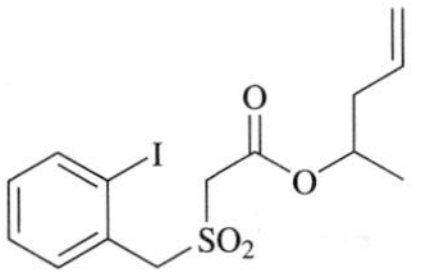

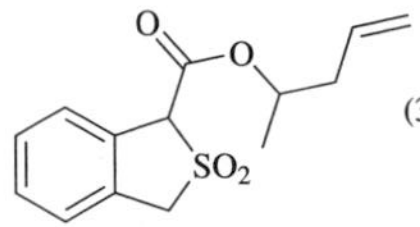

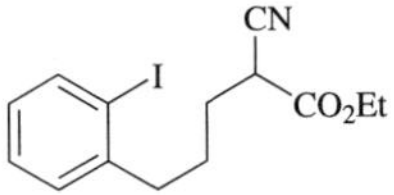

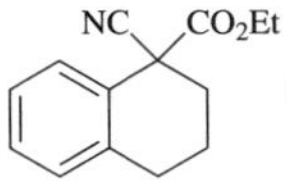

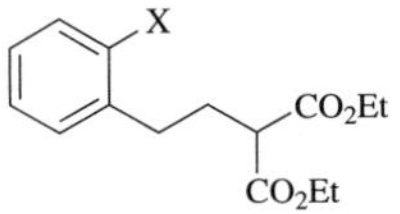

TABLE 3. ARYLATION OF 1,3-DICARBONYLS AND CYANOACETATES (*Continued*)

Substrate	Aryl Compound	Conditions	Product(s) and Yield(s) (%)	Refs.
C_{16} 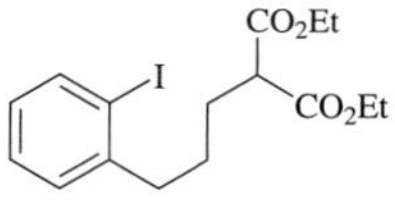		Pd(PPh$_3$)$_4$, NaH, DMF, 130–140°	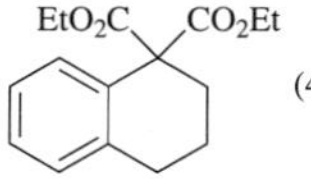 (41)	85
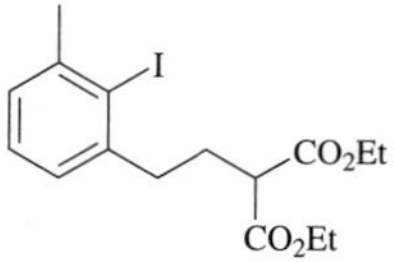		Pd(PPh$_3$)$_4$, NaH, DMF, 130–140°	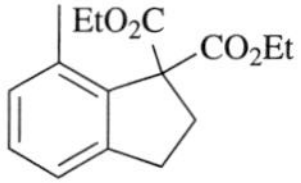 (75)	85
C_{17} 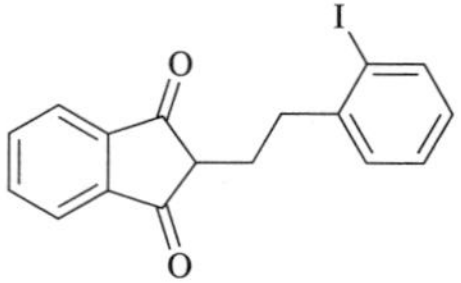		Pd(PPh$_3$)$_4$ (10 mol %), NaH, DMF, 130°, 6 h	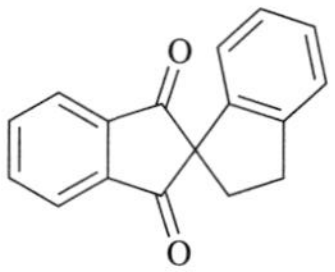 (68)	15
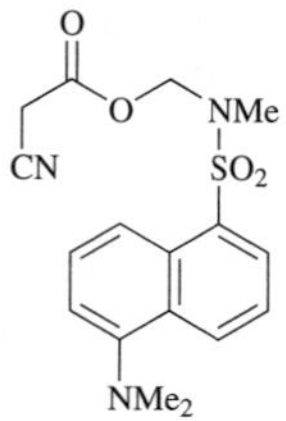	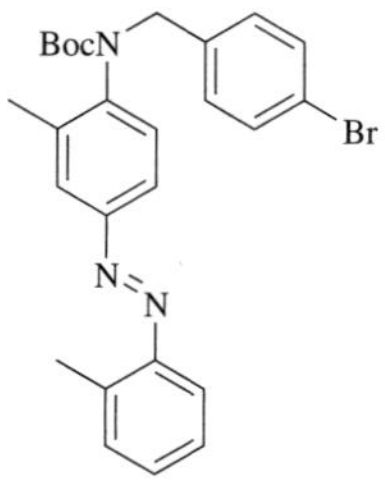	Catalyst	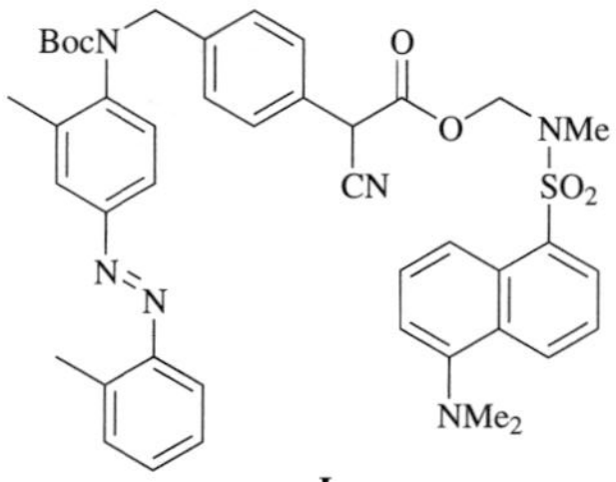 I (—)	90

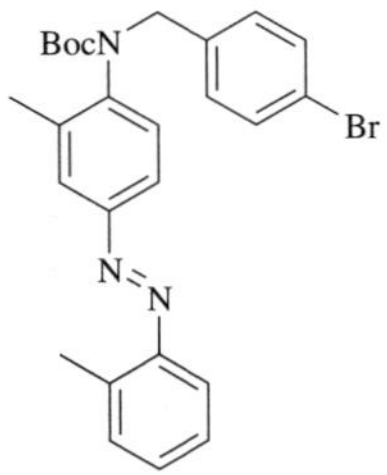

	Pd catalyst, ligand		
		Ligand	
		P(t-Bu)₃	1
	I (—)	**L33**	86
		L34	85
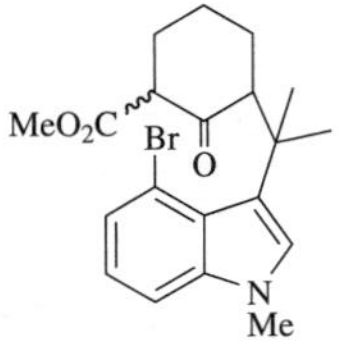	Pd(OAc)₂ (0.3 eq), P(t-Bu)₃ (0.6 eq), KOt-Bu (2 eq), toluene, 70°		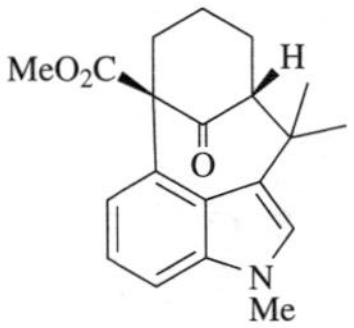 (74)
	Pd(PPh₃)₄, NaH, DMF, 135°		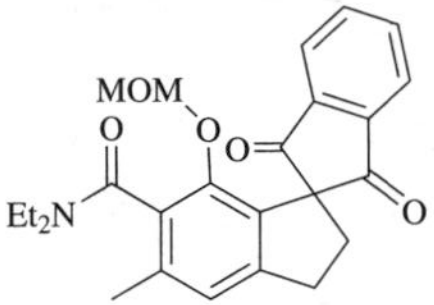 (76)

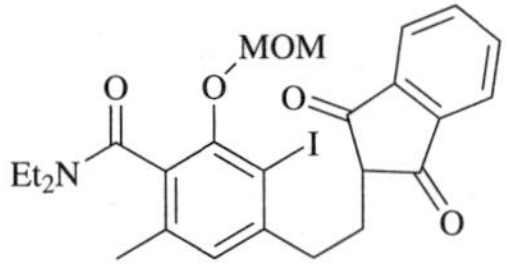

[a] The reaction was performed in 2 hours.
[b] CuI (1 mol %) was used.
[c] CuI (10 mol %) was used.
[d] The ester reactant was used as the solvent.

REFERENCES

[1] Prim, D.; Campagne, J.-M.; Joseph, D.; Andrioletti, B. *Tetrahedron* **2002**, *58*, 2041.
[2] Takeda, T.; Gonda, R.; Hatano, K. *Chem. Pharm. Bull.* **1997**, *45*, 697.
[3] Stratmann, T.; Moore, R. E.; Bonjouklian, R.; Deeter, J. B.; Patterson, G. M. L.; Shaffer, S.; Smith, C. D.; Smitka, T. A. *J. Am. Chem. Soc.* **1994**, *116*, 9935.
[4] Nicolaou, K. C.; Snyder, S. A. *Classics in Total Synthesis II: More Targets, Strategies, Methods*; Wiley-VCH: Weinheim, 2003.
[5] Rieu, J. P.; Boucherle, A.; Cousse, H.; Mouzin, G. *Tetrahedron* **1986**, *42*, 4095.
[6] Glasby, J. S. *Encyclopedia of the Alkaloids*; Plenum: New York, 1975; Vol. 1.
[7] Gao, Y.; Yao, Z. J.; Breckenridge, D.; Soon, L.; Soriano, J. V.; Burke, T. R.; Bottaro, D. P., Jr *J. Biol. Chem.* **2001**, *276*, 14308.
[8] Semmelhack, M. F.; Stauffer, R. D.; Rogerson, T. D. *Tetrahedron Lett.* **1973**, *14*, 4519.
[9] Tamao, K.; Zembayashi, M.; Kumada, M. *Chem. Lett.* **1976**, 1239.
[10] Kosugi, M.; Suzuki, M.; Hagiwara, I.; Goto, K.; Saitoh, K.; Migita, T. *Chem. Lett.* **1982**, 939.
[11] Sakurai, H.; Shirahata, A.; Araki, Y.; Hosomi, A. *Tetrahedron Lett.* **1980**, *21*, 2325.
[12] Kuwajima, I.; Nakamura, E. *Acc. Chem. Res.* **1985**, *18*, 181.
[13] Kuwajima, I.; Urabe, H. *J. Am. Chem. Soc.* **1982**, *104*, 6831.
[14] Uno, M.; Seto, K.; Takahashi, S. *J. Chem. Soc., Chem. Commun.* **1984**, 932.
[15] Ciufolini, M. A.; Browne, M. E. *Tetrahedron Lett.* **1987**, *28*, 171.
[16] Palucki, M.; Buchwald, S. L. *J. Am. Chem. Soc.* **1997**, *119*, 11108.
[17] Hamann, B. C.; Hartwig, J. F. *J. Am. Chem. Soc.* **1997**, *119*, 12382.
[18] Satoh, T.; Kawamura, Y.; Miura, M.; Nomura, M. *Angew. Chem., Int. Ed. Engl.* **1997**, *36*, 1740.
[19] Muratake, H.; Hayakawa, A.; Natsume, M. *Tetrahedron Lett.* **1997**, *38*, 7577.
[20] Muratake, H.; Natsume, M. *Tetrahedron Lett.* **1997**, *38*, 7581.
[21] Lloyd-Jones, G. C. *Angew. Chem., Int. Ed.* **2002**, *41*, 953.
[22] Hartwig, J. F. *Synlett* **2006**, 1283.
[23] Culkin, D. A.; Hartwig, J. F. *Acc. Chem. Res.* **2003**, *36*, 234.
[24] Christmann, U.; Vilar, R. *Angew. Chem., Int. Ed.* **2005**, *44*, 366.
[25] Galardon, E.; Ramdeehul, S.; Brown, J. M.; Cowley, A.; Hii, K. K.; Jutand, A. *Angew. Chem., Int. Ed.* **2002**, *41*, 1760.
[26] Culkin, D. A.; Hartwig, J. F. *J. Am. Chem. Soc.* **2002**, *124*, 9330.
[27] Wolkowski, J. P.; Hartwig, J. F. *Angew. Chem., Int. Ed.* **2002**, *41*, 4289.
[28] Hartwig, J. F. *Acc. Chem. Res.* **1998**, *31*, 852.
[29] Culkin, D. A.; Hartwig, J. F. *J. Am. Chem. Soc.* **2001**, *123*, 5816.
[30] Ahman, J.; Wolfe, J. F.; Troutman, M. V.; Palucki, M.; Buchwald, S. L. *J. Am. Chem. Soc.* **1998**, *120*, 1918.
[30a] Hamada, T.; Chieffi, A.; Ahman, J.; Buchwald, S.L. *J. Amer. Chem. Soc.* **2002**, *124*, 1261.
[31] Hamada, T.; Buchwald, S. L. *Org. Lett.* **2002**, *4*, 999.
[32] Lee, S.; Hartwig, J. F. *J. Org. Chem.* **2001**, *66*, 3402.
[33] Spielvogel, D. J.; Buchwald, S. L. *J. Am. Chem. Soc.* **2002**, *124*, 3500.
[34] Liao, X.; Weng, Z.; Hartwig, J. F. *J. Am. Chem. Soc.* **2008**, *130*, 195.
[35] Chen, G.; Kwong, F. Y.; Chan, H. O.; Yu, W.-Y.; Chan, A. S. C. *Chem. Commun.* **2006**, 1413.
[36] Xie, X.; Chen, Y.; Ma, D. *J. Am. Chem. Soc.* **2006**, *128*, 16050.
[37] Ooi, T.; Goto, R.; Maruoka, K. *J. Am. Chem. Soc.* **2003**, *125*, 10494.
[38] Koech, P. K.; Krische, M. J. *J. Am. Chem. Soc.* **2004**, *126*, 5350.
[39] Kelly, L. F.; Narula, A. S.; Birch, A. J. *Tetrahedron Lett.* **1980**, *21*, 2455.
[40] Kawatsura, M.; Hartwig, J. F. *J. Am. Chem. Soc.* **1999**, *121*, 1473.
[41] Hamada, T.; Chieffi, A.; Ahman, J.; Buchwald, S. L. *J. Am. Chem. Soc.* **2002**, *124*, 1261.
[42] Aranyos, A.; Old, D. W.; Kiyomori, A.; Wolfe, J. P.; Sadighi, J. P.; Buchwald, S. L. *J. Am. Chem. Soc.* **1999**, *121*, 4369.
[43] Fox, J. M.; Huang, X.; Chieffi, A.; Buchwald, S. L. *J. Am. Chem. Soc.* **2000**, *122*, 1360.
[44] Chae, J.; Yun, J.; Buchwald, S. L. *Org. Lett.* **2004**, *6*, 4809.
[45] Rutherford, J. L.; Rainka, M. P.; Buchwald, S. L. *J. Am. Chem. Soc.* **2002**, *124*, 15168.
[46] Satoh, T.; Kametani, Y.; Terao, Y.; Miura, M.; Nomura, M. *Tetrahedron Lett.* **1999**, *40*, 5345.

[47] Terao, Y.; Kametani, Y.; Wakui, H.; Satoh, T.; Miura, M.; Nomura, M. *Tetrahedron* **2001**, *57*, 5967.
[48] Ehrentraut, A.; Zapf, A.; Beller, M. *Adv. Synth. Catal.* **2002**, *344*, 209.
[49] Schnyder, A.; Indolese, A. F.; Studer, M.; Blaser, H.-U. *Angew. Chem., Int. Ed.* **2002**, *41*, 3668.
[50] Viciu, M. S.; Germaneau, R. F.; Nolan, S. P. *Org. Lett.* **2002**, *4*, 4053.
[51] Singh, R.; Nolan, S. P. *J. Organomet. Chem.* **2005**, *690*, 5832.
[52] Viciu, M. S.; Kelly, R. A.; Stevens, E. D.; Naud, F.; Studer, M.; Nolan, S. P. *Org. Lett.* **2003**, *5*, 1479.
[53] Churruca, F.; San Martin, R.; Tellitu, I.; Dominguez, E. *Org. Lett.* **2002**, *4*, 1591.
[54] Churruca, F.; San Martin, R.; Tellitu, I.; Dominguez, E. *Tetrahedron Lett.* **2003**, *44*, 5925.
[55] Churruca, F.; San Martin, R.; Carril, M.; Tellitu, I.; Dominguez, E. *Tetrahedron* **2004**, *60*, 2393.
[56] Churruca, F.; San Martin, R.; Tellitu, I.; Dominguez, E. *Tetrahedron Lett.* **2006**, *47*, 3233.
[57] Limbeck, M.; Wamhoff, H.; Rölle, T.; Griebenow, N. *Tetrahedron Lett.* **2006**, *47*, 2945.
[58] Khan, F. A.; Czerwonka, R.; Reissig, H.-U. *Eur. J. Org. Chem.* **2000**, *2000*, 3607.
[59] Satoh, T.; Jones, W. D. *Organometallics* **2001**, *20*, 2916.
[60] Terao, Y.; Satoh, T.; Miura, M.; Nomura, M. *Bull. Chem. Soc. Jpn.* **1999**, *72*, 2345.
[61] Willis, M. C.; Taylor, D.; Gillmore, A. T. *Org. Lett.* **2004**, *6*, 4755.
[62] Willis, M. C.; Taylor, D.; Gillmore, A. T. *Tetrahedron* **2006**, *62*, 11513.
[63] Mutter, R.; Campbell, I. B.; Martin de la Nava, E. M.; Merritt, A. T.; Wills, M. *J. Org. Chem.* **2001**, *66*, 3284.
[64] Muratake, H.; Nakai, H. *Tetrahedron Lett.* **1999**, *40*, 2355.
[65] Terao, Y.; Fukuoka, Y.; Satoh, T.; Miura, M.; Nomura, M. *Tetrahedron Lett.* **2002**, *43*, 101.
[66] Bordwell, F. G. *Acc. Chem. Res.* **1988**, *21*, 456.
[67] Shaughnessy, K. H.; Hamann, B. C.; Hartwig, J. F. *J. Org. Chem.* **1998**, *63*, 6546.
[68] Hama, T.; Culkin, D. A.; Hartwig, J. F. *J. Am. Chem. Soc.* **2006**, *128*, 4976.
[69] Zhang, T. Y.; Zhang, H. *Tetrahedron Lett.* **2002**, *43*, 193.
[70] Zhang, T. Y.; Zhang, H. *Tetrahedron Lett.* **2002**, *43*, 1363.
[71] Honda, T.; Namiki, H.; Satoh, F. *Org. Lett.* **2001**, *3*, 631.
[72] Cossy, J.; deFilippis, A.; Pardo, D. G. *Org. Lett.* **2003**, *5*, 3037.
[73] Satoh, T.; Inoh, J.-I.; Kawamura, Y.; Kawamura, Y.; Miura, M.; Nomura, M. *Bull. Chem. Soc. Jpn.* **1998**, *71*, 2239.
[74] Moradi, W. A.; Buchwald, S. L. *J. Am. Chem. Soc.* **2001**, *123*, 7996.
[75] Jorgensen, M.; Lee, S.; Liu, X.; Wolkowski, J. P.; Hartwig, J. F. *J. Am. Chem. Soc.* **2002**, *124*, 12557.
[76] Hama, T.; Liu, X.; Culkin, D. A.; Hartwig, J. F. *J. Am. Chem. Soc.* **2003**, *125*, 11176.
[77] Lee, S.; Beare, N. A.; Hartwig, J. F. *J. Am. Chem. Soc.* **2001**, *123*, 8410.
[78] Gaertzen, O.; Buchwald, S. L. *J. Org. Chem.* **2002**, *67*, 465.
[79] Liu, X.; Hartwig, J. F. *Org. Lett.* **2003**, *5*, 1915.
[80] Wu, L.; Hartwig, J. F. *J. Am. Chem. Soc.* **2005**, *127*, 15824.
[81] Uno, M.; Masuda, M.; Ueda, W.; Takahashi, S. *Tetrahedron Lett.* **1985**, *26*, 1553.
[82] Beare, N. A.; Hartwig, J. F. *J. Org. Chem.* **2002**, *67*, 541.
[83] Djakovitch, L.; Kohler, K. *J. Organomet. Chem.* **2000**, *606*, 101.
[84] Aramendia, M. A.; Borau, V.; Jimenez, C.; Marinas, J. M.; Ruiz, J. R.; Urbano, F. J. *Tetrahedron Lett.* **2002**, *43*, 2847.
[85] Ciufolini, M. A.; Qi, H. B.; Browne, M. E. *J. Org. Chem.* **1988**, *53*, 4149.
[86] MacKay, J. A.; Bishop, R. L.; Rawal, V. H. *Org. Lett.* **2005**, *7*, 3421.
[87] Zeevaart, J. G.; Parkinson, C. J.; de Koning, C. B. *Tetrahedron Lett.* **2004**, *45*, 4261.
[88] Uno, M.; Seto, K.; Ueda, W.; Masuda, M.; Takahashi, S. *Synthesis* **1985**, 506.
[89] Matsubara, H.; Seto, K.; Tahara, T.; Takahashi, S. *Bull. Chem. Soc. Jpn.* **1989**, *62*, 3896.
[90] Stauffer, S. R.; Beare, N. A.; Stambuli, J. P.; Hartwig, J. F. *J. Am. Chem. Soc.* **2001**, *123*, 4641.
[91] Fuson, R. C. *Chem. Rev.* **1935**, *16*, 1.
[92] Terao, Y.; Satoh, T.; Miura, M.; Nomura, M. *Tetrahedron Lett.* **1998**, *39*, 6203.
[93] Inoh, J.-I.; Satoh, T.; Pivsa-Art, S.; Miura, M.; Nomura, M. *Tetrahedron Lett.* **1998**, *39*, 4673.
[94] Yamamoto, Y.; Hatsuya, S.; Yamada, J. *J. Chem. Soc., Chem. Commun.* **1988**, 86.

[95] Yamamoto, Y.; Hatsuya, S.; Yamada, J. *J. Org. Chem* **1990**, *55*, 3318.
[96] Fraboni, A.; Fagnoni, M.; Albini, A. *J. Org. Chem.* **2003**, *68*, 4886.
[97] Heckrodt, T. J.; Mulzer, J. *J. Am. Chem. Soc.* **2003**, *125*, 4680.
[98] Agnelli, F.; Sulikowski, G. A. *Tetrahedron Lett.* **1998**, *39*, 8807.
[99] Liu, X.; Hartwig, J. F. *J. Am. Chem. Soc.* **2004**, *126*, 5182.
[100] Carfagna, C.; Musco, A.; Sallese, G.; Santi, R.; Fiorani, T. *J. Org. Chem.* **1991**, *56*, 264.
[101] Bruggink, A.; McKillop, A. *Tetrahedron* **1975**, *31*, 2607.
[102] Okuro, K.; Furuune, M.; Miura, M.; Nomura, M. *J. Org. Chem.* **1993**, *58*, 7606.
[103] Hang, H. C.; Drotleff, E.; Elliott, G. I.; Ritsema, T. A.; Konopelski, J. P. *Synthesis* **1999**, 398.
[104] Lu, B.; Wang, B.; Zhang, Y.; Ma, D. *J. Org. Chem.* **2007**, *72*, 5337.
[105] Hennessy, E. J.; Buchwald, S. L. *Org. Lett.* **2002**, *4*, 269.
[106] Yip, S. F.; Cheung, H. Y.; Zhou, Z.; Kwong, F. Y. *Org. Lett.* **2007**, *9*, 3469.
[107] Matsubara, K.; Ueno, K.; Koga, Y.; Hara, K. *J. Org. Chem.* **2007**, *72*, 5069.
[108] Millard, A. A.; Rathke, M. W. *J. Am. Chem. Soc.* **1977**, *99*, 4833.
[109] Fauvarque, J. F.; Jutand, A. *J. Organomet. Chem.* **1977**, *132*, C17.
[110] Cristau, H. J.; Vogel, R.; Taillefer, M.; Gadras, A. *Tetrahedron Lett.* **2000**, *41*, 8457.
[111] Abramovitch, R. A.; Barton, D. H. R.; Finet, J.-P. *Tetrahedron* **1988**, *44*, 3039.
[112] Sole, D.; Peidro, E.; Bonjoch, J. *Org. Lett.* **2000**, *2*, 2225.
[113] Sole, D.; Vallverdu, L.; Solans, X.; Font-Bardia, M.; Bonjoch, J. *J. Am. Chem. Soc.* **2003**, *125*, 1587.
[114] Carril, M.; SanMartin, R.; Churruca, F.; Tellitu, I.; Dominguez, E. *Org. Lett.* **2005**, *7*, 4787.
[115] Honda, T.; Sakamaki, Y. *Tetrahedron Lett.* **2005**, *46*, 6823.
[116] Veya, P.; Floriani, C.; Chiesi-Villa, A.; Rizzoli, C. *Organometallics* **1993**, *12*, 4899.
[117] Goossen, L. J. *Chem. Commun.* **2001**, 669.
[118] Rosenmund, K. W.; Harms, H. *Chem. Ber.* **1920**, *53*, 2226.
[119] Hurtley, W. R. H. *J. Chem. Soc.* **1929**, 1870.
[120] Ames, D. E.; Dodds, W. D. *J. Chem. Soc., Perkin Trans. 1* **1972**, 705.
[121] Sacks, C. E.; Fuchs, P. L. *J. Am. Chem. Soc.* **1975**, *97*, 7372.
[122] Rathke, M. W.; Vogiazoglou, D. *J. Org. Chem.* **1987**, *52*, 3697.
[123] Setsune, J.-I.; Ueda, T.; Shikata, K.; Matsukawa, K. *Tetrahedron* **1986**, *42*, 2647.
[124] Ugo, R.; Nardi, P.; Psaro, R.; Roberto, D. *Gazz. Chim. Ital.* **1992**, *122*, 511.
[125] Minami, T.; Isonaka, T.; Okada, Y.; Ichikawa, J. *J. Org. Chem.* **1993**, *58*, 7009.
[126] Newman, M. S.; Farbman, M. D. *J. Am. Chem. Soc.* **1944**, *66*, 1550.
[127] Periasamy, M.; KishoreBabu, N.; Jayakumar, K. N. *Tetrahedron Lett.* **2003**, *44*, 8939.
[128] Finet, J.-P. *Chem. Rev.* **1989**, *89*, 1487.
[129] Barton, D. H. R.; Blazejewski, J.-C.; Charpiot, B.; Finet, J.-P.; Motherwell, W. B.; Papoula, M. T. B.; Stanforth, S. P. *J. Chem. Soc., Perkin Trans. 1* **1985**, 2667.
[130] Arnauld, T.; Barton, D. H. R.; Normant, J.-F.; Doris, E. *J. Org. Chem.* **1999**, *64*, 6915.
[131] Pinhey, J. T.; Rowe, B. A. *Tetrahedron Lett.* **1980**, *21*, 965.
[132] Kozyrod, R. P.; Pinhey, J. T. *Tetrahedron Lett.* **1981**, *22*, 783.
[133] Elliott, G. I.; Konopelski, J. P. *Tetrahedron* **2001**, *57*, 5683.
[134] Kozyrod, R. P.; Pinhey, J. T. *Tetrahedron Lett.* **1982**, *23*, 5365.
[135] Donnelly, D. M. X.; Finet, J.-P.; Guiry, P. J.; Nesbitt, K. *Tetrahedron* **2001**, *57*, 413.
[136] Buston, J. E. H.; Moloney, M. G.; Parry, A. V. L.; Wood, P. *Tetrahedron Lett.* **2002**, *43*, 3407.
[137] Moloney, M. G.; Paul, D. R.; Thompson, R. M.; Wright, E. *Tetrahedron: Asymmetry* **1996**, *7*, 2551.
[138] Mino, T.; Matsuda, T.; Maruhashi, K.; Yamashita, M. *Organometallics* **1997**, *16*, 3241.
[139] Barroso, S.; Blay, G.; Cardona, L.; Fernandez, I.; Garcia, B.; Pedro, J. R. *J. Org. Chem.* **2004**, *69*, 6821.
[140] Baumgartner, M. T.; Gallego, M. H.; Pierini, A. B. *J. Org. Chem.* **1998**, *63*, 6394.
[141] Rossi, R. A.; Alonso, R. A. *J. Org. Chem.* **1980**, *45*, 1239.
[142] Ochiai, M.; Shu, T.; Nagaoka, T.; Kitagawa, Y. *J. Org. Chem.* **1997**, *62*, 2130.
[143] Chen, K.; Koser, G. F. *J. Org. Chem.* **1991**, *56*, 5764.
[144] Iwama, T.; Birman, V. B.; Kozmin, S. A.; Rawal, V. H. *Org. Lett.* **1999**, *1*, 673.

[145] Ryan, J. H.; Stang, P. J. *Tetrahedron Lett.* **1997**, *38*, 5061.
[146] Aggarwal, V. K.; Olofsson, B. *Angew. Chem., Int. Ed.* **2005**, *44*, 5516.
[147] Vogl, E. M.; Buchwald, S. L. *J. Org. Chem.* **2002**, *67*, 106.
[148] Martin, R.; Buchwald, S. L. *Org. Lett.* **2008**, *10*, 4561.
[149] Kosugi, M.; Hagiwara, I.; Sumiya, T.; Migita, T. *Bull. Chem. Soc. Jpn.* **1984**, *57*, 242.
[150] Su, W.; Raders, S.; Verkade, J. G.; Liao, X.; Hartwig, J. F. *Angew. Chem., Int. Ed.* **2006**, *45*, 1.
[151] Tetsuo, I.; Rawal, V. H. *Org. Lett.* **2006**, *8*, 5715.
[152] Eidamshaus, C.; Burch, J. D. *Org. Lett.* **2008**, *10*, 4211.
[153] Diedrichs, N.; Ragot, J. P.; Thede, K. *Eur. J. Org. Chem.* **2005**, 1731.
[154] Nilsson, P.; Larhed, M.; Hallberg, A. *J. Am. Chem. Soc.* **2003**, *125*, 3430.
[155] Sole, D.; Vallverdu, L.; Peidro, E.; Bonjoch, J. *Chem. Commun.* **2001**, 1888.
[156] Hama, T.; Hartwig, J. F. *Org. Lett.* **2008**, *10*, 1549.
[157] Hama, T.; Hartwig, J. F. *Org. Lett.* **2008**, *10*, 1545.
[158] Orsini, F.; Pelizzoni, F. *Synth. Commun.* **1987**, *17*, 1389.
[159] Orsini, F.; Pelizzoni, F.; Vallarino, L. M. *J. Organomet. Chem.* **1989**, *367*, 375.
[160] Durbin, M. J.; Willis, M. W. *Org. Lett.* **2008**, *10*, 1413.
[161] Koch, K.; Melvin, L. S., Jr.; Reiter, L. A.; Biggers, M. S.; Showell, H. J.; Griffiths, R. J.; Pettipher, E. R.; Cheng, J. B.; Milici, A. J.; Breslow, R.; Conklyn, M. J.; Smith, M. A.; Hackman, B. C.; Doherty, N. S.; Salter, E.; Farrell, C. A.; Schulte, G. *J. Med. Chem.* **1994**, *37*, 3197.
[162] Chen, Y.; Wang, Y.; Sun, Z.; Ma, D. *Org. Lett.* **2008**, *10*, 625.
[163] Siliphaivanh, P.; Harrington, P.; Witter, D. J.; Otte, K.; Tempest, P.; Kattar, S.; Kral, A. M.; Fleming, J. C.; Desmukh, S. V.; Harsch, A.; Secrist, P. J.; Miller, T. A. *Bioorg. Med. Chem. Lett.* **2007**, *17*, 4619.
[164] Wang, X.; Guram, A.; Bunel, E.; Cao, G.-Q.; Allen, R. A.; Faul, M. M. *J. Org. Chem.* **2008**, *73*, 1643.
[165] Zeevaart, J. G.; Parkinson, C. J.; Koning, C. B. *Tetrahedron Lett.* **2007**, *48*, 3289.
[166] McKillop, A.; Rao, D. P. *Synthesis* **1977**, 759.

CHAPTER 3

INDOLES VIA PALLADIUM-CATALYZED CYCLIZATION

SANDRO CACCHI, GIANCARLO FABRIZI, AND ANTONELLA GOGGIAMANI

Department of Drug Chemistry and Technologies, Sapienza, University of Rome, 00185 Rome, Italy

CONTENTS

	PAGE
INTRODUCTION	284
MECHANISMS	285
Palladium(II)-Catalyzed Cyclizations	287
Palladium(0)-Catalyzed Cyclizations	288
SCOPE AND LIMITATIONS	292
Indole Formation from Alkynes	292
2-Substituted Indoles	292
From 2-Alkynylanilid(n)es	292
From 1,2-Dihaloarenes	293
Under Copper- and/or Phosphine-Free Conditions	294
Via Coupling/Cyclization Methods with Supported Palladium Catalysts	295
From 2-Ethynylaniline	295
From 3-(2-Trifluoroacetamidophenyl)-1-propargyl Carbonate Esters	296
From 2-Halo-N-alkynylanilides	297
3-Substituted Indoles	297
2,3-Disubstituted Indoles	298
From Internal Alkynes and 2-Haloanilid(n)es	299
From 2-Alkynyltrifluoroacetanilides and C_{sp^3}, C_{sp^2}, and C_{sp} Donors	301
From 2-Alkynylanilid(n)es and Allylic Halides, Alkenes, and CO/MeOH	304
From N-Alkynyl-2-haloanilides	306
From 2-Alkynyl-N-alkylideneanilines	307
From 2-Alkynylisocyanobenzenes	307
From 2-(Alkynyl)phenylisocyanates	308
From 2-Alkynylphenyl N,O-Acetals and from 2-Iodoanilides and 1-(Tributylstannyl)-1-substituted Allenes	308
Indole Formation from Alkenes	309
Unsubstituted Indoles	309
2-Substituted Indoles	309
3-Substituted Indoles	312
2,3-Disubstituted Indoles	313

---

sandro.cacchi@uniroma1.it
Organic Reactions, Vol. 76, Edited by Scott E. Denmark et al.
© 2012 Organic Reactions, Inc. Published 2012 by John Wiley & Sons, Inc.

Indoles via Arene Vinylation 315
Indoles via N-Vinylation and N-Arylation 316
   Unsubstituted Indoles 316
   2-Substituted Indoles 318
   3-Substituted Indoles 321
   2,3-Disubstituted Indoles 322
Solid-Phase Synthesis 323
   Indole Formation from Alkynes 323
   Indole Formation from Alkenes 326
   Indole Formation via N-Vinylation and N-Arylation 327
COMPARISON WITH OTHER METHODS 329
   Copper-Catalyzed Indole Formation 329
   Indole Formation from Alkynes 329
   Indole Formation from Alkenes 332
   Indole Formation via N-Vinylation and N-Arylation 332
   Indole Formation via Arene Vinylation 334
   Gold-Catalyzed Indole Formation 334
   Indium-Catalyzed Indole Formation 336
   Iridium-Catalyzed Indole Formation 337
   Molybdenum-Catalyzed Indole Formation 337
   Platinum-Catalyzed Indole Formation 338
   Rhodium-Catalyzed Indole Formation 340
   Ruthenium-Catalyzed Indole Formation 342
   Titanium-Catalyzed Indole Formation 344
   Zinc-Catalyzed Indole Formation 344
EXPERIMENTAL CONDITIONS 344
EXPERIMENTAL PROCEDURES 346
   2-(3α-Acetoxyandrost-16-en-17-yl)-1H-indole [One-Flask Synthesis of a 2-Substituted
      Indole from 2-Ethynylaniline] 346
   N-Acetyl-2-isopropyl-6-carbomethoxyindole [Preparation of a 2-Substituted Indole
      from a 2-Alkynylacetanilide] 346
   2-[(4-Ethylpiperazin-1-yl)methyl]indole [Synthesis of a 2-Substituted Indole through an
      Intramolecular Heterocyclization/Intermolecular Nucleophilic Attack on a
      π-Allylpalladium Intermediate] 347
   3-(4-Acetylphenyl)indole [Synthesis of a 2-Unsubstituted 3-Arylindole via the
      Aminopalladation/Reductive Elimination Pathway] 347
   2-Phenyl-3-(phenylethynyl)indole [Synthesis of a 2,3-Disubstituted Indole from a
      2-Alkynyltrifluoroacetanilide and a 1-Bromoalkyne] 348
   2-(Cyclooct-1-enyl)-3-(4-methoxybenzoyl)indole [Synthesis of a
      2-Substituted-3-Carbonylated Indole via a Carbonylative Three-Component
      Cyclization] 348
   2,3-Diphenylindole [Synthesis of a 2,3-Disubstituted Indole via a One-Pot Tandem
      Cross-Coupling/Aminopalladation/Reductive Elimination Process] . . 349
   (2R, 5S)-3,6-Diethoxy-2-isopropyl-5-[2-(trimethylsilyl)-3-indolyl]methyl-
      2,5-dihydropyrazine [Synthesis of a 2,3-Disubstituted Indole via Heteroannulation of
      an Internal Alkyne with 2-Iodoaniline] 350
   N-Tosylindole [Synthesis of a 2,3-Unsubstituted Indole via Cylization of a
      2-Vinylanilide] 350
   N-(4-Bromobenzyl)-2-ethyl-3-(tert-butyldimethylsilyloxy)-5-methoxyindole [Synthesis
      of a 2,3-Substituted Indole via Cyclization of a 2-Allylaniline] . . . 351
   Indole [Cyclization of 2-Nitrostyrene] 351
   (L)-N,N-Di-tert-butoxycarbonyltryptophan Methyl Ester [Synthesis of a 3-Substituted
      Indole via Cyclization of an in Situ Generated 2-Haloanilinoenamine] . . 352

2,3-Diphenylindole [Synthesis of a 2,3-Disubstituted Indole through a One-Pot
  Hydroamination/Cyclization Process] 352
N-(4-Ethoxycarbonylphenyl)-2-ethoxycarbonyl-5-methoxyindole [Synthesis of a
  2-Substituted Indole Based on an Intramolecular N-Arylation Process] . . 353
Methyl 2-(2-Methoxyquinolin-3-yl)indole-5-carboxylate [Synthesis of a 2-Substituted
  Indole through a Tandem Carbon–Nitrogen/Suzuki–Miyaura Coupling] . . 353
2-[1-[4-(Trifluoromethyl)benzyl]indol-3-yl]acetamide [A Solid-Phase Synthesis of a
  3-Substituted Indole via Cyclization of a 2-Iodo-N-allylaniline] . . . 354
Methyl 2-Indolecarboxylate [A Solid-Phase Synthesis of a 2-Substituted Indole via
  Tandem Heck Reaction/N-Arylation] 355
TABULAR SURVEY 356
  Table 1A. 2-Substituted Indoles from 2-Haloanilines and Alkynes . . . 358
  Table 1B. 2-Substituted Indoles from 2-Haloanilides and Alkynes . . . 362
  Table 1C. 2-Substituted Indoles from 1,2-Dihaloarenes and Alkynes . . 372
  Table 1D. 2-Substituted Indoles from 2-Alkynylanilines 373
  Table 1E. 2-Substituted Indoles from 2-Alkynylanilides 379
  Table 1F. 2-Substituted Indoles from 2-Alkynylhaloarenes 386
  Table 1G. 2-Substituted Indoles from 2-Halo-N-alkynylanilides . . . 391
  Table 1H. 2-Substituted Indoles from 2-Alkynylisocyanatobenzenes . . 392
  Table 2A. 3-Substituted Indoles from 2-Haloanilines and Alkynes . . . 393
  Table 2B. 3-Substituted Indoles from 2-Alkynylanilides 394
  Table 2C. 3-Substituted Indoles from 3-Iodo-N-allylaniline and Internal
    Alkynes 395
  Table 2D. 3-Substituted Indoles from 2-Halo-N-alkylanilines 396
  Table 2E. 3-Substituted Indoles from 2-Iodo-N-propargylanilides and
    N-2-(Halophenyl)allenamides 398
  Table 3A. 2,3-Disubstituted Indoles from 2-Haloanilines, 2-Iodobenzoic Acids, or
    Anilines and Alkynes 400
  Table 3B. 2,3-Disubstituted Indoles from 2-Haloanilides or N-Acyl Benzotriazoles and
    Alkynes 415
  Table 3C. 2,3-Disubstituted Indoles from 2-Alkynylanilines 425
  Table 3D. 2,3-Disubstituted Indoles from 2-Alkynylanilides 427
  Table 3E. 2,3-Disubstituted Indoles from 2-Halo-N-alkynylanilides and
    2-Halo-N-alkylanilines 446
  Table 3F. 2,3-Disubstituted Indoles from 2-Alkynylisocyanobenzenes,
    -isocyanatobenzenes, and -N-alkylideneanilines 447
  Table 3G. 2,3-Disubstituted Indoles from N-(2-Halophenyl)allenamides . . 449
  Table 3H. 2,3-Disubstituted Indoles from 2-Allenylanilides Prepared in Situ . 450
  Table 4A. 2,3-Unsubstituted Indoles from 2-Vinylanilines and -anilides . . 451
  Table 4B. 2,3-Unsubstituted Indoles from 2-Nitrostyrenes 454
  Table 5A. 2-Substituted Indoles from 2-Allylanilines and -anilides . . . 455
  Table 5B. 2-Substituted Indoles from 2-Haloarylenamines and -imines . . 456
  Table 5C. 2-Substituted Indoles from 2-Haloarylenamines and -imines Prepared in Situ 457
  Table 5D. 2-Substituted Indoles from 2-Nitrostyrenes 463
  Table 6A. 3-Substituted Indoles from 2-Halo- and 2-Pseudohalo-N-allylanilines and
    -anilides 467
  Table 6B. 3-Substituted Indoles from 2-Halo-N-allylanilines and -anilides Prepared
    in Situ 471
  Table 6C. 3-Substituted Indoles from 2-Haloarylenamines 474
  Table 6D. 3-Substituted Indoles from 2-Haloarylenamines and -imines Prepared in Situ 475
  Table 6E. 3-Substituted Indoles from Arylenamines 480
  Table 6F. 3-Substituted Indoles from 2-Nitrostyrenes, Nitroalkenes, and Nitroarenes 481
  Table 7A. 2,3-Disubstituted Indoles from 2-Haloarylenamines and -imines . . 483

Table 7B. 2,3-Disubstituted Indoles from 2-Haloarylenamines and -imines Prepared in
Situ 485
Table 7C. 2,3-Disubstituted Indoles from Arylenamines and -imines . . . 488
Table 7D. 2,3-Disubstituted Indoles from 2-Nitrostyrenes, 2-Isocyanostyrene, and
2-Allylanilines 491
Table 8. Indoles via Arene Vinylation 493
Table 9. 2,3-Unsubstituted Indoles via N-Vinylation and N-Arylation . . . 494
Table 10. 2-Substituted Indoles via N-Vinylation and N-Arylation . . . 499
Table 11. 3-Substituted Indoles via N-Vinylation and N-Arylation . . . 509
Table 12. 2,3-Disubstituted Indoles via N-Vinylation and N-Arylation . . . 510
Table 13. Solid-Phase Synthesis of Indoles from Alkynes 512
Table 14. Solid-Phase Synthesis of Indoles from Alkenes 519
Table 15. Solid-Phase Synthesis via N-Arylation 521
Table 16. Miscellaneous 523
REFERENCES 526

INTRODUCTION

The palladium-catalyzed assembly of the functionalized pyrrole nucleus on a benzenoid scaffold is a widely used synthetic tool for the preparation of indole derivatives.[1-10] This construction can be categorized into four main types: (1) cyclization of alkynes, (2) cyclization of alkenes, (3) cyclization via C-vinylation reactions, and (4) cyclization via N-arylation or N-vinylation reactions. The first approach is by far the most versatile in terms of the range of the added functional groups and of the bonds that can be created in the construction of the pyrrole ring. This method is based on the utilization of precursors containing nitrogen nucleophiles and carbon–carbon triple bonds. The nitrogen nucleophile and alkyne moiety may be part of the same molecule or belong to two different molecules. Some of the most general and versatile alkyne-based cyclizations to indoles are summarized in Fig. 1.

Assembly of the pyrrole nucleus from precursors containing nitrogen nucleophiles and carbon–carbon double bonds entails only intramolecular cyclizations and, considering the bonds that can be created in the cyclization step, appears less versatile than the alkyne-based approach. Alkene-based cyclizations to give indoles are summarized in Fig. 2.

Cyclization to indoles via arene vinylation has limited synthetic scope. However, it is interesting that, unlike the above alkyne- and alkene-based procedures where the site of the oxidative addition of the carbon–X bond to the palladium(0) species is located on the benzenoid ring (Figures 1 and 2), the oxidative addition site is located in a vinylic fragment tethered to the benzenoid ring in this type of cyclization. Furthermore, it is the sole example of the construction of the pyrrole ring via palladium-catalyzed vinylation of an *ortho*-unfunctionalized aromatic ring (Fig. 3). Such direct arene vinylation and arylation processes are of great current interest.[11-14]

Finally, indoles can be prepared via cyclizations proceeding through N-arylation and N-vinylation reactions (Fig. 4) that are based on the pioneering work[15-22]

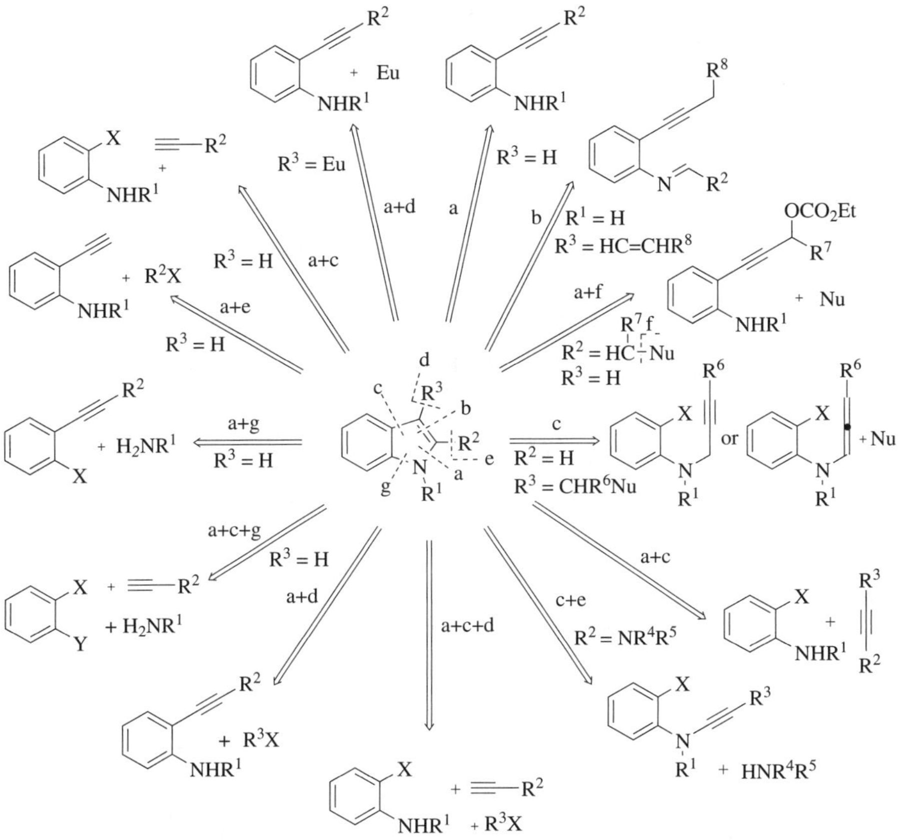

Figure 1. Retrosynthetic representation of the main alkyne-based, palladium-catalyzed constructions of the pyrrole ring. (Nu = nucleophile, E = electrophile).

on palladium-catalyzed carbon–nitrogen bond forming reactions from aryl halides or triflates with amines, amides, and carbamates.

In general, only synthetic procedures where palladium catalysis is involved in the pyrrole ring construction event are discussed herein. Palladium-catalyzed reactions producing indole-related compounds, such as azaindoles, indazoles, indolines, oxindoles, bis(indolyl)methanes, and related systems, or condensed polycyclic compounds, such as carbolines, carbazoles, indoloquinolines, indoloquinazolines, and related systems, are not discussed. Indoles are classified as 2-substituted, 3-substituted, and 2,3-disubstituted derivatives without considering the functionalization of the nitrogen atom.

MECHANISMS

A variety of reaction parameters such as solvents, temperature, the nature of the substrates and ligands, bases, and additives, and sometimes even their

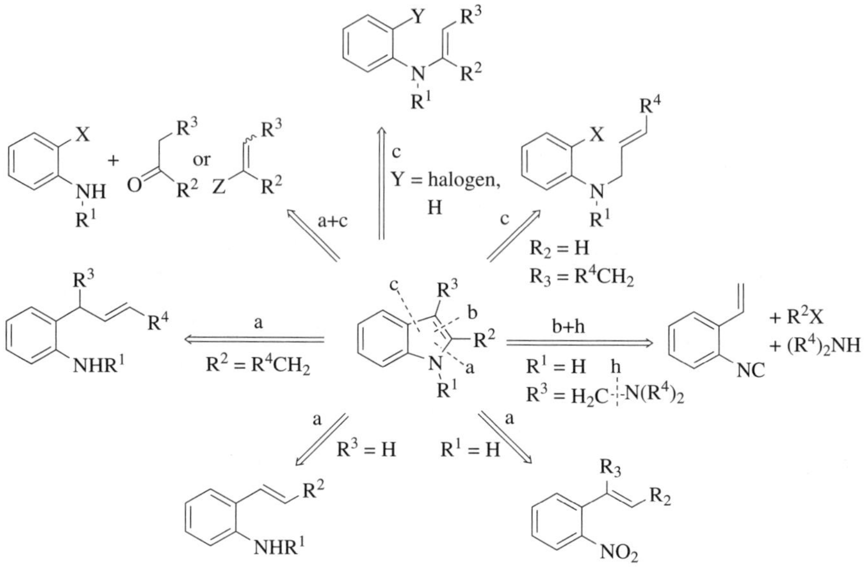

Figure 2. Retrosynthetic representation of the main alkene-based, palladium-catalyzed constructions of the pyrrole ring.

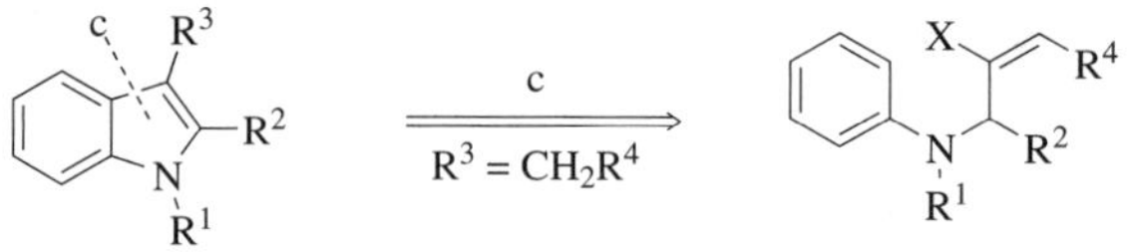

Figure 3. Retrosynthetic representation of the palladium-catalyzed construction of the pyrrole ring via intramolecular arene vinylation.

combination can influence the mechanism operating in this reaction. In addition, catalytic cycles usually consist of several consecutive steps and the chemical nature as well as the reactivity of each intermediate can differ depending on reaction conditions. Some reaction parameters can also exhibit opposing effects on different steps of a catalytic cycle. In view of this complexity, it is not surprising that the literature contains few detailed mechanistic studies. Therefore, the word mechanism is used in this section to indicate a plausible rationalization of how products are formed rather than an experimentally supported mechanism. These plausible rationalizations are categorized into two main types corresponding to two main sections: palladium(II)- and palladium(0)-catalyzed reactions. The two main sections are subclassified by the proposed reaction mechanisms. Since the palladium-catalyzed cyclization to indoles is an extremely diverse class of reactions from a mechanistic point of view, only the main mechanistic proposals are discussed below.

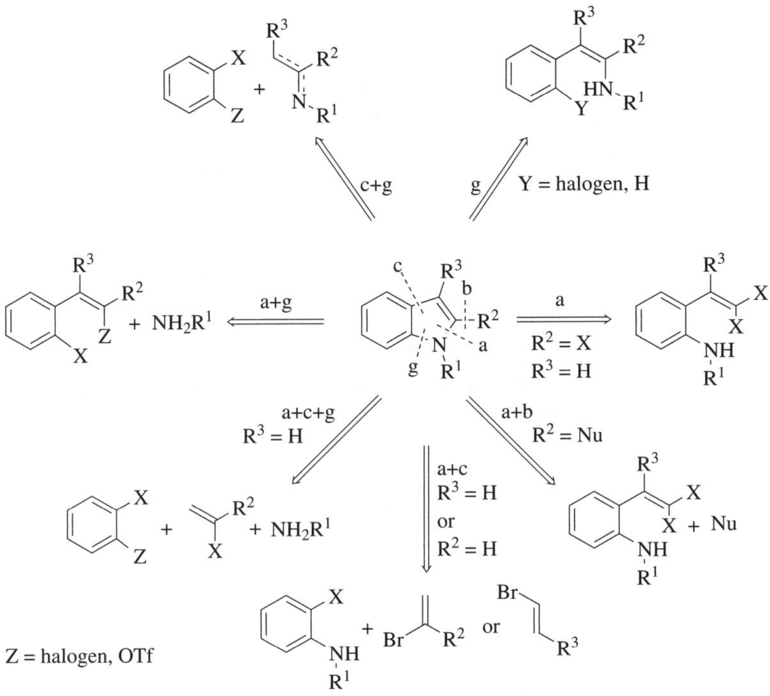

Figure 4. Retrosynthetic representation of the main palladium-catalyzed constructions of the pyrrole ring via N-vinylation and N-arylation reactions.

Palladium(II)-Catalyzed Cyclizations

Most of the syntheses of indoles catalyzed by Pd(II) salts involve cyclizations of aryl alkynes containing *ortho*-nitrogen nucleophiles (Fig. 1, disconnections a and a+d) or allylic and vinylic arenes containing *ortho*-nitrogen nucleophiles (Fig. 2, disconnection a)."

Palladium(II) salts are fairly electrophilic species. For that reason, the first event leading to cyclization in palladium(II)-catalyzed reactions is usually considered to be the coordination of acetylenic or olefinic π-electrons to a palladium(II) species. As shown in Schemes 1 and 2 for 2-alkynylanilides[23] and 2-allylanilines,[24] the resultant π-palladium complexes **1** and **3** subsequently undergo an intramolecular nucleophilic attack of a nitrogen nucleophile across the activated carbon–carbon multiple bond to give the aminopalladation adducts **2** and **4**, respectively. With acetylenic precursors, protonolysis of the carbon–palladium bond of **2** forms 2-substituted indoles and regenerates the active catalytic species. This approach to the construction of the pyrrole ring, which ultimately allows for the addition of nitrogen–hydrogen bonds across carbon–carbon multiple bonds, is frequently described as a hydroamination reaction. With alkene precursors, the conversion of aminopalladation adducts **4** into indole derivatives

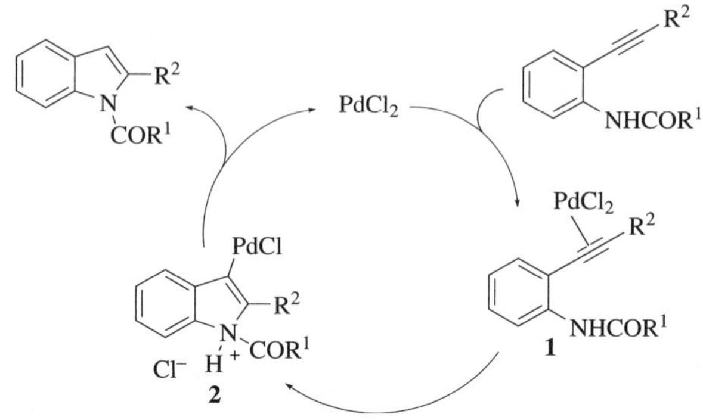

Scheme 1. Catalytic cycle of the palladium(II)-catalyzed cyclization of 2-alkynylanilides via the hydroamination pathway.

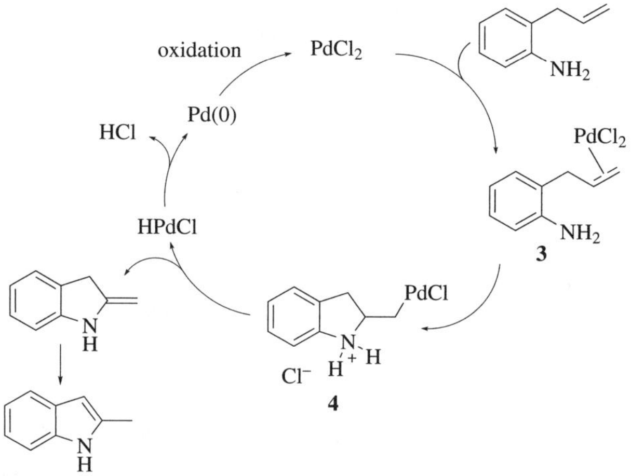

Scheme 2. Catalytic cycle of the palladium(II)-catalyzed cyclization of 2-allylanilines.

involves a β-elimination step that ultimately leads to the formation of palladium(0) species. Consequently, for the reaction to be catalytic with respect to palladium(II), the presence of stoichiometric amounts of oxidants such as $CuCl_2$, $Cu(OAc)_2$, benzoquinone, *tert*-butyl hydroperoxide (TBHP), or MnO_2 is required to allow for the in situ conversion of palladium(0) into palladium(II).

Palladium(0)-Catalyzed Cyclizations

Cyclizations to indoles catalyzed by a palladium(0) species provide a wider variety of applications than palladium(II)-catalyzed cyclizations, and some are among the most efficient and generally applicable methods. They include the great majority of alkyne-based syntheses described in Figure 1, a variety of alkene-based syntheses (Fig. 2, disconnection c, a + c, b + h), cyclization to indoles via arene vinylation (Fig. 3), and cyclizations based on *N*-vinylation and *N*-arylation reactions (Fig. 4).

Palladium(0) complexes are usually nucleophilic and the initial step of the vast majority of palladium(0)-catalyzed cyclizations to indoles involves an oxidative addition of carbon–X bonds (X = I, Br, Cl, OTf) to coordinatively unsaturated palladium(0) species to give carbon–palladium(II)–X intermediates that contain an electrophilic palladium. In general, the oxidative addition step is favored by increasing the electron density on palladium. The observed rate of oxidative addition with carbon$_{aryl}$–halogen bonds increases in the order C–F < C–Cl < C–Br < C–I (aryl fluorides are almost inert).[25] The reactivity of aryl triflates is approximately between that of aryl iodides and aryl bromides. In the presence of monodentate ligands, a *cis*-complex is likely to be the initial product of the oxidative addition. Subsequently, isomerization gives rise to the thermodynamically more stable *trans*-complex. With bidentate ligands, the *cis*-complex is the usual intermediate.

The aminopalladation/reductive elimination mechanism has been suggested to account for the cyclization to indoles of 2-alkynyltrifluoroacetanilides,[7] 2-alkynylisocyanobenzenes,[26] 2-alkynylisocyanatobenzenes[27,28] (Fig. 1, disconnection a+d) and 2-halo-*N*-alkynylanilides[29] (Fig. 1, disconnection c+e). Although some differences exist in the details of the mechanistic proposals for the cyclization of these compounds, the general features of the aminopalladation/reductive elimination pathway are well described by the example shown in Scheme 3 for the synthesis of free (NH) 2,3-disubstituted indoles from 2-alkynyltrifluoroacetanilides **5**. In this mechanism, coordination of π-acetylenic electrons to organopalladium complexes, generated in situ through oxidative addition of organic precursors to palladium(0) species, affords π-alkyne-organopalladium complexes **6** that subsequently undergo nucleophilic attack of the nitrogen atom across the activated carbon–carbon triple bond to give the σ-indolylpalladium intermediates **7** (the aminopalladation adduct). The free

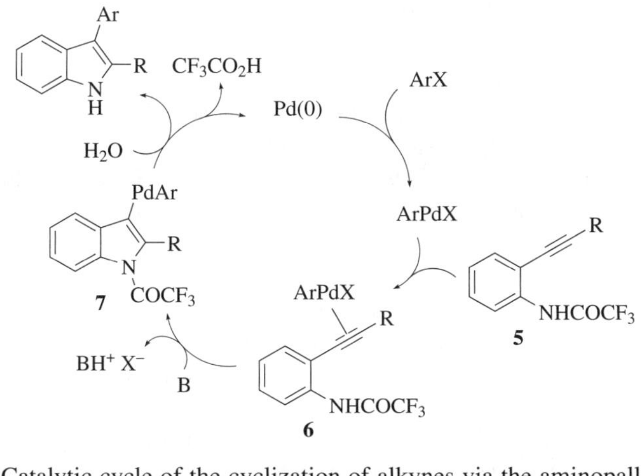

Scheme 3. Catalytic cycle of the cyclization of alkynes via the aminopalladation/reductive elimination pathway.

indole product (NH) is formed by hydrolysis of the amide bond and a reductive elimination step (not necessarily in this order) that produces a new carbon–carbon bond and regenerates the active palladium(0) catalyst.

The palladium-catalyzed reaction of aryl halides with alkynes not containing nucleophiles close to the carbon–carbon triple bond may form π-alkyne-σ-arylpalladium complexes that, unable to undergo an intramolecular nucleophilic attack across the carbon–carbon triple bond, afford carbopalladation adducts **8** (Eq. 1). These adducts, depending on reaction conditions, can be converted into a variety of products via an intermolecular process, as exemplified in Eq. 1.

$$R\text{—}\!\!\!\equiv\!\!\!\text{—}R \;+\; ArX \;\xrightarrow{Pd(0)}\; \left[\begin{array}{c} ArPdX \\ | \\ R\text{—}\!\!\!=\!\!\!\text{—}R \end{array}\right]$$

$$\longrightarrow \left[\begin{array}{c} Ar\quad PdX \\ \diagdown\;\;\diagup \\ \text{C}=\text{C} \\ \diagup\;\;\diagdown \\ R\quad\;\; R \\ \mathbf{8} \end{array}\right] \xrightarrow[-X^-, -Pd(0)]{Nu^-} \begin{array}{c} Ar\quad Nu \\ \diagdown\;\;\diagup \\ \text{C}=\text{C} \\ \diagup\;\;\diagdown \\ R\quad\;\; R \end{array}$$

(Eq. 1)

When the aryl moiety added to palladium(0) contains a nitrogen nucleophile adjacent to the oxidative addition site, as shown in Scheme 4, the carbopalladation adduct **10** can undergo an intramolecular halide displacement from the palladium to give a nitrogen-containing palladacycle **11** that subsequently affords the indole product via a reductive elimination step.[30,31] This carbopalladation route (Fig. 1, disconnection a+c) is one of the most versatile and efficient indole syntheses.

Alkynes or allenes containing a tethered aryl halide fragment such as 2-iodo-N-propargylanilides or N-2-halophenylallenamides can form carbopalladation adducts intramolecularly (Fig. 1, disconnection c). The addition intermediates derived from alkynes have been trapped by norbornene to give indoles containing polycyclic substituents at C(3).[32] Palladium acetate and (n-Bu)$_3$P catalyze the cyclization of 2-alkynyl-N-alkylidene-anilines to indoles (Fig. 1, disconnection b).[33]

The formation of 2-aminomethylindoles **17** from 3-(2-trifluoroacetamidophenyl)-1-propargyl carbonate ester **12**[34] (Fig. 1, disconnection a+f) is likely to proceed through the following basic steps (Scheme 5): (a) initial formation of the σ-allenylpalladium complex **13**—via an S$_N$2' reaction of the palladium complex with ester **12**—that is in equilibrium with the π-propargylpalladium intermediate **14**;[35] (b) intramolecular nucleophilic attack of the nitrogen at the central carbon of the allenyl/propargylpalladium complex;[36–42] (c) protonation of the resultant carbene complex **15** to give the π-allylpalladium complex **16**; (d) site selective intermolecular nucleophilic attack of the nitrogen nucleophile at the less-hindered allylic terminus of **16**. A similar mechanism is most probably operating for the conversion of ester **12** into 2-alkylindoles.[43]

The intramolecular version of the Heck reaction has been used for the construction of the indole ring[44] (Fig. 2, disconnection c) and an intramolecular halide displacement within arylpalladium intermediates by carbon nucleophiles has been

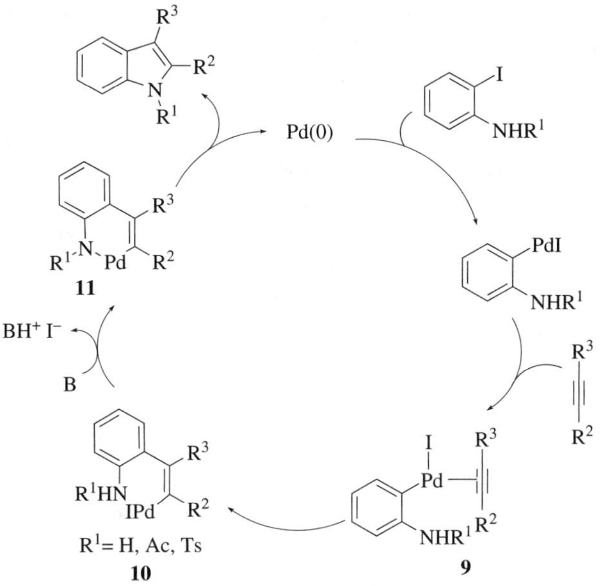

Scheme 4. Catalytic cycle of the palladium(0)-catalyzed annulation of 2-iodoanilines or 2-iodoanilides with internal alkynes.

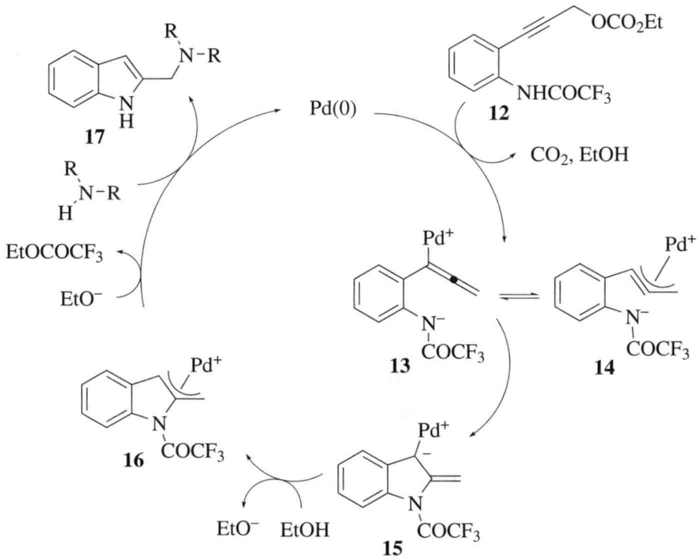

Scheme 5. Catalytic cycle of the palladium(0)-catalyzed cyclization of 3-(2-trifluoroacetamidophenyl)-1-propargyl carbonate esters.

proposed to account for the cyclization of 2-haloanilino enamines to indoles (Fig. 2, disconnection c).[45,46] Phenolic carbamates containing a bromovinylic fragment bound to the nitrogen atom are thought to give indole carbamates through an arene vinylation mechanism (Fig. 3).[47]

The general features of the N-arylation and N-vinylation method (Fig. 4) are shown in Scheme 6 for the cyclization of 2-chlorophenylacetaldehyde N,N-dimethylhydrazone (**18**) to 1-dimethylaminoindole (**21**)[48] and entail (a) an oxidative addition of the aryl chloride fragment to a palladium(0) species to afford the σ-arylpalladium intermediate **19**, (b) an intramolecular chloride displacement by nitrogen to give the palladacycle **20**, and (c) a subsequent reductive elimination leading to the formation of 1-dimethylaminoindole (**21**).

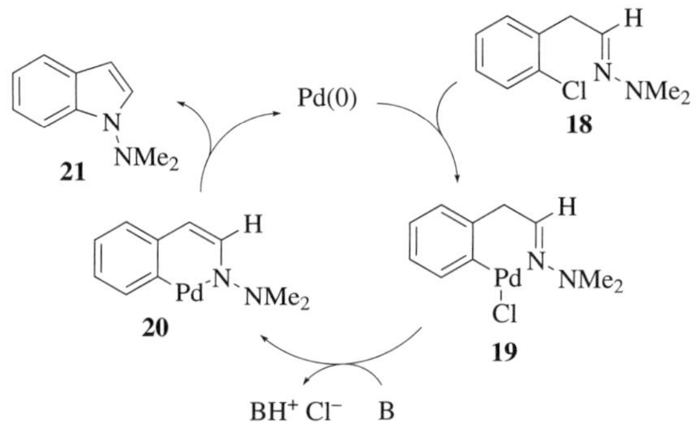

Scheme 6. Catalytic cycle of the palladium(0)-catalyzed cyclization via intramolecular N-arylation.

SCOPE AND LIMITATIONS

Indole Formation from Alkynes

2-Substituted Indoles. *From 2-Alkynylanilid(n)es.* The majority of the examples describing the alkyne-based synthesis of 2-substituted indoles originate from the observation that these indole derivatives can be prepared via palladium (II)-catalyzed cyclization of 2-alkynylanilides (Fig. 1, disconnection a; Scheme 1). The main variations of this method involve the synthesis of the starting 2-alkynylanilides. In early examples, the preparation of 2-alkynylanilides features the coupling of preformed copper(I) salts of terminal alkynes with 2-thallated anilides in acetonitrile.[23] This procedure has rarely found applications in indole synthesis, very likely because of the toxicity of the metal used. 2-Bromoacetanilides have subsequently been used in the coupling reaction,[49] but the 2-alkynylacetanilides are prepared through palladium(0)-catalyzed reaction of 2-bromoacetanilides with alkynylstannanes, a procedure that still uses toxic reagents. A significant improvement came with the discovery[50] that treatment of terminal alkynes with 2-haloanilides under Sonogashira conditions[51,52] can directly afford indole products in a single step through a tandem coupling–cyclization process (Eq. 2; Fig. 1, disconnection a+c).

$$\text{o-Br-C}_6\text{H}_4\text{-NHMs} + \text{HC}\equiv\text{C-SiMe}_3 \xrightarrow[120°, 24\text{ h}]{\text{PdCl}_2(\text{PPh}_3)_2, \text{CuI, Et}_3\text{N}} \left[\text{o-(Me}_3\text{SiC}\equiv\text{C)-C}_6\text{H}_4\text{-NHMs}\right] \longrightarrow \text{2-SiMe}_3\text{-N(Ms)-indole} \quad (58\%)$$

(Eq. 2)

The palladium-catalyzed coupling of terminal alkynes with aryl halides or triflates containing a nitrogen nucleophile in the *ortho* position followed by a palladium-catalyzed cyclization step has been extensively applied, providing stepwise and tandem syntheses of 2-substituted indoles. The cyclization of the coupling products can also be performed using base-mediated[50,53–64] and copper-catalyzed protocols.[50,65–69] The involvement of both palladium and copper catalysis in the cyclization of 2-alkynylanilines or their *N*-substituted derivatives has also been reported.[50,68] In some cases, particularly when indole products are obtained through tandem processes based on Sonogashira cross-coupling followed by a cyclization reaction, the specific role of the palladium catalyst and/or the base and/or copper in the formation of the pyrrole ring are not clearly established.

Cyclizations of 2-alkynylanilines or 2-alkynylanilides in the presence of palladium(II) are performed using $PdCl_2$, $PdCl_2(MeCN)_2$, or $PdCl_2(PPh_3)_2$ in acetonitrile, $PdCl_2/Bu_4NCl$ or Bu_4NBr in a biphasic aqueous HCl/CH_2Cl_2 system, or $PdBr_2$ in toluene.[23,49,70–76] Sodium tetrachloropalladate in dichloroethane at 100° is used in the related cyclization of 2-alkynylisocyanatobenzenes.[28] The combination of $FeCl_3$ and $PdCl_2$ in dichloroethane can give good results for the cyclization of 2-alkynylanilines, where iron may facilitate the in situ reoxidation of palladium(0) to palladium(II).[77]

Tandem coupling/cyclization processes are typically carried out in the presence of copper(I) salts (in general CuI), with $PdCl_2(PPh)_3$ or $Pd(PPh_3)_4$ as precatalyst, i-Pr_2NH or Et_3N as nitrogen base, and MeCN, DMF or DMA/H_2O as the solvent.[50,78–85]

From 1,2-Dihaloarenes. The recently reported use of 1,2-dihaloarenes as the arene partners in the synthesis of 2-substituted indoles is an alternative to the classical methods based on 2-haloanilines or their derivatives.[86–88] 1,2-Dihaloarenes can engage in a cross-coupling reaction with terminal alkynes to give 2-alkynylhaloarenes that subsequently undergo a palladium-catalyzed *N*-arylation/cyclization reaction to give the corresponding indoles. A palladium complex generated from the commercially available imidazolium salt HIPrCl in combination with *t*-BuOK is an efficient catalyst for the conversion of 2-alkynylhaloarenes into indoles (Eq. 3; Fig. 1, disconnection a+g).[86] Mild bases such as Cs_2CO_3 or K_3PO_4 can also be used; however, longer reaction times and, in some cases, incomplete cyclization of the coupling intermediates are observed using these bases. These problems can be circumvented by adding CuI to the reaction mixture. 2-Alkynylhaloarenes can also be cyclized to indoles using Pd(OAc)$_2$ and (t-Bu)$_3$P, with t-BuOK (in toluene) or K$_3$PO$_4$ (in DMA) as the bases.[87]

$$\text{(Eq. 3)}$$

The entire process (cross-coupling/N-arylation/cyclization) can be conducted as a one-pot protocol (Fig. 1, disconnection a+c+g).[86,87] An example is shown in Eq. 4.[86,89]

$$\text{(Eq. 4)}$$

Under Copper- and/or Phosphine-Free Conditions. Although the coupling/cyclization methods mentioned above are usually efficient and versatile, and their synthetic scope is quite large, they suffer from some drawbacks, such as terminal alkyne homocoupling in the presence of copper(I) cocatalysts[90] and/or the use of oxygen-sensitive phosphine ligands. Because of these limitations, some recent studies feature examples of copper- and/or phosphine-free protocols. Thus, 2-iodoanilides and terminal alkynes are converted into 2-substituted indoles via a one-pot coupling/cyclization process with Pd(OAc)$_2$ in the presence of Bu$_4$NOAc as the base under ultrasonic irradiation or standard conditions (Eq. 5).[84] Palladium nanoparticles stabilized in micelles formed by polystyrene-co-poly(ethylene oxide) and cetylpyridinium chloride as a surfactant (PS-PEO-CPC-Pd) are used in the tandem cross-coupling/cyclization of N-mesyl-2-iodoaniline with phenylacetylene (Eq. 6).[85] Both 2-iodoaniline and 2-iodotrifluoroacetanilide give lower yields. The activity of the colloidal catalyst is slightly lower than that of PdCl$_2$(MeCN)$_2$. Indoles can be prepared from 2-iodoaniline and a terminal alkyne under copper-free conditions in the presence of Pd(OAc)$_2$, Ph$_3$P, K$_2$CO$_3$, Bu$_4$NCl in DMF at 100°.[91] Access to 2-substituted indoles from 2-iodoaniline or 2-iodotrifluoroacetanilide and terminal alkynes without any copper promoter is also known in the presence of Pd(OAc)$_2$, TPPTS, and Et$_3$N in an MeCN/H$_2$O mixture.[92]

$$\text{(Eq. 5)}$$

Reagents: Pd(OAc)$_2$, Bu$_4$NOAc, MeCN, 90°; ultrasound, 5 h (90%); rt, 36 h (74%).

$$\text{(Eq. 6)}$$

Reagents: PS-PEO-CPC-Pd, CuI, DMA, H$_2$N(CH$_2$)$_2$OH, 80°, 4.5 h (79%).

Via Coupling/Cyclization Methods with Supported Palladium Catalysts. In addition to soluble palladium complexes, supported palladium precatalysts may be used in the coupling/cyclization protocol. 2-Phenylindole can be prepared through a tandem process from 2-iodoaniline and phenylacetylene in 72% yield by employing palladium on activated carbon in the presence of CuI in DMF/H$_2$O (120°, 6 hours).[93] Palladium on activated carbon can also provide access to a variety of 2-substituted indoles from terminal alkynes and 2-iodoanilides in water.[94] Reactions are carried out in the presence of Ph$_3$P, CuI, and 2-aminoethanol as the base at 80°. Potassium-fluoride-doped alumina in the presence of palladium powder, CuI, and Ph$_3$P is another precatalyst system that furnishes 2-phenylindole in 80% yield from *N*-mesyl-2-iodoaniline under solvent-free and microwave-assisted conditions.[95] The use of a Pd(II)-NaY zeolite precatalyst in DMF at 140° in the presence of LiCl and Cs$_2$CO$_3$ allows for the conversion of 2-iodoanilides and terminal alkynes into the corresponding 2-substituted indoles.[96] The catalyst can be recycled up to five times by adding LiCl and Cs$_2$CO$_3$ to each reaction. However, slightly lower yields are obtained and a longer reaction time is necessary with each recycle.

From 2-Ethynylaniline. All the procedures mentioned above require specific 1-alkynes for each indole, and this can limit their substrate scope. Furthermore, 2-haloanilides are used as starting materials in many cases and this usually requires an additional step to liberate the free indole (NH) derivatives. To circumvent this problem, an alternative approach has been developed in which free 2-substituted indoles (NH) can be synthesized from the same acetylenic building block (Fig. 1, disconnection a+e). Thus, 2-ethynylaniline (**22**) is used as the acetylenic building block. This compound can be prepared in 81% overall yield through a straightforward palladium-catalyzed coupling of 2-iodoaniline with ethynyltrimethylsilane, followed by desilylation with KF.[74] This indole synthesis features a palladium-catalyzed reaction of 2-ethynylaniline with vinylic or aryl triflates or halides followed by a palladium-catalyzed cyclization of the resultant coupling product.[74] The cyclization step can be performed in an acidic two-phase system at room temperature to give yields comparable with or higher than those obtained with PdCl$_2$ in MeCN at 60–80°. The acidic medium does not prevent the unprotected amino group from attacking the activated carbon–carbon triple bond. The entire coupling/cyclization process can be conducted as a one-pot process, as shown in Eq. 7.[71]

$$\text{22} + \text{(cyclohexenyl-Ph-OTf)} \xrightarrow[\text{2. PdCl}_2, \text{Bu}_4\text{NCl}, \text{CH}_2\text{Cl}_2/0.5 \text{ N HCl, rt, 14 h}]{\text{1. Pd(PPh}_3)_4, \text{CuI, Et}_2\text{NH, rt, 5 h}} \text{indole-Ph (98\%)} \quad \text{(Eq. 7)}$$

From 3-(2-Trifluoroacetamidophenyl)-1-propargyl Carbonate Esters. A recent alkyne-based cyclization to 2-substituted indoles uses 3-(2-trifluoroacetamidophenyl)-1-propargyl carbonate esters as the starting materials and involves an intramolecular palladium-catalyzed heterocyclization followed by an intermolecular nucleophilic attack on a π-allylpalladium intermediate (Scheme 5; Fig. 1, disconnection a+f). The trifluoroacetyl group plays a key role in this process by increasing the acidity of the amide, thus aiding the formation of a strong, anionic nucleophile for the intramolecular attack. In addition, the trifluoroacetyl group is easily removed under the reaction conditions and/or upon workup so that free indoles (NH) can be obtained directly. Secondary amines and formate anions can be used as external nucleophiles in this process to prepare 2-aminomethylindoles (Eq. 8)[34] and 2-alkylindoles (Eq. 9),[43] respectively. Steric effects appear to influence the reaction outcome with secondary amines; for example, a moderate yield is obtained with diisopropylamine. With primary amines, the efficiency of the reaction is reduced by the occurrence of side reactions producing complex reaction mixtures. Interestingly, these indole syntheses give excellent results using a monodentate phosphine ligand such as Ph$_3$P, although it has been reported that bidentate ligands afford more stable π-propargylpalladium complexes,[97,98] the suggested intermediates of this cyclization, and that the best ligands for the palladium-catalyzed reaction of propargyl halides[99] and carbonates[36] with soft nucleophiles are the bidentate ligands.

$$\xrightarrow[\text{THF, 80°, 3 h}]{\text{Pd(PPh}_3)_4} \quad (98\%) \quad \text{(Eq. 8)}$$

$$\xrightarrow[\text{MeCN, 80°, 0.5 h}]{\text{Pd(PPh}_3)_4, \text{HCO}_2\text{H, Et}_3\text{N}} \quad (85\%) \quad \text{(Eq. 9)}$$

From 2-Halo-N-alkynylanilides. A common feature of all the syntheses of 2-substituted indoles described above is that the cyclization event involves aniline derivatives bearing an acetylenic moiety adjacent to the nitrogen functionality. An alternative approach uses acetylenic precursors in which the alkyne fragment is directly bound to the nitrogen atom. In particular, 2-halo-*N*-alkynylanilides are converted into 2-aminoindoles via an aminopalladation/reductive elimination process in the presence of primary and secondary amines (Eq. 10; Fig. 1, disconnection c+e).[29] Among the bases investigated, K_2CO_3 and Cs_2CO_3 are the most efficient; $PdCl_2(PPh_3)_2$ as the precatalyst gives higher yields than $Pd(PPh_3)_4$, and THF is more suitable than DMF or toluene as the solvent.

$$\text{(Eq. 10)}$$

3-Substituted Indoles. Only a few studies have been reported that describe the direct synthesis of 2-unsubstituted, 3-substituted indoles from acetylenic precursors. One of the most efficient procedures is the cyclization of 2-ethynyltrifluoroacetanilide with aryl iodides, a reaction that is based on the aminopalladation/reductive elimination protocol (Eq. 11,[100] Fig. 1, disconnection a+d, Scheme 3). The major problem in this cyclization process is the formation of coupling derivative **23**, which is a significant side product or even the main product observed when using a variety of phosphine ligands. In addition, depending on reaction conditions, a competing cyclization to give product **24** is observed. Very likely, this cyclization arises from the nucleophilic attack of the carbonyl oxygen at the internal carbon atom of the acetylenic fragment. Both the nature of the solvent and the catalyst system have a strong influence on the *N-/O*-cyclization ratio. The highest yields of the desired 3-substituted indoles are obtained by using $Pd_2(dba)_3$ as the palladium(0) source, DMSO as the solvent, K_2CO_3 (or Cs_2CO_3) as the base, and omitting phosphine ligands.

$$\text{(Eq. 11)}$$

Indoles containing polycyclic substituents at C(3) are prepared from 2-iodo-*N*-propargylanilides through a cascade process (Fig. 1, disconnection c).[32,101] An example of this construction is shown in Eq. 12.[32] The reaction proceeds through an intramolecular carbopalladation step, followed by capture of norbornene to give a σ-alkylpalladium(II) intermediate that does not contain β-hydrogens aligned for the elimination of HPd species and undergoes an intramolecular Mizoroki–Heck reaction.

$$\text{(Eq. 12)} \quad (45\%)$$

Examples of acetylenic precursors that afford 3-substituted indoles have also been described starting from N-allylic 2-ethynyltrifluoroacetanilides[102] (Eq. 13; Fig. 1, disconnection a+d) and N-mesyl-2-ethynylaniline.[103] N-allylic 2-ethynyltrifluoroacetanilides are converted into the corresponding 3-allylic indoles through the aminopalladation/reductive elimination mechanism. N-mesyl-2-ethynylaniline affords a 3-substituted indole through a tandem process that entails an intramolecular aminopalladation, an olefin insertion, and a protonolysis step. The indole product, however, is isolated in low yield.

$$\text{(Eq. 13)} \quad (49\%)$$

Recently, an additional approach to 3-substituted indoles starting from 3-iodo-N-allylaniline and internal alkynes has been described.[104] The reaction proceeds through carbopalladation, vinylic to aryl palladium migration, and intramolecular Heck reaction. Indole products are isolated in moderate yields. However, this indole synthesis is interesting in that the formation of the pyrrole ring containing a substituent at the C(3) position is accompanied by the introduction of a substituted olefinic system at the C(4) position (Eq. 14).

dppm = bis(diphenylphosphino)methane   (45%) 10:1   (Eq. 14)

2,3-Disubstituted Indoles. Indole syntheses based on the annulation of internal alkynes with 2-haloanilines or their N-substituted derivatives[30,31] (Fig. 1, disconnection a+c; Scheme 4) and those based on the aminopalladation/reductive elimination of 2-alkynyltrifluoroacetanilides with aryl or vinyl halides or triflates, alkyl halides, and allylic carbonates[7,105–108] (Fig. 1, disconnection a+d; Scheme 3) are some of the most powerful methods for the construction of this class of compounds.

From Internal Alkynes and 2-Haloanilid(n)es. The annulation of internal alkynes with 2-haloanilines was initially developed using 2-iodoanilines and their *N*-substituted derivatives.[30,31] The best results were obtained using an excess of the internal alkyne in the presence of sodium or potassium carbonate as bases, LiCl or Bu$_4$NCl as additives, and occasionally adding Ph$_3$P at 100° in DMF. Under these conditions 2,3-disubstituted indoles are isolated in good to excellent yields (Eq. 15).[31] An oxime-derived, chloro-bridged palladacycle, which is thermally stable and insensitive to air or moisture, has also been employed.[109]

$$\text{2-iodoaniline} + \text{Pr-C≡C-Pr} \xrightarrow[\text{DMF, 100°, 20 h}]{\text{Pd(OAc)}_2, \text{LiCl, K}_2\text{CO}_3} \text{2,3-dipropylindole} \quad (80\%) \quad \text{(Eq. 15)}$$

With unsymmetrical alkynes, the regiochemical outcome is controlled by the carbopalladation step. Steric and coordinating effects are the main controlling factors. These effects follow the general trend observed in related reactions involving a carbopalladation step.[110,111] Steric effects tend to control the conversion of the π-alkyne–σ-organopalladium intermediate formed initially (**9**, Scheme 4) into the carbopalladation adduct **10**, preferentially directing the palladium moiety to the more hindered end of the carbon–carbon triple bond (Eq. 16).[31] Coordinating effects influence the formation of vinylic adducts in a way that the added palladium ends up close to the coordinating group (Eq. 17).[31] Electronic factors are believed to play a minor role; however, they must have some influence on the catalytic process. For example, the site selectivity decreases significantly for the annulation with electron-deficient anilines.[112]

$$\text{2-iodoaniline} + \text{HC≡C-}t\text{-Bu} \xrightarrow[\text{DMF, 100°, 20 h}]{\text{Pd(OAc)}_2, \text{Ph}_3\text{P, LiCl, Na}_2\text{CO}_3} \text{3-}t\text{-Bu-indole} \quad (82\%) \quad \text{(Eq. 16)}$$

$$\text{2-iodo-NHAc} + \text{HO-C≡C-} \xrightarrow[\text{DMF, 100°, 12 h}]{\text{Pd(OAc)}_2, \text{Ph}_3\text{P,} \atop \text{Bu}_4\text{NCl, Na}_2\text{CO}_3} \text{product A} + \text{product B}$$

(60%)   (16%)

(Eq. 17)

The major drawback of this procedure is that it relies on substantially different steric bulk between the two substituents on the ends of the alkyne. For example, the direct annulation of a diaryl acetylene usually results in a mixture of two products. Consequently, some of the most successful syntheses based on this annulation protocol involve the reaction of 2-iodoanilines or 2-iodoanilides with alkynes containing a bulky silyl group at one of the acetylenic termini. Indeed, the reaction is highly site selective, with silylalkynes providing 3-substituted-2-silylindoles (Eq. 18),[31] which are versatile intermediates for the synthesis of a vast array of indole derivatives (Eq. 19).[112–121]

$$\text{2-iodoaniline} + \text{SiMe}_3\text{-alkyne} \xrightarrow[\text{DMF, 100°, 24 h}]{\text{Pd(OAc)}_2, \text{Ph}_3\text{P}, \text{Bu}_4\text{NCl}, \text{Na}_2\text{CO}_3} \text{2-SiMe}_3\text{-3-methylindole} \quad (98\%) \qquad \text{(Eq. 18)}$$

(Eq. 19)

1. AlCl$_3$, CH$_2$Cl$_2$, 0°, 0.5 h
2. hydrolysis → (88%)

NBS, CH$_2$Cl$_2$, reflux, 0.5 h → (70%)

CH$_2$=CHCO$_2$Et, Pd(OAc)$_2$, DMF, 100°, 2 d → (75%)

To overcome drawbacks associated with this process, such as the high cost and low stability of 2-iodoanilines, a procedure has been developed in which they are replaced by the much cheaper bromo or chloro derivatives.[122] As the oxidative insertion usually requires an electron-rich palladium species with these substrates, the phosphine-free protocol frequently adopted in the reaction of 2-iodoanilines is not applicable. Several types of highly active phosphine ligands have been examined. The best results, even in terms of site selectivity, are obtained using dtbpf under the conditions shown in Eq. 20 for the reaction of a 2-chloroaniline.[122]

$$\text{2-chloroaniline} + \text{Ph-alkyne-Pr} \xrightarrow[\text{NMP, 130°, 4 h}]{\text{Pd(OAc)}_2, \text{dtbpf}, \text{K}_2\text{CO}_3} \text{3-Pr-2-Ph-indole} \ (76\%) + \text{3-Ph-2-Pr-indole} \ (7\%)$$

dtbpf = 1,1'-bis(di-tert-butylphosphino)ferrocene

(Eq. 20)

From 2-Alkynyltrifluoroacetanilides and C_{sp^3}, C_{sp^2}, and C_{sp} Donors. The aminopalladation/reductive elimination protocol allows for the preparation of a broad range of symmetrical and unsymmetrical 2,3-disubstituted indoles from 2-alkynyltrifluoroacetanilides and a wide range of cyclization partners such as aryl or heteroaryl iodides,[106,123,124] bromides,[105,106] chlorides[107] and triflates,[105,106,123] vinylic triflates,[106,123,124] allylic esters,[102,106] alkyl iodides and bromides,[106,125] and alkynyl bromides.[108] When unsymmetrical 2,3-disubstituted indoles are the target products, this method provides the remarkable advantage that formation of constitutional isomers is excluded. The site selectivity follows from the sequence of events and is unambiguous. Substituents close to the oxidative addition site usually do not hamper the reaction. 2-Alkynyltrifluoroacetanilides containing alkyl, vinylic, electron-withdrawing, and electron-donating substituents on the alkyne moiety have been successfully employed. 2-Alkynylanilines and 2-alkynylacetanilides provides unsatisfactory results, suggesting that the acidity of the nitrogen–hydrogen bond plays a key role in this cyclization reaction and that organopalladium complexes are less effective than $PdCl_2$ in activating the carbon–carbon triple bond toward intramolecular nucleophilic attack. Indeed, a variety of cyclizations of alkynes containing proximal amino[123–125] and amido[23,49,50,70,126–132] groups under catalysis by palladium(II) salts have been described. Notably, the trifluoroacetyl group is readily removed during the reaction and/or the workup to allow for the formation of free indoles (NH) thus avoiding deprotection steps.

Carbonates (Na_2CO_3, K_2CO_3, Cs_2CO_3) are better bases than Et_3N, and MeCN or THF is typically used as the solvent for these aminopalladation/reductive elimination reactions. Depending on the cyclization partner, reaction temperatures range from room temperature for vinyl triflates to 120° for aryl chlorides. As to the phosphine ligands, the popular $Pd(PPh_3)_4$ is an efficient precatalyst with aryl or heteroaryl iodides,[106,123,124] bromides,[105,106] and triflates,[105,106,123] vinylic triflates,[106,123,124] allylic carbonates[102,106] (in stepwise and one-pot protocols), and 1-bromo-2-arylalkynes.[108] Its use is exemplified in Eq. 21[108] with the preparation of 2-aryl-3-alkynylindoles.

(Eq. 21)

The aminopalladation/reductive elimination reaction performed under an atmosphere of carbon monoxide affords 2-substituted 3-acylindoles in good to high yields. With neutral, electron-rich, and slightly electron-poor aryl iodides and vinylic triflates, this three-component reaction may be carried out using Pd(PPh$_3$)$_4$ as the palladium(0) source under a balloon of carbon monoxide (Eq. 22).[106,124] With electron-deficient aryl iodides, good results can be obtained with (2-tol)$_3$P and Pd(dba)$_2$ or by using anhydrous acetonitrile and a higher pressure of carbon monoxide.

(Eq. 22)

The reaction of 1-bromoalkynes with 2-alkynyltrifluoroacetanilides affords 2-substituted 3-alkynylindoles.[108] Reactions are usually carried out using Pd(PPh$_3$)$_4$ and Cs$_2$CO$_3$ or K$_2$CO$_3$ in MeCN. With 1-bromoalkynes containing alkyl groups bound to the alkyne fragment, the use of (t-Bu)$_3$P (added to the reaction mixture as the tetrafluoroborate salt)[133] and Pd$_2$(dba)$_3$ can produce a significant increase in the yields. 2-Substituted 3-alkynylindoles represent useful intermediates for the synthesis of 2-substituted 3-acylindoles. The latter can be prepared from 2-alkynyltrifluoroacetanilides and 1-bromoalkynes via a one-pot cyclization/hydration protocol, omitting the isolation of 2-substituted 3-alkynylindoles (Eq. 23).[108]

(Eq. 23)

Aryl chlorides are more reluctant to undergo oxidative addition to palladium(0) and require the use of XPhos (Eq. 24), one of the biaryl monophosphines that enhance the rate of the oxidative addition of aryl chlorides to palladium(0) species.[134–136] The use of this ligand solves one of the major problems in realizing this type of indole synthesis with relatively unreactive precursors of organopalladium complexes, namely the competitive formation of simple 2-substituted indoles, the formation of which does not involve the aryl halide partner.

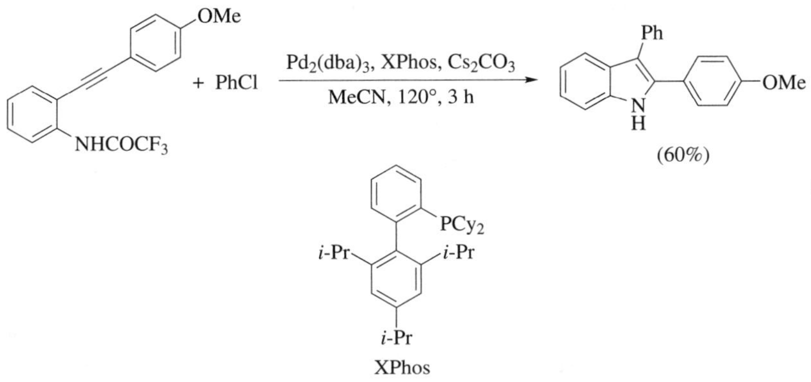

(Eq. 24)

Even with ethyl iodoacetate and benzyl bromides[106,125] a more active phosphine ligand is required for the oxidative addition to palladium(0). Indolylcarboxylate esters and 2-substituted 3-benzylindoles are isolated in good to high yields in the presence of the electron-rich, sterically-encumbered ligand tris(2,4,6-trimethoxyphenyl)phosphine (ttmpp) and $Pd_2(dba)_3$ (Eq. 25).[125]

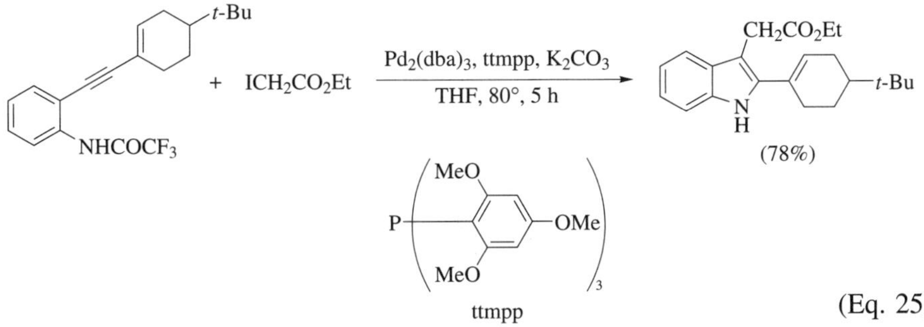

(Eq. 25)

The ttmpp ligand also provides remarkable site selectivity in the reaction of 2-alkynyltrifluoroacetanilides with allylic carbonates where steric differences between the two allylic termini are small.[102] The use of ttmpp in these reactions leads to the formation of products that bear the indole unit located almost exclusively on the less substituted terminus of the allylic system (Eq. 26).[102] The process is accompanied by some isomerization of the olefin. 2-Substituted 3-allylindoles can also be prepared through stepwise and one-pot protocols.[102] The stepwise protocol involves the N-allylation of 2-alkynyltrifluoroacetanilides [$Pd_2(dba)_3$, dppb, THF, 60°] followed by an aminopalladation/reductive elimination step [$Pd(PPh_3)_4$, K_2CO_3, MeCN, 90° or $Pd_2(dba)_3$, ttmpp, DME, 100°]; the one-pot synthesis is carried out by treating 2-alkynyltrifluoroacetanilides with allylic carbonates and $Pd(PPh_3)_4$ in THF at 60° until their disappearance and then adding K_2CO_3 and raising the reaction temperature to 80°.

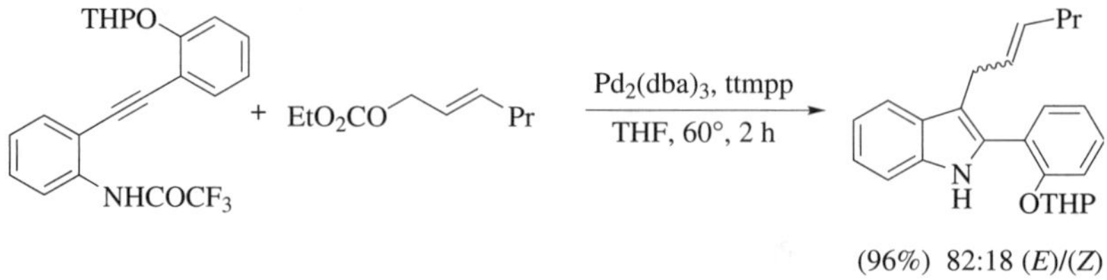

(Eq. 26)

(96%) 82:18 (E)/(Z)

On the whole, the aminopalladation/reductive elimination route to indoles entails three basic steps: (1) acylation of 2-haloanilines with trifluoroacetic anhydride, (2) cross-coupling of terminal alkynes with 2-halotrifluoroacetanilides, (3) indole formation by aminopalladation/reductive elimination. To make this process more practical, one-pot (Eq. 27)[137,138] and one-pot tandem (Eq. 28)[139] protocols (Fig. 1, disconnection a+c+d) have been developed.

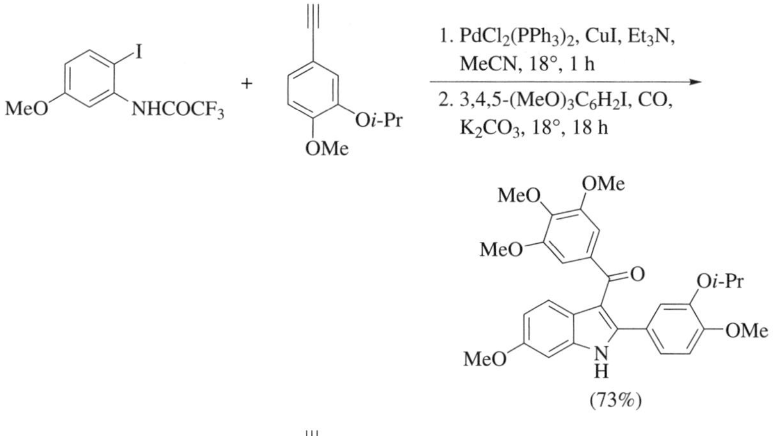

(Eq. 27)

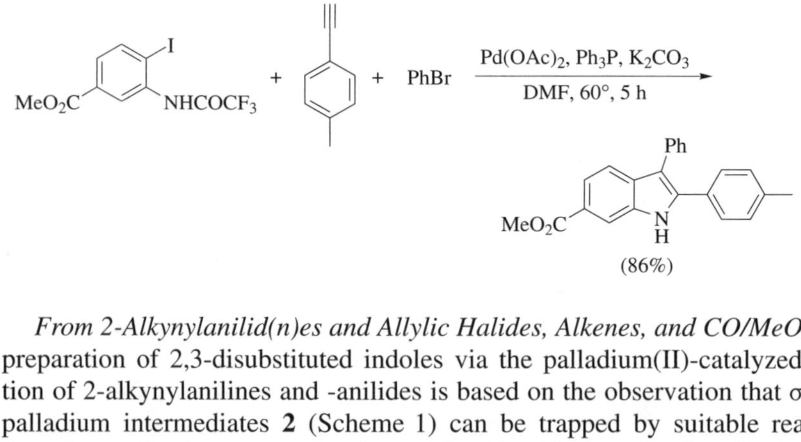

(73%)

(Eq. 28)

(86%)

From 2-Alkynylanilid(n)es and Allylic Halides, Alkenes, and CO/MeOH. The preparation of 2,3-disubstituted indoles via the palladium(II)-catalyzed cyclization of 2-alkynylanilines and -anilides is based on the observation that σ-indolylpalladium intermediates **2** (Scheme 1) can be trapped by suitable reagents so that the cyclization step may be combined with the functionalization of the indole nucleus at C(3) (Fig. 1, disconnection a+d). The potential of this trapping approach to the synthesis of indole derivatives has not gone unnoticed and tandem processes that employ this strategy have been developed.

In the tandem allylative cyclization of 2-alkynyl-N-methoxycarbonylanilides (Eq. 29),[70] the reaction proceeds through a site-selective attack of the σ-indolylpalladium intermediate on the γ position of allyl chlorides. The use of the unprotected amine or the acetamido derivative give unsatisfactory results and lack of control of the olefin geometry is observed in reactions using a substituted allylic chloride. A large excess of the allyl chloride (allyl chloride/alkyne 10:1) is needed to obtain the best results. The presence of methyloxirane as the proton scavenger is crucial for minimizing the competitive protonation of the σ-indolylpalladium intermediate leading to 3-unsubstituted, 2-substituted indoles.

(Eq. 29)

σ-Indolylpalladium intermediates can be trapped by carbon monoxide or alkenes to give indole products incorporating carbon monoxide,[128,129] vinylic,[130,131] or alkyl groups at the C(3) position.[103] In the first case, treatment of a 2-alkynylaniline with $PdCl_2$ in methanol under an atmosphere of carbon monoxide affords a σ-acylpalladium derivative which reacts with methanol to give an indolylcarboxylate ester (Eq. 30).[128,129] Palladium(0) species formed in this step are oxidized to the active palladium(II) species by $CuCl_2$; the use of 1,4-benzoquinone, disodium peroxysulfate, or molecular oxygen met with failure. Similar conditions are used to develop a domino cyclization/Heck reaction producing 2-substituted 3-vinylic indoles with alkenes containing electron-withdrawing groups.[130] Modified conditions ($PdCl_2$, excess amounts of Bu_4NF and $CuCl_2 \cdot H_2O$ as a reoxidant) are necessary to extend the reaction to alkenes lacking the activation of an electron-withdrawing group (Eq. 31).[131] $Cu(OAc)_2$, 2,3-dichloro-5,6-dicyanobenzoquinone (DDQ), and pyridine 1-oxide fail to give the desired products.

(Eq. 30)

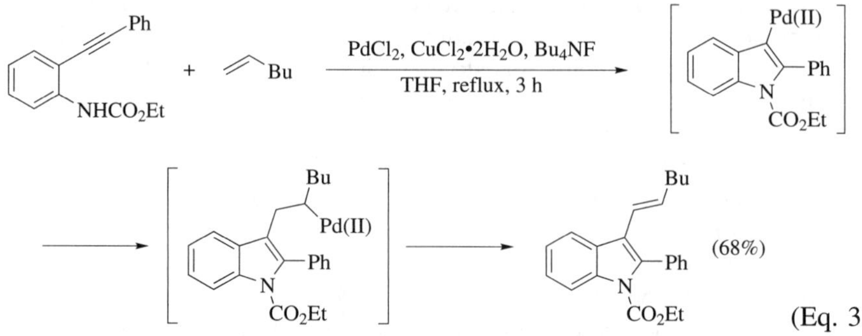

[Eq. 31 scheme]

The reaction of 2-alkynylanilides with α,β-enals and -enones in the presence of LiBr affords 2-substituted, 3-alkylindoles via a tandem palladium(II)-catalyzed aminopalladation that entails addition of the resultant σ-indolylpalladium(II) intermediate to the α,β-unsaturated carbonyl compound, followed by protonolysis of the carbon–palladium bond with regeneration of the palladium(II) species (Eq. 32).[103] The addition of LiBr is crucial to inhibit β-elimination in the carbopalladation intermediate **25**. This remarkable halide effect is accounted for by assuming that the bromide anion inhibits the β-hydride elimination by occupancy of the free coordination sites. Also, electron donation from the bromide anion to palladium results in a highly polarized carbon–palladium bond that is readily cleaved via protonolysis.

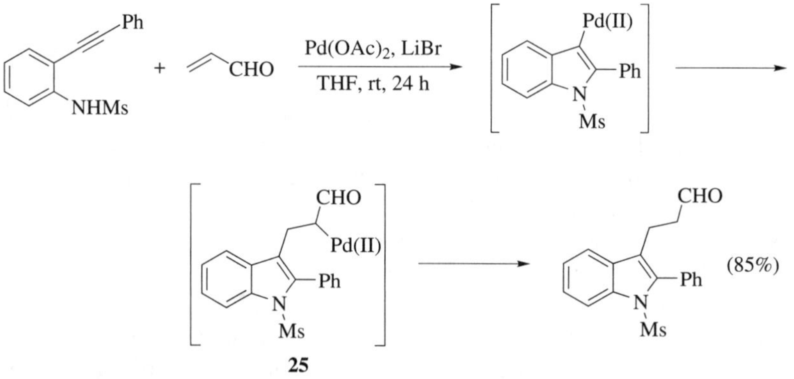

[Eq. 32 scheme]

From N-Alkynyl-2-haloanilides. The cyclization of *N*-alkynyl-2-haloanilides with primary and secondary amines (Fig. 1, disconnection c+e) provides a convenient entry to 2-amino-3-substituted indoles, a class of compounds that is otherwise difficult to obtain. Typical reaction conditions are shown in Eq. 33,[29] although Cs_2CO_3 can also be used. THF is more suitable than DMF or toluene as the solvent. Higher yields are obtained when $PdCl_2(PPh_3)_2$ is used as the precatalyst; $Pd(PPh_3)_4$ is less effective, most probably because of the higher phosphine content, which reduces the activity of the actual palladium(0) catalyst.

(Eq. 33) (95%)

From 2-Alkynyl-N-alkylideneanilines. The cyclization of 2-alkynyl-N-alkylideneanilines bearing an aryl substituent on the alkylidene fragment (Fig. 1, disconnection b) affords 2-aryl- and 2-heteroaryl-3-(1-alkenyl)indoles in good yields (Eq. 34).[33] The reaction involves the addition of a HPdOAc species to the carbon–carbon triple bond followed by a cyclization step. HPdOAc is formed by the oxidative addition of AcOH to palladium(0). Reaction of $(n\text{-Bu})_3\text{P}$ with Pd(OAc)_2 forms Ac_2O and palladium(0), and in situ hydrolysis of the Ac_2O provides the AcOH. The preparation of alkyl-substituted imines tends to fail due to their instability, hence, 2-alkylindoles are best prepared in a one-pot procedure. The formation of imines from 2-alkynylanilines and benzaldehyde or secondary aliphatic aldehydes followed by in situ cyclization proceeds without problems.

(Eq. 34) (59%)

From 2-Alkynylisocyanobenzenes. 2-Alkynylisocyanobenzenes are converted into 2-substituted, 3-allyl-N-cyanoindoles in good to acceptable yields by a three-component reaction with allyl methyl carbonate and trimethylsilyl azide in the presence of $\text{Pd}_2(\text{dba})_3\cdot\text{CHCl}_3$ and $(2\text{-furyl})_3\text{P}$ at 100° (Eq. 35; Fig. 1, disconnection a+d).[26] At lower temperature (up to 40°) the reaction affords N-allyl cyanamides **26**. (2-Furyl)$_3$P gives the best results when combined with the $\text{Pd}_2(\text{dba})_3\cdot\text{CHCl}_3$ complex but other monodentate phosphine ligands such as Ph$_3$P, (2-tol)$_3$P, and (4-FC$_6$H$_4$)$_3$P can afford satisfactory results. In contrast, bidentate phosphine ligands such as 1,2-bis(diphenylphosphino)ethane (dppe), 1,3-bis(diphenylphosphino)propane (dppp), and 1,4-bis(diphenylphosphino)butane (dppb) are ineffective. Toluene or THF can also be used as solvent. However, in polar solvents such as DCE, MeCN, or DMF only small amounts of the indole product are formed.

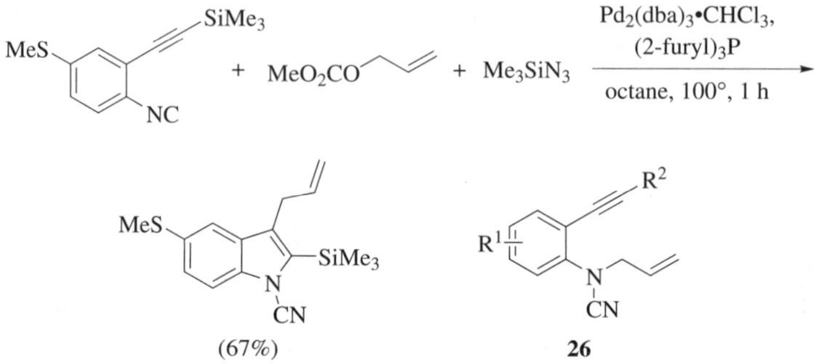

$$\text{(Eq. 35)}$$

From 2-(Alkynyl)phenylisocyanates. The reaction of 2-(alkynyl)phenylisocyanates with allyl carbonates gives 2-substituted 3-allyl-*N*-(alkoxycarbonyl) indoles in the presence of Pd(PPh$_3$)$_4$ and CuCl (Eq. 36; Fig. 1, disconnection a+d).[27,28] Copper(I) chloride affords higher yields than CuBr and is far superior to other copper salts such as CuI, CuOAc, (CuOTf)$_2$·benzene, or CuCl$_2$. Zinc chloride is also usable as a partner for palladium. The combinations Pd(OAc)$_2$/ Ph$_3$P, Pd$_2$(dba)$_3$·CHCl$_3$/dppe, and Pd$_2$(dba)$_3$·CHCl$_3$/(2-furyl)$_3$P are less effective than Pd(PPh$_3$)$_4$. THF is the solvent of choice whereas toluene, MeCN, and DMF give the desired indole product in lower yield. Longer reaction times are required when a bulky substituent is bound to one of the alkyne termini. With an alkynyl *tert*-butyl group, no allylindole is obtained and the sole product is a 2-alkynyl-*N*-allylaniline derivative. Electronic effects of the substituents *para* to the isocyanate group and the bulk of the alcoholic fragment of the allylic carbonates do not seem to exert a significant influence on the reaction outcome.

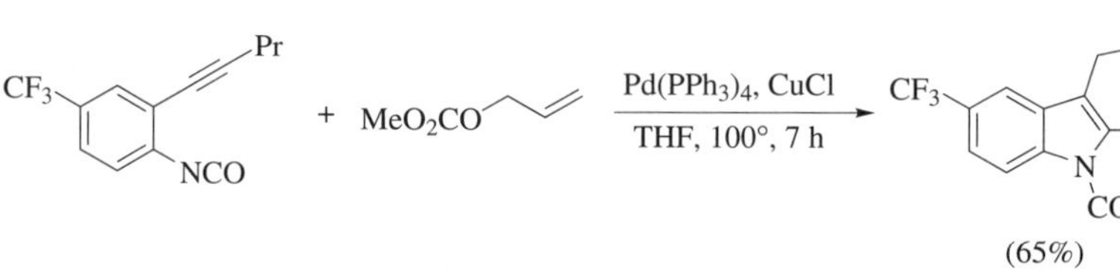

$$\text{(Eq. 36)}$$

From 2-Alkynylphenyl N,O-Acetals and from 2-Iodoanilides and 1-(Tributylstannyl)-1-substituted Allenes. A few examples of intramolecular cyclizations of 2-alkynylphenyl *N,O*-acetals (Eq. 37)[140] and of one-step syntheses of 2-methyl-3-substituted indoles from *N*-acyl-2-iodoanilines and 1-(tributylstannyl)-1-substituted allenes (Eq. 38)[141] are also known.

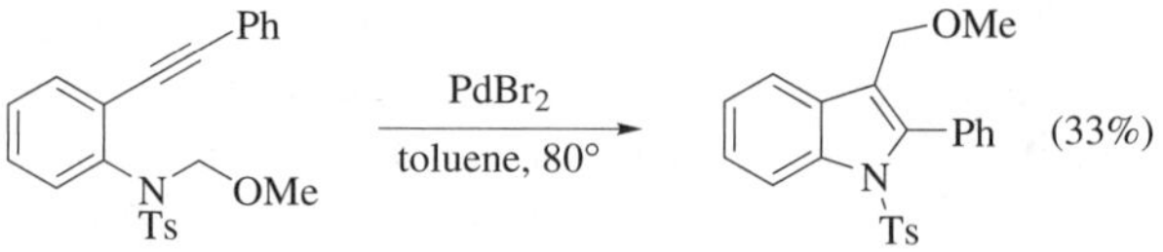

$$\text{(Eq. 37)}$$

$$\text{o-I-C}_6\text{H}_4\text{NHBoc} + \text{Bu}_3\text{Sn-C(=CH}_2)\text{-(CH}_2)_2\text{OTBS} \xrightarrow[\text{DMF, rt, 4 h}]{\text{Pd}_2(\text{dba})_3,\ \text{P(2-furyl)}_3,\ \text{CuI, Bu}_4\text{NCl}} \text{3-(CH}_2)_2\text{OTBS-}N\text{-Boc-indole} \quad (57\%)$$

(Eq. 38)

Indole Formation from Alkenes

Unsubstituted Indoles. 2-Vinylaniline[142] and its N-substituted derivatives, including 2-vinyl-N-tosylanilides,[143–146] 2-vinylacetanilides,[147] and 2-vinyl-N-alkylanilines,[148] undergo palladium-catalyzed cyclization to give unsubstituted indoles (Fig. 2, disconnection a). Palladium chloride and PdCl$_2$(MeCN)$_2$ are typically employed as precatalysts in the presence of benzoquinone or CuCl/O$_2$ as reoxidants and LiCl as an additive. An example is depicted in Eq. 39.[142] Perhaps the main limitation of this approach is that the requisite 2-vinylanilines require lengthy syntheses[145,149–152] and that more direct syntheses often proceed in only moderate yield.[145,153]

$$\text{2-vinylaniline} \xrightarrow[\text{THF, reflux, 18 h}]{\text{PdCl}_2(\text{MeCN})_2,\ \text{LiCl, benzoquinone}} \text{indole} \quad (74\%) \quad (\text{Eq. 39})$$

2-Nitrostyrenes have been used as precursors for unsubstituted indoles through a reductive N-heterocyclization process (Fig. 2, disconnection a). The involvement of 2-nitrostyrenes as substrates in the palladium-catalyzed synthesis of indoles was first observed as a side reaction of treating 2-bromonitrobenzenes with ethylene in the presence of palladium acetate to prepare 2-nitrostyrenes.[147] In some cases, significant amounts of indole products were formed in addition to the expected Mizoroki–Heck products, most probably via the in situ reduction of the nitro group of the 2-nitrostyrenes. Subsequent studies developed this side reaction into a new indole synthesis. In the presence of 20 atm of carbon monoxide, PdCl$_2$(PPh$_3$)$_2$, and an excess of SnCl$_2$ at 100°, 2-nitrostyrene gives indole in 50% yield.[154,155] Other additives such as BF$_3$·Et$_2$O, CuCl$_2$, FeCl$_3$, or SnCl$_4$ are ineffective. Further improvement on these conditions have led to a protocol which involves lower temperature and pressure and does not require an added Lewis acid (Eq. 40).[156] In general, the reaction appears to be unaffected by substituents on the aromatic ring. 2-Nitrostyrenes containing either electron-withdrawing or electron-donating substituents give indoles in moderate to excellent yields.

$$\text{2-OH-3-NO}_2\text{-vinylbenzene} \xrightarrow[\text{MeCN, 70°, overnight}]{\text{Pd(OAc)}_2,\ \text{Ph}_3\text{P, CO (4 atm)}} \text{4-OH-indole} \quad (96\%) \quad (\text{Eq. 40})$$

2-Substituted Indoles. 2-Substituted indoles are prepared from 2-allylanilines, 2-nitrostyrenes (Fig. 2, disconnection a) and 2-haloanilino enamines (Fig. 2, disconnection c). 2-Allylanilines undergo palladium-catalyzed cyclization to 2-substituted indoles in the presence of $PdCl_2(MeCN)_2$ as the source of palladium(II) and benzoquinone to reoxidize palladium(0) to palladium(II) (Eq. 41).[142] Neither palladium acetate nor lithium chloropalladate is as effective. The reaction is rarely applied to the synthesis of 2,3-substituted indoles. In one of the few examples, the indole formation from properly substituted 2-allylanilines is used to prepare 2-substituted 3-alkoxyindoles.[157]

$$\underset{NH_2}{\text{2-allylaniline}} \xrightarrow[\text{THF, reflux, 5 h}]{PdCl_2(MeCN)_2, LiCl, benzoquinone} \text{indole} \quad (86\%) \quad \text{(Eq. 41)}$$

The reductive cyclization of 2-nitrostyrenes to 2-substituted indoles can be carried out using carbon monoxide (20 atm), $PdCl_2(PPh_3)_2$, and an excess of $SnCl_2$ in dioxane at 100°.[154,155] According to an improved protocol, carbon monoxide (4 atm), $Pd(OAc)_2$, and Ph_3P in MeCN at 70° can be successfully employed (Eq. 42).[156,158–161] The configuration of the alkene moiety does not affect the reaction outcome. These improved conditions, however, require a relatively high catalyst loading [6 mol % of $Pd(OAc)_2$] and 24 mol % of Ph_3P. In some cases, chromatography may be necessary to remove both triphenylphosphine oxide and a 3,3′-bisindole derivative that can form under the reaction conditions. Further optimization with regard to catalyst, ligand, solvent, temperature, and carbon monoxide pressure has led to the following conditions: 0.1 mol % of $Pd(OCOCF_3)_2$, 0.7 mol % of tmphen, DMF, 80°, 1 atm of carbon monoxide (Eq. 43).[161] For some reactions, the combination of $Pd(OAc)_2$/phenanthroline or the preformed catalyst $phen_2Pd(BF_4)_2$ give similar results.[161]

$$\xrightarrow[\text{MeCN, 70°, 15 h}]{Pd(OAc)_2, Ph_3P, CO} \quad (95\%) \quad \text{(Eq. 42)}$$

$$\xrightarrow[\text{DMF, 80°, 16 h}]{Pd(OCOCF_3)_2, \text{tmphen, CO}} \quad (78\%)$$

tmphen

(Eq. 43)

N-Boc derivatives **28** of 2-haloanilino enamines, readily available via Suzuki–Miyaura cross-coupling of arylboronic acids with enamine derivatives **27**, provide access to 2-substituted indoles in the presence of Pd(PPh$_3$)$_4$ and Et$_3$N in DMF at 100° (Eq. 44).[162] Changing the base to (i-Pr)$_2$NEt or 1,2,2,6,6-pentamethylpiperidine (PMP) leads to dehalogenation as a significant side reaction. The entire process can also be conducted as a tandem process using Pd(PPh$_3$)$_4$, Cs$_2$CO$_3$, arylboronic acid, and Bu$_4$NBr in DMF/H$_2$O at 50–70°.[162]

More frequently, however, the indole formation from enamines not stabilized by conjugation with carbonyl groups is performed by processes that involve their preparation in situ followed by a cyclization step (Fig. 2, disconnection a+c). One of these procedures is based on the reaction of 2-iodo-[163] or 2-chloroanilines[164] with ketones. The latter is best conducted under the conditions shown in Eq. 45.[164] The reaction can be performed even in the presence of Cs$_2$CO$_3$ or KOAc as base, but variable amounts of side products are formed. Magnesium sulfate, presumably acting as a dehydrating agent, plays an important role in promoting the reaction.

A more recent approach based on the in situ preparation of enamines takes advantage of the palladium-catalyzed reaction of 2-bromoanilines with vinyl bromides (Eq. 46).[165] The reaction is strongly dependent on the structure of the ligand. Among the ligands that have been studied—(2-tol)$_3$P, BINAP, XantPhos (9,9-dimethyl-4,5-bis(diphenylphosphino)xanthene), JohnPhos (2-(biphenyl)di-*tert*-butylphosphine), DavePhos, XPhos, and the imidazolium salt HIMeCl (precursor of a carbene ligand)—DavePhos and XPhos give the best results. In particular, DavePhos is the best ligand with 2-bromoanilines, but XPhos is the ligand of choice with 2-chloroanilines. Even preformed imines have been used as precursors of indoles.[160] Very likely these cyclizations involve an isomerization step to the corresponding enamines.

[Scheme: Cl-C6H3(Br)-NH2 + Br-C(=CH2)-Ph → Pd2(dba)3, DavePhos, t-BuONa, toluene, 100°, 20 h → [Cl-C6H3(Br)-N(H)-C(=CH2)-Ph] → 5-Cl-2-Ph-indole (55%)]

DavePhos = 2-dicyclohexylphosphino-2'-(N,N-dimethylamino)biphenyl

(Eq. 46)

3-Substituted Indoles. Unlike the alkyne-based cyclizations to 3-substituted indoles for which there are only a few examples, the alkene-based cylizations to 3-substituted indoles are relatively abundant. 3-Substituted indoles have been synthesized from 2-halo-N-allylanilines and -anilides (Fig. 2, disconnection c), 2-nitrostyrenes (Fig. 2, disconnection a), and 2-haloanilino enamines (Fig. 2, disconnection c).

The cyclization of 2-halo-N-allylanilines and -anilides to indoles (Fig. 2, disconnection c) is based on the intramolecular Mizoroki–Heck reaction. Initial studies investigated indole formation from 2-iodo- and 2-bromo-N-allylacetanilides bearing an olefinic fragment conjugated to a carbonyl group in the presence of Pd(OAc)$_2$, Ph$_3$P, and usually TMEDA.[166] Formation of the desired indole products is accompanied by the formation of the deallylated 2-bromoacetanilide and the deallylated acetanilide (the latter derived from the reduction of the carbon–bromine bond), which in some examples are the major components of the reaction mixtures. Reaction conditions have subsequently been improved and the cyclization has been extended to a variety of activated and unactivated 2-halo-N-allylanilines and -anilides.[167–179] An example is shown in Eq. 47.[178] Pd(OAc)$_2$ is typically used as the source of palladium(0). The reaction can be carried out under phosphine-free conditions[167,168,171,174,175,177,179] or using Ph$_3$P,[166,172,173] (2-tol)$_3$P,[167,176,178] triphenylphosphine trisulfonate sodium salt (TPTTS, in MeCN/H$_2$O),[169] (C$_6$F$_{13}$CH$_2$CH$_2$)$_2$PPh (in supercritical carbon dioxide),[170] Pd(PPh$_3$)$_4$[177] in the presence of Et$_3$N,[167–171,174,175,177,178] Bu$_3$N,[176] NaHCO$_3$,[166,179] Na$_2$CO$_3$,[168,179] K$_2$CO$_3$,[172,173] Cs$_2$CO$_3$,[173] or NaOAc[168] as bases. Tetrabutylammonium chloride[168,173,175,179] or bromide[171,174] are employed as additives, particularly under phosphine-free conditions.

[Scheme: F,Br-substituted N-allylaniline → Pd(OAc)$_2$, (2-tol)$_3$P, Et$_3$N, MeCN, reflux, 5 h → F,Br-substituted 3-methylindole (65%)] (Eq. 47)

Only a few 3-substituted indoles have been prepared from 2-nitrostyrenes. Indole derivatives containing a C(3)-methyl substituent are obtained by reaction of 2-nitro-α-methylstyrenes with carbon monoxide (20 atm), PdCl$_2$(PPh$_3$)$_2$, and SnCl$_2$ in dioxane[154,155] at 100° or carbon monoxide (4 atm), Pd(OAc)$_2$, and Ph$_3$P in MeCN at 70°.[156] 3-Alkoxy-substituted indoles are synthesized from 2-nitro-α-alkoxystyrenes under the conditions shown in Eq. 48.[180]

$$\text{O}_2\text{N}\text{-}\underset{\text{NO}_2}{\bigcirc}\text{-C(OEt)=CH}_2 \xrightarrow[\text{DMF, 120°, 72 h}]{\text{Pd(dba)}_2\text{, phen, dppp, CO (6 atm)}} \text{O}_2\text{N}\text{-indole-3-OEt} \quad (85\%) \quad \text{(Eq. 48)}$$

phen = 1,10-phenanthroline

The cyclization of 2-haloanilino enamines to 3-substituted indoles has received more attention than their formation from 2-nitrostyrenes. A number of 3-substituted indoles are prepared from preformed 2-haloanilino enamines, stabilized by conjugation with keto or ester groups (Eq. 49).[181,182] Enamines not stabilized by conjugation with carbonyl groups are generated in situ from *trans*-1,2-disubstituted bromoalkenes and 2-bromoanilines (Eq. 50)[165] and from 2-haloanilines and aldehydes.[183] In the latter process, coupling of 2-iodoanilines with aldehydes is realized under mild, phosphine-free conditions (Pd(OAc)$_2$, DABCO, DMF, 85°), whereas XPhos is found to be the ligand of choice with 2-bromo- and 2-chloroanilines (Eq. 51).[183] As shown in Eq. 51, chiral aldehydes can participate in this process without racemization. The reaction also tolerates a variety of 2-haloanilines with different electronic properties.

$$\text{MeO-aniline(Br)-NH-CH=CH-C(O)OEt} \xrightarrow[\text{MeCN, 100°, 20 h}]{\text{Pd(OAc)}_2\text{, (2-tol)}_3\text{P, Et}_3\text{N}} \text{MeO-indole-3-C(O)OEt} \quad (82\%) \quad \text{(Eq. 49)}$$

$$\underset{n\text{-C}_8\text{H}_{17}}{\text{aniline(Br)-NH}} + \text{Br-CH=CH-Ph} \xrightarrow[\text{toluene, 100°, 20 h}]{\text{Pd}_2(\text{dba})_3\text{, DavePhos, }t\text{-BuONa}} \left[\underset{n\text{-C}_8\text{H}_{17}}{\text{aniline(Br)-N-CH=CH-Ph}}\right]$$

$$\longrightarrow \underset{n\text{-C}_8\text{H}_{17}}{\text{3-Ph-indole}} \quad (68\%) \quad \text{(Eq. 50)}$$

$$\underset{\text{MeO}}{\text{aniline(Cl)-NH}_2} + \text{OHC-CH}_2\text{-CH(N(Boc)}_2\text{)-CO}_2\text{Me} \xrightarrow[\text{DMA, 120°}]{\text{Pd}_2(\text{dba})_3\text{, XPhos, KOAc}} \text{MeO-indole-3-CH}_2\text{-CH(N(Boc)}_2\text{)CO}_2\text{Me}$$

(73%) (Eq. 51)

2,3-Disubstituted Indoles. The formation of 2,3-disubstituted indoles from 2-nitrostyrenes (Fig. 2, disconnection a), 2-allylanilines (Fig. 2, disconnection a), and 2-haloanilino enamines (Fig. 2, disconnection c) is performed under the same conditions described for the formation of other substituted indoles from the same substrates. Thus, 2-nitrostyrenes are cyclized in the presence of carbon monoxide (20 atm), $PdCl_2(PPh_3)_2$, and $SnCl_2$ in dioxane[155] or carbon monoxide (4 atm), $Pd(OAc)_2$, and Ph_3P in MeCN.[156] The cyclization of 2-allylanilines is typically performed using $PdCl_2(MeCN)_2$ and Na_2CO_3 or K_2CO_3, with or without LiCl in THF.[157] 2-Haloanilino enamines give indoles in the presence of $Pd(OAc)_2$ and Ph_3P in DMF using Pr_3N or $NaHCO_3$ as bases.[45,46]

Recently, 2-haloanilino enamines have been prepared in situ from 2-bromo- or 2-chloroanilines and internal alkynes via site selective titanium-catalyzed intermolecular hydroamination. The enamine formation is followed by a palladium-catalyzed cyclization step (Eq. 52).[184] The high site selectivity observed in the hydroamination step enables the synthesis of a variety of functionalized indoles with a site selectivity that is complementary to that obtained when using the palladium-catalyzed reaction of 2-haloanilines or their N-substituted derivatives with internal alkynes (Eqs. 16–18).[30,31]

(Eq. 52)

Functionalized indoles have been prepared by palladium-catalyzed intramolecular oxidative coupling from simple arylenamines, without the need to activate the arene fragment through the introduction of a carbon–halogen bond (Eq. 53; Fig. 2, disconnection c).[185] This indole synthesis can also be carried out in a one-pot sequence from anilines and 3-oxo esters.[185]

(Eq. 53)

Other novel alkene-based routes to 2,3-disubstituted indoles include a palladium-catalyzed three-component coupling of aryl iodides, 2-isocyanovinylbenzene and Et_2NH (Eq. 54; Fig. 2, disconnection b+h),[186] and a one-pot, two-step, four-component reaction of cinnamaldehydes, bromoanilines, formic acid, and isocyanides, a process that relies on an Ugi reaction followed by an intramolecular Heck reaction (Eq. 55).[187] Both reactions proceed with low to moderate yields.

(Eq. 54) — reaction scheme: 2-vinylbenzonitrile + 4-iodoanisole + Et₂NH, Pd(OAc)₂, dppp, THF, 40°, 21 h → 3-(NEt₂-methyl)-2-(4-methoxyphenyl)indole (39%)

(Eq. 55) — 2-bromoaniline + Ph–CH=CH–CHO + HCO₂H + BnNC:
1. 2,2,2-trifluoroethanol, rt
2. evaporation of the solvent
Ugi reaction

→ [2-bromo-N-CHO-N-(CH(Ph)CH=CH–)-aniline with CONHBn] → Pd(OAc)₂, Ph₃P, MeCN; 1. Heck reaction; 2. isomerization → 3-benzyl-2-(CONHBn)indole (21%)

Indoles via Arene Vinylation

The palladium-catalyzed intramolecular reaction of a vinylic halide fragment with an arene unit has a limited synthetic scope. This strategy has been applied to the preparation of a few 3-methylindoles from aniline carbamates containing a bromopropenyl fragment bound to the nitrogen atom (Fig. 3).[47] However, it is interesting that the oxidative addition site is located in a vinylic fragment tethered to the benzenoid ring unlike the majority of the cyclization procedures described in this chapter, where the site of the oxidative addition of the carbon–X bond to the palladium(0) species is located on the benzenoid ring. Optimum yields are obtained by using Herrmann's catalyst. Cleavage of the allylic side chain is observed as a side reaction. Cyclization at both the *ortho* and *para* positions with respect to the hydroxy group can occur to generate mixtures of indole derivatives. This can be avoided by blocking one of the positions with a substituent or using a symmetrical phenol such as shown in Eq. 56.[47]

(Eq. 56) — 3,5-dihydroxy-N-(2-bromoallyl)-N-(CO₂Me)aniline, Herrmann's catalyst, Cs₂CO₃, DMA, 70°, 1 d → 4,6-dihydroxy-3-methyl-N-CO₂Me-indole (71%)

Herrmann's catalyst (structure: dimeric Pd with 2-tol, 2-tol phosphine and bridging acetates)

Herrmann's catalyst

Recently, the arene vinylation based indole synthesis has been applied to *N*-methylanilines bearing bromopropenyl fragments bound to the nitrogen atom.[188] However, best results have been obtained when the propenyl moiety is part of a cyclic system. The indole product was isolated in low yield with an acyclic bromopropenyl fragment.

Indoles via N-Vinylation and N-Arylation

Unsubstituted Indoles. The first extension of the carbon–nitrogen bond forming reaction[15–22] to the direct formation of indole rings involves the palladium-catalyzed cyclization of 2-chloroarylacetaldehyde N,N-dimethylhydrazones to give 1-aminoindole derivatives (Fig. 4, disconnection g). The best results are obtained under the conditions shown in Eq. 57 for the synthesis of a fluoroindole derivative.[189] In some reactions, the use of the bulky, electron-rich ligand $(t\text{-Bu})_3\text{P}$ gives satisfactory results; Cs_2CO_3 and Rb_2CO_3 can also be used as bases. Yields of chloroindoles are lower than those of unsubstituted indoles or fluoroindoles because competitive oxidative addition of the product chloroindoles to palladium(0) species takes place under the reaction conditions. Because indole derivatives bearing chloro substituents could be useful substrates for increasing the molecular complexity of the indole products, a tandem process has been developed that involves a palladium-catalyzed intramolecular cyclization to chloroindoles followed by palladium-catalyzed functionalization of their carbocyclic rings (Eq. 58).[189]

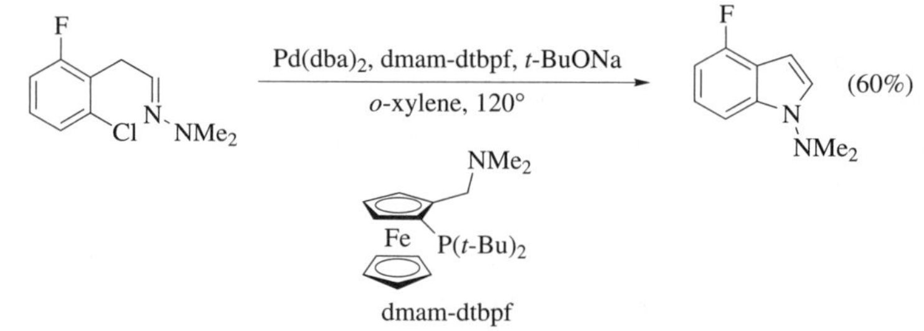

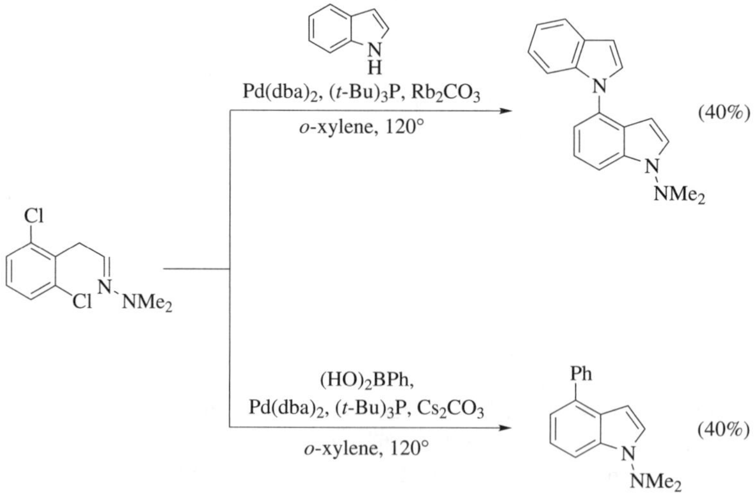

Unsubstituted indoles have also been prepared from benzenoid precursors that do not incorporate the nitrogen atom required for the final ring-closing indole-forming event. Thus, 2-(2-haloalkenyl)aryl halides are converted into indoles via palladium-catalyzed reaction with nitrogen nucleophiles (Fig. 4, disconnection a+g).[190–192] Introducing an external nitrogen unit is particularly useful for the preparation of N-functionalized indoles, including indoles with sterically demanding N-substituents.[191] A palladium(0) source (Pd$_2$(dba)$_3$) along with a variety of phosphine ligands such as DPEPhos, PhXPhos, SPhos (Fig. 5), DavePhos (see Eq. 46), and HP(t-Bu)$_3$BF$_4$ are used. Sodium $tert$-butoxide is usually employed as the base (Eq. 59), although Cs$_2$CO$_3$ gives good results in some examples.[190] Alkene partners have been used successfully as a (Z) and (E) mixture of isomers. In several examples, the $(Z)/(E)$ ratio is low and sometimes the (E) isomer dominates, suggesting that both geometrical isomers can be converted into the indole products. As control experiments revealed that no isomerization of the vinyl halide substrates takes place, it is likely that the funneling of the isomer mixtures to a single indole product is due to the isomerization of an initially formed enamine. By careful choice of the substrate, i.e., introducing a suitable second halide leaving group, it is possible to develop one-pot (Eq. 60)[190] or tandem processes to introduce further amine functionality via a third carbon–nitrogen bond formation in the benzenoid ring.[190]

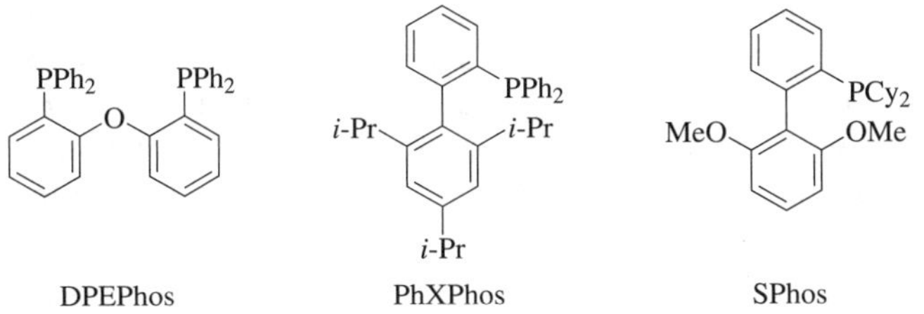

Figure 5. Three of the phosphine ligands examined in the synthesis of indoles from 2-(2-haloalkenyl)aryl halides and amines.

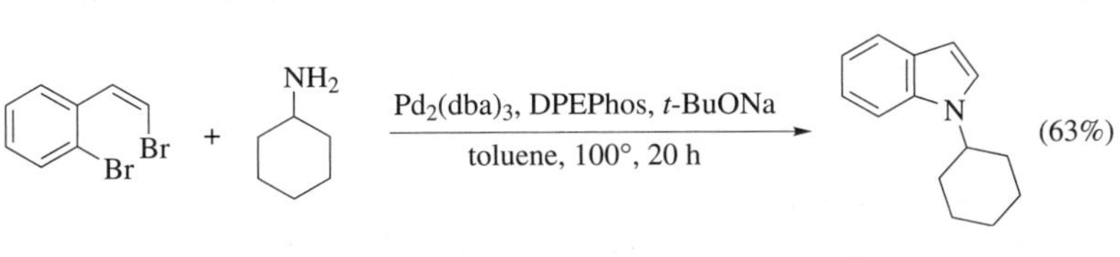

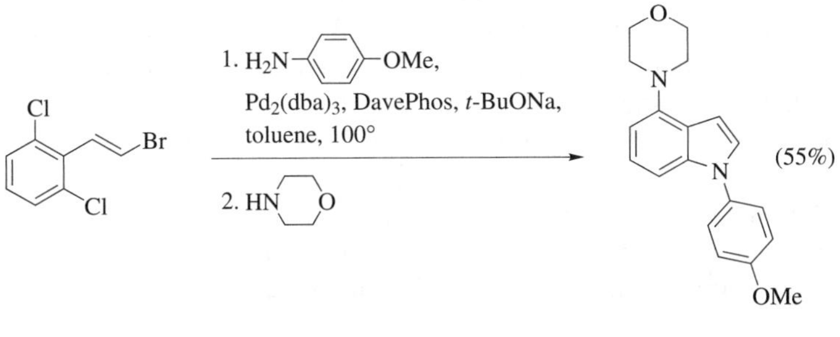

Formation of unsubstituted indoles from 3-nitro-2-methyliodobenzene via a process that usually leads to indoline products entails a tandem intermolecular alkylation/intramolecular amination (Eq. 61; Fig. 4, disconnection c+g).[193] Norbornene is suggested to enter the catalytic cycle favoring the alkylation of the aromatic ring. This alkylation involves carbopalladation, C_{Ar}–H activation, oxidative addition, and reductive elimination steps followed by the extrusion of norbornene. The condensed five-membered ring is formed via an intramolecular Buchwald–Hartwig reaction. Nevertheless, the detailed mechanism of formation of the pyrrole ring is uncertain. The palladium-catalyzed dehydrogenation of an indoline intermediate seems unlikely in view of the fact that indole formation is not observed with any of the other iodoarenes tested. The process is protecting-group dependent: the phenyl-protected amine also gives the corresponding indole in 53% yield, whereas the CO_2Et carbamate gives only the indoline in 20% yield.

$$O_2N\text{-Ar-I} + \text{Br-CH}_2\text{-HN-Ar-NO}_2 \xrightarrow[\text{MeCN, 135°, 20 h}]{\text{Pd(OAc)}_2, \text{(2-furyl)}_3\text{P,} \\ \text{norbornene, Cs}_2\text{CO}_3} \text{indole product} \quad (70\%) \quad \text{(Eq. 61)}$$

2-Substituted Indoles. 2-Substituted indoles may be constructed through N-arylation (Fig. 4, disconnection g) and N-vinylation (Fig. 4, disconnection a) processes. In the N-arylation approach, 2-haloarylenamines are usually the direct precursors of indole products. An example is shown in Eq. 62 (Fig. 4, disconnection g) for the preparation of an indole-2-carboxylate from a preformed 2-iodoarylenamine.[194]

$$\xrightarrow[\text{DMF, 90°}]{\text{PdCl}_2\text{(dppf), KOAc}} \quad (89\%) \quad \text{(Eq. 62)}$$

Preferably, 2-haloarylenamines are generated in situ from suitable precursors. Very likely, they are intermediates in the formation of indoles from 2-(2-haloalkenyl)aryl halides.[190] In the synthesis based on the one-pot titanium-catalyzed hydroamination of 2-(alkynyl)aryl chlorides followed by a palladium-catalyzed N-arylation, 2-haloarylenamines are generated via base-catalyzed isomerization of the initially formed imines (Eq. 63; Fig. 4, disconnection a+g).[195] The reaction usually affords indoles in good yields. However, the reaction of a 2-alkenyl alkyne with tert-butylamine gives the corresponding indole in modest yield (39%), most probably because the site selectivity of the titanium-catalyzed hydroamination of 2-alkenyl alkynes is poorer than that of 2-alkyl alkynes.

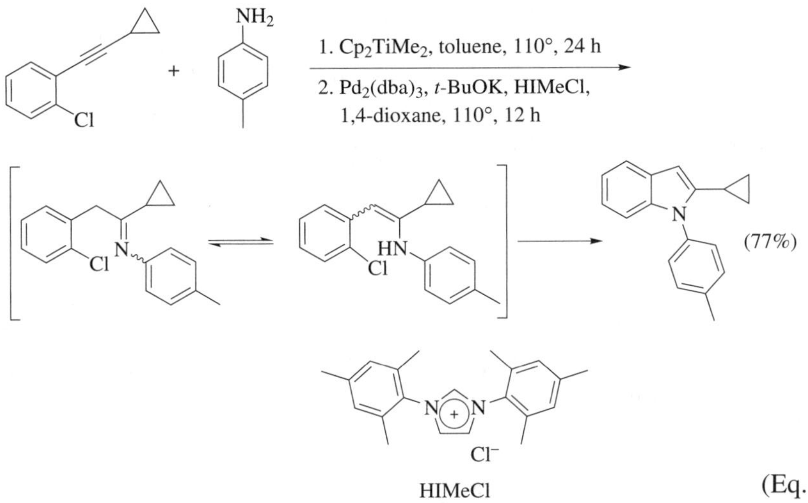

INDOLES VIA PALLADIUM-CATALYZED CYCLIZATION 319

(Eq. 63)

2-Haloarylenamines may also be generated in situ from imines and 1,2-dihalobenzenes or 2-chlorosulfonates to provide an indole synthesis that involves a tandem intermolecular *C*-arylation followed by an intramolecular *N*-arylation (Eq. 64; Fig. 4, disconnection c+g).[196,197] Under the basic conditions used, imines are converted into azaallylic anions (Eq. 65). Control experiments have shown that azaallylic anions are selectively arylated on carbon. With imines containing two different acidic carbon–hydrogen bonds, mixtures of constitutionally isomeric indoles are obtained. For example, the imine derived from 2-heptanone gives a 5:1 mixture of 2-substituted indole (formed via *C*-arylation of the less substituted position of the imine) and 2,3-disubstituted indole (formed via *C*-arylation of the more substituted position of the imine). The reaction has also been developed into a three-component process that provides indoles from 1,2-dihaloarenes, amines, and bromoalkenes (Eq. 66; Fig. 4, disconnection a+c+g).[196]

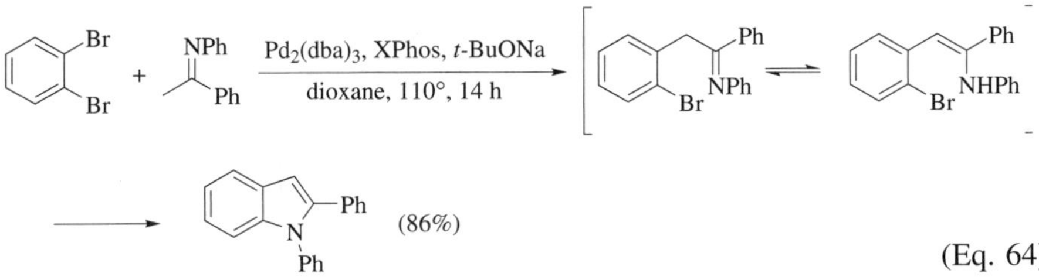

(Eq. 64)

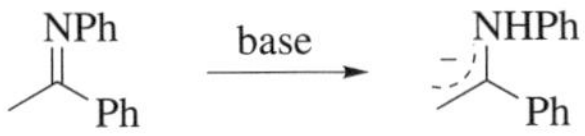

(Eq. 65)

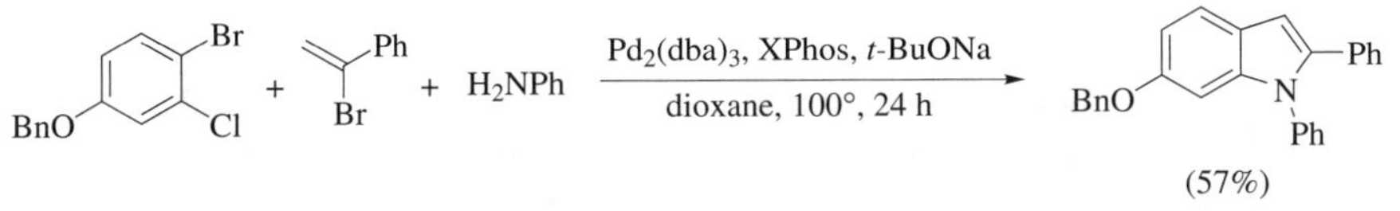

(Eq. 66)

The *N*-vinylation approach has been used to prepare 2-substituted indoles from *ortho-gem*-dihalovinylanilines or -anilides and involves tandem carbon$_{alkenyl}$–nitrogen bond forming processes followed by phosphonylation reactions,[198] Mizoroki–Heck reactions,[199] Suzuki–Miyaura cross-couplings,[200–202] carbonylation/Suzuki–Miyaura cross-couplings,[203] carboalkoxylation reactions,[204] or Sonogashira cross-couplings[205] (Eq. 67; Fig. 4, disconnection a+b). Pd(PPh$_3$)$_4$,[203] PdCl$_2$(PPh$_3$)$_2$,[204] Pd(OAc)$_2$,[198–201] Pd$_2$(dba)$_3$,[198] or Pd/C,[205] the latter four along with Ph$_3$P,[204] 1,1'-bis(diphenylphosphino)ferrocene (dppf),[198] (2-tol)$_3$P,[199] SPhos,[200,201] and (4-MeOC$_6$H$_4$)$_3$P[205] as phosphine ligands, are commonly used as the source of the palladium(0) species. EtN(*i*-Pr)$_2$,[204] HN(*i*-Pr)$_2$,[205] K$_3$PO$_4$·H$_2$O,[199–201] K$_2$CO$_3$,[203] or Et$_3$N[198,199] are used as bases and THF/MeOH,[204] toluene,[198–201,205] or dioxane[203] as solvents. In some cases, beneficial effects are obtained using additives such as Me$_4$NCl.[199]

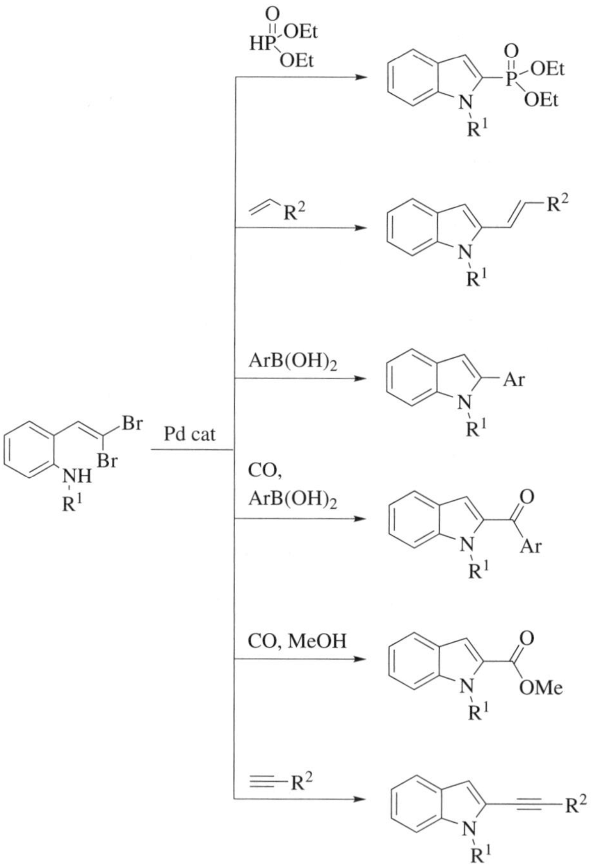

(Eq. 67)

A tandem intermolecular *N*-vinylation followed by an intramolecular *C*-arylation (Fig. 4, disconnection a+c)[165,206] is involved in the reaction of 2-bromoanilines with vinylbromides.[165] The *N*-vinylation approach (Fig. 4, disconnection a) has also been used to prepare 2-bromoindoles from 2-(2,2-dihaloalkenyl)anilines in the presence of Pd(OAc)$_2$, HP(t-Bu)$_3$BF$_4$, and K$_2$CO$_3$ in toluene at 100° (Eq. 68).[207] The use of P(t-Bu)$_3$, generated in situ from the HP(t-Bu)$_3$BF$_4$ salt, appears to be necessary to prevent inhibition of the catalyst by facilitating reversible oxidative addition into the product carbon–bromine bond.

MeO, OBn, Br, Br, NH$_2$ → [Pd(OAc)$_2$, HP(t-Bu)$_3$BF$_4$, K$_2$CO$_3$, toluene, 100°, 14 h] → MeO, OBn, indole-Br, NH (73%)   (Eq. 68)

2-Substituted monochloroindoles have been prepared from trihalogenated alkenylbenzenes and carbamates through a process that is based on two sequential palladium-catalyzed amination reactions, the first intermolecular, the second intramolecular (Eq. 69; Fig. 4, disconnection a+g).[192]

Cl, OMe, Cl, Br + NH$_2$CO$_2$t-Bu → [Pd$_2$(dba)$_3$, SPhos, Cs$_2$CO$_3$, toluene, 110°, 6 h] → Cl-indole-OMe, CO$_2$t-Bu (71%)

(Eq. 69)

3-Substituted Indoles. The preparation of 3-substituted indoles via carbon–nitrogen bond forming reactions has received much less attention than that of 2-substituted indoles. These materials are prepared from 1-bromoalkenes and 2-haloanilines (Fig. 4; disconnection a+c)[165] and from properly substituted 2-(2-haloalkenyl)aryl halides and amines (Eq. 70; Fig. 4, disconnection a+g).[190,191] Reactions are carried out using Pd$_2$(dba)$_3$/DPEPhos or Pd(OAc)$_2$/HP(t-Bu)$_3$BF$_4$ in the presence of t-BuONa in toluene.

Ph, Br, Br + H$_2$NPh → [Pd$_2$(dba)$_3$, DPEPhos, t-BuONa, toluene, 100°] → 3-Ph-indole, N-Ph (94%)   (Eq. 70)

A new approach to 3-substituted indoles from arylenamines that do not contain an aryl halide fragment was subsequently developed (Eq. 71; Fig. 4, disconnection g).[208] The reaction proceeds through a palladium-catalyzed carbon–hydrogen activation followed by an intramolecular amination. Mixtures of indole products are isolated from enamines containing different aromatic rings, suggesting that

an E/Z isomerization occurs rapidly during this process.

$$\text{(Ar)}_2\text{C=CH-NHTs} \xrightarrow[\text{DMSO, 120°, 6 h}]{\text{Pd(OAc)}_2, \text{Cu(OAc)}_2} \text{3-aryl-indole} \quad (51\%) \quad \text{(Eq. 71)}$$

Recently, 2,2-diaryl nitroalkenes have been converted into 3-arylindoles via reductive cyclization under 1 atmosphere of carbon monoxide in the presence of Pd(OAc)$_2$ and 1,10-phenanthroline in DMF at 110° (Eq. 72).[209] The reaction proceeds through direct amination of aromatic C_{sp^2}–H bonds without requiring a functionalized coupling fragment (e.g., an aryl halide or triflate fragment). Both regioisomers of the indole products are obtained with *meta*-substituted 2,2-nitroalkenes. However, a 2,2-dinaphthyl nitroalkene derivative produce only one regioisomer.

$$\text{(Ar)}_2\text{C=CH-NO}_2 \xrightarrow[\text{DMF, 110°, 6 h}]{\text{Pd(OAc)}_2, \text{phen, CO}} \text{3-aryl-indole} \quad (93\%) \quad \text{(Eq. 72)}$$

2,3-Disubstituted Indoles. The pyrrole ring of 2,3-disubstituted indoles can be generated in many ways: (1) from *ortho-gem*-dihalovinylanilines via a formal tandem intramolecular carbon–nitrogen bond forming reaction followed by an intermolecular carbon–carbon cross-coupling reaction with boronic acids (Eq. 73; Fig. 4, disconnection a+b),[200,202] (2) from (2-haloaryl)vinylic triflates via a tandem *N*-vinylation and *N*-arylation process (Eq. 74; Fig. 4, disconnection a+g),[210] (3) from ketimines and 2-dihalobenzenes via the intermediacy of azaallylic anions (Eq. 75; Fig. 4, disconnection c+g),[196,197] and (4) from α-aryloxime *O*-pentafluorobenzoates[211] or substituted enamines[212] (Fig. 4, disconnection g). The cyclization of α-aryloxime *O*-pentafluorobenzoates[211] is supposed to proceed via intramolecular aromatic carbon–hydrogen amination of a vinyl nitrene–palladium intermediate. In the synthesis of indoles from enamines (Eq. 76),[212] the cyclization step is suggested to proceed through an electrophilic attack of palladium(II) on an aromatic ring. Most probably, the role of Cu(OAc)$_2$ is to reoxidize palladium(0).

$$\text{2-CF}_3\text{-3,3-dibromo-aniline} + \text{PhB(OH)}_2 \xrightarrow[\text{toluene, 100°, 1 h}]{\text{Pd(OAc)}_2, \text{SPhos}, \text{K}_3\text{PO}_4 \cdot \text{H}_2\text{O}} \text{2-CF}_3\text{-3-Ph-indole} \quad (79\%) \quad \text{(Eq. 73)}$$

[Eq. 74]

[Eq. 75]

[Eq. 76]

Solid-Phase Synthesis

Solid-phase synthesis is particularly attractive for the generation of libraries of small organic molecules[213–220] and a few very efficient applications of this method to the de novo construction of the indole ring via palladium-catalyzed processes have been introduced.[221] These processes rely on palladium-catalyzed cyclization of properly substituted polymer-bound benzenoid precursors. Described below are indole formations from alkyne and alkene derivatives through carbon–carbon and carbon–carbon/carbon–nitrogen bond forming reactions and the indolization of enamine derivatives via carbon–nitrogen bond forming reactions.

Indole Formation from Alkynes. Palladium-catalyzed, solid-phase syntheses of indoles from alkynes are based on the cyclization of 2-alkynylanilines or -anilides (typically prepared from terminal alkynes and 2-iodoanilines or -anilides via Sonogashira cross-coupling) (Fig. 1, disconnection a), on the intermolecular annulation of 2-iodoanilides with internal alkynes[30,31] (Fig. 1, disconnection a+c), and on the aminopalladation/reductive elimination process[7,9] (Fig. 1, disconnection a+d).

Ester[222,223] and amide[224–226] linkers at the benzene moiety are frequently used as resin attachment points, as shown in the examples in Schemes 7 and 8. In Scheme 7,[222] 1,1,3,3-tetramethylguanidine (TMG) plays a key role in promoting the one-pot coupling/cyclization reaction, with the only byproducts observed arising from an incomplete cyclization step. For the Suzuki–Miyaura cross-coupling shown in Scheme 8,[225] it was found that $Pd(PPh_3)_4$ in DMF/H_2O provides better yields than $Pd_2(dba)_3$ in DMF. Solution-phase conditions can often be successfully applied to solid-phase synthesis. It is interesting to note that K_2CO_3 gives

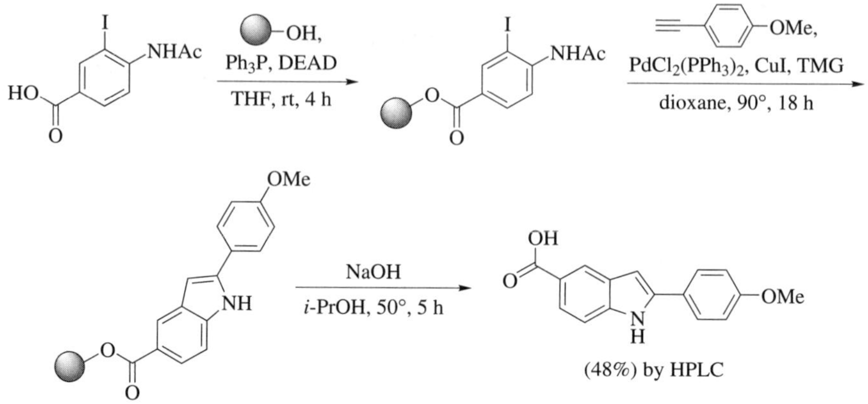

Scheme 7. Synthesis of 2-substituted indoles via tandem Sonogashira cross-coupling/cyclization on the solid phase.

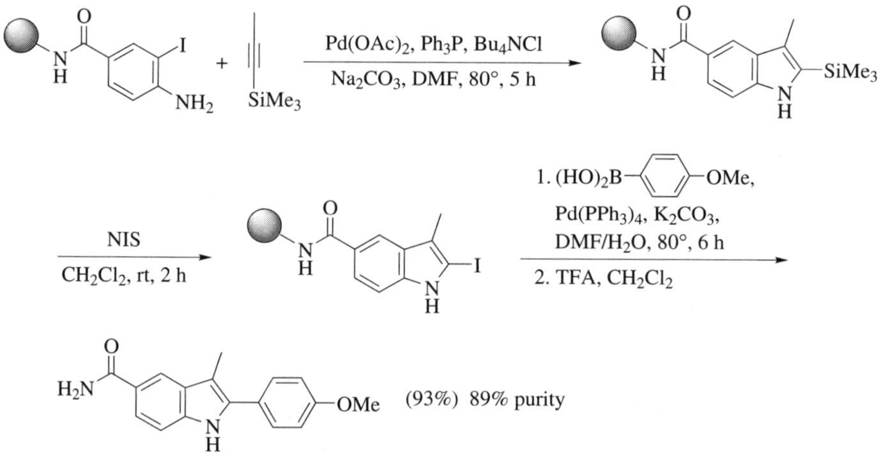

Scheme 8. Synthesis of 3-substituted 2-arylindoles via annulation of internal alkynes with 2-iodoanilines and Suzuki-Miyaura cross-coupling reactions on the solid phase.

better results than Et$_3$N in both the solution-[123] and solid-phase[223] synthesis of 2,3-disubstituted indoles according to the aminopalladation/reductive elimination protocol, despite the expectation that a soluble base would be necessary for a solid-phase reaction (Scheme 9).[223] In fact, little product is obtained when Et$_3$N is used as the base.[223]

Amide linkers at the pyrrole nucleus are also employed. This approach is exemplified by the palladium-catalyzed cyclization of resin-bound 2-alkynylanilides performed under microwave-assisted conditions (Scheme 10).[227] Because

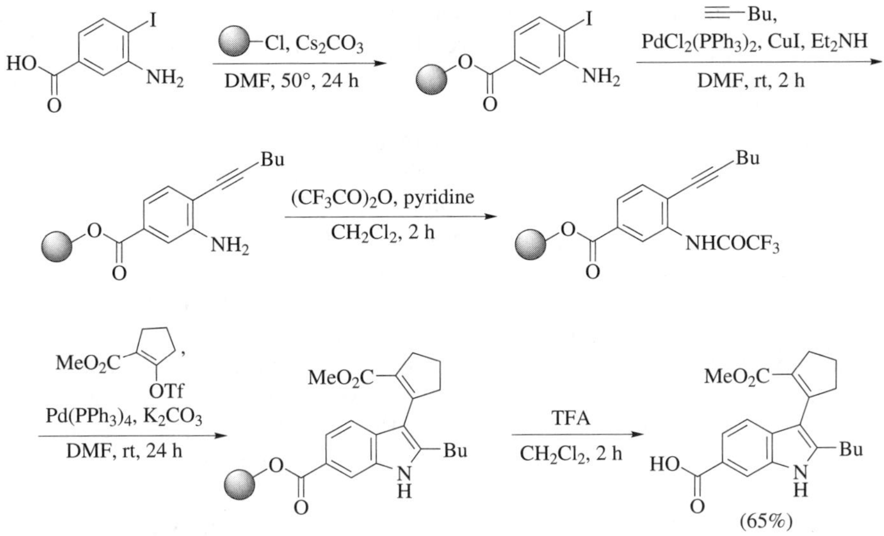

Scheme 9. Synthesis of 2,3-disubstituted indoles via aminopalladation/reductive elimination on the solid phase.

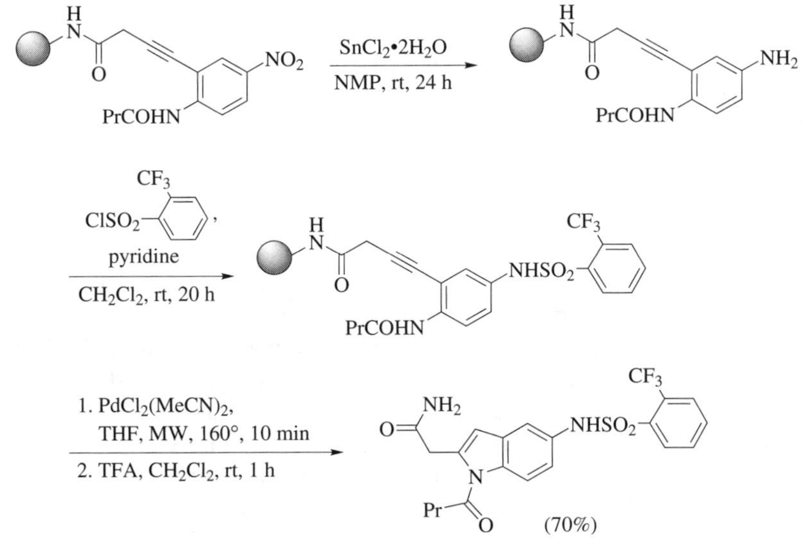

Scheme 10. Synthesis of 2-substituted indoles via cyclization of 2-alkynylanilides on the solid phase.

of their heterogeneous nature, solid-phase syntheses often suffer from long reaction times and/or incomplete conversion of the starting materials. In the latter case, impurities may accumulate on the polymeric surface and lower the purity of compound libraries. Thus, accelerating organic reactions by using microwave conditions appears ideally suited for solid-phase combinatorial synthesis. Indole products are obtained in 65–82% overall yields and with 95–99% purities. Replacement of THF with NMP in the cyclization step decreases both the yield and the purity of the indole products.

All the linkers mentioned above remain as substituents in the final indole derivatives and extraneous substituents such as CO_2H and $CONH_2$ remaining in the final product after cleavage may be undesirable. This may represent a limitation to the scope of the solid-phase approach to the synthesis of indole products and has led to the development of procedures that use traceless linkers. This approach uses the nitrogen–hydrogen bond that will be incorporated into the pyrrole ring to graft the indole precursor on the resin. Cleavage at the end of the synthetic process gives the free indole. In one of the procedures that is based on this strategy, the NH group is attached to the resin via an aminal linkage with a resin-bound 3,4-dihydro-2H-pyran residue (Eq. 77).[228] Resin cleavage with trifluoroacetic acid gives the free indole products. Solution-phase conditions are not particularly successful in this case, with incomplete reaction and large quantities of multiple acetylene insertion products being observed. Optimum yields are obtained with $PdCl_2(PPh_3)_2$, TMG, and resubjection of the reaction mixture to the reaction conditions to drive reactions to completion.

(Eq. 77)

N-Sulfonyl linkers represent a convenient alternative.[225,229,230] The sulfonyl group plays two significant roles: it serves as an activating group to facilitate the cyclization step that occurs under relatively mild conditions and, after indole formation, facilitates the cleavage step, which can be performed under mild conditions (Scheme 11).[229] This should allow the synthesis of diverse indole derivatives bearing either base- or acid-sensitive functional groups. Potassium *tert*-butoxide as the base provides excellent results in some cases.[229,230]

Indole Formation from Alkenes. Palladium-catalyzed solid-phase syntheses of indoles from alkenes are based on the use of an ether linker at the benzene moiety[231] and amide[232] and ester[233–235] linkers at the pyrrole moiety. In these examples, functionalized pyrrole rings are constructed via carbon–carbon bond forming reactions (Fig. 2, disconnection c). Cyclizations have been performed with polymer-bound 2-bromo-N-allylanilides (Eq. 78),[231] 2-haloanilino enamines (Eq. 79),[233,234] and 3-(2-iodoanilino)crotonic acid amides (Scheme 12).[232]

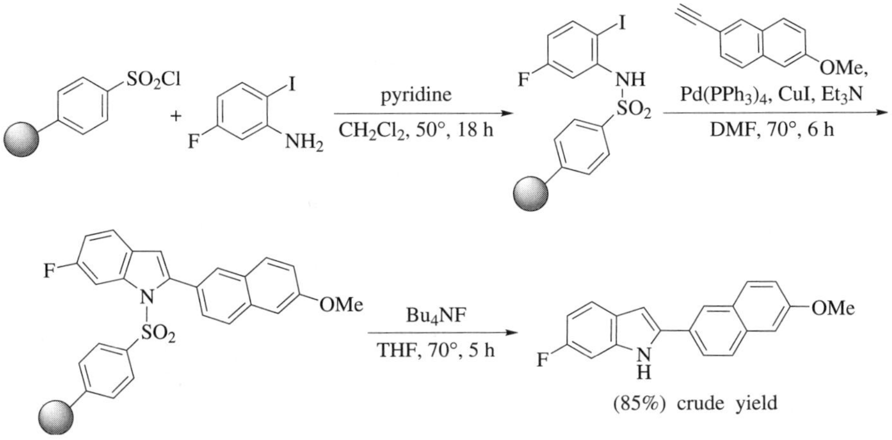

Scheme 11. Traceless linker based synthesis of 2-substituted indoles via tandem Sonogashira cross-coupling/cyclization on the solid phase.

In the latter synthesis, N-alkylated indoles are used because of their higher stability under TFA-cleavage conditions as compared to the free NH counterparts; TFA is reported[236] to induce dimerization of indole-3-acetic acids or their methyl esters.

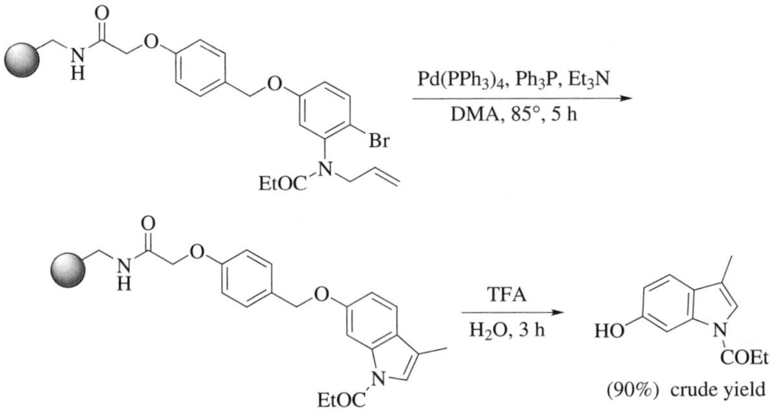

(Eq. 78)

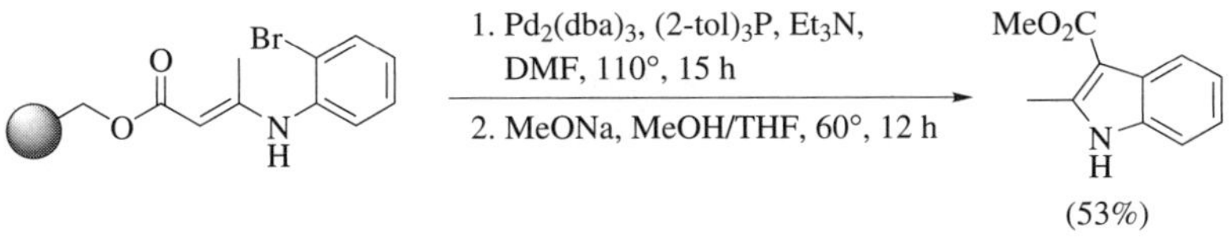

(Eq. 79)

Indole Formation via N-Vinylation and N-Arylation. The cyclization of 2-(2-halophenyl)amino acrylates (immobilized via ester linkers) to methyl 2-indolecarboxylates[234,235] provides examples of solid-phase syntheses of indoles

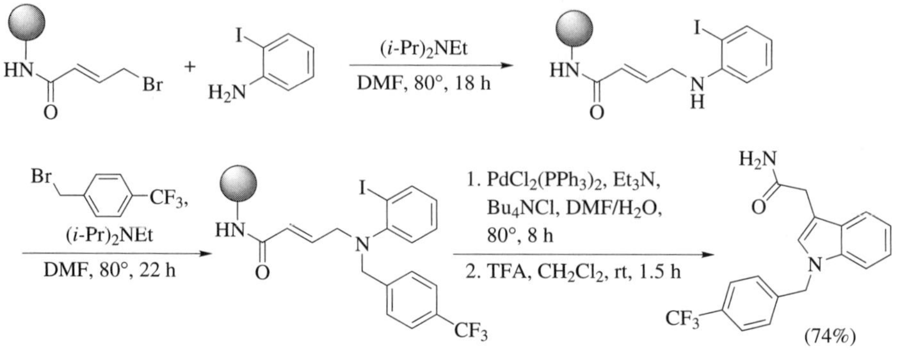

Scheme 12. Indole synthesis via cyclization of 3-(2-iodoanilino)crotonic acid amides on the solid phase.

via the carbon–nitrogen bond forming reaction. The N-arylation is carried out in the presence of $Pd_2(dba)_3$, $(t\text{-}Bu)_3P$, and $(c\text{-}C_6H_{11})_2NMe$ in toluene at 80° when it involves the substitution of the carbon–nitrogen bond for a carbon–bromine bond (Eq. 80)[234] and in the presence of $Pd_2(dba)_3$, the air stable $HP(Bu\text{-}t)_3BF_4$, and $(c\text{-}C_6H_{11})_2NMe$ in DME at 100° when the reaction involves the substitution of the carbon–nitrogen bond for a carbon–triflate bond. This reaction can be conducted as a tandem process that relies on the Mizoroki–Heck reaction of solid-supported N-acetyldehydroalanine with 1,2-dibromobenzenes, followed by in situ intramolecular cyclization of the 2-acetamido-3-(2-bromophenyl)acrylate intermediates (Eq. 81).[234] 1,2-Dibromobenzene gives better results than 1-bromo-2-iodobenzene and 2-bromophenyltriflate.

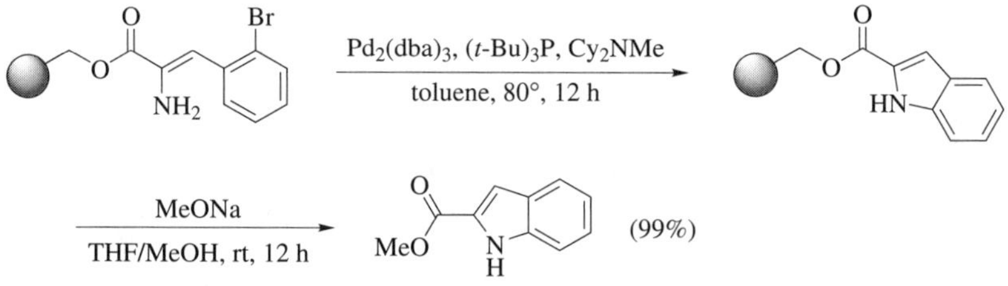

(Eq. 80)

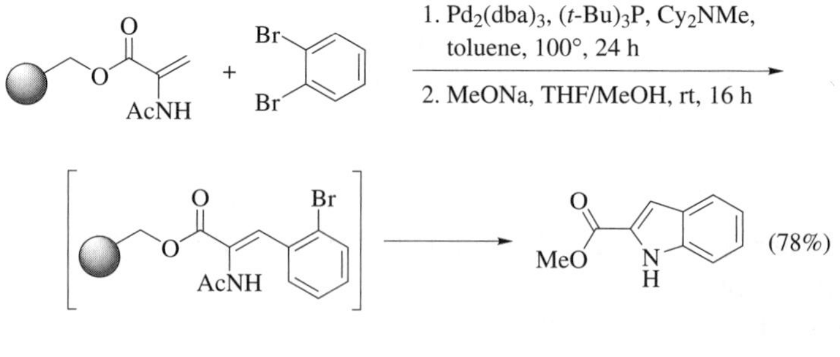

(Eq. 81)

COMPARISON WITH OTHER METHODS

The construction of the indole ring from benzenoid precursors has been performed using a variety of other transition metals. Copper-, gold-, indium-, iridium-, molybdenum-, platinum-, rhodium-, ruthenium-, titanium-, and zinc-catalyzed cyclizations are the most synthetically useful. However, as indicated by the number of approaches developed, the impact of palladium chemistry on indole synthesis has been extraordinary. Although other transition metals can provide better results than palladium in some specific applications, the versatility, flexibility, and substrate scope of palladium-catalyzed reactions are unique. Indole syntheses based on palladium catalysis have incorporated the many advances in catalyst efficiency and allow significant variations in reaction conditions. Despite their importance and utility for the field, indole syntheses based on the other transition metals play a secondary role.

In some cases, and although the same class of indole products is formed, the functionalized pyrrole ring is constructed via bond-forming sequences that are different from those involved in palladium-catalyzed cyclizations. Often, only a few examples are reported so that the substrate scope of the method cannot be critically evaluated. In view of the limited data available, comparisons could be made only on a speculative basis, making it preferable to simply describe the main attributes of the other methods presented in this section. In general, and like the palladium-catalyzed cyclizations to indoles, methods based on the use of stoichiometric amounts of transition metals are not treated and only synthetic procedures where transition metal catalysis is directly involved in the pyrrole ring construction event are discussed. Furthermore, transition-metal-catalyzed reactions producing indole-related compounds such as azaindoles, indazoles, indolines, oxindoles, bis(indolyl)methanes, and related systems or condensed polycyclic compounds such as carbolines, carbazoles, indoloquinolines, indoloquinazolines, and related systems are not treated.

Copper-Catalyzed Indole Formation

The use of copper catalysis is attractive in comparison to palladium-based methods because of its economic advantages and its potential in large-scale reactions. As with palladium, alkynes are the substrates that have been most used with copper to perform cyclizations to indoles. The synthesis of indoles from alkenes, particularly from 2-alkenylphenyl isocyanides, have also been explored. Other synthetic approaches to indoles rely on copper-catalyzed N-arylations or N-vinylations and carbon–carbon bond forming reactions.

Indole Formation from Alkynes. Typically, alkyne-based, copper-catalyzed indole syntheses rely on 2-alkynylanilines and their N-substituted derivatives as direct precursors of 2-substituted indoles. In general, these substrates are prepared via Sonogashira cross-coupling of terminal alkynes with 2-haloanilines or -anilides. Their cyclization into the corresponding indole ring is carried out in the presence of both copper(I) and copper(II) salts. In particular, the cyclization of 2-alkynylanilines and their N-substituted derivatives to indoles has been performed in the presence of CuCl in DMF at 70°,[237] Cu(OAc)$_2$ or Cu(OTf)$_2$ in refluxing 1,2-dichloroethane,[66,238] and Cu(OCOCF$_3$)$_2$·xH$_2$O in MeOH/H$_2$O at room temperature (Eq. 82).[69] Free NH indoles can be obtained from 2-alkynyltrifluoroacetanilides in the presence of CuI and 1,2-*trans*-cyclohexanediamine or PPh$_3$.[239] When the reaction of 2-alkynyltrifluoroacetanilides is carried out in the presence of CuCN, free NH 2-substituted 3-cyanoindoles are obtained through a direct cyclization/cyanation reaction (Eq. 83).[240] With their *N*-tosyl analogues, *N*-tosyl or free NH or a mixture of free NH and *N*-tosyl indole derivatives have been obtained depending on the nature of the substrates (Eq. 83).[240] Tandem Sonogashira cross-coupling of terminal alkynes with 2-haloanilines followed by cyclization to indoles that have been suggested to involve copper catalysis in the cyclization step have also been described.[65,68]

R^1 = COCF$_3$; R^2 = *n*-C$_5$H$_{11}$
R^1 = Ts; R^2 = Ph

R^1 = H; R^2 = *n*-C$_5$H$_{11}$ (73%)
R^1 = Ts; R^2 = Ph (74%)

The copper-catalyzed construction of indole rings from 2-alkynylanilid(n)es has also found its place in solid-phase synthesis. Taking advantage of microwave irradiation, *N*-acyl-[241] and free NH[242] 2-substituted 5-arenesulfamoylindoles have been prepared from resin-bound 2-alkynylanilides.

A notable advance in the alkyne-based, copper-catalyzed route to indoles is the demonstration that copper catalysis can be used both in the formation of 2-alkynylanilides via reaction of terminal alkynes with 2-haloanilides and in the subsequent cyclization to indoles.[239] 2-Iodotrifluoroacetanilide and terminal alkynes can be converted into the corresponding free 2-substituted indoles (NH) through a tandem process that gives the best results in the presence of [Cu(phen)(PPh$_3$)$_2$]NO$_3$ and K$_3$PO$_4$ in toluene or dioxane at 110° (Eq. 84).[239] In some cases, the tandem coupling/cyclization process can be carried out successfully using a CuI/Ph$_3$P combination. Like the related palladium-based tandem processes, the reaction tolerates a wide range of functionalized 1-alkynes, including those containing ether, amide, aldehyde, ester, nitro, and heterocyclic groups. Among the alkynes that have been investigated, a sluggish coupling step that limits the efficiency of the tandem process is observed only with 1-hexyne. No such limitation is observed in the cross-coupling of 1-hexyne with 2-iodotrifluoroacetanilide under Sonogashira conditions.[123,124] The tandem coupling/cyclization procedure was also performed using a catalytic system made of Cu(PPh$_3$)NO$_3$ as the copper source and a 1,10-phenanthroline immobilized on a polystyrene/divinylbenzene solid support.[243] The cyclization step was not as efficient as with [Cu(phen)(PPh$_3$)$_2$]NO$_3$. However, the catalytic system could be reused three times.

(Eq. 84)

The tandem copper-catalyzed coupling/cyclization process has been subsequently extended to 2-bromoalkynyltrifluoroacetanilides using CuI and L-proline as the ligand.[244] Notably, the amino acid ligand allows for running the reaction of the less reactive 2-bromoalkynyltrifluoroacetanilides under conditions milder than those employed with the corresponding iodo derivatives.

The reaction of 2-alkynyl-N-arylideneanilines with alcohols in the presence of catalytic amounts of CuCl affords N-(alkoxybenzyl)indoles (Eq. 85).[245] Other transition metal complexes such as [(η^3-C$_3$H$_5$)PdCl]$_2$, [IrCl(cod)]$_2$, and [RuCl(cod)]$_2$ exhibit catalytic activity, but copper catalysts are the most convenient to use and CuCl gives the best results. Since the N-(alkoxybenzyl)indoles are formed from aldehydes, 2-iodoanilines, terminal alkynes, and alcohols, a wide variety of indole derivatives can be prepared using this protocol.

(Eq. 85)

An alternative to the synthesis of indoles from 2-alkynylanilines and their N-substituted derivatives is the tandem copper-catalyzed reaction of 2-alkynylhaloarenes with primary amines (Eq. 86).[86] Indoles are formed via a carbon–nitrogen bond forming reaction followed by a cyclization step. This reaction is similar to the palladium-catalyzed reaction (Eq. 3),[86,246] with the main differences being the use of K$_3$PO$_4$ or an imidazolium salt (HIPrCl) as a precursor to a carbene ligand for palladium.

(Eq. 86)

The recently described synthesis of 2-(aminomethyl)indoles through copper(I)-catalyzed, tandem three-component coupling/cyclization (Eq. 87),[247,248] which has been applied to the synthesis of a variety of indole derivatives,[249–251] has its counterpart in the palladium-catalyzed reaction of 3-(2-trifluoroacetamidophenyl)-1-propargyl carbonate esters with amines (Eq. 8).[34] The copper-catalyzed process, however, allows for the construction of the functionalized pyrrole ring through the formation of carbon–nitrogen and carbon–carbon bonds that are different from those involved in the related palladium-catalyzed process (compare with Fig. 1, disconnection a+f). Steric effects influence the reaction outcome with secondary amines in the palladium-catalyzed cyclization (for example, a moderate yield is obtained with diisopropylamine), whereas they appear to play a minor role, if any, in the copper-catalyzed process. In addition, the palladium-catalyzed process forms free indoles (NH) while the copper-catalyzed reaction forms N-tosyl indoles.

$$\text{o-alkynyl-NHTs} + (\text{HCHO})_n + (i\text{-Pr})_2\text{NH} \xrightarrow[\text{dioxane, 80°, 15 min}]{\text{CuBr}} \text{2-((i-Pr)_2N-methyl)-1-tosylindole}$$
(81%)

(Eq. 87)

Indole Formation from Alkenes. The use of alkene-based copper-catalyzed synthesis of indoles is still rare. An example of this chemistry has recently been reported and involves the preparation of 2-boryl- (Eq. 88) and 2-silylindoles by copper-catalyzed borylative and silylative cyclization of 2-alkenylaryl isocyanides.[252]

$$\text{alkene-aryl isocyanide} + \text{B}_2(\text{pin})_2 \xrightarrow[\text{MeOH, THF, rt}]{\text{Cu(OAc)}_2, \text{PPh}_3} \text{2-B(pin)-3-CO}_2\text{Me-indole}$$
(58%)

(Eq. 88)

Indole Formation via N-Vinylation and N-Arylation. The potential of the palladium-catalyzed N-arylation and N-vinylation approach to the construction of the pyrrole ring has been demonstrated through several applications. This synthetic approach has been quickly applied to the copper-catalyzed construction of the pyrrole ring incorporated into the indole system. Basically, indoles have been prepared through two main synthetic strategies: the cyclization of 2-haloarylenamid(n)es, and the cyclization of 2-(bromovinyl)anilid(n)es.

Preformed 2-haloarylenamid(n)es have been converted into the corresponding indoles by using the CuI/L-proline precatalyst system (Eq. 89)[253] or under the conditions shown in Eq. 90.[254] Similarly, enehydrazid(n)es and enehydroxylamines have been converted into N-aminoindole and N-alkoxyindole derivatives, respectively.[255]

[Scheme for Eq. 89: F-substituted aryl with Br, NHCbz, and CO2Me vinyl group + CuI, L-proline, K2CO3, dioxane, 100°, 24 h → F-indole-2-CO2Me with N-Cbz (62%) + F-indole-2-CO2Me with N-H (26%)]

(Eq. 89)

[Scheme for Eq. 90: 2-Br-aryl with CH=N-R and CH-CO2Me substituent + CuI, K3PO4, 75° → 3-CO2Me indole with N-R]

R	Conditions	
4-MeOC$_6$H$_4$	ethylene glycol, 2-propanol, 6 h	(89%)
Bn	ethylene glycol, DMF, 15 h	(91%)
c-C$_6$H$_{11}$	DMF, 20 h	(57%)

(Eq. 90)

Indoles have also been synthesized from 2-haloarylenamid(n)es prepared in situ. An example of this chemistry is provided by the synthesis of indole-2-carboxylates from 2-haloaryl aldehydes or ketones and ethyl isocyanoacetate (Eq. 91).[256] The reaction is suggested to proceed through a tandem condensation/coupling/deformylation process. It is performed at room temperature or 50° with iodo- and bromo-containing substrates. With chloride-substituted substrates a higher reaction temperature (80°) is required. CuI and CuBr display a similar catalytic activity whereas CuCl, Cu$_2$O, and CuSO$_4$ are less active. 2-Haloarylenamid(n)es are also generated in situ in the synthesis of indole-2-carboxylic esters from 2-bromoaryl aldehydes and ethyl acetamidoacetate in the presence of CuI and Cs$_2$CO$_3$ in DMSO at 80°.[257]

[Scheme for Eq. 91: 2-Br-aryl with acetyl group (Br-substituted) + CNCH2CO2Et → CuI, Cs2CO3, DMSO, 50°, 4 h → Br-substituted 3-methylindole-2-CO2Et (90%)]

(Eq. 91)

The copper-catalyzed reaction of 2-(2-bromoalkenyl)bromoarenes with carbamates, amides, and anilines allows the preparation of N-functionalized indoles through a tandem amination/cyclization process.[258] In the presence of N,N-dimethylethylenediamine, CuI, CuOAc, and CuTC (copper thiophene-2-carboxylate) can be successfully employed, and K$_2$CO$_3$, K$_3$PO$_4$, and Cs$_2$CO$_3$ are effective bases (Eq. 92).[258] The range of N-coupling partners that can be used complements that achievable using palladium catalysis. The major advantage of employing the copper system is the successful preparation of N-acyl indoles, which could not be effectively prepared using the palladium-catalyzed process.[259] Conversely, the copper chemistry is less efficient in couplings employing simple amines. A related approach to the construction of the indole ring has been used to prepare 2-bromoindole intermediates in a one-pot synthesis of pyrimido[1,6-a]indol-1(2H)-ones by a nucleophilic addition/Cu-catalyzed N-arylation/Pd-catalyzed

C–H activation sequential process.[260]

$$R^1\text{-arene-Br}(Br) + NH_2R^2 \xrightarrow{\text{TMEDA, Cu source, base}}_{\text{toluene, 110°, 24 h}} R^1\text{-indole-}R_2$$

R^1	R^2	Cu Source	Base	
Br	t-BuCO$_2$	CuOAc	Cs$_2$CO$_3$	(53%)
H	n-C$_5$H$_{11}$	CuTC	K$_2$CO$_3$	(87%)
H	Ph	CuI	K$_2$CO$_3$	(80%)

(Eq. 92)

Indole Formation via Arene Vinylation. Some approaches to the contruction of the indole skeleton have been based on the ability of copper to catalyze the formation of carbon–carbon bonds. In particular, this strategy has been applied to the preparation of indoles from N-(2-haloaryl)-[261,162] and N-(aryl)enaminones.[263] N-(2-Haloaryl)enaminones have been converted into the corresponding 2-substituted 3-acylindoles through a process that involves the copper-catalyzed substitution of the carbon–carbon bond for the carbon–halogen bond (Eq. 93).[261] The synthesis of indoles from N-(aryl)enaminones is based on the formation of carbon–carbon bonds through selective catalytic activation of aryl carbon–hydrogen bonds (Eq. 94).[263] This reaction reflects the current interest in minimizing substrate preactivation in indole synthesis,[264,265] taking advantage of carbon–carbon bond forming processes that do not rely on preactivation of the starting materials, an inherently wasteful requirement since the installation of activating groups (commonly halogens) may require multiple steps while none appear in the final products.

$$\xrightarrow{\text{CuI, phen, K}_2\text{CO}_3}_{\text{DMF, 100°, 6 h}} \quad (92\%)$$

(Eq. 93)

$$\xrightarrow{\text{CuI, phen, Li}_2\text{CO}_3}_{\text{air, DMF, 100°, 12 h}} \quad (66\%)$$

(Eq. 94)

Gold-Catalyzed Indole Formation

Alkynes are the typical substrates even for gold-based indole syntheses. In particular, 2-alkynylanilines[70,266,267] and their N-substituted derivatives[70] are converted into 2-substituted indoles using NaAuCl$_4$ in THF,[70] EtOH, or EtOH–H$_2$O mixtures,[266] and AuCl$_3$ in EtOH.[267] Gold-catalyzed cyclizations to indoles may be carried out using a polystyrene-silica-gel-supported gold(III) catalyst[268] or with water[269] or ionic liquids[270] as the reaction medium. In the latter case, cyclization of 2-alkynylanilines with $NaAuCl_4 \cdot H_2O$ in 1-butyl-3-methylimidazolium tetrafluoroborate ([bmim]BF_4) affords 2-substituted indoles in high yields. The catalyst system is best recycled using Bu_4NAuCl_4.[270] The related synthesis of 2-substituted indoles from 2-alkynylnitroarenes proceeds through a one-pot, one-step (Eq. 95) or one-pot, two-step hydrogenation/hydroamination process catalyzed by gold nanoparticles supported on Fe_2O_3.[271]

$$\text{2-alkynyl-NO}_2\text{-arene} \xrightarrow[\text{toluene, 120°, 1 h}]{H_2,\ Au_{np}/Fe_2O_3} \text{2-Ph-indole} \quad (73\%) \qquad \text{(Eq. 95)}$$

The gold-catalyzed hydroamination of 2-alkynylanilines has been combined with a C-3 functionalization step to provide a general entry into 2,3-disubstituted indoles.[272–274] An example of this approach to 2,3-disubstituted indoles is shown in Eq. 96.[272] The reaction involves the conjugate addition to α,β-enones of indolylgold intermediates formed in situ. Both the cyclization reaction and the conjugate addition reaction are completely inhibited when the nitrogen nucleophilicity is decreased as with 2-alkynylacetanilides. In these cases, a competitive addition of water to the triple bond is observed. Both gold(III)[275–278] and gold(I)[279] species are known to catalyze the hydration of alkynes. A related palladium-catalyzed cyclization of aryl alkynes containing *ortho* nitrogen nucleophiles with α,β-enals and -enones has been described.[103] However, the reaction fails to give the desired 2-substituted 3-alkylindoles using anilines, requiring the use of 2-alkynylanilides to give the best results.

$$\text{(Eq. 96)}$$

Some procedures that involve the in situ preparation and cyclization of 2-alkynylanilines to 2-substituted indoles have been developed. Terminal alkynes and 2-iodoaniline have been converted into 2-substituted indoles through a gold-catalyzed coupling/cyclization sequence (Eq. 97).[280] *N*-Boc, *N*-Ts, *N*-Ms, and *N*-acetyl 2-iodoanilines are also suitable coupling/cyclization partners. However, no indole formation is observed with 2-bromoaniline. Recently, a three-component coupling/cyclization of *N*-Ts ethynylaniline, aldehydes, and amines has been described (Eq. 98).[281] The reaction occurs in the presence of a heterogeneous catalyst based on gold supported on nanocrystalline ZrO_2.

[Structure: 2-iodoaniline] + ≡—Ph $\xrightarrow[\text{toluene, 130°, 24 h}]{\text{AuI, dppf, } K_2CO_3}$ [2-phenylindole]—Ph (99%) (Eq. 97)

[Structure: 2-ethynyl-NHTs aniline] + [cyclohexanecarbaldehyde CHO] + [piperidine] $\xrightarrow[\text{dioxane, 100°, 6 h}]{\text{Au/ZrO}_2}$ [indole product] (75%) (Eq. 98)

In addition to 2-alkynylanilid(n)es, 2-tosylaminophenylprop-1-yn-3-ols have been shown to be useful precursors of indole derivatives (Eq. 99).[282]

[Ph-OH substrate with NHTs] $\xrightarrow[\text{HMPA, toluene, reflux, 2 h}]{\text{AuCl, AgOTf, CaSO}_4}$ [3-Ph, 2-vinyl indole N-Ts] (90%) (Eq. 99)

All the above-mentioned alkyne-based gold-catalyzed indole syntheses involve a hydroamination reaction, that is, the addition of a nitrogen–hydrogen bond across a carbon–carbon triple bond. Recently, a synthetic approach that is based on the quite rare carboamination of alkynes[7,102,140,283] (i.e., the addition of a carbon–nitrogen bond to a carbon–carbon triple bond) has been developed. In particular, 2-substituted 3-methylindoles are formed from 2-alkynyl-N,N-dimethylanilines through an intramolecular methylamination catalyzed by AuCl(CAAC) (CAAC = cyclic (alkyl)(amino)carbene) (Eq. 100).[284] In the same paper, cationic gold(I) complexes supported by CAAC ligands were shown to promote the formation of indole derivatives via an intramolecular hydroammoniumation reaction.

[2-alkynyl-N,N-dimethylaniline with Ph] $\xrightarrow[C_6D_6, 160°, 20 \text{ h}]{\text{AuCl(CAAC)/KB(C}_6\text{F}_5)_4}$ [3-methyl-2-phenyl-N-methylindole]—Ph (90%) (Eq. 100)

Indium-Catalyzed Indole Formation

Indium(III) bromide has been reported to catalyze the intramolecular hydroamination of 2-ethynylanilines having an alkyl or aryl group on the alkyne to selectively afford 2-substituted indole derivatives (Eq. 101).[285,286] Interestingly, using substrates with a trimethylsilyl group or no substituents on the triple bond exclusively gives quinoline derivatives.

[NC-substituted 2-ethynylaniline with Ph] $\xrightarrow[\text{toluene, reflux, 2 h}]{\text{InBr}_3}$ [NC-substituted 2-phenylindole]—Ph (98%) (Eq. 101)

Iridium-Catalyzed Indole Formation

The combination of iridium complex **29** with $NaB[3,5-(CF_3)_2C_6H_3]_4$ provides a catalyst system that can be used for the synthesis of 2-substituted indoles from 2-alkynylanilines (Eq. 102).[287] High to excellent yields are obtained with neutral and electron-donating substituents on the aromatic ring and/or the nitrogen whereas indoles are isolated in very low yields when either the aromatic ring or the nitrogen atom bears electron-withdrawing substituents. The number of examples investigated is relatively limited in comparison with the large number of related palladium(II)-catalyzed hydroaminations and there is room for further improvement. Nevertheless, the substrate scope of the palladium(II)-catalyzed processes is wider. Indeed, a number of successful palladium(II)-catalyzed hydroaminations to indoles have been performed using aryl alkynes containing *ortho* nitrogen nucleophiles with electron-withdrawing substituents both on the aromatic ring and/or the nitrogen atom.

(Eq. 102)

R^1	R^2	R^3	Time (h)	
H	H	H	24	(10%)
H	*n*-Bu	H	6	(96%)
H	*n*-Bu	Cl	24	(84%)
H	*n*-Bu	NO_2	24	traces
H	*n*-Bu	Me	6	(96%)
Ac	*n*-Bu	H	24	(21%)
H	Ph	H	12	(95%)
H	Ph	Cl	24	(27%)
H	Ph	NO_2	24	(16%)
H	Ph	Me	24	(96%)

Several types of 4-acetylindoles have been selectively obtained through directed cyclodehydration of α-arylamino ketones catalyzed by a cationic iridium–BINAP complex (Eq. 103).[288] The acetyl group at the *meta* position plays a key directing role and enables carbon–iridium bond formation at the congested *ortho* position, which is followed by an intramolecular 1,2-addition to a carbonyl moiety and a dehydration step.

(Eq. 103)

BARF = tetrakis[3,5-bis(trifluoromethyl)phenyl]borate

Molybdenum-Catalyzed Indole Formation

Molybdenum catalysis has been applied to a few alkyne-based indole syntheses. In particular, the Boc derivative of 2-ethynylaniline, a terminal alkyne, can be converted into the corresponding indoles in the presence of the Mo(CO)$_5$(Et$_3$N) complex (Eq. 104).[289] Interestingly, the cyclization of 2-ethynylaniline provides the desired product in high yield under molybdenum-catalyzed conditions, whereas a poor yield is obtained using an iridium complex (Eq. 102).[287]

$$\text{Mo(CO)}_5(\text{Et}_3\text{N}), \text{Et}_3\text{N}, \text{Et}_2\text{O}$$

R = H (79%)
R = Boc (50%)

(Eq. 104)

An alkene-based route to indoles has also been investigated using molybdenum complexes as catalysts. In particular, 2-nitrostyrenes provide access to 2-substituted and 2,3-disubstituted indoles by molybdenum-catalyzed reductive cyclization with MoO$_2$Cl$_2$(DMF)$_2$ and Ph$_3$P (Eq. 105).[290] Toluene is the most suitable solvent and the use of an inert atmosphere leads to a better conversion, probably due to the oxidation of Ph$_3$P in air. Both the *cis* and *trans* isomers react, although a slightly higher yield is obtained from the former. In comparison to palladium-catalyzed methods,[154–156,158–161,180] no carbon monoxide is required. To make the procedure more practical, the dioxomolybdenum-catalyzed reductive cyclization of 2-nitrostyrenes to indoles can be carried out using a polymer-bound triphenylphosphine.[290] Under these conditions, reaction times are a bit longer, but the isolation of the product by simple filtration to remove the solid-supported phosphine is much easier.

(E)/(Z)	R^1	R^2	
4:1	Me	H	(77%)
1:0	CO$_2$Et	H	(75%)
0:1	CO$_2$Et	Me	(84%)
1:2.3	n-C$_5$H$_{11}$	H	(80%)
1:2	Ph	H	(73%)
10:1	Ph	H	(64%)
1.5:1	Ph	H	(80%)

(Eq. 105)

Platinum-Catalyzed Indole Formation

2-Alkynylanilides also are the typical indole precursors in the platinum-catalyzed cyclizations. However, some of the alkyne-based, platinum-catalyzed cyclizations provide routes to indoles that do not have a palladium counterpart. Furthermore, some of the acetylenic substrates that afford indoles under platinum-catalyzed conditions do not undergo indole formation using palladium. This divergence is the case with the platinum-catalyzed synthesis of 2-substituted-3-acyl indoles (Eq. 106),[283] where PtCl$_2$ gives the best results. Slightly lower yields are obtained with other platinum(II) precatalysts, such as $PtCl_2(MeCN)_2$ and $PtBr_2$, whereas $Pt(PPh_3)_4$ does not afford the products at all. Palladium catalysts such as $Pd(PPh_3)_4$ and $PdCl_2$ do not exhibit useful catalytic activity. 2-Substituted 3-acyl indoles can be accessed using palladium catalysis by the reaction of 2-alkynyltrifluoroacetanilides with aryl iodides or vinyl triflates under an atmosphere of carbon monoxide.[124] This protocol allows for the synthesis of indoles containing aryl and vinylic units bound to the carbonyl group at C(3) but no alkyl substituents can be introduced. In contrast, the synthesis of 2-substituted 3-acylindoles containing alkyl substituents bound to the carbonyl group at C(3) can be readily accomplished by the platinum-catalyzed process.

$$\underset{\substack{\text{Me}}}{\text{[2-(1-propynyl)-N-benzoyl-N-methylaniline]}} \xrightarrow[\text{anisole, 80°, 0.5 h}]{PtCl_2} \underset{\substack{\text{Me}}}{\text{[2-propyl-3-benzoyl-N-methylindole]}} \quad (99\%) \qquad \text{(Eq. 106)}$$

The platinum-catalyzed cyclization of 2-alkynylanilides to indoles has been combined with the reaction of the latter with electron-poor alkynes such as ethyl propiolate and dimethyl acetylenedicarboxylate to give 2,3-disubstituted indoles.[291] The composition of the products is largely influenced by the substituents on the indoles as well as the amount of alkyne used.

A few 2-(alkynyl)phenylisocyanates have been converted into 2-substituted N-(alkoxycarbonyl)indoles using $PtCl_2$, although most of the 2-(alkynyl)phenyliso-cyanates investigated have been converted into the corresponding indoles with Na_2PdCl_4.[28] In some cases, platinum catalysis affords better results than palladium catalysis. For example, an isocyanate having a terminal acetylenic group gives the corresponding indole derivative in 45% yield with $PtCl_2$ and n-propanol (Eq. 107)[28] whereas the use of Na_2PdCl_4 results in the formation of a complex mixture of unidentified products. Longer reaction times are needed with increasing bulk of the alcohols. With tert-butyl alcohol, $PtCl_2$ shows higher catalytic activity than Na_2PdCl_4, and only the use of $PtCl_2$ allows reaction of an internal alkyne with allyl alcohol for formation of the desired 2-substituted indole.[28] Recently, 2-(alkynyl)phenylisocyanates have been prepared via a Hofmann-type rearrangement of 2-(alkynyl)benzamides promoted by $PhI(OAc)_2$ and cyclized in situ to 2-substituted indoles with $PtCl_2$ through a tandem procedure.[292,293]

$$\underset{NCO}{\text{[2-ethynylphenyl isocyanate]}} + PrOH \xrightarrow[\text{DME, 100°, 2 h}]{PtCl_2} \underset{CO_2Pr}{\text{[N-propoxycarbonylindole]}} \quad (45\%) \qquad \text{(Eq. 107)}$$

The preparation of 2,3-disubstituted indoles and particularly 3-alkoxyindoles from aniline **30** (Eq. 108)[294] is another platinum-catalyzed reaction without a palladium counterpart. It can be carried out even using proton catalysis.

$$\text{30} \xrightarrow{\text{PtCl}_2, \text{ toluene}, 80°, 1\text{ h}} \text{product} \quad (92\%) \quad \text{(Eq. 108)}$$

The cyclization of precursor **31** to give the 2-substituted indole derivative **32** has numerous related palladium-based analogs and the conversion of precursor **33** into indole **34** (Eq. 109)[295] resembles the related palladium-catalyzed reaction of 2-alkynyl-N-allyltrifluoroacetanilides.[102] These are the only examples reported. However, unlike the palladium-based version, the platinum-catalyzed reaction requires the presence of carbon monoxide (its presence has been shown to accelerate certain PtCl$_2$-catalyzed skeletal rearrangements).[296] This reaction is performed with anilides, thus forming N-protected indoles, whereas free indoles (NH) are obtained in the palladium-catalyzed cyclization. Furthermore, the two methods differ mechanistically in that the palladium-based reaction relies on a redox palladium(0)–palladium(II) cycle, whereas the platinum-based one does not. This feature may be of interest when working with substrates that contain additional sites prone to oxidative addition.

$$\xrightarrow{\text{PtCl}_2, \text{CO}, \text{ toluene}, 80°, 0.5\text{ h}} \quad \text{(Eq. 109)}$$

	R	
31	H	32 (93%)
33	CH$_2$CH=CH$_2$	34 (59%)

2-Propargyl anilines can give indoles through a platinum-catalyzed cycloisomerization that can occur under acid-catalyzed or even uncatalyzed conditions.[297]

Rhodium-Catalyzed Indole Formation

The rhodium-catalyzed synthesis of indoles[298] provides interesting alternatives to palladium-based processes. Unprotected 2-ethynylanilines have been converted into parent indoles through a cycloisomerization process catalyzed by [Rh(cod)Cl]$_2$ in the presence of Ph$_3$P (Eq. 110) or (4-FC$_6$H$_4$)$_3$P.[299] The reaction is suggested to involve a rhodium-vinylidene intermediate. Thus, only terminal alkynes can serve as substrates for indole formation. The synthesis of parent indoles from the cyclization of unprotected 2-ethynlanilines distinguishes this process from other metal-catalyzed cyclization methods.

$$\xrightarrow{[\text{Rh(cod)Cl}]_2, \text{Ph}_3\text{P}, \text{DMF}, 85°, 2\text{ h}} \quad (84\%) \quad \text{(Eq. 110)}$$

One of the advantages of using cycloisomerisation to synthesize indoles is that the cyclization step affords metalloindoles that can be trapped using suitable reagents, allowing for the design of processes in which several sequential transformations occur. Such a strategy has been applied to the rhodium-catalyzed synthesis of 2,3-disubstituted indoles from 2-alkynylanilides and alkenes (Eq. 111)[300] or alkynes.[301] The reaction outcome is dependent on the catalyst used. With $Rh(CO)_2acac$, the major pathway is the protodemetallation to generate the corresponding 2-substituted indole product. With $[Rh(cod)OH]_2$, the tandem reaction is favored.

$$\text{2-alkynylanilide (NHMs, Bu)} + \text{CH}_2=\text{CHCOEt} \xrightarrow[\text{dioxane/H}_2\text{O, 90°, 4 h}]{[Rh(cod)OH]_2} \text{indole product} \quad (94\%) \quad \text{(Eq. 111)}$$

The rhodium complex **35** can catalyze the hydroamination of 2-alkynylanilines to indoles. Specifically, 2-ethynylaniline and 2-(phenylethynyl)aniline are converted into indole and 2-phenylindole, respectively, in acetone at 55°.[302]

35

Following a current trend aimed at minimizing substrate preactivation in indole synthesis to reduce cost and increase the breadth of readily available starting materials,[263–265] new approaches based on the rhodium-catalyzed oxidative coupling of alkynes with N-acetyl anilines (Eq. 112)[303,304] and N-aryl-2-aminopyridine (Eq. 113)[305] have been realized.

$$\text{Cl-C}_6\text{H}_4\text{-NHAc} + \text{HC}\equiv\text{C-Ph} \xrightarrow[t\text{-AmOH, 120°, 1 h}]{[Cp^*RhCl_2]_2,\ AgSbF_6,\ Cu(OAc)_2\cdot H_2O} \text{Cl-indole-Ph (N-Ac)} \quad \text{(Eq. 112)}$$

(62%)

$$\text{Ar-NH-(2-pyridyl)} + \text{Ph}\equiv\text{Ph} \xrightarrow[\text{DMF, 120°, 12 h}]{[Cp^*RhCl_2]_2,\ Cu(OAc)_2} \text{2,3-diphenylindole (N-2-pyridyl)} \quad (96\%) \quad \text{(Eq. 113)}$$

N-Propargylanilines have been converted into 2-substituted and 2,3-disubstituted indoles in the presence of RhH(CO)(PPh$_3$)$_3$ or [Rh(cod)$_2$]OTf in hexafluoroisopropyl alcohol (HFIP) (Eq. 114).[306,307] The cyclization proceeds via the corresponding 2-allenylaniline intermediates, which are generated by the rhodium(I)-catalyzed amino-Claisen rearrangement of N-propargylanilines. The reaction was also developed into a one-pot synthesis of indoles by reacting N-alkylaniline with propargyl bromide.

$$\text{(Eq. 114)} \quad (89\%)$$

3-Acetyl-2-hydroxyindoles have been prepared via rhodium(II)-catalyzed decomposition of α-diazoanilides.[308–310] The course of this type of reaction is highly dependent on the substituents surrounding the diazo group. Eq. 115 illustrates an interesting example in which the exclusive alkylation of the nitrophenyl group takes place.[310] Frequently, in similar substrates, insertion of the carbenoid into an aliphatic carbon–hydrogen bond tends to compete with the alkylation of the aryl group. No related palladium-catalyzed reactions have been developed.

$$\text{(Eq. 115)} \quad (65\%)$$

A variety of 2,3-disubstituted indoles have been synthesized by Rh$_2$(O$_2$CCF$_3$)$_4$ catalyzed isomerization of 2-aryl-2H-azirines.[311]

Ruthenium-Catalyzed Indole Formation

A few examples of indole synthesis via ruthenium-catalyzed, intramolecular hydroamination of an acetylenic precursor have been described. By subjecting 2-ethynylaniline to Ru$_3$(CO)$_{12}$ in diglyme for 4 hours at 110° under an argon atmosphere, indole is isolated in 54% yield.[312] 2-Ethynylanilid(n)es have been converted into the corresponding indoles in the presence of [RuL$_2$Cp(MeCN)]PF$_6$ (Eq. 116).[313,314] No reaction is observed with 2-(phenylethynyl)aniline whereas parent indole is isolated in 84% yield after 400 hours using 2-(trimethylsilylethynyl)aniline as the starting alkyne. The reaction has been developed into a one-pot cyclization/hydration process to give indoles containing a C-6 acetaldehyde group.[313,314]

$$\text{(Eq. 116)} \quad (91\%)$$

L = Ph$_2$P-N-t-Bu

Another ruthenium-catalyzed indole formation is based on the functionalization of benzylic carbon–hydrogen bonds of 1,2-disubstituted isocyanates.[315] In one example, heating a solution of 2,6-xylyl isocyanide and $RuH_2(dmpe)_2$ at 140° in benzene-d_6 for 24 hours results in the formation of 7-methylindole in 98% yield as determined by NMR spectroscopy.

More attention has been paid to the preparation of indoles from anilines and alcohol derivatives. Anilines and 1,2-diols are converted into indole products with $RuCl_2(PPh_3)_2$ at 180° in dioxane[316] or $RuCl_3 \cdot xH_2O$ and Ph_3P or XantPhos at 170°.[317] The reaction of anilines with trialkanolamines[318] and trialkanolammonium chorides[319,320] (Eq. 117) also provides access to indoles. 2,3-Unsubstituted,[315,316,318,319] 2-methyl-,[319] and 2,3-dimethylindoles[315] have been prepared using these methods. The alcoholic components act as two-carbon donors in the construction of the pyrrole ring. In this sense, the reaction is reminiscent of the synthesis of indoles via palladium-catalyzed annulation of 2-haloanilines or their derivatives with internal alkynes.[30,31] The palladium-catalyzed reactions, however, are more versatile.

(Eq. 117)

N-Allyl-2-vinylanilides are converted into indoles through a ruthenium-catalyzed isomerization to enamines in the presence of vinyloxytrimethylsilane followed by a ruthenium-catalyzed ring-closing metathesis which is performed on the crude isomerization mixture after evaporation of the volatile materials (Eq. 118).[321,322] The aromatic enamide/ene methatesis has been subsequently applied to the synthesis of indomethacins.[323]

(Eq. 118)

The cyclization reaction of diallylanilines containing an ethynyl group at the *ortho* position of the aromatic ring in the presence of $CpRuCl(PPh)_3$ or $CpRuCl(dppe)$ is accompanied by an aza-Claisen rearrangement, causing an allyl group migration to give substituted indole compounds. This cyclization can also be performed by using the $AuCl(PPh_3)/AgSbF_6$ combination.[324]

Titanium-Catalyzed Indole Formation

Indoles have been obtained through titanium-catalyzed reductive coupling of carbonyl compounds, a reaction that is based on the high reducing ability and pronounced oxophilicity of low-valent titanium (Eq. 119).[325] Heating oxoamides with catalytic amounts of $TiCl_3$, Zn dust as the stoichiometric reducing agent, and an excess of R_3SiCl in MeCN or DME affords indole derivatives in yields comparable to those obtained in stoichiometric reactions.[326,327]

$$\text{(Eq. 119)}$$

Zinc-Catalyzed Indole Formation

Zinc-catalyzed hydroamination of 2-alkynyl-N-tosylanilides (with Et_2Zn)[328] and 2-alkynylanilines (with $ZnBr_2$ or ZnI_2)[329] to the corresponding 2-substituted indole derivatives have been described. A different alkyne-based zinc-catalyzed indole synthesis involves the reaction of propargyl alcohols with anilines in toluene without additives (Eq. 120).[330] The mechanism has been elucidated and the reaction proceeds through a 1,2-nitrogen shift catalyzed by $Zn(OTf)_2$.

$$\text{(Eq. 120)}$$

Zinc catalysis has also been proven to favor the synthesis of 5-hydroxyindoles from benzoquinone and enaminones.[331,332] An example of this chemistry is shown in Eq. 121.[332]

$$\text{(Eq. 121)}$$

Fischer indole synthesis has taken advantage of zinc catalysis. Particularly, triethylene glycol with a catalytic quantity of zinc chloride has been described as an efficient reaction medium for the difficult Fischer synthesis of sensitive indoles.[333]

EXPERIMENTAL CONDITIONS

Both palladium(II) salts and palladium(0) complexes have been used in the construction of the indole ring. Commercial samples are normally used without further purification. PdCl$_2$ and Pd(OAc)$_2$ are the most commonly used palladium(II) salts, but the use of Pd(OCOCF$_3$)$_2$ has also been described. Very often palladium(II) salts (particularly PdCl$_2$, which has a low solubility in water and organic solvents) are used as complexes of the type PdX$_2$L$_2$ such as PdCl$_2$(PPh$_3$)$_2$, Pd(OAc)$_2$(PPh$_3$)$_2$, and PdCl$_2$(MeCN)$_2$. Complexes containing phosphine ligands are frequently formed in situ by combining palladium(II) salts with the phosphine ligands.

The commercially available Pd(PPh$_3$)$_4$ and Pd$_2$(dba)$_3$ are two of the most commonly used sources of palladium(0) species. Pd(PPh$_3$)$_4$ is unstable in air and light sensitive whereas Pd$_2$(dba)$_3$ is much easier to store and manipulate. Palladium on charcoal, or other supported palladium metal catalysts, are also employed as a source of palladium(0). As an alternative to the use of preformed palladium(0) complexes, palladium(0) species can be formed in situ by reduction of palladium(II) species by several reagents such as alkenes, terminal alkynes, carbon monoxide, alcohols, amines, formate anions, metal hydrides, butyllithium, or phosphines. Reactions involving palladium(0) catalysis are usually carried out in an inert atmosphere of argon or nitrogen.

The efficiency of palladium catalysts is dependent on the nature of the ligands and on the ratio of the ligand to palladium. For example, with the coordinatively saturated palladium(0) complex Pd(PPh$_3$)$_4$, the dissociation of two Ph$_3$P is necessary to generate the coordinatively unsaturated Pd(PPh$_3$)$_2$, which allows for the coordination of the reactants to palladium. Although a number of reactions have been carried out under phosphine-free conditions, phosphines are usually required to generate soluble palladium catalysts and to modulate the reactivity of palladium complexes. The recent development of several indole syntheses involving the oxidative addition of carbon–bromine or carbon–chlorine bonds to palladium employ biarylmonophosphines[134–136] because these bonds are usually reluctant to undergo oxidative addition with other commonly used ligands. Carbene ligands have also been employed.[86]

Palladium(II) salts reduced in situ to palladium(0) species or commercially available palladium(0) compounds (particularly Pd$_2$(dba)$_3$) are frequently used to prepare palladium–phosphine complexes in situ via a ligand exchange reaction. Such an exchange reaction has been carried out with a vast range of monodentate and bidentate phosphines and some carbene ligands and represents a convenient entry into the generation of "tailor-made" catalyst systems.

In addition to phosphine and carbene ligands, additives (mostly halide additives such as LiCl, LiBr, Bu$_4$NCl, or BuN$_4$Br), bases, and solvents play an important role in controlling the outcome of palladium-catalyzed reactions. Chloride anions stabilize palladium species and provide more efficient catalytic cycles.[31,334,335] Bromide anions control the vinylic substitution/conjugate addition-type ratio in the reaction of 2-alkynylanilides with α,β-enals and -enones.[103] In general, and apart from some important rationalizations, the specific role of all these factors, which may change from one type of reaction to another, is not always well understood. They combine to afford a toolbox of tunable reaction conditions that make

EXPERIMENTAL PROCEDURES

2-(3α-Acetoxyandrost-16-en-17-yl)-1H-indole [One-Flask Synthesis of a 2-Substituted Indole from 2-Ethynylaniline].[71] To a stirred solution of 3α-acetoxyandrost-16-en-17-yl triflate (0.230 g, 0.49 mmol) in DMF (0.5 mL) and Et$_2$NH (2 mL) were added 2-ethynylaniline (0.058 g, 0.49 mmol), Pd(PPh$_3$)$_4$ (0.011 g, 0.009 mmol), and CuI (0.004 g, 0.020 mmol). The reaction mixture was stirred for 6 h at rt under a nitrogen atmosphere, and then evaporated under reduced pressure. The residue was dissolved in CH$_2$Cl$_2$ (13 mL) and 0.5 N HCl (5 mL), and PdCl$_2$ (0.05 g, 0.028 mmol) and Bu$_4$NCl (0.015 g, 0.051 mmol) were added. The reaction mixture was stirred at rt for 48 h under nitrogen, then poured into a separatory funnel containing Et$_2$O and saturated, aqueous NaHCO$_3$ solution. The organic layer was separated and the aqueous layer was extracted twice with Et$_2$O. The combined organic layers were dried over Na$_2$SO$_4$ and evaporated under vacuum. The residue was purified by silica gel chromatography, eluting with 20% EtOAc/n-hexane to give 0.205 g (96%) of the title product: mp 119–121°; IR (KBr) 3400, 1740 cm^{-1}; ^{1}H NMR (CDCl$_3$) δ 8.16 (br s, 1H), 7.55 (d, J = 8.2 Hz, 1H), 7.31–7.01 (m, 3H), 6.53 (d, J = 1.6 Hz, 1H), 5.95 (br s, 1H), 5.03 (br s, 1H), 2.05 (s, 3H), 1.03 (s, 3H), 0.86 (s, 3H); ^{13}C NMR (CDCl$_3$) δ 170.8, 146.7, 136.0, 134.1, 129.0, 124.9, 122.1, 120.4, 119.8, 110.3, 99.9, 70.1; EIMS m/z (relative intensity): M$^+$ 431 (100), 372 (46).

N-Acetyl-2-isopropyl-6-carbomethoxyindole [Preparation of a 2-Substituted Indole from a 2-Alkynylacetanilide].[49] To a solution of N-acetyl-2-(3-methylbut-1-yn-1-yl)-5-carbomethoxyaniline (0.107 g, 0.413 mmol) in MeCN (4 mL) was added PdCl$_2$(MeCN)$_2$ (11 mg, 0.041 mmol) and the mixture was heated at 80° for 1.5 h. The solvent was removed in vacuo and the resulting oil was purified by column chromatography on silica gel (17% EtOAc/n-hexane) to yield 0.088 g (82%) of the title product as a white, crystalline solid: mp 67.5–68.5°; IR (CHCl$_3$) 1711, 1554, 1462, 1313, 1304, 1297, 1255, 1108 cm^{-1}; ^{1}H NMR (CDCl$_3$) δ 8.41 (d, J = 1.4 Hz, 1H), 7.89 (dd, J = 1.4, 8.1 Hz, 1H), 7.49 (d, J = 8.1 Hz, 1H), 6.50 (s, 1H), 3.92 (s, 3H), 3.72 (hept, 1H), 2.84 (s, 3H), 1.30 (d, J = 6.8 Hz, 6H). Anal. Calcd for C$_{15}$H$_{17}$NO$_3$: C, 69.48; H, 6.61. Found: C, 69.40; H, 6.61.

2-[(4-Ethylpiperazin-1-yl)methyl]indole [Synthesis of a 2-Substituted Indole through an Intramolecular Heterocyclization/Intermolecular Nucleophilic Attack on a π-Allylpalladium Intermediate].[34] A Carousel Tube Reactor (Radley Discovery Technology) equipped with a magnetic stirrer was charged with ethyl 3-(2-trifluoroacetamidophenyl)-1-propargyl carbonate (0.050 g, 0.159 mmol), N-ethylpiperazine (0.055 g, 0.477 mmol), and Pd(PPh$_3$)$_4$ (0.009 g, 0.00795 mmol) in 1.0 mL of anhydrous THF under argon. The mixture was warmed at 80° and stirred for 1.5 h. After cooling, the reaction mixture was concentrated under reduced pressure and the residue was purified by chromatography (Al$_2$O$_3$, 50 g; 30% EtOAc/n-hexane) to give 0.035 g (90%) of the title product as an oil: IR (neat) 3404, 2935, 2816, 1454 cm^{-1}; ^{1}H NMR (CDCl$_3$) δ 8.64 (br s, 1H), 7.56 (d, J = 8.3 Hz, 1H), 7.33 (d, J = 8.3 Hz, 1H), 7.16–7.07 (m, 2H), 6.37 (s, 1H), 3.67 (s, 2H), 2.54–2.41 (m, 10H), 1.09 (t, J = 8.3 Hz, 3H); ^{13}C NMR (CDCl$_3$) δ 136.2, 135.8, 128.4, 121.6, 120.2, 119.6, 110.7, 101.7, 55.9, 53.3, 52.8, 52.3, 12.0. Anal. Calcd for C$_{15}$H$_{21}$N$_3$: C, 74.03; H, 8.70; N, 17.27. Found: C, 74.01; H, 8.68; N, 17.25.

3-(4-Acetylphenyl)indole [Synthesis of a 2-Unsubstituted 3-Arylindole via the Aminopalladation/Reductive Elimination Pathway].[100] To a stirred solution of 2-ethynyltrifluoroacetanilide (0.260 g, 1.22 mmol) and 4-iodoacetophenone (0.200 g, 0.81 mmol) in DMSO (3.0 mL) was added Pd$_2$(dba)$_3$ (0.019 g, 0.020 mmol) and K$_2$CO$_3$ (0.168 g, 1.22 mmol) under argon. The reaction mixture was heated at 40° for 1.25 h. Ethyl acetate was added and the resulting solution was washed with a saturated aqueous NaCl solution, dried over

Na$_2$SO$_4$, and concentrated under reduced pressure. The residue was purified by chromatography (silica gel, 40 g; 30% EtOAc/n-hexane) to give 0.120 g (64%) of 3-(4-acetylphenyl)indole: mp 127–128°; IR 3345, 1663, 744 cm^{-1}; ^{1}H NMR δ 8.73 (br s, 1H), 8.05–7.96 (m, 3H), 7.75 (d, $J = 8.2$ Hz, 2H), 7.43–7.40 (m, 2H), 7.3–7.23 (m, 2H) 2.63 (s, 3H); ^{13}C NMR δ 198.2, 141.0, 136.8, 134.3, 129.1, 126.8, 125.3, 123.2, 122.7, 120.8, 119.7, 116.9, 111.8, 26.6; MS m/z (relative intensity): M$^+$ 235 (88), 220 (100), 192 (44), 165 (30). Anal. Calcd for C$_{16}$H$_{13}$NO: C, 81.67; H, 5.57; N, 5.96. Found: C, 81.57; H, 5.59; N, 5.95.

2-Phenyl-3-(phenylethynyl)indole [Synthesis of a 2,3-Disubstituted Indole from a 2-Alkynyltrifluoroacetanilide and a 1-Bromoalkyne].[108] In a Carousel Tube Reactor (Radley Discovery Technology), a solution of 2-phenylethynyltrifluoroacetanilide (0.100 g, 0.346 mmol) in 2 mL of MeCN was treated with 1-bromophenylacetylene (0.075 g, 0.415 mmol), Pd(PPh$_3$)$_4$ (0.020 g, 0.017 mmol), and Cs$_2$CO$_3$ (0.169 g, 0.519 mmol). The reaction mixture was stirred at 60° for 6 h. After cooling, the reaction mixture was diluted with EtOAc, washed with water, dried over Na$_2$SO$_4$, and concentrated under reduced pressure. The residue was purified by chromatography (silica gel, 35 g; 10% EtOAc/n-hexane) to give 0.077 g (76%) of 2-phenyl-3-(phenylethynyl)indole: mp 81–83°; IR (KBr) 3407, 3057, 2201 cm^{-1}; ^{1}H NMR (CDCl$_3$) δ 8.41 (s, 1H), 8.08 (d, $J = 7.4$ Hz, 2H), 7.88–7.84 (m, 1H), 7.63–7.51 (m, 4H), 7.43–7.26 (m, 7H); ^{13}C NMR (CDCl$_3$) δ 139.5, 135.4, 131.7, 131.3, 130.4, 129.0, 128.5, 128.3, 127.6, 126.6, 124.4, 123.6, 121.0, 120.2, 111.0, 96.1, 93.6, 84.1. Anal. Calcd for C$_{22}$H$_{15}$N: C, 90.07; H, 5.15; N, 4.77. Found: C, 89.91; H, 5.17; N, 4.74.

2-(Cyclooct-1-enyl)-3-(4-methoxybenzoyl)indole [Synthesis of a 2-Substituted-3-Carbonylated Indole via a Carbonylative Three-Component Cyclization].[124] To a solution of 2-(cyclooct-1-enyl)ethynyltrifluoroacetanilide (0.180 g, 0.56 mmol) in MeCN (6 mL) were added 4-iodoanisole (0.157 g, 0.67 mmol), K$_2$CO$_3$ (0.387 g, 2.80 mmol), and Pd(PPh$_3$)$_4$ (0.032 g, 0.028 mmol). The flask was purged with carbon monoxide for a few seconds and connected to a balloon of carbon monoxide. The reaction mixture was stirred at 45° overnight and poured into a separatory funnel containing 0.1 N HCl and EtOAc. The organic layer was separated and the aqueous layer was extracted twice with EtOAc. The combined organic layers were dried (Na_2SO_4) and evaporated under vacuum. The residue was purified by silica gel chromatography, eluting with 20% EtOAc/n-hexane to give 0.155 g (77%) of the title product: mp 72–76°; IR 3250, 1590 cm^{-1}; ^{1}H NMR ($CDCl_3$) δ 8.45 (br s, 1H), 7.82 (AA' part of an AA'BB' system, $J = 8.9$ Hz, 2H), 7.71–7.64 (m, 1H), 7.40–7.34 (m, 2H), 7.25–7.08 (m, 2H), 6.88 (BB' part of an AA'BB' system, $J = 8.9$ Hz, 2H), 6.06 (t, $J = 8.2$ Hz, 1H), 3.87 (s, 3H), 2.38–2.27 (m, 2H), 2.21–2.08 (m, 2H), 1.46 (br s, 8H); ^{13}C NMR ($CDCl_3$) δ 192.5, 162.7, 145.9, 134.8, 134.4, 133.6, 133.2, 131.9, 128.5, 122.6, 121.4, 120.9, 113.2, 112.6, 111.0, 54.4; MS m/z (relative intensity): M^+ 359 (51), 135 (60). Anal. Calcd for $C_{24}H_{25}O_2N$: C, 80.19; H, 7.01; N, 3.90. Found: C, 80.77; H, 7.12; N, 4.56.

2,3-Diphenylindole [Synthesis of a 2,3-Disubstituted Indole via a One-Pot Tandem Cross-Coupling/Aminopalladation/Reductive Elimination Process].[139] A 10 mL 3-neck flask equipped with a magnetic stirring bar, a thermocouple, and an argon inlet was charged with 2-iodotrifluoroacetanilide (0.5 g, 1.54 mmol), Pd(OAc)$_2$ (17.3 mg, 0.08 mmol), Ph$_3$P (80.9 mg, 0.154 mmol), and K$_2$CO$_3$ (0.851 g, 6.16 mmol), followed by addition of 5 mL of anhydrous DMF. Phenylacetylene (0.189 g, 1.85 mmol) and bromobenzene (0.290 g, 1.85 mmol) were added to the reaction mixture with stirring at rt. The reaction mixture was heated at 60° for 0.5 h. The mixture was quenched with water, and the aqueous solution was extracted three times with EtOAc. The organic solution was washed with saturated aqueous NaCl solution, and dried over Na$_2$SO$_4$. The product was purified by column chromatography to give 0.453 g (91%) of 2,3-diphenylindole as an off-white solid: mp 108–110°; ^{1}H NMR (400 MHz, DMSO-d_6) δ 11.55 (s, 1H), 7.46–6.90 (m, 14H); ^{13}C NMR (400 MHz, CDCl$_3$) δ 135.9, 135.1, 134.1, 132.7, 130.2, 128.8, 128.7, 128.5, 128.2, 127.7, 126.2, 122.7, 120.4, 119.7, 115.1, 110.9; LC-MSD (API-ES, positive) m/z: (M + H$^+$) 270.

(2R,5S)-3,6-Diethoxy-2-isopropyl-5-[2-(trimethylsilyl)-3-indolyl]methyl-2,5-dihydropyrazine [Synthesis of a 2,3-Disubstituted Indole via

Heteroannulation of an Internal Alkyne with 2-Iodoaniline].[112] In a 100 mL round-bottom flask equipped with a stirring bar were placed 2-iodoaniline (200 mg, 0.91 mmol), compound **36** (322 mg, 1 mmol), Pd(OAc)$_2$ (8 mg, 0.036 mmol), LiCl (39 mg, 0.91 mmol), Na$_2$CO$_3$ (193 mg, 1.8 mmol), and DMF (12 mL). The reaction mixture was degassed and then heated at 100° under argon until the starting iodoaniline was no longer detected on analysis by TLC (30 h). The DMF was removed under reduced pressure, and the residue was taken up in CH$_2$Cl$_2$ (50 mL). The suspension that resulted was passed through a Celite pad to remove the insoluble solids. The solution was concentrated under vacuum and the product was purified by silica gel column chromatography (2% EtOAc/n-hexane) to afford 301 mg (81%) of the title product as an oil: IR (NaCl) 3415, 2952, 1687 cm^{-1}; ^{1}H NMR (300 MHz, CDCl$_3$) δ 7.93 (br s, 1H), 7.72 (d, $J = 7.9$ Hz, 1H), 7.32 (d, $J = 8.1$ Hz, 1H), 7.13 (t, $J = 8.0$ Hz, 1H), 7.05 (t, $J = 7.9$ Hz, 1H), 4.14 (m, 5H), 3.87 (t, 1H), 3.54 (dd, $J = 3.6, 14.2$ Hz, 1H), 2.88 (dd, $J = 9.6, 14.2$ Hz, 1H), 2.27 (m, 1H), 1.30 (t, $J = 7.1$ Hz, 3H), 1.19 (t, $J = 7.1$ Hz, 3H), 1.03 (d, $J = 6.8$ Hz, 3H), 0.67 (d, $J = 6.8$ Hz, 3H), 0.41 (s, 9H); ^{13}C NMR (75.5 MHz, CDCl$_3$) δ 164.2, 163.3, 138.7, 134.4, 130.1, 123.4, 122.6, 121.0, 119.1, 111.1, 61.2, 61.0, 59.1, 32.4, 19.7, 17.1, 14.9, 14.8, 14.6; EIMS m/z (relative intensity): M$^+$ 413 (4), 202 (100), 186 (18), 169 (36), 160 (11); exact mass calcd for C$_{23}$H$_{35}$N$_3$O$_2$Si, 413.2499; found, 413.2473.

N-Tosylindole [Synthesis of a 2,3-Unsubstituted Indole via Cylization of a 2-Vinylanilide].[145] 2-Vinyl-N-tosylaniline (273 mg, 1.00 mmol) was dissolved in 5 mL of DMF. The system was flushed with argon, and PdCl$_2$(MeCN)$_2$ (26 mg, 0.10 mmol), benzoquinone (216 mg, 2.0 mmol), and LiCl (445 mg, 10 mmol) were added. The mixture was heated at 100–110° for 28 h, cooled, diluted with 25 mL each of Et$_2$O and water, and filtered through Florisil. The Florisil was washed with 100 mL of Et$_2$O and the combined filtrates were washed with 50 mL each of water and saturated aqueous NaCl. After drying (Na$_2$SO$_4$), the solution was concentrated in vacuo and the residue was purified by silica gel chromatography to yield N-tosylindole as a colorless solid that rapidly decomposed to a red oil upon exposure to air: IR (CDCl$_3$) 3010, 2960, 1585, 1435, 1370 cm^{-1}; ^{1}H NMR (CDCl$_3$) δ 7.95 (br d, $J = 9$ Hz, 1H), 7.78 (d, $J = 9$ Hz, 2H), 7.58 (d, $J = 4$ Hz, 1H), 7.50 (d, $J = 2$ Hz, 1H), 7.29 (d, $J = 9$ Hz, 1H), 7.27 (d, $J = 9$ Hz, 2H), 6.58 (d, $J = 4$ Hz, 1H), 2.35 (s, 3H).

***N*-(4-Bromobenzyl)-2-ethyl-3-(*tert*-butyldimethylsilyloxy)-5-methoxyindole [Synthesis of a 2,3-Substituted Indole via Cyclization of a 2-Allylaniline].**[157] To a degassed suspension of K_2CO_3 (414 mg, 3 mmol), benzoquinone (162 mg, 1.5 mmol), and $PdCl_2(MeCN)_2$ (52 mg, 0.2 mmol) in THF (10 mL) was added a degassed solution of precursor **37** (462 mg, 10 mmol) in THF (5 mL) under nitrogen. The mixture was stirred at rt for 22 h. The THF was evaporated under vacuum and the residue was dissolved in Et_2O and purified by silica gel chromatography to give 389 mg (84%) of the title product: ^{1}H NMR (400 MHz, CD_3COCD_3) δ 7.44 (dd, J = 6.6, 1.8 Hz, 2H), 7.15 (d, J = 8.8 Hz, 1H), 6.93 (d, J = 2.4 Hz, 1H), 6.87 (d, 1H), 6.68 (dd, J = 8.8, 2.5 Hz, 1H), 5.30 (s, 2H, C$\underline{H}_2$Ph), 3.78 (s, 3H, OMe), 2.27 (s, 3H), 1.09 (s, 9H), 0.18 (s, 6H); ^{13}C NMR (100.6 MHz, CD_3COCD_3) δ 158.7, 143.5, 136.5, 135.7, 134.1, 133.0, 127.8, 127.0, 125.2, 115.7, 114.9, 104.0, 59.8, 50.4, 30.4, 22.9, 13.5, 5.8, 0.2. Anal. Calcd for $C_{23}H_{30}BrNO_2Si$: C, 59.99; H, 6.57; N, 3.04. Found: C, 59.89; H, 6.73; N, 2.99.

Indole [Cyclization of 2-Nitrostyrene].[156] To an oven-dried, threaded ACE glass pressure tube was added 2-nitrostyrene (298 mg, 2.00 mmol), $Pd(OAc)_2$ (26 mg, 0.12 mmol), Ph_3P (124 mg, 0.48 mmol), and 4 mL of MeCN. The tube was fitted with a pressure head, the solution was saturated with CO (four cycles to 4 atm of CO), and the reaction mixture was heated to 70° (oil bath temperature) under CO (4 atm) until all starting material was consumed (15 h) as judged by TLC. The reaction mixture was diluted with 10% aqueous HCl (10 mL) and extracted with Et_2O (3 × 10 mL). The combined organic phases were washed with 10% aqueous HCl (10 mL) and dried ($MgSO_4$), and the solvent was removed to give the crude product which was purified by chromatography (10% EtOAc/*n*-hexanes) to give 203 mg (87%) of indole as white crystals. The spectroscopic data matched those found in the *Aldrich Library of Spectra*: FT-IR spectra **2**, 653 A; ^{1}H and ^{13}C NMR spectra **3**, 121 A.

(L)-*N*,*N*-Di-*tert*-butoxycarbonyltryptophan Methyl Ester [Synthesis of a 3-Substituted Indole via Cyclization of an in Situ Generated 2-Haloanilinoenamine].[336] A solution of 2-iodoaniline (73.0 mg, 0.33 mmol),

(S)-methyl 2-(bis(*tert*-butoxycarbonyl)amino)-5-oxopentanoate (104.0 mg, 0.30 mmol), DABCO (101.0 mg, 0.9 mmol), and Pd(OAc)$_2$ (3.4 mg, 0.015 mmol) in anhydrous DMF (1.5 mL) was degassed. The reaction mixture was heated to 85° until the reaction was complete (usually 8–12 h). The reaction mixture was cooled to rt and diluted with H$_2$O. The aqueous phase was extracted with EtOAc and the combined organic phase was washed with saturated aqueous NaCl solution, dried (Na$_2$SO$_4$), and evaporated to dryness under reduced pressure. Purification of the crude product by silica gel chromatography (20% EtOAc/heptane) provided 101 mg (81%) of the title product as a yellow oil: $[\alpha]_D^{23}$ −60.0 (*c* 1.0, CHCl$_3$); IR (CHCl$_3$) 3348, 2980, 2359, 1782, 1741, 1457, 1369, 1273, 1140, 1092, 852 cm^{-1}; ^{1}H NMR (300 MHz, CDCl$_3$) δ 8.45 (br s, 1H), 7.58 (d, $J = 7.7$ Hz, 1H), 7.34 (d, $J = 7.9$ Hz, 1H), 7.15 (dt, $J = 1.2, 7.7$ Hz, 1H), 7.09 (dt, $J = 1.2$, 7.9 Hz, 1H), 6.98 (d, $J = 2.1$ Hz, 1H), 5.20 (dd, $J = 4.7, 10.3$ Hz, 1H), 3.77 (s, 3H), 3.62 (dd, $J = 4.7, 14.9$ Hz, 1H), 3.40 (dd, $J = 10.3, 14.9$ Hz, 1H), 1.28 (s, 18H); ^{13}C NMR (75 MHz, CDCl$_3$) δ 171.1, 151.5, 136.3, 127.5, 123.2, 121.7, 119.2, 118.5, 111.2, 82.8, 58.9, 52.1, 27.6 (6C), 25.8; MS (ESI) *m/z*: [M + Na] 441; HRMS (ESI) *m/z*: [M + Na] calcd for C$_{22}$H$_{30}$N$_2$O$_6$Na, 441.2002; found, 441.1975.

2,3-Diphenylindole [Synthesis of a 2,3-Disubstituted Indole through a One-Pot Hydroamination/Cyclization Process].[184] 2-Chloroaniline (610 mg, 4.76 mmol) and diphenylacetylene (1.02 g, 5.70 mmol) were added to a solution of TiCl$_4$ (0.05 mL, 0.47 mmol) and *t*-BuNH$_2$ (0.30 mL, 2.86 mmol) in toluene (5 mL) and the resulting mixture was stirred for 20 h at 105°. The solvent was partially removed and *t*-BuOK (1.60 g, 14.0 mmol), HIPrCl (202 mg, 0.48 mmol), and Pd(OAc)$_2$ (106 mg, 0.48 mmol) were added. The mixture was stirred at 105° for 24 h. CH$_2$Cl$_2$ (75 mL) and aqueous 2 M HCl (50 mL) were added to the cold suspension. The separated aqueous phase was washed with CH$_2$Cl$_2$ (2 × 75 mL) and the combined organic phases were washed with saturated aqueous NaHCO$_3$ (50 mL) and saturated aqueous NaCl (50 mL). Drying with MgSO$_4$ and purification by silica gel chromatography (5% → 10% → 20% Et$_2$O/*n*-pentane) yielded 974 mg (76%) of 2,3-diphenylindole as an off-white solid: ^{1}H NMR (CDCl$_3$, 300 MHz) δ 8.22 (br s, 1H), 7.67 (d, $J = 7.8$ Hz, 1H), 7.45–7.12 (m, 13H); ^{13}C NMR (CDCl$_3$, 75 MHz) δ 135.9, 135.1, 134.1, 132.7, 130.2, 128.8, 128.7, 128.5, 128.2, 127.7, 126.2, 122.7, 120.4, 119.7, 115.2, 110.9; EIMS *m/z* (relative intensity): M$^+$ 269 (100), 254 (4), 239 (5), 165 (11), 134 (6), 127 (4); HRMS (EI) m/z: calcd for $C_{20}H_{15}N$, 269.1204; found, 269.1198.

N-(4-Ethoxycarbonylphenyl)-2-ethoxycarbonyl-5-methoxyindole [Synthesis of a 2-Substituted Indole Based on an Intramolecular N-Arylation Process].[194] To a solution of precursor **38** (148 mg, 0.3 mmol) in DMF (5 ml) under nitrogen at rt was added KOAc (95 mg, 1 mmol) and PdCl$_2$(dppf) (14 mg, 6 mol %). The mixture was heated to 90° for 30 min, and then partitioned between EtOAc (50 mL) and water (50 mL). The aqueous layer was separated and the organic phase was washed with water (4 × 25 mL), saturated aqueous NaCl (30 mL), dried (MgSO$_4$), filtered, and the solvent was removed in vacuo to give a brown oil. Column chromatography (20% EtOAc/n-hexane) gave 111 mg (94%) of the title product as a clear oil: IR (film) 2982, 1710, 1610 cm^{-1}; ^{1}H NMR (300 MHz, CDCl$_3$) δ 8.20 (d, $J = 8.2$ Hz, 2H), 7.42 (s, 1H), 7.39 (d, $J = 8.2$ Hz, 2H), 7.11 (d, $J = 2.1$ Hz, 1H), 7.02 (d, $J = 9.1$ Hz, 1H), 6.95 (dd, $J = 2.2, 9.1$ Hz, 1H), 4.43 (q, $J = 7.1$ Hz, 2H), 4.23 (q, $J = 7.1$ Hz, 2H), 3.86 (s, 3H), 1.43 (t, $J = 7.2$ Hz, 3H), 1.24 (t, $J = 7.2$ Hz, 3H); ^{13}C NMR (100 MHz, CDCl$_3$) δ 166.3, 161.5, 155.7, 143.1, 136.1, 130.8, 130.5, 130.3, 129.6, 128.3, 127.1, 117.5, 112.5, 112.2, 103.0, 61.6, 61.0, 56.1, 14.8, 14.5; MS (ESI$^+$, 70 V) m/z: [MH$^+$] 368.

Methyl 2-(2-Methoxyquinolin-3-yl)indole-5-carboxylate [Synthesis of a 2-Substituted Indole through a Tandem Carbon–Nitrogen/Suzuki–Miyaura Coupling].[201] A 5 mL round-bottomed flask was charged with **39** (0.1675 g, 0.5 mmol), **40** (0.1523 g, 0.75 mmol), Pd(OAc)$_2$ (3.4 mg, 0.015 mmol), SPhos (12.3 mg, 0.03 mmol), and K$_3$PO$_4$·H$_2$O (0.58 g, 2.5 mmol). The solid mixture was purged with argon for 10 min followed by addition of toluene (2.5 mL). The resulting mixture was stirred at rt for 2 min, then heated at 100° for 1.5 h. The mixture was diluted with EtOAc (10 mL) and H$_2$O, and the organic phase was separated and dried over Na$_2$SO$_4$. The crude material was purified by chromatography with 20% EtOAc/n-hexane to afford 0.143 g (86%) of the title product as a white solid: ^{1}H NMR (300 MHz, DMSO-d_6) δ 11.89 (s, 1H), 8.74 (s, 1H), 8.31 (s, 1H), 7.94 (d, J = 7.2 Hz, 1H), 7.84–7.77 (m, 2H), 7.70 (dd, J = 7.0, 1.3 Hz, 1H), 7.56 (d, J = 8.5 Hz, 1H), 7.50 (dd, J = 6.9, 1.2 Hz, 1H), 7.32 (d, J = 1.3 Hz, 1H), 4.18 (s, 3H), 3.86 (s, 3H); ^{13}C NMR (100 MHz, DMSO-d_6) δ 167.2, 158.3, 144.7, 139.4, 135.5, 134.2, 130.0, 127.8, 127.7, 126.4, 124.9, 124.8, 123.0, 122.9, 120.9, 116.5, 111.4, 104.7, 53.8, 51.7; HRMS (EI) m/z: [M]$^+$ calcd for C$_{20}$H$_{16}$N$_2$O$_3$, 332.1161; found, 332.1161.

2-[1-[4-(Trifluoromethyl)benzyl]indol-3-yl]acetamide [A Solid-Phase Synthesis of a 3-Substituted Indole via Cyclization of a 2-Iodo-N-allylaniline].[232] Rink amide resin (7.5 g, 0.48 mmol/g, 3.6 mmol) was deprotected with 20% piperidine in DMF (100 mL) at rt for 1.5 h and then filtered and washed with DMF, MeOH, and CH$_2$Cl$_2$. The deprotected resin was suspended in DMF (36 mL) and treated with 1,3-diisopropylcarbodiimide (2.73 g, 21.6 mmol), followed by 4-bromocrotonic acid (3.56 g, 21.6 mmol). The mixture was stirred at rt for 30 min, and then filtered, washed with CH$_2$Cl$_2$ and DMF. The resulting resin was retreated with DMF (36 mL), 1,3-diisopropylcarbodiimide (21.6 mmol), and 4-bromocrotonic acid (21.6 mmol) at rt for 30 min and then washed with DMF, MeOH, CH$_2$Cl$_2$, and Et$_2$O, and dried in vacuo to give 7.41 g of resin **41** with a loading level of 0.32 mmol/g, which was determined by cleaving an aliquot with 30% TFA in CH$_2$Cl$_2$ at rt for 80 min. Resin **41** (1.2 g, 0.38 mmol) was suspended in DMF (10 mL) and treated with (i-Pr)$_2$NEt (387 mg, 3.0 mmol) followed by 2-iodoaniline (420 mg, 1.9 mmol). The reaction mixture was stirred at 80° for 18 h and then filtered, washed with CH$_2$Cl$_2$, MeOH, and CH$_2$Cl$_2$, and dried in vacuo to give 1.25 g of resin **42**. A mixture of resin **42** (230 mg, 0.070 mmol), (*i*-Pr)$_2$NEt (90 mg, 0.70 mmol), and 4-(trifluoromethyl)benzyl bromide (167 mg, 0.70 mmol) in DMF (2.5 mL) was stirred at 80° for 22 h and then filtered, washed sequentially with MeOH and CH$_2$Cl$_2$, and dried in vacuo to give resin **43**. The resulting resin was then suspended in DMF/H$_2$O (9:1, 4 mL) and treated with Bu$_4$NCl (29 mg, 0.11 mmol), Et$_3$N (21 mg, 0.21 mmol), and PdCl$_2$(PPh$_3$)$_4$ (4.9 mg, 0.007 mmol). The suspension was stirred at 80° for 8 h, at which time TLC indicated that the reaction was complete. The dark-brown reaction mixture was filtered and the solid was washed sequentially with CH$_2$Cl$_2$, MeOH, and CH$_2$Cl$_2$, and then dried in vacuo. The resulting resin was cleaved with 30% TFA in CH$_2$Cl$_2$ (8 mL) at rt for 1.5 h. The crude cleaved product obtained was dissolved in EtOAc (25 mL), and the solution was washed with H$_2$O (5 mL, to remove contaminated Et$_3$N–TFA salt), and saturated aqueous NaCl (5 mL), dried over anhydrous Na$_2$SO$_4$, and concentrated under reduced pressure. The resulting product showed 85% purity by reversed-phase HPLC [2 mL/ min, 30% H$_2$O/MeCN (0.2% TFA), linear gradient to 5:95 in 30 min; Rf = 18.5 min]. After purification by preparative TLC using 5% MeOH/EtOAc as the eluent, 17.2 mg (74% yield for four steps, based on the loading level of resin **41**) of the title product was obtained as a colorless solid: ^{1}H NMR (CD$_3$OD) δ 7.61–7.56 (m, 3H), 7.31–7.25 (m, 4H), 7.13 (t, J = 7.2 Hz, 1H), 7.06 (t, J = 7.0 Hz, 1H), 5.46 (s, 2H), 3.67 (s, 2H); ^{13}C NMR (CD$_3$OD) δ 177.7, 144.4, 138.2, 130.8 (q, $^2J_{CF}$ = 32.3 Hz), 129.6, 128.9, 128.6, 126.7, 125.8 (q, J_{CF} = 271.9 Hz), 123.2, 120.7, 120.1, 111.0, 110.3, 50.3, 33.4; MS m/z: [MH$^+$] 333; HRMS-FAB m/z: [M + H]$^+$ calcd for C$_{18}$H$_{15}$F$_3$N$_2$O, 333.1215; found, 333.1165. Anal. Calcd for C$_{18}$H$_{15}$F$_3$N$_2$O•1.3 H$_2$O: C, 60.77; H, 4.99; N, 7.87; F, 16.02. Found: C, 60.45; H, 4.22; N, 7.79; F, 16.57.

Methyl 2-Indolecarboxylate [A Solid-Phase Synthesis of a 2-Substituted Indole via Tandem Heck Reaction/N-Arylation].[234] To a mixture of solid-supported N-acetyl dehydroalanine (300 mg, 0.285 mmol), 1,2-dibromobenzene (0.051 mL, 0.428 mmol), Pd$_2$(dba)$_3$•CHCl$_3$ (39 mg, 0.043 mmol), and (*c*-C$_6$H$_{11}$)$_2$NMe (0.18 mL, 0.855 mmol) in toluene (3 mL) was added a 0.5 M toluene solution of (*t*-Bu)$_3$P (0.34 mL, 0.17 mmol) and the mixture was then heated at 100° for 24 h. The resin was collected by filtration and washed with DMF (three times), DMF/H$_2$O 1:1 (three times), DMF (three times), THF (three times), and MeOH (three times), and the resin was dried under reduced pressure at 40°. A mixture of the above resin and NaOMe (15 mg, 0.285 mmol) in THF (3 mL) and MeOH (1.5 mL) was agitated at rt for 16 h. The resin was separated by filtration and washed with EtOAc; the filtrate was washed with saturated aqueous NH$_4$Cl, H$_2$O, and saturated aqueous NaCl, dried over Na$_2$SO$_4$, and evaporated to afford the crude product which was purified by silica gel chromatography using 20% EtOAc/*n*-hexane to afford 39 mg (78%) of methyl 2-indolecarboxylate as a colorless solid: mp 150–151° (EtOAc/*n*-hexane); IR 3330, 1696, 1684 cm^{-1}; ^{1}H NMR (CDCl$_3$) δ 8.89 (br s, 1H), 7.70 (d, $J = 8.0$ Hz, 1H), 7.45–7.14 (m, 4H), 3.95 (s, 3H); MS *m/z*: M$^+$ 175; HRMS calcd for C$_{10}$H$_9$HNO$_2$, 175.0633; found, 175.0609.

TABULAR SURVEY

The literature has been surveyed up to the end of 2010. No attempts have been made to cover the patent literature. In general, Tables 1–15 are organized according to the sequence used in the "Scope and Limitations" section. Failed reactions have not been included in the tables. Entries in the tables are ordered by increasing carbon count of the substrates, including protecting groups. The carbon count of Tables 13–15 (solid-phase syntheses) applies only to polymer-bound benzenoid fragments, including the functional groups involved in the linkage to the solid support that remain in the indole product. Yields given for solid-phase syntheses refer to the entire synthetic process; conditions are given for the indole formation step and for the reactions that follow the indole formation step leading to the isolated products. When the numbering is different for an R group in the starting material and the product, the numbering is based on the product.

The following abbreviations are used in the tables:

addn	addition
BINAP	2,2'-bis(diphenylphosphino)-1,1'-binaphthyl
bmim	1-butyl-3-methylimidazolium
CPC	cetylpyridinium chloride
DavePhos	2-(2'-*N*,*N*-dimethylaminobiphenyl)dicyclohexylphosphine
dba	dibenzylideneacetone
DIC	*N*,*N*'-diisopropylcarbodiimide
dipf	1,1'-bis(di-*iso*-propylphosphino)ferrocene
dmam-dtbpf	2-(dimethylaminomethyl)-1-(di-*tert*-butylphosphanyl)ferrocene
dmpe	1,2-bis(dimethylphosphino)ethane
DPEPhos	bis[(2-diphenylphosphino)phenyl]ether
dppb	1,4-bis(diphenylphosphino)butane
dppe	1,2-bis(diphenylphosphino)ethane
dppf	1,1'-bis(diphenylphoshino)ferrocene
dppm	bis(diphenylphosphino)methane
dppp	1,3-bis(diphenylphosphino)propane
dtbpf	1,1'-bis(di-*tert*-butylphosphino)ferrocene
HIPrCl	1,3-bis(2,6-diisopropylphenyl)imidazolium chloride

JohnPhos	2-(biphenyl)di-*tert*-butylphosphine
MW	microwave irradiation
NfO	nonafluorobutanesulfonate
NIS	*N*-iodosuccinimide
Np	naphthyl
phen	1,10-phenanthroline
PhXPhos	2-(2',4',6'-triisopropylbiphenyl)diphenylphosphine
PMP	1,2,2,6,6-pentamethylpiperidine
PS-PEO	polystyrene–polyethylene oxide copolymer
SBA-15	silica mesophases
$scCO_2$	supercritical carbon dioxide
SPhos	2-(2',6'-dimethoxybiphenyl)dicyclohexylphosphine
TES	triethylsilyl
TMG	1,1,3,3-tetramethylguanidine
tmphen	3,4,7,8-tetramethyl-1,10-phenanthroline
tol	tolyl, methylphenyl
TPPTS	triphenylphosphine-3,3',3"-trisulfonate sodium salt
ttmpp	tris(2,4,6-trimethoxyphenyl)phosphine
XantPhos	9,9-dimethyl-4,5-bis(diphenylphosphino)xanthene
XPhos	2-(2',4',6'-triisopropylbiphenyl)dicyclohexylphosphine
))))	ultrasound irradiation

TABLE 1A. 2-SUBSTITUTED INDOLES FROM 2-HALOANILINES AND ALKYNES

2-Haloaniline	Alkyne	Conditions	Product(s) and Yield(s) (%)	Refs.
C_6 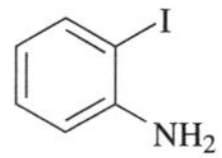 2-iodoaniline	 ≡—R	[Pd(NH$_3$)$_4$]/NaY zeolite, Et$_3$N, DMF/H$_2$O, 80°	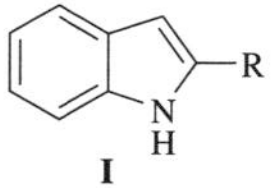 2-R-indole (**I**) R / Time (d) / Yield CMe$_2$OH / 1 / (64) Bu / 1 / (70) (CH$_2$)$_3$CO$_2$Me / 1 / (51) Ph / 0.4 / (91)[a]	78
	 ≡—R	[PdI]/SBA-15, Et$_3$N, DMF/H$_2$O (4:1), 80°	**I** R / Time (d) / Yield C(Me)$_2$OH / 1 / (93)[a] Bu / 1 / (62) (CH$_2$)$_3$CO$_2$Me / 1 / (89) Ph / 0.1 / (72)	78
	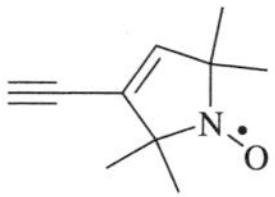 ≡— (2,2,5,5-tetramethyl-pyrrolinyl-N-oxide)	Pd(OAc)$_2$, PPh$_3$, K$_2$CO$_3$, Bu$_4$NCl, DMF, 100°, 3 h	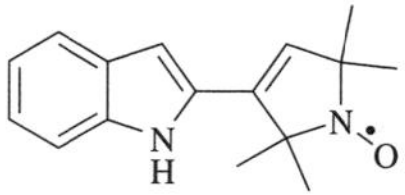 2-(pyrrolinyl-N-oxide)indole (50)	91
	 ≡—Ph	Pd/C, CuI, Et$_3$N, DMF/H$_2$O (1:1), 120°, 6 h	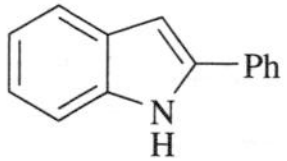 2-Ph-indole (72)	93
	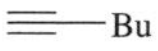 ≡—Bu	Pd(OAc)$_2$, TPPTS, Et$_3$N, MeCN/H$_2$O, 65°, 72 h	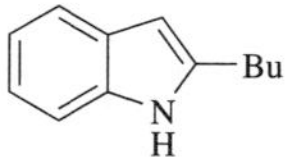 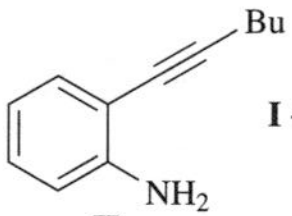 2-Bu-indole (**I**) + ortho-alkynyl aniline (**II**) **I** + **II** (75), **I:II** = 75:25	92
	 ≡—R	Pd(OAc)$_2$, Et$_3$N, DMF/H$_2$O, 80°, 2 d	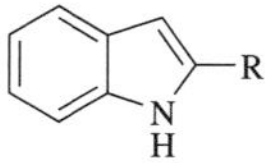 2-R-indole R / Yield CMe$_2$OH / (78)[a] (CH$_2$)$_3$CO$_2$Me / (80)[a] Ph / (75)[a]	78

Substrate	Alkyne	Conditions	Product	Results	Refs.
2-X-aniline (X=Br/I)	≡—R	PdCl₂(PPh₃)₂, CuI, Et₃N, DMF, 24 h	2-R-indole	X=Br, R=Ph, rt (78); X=I, R=n-C₅H₁₁, reflux (68); X=I, R=Ph, rt (72)	79
4-R¹-2-iodoaniline	≡—N(R²)Bn	Pd(OAc)₂, PPh₃, Bu₄NOAc, DMF, 60°	5-R¹-2-N(R²)Bn-indole	R¹=H, R²=Boc (40); R¹=H, R²=Ts (64); R¹=O₂N, R²=Ts (26)	337
2-Br-6-NO₂-4-X-aniline	≡—R	1. PdCl₂(PPh₃)₂, CuI, Et₂NH, DMF, 70°; 2. NaOH, 140°, 2–4 h	7-NH₂-indole (2-R, 4-NO₂)	R=Bu (40); R=n-C₅H₁₁ (50); R=c-C₆H₁₁ (36); R=Ph (46); R=3-ClC₆H₄ (44); R=4-MeC₆H₄ (43)	338
2-X-aniline (R¹,R²,R³ substituted)	≡—R⁴	1. PdCl₂(PPh₃)₂, CuI, Et₂NH, DMA, rt; 2. NaOH, 140°, 2–4 h	2-R⁴-indole (R¹,R²,R³ on ring)	see sub-table below	338

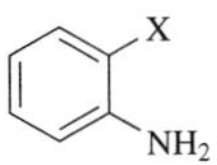

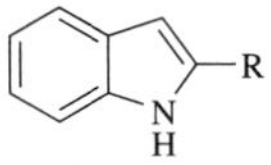

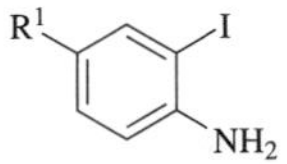

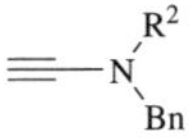

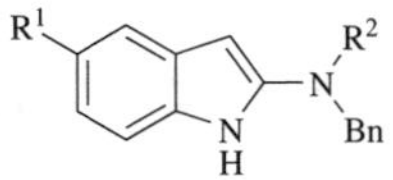

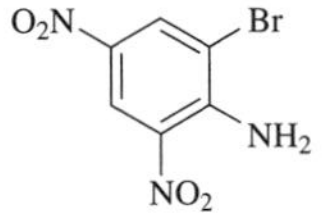

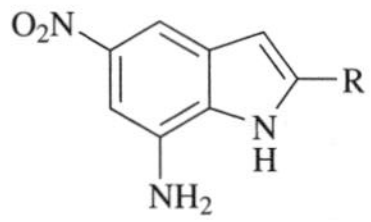

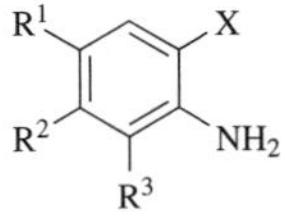

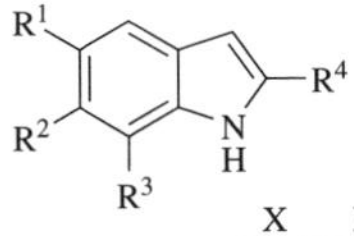

X	R¹	R²	R³	R⁴	(Yield)
I	O₂N	H	H	Bu	(89)[b]
I	Cl	H	O₂N	Bu	(55)[b]
I	O₂N	H	H	n-C₅H₁₁	(81)
I	O₂N	H	H	Ph	(84)
I	O₂N	H	H	1-cyclohexenyl	(76)
I	Cl	H	O₂N	3-ClC₆H₄	(76)[b]
Br	H	O₂N	H	n-C₅H₁₁	(47)
Br	H	O₂N	H	Ph	(60)
I	Me	H	H	n-C₅H₁₁	(52)[b]
I	Me	H	H	Ph	(69)

TABLE 1A. 2-SUBSTITUTED INDOLES FROM 2-HALOANILINES AND ALKYNES (*Continued*)

2-Haloaniline	Alkyne	Conditions	Product(s) and Yield(s) (%)			Refs.
C$_{6-8}$						
R^1–C$_6$H$_3$(I)NH$_2$	≡—TMS	1. PdCl$_2$dpp—Si, CuI, *i*-Pr$_2$NH, MeCN, 60°, 12 h 2. Bu$_4$NF on silica, 90°, 2 h 3. R^{2}X, 60°, time 1 4. HCl (1 N), PdCl$_2$, reflux, time 2	R^1–indole–R^2 (N-H)			339

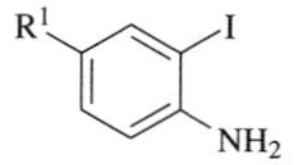

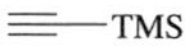

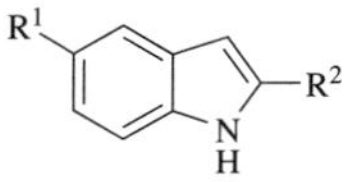

R^1	R^2	Time 1 (h)	Time 2 (h)
H	2-IC$_4$H$_3$S	2	5 (57)
H	3-IC$_5$H$_4$N	2	45 (40)
5-Cl	PhI	2	7 (52)
H	4-IC$_6$H$_4$Br	3	5 (28)
H	3-IC$_6$H$_4$F	2	7 (59)
5-O$_2$N	PhI	2	7 (11)
H	3-(TfO)C$_{10}$H$_7$	3	6 (38)
H	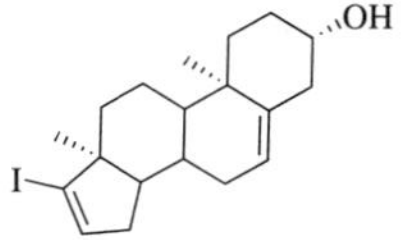 steroid-OH	20	24 (21)
5-NC	PhI	2	24 (47)
5-MeO$_2$C	PhI	2	7 (20)

| R^1–C$_6$H$_2$(I)(R^2)(R^3)NH$_2$ | ≡—oxazolidinone-Bn | Pd(OAc)$_2$, PPh$_3$, Bu$_4$NOAc, DMF, 60° | indole–oxazolidinone-Bn | | | 337 |

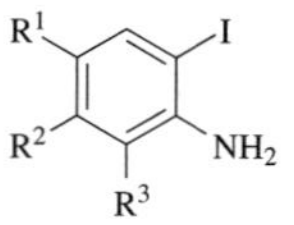

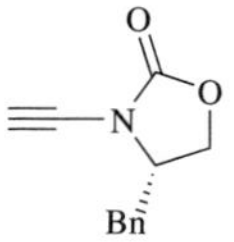

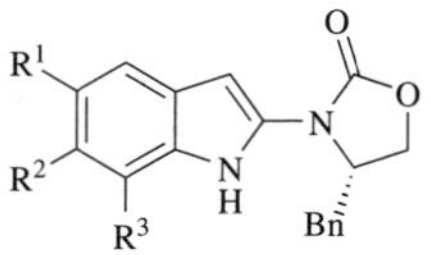

R^1	R^2	R^3	
H	H	H	(81)
Cl	H	H	(87)
H	F	H	(69)
Cl	H	F	(87)
O$_2$N	H	H	(31)
CF$_3$	H	H	(48)
NC	H	H	(38)
H	MeO$_2$C	H	(42)

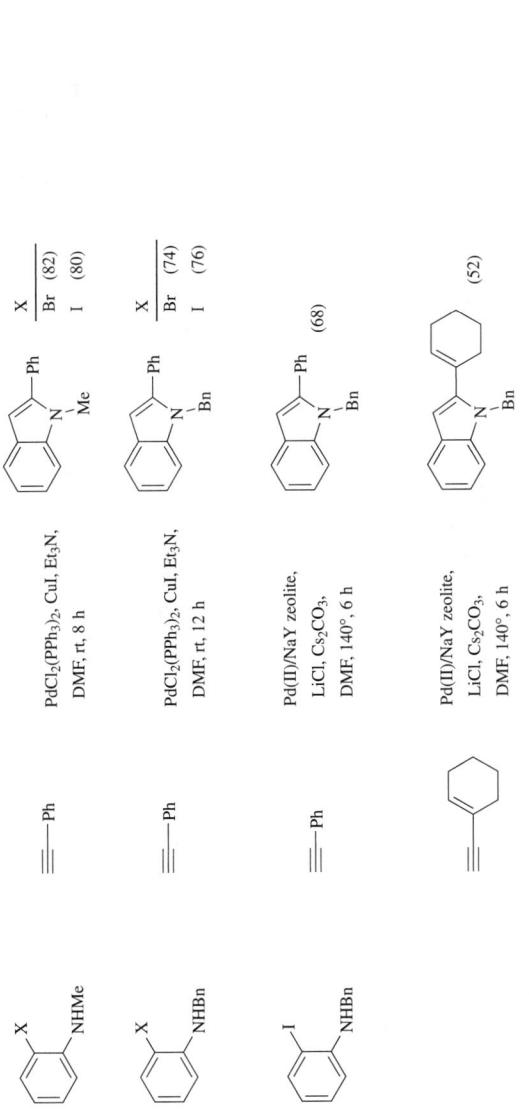

[a] The yield was determined using gas chromatography.
[b] The reaction was carried out under microwave irradiation.

TABLE 1B. 2-SUBSTITUTED INDOLES FROM 2-HALOANILIDES AND ALKYNES

2-Haloanilide	Alkyne	Conditions	Product(s) and Yield(s) (%)	Refs.
C₇				
2-I-C₆H₄-NHMs	≡—Ph	PS-PEO-CPC-Pd, ligand, H₂N(CH₂)₂OH, CuI, 80°	2-Ph-1-Ms-indole Solvent Ligand Time (h) DMA — 4.5 (79) DMA PPh₃ 3 (70) H₂O — 3 (20) H₂O PPh₃ 3 (42)	85
	≡—Ph	Pd, CuI, PPh₃/KF–Al₂O₃, no solvent, MW	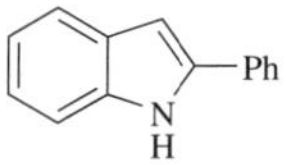 2-Ph-1H-indole (80)	95
	≡—Ph	PdCl₂(MeCN)₂, PPh₃, H₂N(CH₂)₂OH, CuI, 80°, 4 h	2-Ph-1-Ms-indole (91)	85
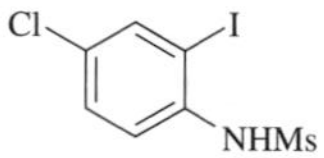 4-Cl-2-I-C₆H₃-NHMs	≡—R	Pd/C, PPh₃, CuI, 2-aminoethanol, H₂O, 80°	5-Cl-2-R-1-Ms-indole R Time (h) CH₂OH 12 (65) (CH₂)₂OH 12 (60) C(OH)Me₂ 12 (89) Ph 4 (62)	94
2-X-C₆H₄-NHMs	≡—R	PdCl₂(PPh₃)₂, CuI, Et₃N, DMF, 80° (X = I) or 120° (X = Br), 24 h	2-R-1-Ms-indole R X = I X = Br CH₂OH (53) (—) (CH₂)₂OH (59) (47) CH₂OMe (71) (12) TMS (31) (58) Bu (64) (43) CH(OEt)₂ (63) (—) (CH₂)₂CO₂Et (43) (49) Ph (66) (20)	50

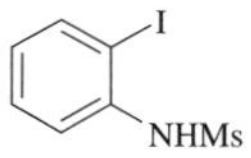

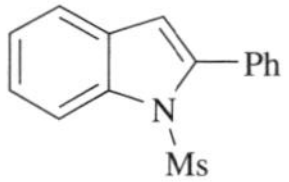

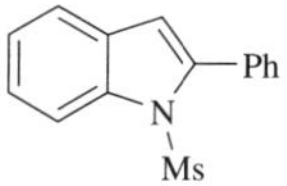

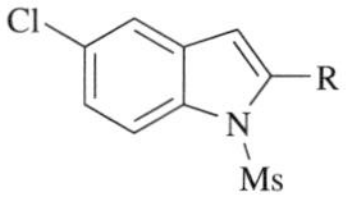

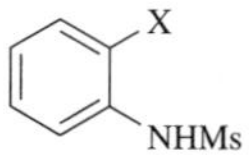

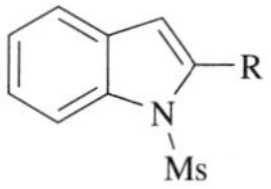

C_{7-9}: aryl iodide with R^1, NHMs, and alkyne R^2

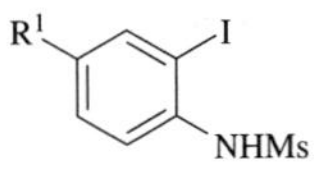

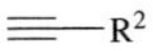

Pd/C, CuI, PPh$_3$, 2-aminoethanol, H$_2$O, 80°

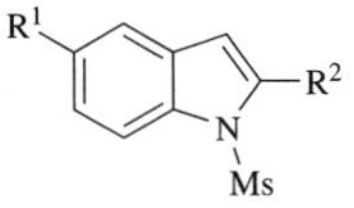

Product: indole with R^1, R^2, N-Ms

R^1	R^2	Time (h)	
F	(CH$_2$)$_3$Cl	5	(74)
F	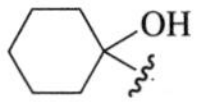 1-hydroxycyclohexyl	3	(85)
F	2-O$_2$NC$_6$H$_4$	24	(78)
F	2-MeC$_6$H$_4$	12	(78)
F	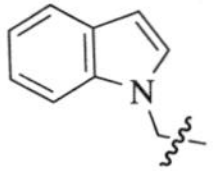 N-indolyl	14	(75)
F	6-MeO-2-naphthyl	12	(78)
F	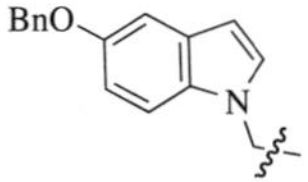 5-BnO-1-methylindol-2-yl	14	(90)
Cl	MeCH(OH)CH$_2$	10	(79)
Cl	4-MeC$_6$H$_4$	4	(86)
Et	4-MeC$_6$H$_4$	6	(90)
Et	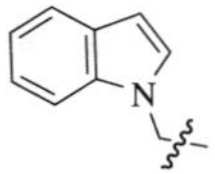 N-indolyl	4	(86)

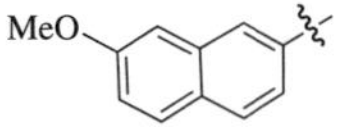

340

C$_8$: aryl iodide with X, NHCOCF$_3$, and phenylacetylene

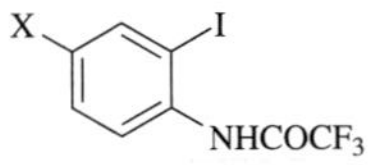

Pd/C, PPh$_3$, CuI, 2-aminoethanol, H$_2$O, 80°

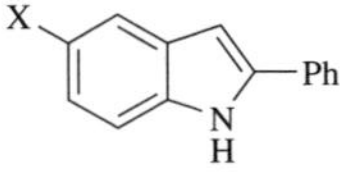

Product: 2-phenyl-1H-indole with X

X	Time (h)	
F	12	(78)
Cl	4	(40)

94

TABLE 1B. 2-SUBSTITUTED INDOLES FROM 2-HALOANILIDES AND ALKYNES (*Continued*)

2-Haloanilide	Alkyne	Conditions	Product(s) and Yield(s) (%)	Refs.
C_8 2-X-C₆H₄-NHAc 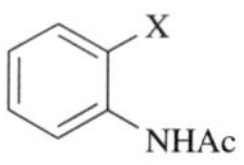 	≡—Ph	PdCl₂(PPh₃)₂, CuI, Et₃N, DMF, rt, 24 h	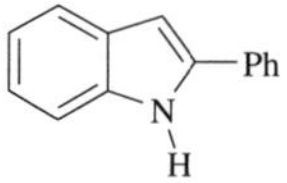 2-Ph-indole: X = Br (75), I (69)	79
2-I-C₆H₄-NHAc	≡—R	Pd(II)/NaY zeolite, LiCl, Cs₂CO₃, DMF, 140°, 6 h	2-R-indole: R = H (40), CH₂OMe (48), CH₂OTHP (69), n-C₅H₁₁ (72), 6-methylpyridin-3-yl (51); R = Ph (82), 1-cyclohexenyl (78), 5-acetylthien-2-yl (52), quinolin-3-yl (50)	96
2-I-4-Me-C₆H₃-NHMs	≡—R	Pd/C, PPh₃, Cu, 2-aminoethanol, H₂O, 80°	2-R-5-Me-1-Ms-indole; R / Time (h) / Yield: CH₂OH 8 (70); (CH₂)₂OH 12 (82); CH(OH)Me 12 (78); (CH₂)₂Me 48 (40); C(OH)Me₂ 12 (85); CH(OH)Et 12 (80); Ph 3 (70); CH(OH)Ph 12 (80)	94
2-I-C₆H₄-NHCOCF₃	≡—Bu	Pd(OAc)₂, TPPTS, Et₃N, MeCN/H₂O, rt, 72 h	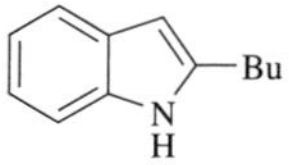 I 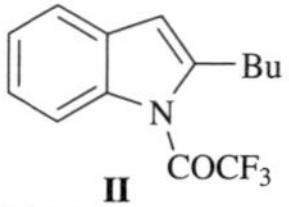 (2-Bu-indole) + II (2-Bu-1-COCF₃-indole); I + II (70), I:II = 74:26	92

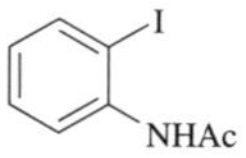

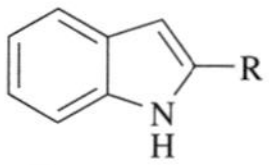

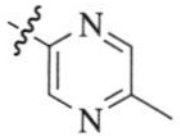

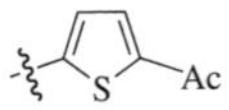

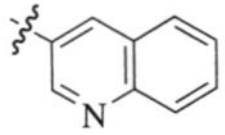

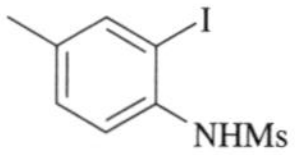

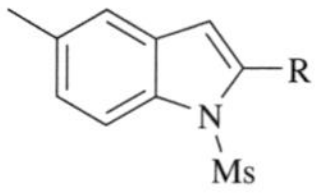

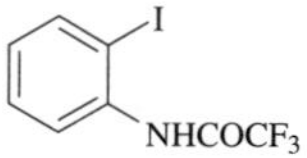

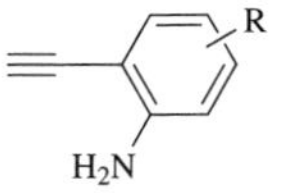

Starting material	Alkyne	Conditions	Product	Yield (%)	Refs.
2-Br-6-F-4-F-C₆H₂(NHCOCF₃)	4-R-2-ethynylaniline	PdCl₂(PPh₃)₂, Et₃N, DMF, 90°	CF₃COHN-(F,F-indol-2-yl)-C₆H₃-R	R: H (75), 4h; 3,5-Cl₂ (54), 4h; 4-CF₃ (56), 8h; 3,5-Me₂ (58), 4.5h	83
2-Br-4-Cl-6-Cl·C₆H₂(NHCOCF₃)	4-R-2-ethynylaniline	PdCl₂(PPh₃)₂, Et₃N, DMF, 90°, 4 h	CF₃COHN-(F,F-indol-2-yl)-C₆H₃-R	R: H (58); 3,5-Cl₂ (54)	83
2-Br-4-Cl-6-Cl·C₆H₂(NHCOCF₃)	2-ethynylaniline	PdCl₂(PPh₃)₂, Et₃N, DMF, 90°, 4 h	CF₃COHN-(Cl,Cl-indol-2-yl)-C₆H₄	(55)	83
2-Br-4-O₂N-6-Cl·C₆H₂(NHCOCF₃)	4-R-2-ethynylaniline	PdCl₂(PPh₃)₂, Et₃N, DMF, 90°	CF₃COHN-(O₂N,Cl-indol-2-yl)-C₆H₃-R	R: H (59), 4h; CF₃ (45), 8h	83
2-I-4-MeOC-C₆H₃(NHMs)	R-C≡CH	Pd(OAc)₂, Bu₄NOAc, MeCN,))), 90°, 6 h	MeOC-(2-R-N-Ms-indole) **I**	R: Ph (52); 4-MeC₆H₄ (65); 4-MeOC₆H₄ (66)	84
C₉	R-C≡CH	Pd(OAc)₂, Bu₄NOAc, MeCN, 90°, 12 h	**I**	R: Ph (45); 4-MeC₆H₄ (71); 4-MeOC₆H₄ (67)	84

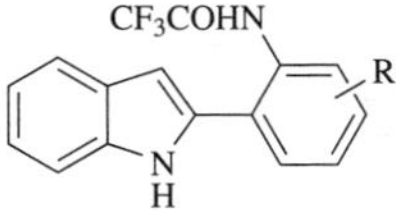

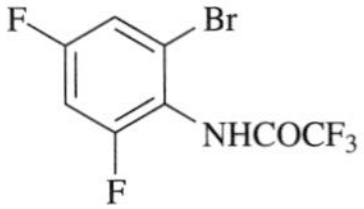

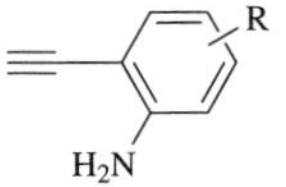

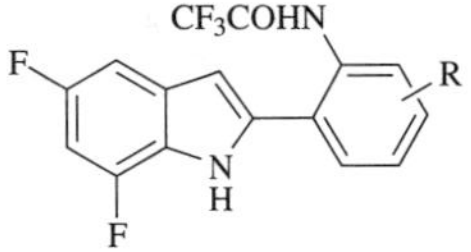

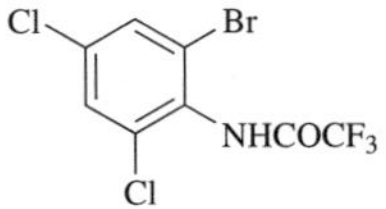

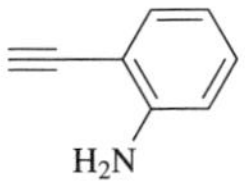

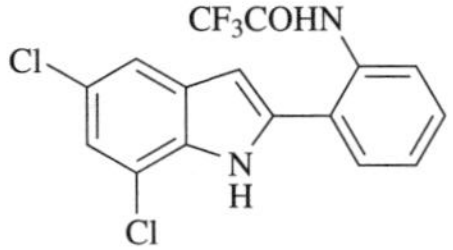

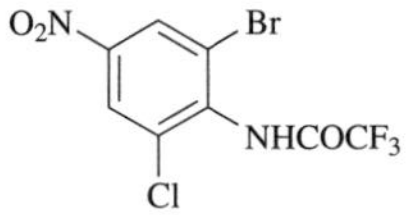

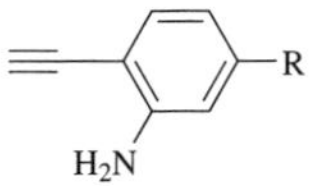

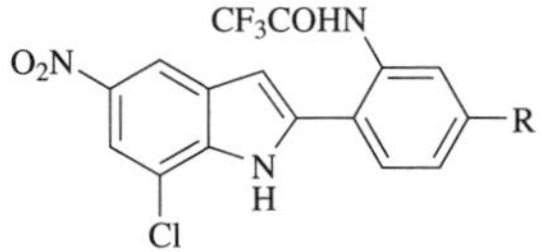

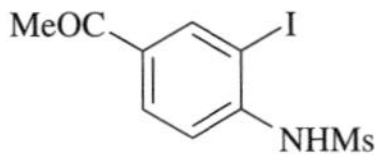

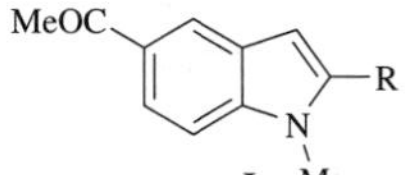

TABLE 1B. 2-SUBSTITUTED INDOLES FROM 2-HALOANILIDES AND ALKYNES (Continued)

2-Haloanilide	Alkyne	Conditions	Product(s) and Yield(s) (%)		Refs.
C9 MeO2C–C6H3(I)–NHMs	≡–R	Pd(OAc)2, Bu4NOAc, MeCN,))), 90°, 6 h	MeO2C–[indole-N(Ms)]–R **I** R: CHMeOH (56); Ph (43); 4-MeC6H4 (58); 4-MeOC6H4 (54); 1-Np (60)		84
	≡–R	Pd(OAc)2, Bu4NOAc, MeCN, 90°, 12 h	**I** R: CHMeOH (56); Ph (43); 4-MeC6H4 (58); 4-MeOC6H4 (54); 1-Np (60)		84
2-iodo-4-methyl-NHCOCF3	≡–Ph	Pd/C, PPh3, CuI, 2-aminoethanol, H2O, 80°, 4 h	4-Me-[indole-NH]-Ph (20) + 2-alkynyl-NHCOCF3 with Ph (35)		94
4-OTf-2-NO2-NHCOCF3	≡–R	Pd(PPh3)4, CuI, Bu4NI, Et3N, DMF, 80°	O2N–[indole-NH]–R		80

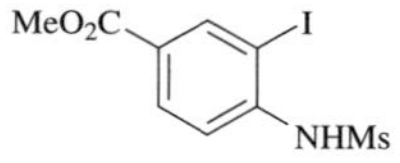

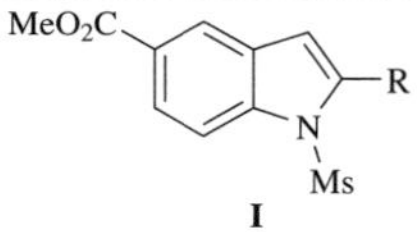

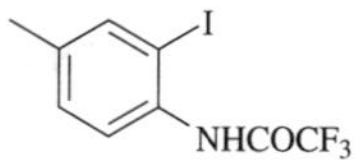

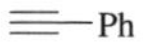

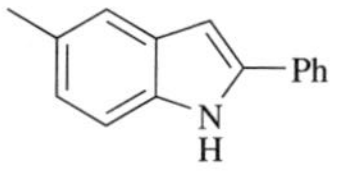

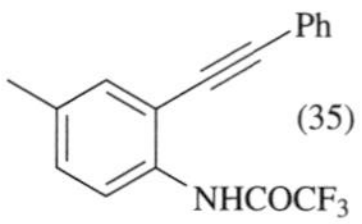

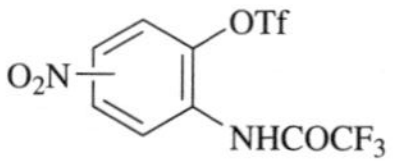

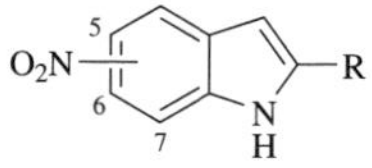

NO2 isomer	R	Time (h)	(%)	NO2 isomer	R	Time (h)	(%)
5	(CH2)2OH	42	(68)	5	(CH2)3Cl	24	(45)
6	(CH2)2OH	19	(69)	6	(CH2)3Cl	20	(66)
7	(CH2)2OH	12	(48)	5	(CH2)3CN	25	(75)
5	Pr	24	(90)	6	(CH2)3CN	8	(73)
6	Pr	21	(84)	5	Ph	41	(85)
7	Pr	7.5	(39)	6	Ph	17	(88)
				7	Ph	3.5	(52)

| 2-iodo-4-NO2–NHCO2Et | ≡–N(Ts)(Bn) | Pd(OAc)2, PPh3, Bu4NOAc, DMF, 60° | O2N–[indole-N(CO2Et)]–N(Ts)(Bn) (55) | | 337 |

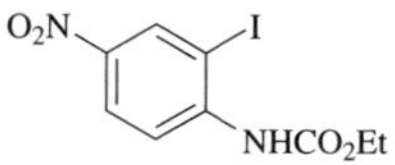

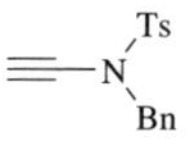

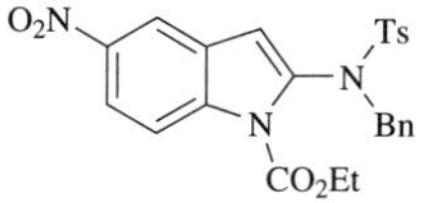

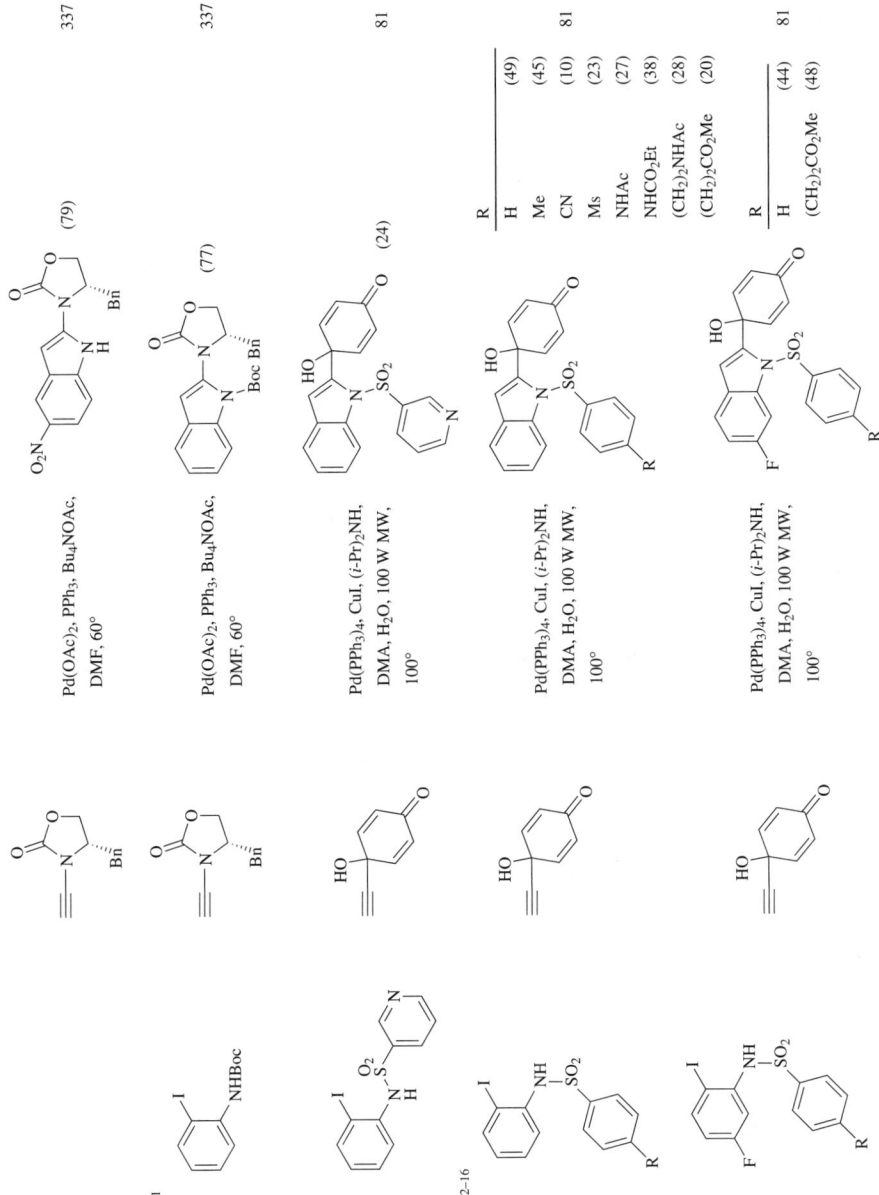

TABLE 1B. 2-SUBSTITUTED INDOLES FROM 2-HALOANILIDES AND ALKYNES (*Continued*)

2-Haloaniide	Alkyne	Conditions	Product(s) and Yield(s) (%)	Refs.
C_{13} 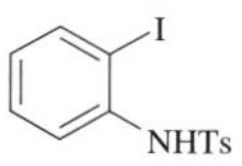 2-I-C$_6$H$_4$-NHTs	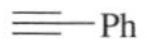 ≡—Ph	Pd(II)/NaY zeolite, LiCl, Cs$_2$CO$_3$, DMF, 140°, 6 h	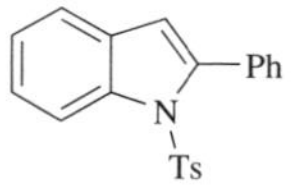 2-Ph-N-Ts-indole (68)	95
	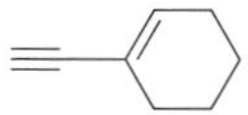 cyclohexenyl-C≡CH	Pd(II)/NaY zeolite, LiCl, Cs$_2$CO$_3$, DMF, 140°, 6 h	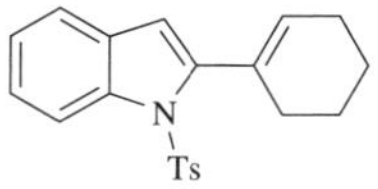 2-(cyclohexenyl)-N-Ts-indole (52)	95
	 ≡—R	Pd(OAc)$_2$, Et$_3$N, DMF/H$_2$O, 80°	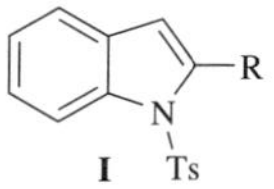 2-R-N-Ts-indole (**I**) R / Time (d) C(Me)$_2$OH 7 (35)[a] Bu 4 (47) (CH$_2$)$_3$CO$_2$Me 3 (76)[a] Ph 2 (42)	78
	 ≡—R	[Pd(NH$_3$)$_4$]/NaY, Et$_3$N, DMF/H$_2$O, 80°	**I** R / Time (d) C(Me)$_2$OH 8 (56) Bu 1 (52) (CH$_2$)$_3$CO$_2$Me 1 (91)[a] Ph 6 (76)	78
	 ≡—R	[Pd]/SBA-15, Et$_3$N, DMF/H$_2$O, 80°	**I** R / Time (d) C(Me)$_2$OH 7 (67) Bu 1 (48) (CH$_2$)$_3$CO$_2$Me 1 (61) Ph 6 (65)	78
	 ≡—R	Pd(OAc)$_2$, Bu$_4$NOAc, MeCN,))), 90°	**I** R / Time (h) C(Me)$_2$OH 6 (44) Ph 4 (82) 4-MeC$_6$H$_4$ 5 (71) 4-MeOC$_6$H$_4$ 6 (72) 1-Np 6 (63)	84

Alkyne	Conditions	Product			Refs.
≡—R	Pd(OAc)₂, Bu₄NOAc, MeCN, 90°	**I** R-indole (2-R)	R, Time (h) C(Me)₂OH, 24 (41) Ph, 24 (80) 4-MeC₆H₄, 30 (69) 4-MeOC₆H₄, 3 (76) 1-Np, 24 (67)		84
≡—Bu	Pd/C, additive, NaOAc, NMP, 120°, 2 h	**I** 2-Bu-indole + **II** 3-Bu-indole	Additive, I+II, I:II LiCl, (92), 89:11 —, (84), 77:23		341
≡—Ph	Pd/C, additive, NaOAc, NMP, 120°, 12 h	2-Ph-indole	Additive LiCl (69) — (70)		341
≡—R	Pd/C, NaOAc, NMP, 120°	2-R-indole	R, Time (h) c-C₆H₁₁, 24 (77) 4-MeOC₆H₄, 24 (64) 3-MeC₆H₄, 24 (72) 4-MeC₆H₄, 24 (63) 1-Np, 24 (52) 6-MeO-2-Np, 5 (66)		341
Br-CH₂-C≡CH	PdCl₂(PPh₃)₂, CuI, piperidine, 50°	1-Ts-2-(piperidin-1-ylmethyl)indole (97)			82

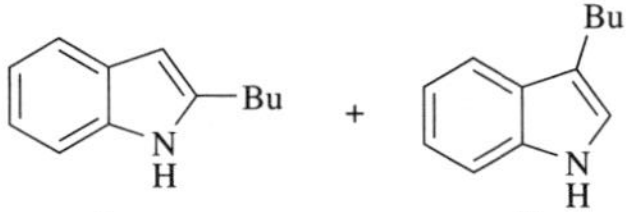

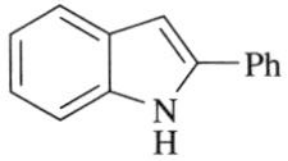

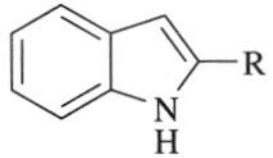

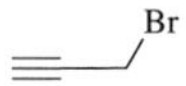

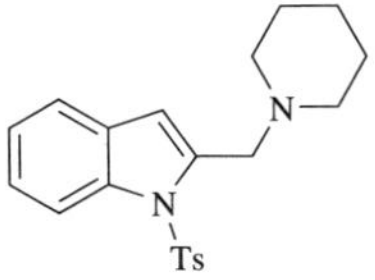

TABLE 1B. 2-SUBSTITUTED INDOLES FROM 2-HALOANILIDES AND ALKYNES (Continued)

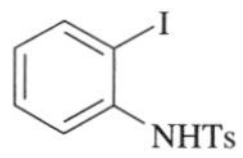

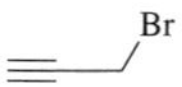

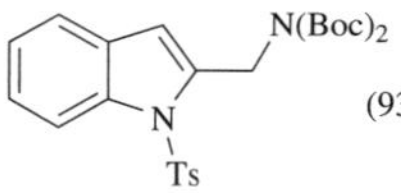

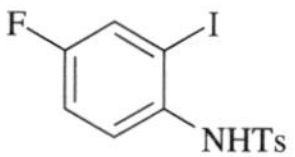

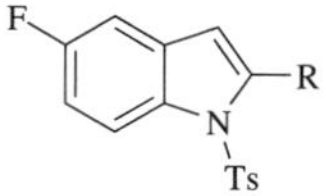

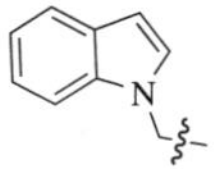

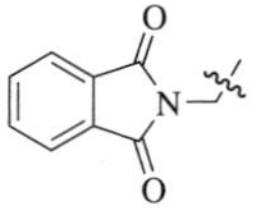

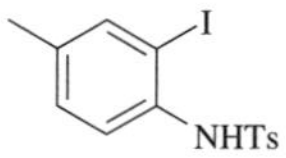

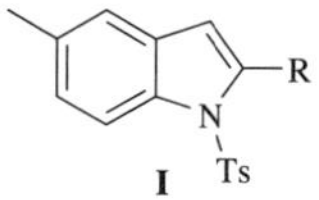

2-Haloanilide	Alkyne	Conditions	Product(s) and Yield(s) (%)	Refs.
C_{13} 2-I-C$_6$H$_4$-NHTs	≡—CH$_2$Br	PdCl$_2$(PPh$_3$)$_2$, CuI, KN(Boc)$_2$, Et$_3$N, DMF, rt, 80°	2-(CH$_2$N(Boc)$_2$)-N-Ts-indole (93)	82
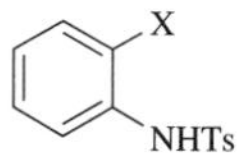 2-X-C$_6$H$_4$-NHTs	 ≡—Ph	PdCl$_2$(PPh$_3$)$_2$, CuI, Et$_3$N, DMF, rt, 12 h	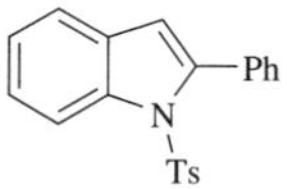 2-Ph-N-Ts-indole X: Br (68); I (70)	79
2-I-4-F-C$_6$H$_3$-NHTs	≡—R	Pd/C, CuI, PPh$_3$, 2-aminoethanol, H$_2$O, 80°	5-F-2-R-N-Ts-indole R / Time (h) / Yield HO(CH$_2$)$_3$ / 14 / (80) MeCH(OH)CH$_2$ / 12 / (65) HOCMe$_2$ / 10 / (70) Me(CH$_2$)$_3$ / 12 / (74) Ph / 8 / (68) 4-MeC$_6$H$_4$ / 10 / (78) CH$_2$-(N-indolyl) / 10 / (76) CH$_2$-(phthalimidyl) / 12 / (70)	340
C_{14} 2-I-4-Me-C$_6$H$_3$-NHTs	≡—R	Pd(OAc)$_2$, Bu$_4$NOAc, MeCN,))), 90°	5-Me-2-R-N-Ts-indole (**I**) R / Time (h) / Yield Ph / 5 / (82) 3-FC$_6$H$_4$ / 6 / (65) 4-MeC$_6$H$_4$ / 5 / (90) 4-MeOC$_6$H$_4$ / 5 / (90) 1-Np / 6 / (42)	84

| | ≡—R | Pd(OAc)$_2$, Bu$_4$NOAc, MeCN, 90° | [cyclohexadienone-alkyne product] | 84 |

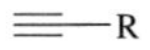

R	Time (h)	
Ph	30	(71)
3-FC$_6$H$_4$	36	(56)
4-MeC$_6$H$_4$	35	(87)
4-MeOC$_6$H$_4$	36	(74)
1-Np	6	(46)

C$_{18}$ [2-iodo-N-(5-(dimethylamino)naphthalen-1-ylsulfonyl)aniline]

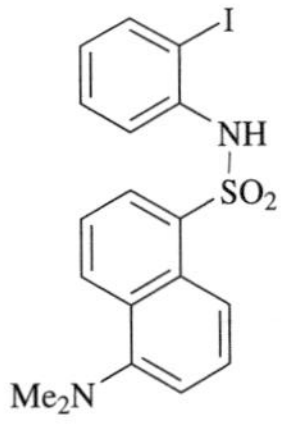

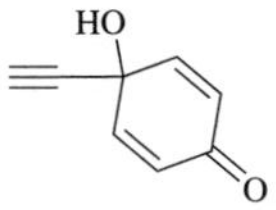

| | | Pd(PPh$_3$)$_4$, CuI, (i-Pr)$_2$NH, DMA, H$_2$O, 100 W MW, 100° | **I** [indole product with HO-cyclohexadienone and N-sulfonyl-5-(dimethylamino)naphthalene] (70) | 81 |

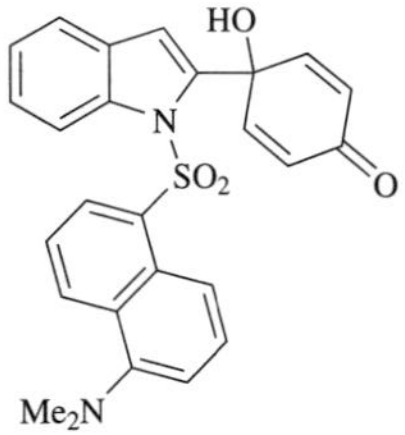

---

[a] The reported yield was determined by GC.

TABLE 1C. 2-SUBSTITUTED INDOLES FROM 1,2-DIHALOARENES AND ALKYNES

1,2-Dihaloarene	Alkyne	Conditions	Product(s) and Yield(s) (%)	Refs.
C_6				
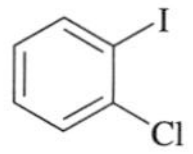 2-iodo-1-chlorobenzene	 ≡—Ph	1. Pd(OAc)$_2$, HIPrCl, CuI, Cs$_2$CO$_3$, toluene, 105°, 1 h 2. 4-MeC$_6$H$_4$NH$_2$, 18 h 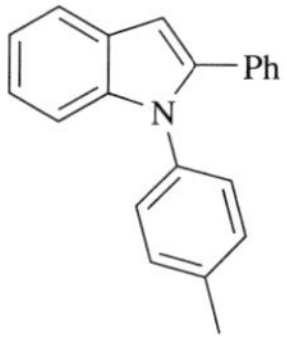	2-Ph, N-(4-MeC$_6$H$_4$)-indole (64)	86
	≡—Ph	1. Pd(OAc)$_2$, HIPrCl, CuI, Cs$_2$CO$_3$, toluene, 105°, 1 h 2. RNH$_2$, t-BuOK, 22 h	2-Ph-N-R-indole R Ph (52) 2-FC$_6$H$_4$ (61) 4-FC$_6$H$_4$ (54) 3-MeC$_6$H$_4$ (66) 4-MeC$_6$H$_4$ (65) Bn (50) 3,5-Me$_2$C$_6$H$_3$ (67) n-C$_8$H$_{17}$ (58) 2,4,6-Me$_3$C$_6$H$_2$ (58)	89
C_7				
4-CF$_3$-2-iodo-1-chlorobenzene	≡—Ph	1. Pd(OAc)$_2$, HIPrCl, CuI, Cs$_2$CO$_3$, toluene, 105°, 1 h 2. RNH$_2$, t-BuOK, 22 h	5-CF$_3$-2-Ph-N-R-indole R 2-FC$_6$H$_4$ (74) 4-FC$_6$H$_4$ (79) 4-MeC$_6$H$_4$ (63) 2-MeOC$_6$H$_4$ (76) 4-MeOC$_6$H$_4$ (65) 2,4,6-Me$_3$C$_6$H$_2$ (68)	87

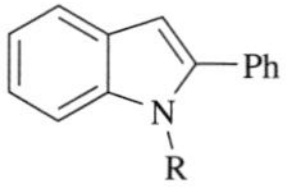

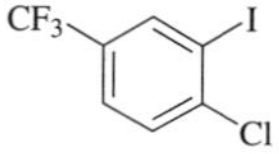

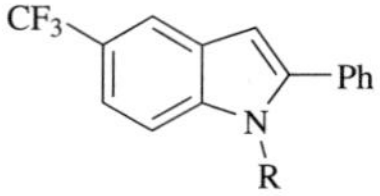

TABLE 1D. 2-SUBSTITUTED INDOLES FROM 2-ALKYNYLANILINES

2-Alkynylaniline	Conditions	Product(s) and Yield(s) (%)	Refs.

C_8

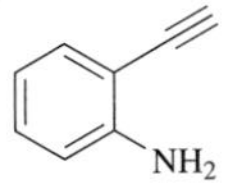

2-Alkynylaniline: 2-ethynylaniline (NH_2, C≡CH)

Conditions:
1. ROTf, Pd(PPh$_3$)$_4$, CuI, Et$_2$NH, rt, time 1
2. PdCl$_2$, Bu$_4$NCl, CH$_2$Cl$_2$/0.5 N HCl, rt, time 2

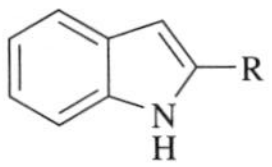

Product: 2-R-indole

R	Time 1 (h)	Time 2 (h)		Refs.
1-cyclooctenyl	5.5	14	(62)	71
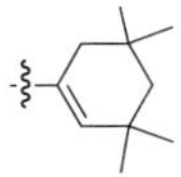	5.5	40	(81)	
4-phenyl-1-cyclohexenyl	5	14	(98)	
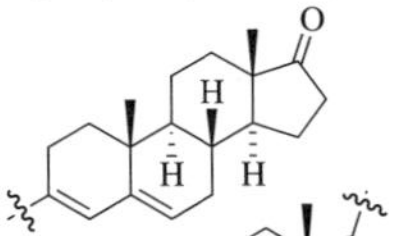 (steroid ketone)	5.5	14	(65)	
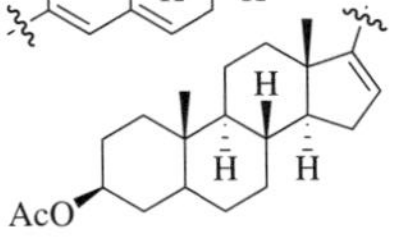 (AcO-steroid)	6	48	(96)	

C_{11-15}

2-Alkynylaniline: R^1-substituted 2-(C≡C-R^2)aniline

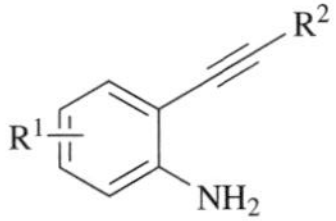

Conditions: R^3NCO, PdCl$_2$, THF, 80°

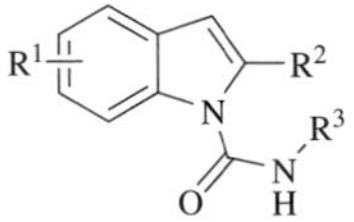

Product: N-carbamoyl indole (R^1, R^2, R^3 substituted)

R^1	R^2	R^3		Refs.
H	c-C$_3$H$_5$	Ph	(90)	
H	c-C$_3$H$_5$	4-FC$_6$H$_4$	(72)	
H	c-C$_3$H$_5$	4-MeOC$_6$H$_4$	(88)	342
H	Ph	Ph	(87)	
H	Ph	4-FC$_6$H$_4$	(75)	
H	Ph	4-MeOC$_6$H$_4$	(86)	
H	4-MeOC$_6$H$_4$	Ph	(53)	
H	4-MeOC$_6$H$_4$	4-FC$_6$H$_4$	(65)	
5-Me	Ph	Ph	(74)	
5-Me	Ph	4-FC$_6$H$_4$	(72)	
5-CF$_3$	Ph	Ph	(46)	
5-CF$_3$	Ph	4-FC$_6$H$_4$	(60)	
H	4-MeOC$_6$H$_4$	4-MeOC$_6$H$_4$	(73)	
5-Me	Ph	4-MeOC$_6$H$_4$	(83)	
5-CF$_3$	Ph	4-MeOC$_6$H$_4$	(43)	

TABLE 1D. 2-SUBSTITUTED INDOLES FROM 2-ALKYNYLANILINES (*Continued*)

2-Alkynylaniline	Conditions	Product(s) and Yield(s) (%)	Refs.
C_{12} 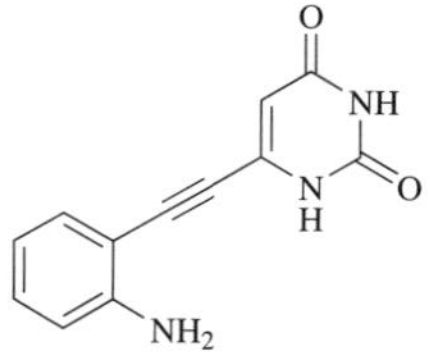 — 2-alkynylaniline with uracil substituent	$PdCl_2$, DMF, 120°, 4.5 h	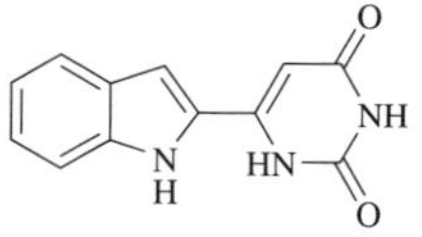 2-substituted indole with uracil (28)	74
C_{12-15} 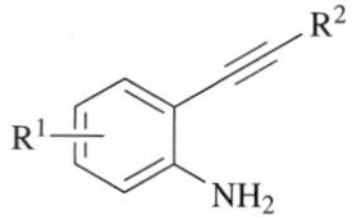 — R^1-substituted 2-(R^2-ethynyl)aniline	$PdCl_2$, $FeCl_3$, DCE, 80°	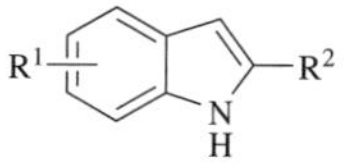 R^1-indole with 2-R^2 substituent R^1 — R^2 — Time (h) H — 2-thienyl — 4 (50) H — Ph — 4 (88) 5-Cl — Ph — 4 (83) 5-NO_2 — Ph — 4 (82) H — 4-$O_2NC_6H_4$ — 20 (40) 4-Me — Ph — 4 (80) H — 4-$MeOC_6H_4$ — 20 (76)	77
C_{12-16} — 2-(R-ethynyl)aniline	$PdCl_2$, MeCN, reflux	2-R-indole R — Time (h) Bu — 1 (81) t-Bu — 0.5 (77) Ph — 0.5 (52) $MeCH(CH_2)_5Me$ — 0.5 (83)	70
C_{12-34} — 2-(R-ethynyl)aniline 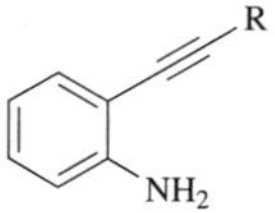	$PdCl_2$, MeCN, reflux	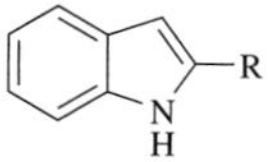 2-R-indole	74

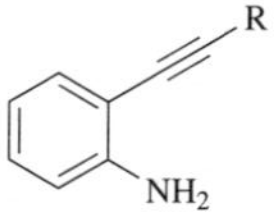

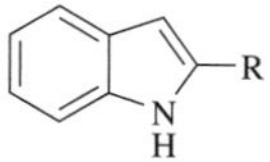

R	Time (h)		R	Time (h)	
2-thienyl	3	(82)	Ph-chromene	4	(60)
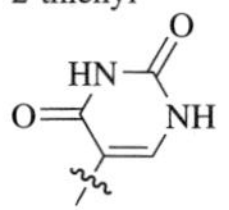 thymine-like	4.5	(28)[a]	MeO-steroid	4	(75)
3-pyridyl	36	(38)	HO-steroid	4	(74)
3-FC$_6$H$_4$	6	(51)	steroid ketone	4.5	(82)
4-MeC$_6$H$_4$	3	(80)	O$_2$CPh-steroid	4.5	(77)
4-PhCOC$_6$H$_4$	3	(39)			
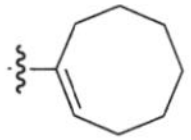 cyclooctenyl	9	(80)			
2-Np	3.5	(76)			
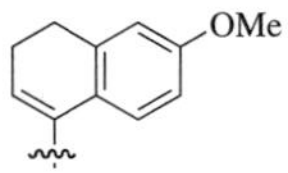 6-OMe-tetrahydronaphthyl	3.5	(64)			
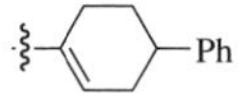 4-Ph-cyclohexenyl	3	(87)			

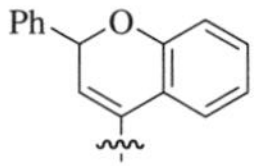

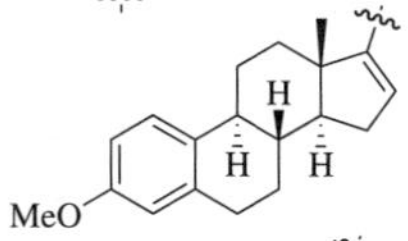

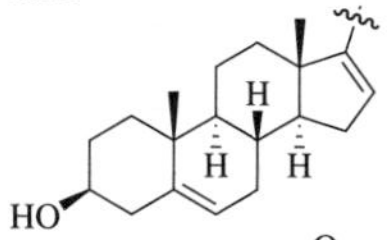

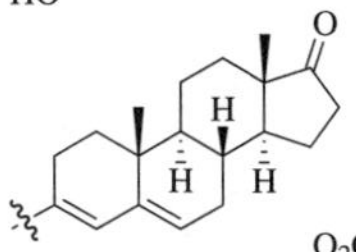

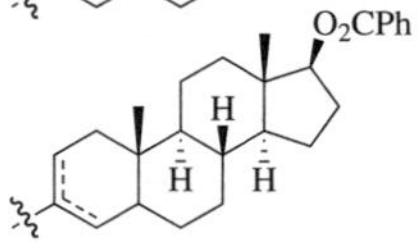

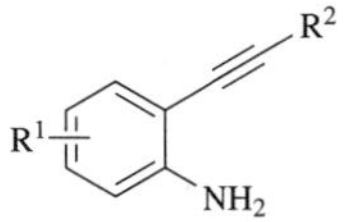

C1, DMF, 100°, 1 h

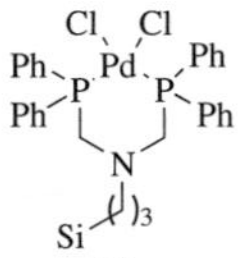

Catalyst C1: [Cl$_2$Pd(Ph$_2$PCH$_2$N(CH$_2$SiCl$_3$))]

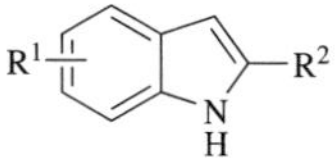

R^1	R^2	
H	Ph	(96)
H	2-thienyl	(90)
6-Cl	3-thienyl	(89)
H	5-indolyl	(65)

C14–16

343

TABLE 1D. 2-SUBSTITUTED INDOLES FROM 2-ALKYNYLANILINES (Continued)

2-Alkynylaniline	Conditions	Product(s) and Yield(s) (%)	Refs.
C_{15} 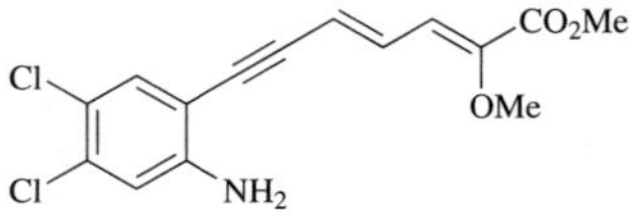	PdCl$_2$(MeCN)$_2$, DMF, 70°	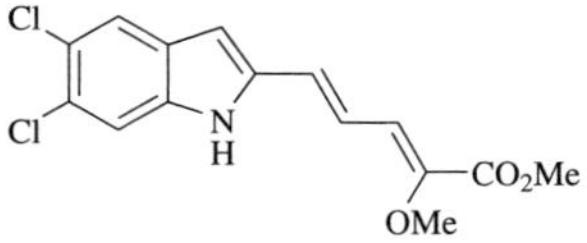 (90)	75
C_{15-35}	PdCl$_2$, Bu$_4$NCl, CH$_2$Cl$_2$/HCl (x N), rt		71

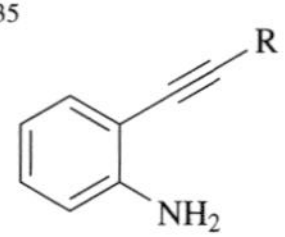

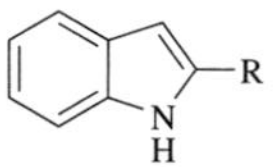

R	x	Time (h)	
4-MeC$_6$H$_4$	3	48	(82)
4-MeOC$_6$H$_4$	3	21	(68)
3-FC$_6$H$_4$	3	72	(45)
CH=CHPh	3	72	(57)
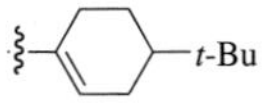 t-Bu	0.5	16	(97)
(tetrahydronaphthyl)	—	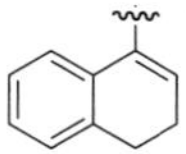 36	(92)
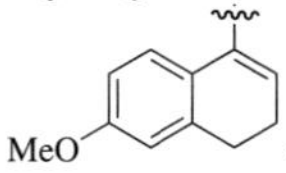 (MeO-naphthyl)	2	72	(70)
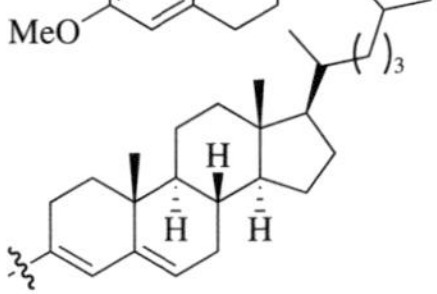 (steroid)	0.5	10	(89)

| C_{16} 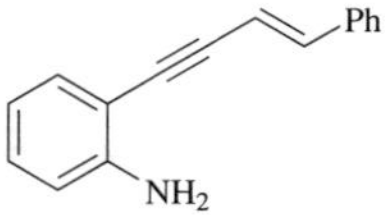 | PdCl$_2$(MeCN)$_2$, DMF, 70° 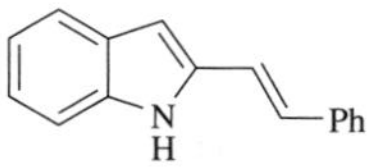 | (83) **I** | 75 |

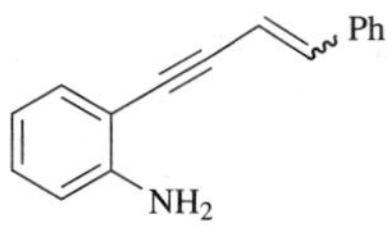

Substrate	Conditions	Product	Refs.
C₁₈ (2-alkynylaniline with Ph)	PdCl₂, MeCN, 70°, 2.5 h	**I** (74)[b]	74
C₁₈ (ferrocenyl, O₂N)	PdCl₂, Bu₄NBr, HCl/CH₂Cl₂, rt, 24 h	(58) 5-nitro-2-ferrocenylindole	76
C₁₈₋₂₂ (MeO₂C-CH(N(Boc)₂)-(CH₂)ₙ-alkyne)	PdCl₂(MeCN)₂, MeCN, reflux	indole product, see table	73, 344

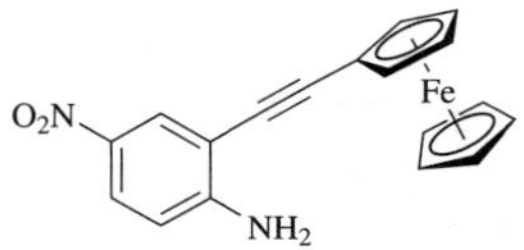

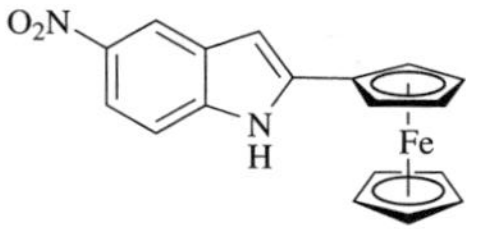

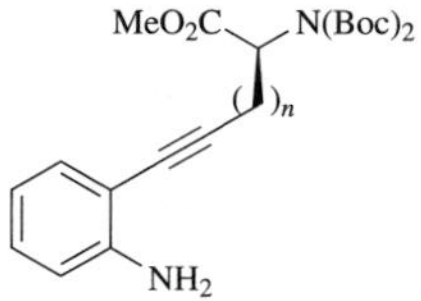

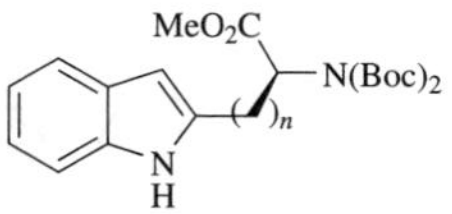

R	n	Time (h)	
H	2	0.5	(60)
H	3	0.5	(55)
Boc	1	3	(52)

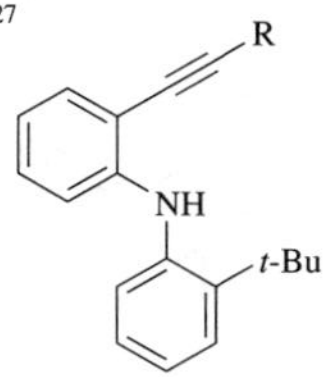

C₂₂₋₂₇ (2-alkynyl-N-(2-t-Bu-phenyl)aniline)

PdCl₂, **L1**, EtOH, 80°

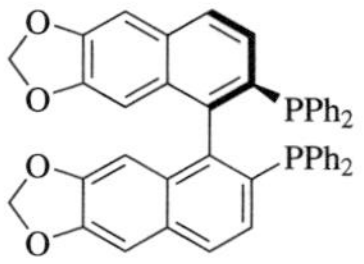

2-R-1-(2-t-Bu-phenyl)indole

345

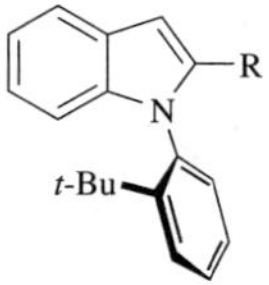

R	Time (h)		% ee
Bu	24	(84)	35
Ph	4	(93)	60
2-BrC₆H₄	23	(90)	83
2-ClC₆H₄	13	(85)	83
2-O₂NC₆H₄	24	(67)	82
2-MeC₆H₄	9	(95)	67
4-MeC₆H₄	7	(89)	49
2-MOMOCH₂C₆H₄	7	(71)	77
2-i-PrC₆H₄	12	(90)	80

L1: biaryl bis(methylenedioxy)-naphthyl bisphosphine (PPh₂, PPh₂)

TABLE 1D. 2-SUBSTITUTED INDOLES FROM 2-ALKYNYLANILINES (*Continued*)

2-Alkynylaniline	Conditions	Product(s) and Yield(s) (%)	Refs.
C_{27}	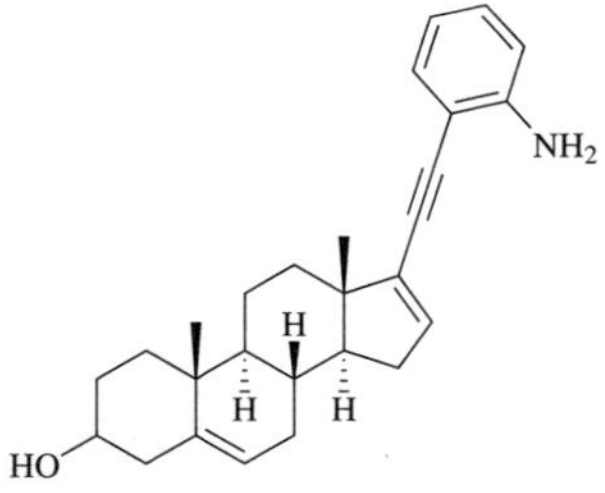 PdCl_2, MeCN, 70°, 4 h	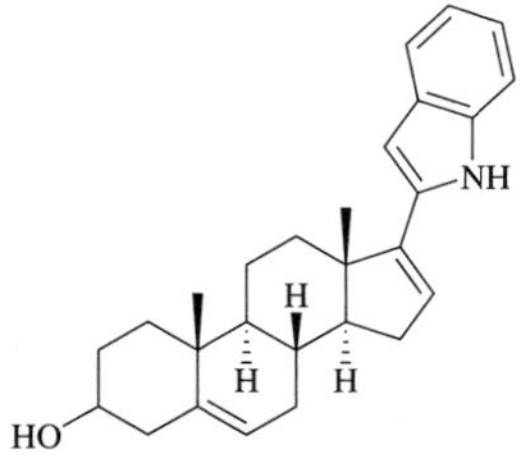 (64)	74
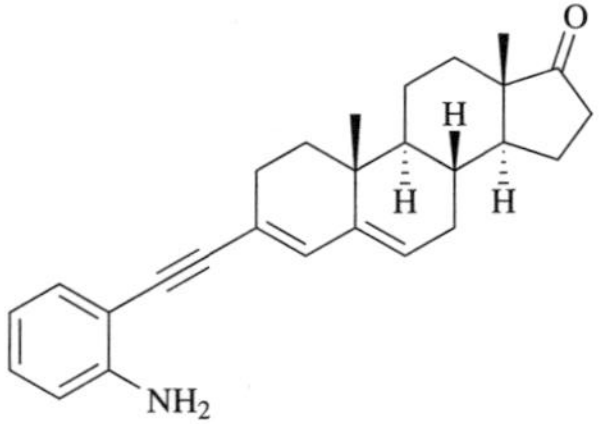	PdCl_2, MeCN, 70°, 4.5 h	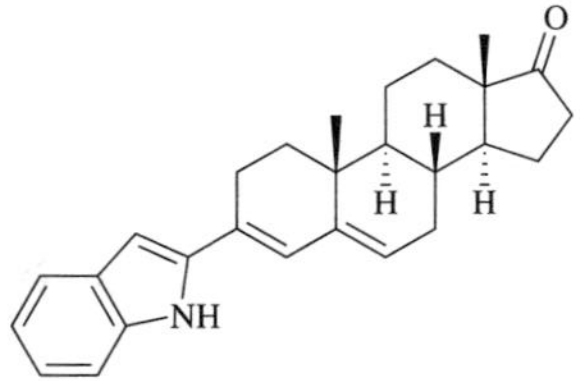 (82)	74
C_{34}	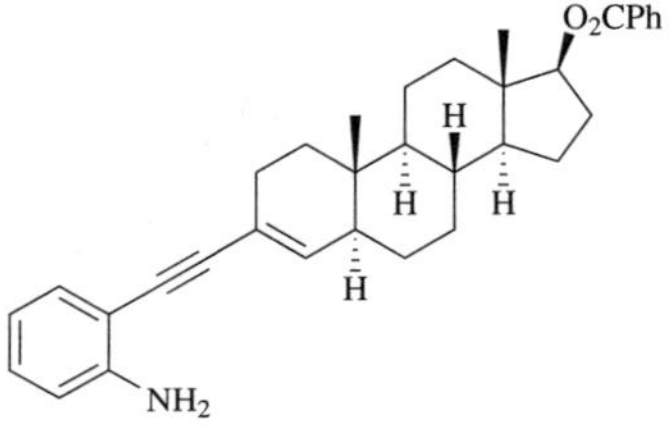 PdCl_2, MeCN, 70°, 4.5 h	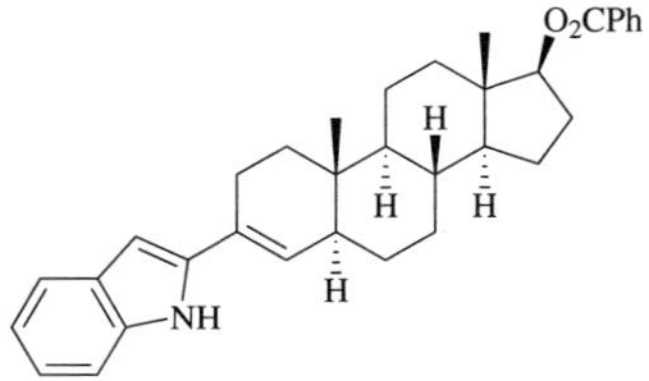 (77)	74

[a] The reaction was carried out at 120° in DMF.

[b] A commercially available (*E*)/(*Z*) mixture was used but the (*E*)-isomer was found to react preferentially.

TABLE 1E. 2-SUBSTITUTED INDOLES FROM 2-ALKYNYLANILIDES

	2-Alkynylanilide	Conditions	Product(s) and Yield(s) (%)	Refs.
C_{11} 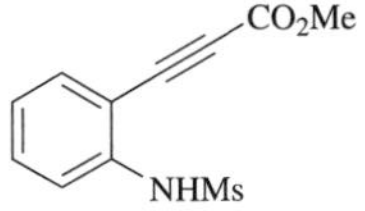	(2-NHMs-C6H4)−C≡C−CO2Me	≡—CO2Me, Pd(PPh3)4, ZnBr2, i-Pr2NEt, THF, reflux	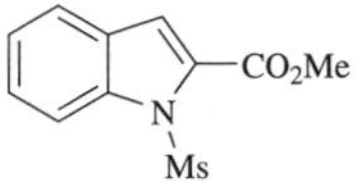 2-CO2Me-N-Ms-indole (94)	346
C_{13} 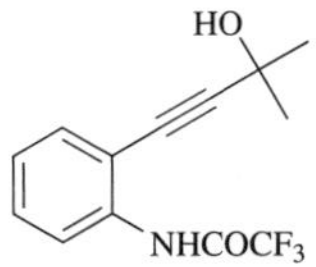	(2-NHCOCF3-C6H4)−C≡C−C(Me)2OH	Pd(OAc)2, LiCl, K2CO3, DMF, 100°, 16 h	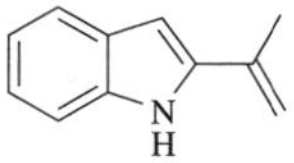 2-isopropenyl-indole (95)	72
C_{13-20} 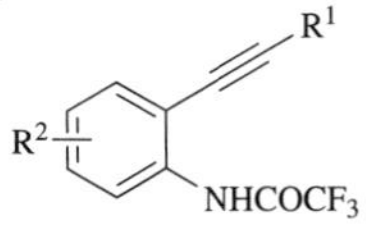	R^2—(2-NHCOCF3-C6H4)−C≡C−R^1	PdCl2(MeCN)2, MeCN, 60°	2-R^1-R^2-indole	49

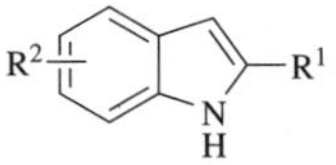

R^1	R^2	Time (h)	(%)
Pr	5-Cl	2.75	(76)
i-Pr	5-Cl	1.5	(80)
Bu	5-Cl	2	(83)
i-Pr	5-OTf	3	(40)
Pr	5-Me	1	(81)
Pr	6-OMe	0.5	(75)
i-Pr	6-OMe	1	(78)
i-Pr	6-CO2Me	1.5	(82)
Bu	5-OTf	2.5	(65)
Bu	5-Me	0.5	(77)

R^1	R^2	Time (h)	(%)
Bu	6-OMe	1.5	(66)
Ph	6-Cl	4	(48)
Bu	6-CO2Me	1.75	(71)
Ph	5-OTf	3.5	(53)
Ph	5-Me	1.25	(80)
Ph	6-OMe	3	(35)
Ph	6-CO2Me	1.5	(76)
(CH2)2OTBS	5-Me	1.5	(37)
(CH2)2OTBS	5-Me	2.5	(35)
(CH2)4C≡CTMS	5-Me	3.5	(53)

TABLE 1E. 2-SUBSTITUTED INDOLES FROM 2-ALKYNYLANILIDES (*Continued*)

2-Alkynylanilide	Conditions	Product(s) and Yield(s) (%)			Refs.
C14					
(structure: 2-NHCOCF₃ phenyl–C≡C–CH₂OCO₂Et)	HN⌒NR, Pd(PPh₃)₄, THF, 80°	(indole-2-CH₂-piperazine-N-R)			34
		R	Time (h)		
		Et	1.5	(91)	
		CO₂Et	20	(88)	
		2-FC₆H₄	2.5	(97)	
		4-FC₆H₄	6	(96)	
		3,4-Cl₂C₆H₃	2	(96)	
		2-NCC₆H₄	3	(98)	
		2-FC₆H₄CH₂	6	(98)	
		4-ClC₆H₄CH₂	3	(92)	
		4-BrC₆H₄CH₂	3	(94)	
		3,4-Cl₂C₆H₃CH₂	24	(80)	
		2-Cl-6-FC₆H₃CH₂	12	(78)	
		4-MeC₆H₄CH₂	4	(85)	
		4-MeOC₆H₄CH₂	4	(81)	
		2,4,6-Me₃C₆H₂CH₂	3	(80)	
		4-ClC₆H₄CH(Ph)	8	(92)	
	HN⌒N–R¹, Pd(PPh₃)₄, THF, 80° (with R² substituent)	(indole-2-CH₂-piperazine with R¹, R²)			
		R¹	R²	Time (h)	
		Et	H	1.5 (91)	347
		EtO₂C	H	20 (88)	347
		3,4-Cl₂C₆H₃	H	2 (96)	347
		4-FC₆H₄	H	6 (96)	347
		2-FC₆H₄	H	2.5 (97)	347
		2-NCC₆H₄	H	3 (98)	347
		Me	Ph	6 (58)	347, 34

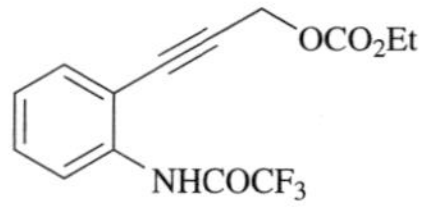

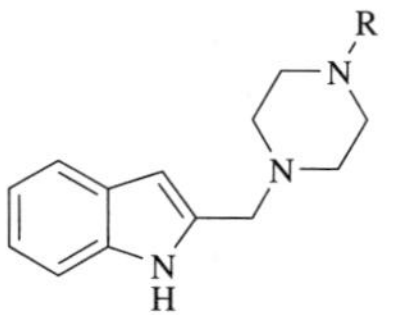

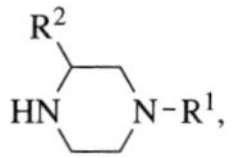

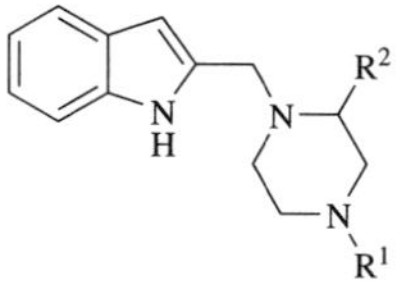

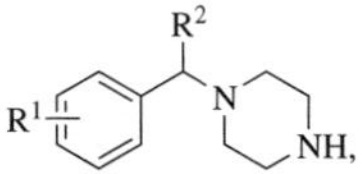

Scheme 1

Indole-CH2-piperazine-N-CH(R²)-C6H4R¹ + piperazine-NH reagent, Pd(PPh3)4, THF, 80°

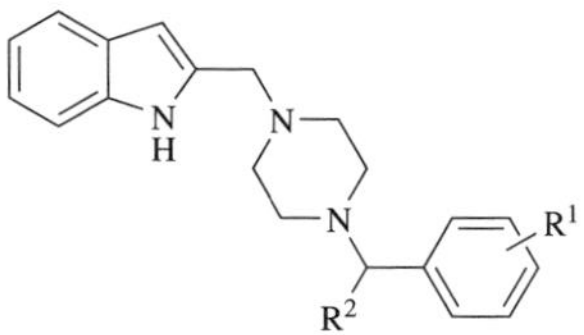

R¹	R²	Time (h)		
4-MeO	H	4	(81)	347
2-FC6H4	H	6	(98)	
4-ClC6H4	H	3	(92)	
4-BrC6H4	H	3	(94)	
4-MeC6H4	H	4	(85)	
4-ClC6H4	Ph	8	(92)	

Scheme 2

Indole-CH2-piperazine-N-CH2-C6H2(R¹)(R²)(R³)(R⁴), Pd(PPh3)4, THF, 80°

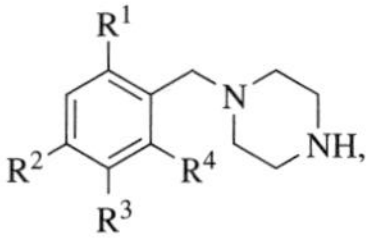

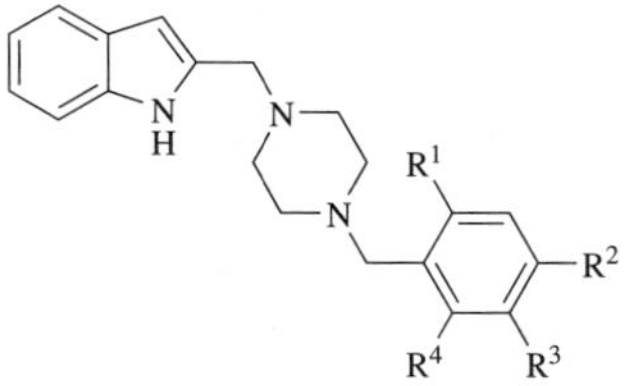

R¹	R²	R³	R⁴	Time (h)		
H	Cl	Cl	H	24	(80)	347
Cl	H	H	F	12	(78)	
Me	H	H	Me	4	(80)	

Scheme 3

Indole-CH2-NR2, R2NH, Pd(PPh3)4, THF, 80°

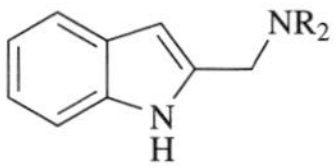

R2NH	Time (h)		
Et	2	(60)	34
i-Pr	4	(45)	
morpholine	1	(98)	
piperidine	1	(94)	

Scheme 4

2-(BuC≡C)-C6H4-NHR → 2-Bu-indole-N-R, PdCl2, MeCN, reflux

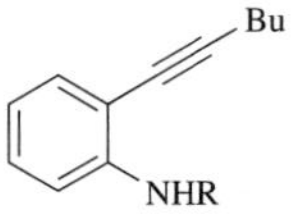

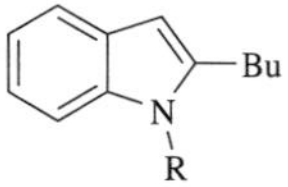

R	Time (h)		
Ac	4	(74)	70
CO2Me	2	(78)	

TABLE 1E. 2-SUBSTITUTED INDOLES FROM 2-ALKYNYLANILIDES (*Continued*)

2-Alkynylanilide	Conditions	Product(s) and Yield(s) (%)	Refs.

C_{14-16}

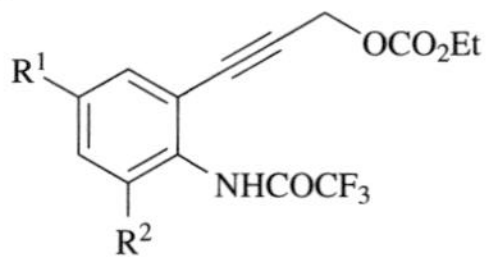

2-Alkynylanilide: structure with R^1, R^2, OCO$_2$Et, NHCOCF$_3$

Conditions: HCO$_2$H, Et$_3$N, MeCN, 80°

Product: 2-methylindole with R^1, R^2 substituents

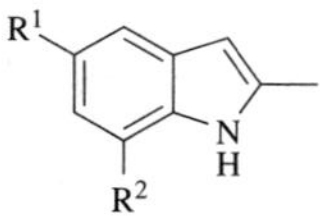

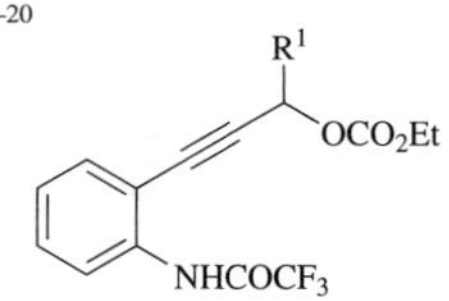

R^1	R^2	Time (h)	
H	H	1	(91)
Cl	H	0.5	(60)
Me	H	0.5	(80)
Me	O$_2$N	2	(78)
Me	Cl	2	(51)
F	Me	1	(65)
Cl	CF$_3$	1	(86)
Me	Me	1	(99)
MeCO	H	0.66	(73)
MeO$_2$C	H	1	(70)

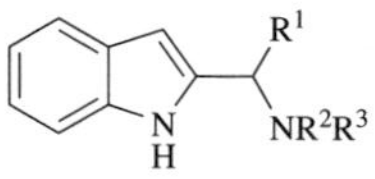

Refs. 347

C_{14-20}

2-Alkynylanilide: structure with R^1, OCO$_2$Et, NHCOCF$_3$

Conditions: HNR^{2}R^3, Pd(PPh$_3$)$_4$, THF, 80°

Product: indole with CH(R^1)NR^{2}R^3 at 2-position

R^1	HNR^{2}R^3	Time (h)	
H	morpholine	1	(98)
H	Et$_2$NH	2	(60)
H	piperidine	1	(94)
H	(*i*-Pr)$_2$NH	4	(45)
Ph	morpholine	1	(98)
Ph	Et$_2$NH	2	(60)
Ph	BuNH$_2$	5	(80)
Ph	piperidine	1	(94)
Ph	1-ethylpiperazine	1	(92)
Ph	(*i*-Pr)$_2$NH	4	(45)
Ph	1-ethylpiperazine	2	(90)
Ph	BnNH$_2$	4	(54)

Refs. 347

C_{14-23}

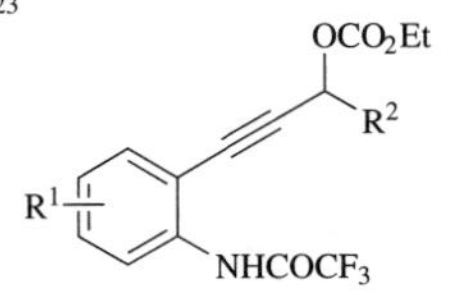

[Structure: R¹-substituted phenyl with NHCOCF₃ and C≡C-CR²(OCO₂Et)]

Pd(PPh₃)₄, HCO₂H, Et₃N, MeCN, 80°

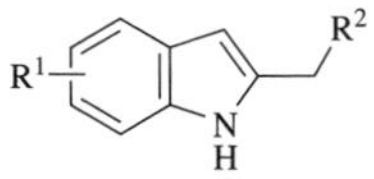

[Product: 5-R¹-2-(CH₂R²)-indole]

R¹	R²	Time (h)	
H	H	1	(91)
4-Cl	H	0.5	(60)
H	Me	1	(70)
4-Me	H	0.5	(80)
4-Me-6-Cl	H	2	(51)
4-F-6-Me	H	1	(86)
4-Me-6-O₂N	H	2	(78)
4-Cl-6-CF₃	H	1	(86)
H	Et	2	(70)
4,6-Me₂	H	1	(99)

R¹	R²	Time (h)	
4-MeCO	H	0.66	(73)
4-MeO₂C	H	1	(70)
H	Pr	1	(67)
H	Ph	3	(75)
H	4-FC₆H₄	1	(85)
H	3-MeOC₆H₄	1	(75)
4,6-Me₂	Ph	1	(50)
4-Cl-6-CF₃	3-MeOC₆H₄	0.5	(95)
4,6-Me₂	3-MeOC₆H₄	2	(72)

43

C_{15-20}

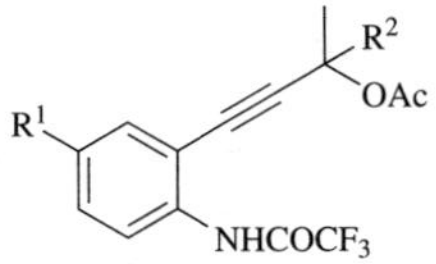

[Structure: R¹-substituted phenyl with NHCOCF₃ and C≡C-CR²(Me)(OAc)]

Pd(OAc)₂, PPh₃, Et₃N, THF, 80°

[Product: 5-R¹-2-(1-R²-vinyl)-indole]

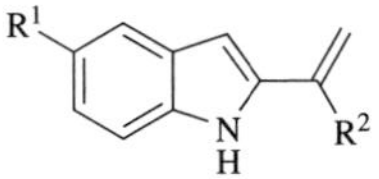

R¹	R²	Time (h)	
H	Me	1	(80)
Cl	Me	2	(85)
NC	Me	1	(83)
Me	Me	3	(84)
MeCO	Me	1	(91)
MeO₂C	Me	2	(87)
H	Ph	6	(78)
F	Ph	1	(68)

347

TABLE 1E. 2-SUBSTITUTED INDOLES FROM 2-ALKYNYLANILIDES (*Continued*)

2-Alkynylanilide	Conditions	Product(s) and Yield(s) (%)	Refs.

C_{15-22}

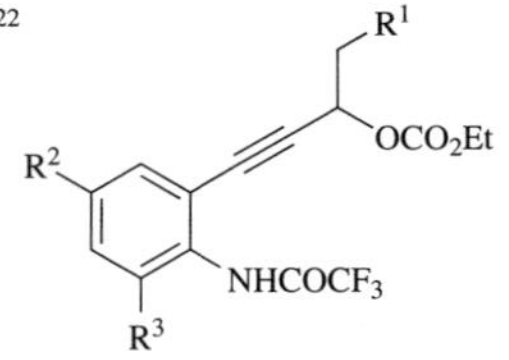

Alkynylanilide with R^1, R^2, R^3, OCO$_2$Et, NHCOCF$_3$ substituents

Pd(OAc)$_2$, PPh$_3$, Et$_3$N, THF, 80°

2-styrylindole product with R^1, R^2, R^3

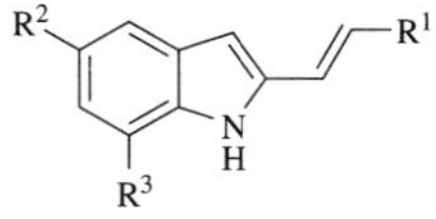

R^1	R^2	R^3	Time (h)	
H	H	H	0.5	(90)
H	F	F	2	(82)
H	Me	H	2	(63)
Me	H	H	1	(95)
H	Me	Me	1	(87)
Et	H	H	1	(83)
Ph	Me	H	0.5	(85)

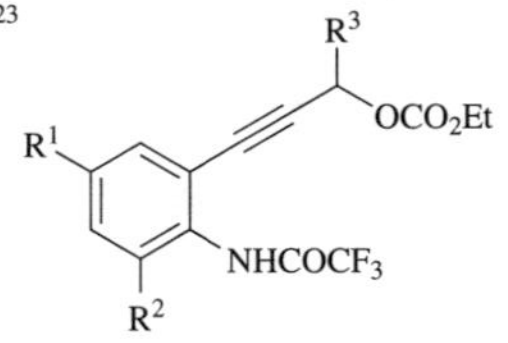

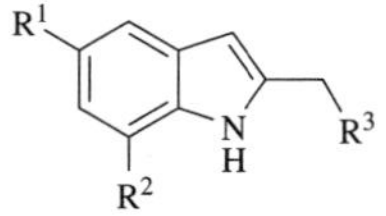

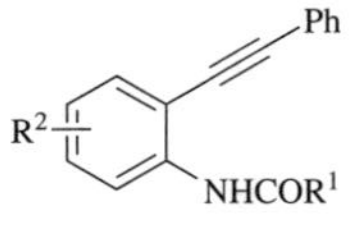

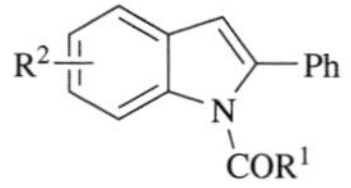

347

C_{15-23}

Alkynylanilide with R^1, R^2, R^3, OCO$_2$Et, NHCOCF$_3$ substituents

HCO$_2$H, Et$_3$N, MeCN, 80°

2-substituted indole product

R^1	R^2	R^3	Time (h)	
H	H	Me	1	(70)
H	H	Et	2	(70)
H	H	Pr	1	(67)
H	H	Ph	3	(75)
H	H	4-FC$_6$H$_4$	1	(85)
H	H	3-MeOC$_6$H$_4$	1	(75)
H	H	4-MeOC$_6$H$_4$	1	(70)
H	H	Bn	1	(50)
Cl	CF$_3$	3-MeOC$_6$H$_4$	0.5	(85)
Me	Me	4-MeOC$_6$H$_4$	2	(72)

347

C_{16-21}

Ph-alkynyl anilide with R^2, NHCOR^1

PdCl$_2$, MeCN, 81°

2-Ph-indole product with COR^1, R^2

R^1	R^2	
Me	6-Cl	(67)
Me	5-Me	(80)
Me	6-Me	(95)
Me	5,6-Me$_2$	(90)
Ph	H	(78)

23

C_20

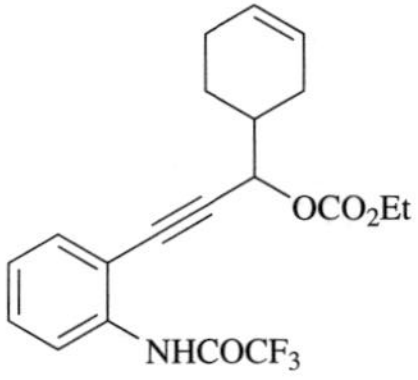

Pd(OAc)$_2$, PPh$_3$,
Et$_3$N, THF, 80°, 1 h

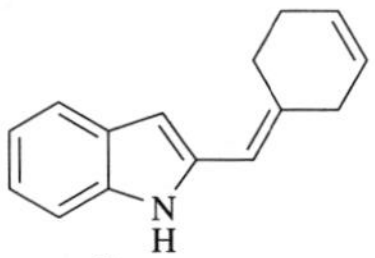

I + **II** (76), **I:II** = 1:1   347

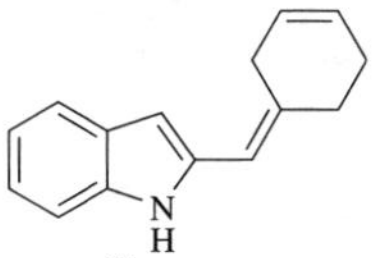

TABLE 1F. 2-SUBSTITUTED INDOLES FROM 2-ALKYNYLHALOARENES

2-Alkynylhaloarene	Amine	Conditions	Product(s) and Yield(s) (%)				Refs.

C_{11-17}

2-Alkynylhaloarene: 2-alkynyl-chloroarene with R^1 and R^2

Amine: $ArNH_2$

Conditions: Pd(OAc)$_2$, **L2**, t-BuOK, toluene, 14 h

L2 = 1,3-bis(2,6-diisopropylphenyl)imidazolium chloride

Product: 1-Ar-2-R^2-indole with R^1

Ar	R^1	R^2	Temp (°)	
2,4,6-Me$_3$C$_6$H$_2$	H	c-C$_3$H$_5$	105	(85)
2,6-(i-Pr)$_2$C$_6$H$_3$	H	c-C$_3$H$_5$	105	(82)
2,4,6-Me$_3$C$_6$H$_2$	H	Ph	120	(91)
2,6-(i-Pr)$_2$C$_6$H$_3$	H	Ph	120	(94)
2,4,6-Me$_3$C$_6$H$_2$	5-OH	Ph	120	(75)
2,6-(i-Pr)$_2$C$_6$H$_3$	5-OH	Ph	120	(71)
2,4,6-Me$_3$C$_6$H$_2$	H	3-MeC$_6$H$_4$	120	(91)
2,4,6-Me$_3$C$_6$H$_2$	6-Me	Ph	120	(56)
2,6-(i-Pr)$_2$C$_6$H$_3$	6-Me	Ph	120	(91)
2,4,6-Me$_3$C$_6$H$_2$	H	4-PrC$_6$H$_4$	120	(95)
2,6-(i-Pr)$_2$C$_6$H$_3$	H	4-PrC$_6$H$_4$	120	(88)

348

C_{11-18}

Amine: 1-adamantylamine (H$_2$N-Ad)

Conditions: Pd(OAc)$_2$, **L2**,[a] t-BuOK, toluene, 14 h

Product: 1-(1-adamantyl)-2-R^2-indole with R^1

R^1	R^2	Temp (°)	
H	c-C$_3$H$_5$	105	(80)
H	n-C$_6$H$_{13}$	105	(59)
H	Ph	120	(83)
H	4-FC$_6$H$_4$	120	(94)
H	4-MeOC$_6$H$_4$	120	(79)
H	4-CF$_3$C$_6$H$_4$	120	(83)
6-Me	Ph	120	(94)
H	4-PrC$_6$H$_4$	120	(88)
5-MeO	4-PrC$_6$H$_4$	120	(55)

348

Substrate	Amine	Conditions	Product				Refs.
C$_{11-19}$ 1-alkynyl-2-chlorobenzene (R^1)	t-BuNH$_2$	Pd(OAc)$_2$, **L2**,^{a} t-BuOK, toluene	R^1-indole, N-t-Bu, 2-R^2	R^1	R^2	Temp (°)	348
				H	c-C$_3$H$_5$	120 (87)	
				H	Bu	120 (63)	
				H	Ph	120 (55)	
				6-Me	Ph	105 (90)	
				H	4-PrC$_6$H$_4$	105 (84)	
				5-MeO	4-PrC$_6$H$_4$	105 (65)	
C$_{11-21}$ 1-alkynyl-2-chloro-3-alkoxybenzene (OR^1, R^2)	R^{2}NH$_2$	1. Cp$_2$TiMe$_2$, toluene, 110°, 24 h 2. Pd$_2$(dba)$_3$, **L3**, t-BuOK, 1,4-dioxane, 110°, 12 h	indole, N-R^2, 2-R^1	R^1	R^2		195
				Pr	CHMePh	(65)	
				c-C$_3$H$_5$	4-MeC$_6$H$_4$	(77)	
				Bu	4-MeOC$_6$H$_4$	(78)	
				1-cyclohexenyl	t-Bu	(39)	
				CH$_2$(CH$_2$)$_3$OBn	t-Bu	(70)	
C$_{11-21}$	BnNH$_2$	Pd(OAc)$_2$, HIPrCl, t-BuOK, toluene, reflux	4-OR^1 indole, N-Bn, 2-R^2	R^1	R^2		88
				Me	Et	(70)	
				Me	Pr	(57)	
				Me	Bu	(79)	
				Me	Ph	(84)	
				Bn	Bu	(75)	
				Bn	Ph	(89)	
C$_{12}$ 1-butynyl-2-chlorobenzene	RNH$_2$	Pd(OAc)$_2$, (t-Bu)$_3$P, t-BuOK, toluene, 110°, 14 h	indole, N-R, 2-Bu	R			87
				Ph		(23)	
				Bn		(85)	
				n-C$_8$H$_{17}$		(90)	

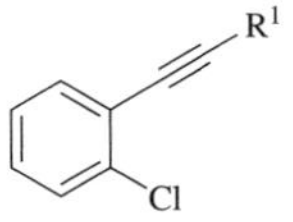

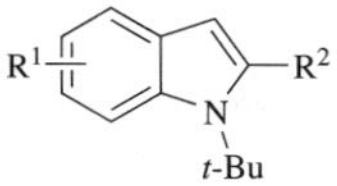

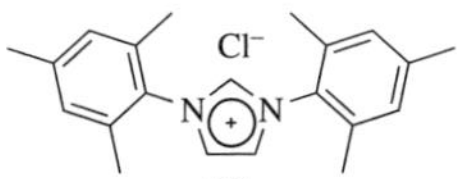

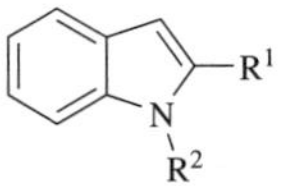

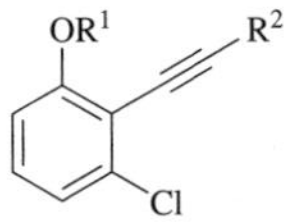

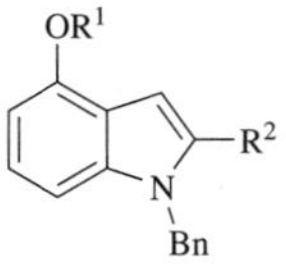

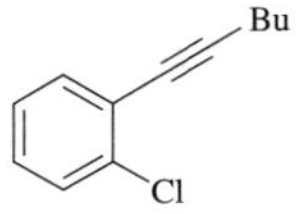

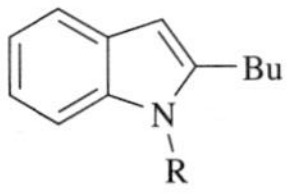

L3: 1,3-dimesitylimidazolium chloride

TABLE 1F. 2-SUBSTITUTED INDOLES FROM 2-ALKYNYLHALOARENES (*Continued*)

2-Alkynylhaloarene	Amine	Conditions	Product(s) and Yield(s) (%)	Refs.
C_{12-15} R^1–(aryl with R^2 alkynyl, Cl)	R^3NH_2	Pd(OAc)$_2$, HIPrCl, base, toluene, 105°, 2 h	(indole product with R^1, R^2, R^3)	86

R^1	R^2	R^3	Base	Yield
H	t-Bu	2-MeC$_6$H$_4$	t-BuOK	(83)
H	t-Bu	Bn	t-BuOK	(74)
H	Ph	n-C$_6$H$_{13}$	t-BuOK	(93)
H	Ph	n-C$_8$H$_{17}$	t-BuOK	(93)
H	Ph	2-MeC$_6$H$_4$	t-BuOK	(99)
H	Ph	PhCH$_2$	t-BuOK	(92)
H	Ph	4-MeOC$_6$H$_4$CH$_2$	t-BuOK	(78)
H	Ph	2,4,6-Me$_3$C$_6$H$_2$	t-BuOK	(99)
CF$_3$	n-C$_6$H$_{13}$	PhCH$_2$	t-BuOK	(66)
CF$_3$	n-C$_6$H$_{13}$	4-EtO$_2$CC$_6$H$_4$	K$_2$CO$_3$/CuI[b]	(67)
CF$_3$	Ph	4-MeC$_6$H$_4$	K$_2$CO$_3$/CuI[b]	(95)
CF$_3$	Ph	4-EtO$_2$CC$_6$H$_4$	K$_2$CO$_3$/CuI[b]	(92)

| C_{12-17} (R-alkynyl-Cl arene) | t-amyl-NH$_2$ | Pd(OAc)$_2$, **L2**,[a] t-BuOK, toluene, 120° | (indole with R, t-amyl N-substituent) | 348 |

R	Yield
Bu	(43)
4-CF$_3$C$_6$H$_4$	(80)
3-MeC$_6$H$_4$	(85)
4-PrC$_6$H$_4$	(92)

| C_{12-20} R^1–(aryl with R^2 alkynyl, Cl) | R^3NH_2 | 1. Cp$_2$TiMe$_2$, toluene, 110°, 24 h; 2. Pd$_2$(dba)$_3$, **L3**,[c] t-BuOK, 1,4-dioxane, 110°, 12 h | (indole product with R^1, R^2, R^3) | 195 |

R^1	R^2	R^3	Yield
OMe	Pr	t-Bu	(65)
CF$_3$	c-C$_3$H$_5$	4-MeC$_6$H$_4$	(77)
OMe	CH$_2$(CH$_2$)$_3$OBn	4-MeOC$_6$H$_4$	(78)

Substrate	Reagent	Conditions	Product	Yield (%)	Refs.
C_{13-21} 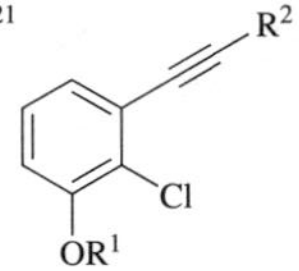 ![alkyne with R^2, Cl, OR^1]	BnNH$_2$	Pd(OAc)$_2$, HIPrCl, t-BuOK, toluene, reflux	2-R^2-7-OR^1-1-Bn-indole	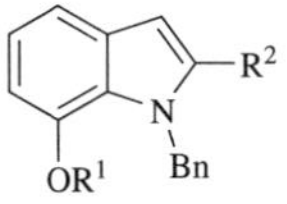 R^1/R^2: Me/Bu (74), Me/Ph (63), Bn/Bu (64), Bn/Ph (70)	88
C_{14} 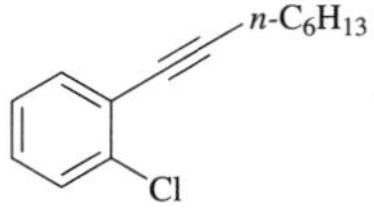 ![alkyne with n-C$_6$H$_{13}$, Cl]	RNH$_2$	Pd(OAc)$_2$, (t-Bu)$_3$P, t-BuOK, toluene, 110–112°, 14 h	2-(n-C$_6$H$_{13}$)-1-R-indole	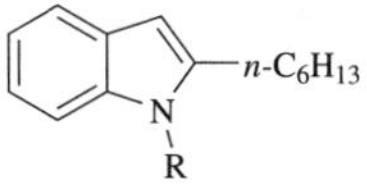 R: Bn (85), n-C$_8$H$_{17}$ (74)	87
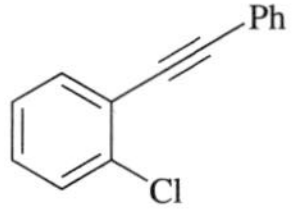	n-C$_8$H$_{17}$NH$_2$	Pd(OAc)$_2$, (t-Bu)$_3$P, t-BuOK, toluene, 110°, 14 h	2-Ph-1-(n-C$_8$H$_{17}$)-indole	(99)	87
![alkyne with Ph, Cl]	RNH$_2$	Pd(OAc)$_2$, (t-Bu)$_3$P, K$_3$PO$_4$, DMA, 130°, 14 h	2-Ph-1-R-indole	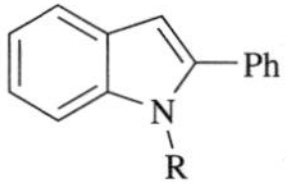 R: Ph (99), Bn (95), 4-MeOC$_6$H$_4$ (95)	87
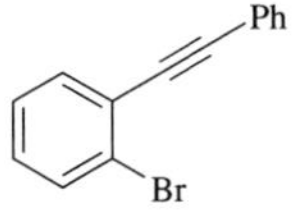	RNH$_2$	Pd(OAc)$_2$, (t-Bu)$_3$P, t-BuOK, toluene, 110°, 14 h	2-Ph-1-R-indole	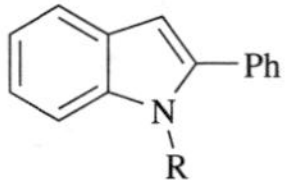 R: Ph (95), Bn (95)	87
C_{17-19} ![alkyne with Ph, Br]	4-MeC$_6$H$_4$NH$_2$				
![alkyne with N(CO$_2$Me)Bn, Br, R]		Pd$_2$(dba)$_3$, XPhos, Cs$_2$CO$_3$, 1,4-dioxane, 110°, 8–24 h	2-N(CO$_2$Me)Bn-1-(4-MeC$_6$H$_4$)-6-R-indole	R: H (82), MeO$_2$C (75)	349

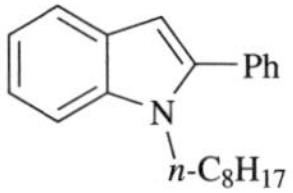

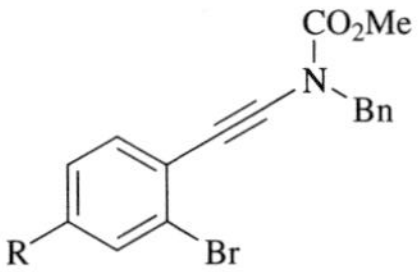

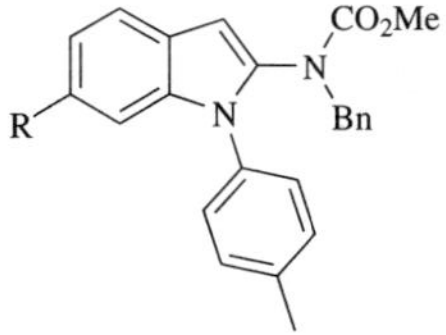

TABLE 1F. 2-SUBSTITUTED INDOLES FROM 2-ALKYNYLHALOARENES (*Continued*)

2-Alkynylhaloarene	Amine	Conditions	Product(s) and Yield(s) (%)							Refs.
C_{17-23}	R^5NH_2	$Pd_2(dba)_3$, XPhos, Cs_2CO_3, 110°, 8–24 h								349
			X	R^1	R^2	R^3	R^4	R^5	Solvent	
			Br	H	H	Ph	H	4-MeC$_6$H$_4$	1,4-dioxane (91)	
			Br	H	H	Ph	H	2-MeC$_6$H$_4$	1,4-dioxane (82)	
			Cl	O$_2$N	H	Ph	H	4-MeC$_6$H$_4$	toluene (64)	
			Cl	H	H	Bn	H	4-MeOC$_6$H$_4$	1,4-dioxane (91)	
			Br	Me	H	Ph	H	4-MeC$_6$H$_4$	toluene (65)	
			Br	H	MeO$_2$C	Ph	H	c-C$_6$H$_{11}$	1,4-dioxane (80)	
			Br	MeO	MeO	Ph	H	4-MeC$_6$H$_4$	toluene (60)	
			Cl	MeO$_2$C	H	Ph	H	Bn	1,4-dioxane (72)	
			Br	H	H	Ph	Ph	Bn	toluene (71)	
C_{23}	4-MeC$_6$H$_4$NH$_2$	$Pd_2(dba)_3$, XPhos, Cs_2CO_3, 1,4-dioxane, 110°, 8–24 h	(88)							349
C_{24-25}	R^2NH_2	1. Cp$_2$TiMe$_2$, toluene, 110°, 24 h 2. Pd$_2$(dba)$_3$, **L3**,[c] t-BuOK, 1,4-dioxane, 110°, 12 h								195
			R^1	R^2						
			c-C$_3$H$_5$	4-MeC$_6$H$_4$					(75)	
			Bu	4-MeOC$_6$H$_4$					(64)	

[a] The structure of ligand **2** is shown in the first entry of the first page of Table 1F.
[b] The reaction time was 5–18 hours.
[c] The structure of ligand **3** is shown in the second entry of the second page of Table 1F.

TABLE 1G. 2-SUBSTITUTED INDOLES FROM 2-HALO-N-ALKYNYLANILIDES

2-Halo-N-alkynylanilide	Amine	Conditions	Product(s) and Yield(s) (%)	Refs.
C$_{18}$				
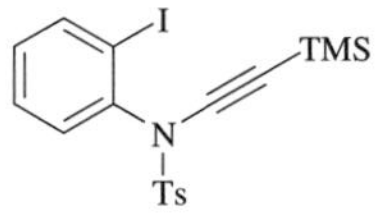 (2-iodo-N-Ts anilide with TMS-alkyne)	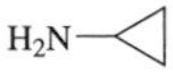 H$_2$N–cyclopropyl	1. PdCl$_2$(PPh$_3$)$_2$, K$_2$CO$_3$, THF, 80° 2. Bu$_4$NF, THF	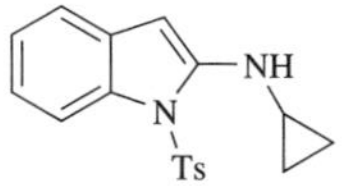 2-(cyclopropyl-NH)-1-Ts-indole (66)	29
(2-iodo-N-Ts anilide with TMS-alkyne)	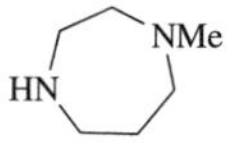 HN(CH$_2$)$_5$NMe (homopiperazine)	1. PdCl$_2$(PPh$_3$)$_2$, K$_2$CO$_3$, THF, 80° 2. Bu$_4$NF, THF	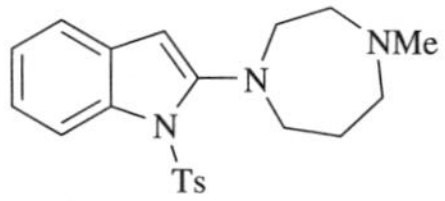 2-(4-Me-homopiperazinyl)-1-Ts-indole (75)	29
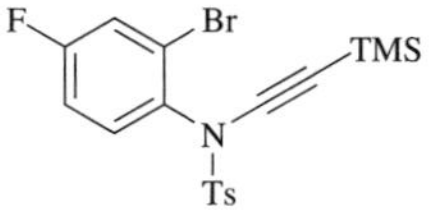 (2-Br, 4-F-N-Ts anilide with TMS-alkyne)	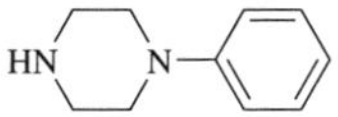 HN(piperazinyl)Ph	1. PdCl$_2$(PPh$_3$)$_2$, K$_2$CO$_3$, THF, 80° 2. Bu$_4$NF, THF	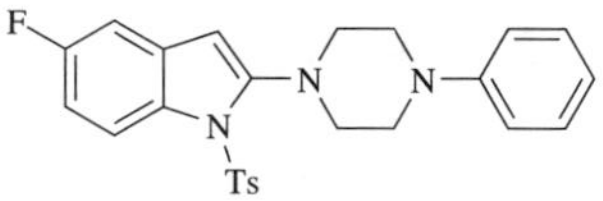 5-F-2-(4-Ph-piperazinyl)-1-Ts-indole (99)	29
C$_{19}$				
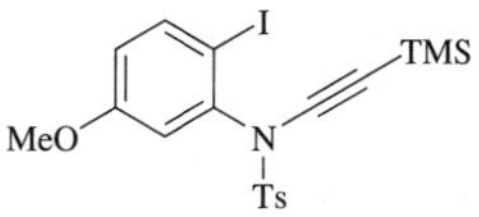 (2-I, 5-MeO-N-Ts anilide with TMS-alkyne)	 HN(piperazinyl)NBoc	1. PdCl$_2$(PPh$_3$)$_2$, K$_2$CO$_3$, THF, 80° 2. Bu$_4$NF, THF	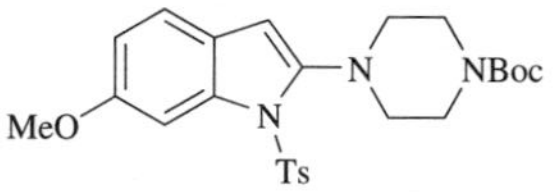 6-MeO-2-(4-Boc-piperazinyl)-1-Ts-indole (89)	29

TABLE 1H. 2-SUBSTITUTED INDOLES FROM 2-ALKYNYLISOCYANATOBENZENES

2-Alkynylisocyanatobenzene	Conditions	Product(s) and Yield(s) (%)	Refs.
C_{12-16}			
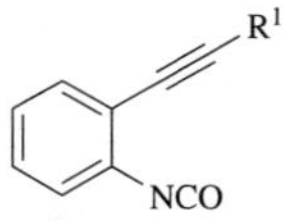 (alkynyl-NCO benzene with R^1)	Na_2PdCl_4, DCE, 100°, R^2OH	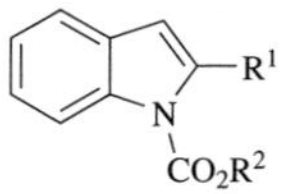 (2-R^1-indole-N-CO_2R^2)	28

R^1	R^2	Time (h)	
Pr	Me	1.5	(69)
Pr	Pr	1.5	(74)
Pr	i-Pr	2	(67)
Pr	Bu	3	(64)
Pr	Bu	24	(56)
Pr	CH_2=$CHCH_2$	4	(85)
t-Bu	Pr	2	(89)
c-C_5H_9	Pr	1.5	(83)
Ph	Pr	1.5	(59)
4-$MeOC_6H_4$	Pr	1.5	(58)
4-$CF_3C_6H_4$	Pr	1.5	(55)

TABLE 2A. 3-SUBSTITUTED INDOLES FROM 2-HALOANILINES AND ALKYNES

2-Haloaniline	Alkyne	Conditions	Product(s) and Yield(s) (%)	Refs.
C$_6$ 2-iodoaniline with R^1	TES—≡—R^2	1. Catalyst, base, DMF, 120° 2. HCl (2 M) or Bu$_4$NF (1 M in THF)	3-R^3-indole with R^1	350

R^1	R^2	R^3	Catalyst	Base	Time (h)	
H	(CH$_2$)$_2$OTES	(CH$_2$)$_2$OH	Pd$_{(ALD)}$[a]	K$_2$CO$_3$	14	(48)
H	(CH$_2$)$_2$OTES	(CH$_2$)$_2$OH	Pd$_{(ALD)}$[a]	Na$_2$CO$_3$	14	(62)
H	(CH$_2$)$_2$OTES	(CH$_2$)$_2$OH	Pd/NaY	Na$_2$CO$_3$	14	(98)
5-O$_2$N	(CH$_2$)$_2$OTES	(CH$_2$)$_2$OH	Pd$_{(ALD)}$[a]	Na$_2$CO$_3$	216	(60)
5-O$_2$N	(CH$_2$)$_2$OTES	(CH$_2$)$_2$OH	Pd/NaY	Na$_2$CO$_3$	144	(70)
H	Ph	Ph	Pd$_{(ALD)}$[a]	Na$_2$CO$_3$	24	(80)
H	Ph	Ph	Pd/NaY	Na$_2$CO$_3$	24	(70)
5-O$_2$N	Ph	Ph	Pd$_{(ALD)}$[a]	Na$_2$CO$_3$	288	(42)
5-O$_2$N	Ph	Ph	Pd/NaY	Na$_2$CO$_3$	84	(80)

2-Haloaniline	Alkyne	Conditions	Product(s) and Yield(s) (%)	Refs.
C$_8$ 4-OMe, 6-OMe, 2-iodoaniline	TES—≡—R^1	1. Catalyst, K$_2$CO$_3$, DMF, 120°, 14 h 2. HCl (2 M) or Bu$_4$NF (1 M in THF)	4-OMe, 6-OMe indole with R^2	350

R^1	R^2	Catalyst	
Ph	Ph	Pd$_{(ALD)}$[a]	(83)
Ph	Ph	Pd/NaY	(88)
(CH$_2$)$_2$OTES	(CH$_2$)$_2$OH	Pd$_{(ALD)}$[a]	(70)
(CH$_2$)$_2$OTES	(CH$_2$)$_2$OH	Pd/NaY	(81)

[a] Pd$_{(ALD)}$ is palladium supported on activated carbon purchased from Aldrich.

TABLE 2B. 3-SUBSTITUTED INDOLES FROM 2-ALKYNYLANILIDES

	2-Alkynylanilide	Conditions	Product(s) and Yield(s) (%)	Refs.
C_9	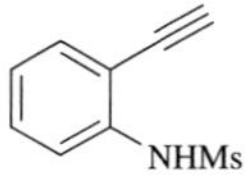 (2-ethynyl-NHMs aniline)	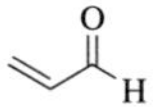 H, Pd(OAc)$_2$, LiBr, THF, rt, 3 d	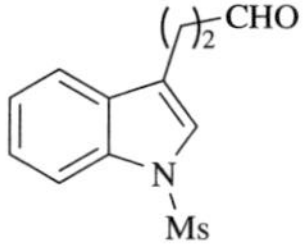 3-((CH$_2$)$_2$CHO)-1-Ms-indole (27)	103
C_{10}	(2-ethynyl-NHCOCF$_3$ aniline)	ArI, Pd$_2$(dba)$_3$, K$_2$CO$_3$, DMSO, 40°	3-Ar-indole Ar / Time (h) Ph / 1.5 (67) 3-FC$_6$H$_4$ / 1.5 (57) 4-ClC$_6$H$_4$ / 1.25 (71) 3-O$_2$NC$_6$H$_4$ / 2 (86) 4-MeC$_6$H$_4$ / 2 (63) 3-CF$_3$C$_6$H$_4$ / 2 (82) 4-MeOC$_6$H$_4$ / 1.75 (56) 3-O$_2$N-4-Me-C$_6$H$_3$ / 1 (69) 4-MeCONHC$_6$H$_4$ / 4 (62) 4-EtO$_2$CC$_6$H$_4$ / 8 (69) 3-EtO$_2$CC$_6$H$_4$ / 1.15 (78)	100
C_{18}	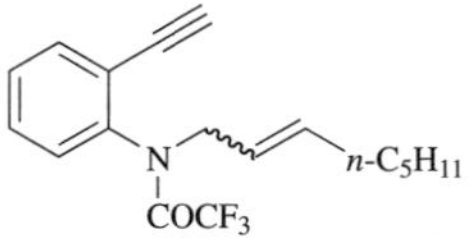 (2-ethynyl aniline with N(COCF$_3$)CH$_2$CH=CH-n-C$_5$H$_{11}$), $(E):(Z) = 90:10$	Pd(PPh$_3$)$_4$, K$_2$CO$_3$, MeCN, 90°, 24 h 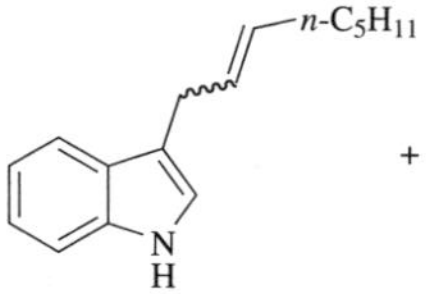	I (3-(CH=CH-n-C$_5$H$_{11}$)-indole) + II (3-(CH(n-C$_5$H$_{11}$)CH=CH$_2$)-indole) I + II (38), I:II = 92:8 I $(E):(Z) = 89:11$	102
C_{19}	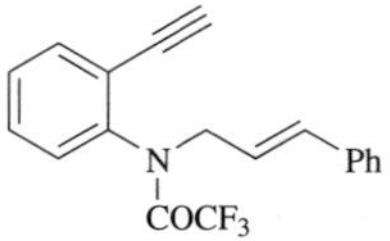 (2-ethynyl aniline with N(COCF$_3$)CH$_2$CH=CH-Ph)	Pd(PPh$_3$)$_4$, K$_2$CO$_3$, MeCN, 90°, 24 h 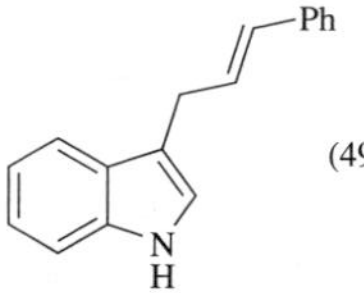	3-(CH$_2$CH=CH-Ph)-indole (49)	102

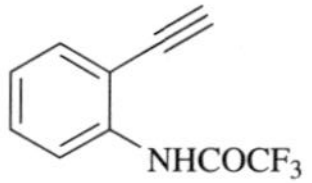

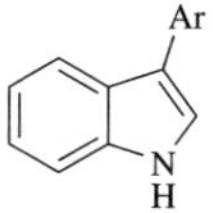

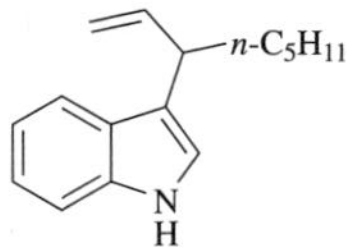

TABLE 2C. 3-SUBSTITUTED INDOLES FROM 3-IODO-N-ALLYLANILINE AND INTERNAL ALKYNES

3-Iodo-N-allylaniline	Alkyne	Conditions	Product(s) and Yield(s) (%)	Refs.
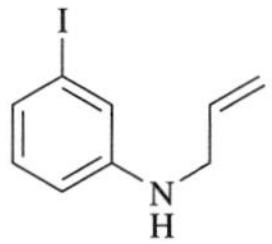 C₉ (3-iodo-N-allylaniline structure)	$R^1\!-\!\!\equiv\!\!-R^2$	Pd(OAc)₂, dppm, CsO₂CCMe₃, DMF, 100°, 3 h	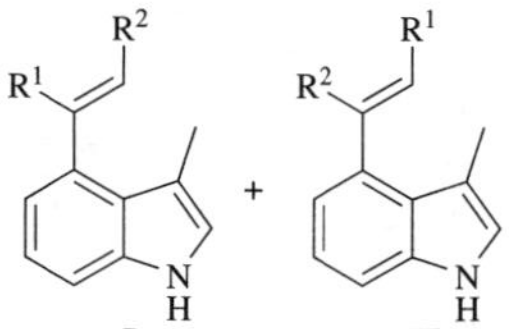 **I** + **II** R¹ R² **I + II** **I:II** Me Ph (26) 15:1 Et Ph (45) 10:1 Ph Ph (31) —	104

TABLE 2D. 3-SUBSTITUTED INDOLES FROM 2-HALO-N-ALKYLANILINES

2-Halo-N-alkylaniline	Conditions	Product(s) and Yield(s) (%)			Refs.

C_{12-13}

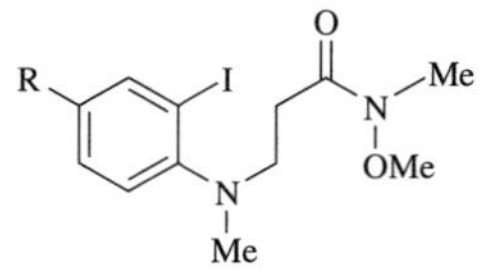

2-Halo-N-alkylaniline: (structure with R, I, N(Me)C(O)CH$_2$N(Me)(OMe))

Conditions: Pd(PPh$_3$)$_4$, phenol, t-BuOK, THF, reflux

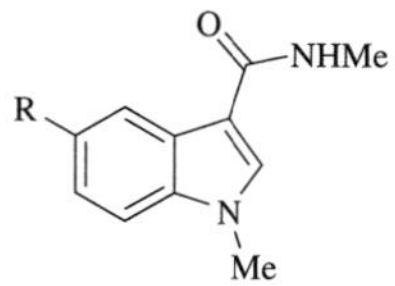

Product: 5-R-1-methyl-3-(N-methylcarboxamido)indole

R	Time (h)	
H	8	(58)
Cl	18	(78)
Me	8	(59)

Refs. 351, 352

C_{12-19}

2-Halo-N-alkylaniline: (structure with R^1, I, N(Me)CH$_2$CH$_2$CO$_2$R^2)

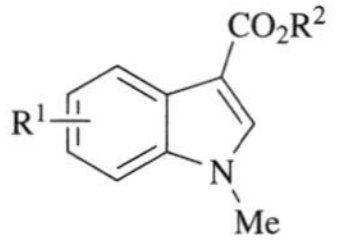

Conditions: Pd(PPh$_3$)$_4$, t-BuOK, THF, reflux

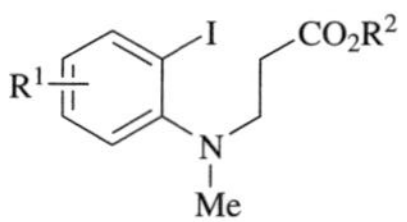

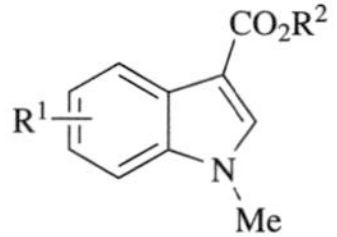

Product: 1-methyl-3-CO$_2$R^2-indole with R^1 substitution

R^1	R^2	Time (h)	[a]
H	Me	8	(20)
5-Cl	Me	24	(6)
5-F	Me	20	(70)
6-F	Me	24	(20)
6-Cl	Me	24	(20)
6-Me	Me	8	(23)
5-MeO	Me	8	(46)
5-MeO$_2$C	Me	16	(24)
5-Me	Bn	8	(7)

Refs. 351, 352

Conditions: Pd(PPh$_3$)$_4$, K$_3$PO$_4$, DMF, 90°

Product: 1-methyl-3-CO$_2$R^2-indole with R^1 substitution

R^1	R^2	Time (h)	[a]
H	Me	24	(51)
5-Cl	Me	24	(32)
5-F	Me	24	(60)
6-F	Me	24	(7)
6-Cl	Me	24	(30)
6-Me	Me	24	(42)
5-MeO	Me	48	(67)
5-MeO$_2$C	Me	36	(29)
5-Me	Bn	48	(36)

Refs. 351, 352

C$_{13}$

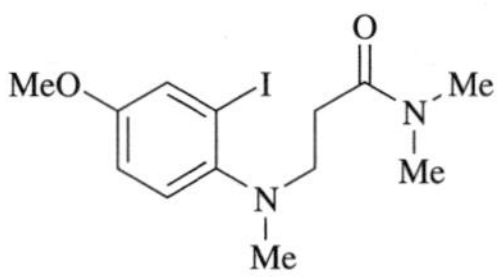

[Structure: MeO-substituted iodoaryl with N(Me)CH$_2$CH$_2$C(O)N(Me)$_2$ side chain]

Pd(PPh$_3$)$_4$, phenol, t-BuOK, THF, reflux, 18 h

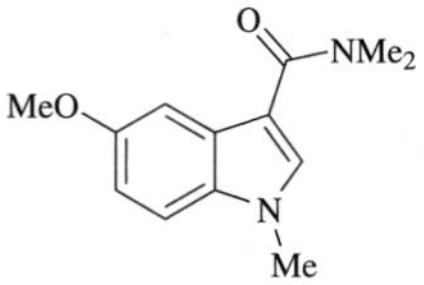

[Structure: 5-MeO-1-Me-indole-3-C(O)NMe$_2$] (58)

351, 352

---

[a] The corresponding indolines were isolated in variable amounts.

TABLE 2E. 3-SUBSTITUTED INDOLES FROM 2-IODO-N-PROPARGYLANILIDES AND N-2-(HALOPHENYL)ALLENAMIDES

Substrate	Conditions	Product(s) and Yield(s) (%)	Refs.
C_{11-12} 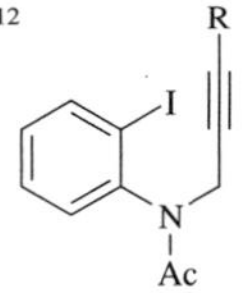	Norbornene, Pd(OAc)$_2$, PPh$_3$, K$_2$CO$_3$, Et$_4$NCl, MeCN, rt, 18 h	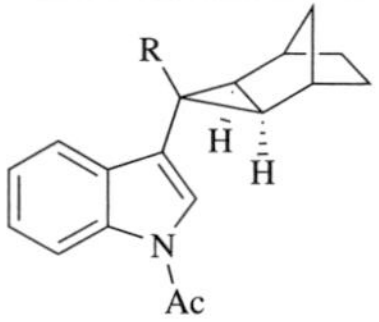 R: H (40); Me (45)	32
C_{14} 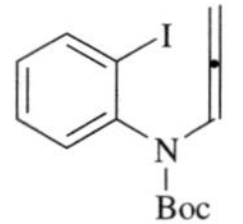	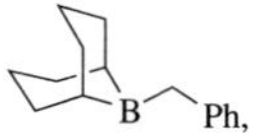 Pd$_2$(dba)$_3$, 9-BBN-CH$_2$CH$_2$Ph, aq Cs$_2$CO$_3$ (3 M), EtOH, 80°	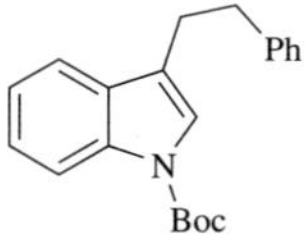 3-(CH$_2$CH$_2$Ph)-N-Boc-indole (64)	353
	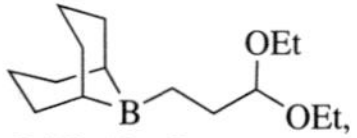 PdCl$_2$(dppf), 9-BBN-CH$_2$CH$_2$CH(OEt)$_2$, aq Cs$_2$CO$_3$ (3 M), DMF, 50°	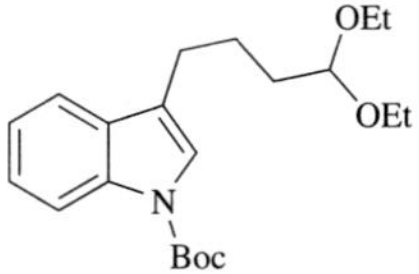 3-(CH$_2$CH$_2$CH(OEt)$_2$)-N-Boc-indole (66)	353
	NaN$_3$, Pd(PPh$_3$), DMF, rt	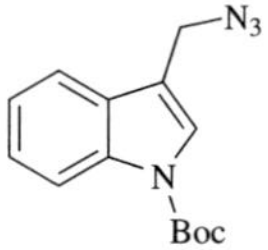 3-(CH$_2$N$_3$)-N-Boc-indole (97)	353
	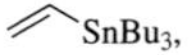 CH$_2$=CHSnBu$_3$, Pd$_2$(dba)$_3$, DMF, 80°	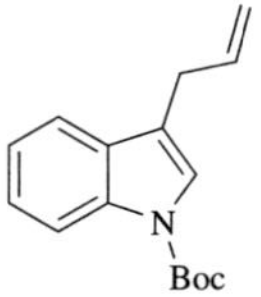 3-allyl-N-Boc-indole (96)	353

Substrate	Conditions	Product(s) and Yield(s) (%)	Refs.

C_{14-16}

Substrate: 2-iodo-N-R¹-N-vinylaniline

CO (1 atm), PdCl₂(dppf), Et₃N, MeOH/DMF, 60°

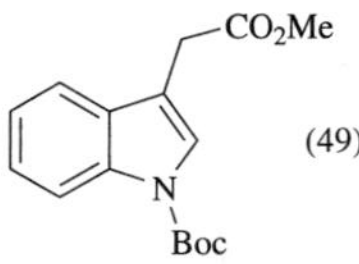

Product: 3-(CH₂CO₂Me)-N-Boc-indole (49)

353

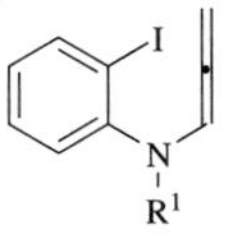

(HO)₂BR², Pd₂(dba)₃, aq Cs₂CO₃ (3 M), EtOH, 80°

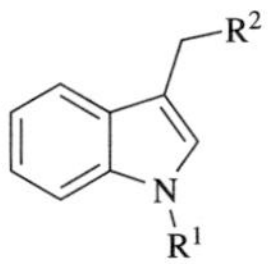

Product: 3-(CH₂R²)-N-R¹-indole

R¹	R²	
Boc	4-MeOC₆H₄	(90)
Boc	2-thienyl	(58)
Ts	Ph	(88)
Ts	4-ClC₆H₄	(71)
Ts	Ph-CH=CH-	(73)

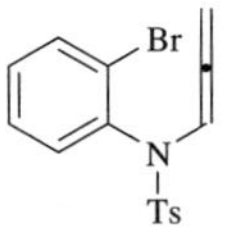

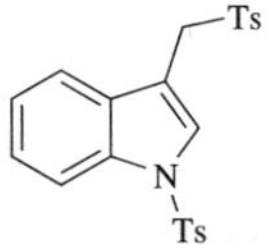

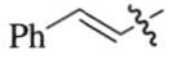

353

C_{16}

Substrate: 2-bromo-N-Ts-N-vinylaniline

NaTs, Pd(PPh₃)₄, DMF, 70°

Product: 3-(CH₂Ts)-N-Ts-indole (83)

353

9-BBN-CH₂CH(OEt)₂ derivative, PdCl₂(dppf), aq Cs₂CO₃ (3 M), DMF

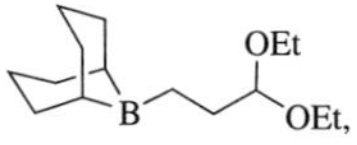

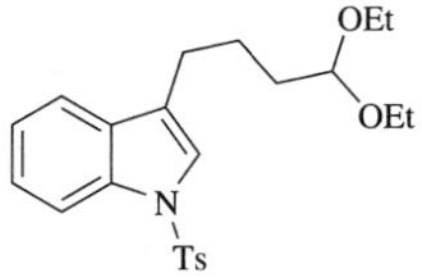

Product: 3-(CH₂CH₂CH(OEt)₂)-N-Ts-indole — (82)

Wait — product shown: 3-(CH₂CH(OEt)OEt...) N-Ts-indole (82)

353

C_{28}

Substrate: bis(2-iodo-N-SO₂Ph-anilino)-2-butyne

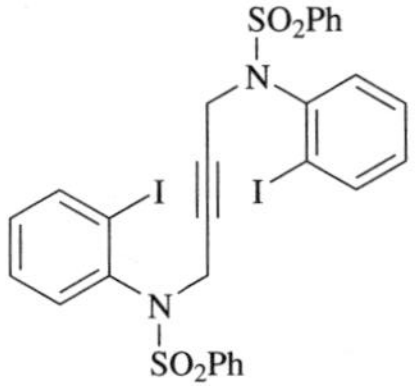

Pd(OAc)₂, PPh₃, (Me₃Sn)₂, anisole, 100°, 16 h

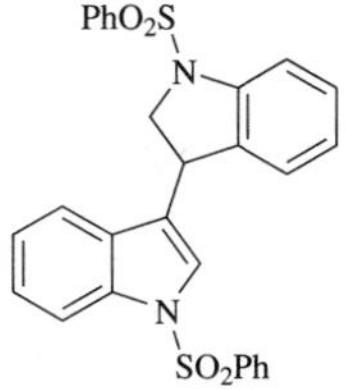

Product: 3-(N-SO₂Ph-indolin-3-yl)-N-SO₂Ph-indole (48)

101

TABLE 3A. 2,3-DISUBSTITUTED INDOLES FROM 2-HALOANILINES, 2-IODOBENZOIC ACIDS, OR ANILINES AND ALKYNES

Substrate	Alkyne	Conditions	Product(s) and Yield(s) (%)	Refs.
C_6				
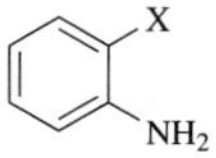 2-X-aniline (X = Br, Cl)	 Ph—≡—Ph	Pd(OAc)$_2$, PCy$_3$, K$_2$CO$_3$, NMP	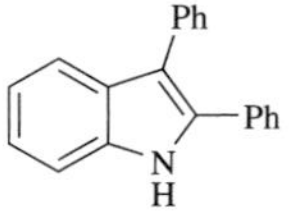 2-Ph-3-Ph-indole: X=Br (99), X=Cl (97)	122
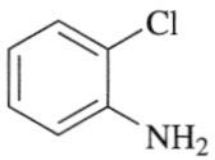 2-Cl-aniline	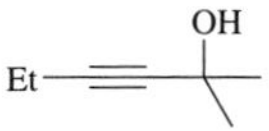 Et—≡—C(OH)Me$_2$	Pd(OAc)$_2$, dtbpf, K$_2$CO$_3$, NMP	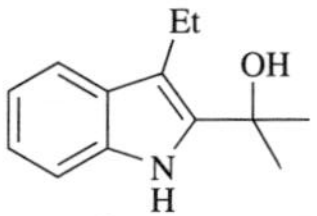 I 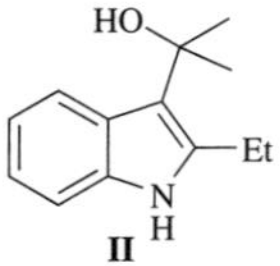 (2-C(OH)Me$_2$-3-Et) + II (2-Et-3-C(OH)Me$_2$), I + II (60), I:II = >99:1	122
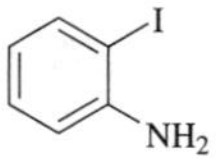 2-I-aniline	 Et—≡—Me	Pd(OAc)$_2$, PPh$_3$, Bu$_4$NCl, K$_2$CO$_3$, DMF, 100°, 24 h	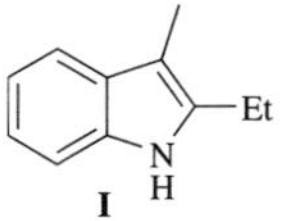 I 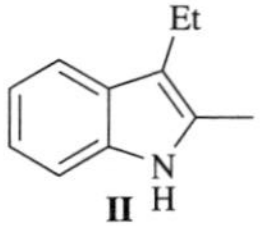 (2-Me-3-Et) + II (2-Et-3-Me), I + II (62), I:II = 60:40	30, 31
2-I-aniline	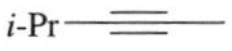 i-Pr—≡—Me	Pd(OAc)$_2$, PPh$_3$, Bu$_4$NCl, K$_2$CO$_3$, DMF, 100°, 24 h	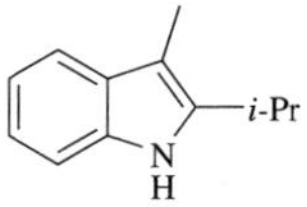 2-Me-3-i-Pr-indole (62) 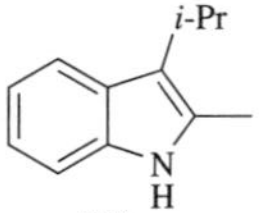 + 2-i-Pr-3-Me-indole (25)	30, 31
2-I-aniline	 R^1—≡—R^2	Pd(OAc)$_2$, LiCl, base, DMF, 100°	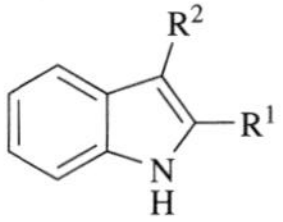 2-R^2-3-R^1-indole; R^1=Pr, R^2=Et, Base=K$_2$CO$_3$, Time=20 h (80); R^1=HO-cyclohexyl, R^2=Et, Base=KOAc, Time=24 h (85)	30, 31

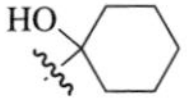

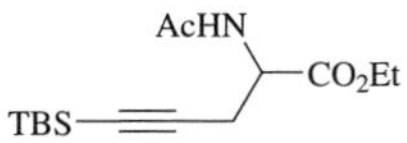

AcHN-CH(CO₂Et)-CH₂-C≡C-TBS

$\xrightarrow{\text{Pd(OAc)}_2,\ \text{Bu}_4\text{NCl},\ \text{KOAc, DMF, 90–100°, 22 h}}$ 3-(2-acetamido-2-ethoxycarbonylethyl)-2-TBS-indole (38)   115

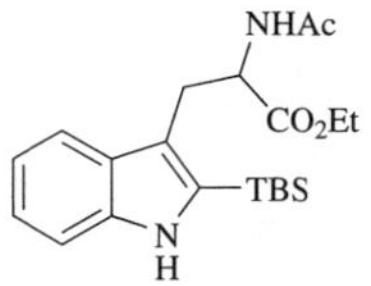

$R^1\!\!-\!\!\equiv\!\!-\!\!R^2$ $\xrightarrow{\text{Pd(OAc)}_2,\ \text{phosphine, Bu}_4\text{NCl, base, DMF, 100°}}$ 2-R^2-3-R^1-indole   30, 31

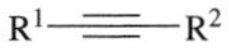

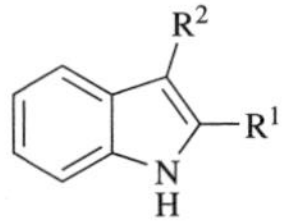

R^1	R^2	Phosphine	Base	Time (h)	
CMe₂OH	Me	PPh₃	Na₂CO₃	12	(52)
TMS	Me	PPh₃	Na₂CO₃	24	(98)
TMS	CH₂OH	PPh₃	Na₂CO₃	24	(60)
t-Bu	Me	PPh₃	Na₂CO₃	24	(82)
CMe₂OH	CMe₂OH	PPh₃	Na₂CO₃	72	(54)
TMS	TMS	—	NaOAc	20	(54)
TMS	Bu	—	NaOAc	12	(81)
TMS	Me	—	KOAc	20	(54)
c-C₆H₁₁	Ph	—	NaOAc	16	(68)
 TMS	MeC=CH₂	—	KOAc	24	(70)
1-hydroxycyclohexyl					

Ph-C≡C-N(oxazolidin-2-one) $\xrightarrow{\text{Pd(OAc)}_2,\ t\text{-Bu}_3\text{P·HBF}_4,\ \text{K}_2\text{CO}_3,\ \text{DMF, 100°, 21 h}}$ **I** (2-Ph-3-oxazolidinonyl-indole) + **II** (3-Ph-2-oxazolidinonyl-indole)   354

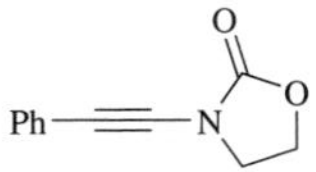

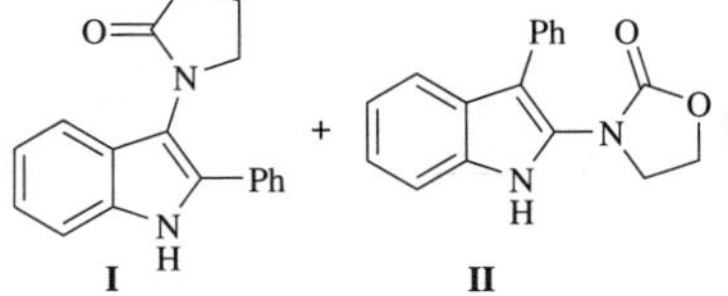

I + II (65), **I:II** = 4:1

TABLE 3A. 2,3-DISUBSTITUTED INDOLES FROM 2-HALOANILINES, 2-IODOBENZOIC ACIDS, OR ANILINES AND ALKYNES (*Continued*)

Substrate	Alkyne	Conditions	Product(s) and Yield(s) (%)	Refs.
C_6				
(4-iodoaniline with R)	BocHN-CH(CO$_2$Et)-CH$_2$-C≡C-TBS	Pd(OAc)$_2$, Bu$_4$NCl, Et$_3$N, DMF, 90–100°	2,3-disubstituted indole (NHBoc, CO$_2$Et, TBS): R / Time (h) / yield — H 23 (62); 5-NO$_2$ 24 (53); 5-Cl 20 (48); 5-F 22 (47); 6-NO$_2$ 24 (46)	115
C_{6-7}				
(2-iodoaniline with R^1, R^2)	TES-C≡C-CH$_2$-piperazinone (iPr, OEt)	Pd(OAc)$_2$, LiCl, Na$_2$CO$_3$, DMF, 100°	**I** + **II**; R^1, R^2, I, II: NO$_2$, H (83) (4); F, H (62) (—); Cl, Cl (80) (—); OMe, H (65) (<5); H, OMe (77) (<5)	112, 117

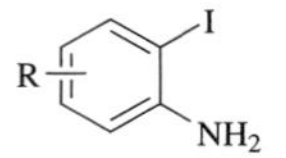

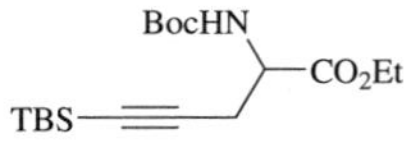

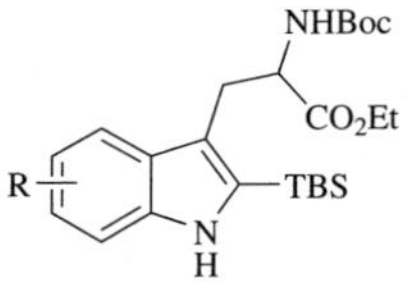

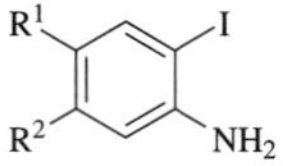

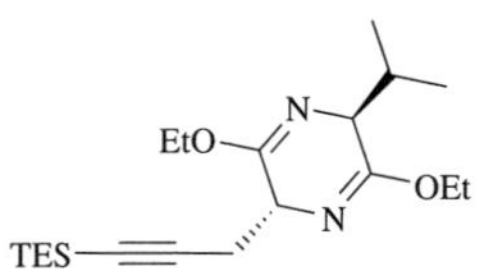

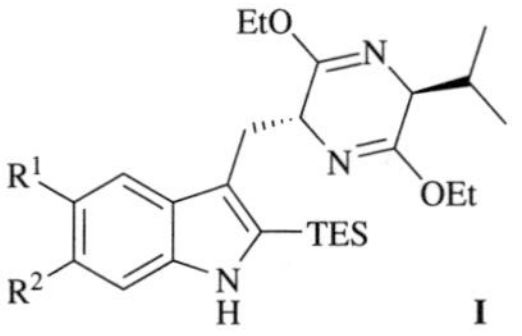

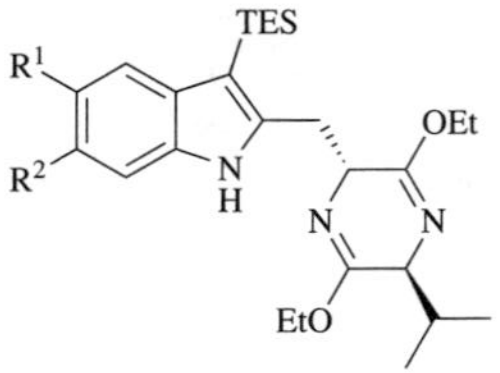

R^3—≡—R^4

Pd(PPh$_3$)$_4$, PPh$_3$,
Et$_3$N, DMF,
80°, 8 h

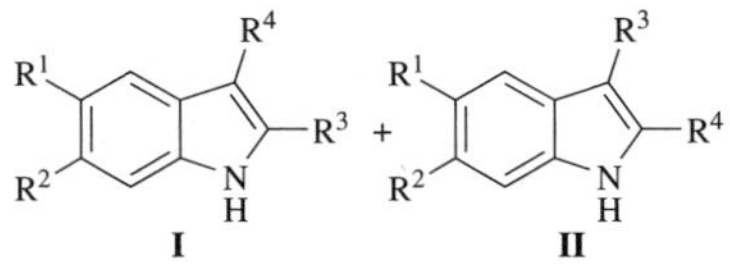

[Scheme showing products I and II: 2,3-disubstituted indoles with R^1, R^2 on benzene ring, R^3, R^4 at 2,3-positions; 355, 356]

R^1	R^2	R^3	R^4	I + II	I:II^a
H	H	CF$_3$	4-ClC$_6$H$_4$	(79)	78:22
H	H	CF$_3$	2-ClC$_6$H$_4$	(59)	76:24
H	H	CF$_3$	3-ClC$_6$H$_4$	(80)	72:28
H	H	CHF$_2$	4-ClC$_6$H$_4$	(70)	100:0
Cl	H	CF$_3$	4-ClC$_6$H$_4$	(79)	84:16
O$_2$N	H	CF$_3$	4-ClC$_6$H$_4$	(57)	84:16
Cl	Cl	CF$_3$	4-ClC$_6$H$_4$	(74)	84:16
H	H	CF$_3$	4-MeC$_6$H$_4$	(81)	88:12
H	H	CF$_3$	4-MeOC$_6$H$_4$	(73)	91:9
H	H	CF$_3$	4-EtO$_2$CC$_6$H$_4$	(92)	68:32
H	H	CF$_3$	4-O$_2$NC$_6$H$_4$	(65)	53:47
H	H	CF$_3$	Ph(CH$_2$)$_2$	(47)	81:19
H	H	CF$_3$	PhCH(Me)CH$_2$	(55)	72:28
Me	H	CF$_3$	4-ClC$_6$H$_4$	(82)	81:19
H	MeO	CF$_3$	4-ClC$_6$H$_4$	(65)	88:12

TABLE 3A. 2,3-DISUBSTITUTED INDOLES FROM 2-HALOANILINES, 2-IODOBENZOIC ACIDS, OR ANILINES AND ALKYNES (*Continued*)

Substrate	Alkyne	Conditions	Product(s) and Yield(s) (%)					Refs.
C_{6-7} 2-iodoaniline with R^1, R^2 substituents	R^3—≡—R^4	$Pd_2(dba)_3 \cdot CHCl_3$, $P(o\text{-tol})_3$, Et_3N, DMF, 80°, 8 h	I (R^1, R^2, R^3, R^4 indole) + II (regioisomer)					356
			R^1	R^2	R^3	R^4	I+II	I:II[a]
			H	H	CF_3	4-ClC_6H_4	(85)	34:66
			H	H	CF_3	2-ClC_6H_4	(56)	16:84
			H	H	CF_3	3-ClC_6H_4	(74)	27:73
			H	H	CHF_2	4-ClC_6H_4	(—)	—
			Cl	H	CF_3	4-ClC_6H_4	(89)	22:78
			O_2N	H	CF_3	4-ClC_6H_4	(81)	36:64
			Cl	Cl	CF_3	4-ClC_6H_4	(86)	11:89
			H	H	CF_3	4-MeC_6H_4	(79)	9:91
			H	H	CF_3	4-MeOC_6H_4	(52)	8:92
			H	H	CF_3	$4\text{-EtO}_2CC_6H_4$	(85)	34:66
			H	H	CF_3	$4\text{-O}_2NC_6H_4$	(51)	48:52
			H	H	CF_3	$Ph(CH_2)_2$	(60)	22:78
			H	H	CF_3	$PhCH(Me)CH_2$	(27)	17:83
			Me	H	CF_3	4-ClC_6H_4	(62)	27:73
			H	MeO	CF_3	4-ClC_6H_4	(66)	17:83
2-iodoaniline NHR^1	R^2—≡—$P(R^3)_2$=O	$Pd(acac)_2$, K_2CO_3, DMSO	I (2-$P(R^3)_2$=O, 3-R^2 indole-R^1) + II (2-R^2, 3-$P(R^3)_2$=O indole-R^1)					357

2-bromoaniline (R¹ on ring, Br, NH₂) + R²—≡—R³ →

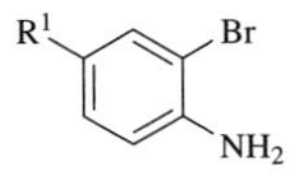

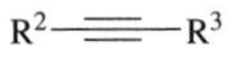

Pd(OAc)₂, phenylurea,
K₂CO₃, DMF, 130°, 30 h

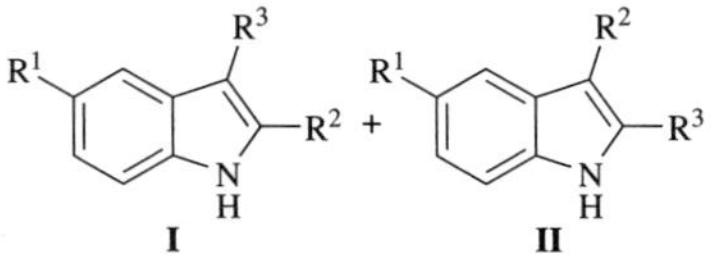

→ I (indole with R¹, R², R³, NH) + II (indole with R¹, R², R³ positions swapped, NH)

R¹	R²	R³	Time (h)	Temp (°)	I	II	358
H	Ph	Ph	11	120	(80)	(—)	
Me	Ph	t-Bu	22	120	(50)	(—)	
Me	Ph	Ph	11	90	(80)	(20)	
Me	Ph₂P(O)	Ph	11	90	(57)	(—)	

R¹	R²	R³	I + II	I:II
H	Ph	Ph	(84)	—
H	Ph	Me	(62)	96:4
H	Ph	Pr	(67)	82:18
H	Pr	Pr	(80)	—
H	Ph	CO₂H	(—)	—
H	Ph	CO₂Ph	(—)	—
H	2-MeOC₆H₄	Ph	(71)	70:30
Me	Ph	Ph	(86)	—
Me	Ph	Me	(65)	88:12
Me	Ph	Pr	(55)	80:20

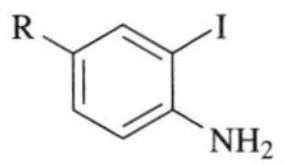

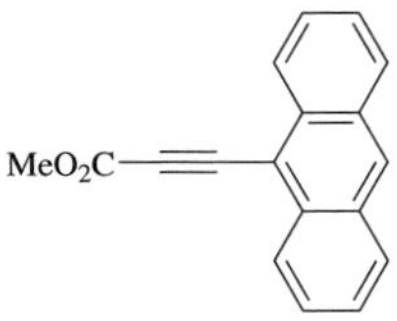

2-iodoaniline (R on ring, I, NH₂) + 9-(methoxycarbonylethynyl)anthracene (MeO₂C—≡—anthracenyl) →

PdCl₂(PPh₃)₂, Bu₄NBr, NaOAc, DMF, 100°, 10 h

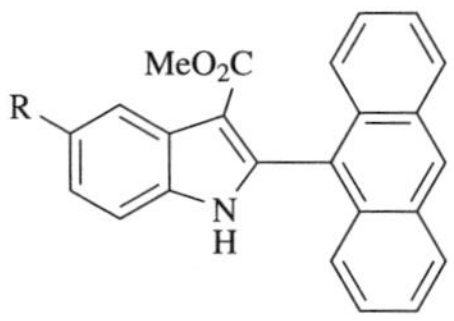

→ indole bearing CO₂Me at C-3 and 9-anthracenyl at C-2, with R on ring, NH

R		359
H	(34)	
Me	(30)	

TABLE 3A. 2,3-DISUBSTITUTED INDOLES FROM 2-HALOANILINES, 2-IODOBENZOIC ACIDS, OR ANILINES AND ALKYNES (*Continued*)

Substrate	Alkyne	Conditions	Product(s) and Yield(s) (%)	Refs.
C_{6-8}				
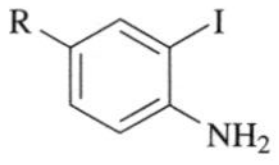 (2-iodo-4-R-aniline)	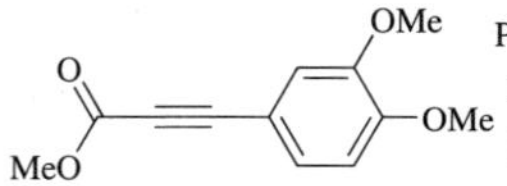 (3,4-dimethoxyphenyl propynone, MeO)	PdCl$_2$(PPh$_3$)$_2$, Bu$_4$NBr, NaOAc, DMF, 100°, 0.5 h	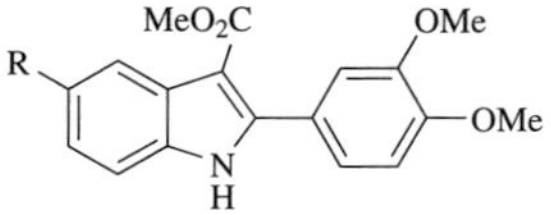 2-(3,4-dimethoxyphenyl)-3-MeO$_2$C-indole, R at 5-position, OMe, OMe R / yield: H (65) Me (40); F (40) NC (38); Cl (47) MeO$_2$C (33); O$_2$N (30)	359
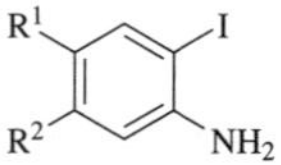 (2-iodo-4-R^1-5-R^2-aniline)	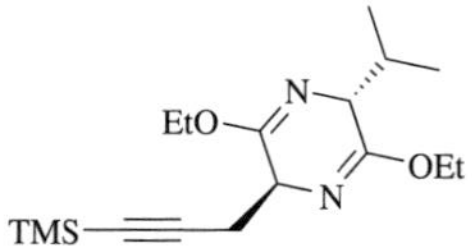 (TMS-alkyne with bis-lactim ether, iPr, OEt, EtO)	Pd(OAc)$_2$, LiCl, Na$_2$CO$_3$, DMF, 100°	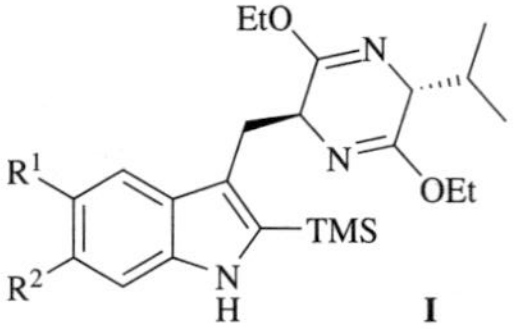 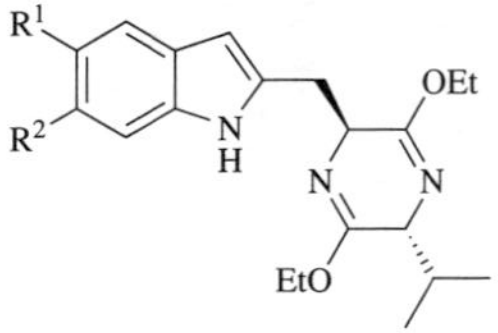 **I** (TMS-substituted indole product) + **II** (regioisomer) R^1 / R^2 / I / II: H, H (81) (—); F, H (50) (15); NO$_2$, H (65) (22); Me, H (63) (—); Me, Me (70) (—)	112

406

C$_{6-12}$			
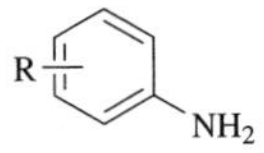 ArNH$_2$ (R-substituted aniline)	 MeO$_2$C—≡—CO$_2$Me	Pd(OAc)$_2$, O$_2$ (1 atm), DMA/pivOH (4:1), 120°, 12 h	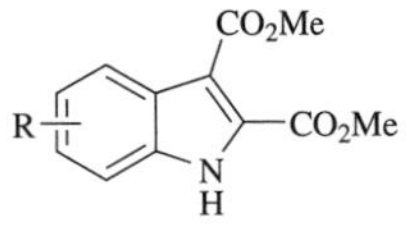 Indole-2,3-dicarboxylate (CO$_2$Me at 2,3; R on ring), NH	360

R		R	
5-Cl	(45)	5-MeO	(99)
5-F	(88)	7-MeO	(81)
5-OH	(38)	5-CF$_3$O	(46)
5-Me	(93)	5-EtO$_2$C	(97)
6-Me	(95)	6-i-Pr	(99)
7-Me	(72)	6-c-C$_6$H$_{11}$	(81)

C$_7$

2-iodo-N-methylaniline + Pr—≡—Pr, Pd(OAc)$_2$, PPh$_3$, Bu$_4$NCl, K$_2$CO$_3$, DMF, 100°, 24 h → 2,3-dipropyl-1-methylindole (71)   30, 31

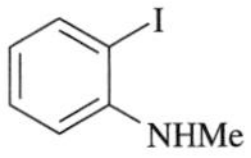

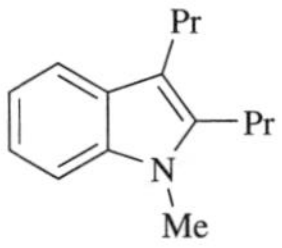

2-bromo-N-methylaniline + R^1—≡—R^2, Pd(OAc)$_2$, phenylurea, K$_2$CO$_3$, DMF, 130°, 30 h →

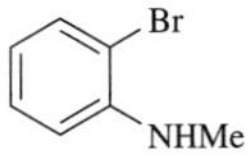

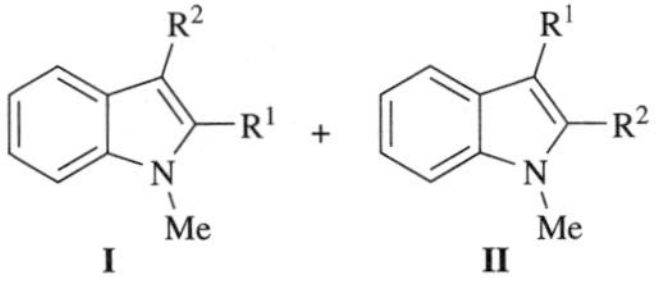

I (3-R^2, 2-R^1, N-Me) + II (3-R^1, 2-R^2, N-Me)   358

R^1	R^2	I + II	I:II
Ph	Me	(51)	78:22
Ph	Ph	(67)	—

2-iodobenzoic acid, 1. NaN$_3$, CbzCl, t-BuONa, DMF, 75°, 5 h; 2. Pd(OAc)$_2$, alkyne, Na$_2$CO$_3$, 120°, 16 h → 2-R^1,3-R^2-indole (NH)   361

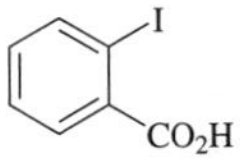

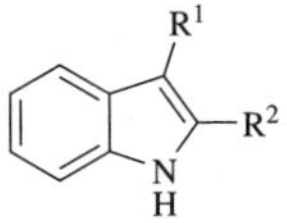

R^1	R^2	
t-Bu	Me	(56)
Ph	TMS	(82)
Ph	Ph	(77)

TABLE 3A. 2,3-DISUBSTITUTED INDOLES FROM 2-HALOANILINES, 2-IODOBENZOIC ACIDS, OR ANILINES AND ALKYNES (*Continued*)

Substrate	Alkyne	Conditions	Product(s) and Yield(s) (%)	Refs.
C_7				
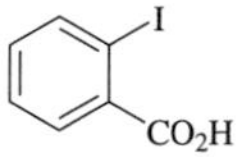 (2-iodobenzoic acid, CO_2H)	 R^1—≡—R^2	1. NaN_3, PhOCOCl, *t*-BuONa, NMP, 75°, 5 h 2. Piperidine, NMP, 75°, 3 h 3. $Pd(OAc)_2$, alkyne, Na_2CO_3, 120°, 16 h	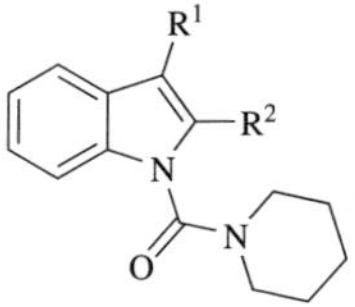 (indole with R^1, R^2, N-C(O)-piperidine) R^1 R^2 Pr Pr (54) Ph Ph (68)	361
	 Ph—≡—Ph	1. NaN_3, PhOCOCl, *t*-BuONa, NMP, 75°, 5 h 2. HNR_2, NMP, 75°, 3 h 3. $Pd(OAc)_2$, alkyne, Na_2CO_3, 120°, 16 h	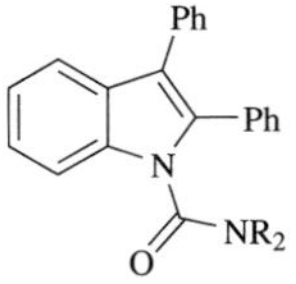 (2,3-diphenyl indole, N-C(O)-NR_2) Amine morpholine (64) pyrrolidine (62)	361
	 Pr—≡—Pr	1. NaN_3, PhOCOCl, *t*-BuONa, NMP, 75°, 5 h 2. RNH_2, NMP, 75°, 3 h 3. $Pd(OAc)_2$, alkyne, Na_2CO_3, 120°, 16 h	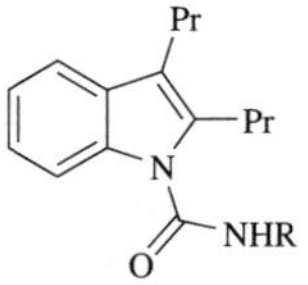 (2,3-dipropyl indole, N-C(O)-NHR) R Ph(Me)CH (59) Ph(CH$_2$)$_2$ (39)	361
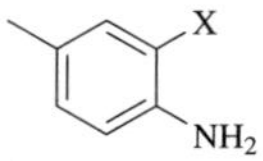 (4-methyl-2-X-aniline, NH_2)	 R^1—≡—R^2	$Pd(OAc)_2$, dtbpf, K_2CO_3, NMP, 110 or 130°	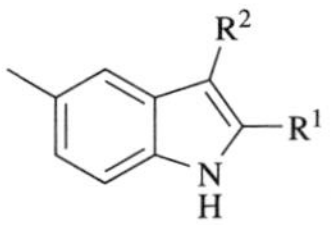 I 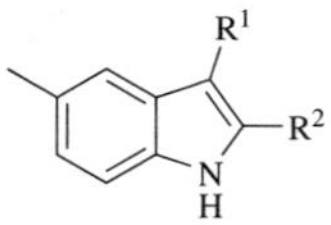 (5-methyl indole with R^1, R^2) + II (6-methyl indole with R^1, R^2)	122

X	R[1]	R[2]	I	I:II
Br	Ph	Pr	(86)	78:22
Cl	MeCH=CH$_2$	Et	(60)	>99:1
Cl	Pr	Pr	(82)	—
Cl	Ph	Pr	(76)	91:9

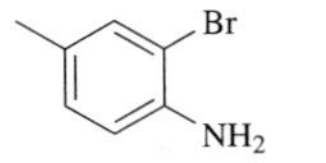

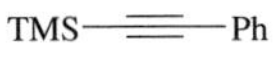

$Pd(OAc)_2$, dtbpf, $KHCO_3$, NMP

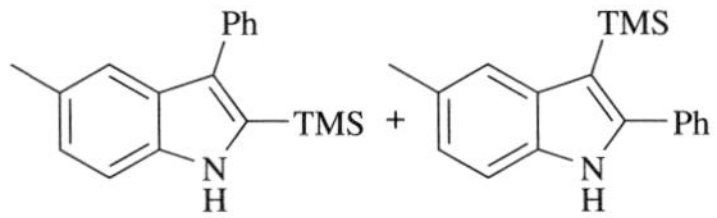

I **II**

I (63), **I:II** = >99:1

122

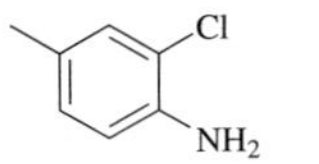

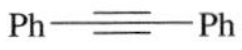

$Pd(OAc)_2$, PCy_3, K_2CO_3, NMP

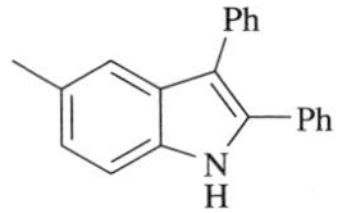

(86)

122

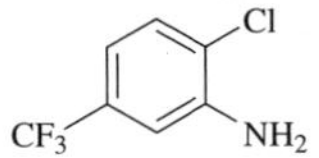

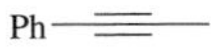

$Pd(OAc)_2$, dtbpf, K_2CO_3, NMP

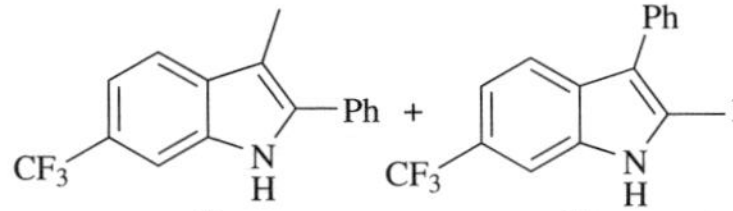

I **II**

I (63), **I:II** = >89:11

122

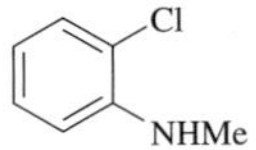

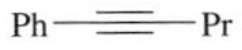

$Pd(OAc)_2$, dtbpf, K_2CO_3, NMP

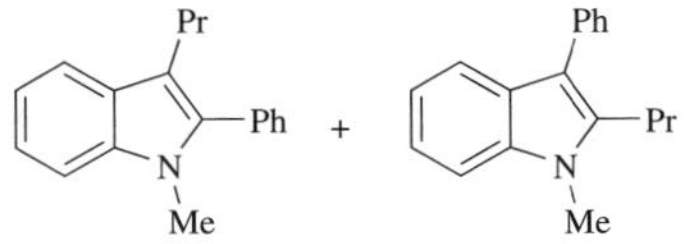

I **II**

I (68), **I:II** = 91:9

122

TABLE 3A. 2,3-DISUBSTITUTED INDOLES FROM 2-HALOANILINES, 2-IODOBENZOIC ACIDS, OR ANILINES AND ALKYNES (*Continued*)

Substrate	Alkyne	Conditions	Product(s) and Yield(s) (%)	Refs.
C₇ 2-iodo-6-methoxyaniline	EtO-isopropyl-OEt alkyne with TES	Pd(OAc)₂, LiCl, Na₂CO₃, DMF, 100°, 36 h	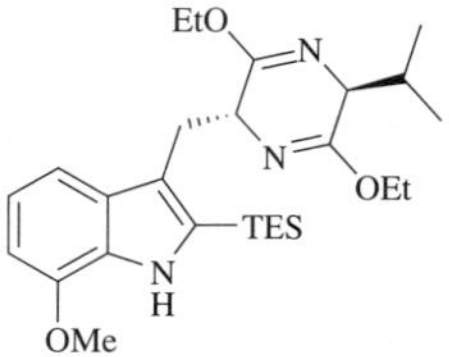 isopropyl-diketopiperazine-CH₂-indole(TES)(OMe) (75)	118, 120
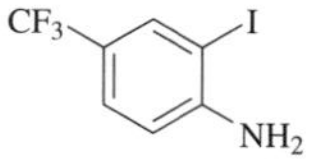 2-iodo-4-(trifluoromethyl)aniline	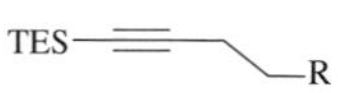 TES─≡─R	PdCl₂(PPh₃)₂, LiCl, Na₂CO₃, DMF, 100°	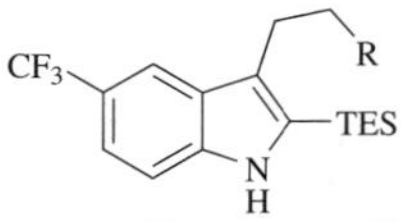 CF₃-indole(TES)(R) R Time (h) TESO 12 (40) 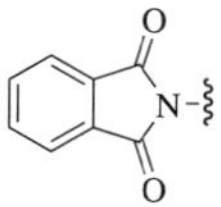 phthalimido 18 (83)	362
C₇₋₁₃ 2-iodo-N-R²-aniline (R¹)	R³─≡─P(=S)Ph₂	Pd(acac)₂, K₂CO₃, DMSO, 90°, 11 h	Indoles I (2-P(=S)Ph₂, 3-R³) + II (3-P(=S)Ph₂, 2-R³) | R¹	R²	R³	I	II | |---|---|---|---|---| | H | Me | 2-thienyl | (57) | (10) | | H | Me | *t*-Bu | (35) | (—) | | H | Me | *c*-C₆H₁₁ | (91) | (—) | | H | Me | *n*-C₆H₁₃ | (67) | (—) | | H | Me | Ph | (74) | (6) |	357

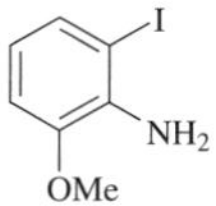

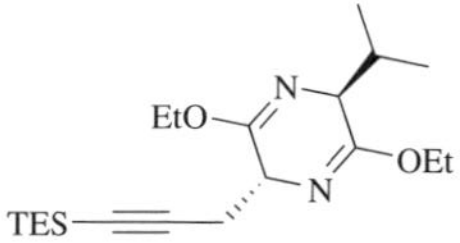

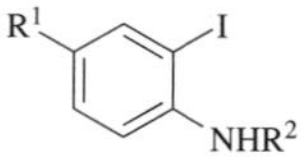

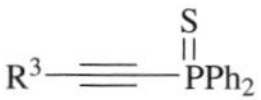

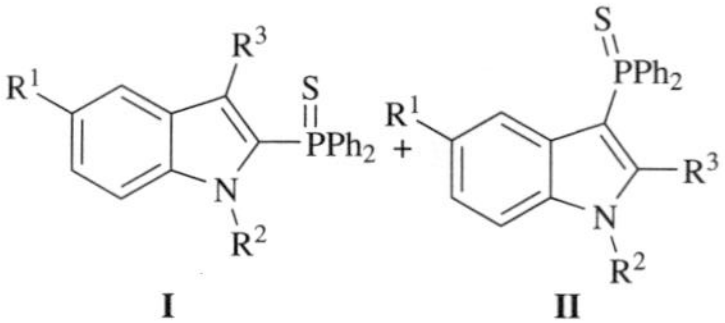

C8

[Structure: 2-iodo-5-(methoxycarbonyl)aniline] + [Structure: R¹–C≡C–pyridine–R²]

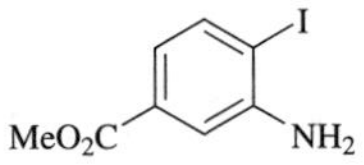

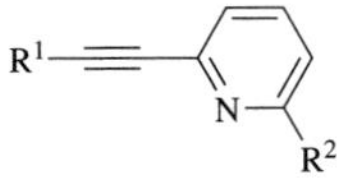

Pd(OAc)₂, dppf, KOAc,
NMP, 140°, 1 h

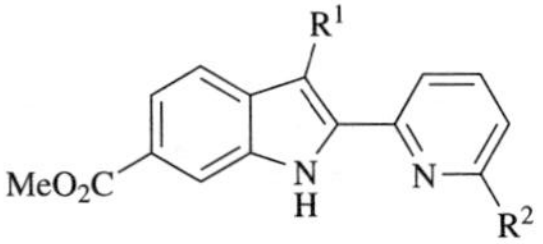

→ I (MeO₂C-indole-2-(pyridine-R²) with R¹ at 3-position) + II (regioisomer)   363

			I + II	
H	Me	4-MeOC₆H₄	(75)	(8)
H	Me	2-MeOC₆H₄	(54)	(17)
H	Me	4-MeO₂CC₆H₄	(58)	(6)
H	Me	4-MeCOC₆H₄	(59)	(6)
Br	Me	Ph	(57)	(11)
Cl	Me	Ph	(56)	(18)
Me	Me	Ph	(71)	(13)
H	Et	Ph	(78)	(6)
H	i-Pr	Ph	(73)	(—)
H	Bn	Ph	(59)	(5)

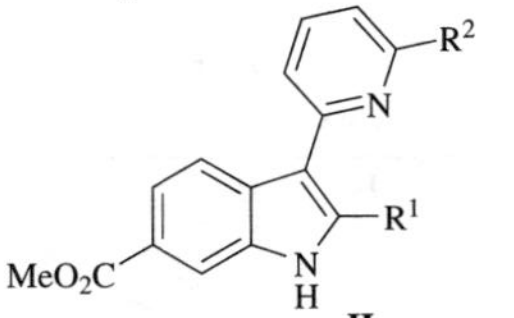

R¹	R²	I + II	I:II[b]
Bu	H	(91)	97:3
t-Bu	H	(78)	31:69
c-C₅H₉	H	(94)	94:6
c-C₅H₉	Me	(88)	87:13
c-C₅H₉	MeO	(84)	80:20
Ph	H	(63)	57:43
4-MeOC₆H₄	H	(63)	41:59

TABLE 3A. 2,3-DISUBSTITUTED INDOLES FROM 2-HALOANILINES, 2-IODOBENZOIC ACIDS, OR ANILINES AND ALKYNES (*Continued*)

Substrate	Alkyne	Conditions	Product(s) and Yield(s) (%)	Refs.

C$_8$

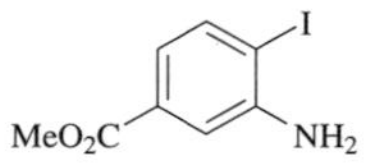

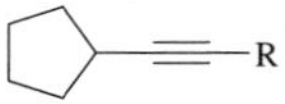

Substrate: 2-iodo-5-(methoxycarbonyl)aniline (MeO$_2$C–C$_6$H$_3$(I)(NH$_2$))
Alkyne: cyclopentylacetylene (R≡, R = cyclopentyl)
Conditions: Pd(OAc)$_2$, dppf, AcOK, NMP, 140°, 1 h

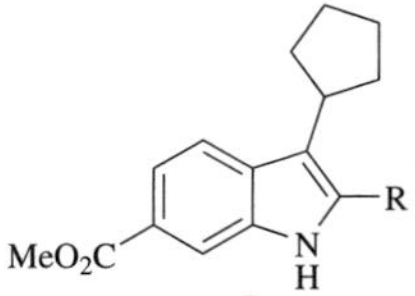

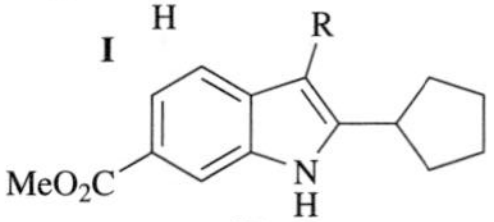

Products I + II (I: 2-cyclopentyl-6-(methoxycarbonyl)indole; II: 3-cyclopentyl-6-(methoxycarbonyl)indole)

R	I + II	I:II^b
3-pyridyl	(93)	68:32
4-pyridyl	(76)	72:28
Ph	(86)	67:33

Ref. 363

Substrate: 2-iodo-N,N-dimethylaniline
Alkyne: R≡
Conditions:
1. PdCl$_2$(PPh$_3$)$_2$, CuI, Et$_3$N, 300 W MW, 60°, time 1
2. ArI, MeCN, 300 W MW, 90°, time 2

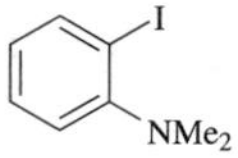

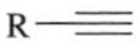

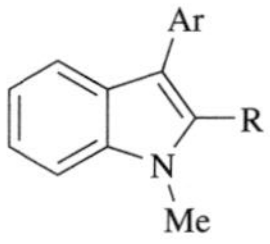

Product: 1-methyl-2-R-3-Ar-indole

R	Ar	Time 1 (h)	Time 2 (h)	
NC(CH$_2$)$_3$	4-O$_2$NC$_6$H$_4$	0.3	0.5	(72)
NC(CH$_2$)$_3$	4-MeO$_2$CC$_6$H$_4$	0.3	0.5	(63)
3-thienyl	3-EtO$_2$CC$_6$H$_4$	0.3	0.5	(78)
Ph	4-EtO$_2$CC$_6$H$_4$	6	4	(82)
Ph	4-EtO$_2$CC$_6$H$_4$	0.3	0.5	(86)
1-cyclohexenyl	4-O$_2$NC$_6$H$_4$	0.3	0.5	(91)
2-MeOC$_6$H$_4$	4-ClC$_6$H$_4$	0.3	0.83	(87)
3-MeOC$_6$H$_4$	Ph	0.3	0.5	(91)
4-MeOC$_6$H$_4$	4-EtO$_2$CC$_6$H$_4$	0.3	0.3	(86)
4-NCC$_6$H$_4$	4-EtO$_2$CC$_6$H$_4$	0.5	0.83	(77)

Ref. 364

412

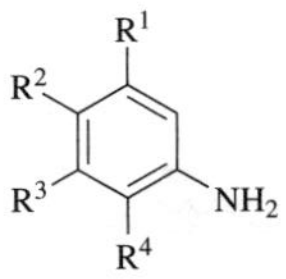

MeO2C—≡—CO2Me   Pd(OAc)$_2$, O$_2$ (1 atm), DMA/pivOH (4:1), 120°, 12 h

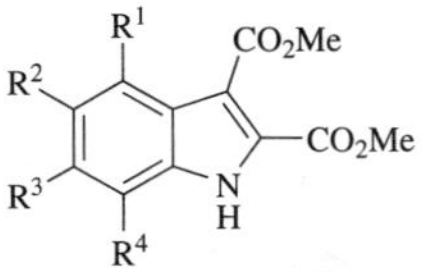

359

R^1	R^2	R^3	R^4
H	MeO	H	MeO (99)
MeO	H	H	MeO (63)
H	MeO	MeO	H (93)

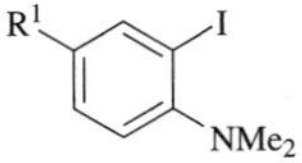

R^2—≡

1. PdCl$_2$(PPh$_3$)$_2$, CuI, Et$_3$N, 300 W MW, 60°, time 1
2. ArI, MeCN, 300 W MW, 90°, time 2

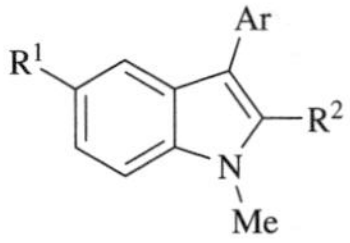

364

R^1	R^2	Ar	Time 1 (h)	Time 2 (h)
Br	3-thienyl	2-thienyl	0.3	0.5 (85)
Br	Ph	2-thienyl	0.3	0.5 (79)
Br	3,5-(MeO)$_2$C$_6$H$_3$	2-thienyl	5	4 (82)
Br	3,5-(MeO)$_2$C$_6$H$_3$	2-thienyl	0.3	0.5 (74)
Br	3,5-(MeO)$_2$C$_6$H$_3$	3-thienyl	0.3	0.5 (76)
Me	HOCH$_2$CH$_2$	3,4-(MeO)$_2$C$_6$H$_3$	0.5	0.5 (33)
Me	4-MeOC$_6$H$_4$	4-ClC$_6$H$_4$	0.5	0.5 (94)
Me	Ph	3-EtO$_2$CC$_6$H$_4$	0.5	0.5 (68)
MeO$_2$C	3-MeOC$_6$H$_4$	3-thienyl	0.5	0.5 (70)

C$_{8-10}$

TABLE 3A. 2,3-DISUBSTITUTED INDOLES FROM 2-HALOANILINES, 2-IODOBENZOIC ACIDS, OR ANILINES AND ALKYNES (*Continued*)

Substrate	Alkyne	Conditions	Product(s) and Yield(s) (%)	Refs.
C₉	TES─≡─OTES	Pd(OAc)₂, Na₂CO₃, DMF, 100°	(indole with OR and TES substituents); I + II (88), I:II = 75:25; R: I=TES, II=H	121
C₁₂	TES─≡─CH(Me)CH₂OBn	Pd(OAc)₂, PPh₃, LiCl, K₂CO₃, DMF, 100°, 16 h	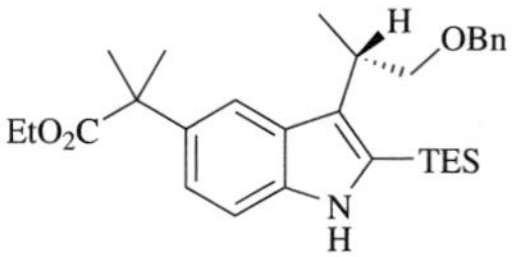 (89)	114
C₁₆	TES─≡─CH(Me)CH₂OBn	Pd(OAc)₂, PPh₃, LiCl, K₂CO₃, DMF, 100°, 16 h	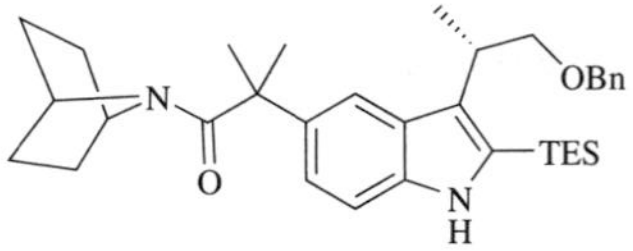 (72)	114
C₄₈	(complex substrate)	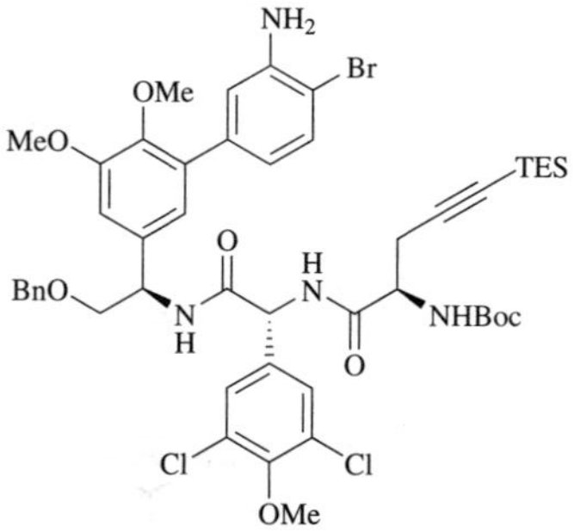 Pd(OAc)₂, dtbpf, Et₃N, toluene/MeCN, 110°, 1 h	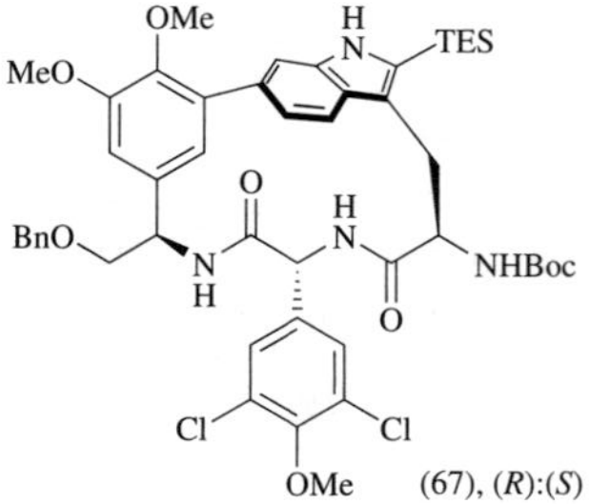 (67), (*R*):(*S*) = 1:1	365

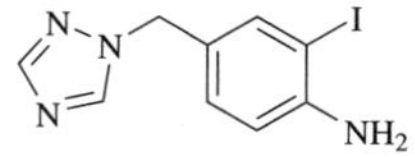

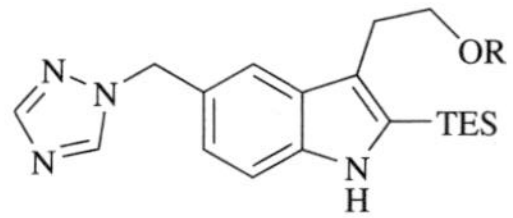

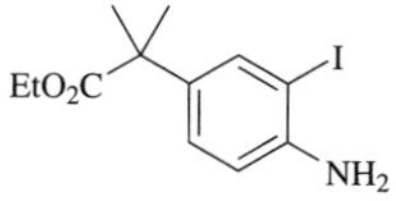

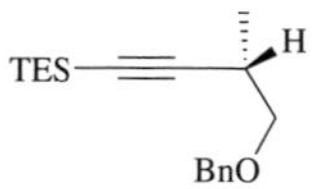

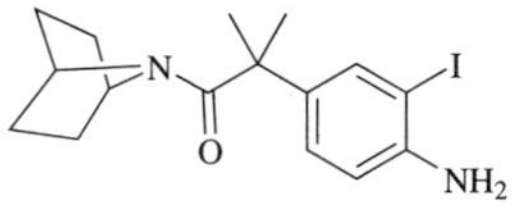

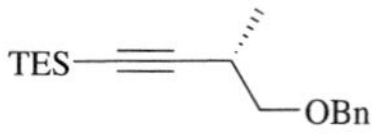

[a] The ratios were determined by ¹⁹F NMR spectroscopy.
[b] The ratios were determined by ¹H NMR spectroscopy.

TABLE 3B. 2,3-DISUBSTITUTED INDOLES FROM 2-HALOANILIDES OR N-ACYL BENZOTRIAZOLES AND ALKYNES

Substrate	Alkyne	Conditions	Product(s) and Yield(s) (%)	Refs.
C$_8$				
2-I-C$_6$H$_4$-NHCOCF$_3$	≡—Ph	ArBr, Pd(OAc)$_2$, PPh$_3$, K$_2$CO$_3$, DMF, 60°	2-Ar-3-Ph-indole (NH) Ar — Time (h) Ph — 0.5 (91) 4-MeOC$_6$H$_4$ — 1 (60)	139
	 ≡—Ph	1. Pd(OAc)$_2$, PPh$_3$, K$_2$CO$_3$, DMF, 60° 2. ArBr	**I** (2-Ar-3-Ph-N-COCF$_3$-indole) Ar — Time (h) Ph — 4.5 (86) 4-O$_2$NC$_6$H$_4$ — 8 (98) 4-MeO$_2$CC$_6$H$_4$ — 2.5 (85)	139
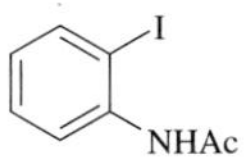 2-I-C$_6$H$_4$-NHAc	 R^1—≡—R^2	Pd(OAc)$_2$, LiCl, KOAc, DMF, 100°	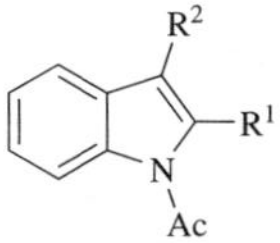 2-R^1-3-R^2-N-Ac-indole R^1 — R^2 — Time (h) TMS — Me — 12 (70) Ph — Me — 24 (75)	30, 31
	 ≡—Et	Pd(OAc)$_2$, LiCl, K$_2$CO$_3$, DMF, 100°, 24 h	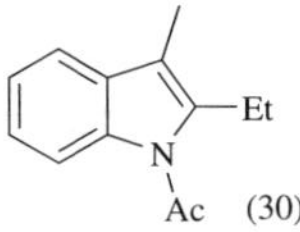 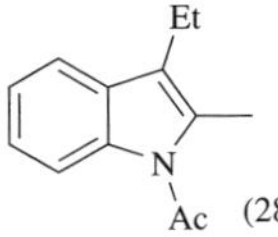 2-Me-3-Et-N-Ac-indole (28) + 2-Et-3-Me-N-Ac-indole (30)	31
	 R^1—≡—R^2	Pd(OAc)$_2$, Bu$_4$NCl, base, DMF, 100°, 24 h	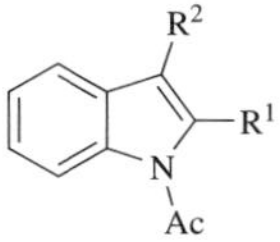 2-R^1-3-R^2-N-Ac-indole R^1 — R^2 — Base CH$_2$OH — MeCH=CH$_2$ — NaOAc (27) Pr — Pr — KOAc (91)	30, 31

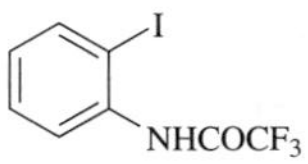

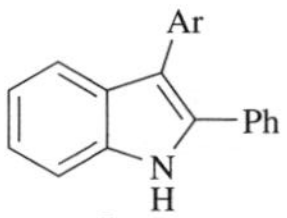

TABLE 3B. 2,3-DISUBSTITUTED INDOLES FROM 2-HALOANILIDES OR N-ACYL BENZOTRIAZOLES AND ALKYNES (*Continued*)

Substrate	Alkyne	Conditions	Product(s) and Yield(s) (%)	Refs.
C_8 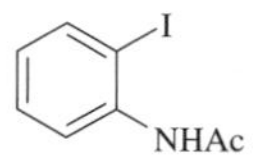 2-iodo-NHAc	 ≡—i-Pr	Pd(OAc)$_2$, Bu$_4$NCl, Na$_2$CO$_3$, DMF, 100°, 24 h	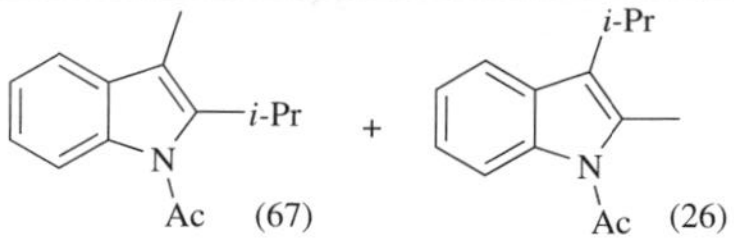 i-Pr indole-Ac (67) + i-Pr indole-Ac (26)	30, 31
	 ≡—CH$_2$OH	Pd(OAc)$_2$, PPh$_3$, Bu$_4$NCl, K$_2$CO$_3$, DMF, 100°, 24 h	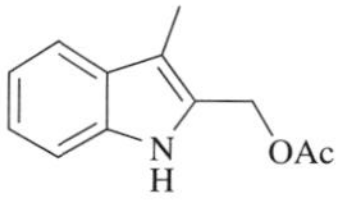 indole-OAc (60)	31
	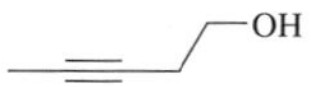 ≡—CH$_2$CH$_2$OH	Pd(OAc)$_2$, Bu$_4$NCl, KOAc, DMF, 100°, 24 h	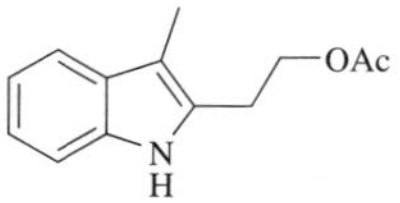 indole-OAc (43)	31
	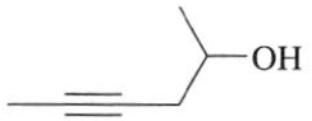 ≡—CH(OH)CH$_3$	Pd(OAc)$_2$, PPh$_3$, Bu$_4$NCl, Na$_2$CO$_3$, DMF, 100°, 12 h	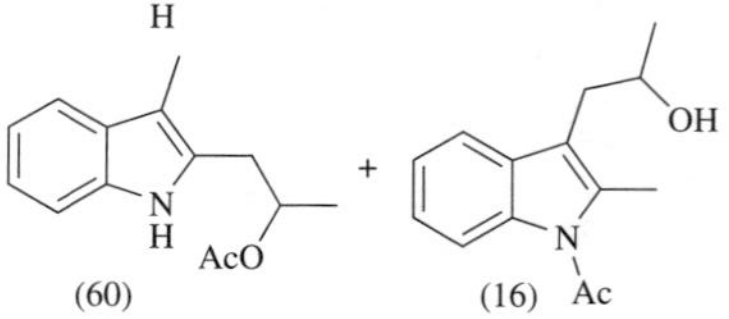 indole-AcO (60) + indole-OH Ac (16)	31
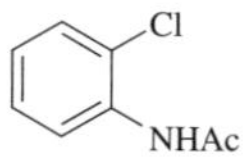 C_8 2-Cl-NHAc	 ≡—t-Bu	Pd(OAc)$_2$, dtbpf, K$_2$CO$_3$, NMP	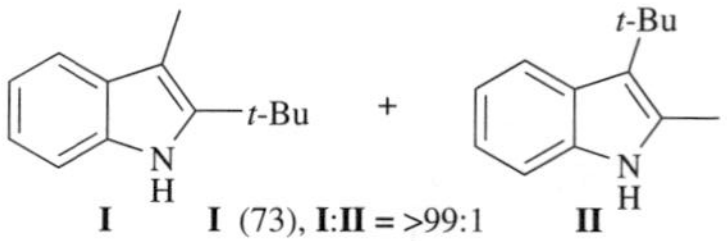 t-Bu indole I (73), I:II = >99:1 + t-Bu indole II	122
C_{8-9} 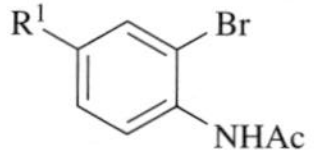 2-Br-R^1-NHAc	 Ph—≡—R^2	Pd(OAc)$_2$, phenylurea, K$_2$CO$_3$, DMF, 130°, 30 h	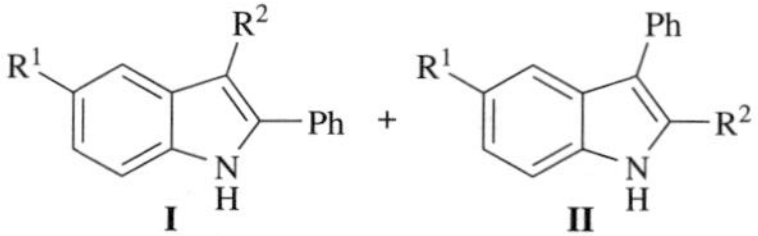 R^2-Ph indole I + Ph-R^2 indole II R^1 R^2 I+II I:II H Ph (74) — H Me (59) 93:7 Me Ph (70) —	358

C8-10

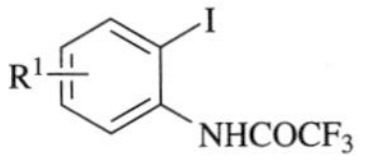

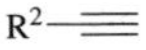

R¹-C6H3(I)(NHCOCF3) + R²−≡ →
1. PdCl₂(PPh₃)₂, CuI, Et₃N, 300 W MW, 60°, 0.5 h
2. ArI, MeCN, 300 W MW, 90°, 0.5 h

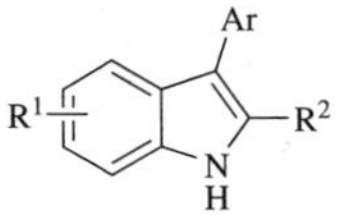

Product: 2-R², 3-Ar indole with R¹ substituent, NH

364

R¹	R²	Ar	(%)
H	3-MeOC₆H₄	4-MeOC₆H₄	(82)
H	Ph	3-EtO₂CC₆H₄	(93)
5-Me	4-NCC₆H₄	Ph	(66)
5-Me	4-MeOC₆H₄	3-O₂NC₆H₄	(88)
5-Me	3-thienyl	4-ClC₆H₄	(67)
5-MeO₂C	Ph	3-MeOC₆H₄	(76)
5-MeO₂C	4-MeOC₆H₄	4-EtO₂CC₆H₄	(60)

C8-16

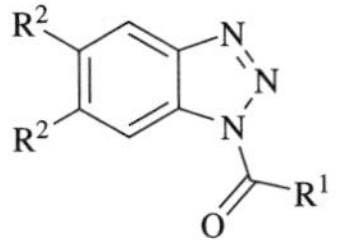

Benzotriazole-N-C(O)R¹ (with R² substituents) + n-C₅H₁₁−≡−n-C₅H₁₁ →
Pd(PPh₃)₄, neat, 130°, 30 h

Product: N-acyl indole with 2,3-di-n-C₅H₁₁ substituents

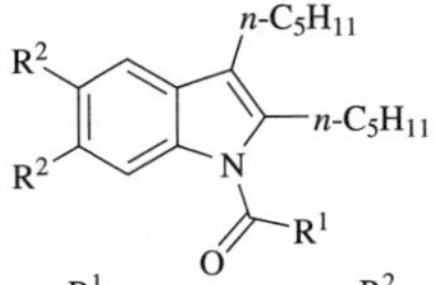

366

R¹	R²	Time (h)	(%)
Me	H	72	(—)
EtO	H	9	traces
Ph	H	18	(42)
3-FC₆H₄	H	8	(54)
3,5-F₂C₆H₃	H	24	(66)
4-CF₃C₆H₄	H	8	(69)
4-MeOC₆H₄	H	72	(47)
3,5-(CF₃)₂C₆H₃	H	12	(39)
4-MeCOC₆H₄	H	18	(64)
4-CF₃C₆H₄	Me	8	(74)
4-CF₃C₆H₄	MeO	11	(41)
4-CF₃C₆H₄	NC	41	(—)

TABLE 3B. 2,3-DISUBSTITUTED INDOLES FROM 2-HALOANILIDES OR N-ACYL BENZOTRIAZOLES AND ALKYNES (*Continued*)

Substrate	Alkyne	Conditions	Product(s) and Yield(s) (%)	Refs.
C₉				
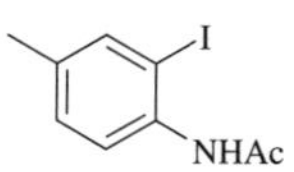 2-I-4-Me-C₆H₃-NHAc	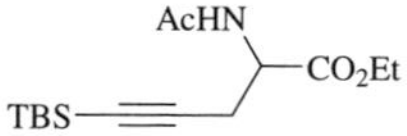 TBS-≡-CH₂-CH(NHAc)CO₂Et	Pd(OAc)₂, Bu₄NCl, KOAc, DMF, 90–100°, 22 h	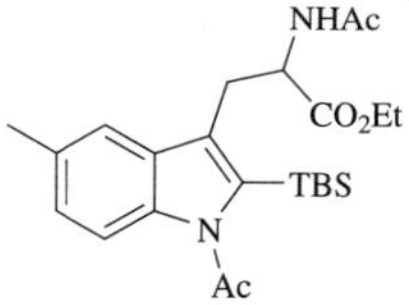 2,3-disubst indole with NHAc, CO₂Et, TBS, N-Ac, 5-Me (27)	115
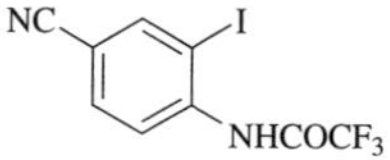 4-NC-2-I-C₆H₃-NHCOCF₃	 ≡-Ph	ArBr, Pd(OAc)₂, PPh₃, K₂CO₃, DMF, 60°	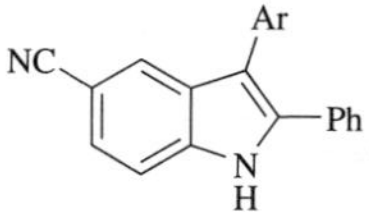 5-CN-2-Ph-3-Ar indole (NH) — Ar: Ph, Time 12.5 h (91); 4-MeOC₆H₄, 1.5 h (78)	139
	 ≡-Ph	1. Pd(OAc)₂, PPh₃, K₂CO₃, DMF, 60° 2. 2-O₂NC₆H₄Br, 12.5 h	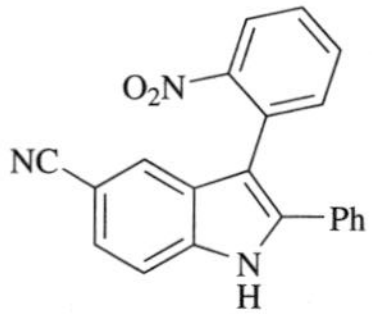 5-CN-2-Ph-3-(2-O₂NC₆H₄) indole (NH) (80)	139
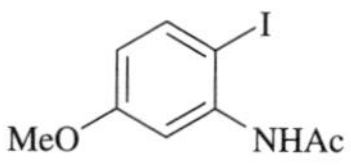 2-I-5-MeO-C₆H₃-NHAc	 ≡-C₆H₄-OMe (4-)	1. MeMgCl, THF, 0° 2. PdCl₂(PPh₃)₂, 65°, 1–2 h 3. 18° 4. 3,4,5-(MeO)₃C₆H₂I, DMSO, 80°, 16–18 h	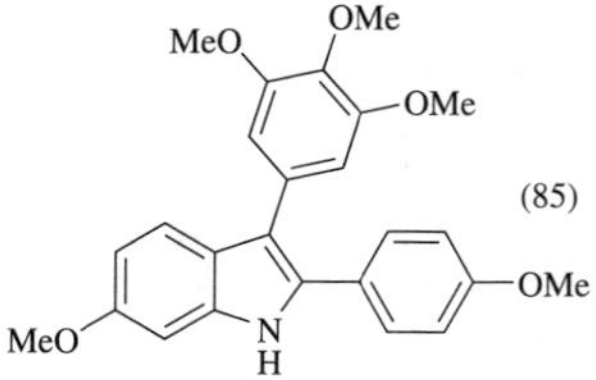 6-MeO-2-(3,4,5-triMeOC₆H₂)-3-(4-MeOC₆H₄) indole (NH) (85)	137, 138
	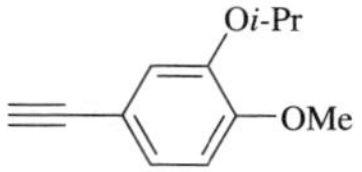 ≡-C₆H₃(Oi-Pr)(OMe)	1. PdCl₂(PPh₃)₂, CuI, Et₃N, MeCN, 18°, 1 h 2. 3,4,5-(MeO)₃C₆H₂I, 18°, 18 h	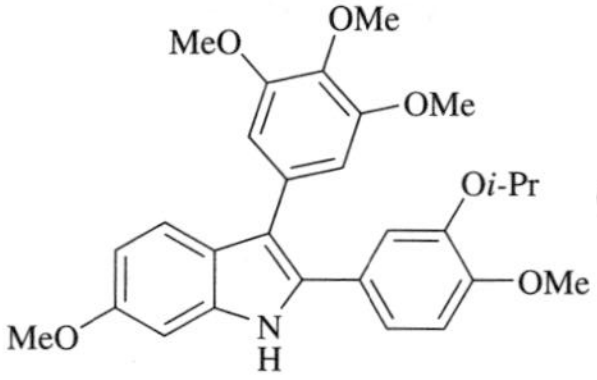 6-MeO-2-(3,4,5-triMeOC₆H₂)-3-(3-Oi-Pr-4-MeOC₆H₃) indole (NH) (77)	138

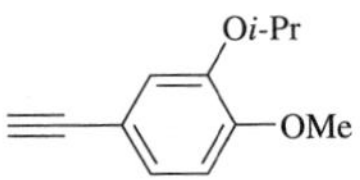

Starting Material	Alkyne	Conditions	Product	Ref.

C10: MeO2C—C6H3(I)(NHCOCF3)

Alkyne: HC≡C—C6H3(Oi-Pr)(OMe)

1. PdCl2(PPh3)2, CuI, CO (balloon), Et3N, MeCN, 18°, 1 h
2. 3,4,5-(MeO)3C6H2I, 18°, 18 h

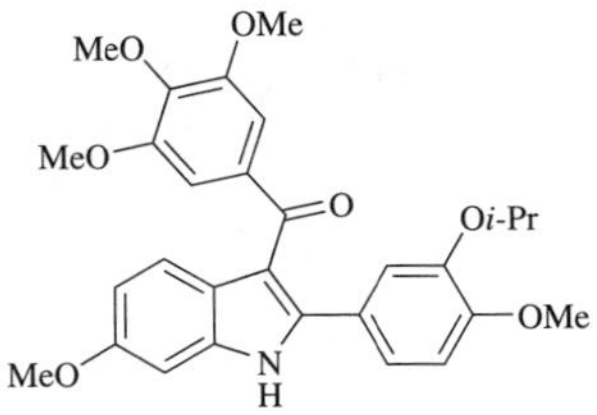

Product: indole with MeO2C, N-H, 6-OMe; C2 = 2-(Oi-Pr,OMe-C6H3); C3 = C(O)–C6H2(OMe)3 (73)

Ref. 138

Alkyne: HC≡C—R

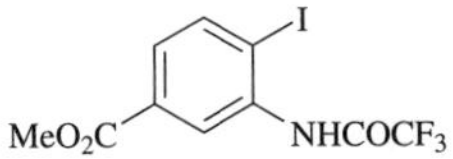

PhBr, Pd(OAc)2, PPh3, K2CO3, DMF, 60°

Product: 6-MeO2C-indole, 2-R, 3-Ph, N-H

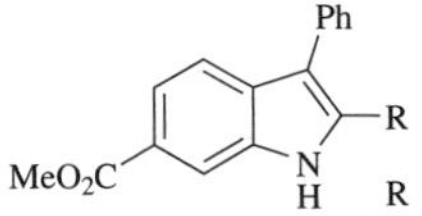

R	Time (h)	
Ph	5	(86)
4-MeC6H4	3.5	(82)

Ref. 139

Alkyne: HC≡C—Ph

1. Pd(OAc)2, PPh3, K2CO3, DMF, 60°
2. 2-O2NC6H4Br, 60°, 3.8 h

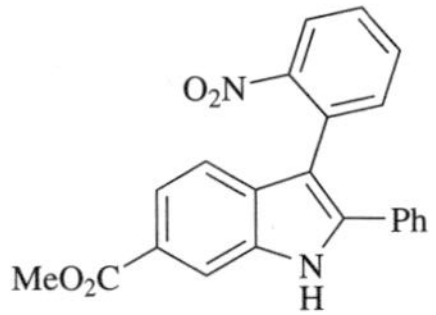

Product: 6-MeO2C-indole, 2-Ph, 3-(2-O2N-C6H4), N-H (94)

Ref. 139

C11-12: 4-R-2-I-C6H3-NHBoc

Alkyne: TMS–C≡C–CH(Me)–CH(OAc)(CO2Me)

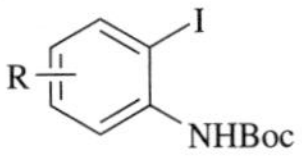

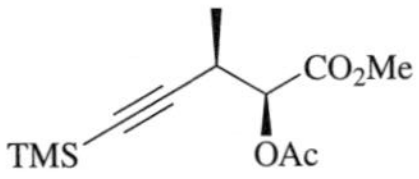

Pd(OAc)2, PPh3, Et4NCl, i-Pr2NEt, DMF, 100°, 1.5 h

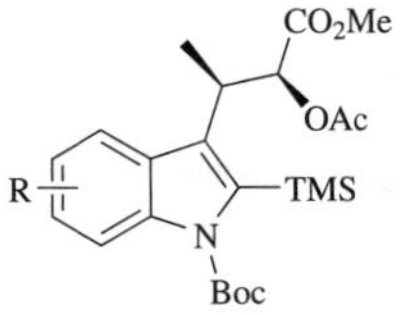

Product: indole with N-Boc, 2-TMS, 3-CH(Me)CH(OAc)CO2Me

R		
H	(85)	
MeO	(88)	

Ref. 367

C12: 2-I-3-OMe-C6H3-NHBoc

Alkyne: TMS–C≡C–CH2CH2–NR2

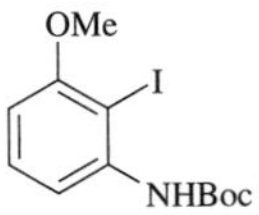

Pd(OAc)2, PPh3, Bu4NCl, i-Pr2EtN, DMF, 80°, 48 h

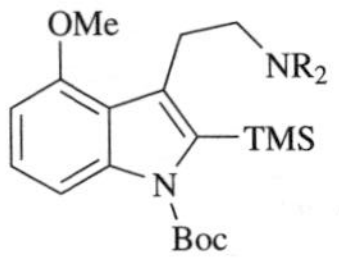

Product: 4-OMe-indole, N-Boc, 2-TMS, 3-CH2CH2NR2

R		
H	(83)	
Me	(69)	
Bn	(77)	

Ref. 368

TABLE 3B. 2,3-DISUBSTITUTED INDOLES FROM 2-HALOANILIDES OR N-ACYL BENZOTRIAZOLES AND ALKYNES (*Continued*)

Substrate	Alkyne	Conditions	Product(s) and Yield(s) (%)	Refs.
C_{13} 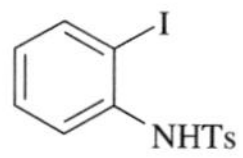 2-I-C₆H₄-NHTs	 R¹—≡—R²	Pd(OAc)₂, additive, base, DMF, 100°	2,3-disubstituted indole (R¹ at C2, R² at C3, N-Ts) | R¹	R²	Additive	Base	Time (h) | |---|---|---|---|---| | *t*-Bu | Me | Bu₄NCl | KOAc | 24 (86) | | Pr | Pr | Bu₄NCl | KOAc | 24 (86) | | CMe₂OH | MeCH=CH₂ | Bu₄NCl | KOAc | 24 (45) | | Ph | Ph | LiCl | K₂CO₃ | 48 (60) |	31, 30
	 ≡—Et	Pd(OAc)₂, LiCl, K₂CO₃, DMF, 100°, 36 h	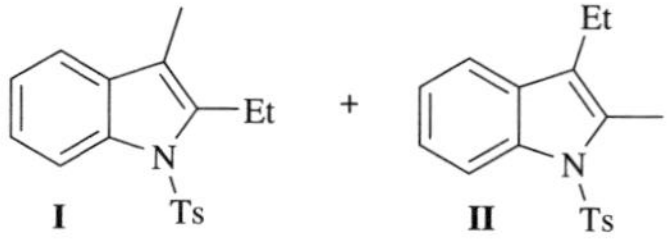 2-Et-3-Me-N-Ts-indole (I) + 2-Me-3-Et-N-Ts-indole (II) I + II (60), I:II = 1:1	31, 30
	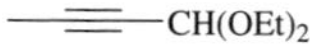 ≡—CH(OEt)₂	Pd(OAc)₂, Bu₄NCl, NaOAc, DMF, 100°, 24 h	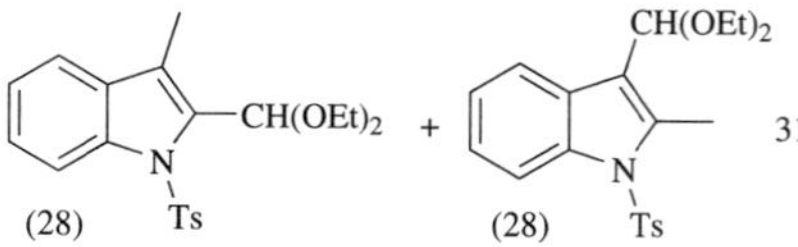 2-Me-3-CH(OEt)₂-N-Ts-indole (28) + 2-CH(OEt)₂-3-Me-N-Ts-indole (28)	31, 30
	BnO—(sugar alkyne with OBn, OBn, OBn, NHBoc, CO₂Me)	Pd(OAc)₂, PPh₃, Bu₄NCl, Na₂CO₃, DMF, 100°, 24 h	indole product with sugar and NHBoc/CO₂Me side chain (89)	113

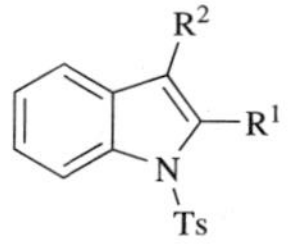

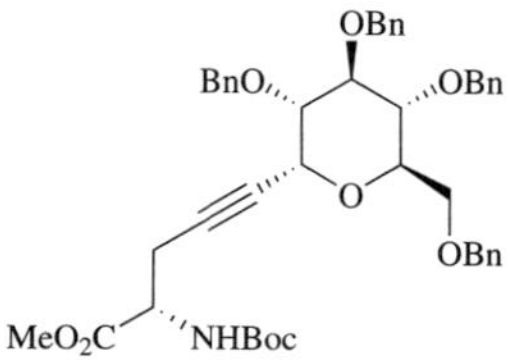

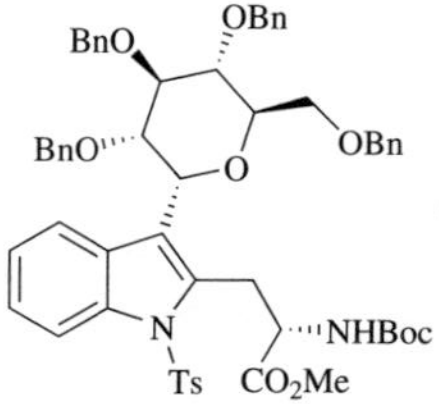

Pd(OAc)₂, PPh₃, Bu₄NCl, Na₂CO₃, 100°

R—≡—Et

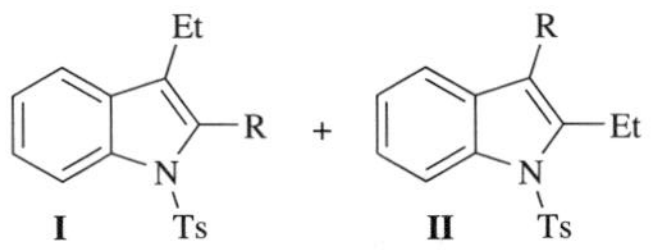

R	Time (h)	I + II	I:II
MeO₂C(CH₂)₂	2	(75)	60:40
BocNH(CH₂)₂	24	(90)	56:44
MeO₂CCH(NHBoc)CH₂	40	(50)	56:44

369

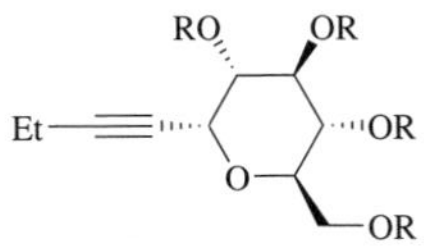

Pd(OAc)₂, PPh₃, Bu₄NCl, Na₂CO₃, 100°

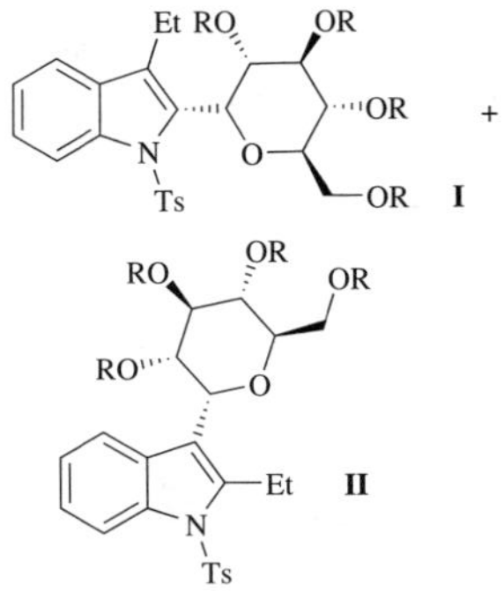

R	Time (h)	I + II	I:II
H	1	(60)	65:35
Bn	48	(16)	69:31

369

Pd/C, additive, NaOAc, NMP, 120°, 24 h

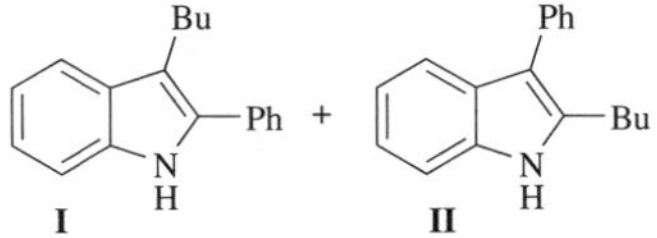

Bu—≡—Ph

Additive	I + II	I:II
LiCl	(90)	72:28
—	(84)	77:23

341

TABLE 3B. 2,3-DISUBSTITUTED INDOLES FROM 2-HALOANILIDES OR N-ACYL BENZOTRIAZOLES AND ALKYNES (Continued)

Substrate	Alkyne	Conditions	Product(s) and Yield(s) (%)	Refs.
C_{13} 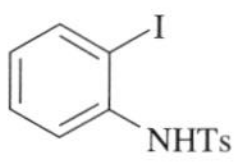 2-I-C$_6$H$_4$-NHTs	 Ph-≡-	Pd/C, additive, NaOAc, NMP, 120°, 24 h	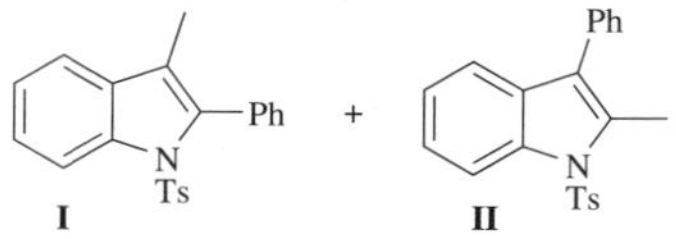 **I** (2-Me-3-Ph-indole-N-Ts) + **II** (2-Ph-3-Me-indole-N-Ts) Additive / Time (h) / I+II / I:II LiCl / 10 / (84) / 64:36 — / 2 / (90) / 68:32	341
	 R-≡-R	Pd/C, additive, NaOAc, NMP, 120°	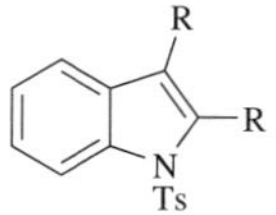 2,3-R,R-indole-N-Ts R / Additive / Time (h) Et / LiCl / 24 (97) Pr / LiCl / 7 (96) Pr / — / 6 (90) n-C$_5$H$_{11}$ / LiCl / 24 (100) n-C$_5$H$_{11}$ / — / 3 (100) Ph / LiCl / 8 (69) Ph / — / 8 (94)	341
	 R-≡-R	Pd/C, NaOAc, NMP	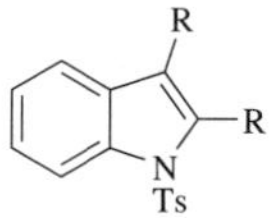 2,3-R,R-indole-N-Ts R / Temp (°) / Time (h) Pr / 120 / 24 (86) Bu / 120 / 6 (86) n-C$_5$H$_{11}$ / 110 / 12 (88) n-C$_5$H$_{11}$ / 120 / 3 (99) Ph / 130 / 24 (87)	341

C15

MeO2C— (aryl with I and NHTs) + n-C5H11—≡—n-C5H11   →   Pd/C, NaOAc, NMP, 120°, 24 h   → MeO2C-indole with n-C5H11 and n-C5H11, N-Ts   (70)   341

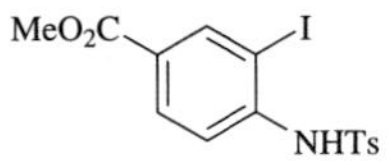

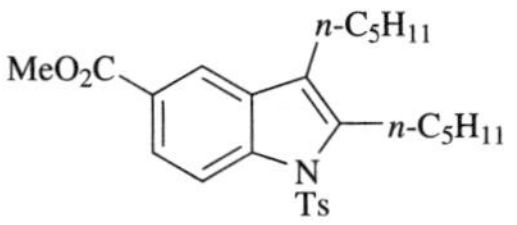

C16

(benzotriazole with dimethyl and N-C(O)-C6H4-CF3) + R—≡—R   →   Pd(PPh3)4, neat, 130°, 8–9 h   →   indole products with R groups, N-C(O)-C6H4-CF3   366

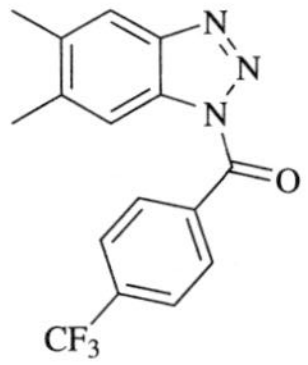

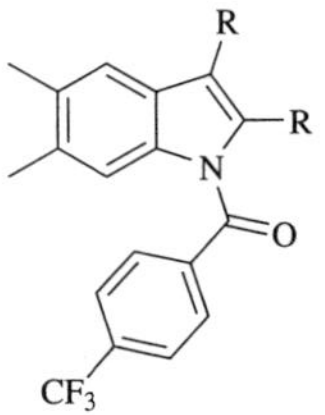

R	(yield)
Pr	(71)
MeCO(CH2)2	(64)
MOMO(CH2)3	(67)
TBSO(CH2)3	(51)
Ph	(41)

(benzotriazole substrate) + R1—≡—R2   →   Pd(PPh3)4, neat, 130°, 8–9 h   →   I + II   366

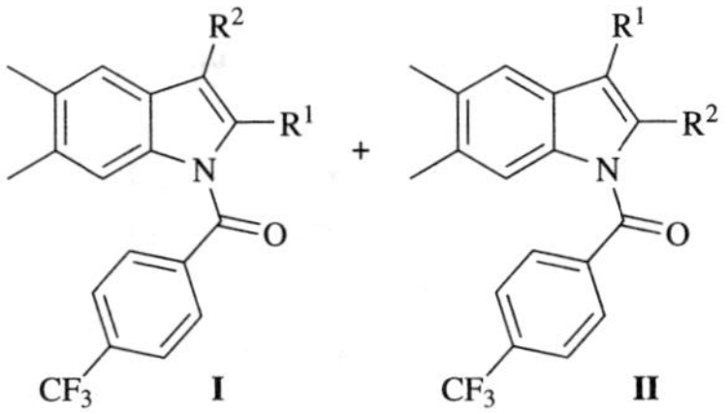

R1	R2	I + II	I:II
t-Bu	Me	(22)	78:22
c-C6H11	Me	(65)	55:45
Ph	Bu	(72)	74:26
4-MeOC6H4	Bu	(66)	52:48
4-CF3C6H4	Bu	(59)	82:18

TABLE 3B. 2,3-DISUBSTITUTED INDOLES FROM 2-HALOANILIDES OR N-ACYL BENZOTRIAZOLES AND ALKYNES (Continued)

Substrate	Alkyne	Conditions	Product(s) and Yield(s) (%)	Refs.
C$_{50-56}$		Pd(OAc)$_2$, dtbpf, Et$_3$N, toluene/MeCN, 110°, 1 h		365

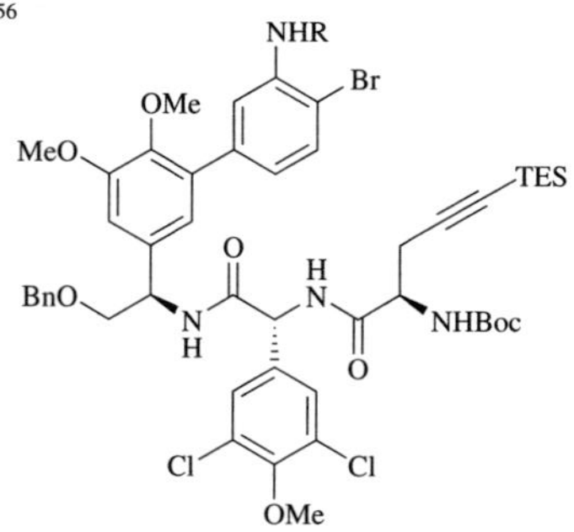

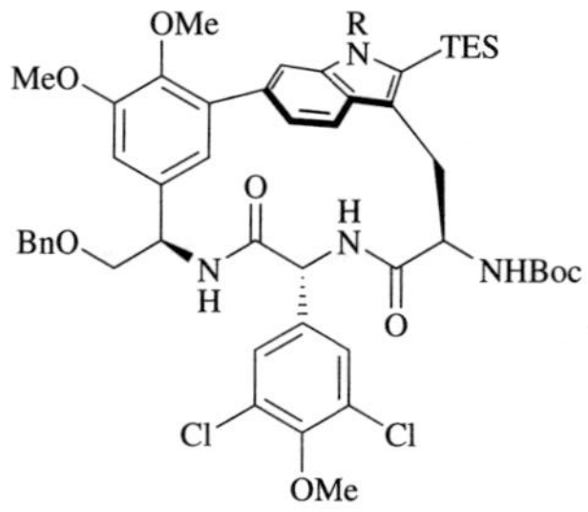

R		(R):(S)
CO$_2$Me	(78)	12.2:1
COMe	(89)	4:1
COPh	(70)	>20:1
CO$_2$Bn	(74)	6.5:1

TABLE 3C. 2,3-DISUBSTITUTED INDOLES FROM 2-ALKYNYLANILINES

2-Alkynylaniline	Conditions	Product(s) and Yield(s) (%)	Refs.
C_{11} (Et-alkynyl aniline with NH$_2$)	RCHO, Pd(OAc)$_2$, Bu$_3$P, THF, 100°	R: i-Pr (57)[a]; c-C$_6$H$_{11}$ (52); Ph (74)[a]; Bn (30); PhMeCH (67)[a]	33
C_{11-17} (R^2-alkynyl aniline with NHR^3, R^1 on ring)	⧸⧸CO$_2$Bu, PdCl$_2$, KI, DMF, 100°, 20 h	R^1 R^2 R^3 H TMS H (72) H HO(CH$_2$)$_4$ H (32) H n-C$_6$H$_{13}$ H (76) H 4-MeC$_6$H$_4$ H (85) H 4-MeC$_6$H$_4$ Me (74) Me 4-MeC$_6$H$_4$ H (57) 5-MeO$_2$C 4-MeC$_6$H$_4$ H (76)	370
C_{12-14} (R-alkynyl aniline with NH$_2$)	CO, PdCl$_2$, CuCl$_2$, AcONa, K$_2$CO$_3$, MeOH, rt, 3 h	R: Bu (30); Ph (51)	128
C_{12-16} (R^3-alkynyl aniline with NHR^2, R^1 on ring)	PhS—SPh, PdCl$_2$, DMSO, 80°	R^1 R^2 R^3 Time (h) H H t-Bu 24 (83) H H 2-thienyl 24 (82) H H 4-pyridyl 96 trace H H n-C$_6$H$_{13}$ 24 (56) Cl H Ph 24 (95) H H 2-BrC$_6$H$_4$ 30 (59) H H 2-MeOC$_6$H$_4$ 24 (35) Me H Ph 24 (61) H Me Ph 24 (65) H H 4-MeCOC$_6$H$_4$ 24 (19)	371

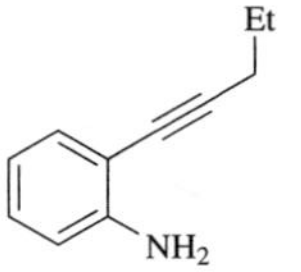

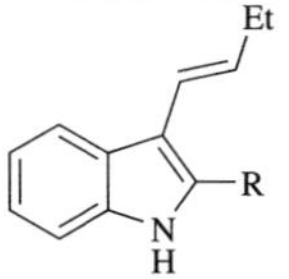

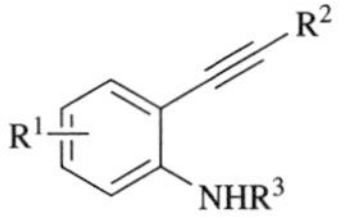

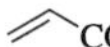

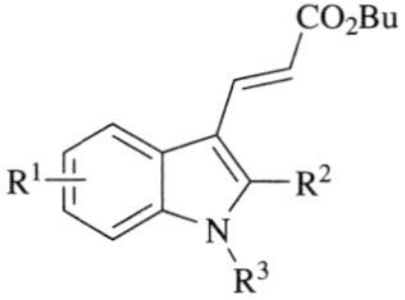

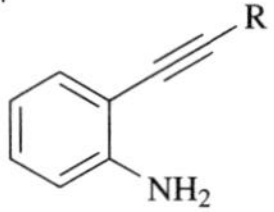

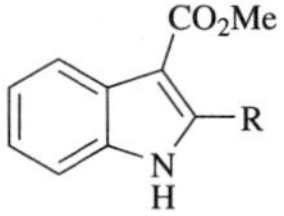

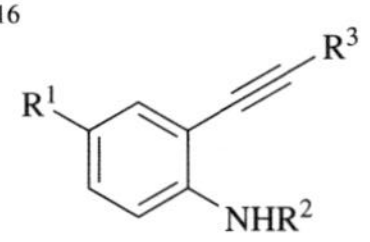

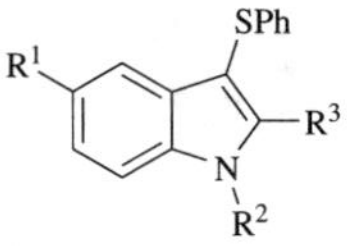

425

TABLE 3C. 2,3-DISUBSTITUTED INDOLES FROM 2-ALKYNYLANILINES (Continued)

2-Alkynylaniline	Conditions	Product(s) and Yield(s) (%)	Refs.
C₁₄ 4-Cl-2-(phenylethynyl)aniline	Methyl vinyl ketone, PdCl₂, FeCl₃, DCE	2,3-disubstituted indole with Ph and two -(CH₂)₂C(O)- chains, 6-Cl (80)	77
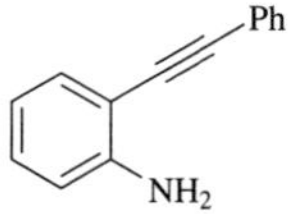 2-(phenylethynyl)aniline	RS—SR, PdCl₂, DMSO, 80°	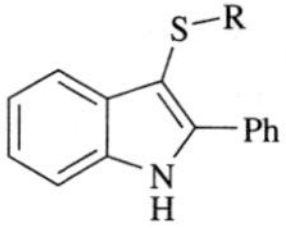 3-(SR)-2-Ph-indole	371

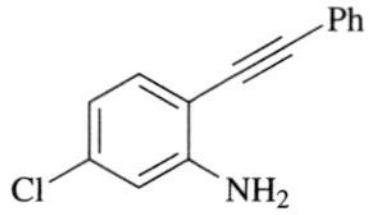

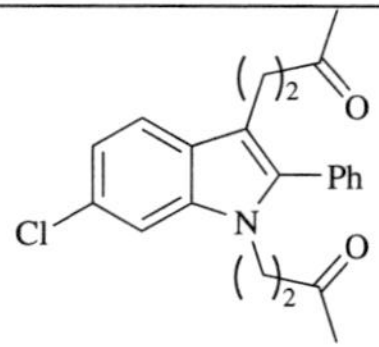

R	Time (h)	
Me	24	(78)
2-pyridyl	24	(74)
2-NH₂C₆H₄	48	(27)
4-ClC₆H₄	24	(92)
4-FC₆H₄	24	(89)
4-O₂NC₆H₄	24	(96)
Bn	24	trace
4-MeOC₆H₄	24	(70)
4-MeC₆H₄	25	(58)
pyrazole-linked (NC, H₂N, CF₃, 2,6-Cl₂C₆H₃)	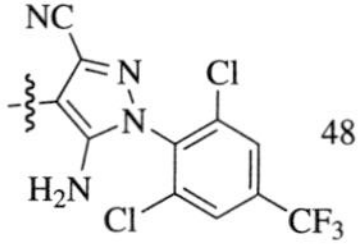 48	(27)

| C₁₅ 2-(p-tolylethynyl)aniline | R¹R²C=CR³, PdCl₂, KI, DMF, 100°, 20 h | 3-vinyl-2-(p-tolyl)indole | 370 |

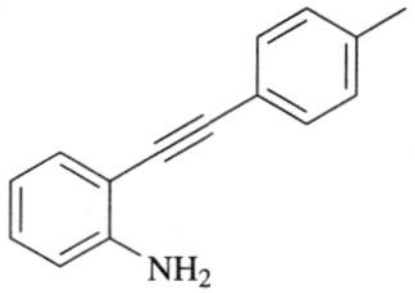

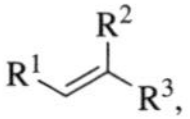

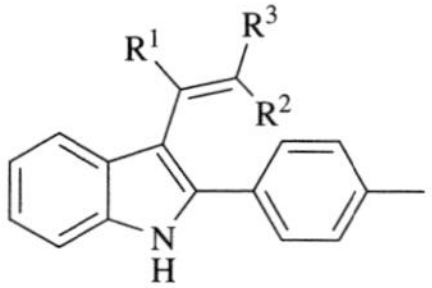

R¹	R²	R³	
H	H	CN	(77)
H	H	CONH₂	(95)
H	H	SO₂Me	(87)
H	H	COMe	(79)
Me	H	CO₂Me	(51)
H	Me	CO₂Me	(67)

[a] The yield was determined by NMR spectroscopy.

TABLE 3D. 2,3-DISUBSTITUTED INDOLES FROM 2-ALKYNYLANILIDES

C_{11-21}

2-Alkynylanilide	Conditions	Product(s) and Yield(s) (%)	Refs.

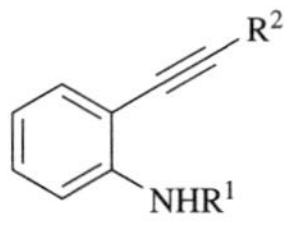

Acrolein, Pd(OAc)$_2$, LiBr, THF, rt

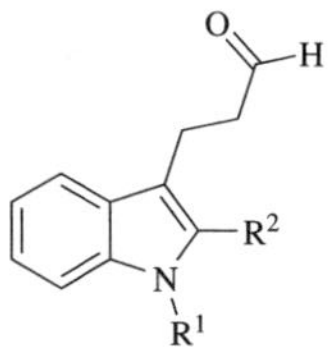

R^1	R^2	Time (d)	
Ms	CH$_2$OMe	1	(89)
Ms	TMS	4	(72)
Ms	Bu	0.5	(75)
Ms	Ph	1	(85)
Ts	CH$_2$OMe	0.5	(94)
Ts	Bu	0.5	(88)
Ts	Ph	0.5	(83)

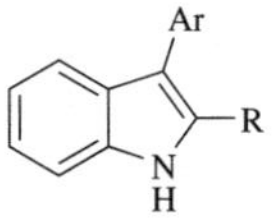

103

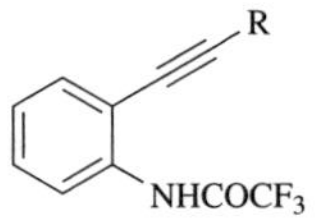

ArI, CO (balloon), Pd(PPh$_3$)$_4$, K$_2$CO$_3$, MeCN, 45°, overnight

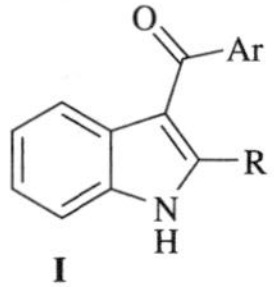

R	Ar	I	II
Me	4-MeOC$_6$H$_4$	(73)	(15)
Bu	4-MeOC$_6$H$_4$	(83)	(—)
Bu	4-MeCONHC$_6$H$_4$	(72)^a	(—)
2-thienyl	4-MeC$_6$H$_4$	(73)	(—)
Ph	3-MeC$_6$H$_4$	(57)	(—)
Ph	4-MeOC$_6$H$_4$	(60)	(—)
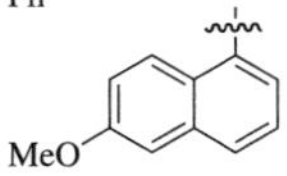 MeO-naphthyl	3-FC$_6$H$_4$	(64)^b	(8)

106, 124

C_{12-15}

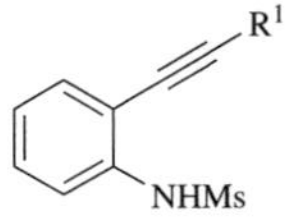

CH$_2$=CHR^2, PdCl$_2$, CuCl$_2$, NaOAc, K$_2$CO$_3$, MeCN, 50°

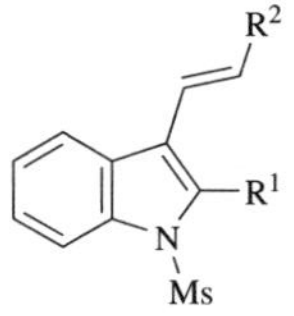

R^1	R^2	Time (h)	
TMS	CO$_2$Et	1	(32)
(CH$_2$)$_5$Me	CO$_2$Et	1	(46)
Ph	CHO	1.5	(48)
Ph	CO$_2$Et	1	(74)
Ph	Ac	1.5	(55)

130

TABLE 3D. 2,3-DISUBSTITUTED INDOLES FROM 2-ALKYNYLANILIDES (*Continued*)

2-Alkynylanilide	Conditions	Product(s) and Yield(s) (%)	Refs.

C_{13-15}

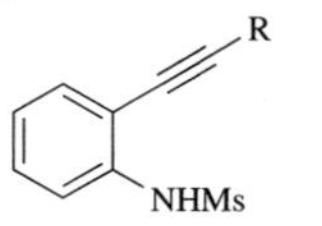

2-Alkynylanilide: aryl with C≡C-R and NHMs

Conditions: CO, $PdCl_2$, $CuCl_2$, AcONa, K_2CO_3, MeOH, rt, 3 h

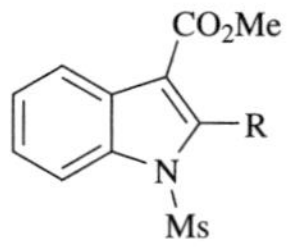

Product: indole with 2-R, 3-CO_2Me, N-Ms

R	
Bu	(67)
Ph	(76)

Ref: 128

C_{13-18}

2-Alkynylanilide: aryl with C≡C-R^1 and $NHCO_2Me$

Conditions: $R^2CH=CR^3CHR^4Cl$, $PdCl_2(MeCN)_2$, methyloxirane, THF

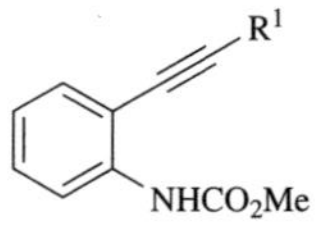

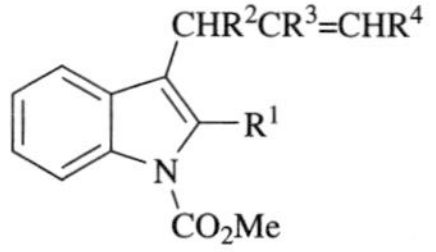

Product: indole with 2-R^1, 3-$CHR^2CR^3=CHR^4$, N-CO_2Me

Ref: 70

R^1	R^2	R^3	R^4	Temp	Time (h)	
$CH_2=CMe$	H	H	H	rt	20	(81)
Bu	H	H	H	rt	20	(73)
Bu	H	H	Me	rt	5.5	(80)[c]
Bu	H	Me	H	rt	2.5	(74)
Bu	Me	H	H	rt	48	(69)
Bu	H	H	Et	rt	72	(31)
Bu	H	H	CH_2Cl	rt	1	(71)[d]
Bu	H	Cl	H	rt	20	(52)
Bu	CH_2Cl	H	H	rt	5	(51)
Bu	H	CH_2Cl	H	rt	1	(72)
t-Bu	H	H	H	reflux	5	(33)
Ph	H	H	H	rt	48	(82)
Ph	H	H	H	reflux	0.5	(82)
$MeCH(CH_2)_5Me$	H	H	H	rt	20	(81)

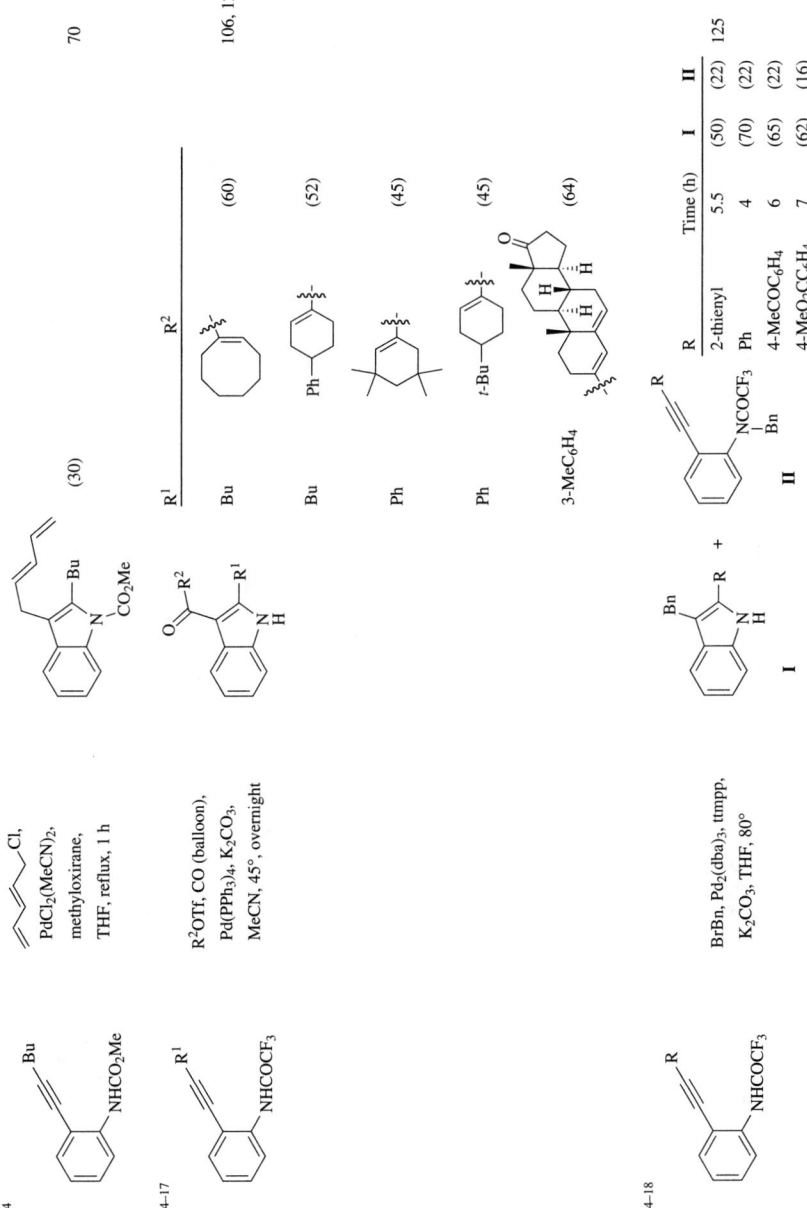

TABLE 3D. 2,3-DISUBSTITUTED INDOLES FROM 2-ALKYNYLANILIDES (*Continued*)

C$_{14-18}$

2-Alkynylanilide	Conditions	Product(s) and Yield(s) (%)					Refs.

PdX$_2$, CuX$_2$, DCE, 40°

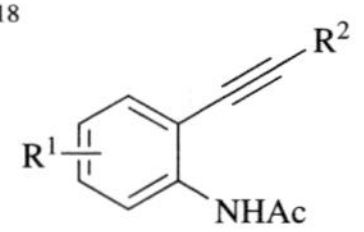

R^1	R^2	X	Time (h)	I	II
H	*t*-Bu	Br	15	(5)	(5)
H	*t*-Bu	Cl	15	(5)	(88)
H	MeC(O)CH$_2$	Br	5	(5)	(57)
H	Ph	Br	12	(18)	(64)
H	Ph	Cl	12	(20)	(59)
H	*n*-C$_8$H$_{17}$	Cl	5	(25)	(56)
5-O$_2$N	*n*-C$_8$H$_{17}$	Br	5	(18)	(58)
5-O$_2$N	*n*-C$_8$H$_{17}$	Cl	5	(13)	(65)
5,6-F$_2$	*n*-C$_8$H$_{17}$	Br	11	(30)	(62)
5,6-F$_2$	*n*-C$_8$H$_{17}$	Cl	11	(40)	(45)
5,7-Cl$_2$	*n*-C$_8$H$_{17}$	Br	5	(50)	(41)
5,7-Cl$_2$	*n*-C$_8$H$_{17}$	Cl	5	(41)	(54)

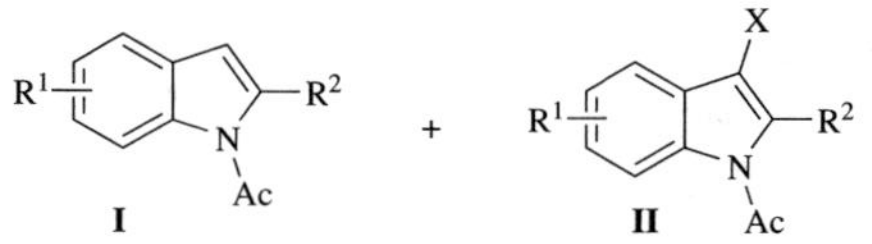

372

C$_{14-21}$

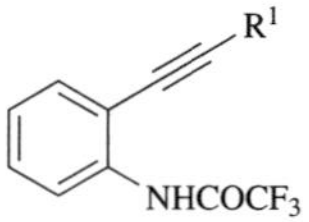

R^{2}OTf, Pd(PPh$_3$)$_4$, K$_2$CO$_3$, MeCN, rt

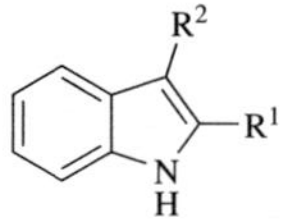

R^1	R^2	Time (h)	
Bu	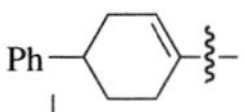 (4-Ph-cyclohexenyl)	35	(81)
Bu	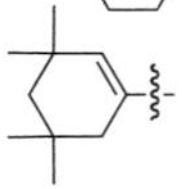 (3,5,5-trimethylcyclohexenyl)	1.5	(74)
Ph	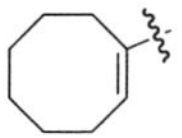 (cyclooctenyl)	4	(74)

123

430

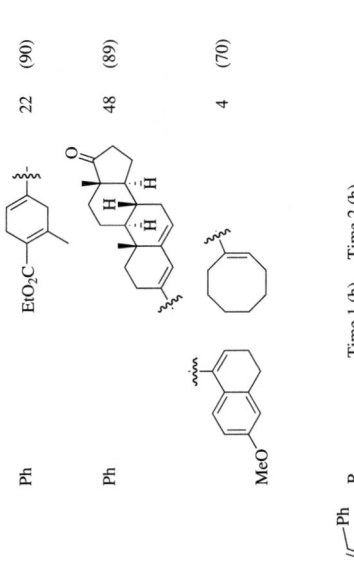

R	Time 1 (h)	Time 2 (h)	
CH$_2$NHCOEt	48	1	(45)
n-C$_5$H$_{11}$	5.5	24	(75)
Ph	3	1.5	(91)
CH$_2$OTHP	1.5	1.5	(77)
4-MeOC$_6$H$_4$	1	2	(74)
4-MeCOC$_6$H$_4$	0.5	8	(91)
2-THPOC$_6$H$_4$	3	24	(64)

102, 106

TABLE 3D. 2,3-DISUBSTITUTED INDOLES FROM 2-ALKYNYLANILIDES (*Continued*)

2-Alkynylanilide	Conditions	Product(s) and Yield(s) (%)	Refs.
C_{14-22} 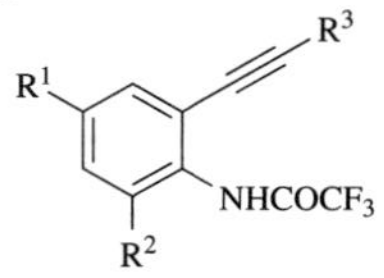	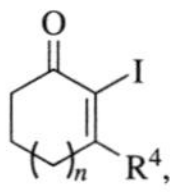	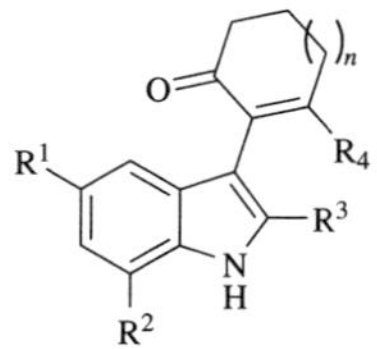	

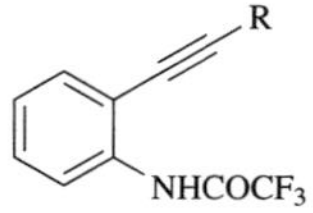

Pd$_2$(dba)$_3$, AsPh$_3$, Cs$_2$CO$_3$, MeCN, 80°

R^1	R^2	R^3	R^4	n	Time (h)	
H	H	Bu	H	1	2	(77)
H	H	2-thienyl	H	1	2	(95)
O$_2$N	NC	Bu	H	1	3.5	(60)
H	H	Ph	H	0	2.5	(76)
H	H	Ph	H	1	2	(95)
H	H	2-BrC$_6$H$_4$	H	1	2	(87)
H	H	Ph	H	2	3	(67)
H	H	Ph	Me	1	2	(58)
H	H	3-MeO$_2$CC$_6$H$_4$	H	1	3.5	(95)
H	H	4-MeOC$_6$H$_4$	H	1	2	(67)
F	F	4-MeOC$_6$H$_4$	H	1	3	(72)
H	H	4-phenyl-1-cyclohexenyl	H	1	2	(40)

373

ArI, Pd(PPh$_3$)$_4$, K$_2$CO$_3$, MeCN, 80°

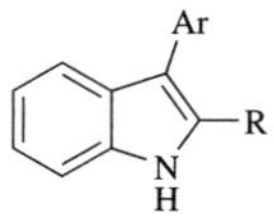

R	Ar	Time (h)	
Bu	4-ClC$_6$H$_4$	2	(82)
Ph	4-MeOC$_6$H$_4$	7	(80)
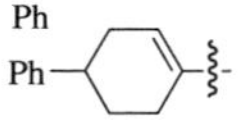 Ph	4-MeO$_2$CC$_6$H$_4$	2	(68)

123

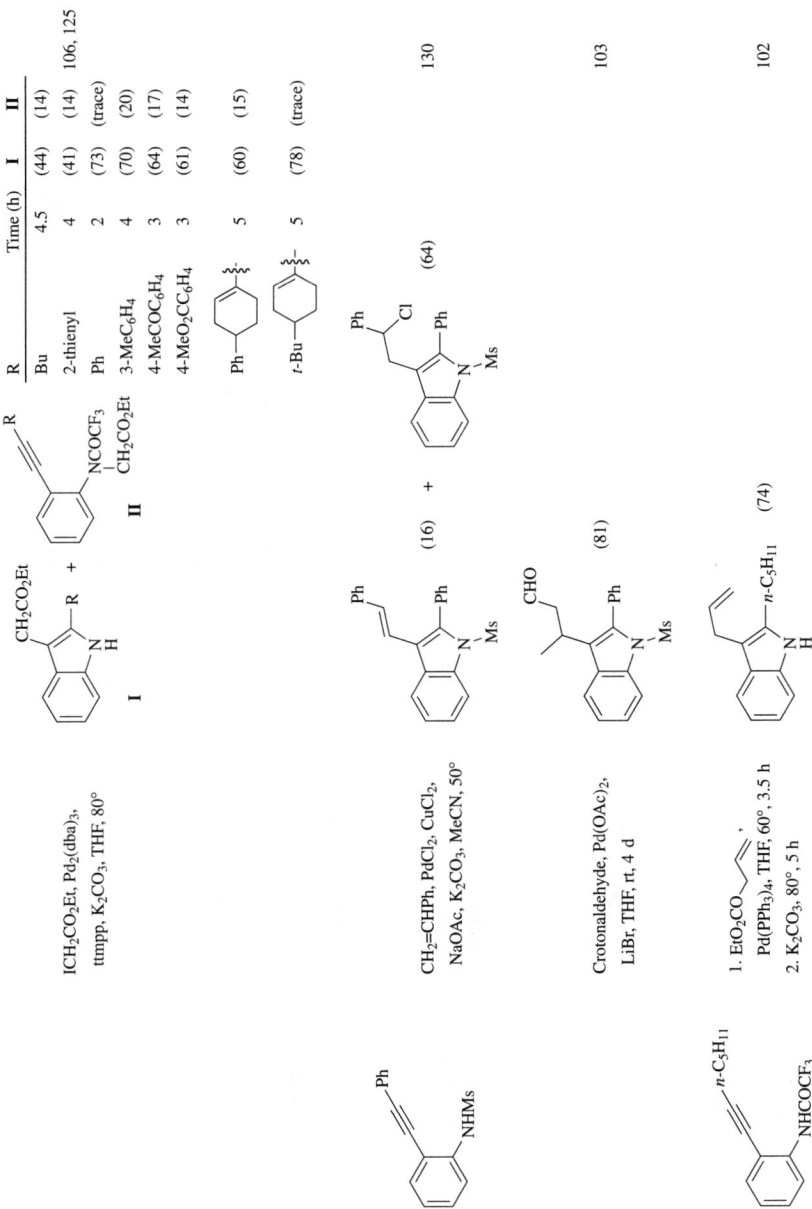

TABLE 3D. 2,3-DISUBSTITUTED INDOLES FROM 2-ALKYNYLANILIDES (*Continued*)

2-Alkynylanilide	Conditions	Product(s) and Yield(s) (%)		Refs.
C₁₅ (n-C₅H₁₁ alkynyl, NHCOCF₃)	ArBr, Pd(PPh₃)₄, Cs₂CO₃, MeCN, 100°, 3 h	Ar-thiazole-N-n-C₅H₁₁ (75); 3-NCC₆H₄ (88)		105, 106
	ArCl, Pd₂(dba)₃, XPhos, Cs₂CO₃, MeCN, 120°, 1.5 h	Ar-indole-n-C₅H₁₁: Ph (40); 3-MeOC₆H₄ (54)		107
(Pr alkynyl, N(OMe)(Ac))	PdBr₂, toluene, 80°	Ar-indole-Pr, N-CH₂OMe (52)		140
C₁₅₋₁₆ (R alkynyl, NHCOCF₃)	ArN₂BF₄, Pd(PPh₃)₄, K₂CO₃, Bu₄NI, MeCN (anhyd), 60°, 1–5.5 h			374

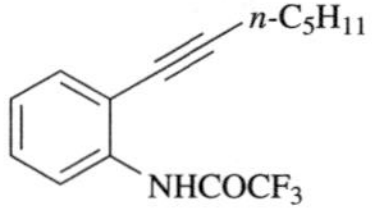

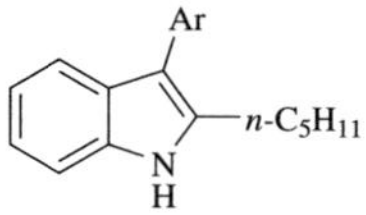

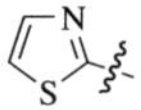

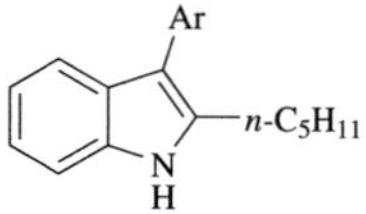

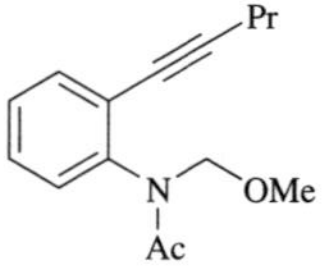

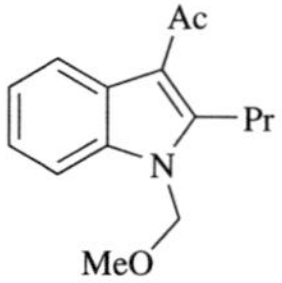

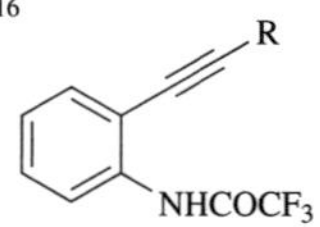

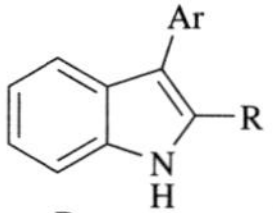

R	Ar		R	Ar	
n-C₅H₁₁	Ph	(63)	Ph	2-MeOC₆H₄	(69)
n-C₅H₁₁	4-ClC₆H₄	(89)	Ph	4-MeC₆H₄	(84)
n-C₅H₁₁	4-MeC₆H₄	(81)	2-BrC₆H₄	4-MeC₆H₄	(66)
Ph	Ph	(76)	Ph	4-MeCOC₆H₄	(96)
Ph	4-ClC₆H₄	(89)	Ph	2-Me-4-MeOC₆H₃	(49)
Ph	4-O₂NC₆H₄	(93)	4-MeOC₆H₄	Ph	(82)
2-BrC₆H₄	4-O₂NC₆H₄	(91)	4-MeOC₆H₄	4-MeC₆H₄	(86)
Ph	3-CF₃C₆H₄	(85)	4-MeOC₆H₄	4-NCC₆H₄	(86)
Ph	4-MeOC₆H₄	(69)	4-MeCOC₆H₄	4-MeOC₆H₄	(65)
Ph	3-MeOC₆H₄	(68)			

C$_{15-17}$

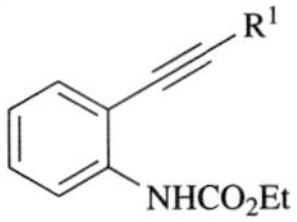

[Structure: 2-R^1-3-(R^2-vinyl)-1-(CO$_2$Et)indole]

from NHCO$_2$Et aryl alkyne with R^1

CH$_2$=CHR^2, PdCl$_2$,
CuCl$_2$, H$_2$O,
Bu$_4$NF, THF, reflux

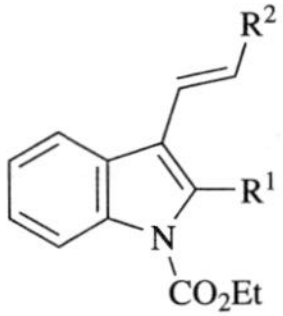

R^1	R^2	Time (h)	
Bu	CO$_2$Me	6	(42)
t-Bu	Ac	2.5	(22)
Ph	CO$_2$Me	4	(64)
Ph	Bu	3	(68)
Ph	Ac	16	(51)

131

C$_{15-22}$

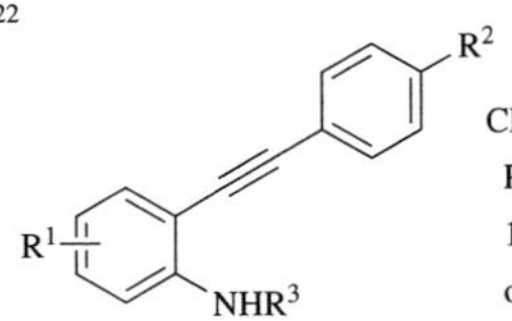

[Structure: 2-(4-R^2-phenyl)-3-(CH(OH)CO$_2$Et)-N-R^3-indole with R^1 on benzene ring]

from aryl alkyne-NHR^3

CHOCO$_2$Et,
Pd(bpy)(H$_2$O)$_2$(OTf)$_2$,
1,4-dioxane, 60°,
overnight

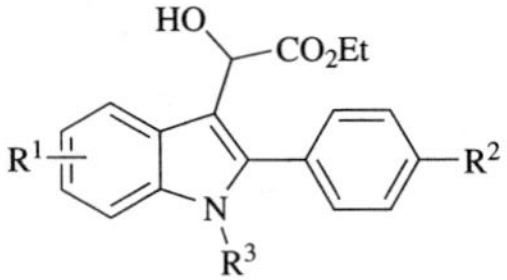

R^1	R^2	R^3	
H	H	Ms	(51)
H	H	COCF$_3$	(—)^e
H	H	Ts	(89)
5-F	H	Ts	(81)
5-Cl	H	Ts	(73)
6-Cl	H	Ts	(67)
H	Br	Ts	(81)
H	Me	Ts	(83)
5-Me	H	Ts	(80)

375

C$_{16}$

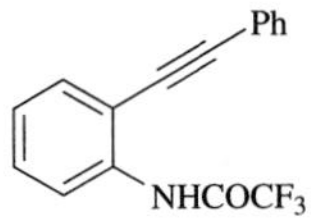

[Structure: 2-Ph-3-(3-methylbut-2-enyl)-1H-indole]

from Ph-alkyne-NHCOCF$_3$

1. 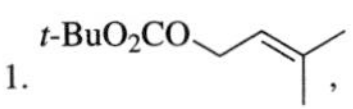 t-BuO$_2$CO-CH$_2$-C(Me)=CH$_2$ (prenyl carbonate),
   Pd(PPh$_3$)$_4$, THF, 60°, 48 h
2. K$_2$CO$_3$, 80°, 2 h

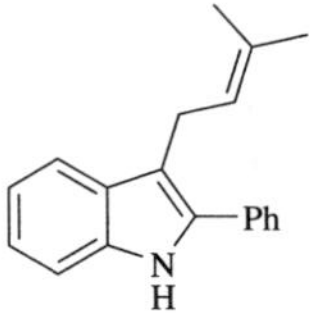

(52)

102

[Structure: 2-Ph-3-(2-methylallyl)-1H-indole]

1. 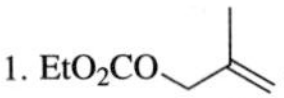 EtO$_2$CO-CH$_2$-C(Me)=CH$_2$,
   Pd(PPh$_3$)$_4$, THF, 60°, 24 h
2. K$_2$CO$_3$, 80°, 10 h

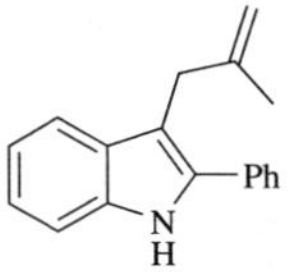

(74)

102

435

TABLE 3D. 2,3-DISUBSTITUTED INDOLES FROM 2-ALKYNYLANILIDES (*Continued*)

2-Alkynylanilide	Conditions	Product(s) and Yield(s) (%)	Refs.
C$_{16}$			
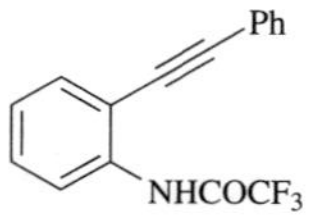 (Ph-alkynyl NHCOCF$_3$ anilide)	 EtO$_2$CO—Ph, Pd$_2$(dba)$_3$, ttmpp, THF, 60°, 3 h	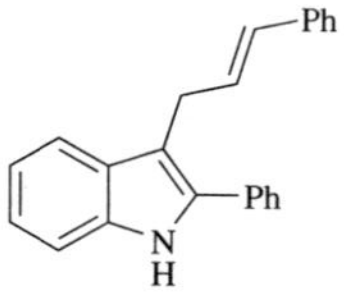 3-cinnamyl-2-phenylindole (94)	102
	4-PhC$_6$H$_4$ONf, Pd(PPh$_3$)$_4$, Cs$_2$CO$_3$, MeCN, 100°, 1 h	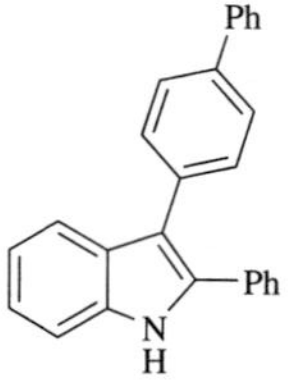 3-(4-biphenylyl)-2-phenylindole (89)	105
	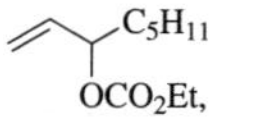 C$_5$H$_{11}$-CH=CH-CH(OCO$_2$Et), Pd$_2$(dba)$_3$, ttmpp, THF, 60°, 3 h	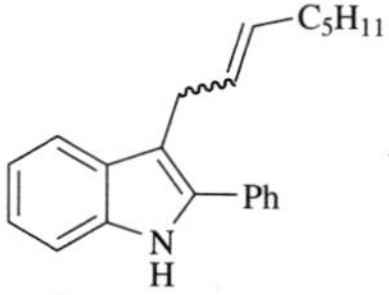 **I** 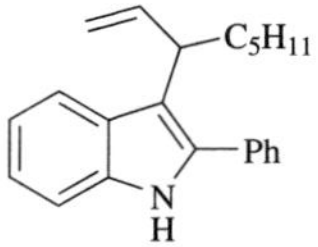 (3-(oct-2-enyl)-2-phenylindole) + **II** (3-(oct-1-en-3-yl)-2-phenylindole); **I + II** (80), **I:II** = 97:3, **I** (*E*):(*Z*) = 88:12	102
	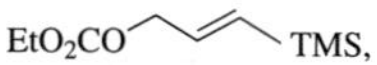 EtO$_2$CO—TMS, Pd$_2$(dba)$_3$, ttmpp, THF, 60°, 3 h	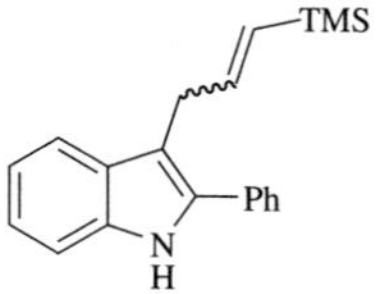 3-(3-TMS-allyl)-2-phenylindole (46), (*E*):(*Z*) = 95:5	102

I

ArOTf, Pd(PPh$_3$)$_4$,
Cs$_2$CO$_3$, MeCN, 100°

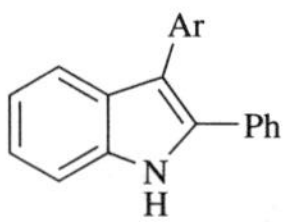

Ar	Time (h)	
Ph	0.5	(97)
4-O$_2$NC$_6$H$_4$	1	(94)
4-MeOC$_6$H$_4$	1	(98)
3,4,5-(MeO)$_3$C$_6$H$_4$	1	(98)
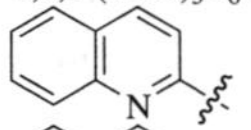 (quinolinyl)	2	(35)
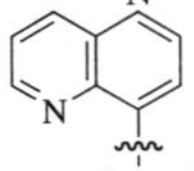 (quinolinyl)	1	(91)
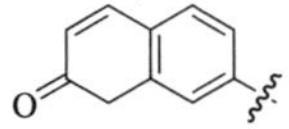 (tetrahydronaphthalenone)	1	(80)
4-PhCOC$_6$H$_4$	1	(99)

105, 106

I

ArCl, Pd$_2$(dba)$_3$, XPhos,
Cs$_2$CO$_3$, MeCN, 120°

Ar	Time (h)	
2-pyridyl	3	(83)
3-pyridyl	2	(80)
Ph	3.5	(90)
4-MeC$_6$H$_4$	5.5	(48)
2-MeC$_6$H$_4$	5.5	(40)
3-CF$_3$C$_6$H$_4$	1.5	(85)
4-MeOC$_6$H$_4$	12	(65)
3-MeOC$_6$H$_4$	3	(88)
3-MeCOC$_6$H$_4$	2	(94)
3-MeO$_2$CC$_6$H$_4$	3	(84)

107

TABLE 3D. 2,3-DISUBSTITUTED INDOLES FROM 2-ALKYNYLANILIDES (Continued)

2-Alkynylanilide	Conditions	Product(s) and Yield(s) (%)					Refs.
C$_{16}$ Ph–C≡C–C$_6$H$_4$–NHCOCF$_3$ (ortho)	ArBr, Pd(PPh$_3$)$_4$, Cs$_2$CO$_3$, MeCN, 100°	[2-Ph-3-Ar-indole]					105, 106
		Ar	Time (h)		Ar	Time (h)	
		Ph	0.5	(98)	3-OHC$_6$H$_4$	0.5	(94)
		3-FC$_6$H$_4$	0.5	(97)	4-OHC$_6$H$_4$	1	(98)
		4-O$_2$NC$_6$H$_4$	1	(98)	3,5-Me$_2$C$_6$H$_3$	0.5	(98)
		4-MeC$_6$H$_4$	1	(98)	4-MeCOC$_6$H$_4$	0.5	(91)
		3-MeC$_6$H$_4$	1	(96)	4-t-BuC$_6$H$_4$	0.5	(98)
		2-MeC$_6$H$_4$	1	(98)	4-PhC$_6$H$_4$	1	(98)
		3-CF$_3$C$_6$H$_4$	0.5	(88)			
		4-MeOC$_6$H$_4$	0.25	(86)			
		3-MeOC$_6$H$_4$	0.5	(99)			
		3-NCC$_6$H$_4$	0.5	(90)			
		4-NCC$_6$H$_4$	0.5	(84)			
		2-thiazolyl	2	(70)			
		5-pyrimidinyl	5	(85)			
		2,4-dichloropyrimidin-5-yl	5	(70)			
		2-pyridyl	2	(94)			
		5-indolyl	5	(56)			

	Br–C≡C–R, Pd(PPh$_3$)$_4$, Cs$_2$CO$_3$, MeCN, 60°	[3-(RC≡C)-2-Ph-indole] **I**					108
		R	Time (h)				
		CMe$_2$OH	8	(50)			
		Ph	6	(76)			
		4-O$_2$NC$_6$H$_4$	8	(58)			
		4-MeOC$_6$H$_4$	55	(57)			
		4-MeCOC$_6$H$_4$	7	(65)			
		2-quinolinyl	48	(52)			
		6-MeO-2-naphthyl	22	(63)			

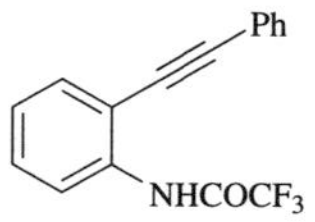

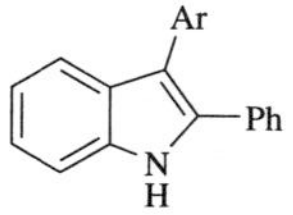

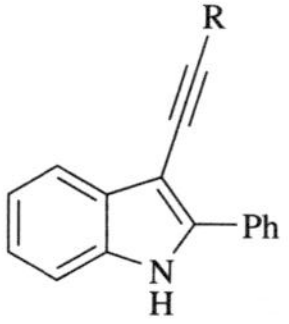

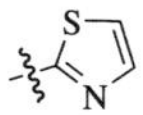

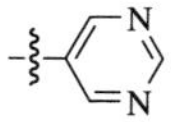

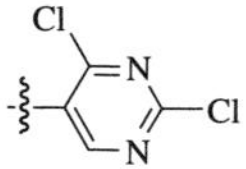

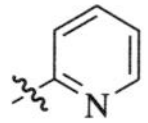

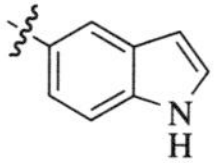

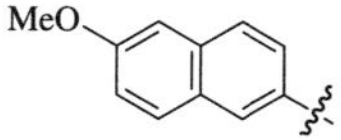

Br———R, Pd(PPh$_3$)$_4$, K$_2$CO$_3$, MeCN, 60° → **I**

R	Time (h)	
CMe$_2$OH	30	(62)
Ph	16	(78)
4-O$_2$NC$_6$H$_4$	6	(66)
4-CHOC$_6$H$_4$	16	(86)
3-O$_2$N-4-MeC$_6$H$_3$	4	(81)
4-MeCOC$_6$H$_4$	9	(50)
n-C$_8$H$_{17}$	30	(40)
2-quinolyl	16	(45)

108

1. Br———n-C$_8$H$_{17}$, Pd$_2$(dba)$_3$, HP(t-Bu)$_3$BF$_4$, Cs$_2$CO$_3$, MeCN, 60°, 6 h
2. TsOH, MeOH/Me$_2$CO, rt, 2 h

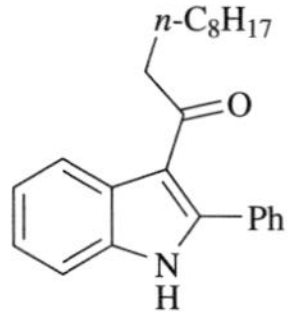

n-C$_8$H$_{17}$–C(O)–CH$_2$–[2-Ph-indol-3-yl] (57)

108

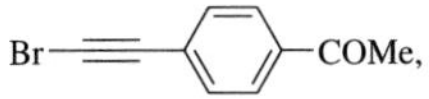

Br—C$_6$H$_4$—COMe, Pd(PPh$_3$)$_4$, K$_2$CO$_3$, MeCN, 60°, 5 h

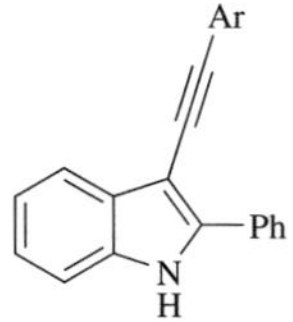

Ar—C≡C—[2-Ph-indol-3-yl], Ar = 4-MeCOC$_6$H$_4$ (56)

108

ArI, supported Pd nanoparticles, K$_2$CO$_3$, MeCN, 110°

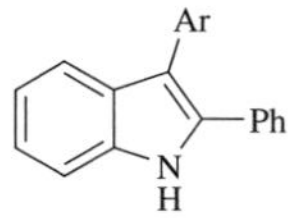

Ar—[2-Ph-indol-3-yl]

Ar	Time (h)		f
Ph	9	(82)	—
4-ClC$_6$H$_4$	5	(96)	(92, 83)
4-O$_2$NC$_6$H$_4$	4	(90)	—
4-NCC$_6$H$_4$	4	(84)	(92, 90, 86)
3-MeC$_6$H$_4$	40	(77)	—
4-MeOC$_6$H$_4$	48	(70)	—
3-CF$_3$C$_6$H$_4$	2	(91)	(90, 86, 87)
4-MeCOC$_6$H$_4$	2	(89)	(91, 87)

376

TABLE 3D. 2,3-DISUBSTITUTED INDOLES FROM 2-ALKYNYLANILIDES (*Continued*)

2-Alkynylanilide	Conditions	Product(s) and Yield(s) (%)	Refs.

C$_{16}$

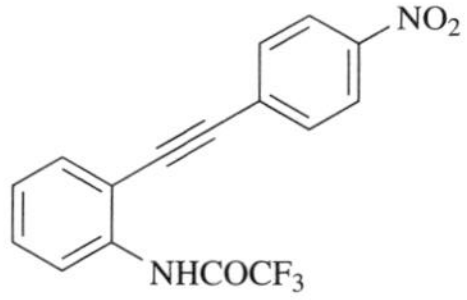

2-Alkynylanilide: 4-NO$_2$-C$_6$H$_4$-C≡C-C$_6$H$_4$-NHCOCF$_3$

Conditions: Br-(thiazole), Pd(PPh$_3$)$_4$, Cs$_2$CO$_3$, MeCN, 100°, 2 h

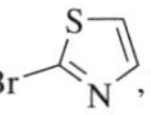

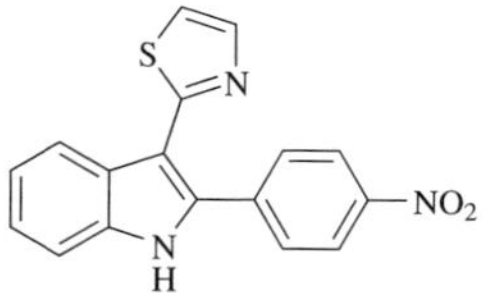

Product: 2-(thiazol-2-yl)-3-(4-nitrophenyl)indole (40)

Ref: 105

C$_{16–18}$

2-Alkynylanilide: R^2-C≡C-C$_6$H$_3$(R^1)-NHCOCF$_3$

Conditions: ArN$_2$BF$_4$, Pd(PPh$_3$)$_4$, K$_2$CO$_3$, Bu$_4$NI, MeCN (anhyd), 60°, 1–5.5 h

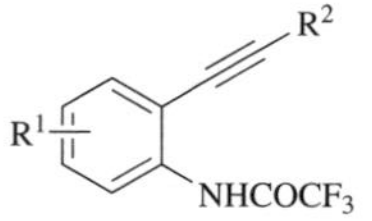

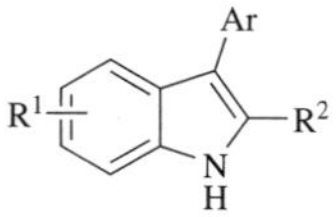

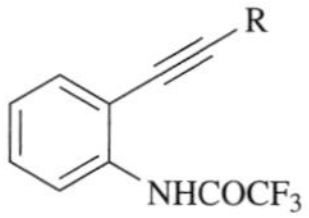

Product: 2-R^2, 3-Ar indole with R^1

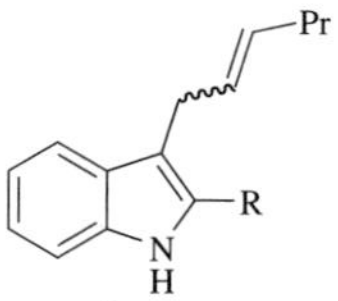

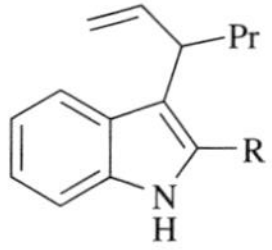

R^1	R^2	Ar	
5-F	Ph	4-NCC$_6$H$_4$	(71)
5-F	Ph	4-MeC$_6$H$_4$	(71)
5-Cl	Ph	4-MeOC$_6$H$_4$	(74)
5-Cl	4-MeC$_6$H$_4$	2-Me-4-MeOC$_6$H$_3$	(66)
5-MeCO	Ph	Ph	(68)
5-Cl	4-MeCOC$_6$H$_4$	4-O$_2$NC$_6$H$_4$	(75)
5,7-Me$_2$	Ph	4-FC$_6$H$_4$	(70)
5,7-Me$_2$	Ph	4-MeOC$_6$H$_4$	(50)

Ref: 374

C$_{16–21}$

2-Alkynylanilide: R-C≡C-C$_6$H$_4$-NHCOCF$_3$

Conditions: EtO$_2$CO-CH=CH-Pr, Pd$_2$(dba)$_3$, ttmpp, THF, 60°

Products: I (3-(1-propylallyl)-2-R-indole) + II (3-(pent-2-enyl)-2-R-indole)

R	Time (h)	I + II	I:II	I (E):(Z)
Ph	3	(76)	97:3	87:13
2-BrC$_6$H$_4$	10	(34)	100	91:9
4-MeOC$_6$H$_4$	4	(72)	97:3	87:13
4-MeCOC$_6$H$_4$	8	(67)	>99:1	82:18
4-THPOC$_6$H$_4$	2	(96)	100	82:18

Ref: 102

C_{17} 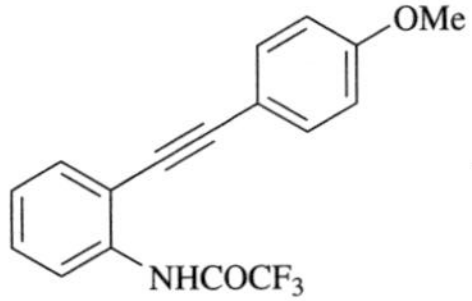	ArBr, Pd(PPh$_3$)$_4$, Cs$_2$CO$_3$, MeCN, 100°	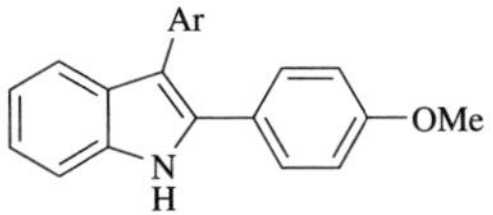 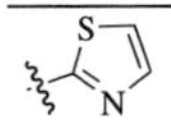 **I** with Ar table: Ph 3 (60); 4-MeOC$_6$H$_4$ 5 (40); 4-MeCOC$_6$H$_4$ 5 (73)	105, 106
	ArCl, Pd$_2$(dba)$_3$, XPhos, Cs$_2$CO$_3$, MeCN, 120°	**I** with Ar table: (thiazolyl) 1 (87); 4-*t*-BuC$_6$H$_4$ 2 (80)	107
	Br—≡—R, Pd(PPh$_3$)$_4$, Cs$_2$CO$_3$, MeCN, 60°	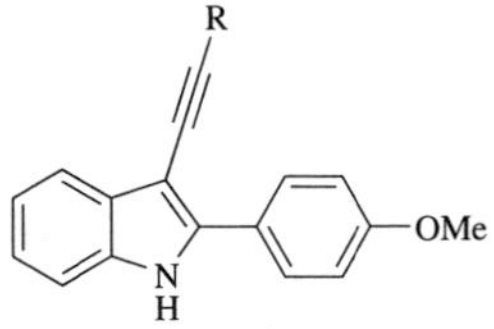 3-R-alkynyl-2-(4-methoxyphenyl)indole; R: Ph 8 (75); 4-MeCOC$_6$H$_4$ 6 (73)	108
	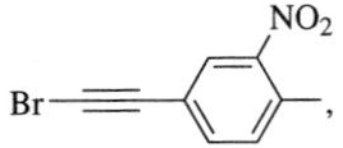 4-Br-2-NO$_2$-toluene, Pd(PPh$_3$)$_4$, K$_2$CO$_3$, MeCN, 60°, 3 h	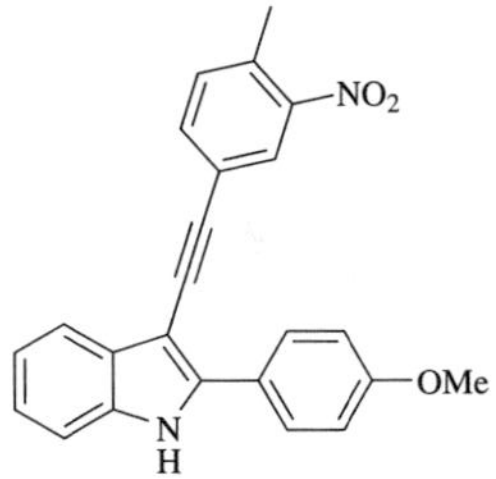 (85)	108
	Br—≡—*n*-C$_8$H$_{17}$, Pd$_2$(dba)$_3$, HP(*t*-Bu)$_3$BF$_4$, Cs$_2$CO$_3$, MeCN, 60°, 2 h	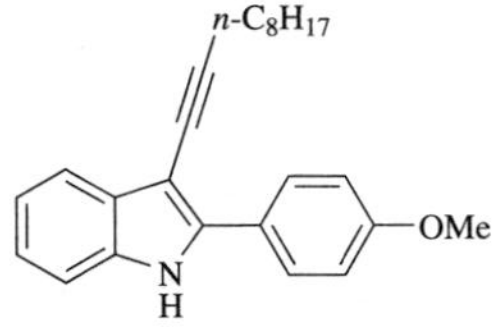 (43)	108

Starting material: 4-MeO-C$_6$H$_4$-C≡C-C$_6$H$_4$-NHCOCF$_3$ (ortho)

TABLE 3D. 2,3-DISUBSTITUTED INDOLES FROM 2-ALKYNYLANILIDES (*Continued*)

2-Alkynylanilide	Conditions	Product(s) and Yield(s) (%)	Refs.

C$_{17}$

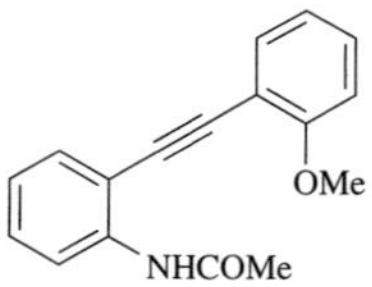

PhS—SPh, PdCl$_2$, DMSO, 80°, 24 h

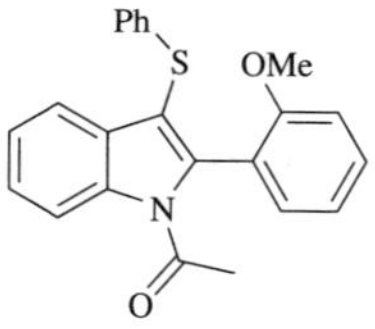

(35)

371

C$_{17-18}$

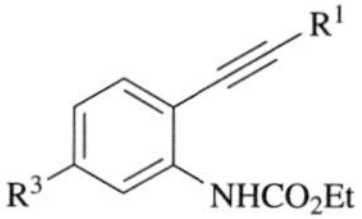

CH$_2$=CHR^2, PdCl$_2$, CuCl$_2$·H$_2$O, Bu$_4$NF, THF, reflux

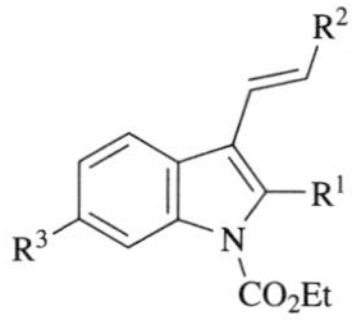

R^1	R^2	R^3	Time (h)	
Ph	CN	Cl	4.5	(31)
Ph	CO$_2$Me	Cl	3	(66)
Ph	Bu	OMe	8	(42)
Ph	CO$_2$Me	CN	22	(66)

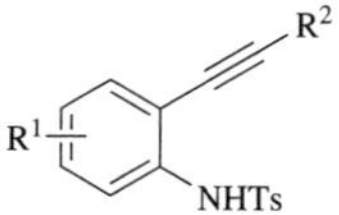

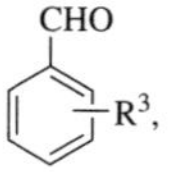

131

C$_{17-22}$

Pd(bpy)(H$_2$O)$_2$(OTf)$_2$, 1,4-dioxane, 60°, overnight

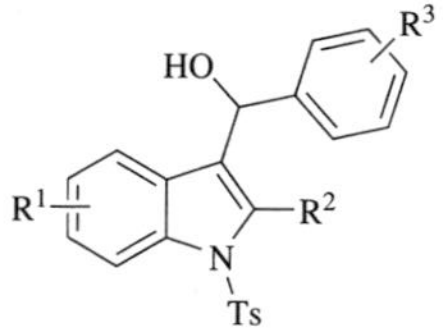

R^1	R^2	R^3	
H	MeOCH$_2$	4-O$_2$N	traces
H	Ph	4-O$_2$N	(75)
H	Ph	3-O$_2$N	(63)
H	Ph	5-Cl-4-O$_2$N	(93)
H	Ph	H	(—)^a
5-F	Ph	4-O$_2$N	(64)
5-Cl	Ph	4-O$_2$N	(62)
6-Cl	Ph	4-O$_2$N	(49)
H	4-BrC$_6$H$_4$	4-O$_2$N	(62)
H	n-C$_6$H$_{13}$	4-O$_2$N	traces
H	Ph	4-NC	traces
H	Ph	4-MeCO	traces
5-Me	Ph	4-O$_2$N	(65)
H	4-MeC$_6$H$_4$	4-O$_2$N	(78)

375

C_{18}	structure: 4-COMe-C6H4-C≡C- attached to aniline with NHCOCF3	ArCl, Pd2(dba)3, XPhos, Cs2CO3, MeCN, 120°	3-Ar-2-(4-acetylphenyl)indole **I**	107

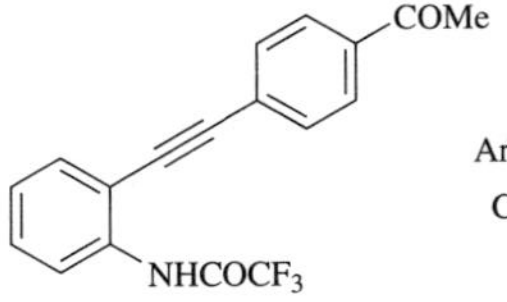

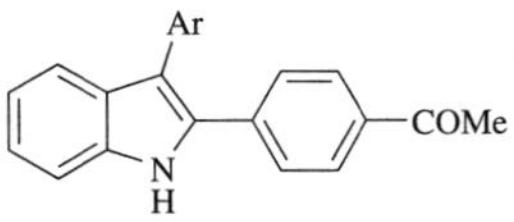

Ar	Time (h)
Ph	5.5 (50)
3-MeOC6H4	2.5 (56)
4-MeCOC6H4	1 (87)

		ArBr, Pd(PPh3)4, Cs2CO3, MeCN, 100°	2-Ar-indole **I**	105, 106

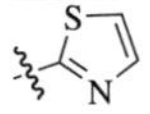

Ar	Time (h)
2-thienyl	2 (74)
4-t-BuC6H4	1 (98)

		Br–≡–R, Pd(PPh3)4, Cs2CO3, MeCN, 60°	3-(R-C≡C)-2-(4-acetylphenyl)indole	108

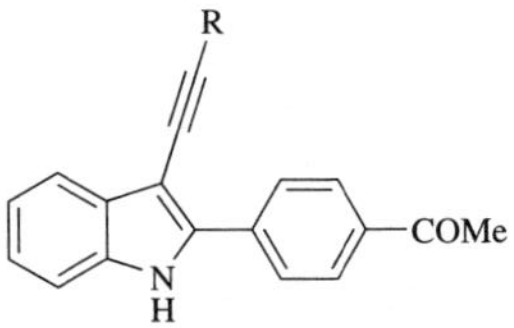

R	Time (h)
Ph	16 (52)
4-MeCOC6H4	8 (59)

C_{19}	2-Bu-C≡C-aniline (NHTs)	methyl vinyl ketone, Pd(OAc)2, LiBr, THF, rt, 2 d	3-(3-oxobutyl)-2-Bu-N-Ts-indole (76)	103

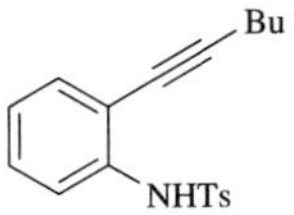

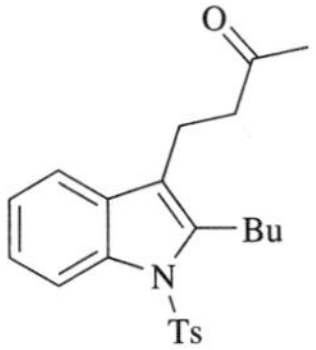

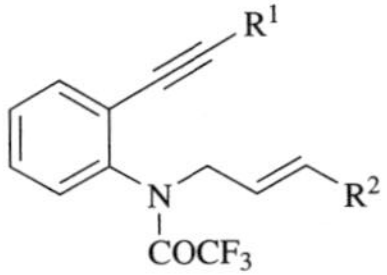

C_{19-27}	R1-C≡C-aniline with N(COCF3)(CH2CH=CHR2)	Pd(PPh3)4, K2CO3, MeCN, 90°	3-(CH2CH=CHR2)-2-R1-indole	102

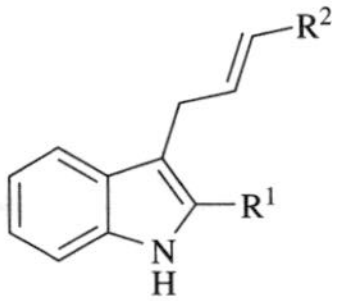

R1	R2	Time (h)
Ph	H	1 (91)
CH2NHCOEt	Ph	1 (76)
Ph	Ph	1.5 (91)
Ph	4-MeOC6H4	1.5 (84)
Ph	4-MeCOC6H4	2 (77)

TABLE 3D. 2,3-DISUBSTITUTED INDOLES FROM 2-ALKYNYLANILIDES (*Continued*)

2-Alkynylanilide	Conditions	Product(s) and Yield(s) (%)	Refs.
C$_{20}$ 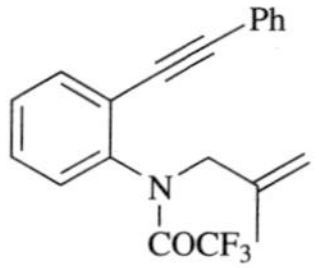	Pd(PPh$_3$)$_4$, K$_2$CO$_3$, MeCN, 90°, 3 h	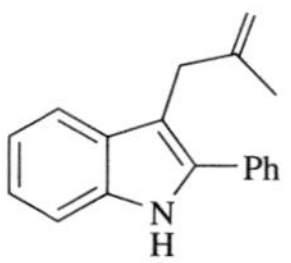 (95) [2-Ph, 3-(2-methylallyl)indole]	102
C$_{20-23}$ 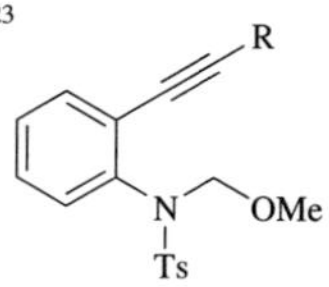	PdBr$_2$, toluene, 80°	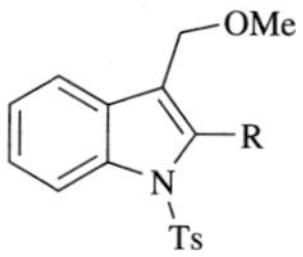 R: Pr (33), Ph (33) [3-(OMe-methyl), N-Ts indole]	140
C$_{21}$ 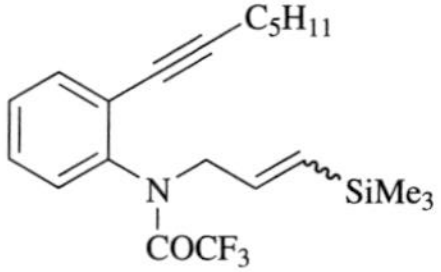 (E):(Z) = 89:11	Pd$_2$(dba)$_3$, ttmpp, DME, 100°, 4 h	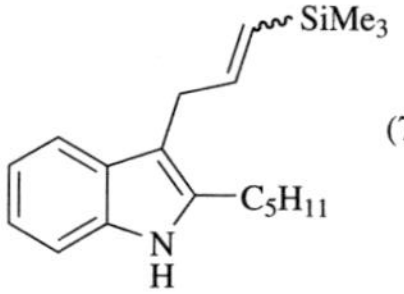 (75), (E):(Z) = 88:12 [2-C$_5$H$_{11}$, 3-allyl-SiMe$_3$ indole]	102
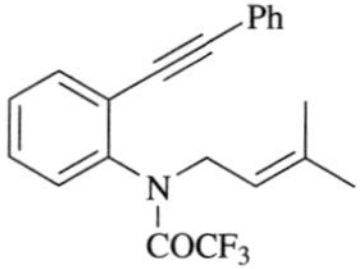 C$_{21-23}$	Pd(PPh$_3$)$_4$, K$_2$CO$_3$, MeCN, 90°, 24 h	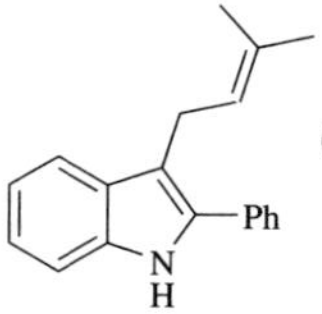 (44) [2-Ph, 3-(3-methylbut-2-enyl)indole]	102
C$_{21-23}$ 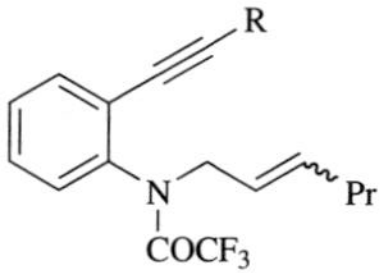	Pd(PPh$_3$)$_4$, K$_2$CO$_3$, MeCN, 90°	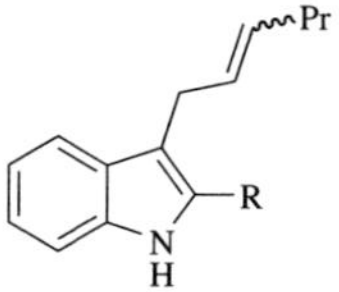 I 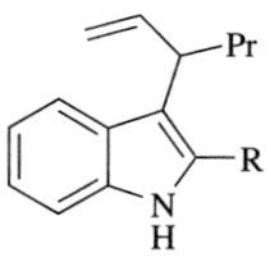 (3-allyl-Pr, 2-R indole) + II (3-(1-Pr-allyl), 2-R indole)	102

R	Time (h)	I + II	I:II	I (E):(Z)
n-C$_5$H$_{11}$	8	(66)	81:18	81:19
CH$_2$OTHP	5	(71)	88:12	88:12
4-MeOC$_6$H$_4$	2	(81)	65:35	88:12
4-MeOC$_6$H$_4$	4	(83)	75:25	78:22

(E):(Z) = 97:3
(E):(Z) = 92:8
(E):(Z) = 92:8
(E):(Z) = 92:8

C$_{21-24}$

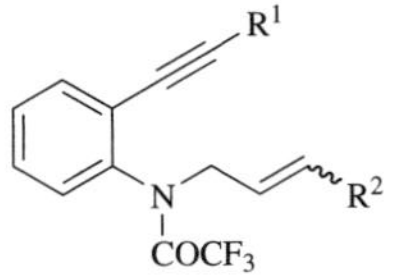

Pd(PPh$_3$)$_4$, K$_2$CO$_3$, DME, 100°

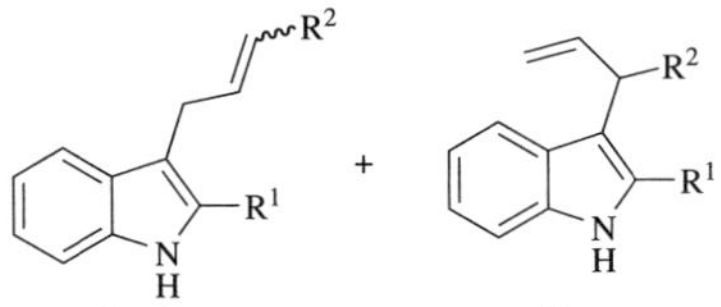

102

I + II

R^1	R^2	Time (h)	I + II	I:II	I (E):(Z)
n-C$_5$H$_{11}$	Pr	24	(69)	97:3	85:15
CH$_2$OTHP	Pr	12	(76)	96:4	82:18
CH$_2$OTHP	C$_5$H$_{11}$	12	(69)	97:3	83:17

(E):(Z) = 97:3
(E):(Z) = 93:7
(E):(Z) = 92:8

C$_{26}$

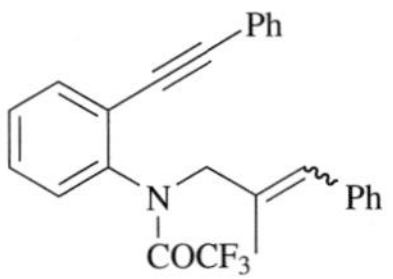

Pd(PPh$_3$)$_4$, K$_2$CO$_3$, MeCN, 90°, 5.5 h

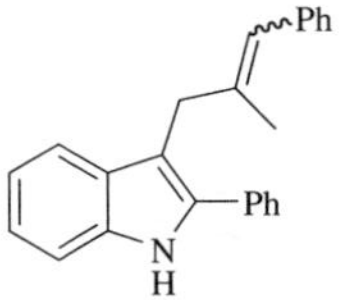

(78), (E):(Z) = 67:33

102

(E):(Z) = 98:2

[a] The reaction time was 40 hours.
[b] The reaction was carried out in a stainless steel bomb, under 7 atm of CO, in anhydrous MeCN.
[c] The product was a mixture of isomers, (E):(Z) = 77:23.
[d] The product was a mixture of isomers.
[e] The starting material was recovered.
[f] Yields are of successive runs carried out with the recovered catalyst

TABLE 3E. 2,3-DISUBSTITUTED INDOLES FROM 2-HALO-N-ALKYNYLANILIDES AND 2-HALO-N-ALKYLANILINES

Substrate	Conditions	Product(s) and Yield(s) (%)	Refs.
C_{13} 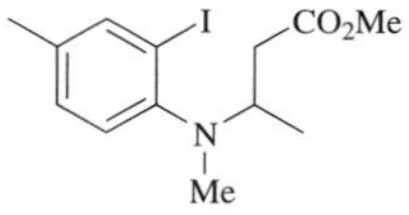 (substrate with I, CO₂Me, N-Me on aryl)	Pd(PPh₃)₄, K₃PO₄, DMF, 90°, 70 h	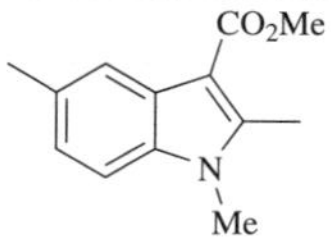 2,3-disubstituted indole with CO₂Me, Me, N-Me (42)	351, 352
C_{14} 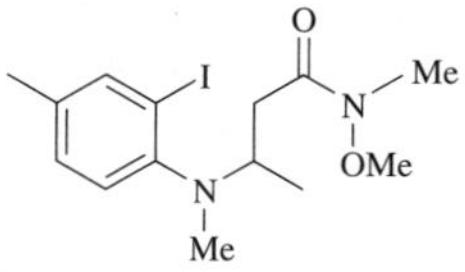 (substrate with I, C(O)N(OMe)Me, N-Me)	Pd(PPh₃)₄, phenol, t-BuOK, THF, reflux, 12 h	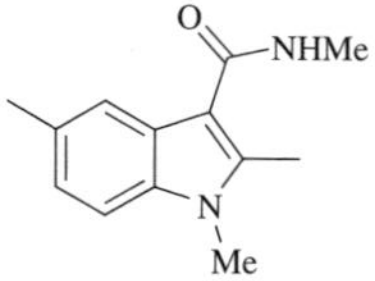 indole-3-C(O)NHMe, 2-Me, N-Me (58)	351, 352
C_{18-20} 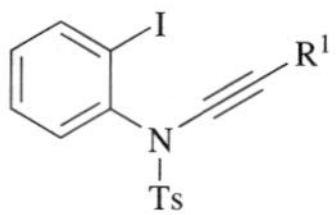 (2-iodo-N-Ts alkynylaniline with R¹)	 HN(R²)-piperazine, PdCl₂(PPh₃)₂, K₂CO₃, THF, 80°, 5–24 h	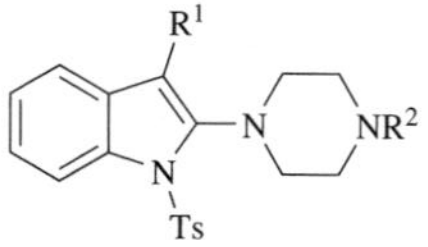 2-(4-R²-piperazinyl)-3-R¹-1-Ts-indole R¹ / R² / yield: TMS / Boc (83) TMS / Me (95) n-C₅H₁₁ / Ph (97)	29
(2-iodo-N-Ts alkynylaniline, R¹ = Ph)	pyrrolidine, PdCl₂(PPh₃)₂, K₂CO₃, THF, 80°	2-pyrrolidinyl-3-Ph-1-Ts-indole (94)	29
C_{21} (2-iodo-N-Ts alkynylaniline, Ph)	 HN(NBoc)-piperazine, PdCl₂(PPh₃)₂, K₂CO₃, THF, 80°, 2–24 h	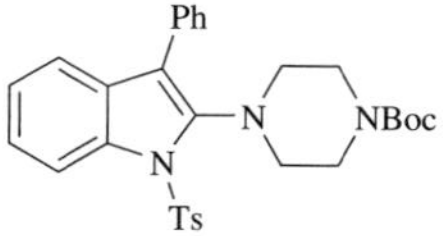 2-(4-Boc-piperazinyl)-3-Ph-1-Ts-indole (95)	29
	RNH₂, PdCl₂(PPh₃)₂, K₂CO₃, THF, 80°, 2–24 h	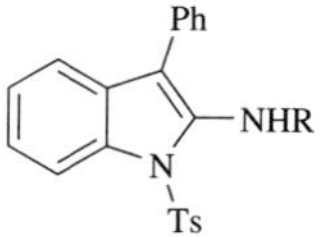 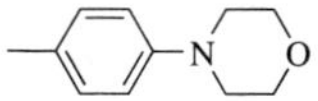 2-NHR-3-Ph-1-Ts-indole R: CH₂CH=CH₂ (98) c-C₃H₅ (94) 4-morpholinyl-C₆H₄-Me (77)	29

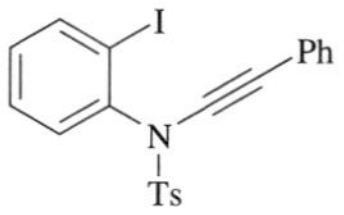

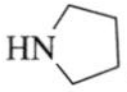

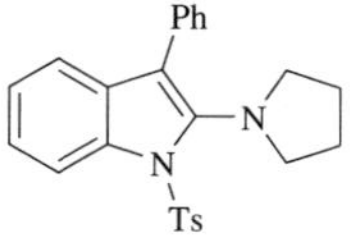

TABLE 3F. 2,3-DISUBSTITUTED INDOLES FROM 2-ALKYNYLISOCYANOBENZENES, -ISOCYANATOBENZENES, AND -N-ALKYLIDENEANILINES

Alkyne	Conditions	Product(s) and Yield(s) (%)				Refs.
C_{12-16}						
(alkyne with R^1, NCO)	$CH_2=CHCH_2OCO_2R^2$, $Pd(PPh_3)_4$, CuCl, THF, 100°	(indole with R^1, CO_2R^2, allyl)				28
		R^1	R^2	Time (h)		
		Pr	Me	1	(81)	
		Pr	i-Pr	1	(69)	
		Pr	t-Bu	1	(72)	
		Pr	Ph	1	(86)	
		Pr	Bn	1	(83)	
		c-C$_5$H$_9$	Me	3	(71)	
		Ph	Me	2	(62)	
		4-MeOC$_6$H$_4$	Me	6	(62)	
		4-CF$_3$C$_6$H$_4$	Me	7	(65)	
C_{12-18}						
(alkyne with TMS, NC, R)	$CH_2=CHCH_2OCO_2Me$, $Pd_2(dba)_3\cdot CHCl_3$, (2-furyl)$_3$P, TMSN$_3$, rt, 10 min; 100°, 1 h	(indole with TMS, CN, allyl, R)				26
		R	Solvent		R	Solvent
		H	octane (59)		7-MeO	octane (63)
		5-F	THF (56)		5-MeS	octane (67)
		5-Cl	THF (54)		5-CF$_3$	octane (65)
		6-Cl	octane (47)		5-CN	THF (30)
		5-Br	octane (47)		5-CO$_2$Me	THF (53)
		5-NO$_2$	THF (34)		5-COMe	THF (37)
		6-NO$_2$	THF (59)		6-COMe	THF (45)
		5-Me	octane (68)		5-i-Pr	octane (77)
		6-Me	octane (65)		5-C≡C-t-Bu	octane (58)
		4-MeO	octane (62)		6-C≡C-t-Bu	THF (61)
		5-MeO	octane (69)		5-Ph-N=N	octane (45)
		6-MeO	octane (57)			

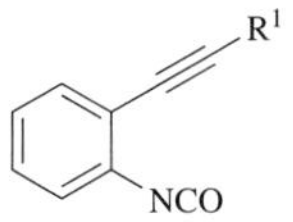

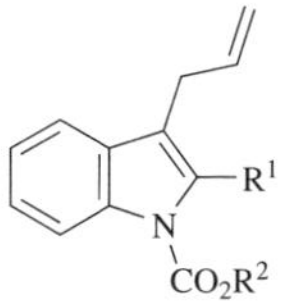

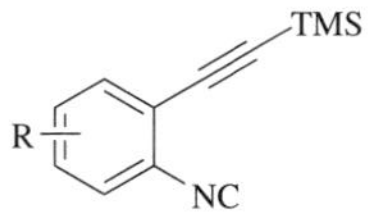

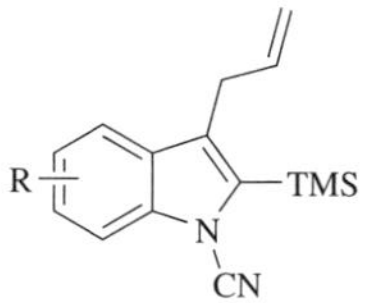

TABLE 3F. 2,3-DISUBSTITUTED INDOLES FROM 2-ALKYNYLISOCYANOBENZENES, -ISOCYANATOBENZENES, AND -N-ALKYLIDENEANILINES (*Continued*)

Alkyne	Conditions	Product(s) and Yield(s) (%)				Refs.
		R^1	R^2	R^3		
C$_{16-21}$						
	Pd(OAc)$_2$, Bu$_3$P, 1,4-dioxane, 100°	Et	H	2-thienyl	(70)	33
		Et	H	5-methyl-2-furyl	(63)	
		Et	H	4-pyridyl	(64)	
		Et	H	Ph	(58)	
		Et	H	4-O$_2$NC$_6$H$_4$	(70)	
		CH$_2$OMOM	H	4-O$_2$NC$_6$H$_4$	(55)	
		—(CH$_2$)$_5$—		4-O$_2$NC$_6$H$_4$	(71)^a	
		OTHP	H	4-O$_2$NC$_6$H$_4$	(56)	
		(CH$_2$)$_2$CO$_2$Et	H	4-O$_2$NC$_6$H$_4$	(59)	

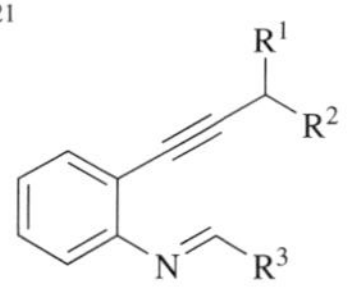

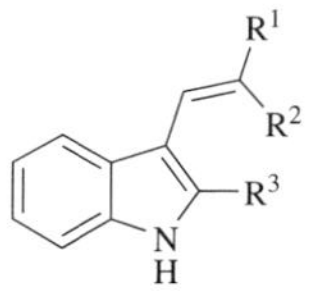

^{a} The reaction was run at 80°.

TABLE 3G. 2,3-DISUBSTITUTED INDOLES FROM N-(2-HALOPHENYL)ALLENAMIDES

N-(2-Halophenyl)allenamide	Reagent	Conditions	Product(s) and Yield(s) (%)	Refs.
C$_{15}$				
(Br, Boc allenamide)	9-BBN-CH$_2$CH$_2$CH(OEt)$_2$	PdCl$_2$(dppf), aq Cs$_2$CO$_3$ (3 M), DMF, 70°	2-Me-3-(CH$_2$CH$_2$CH(OEt)$_2$)-N-Boc-indole (42) + 2,3-diMe-N-Boc-indole (56)	353
C$_{15-21}$				
(I, Boc allenamide with R^1)	(HO)$_2$BR^2	Pd$_2$(dba)$_3$, aq Cs$_2$CO$_3$ (3 M), EtOH, 80°	2-R^1-3-(CH$_2$R^2)-N-Boc-indole R^1 / R^2 / yield: Me / Ph (81); Me$_2$(HO)Si / Ph (61); TMS / Ph (71); Bn / Ph (98); Bn / allyl (84)	353

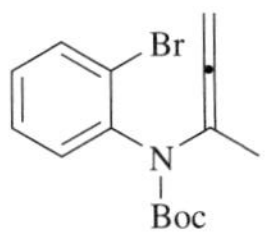

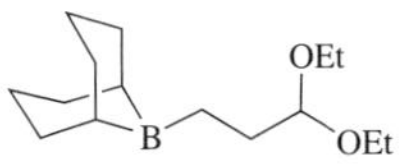

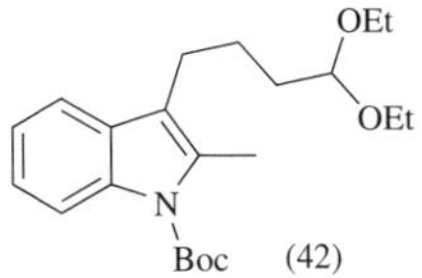

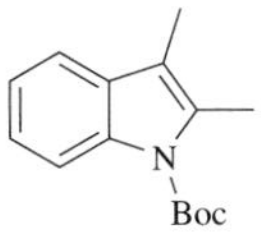

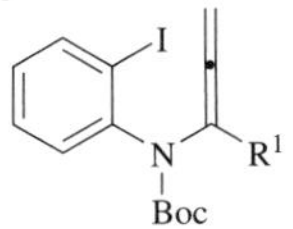

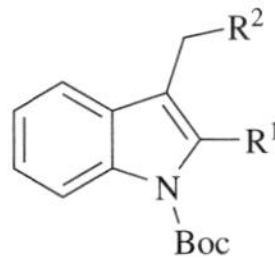

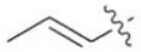

TABLE 3H. 2,3-DISUBSTITUTED INDOLES FROM 2-ALLENYLANILIDES PREPARED IN SITU

Substrate	Reagent	Conditions	Product(s) and Yield(s) (%)	Refs.
C_{11} (aryl iodide with NHBoc, R^1)	Bu_3Sn-allene with R^2, R^3	$Pd_2(dba)_3$, $P(2\text{-furyl})_3$, Bu_4NCl, CuI, DMF	Indole with R^1, R^2, R^3, N-Boc: R^1 / R^2 / R^3 / Temp / Time (h) H / CH_2OPMB / H / rt / 5.5 (65) H / $(CH_2)_2OTBS$ / H / rt / 4 (57) H / $n\text{-}C_6H_{13}$ / H / rt / 0.5 (46) O_2N / $(CH_2)_2OTBS$ / Me / 50° / 6 (71)	377
$C_{11\text{-}12}$ (aryl iodide with NHBoc, R)	Bu_3Sn-allene with OTBS	$Pd_2(dba)_3$, $P(2\text{-furyl})_3$, Bu_4NCl, CuI, DMF	Indole with R, Me, CH_2OTBS, N-Boc: R / Temp / Time (h) O_2N / 50° / 3 (72) [a] MeO / rt / 4 (65)	377
$C_{11\text{-}12}$ (aryl iodide with NHBoc, R^1)	Bu_3Sn-allene with R^2, R^3	1. $Pd_2(dba)_3$, $P(2\text{-furyl})_3$, Bu_4NCl, CuI, DMF, temp 2. Bu_4NF, 0°	Indole with R^1, R^2, R^3, R^4, N-Boc: R^1 / R^2 / R^3 / R^4 / Temp H / CH_2OTBS / Me / CH_2OH / rt (70) H / CH_2OTBS / Pr / CH_2OH / rt (69) H / $n\text{-}C_7H_{15}$ / Me / $n\text{-}C_7H_{15}$ / rt (56) H / H / $n\text{-}C_5H_{11}$ / H / rt (54) MeO / CH_2OPMB / Me / CH_2OH / 50° (62)	377

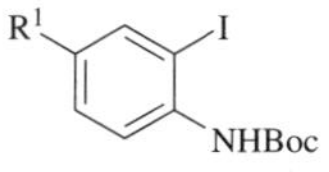

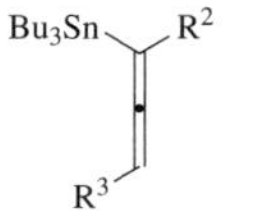

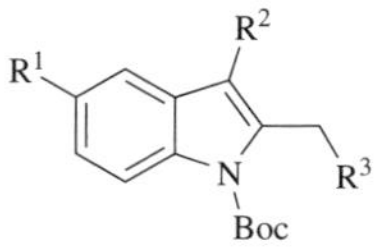

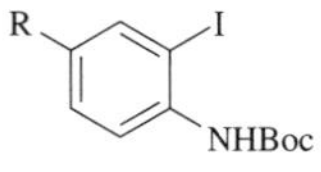

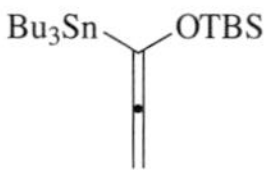

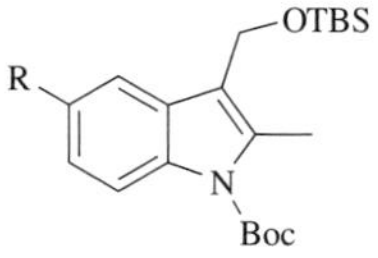

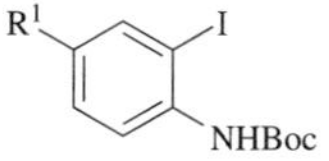

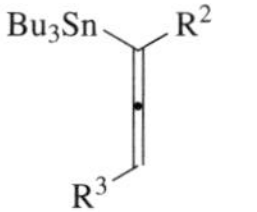

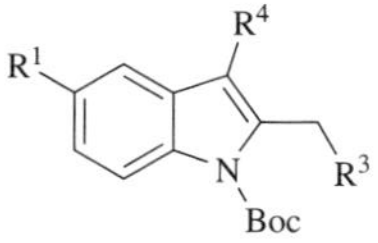

[a] The coupling products were isolated in variable amount.

TABLE 4A. 2,3-UNSUBSTITUTED INDOLES FROM 2-VINYLANILINES AND -ANILIDES

2-Vinylaniline or -anilide	Conditions	Product(s) and Yield(s) (%)	Refs.
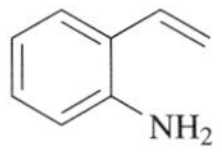 C₈ (2-vinylaniline, NH₂)	PdCl₂(MeCN)₂, LiCl, benzoquinone, THF, reflux, 18 h	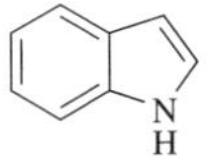 indole (74)	142
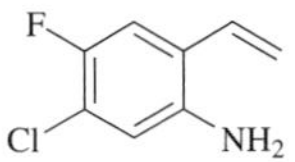 (4-F, 5-Cl vinylaniline, NH₂)	PdCl₂, LiCl, benzoquinone, THF, reflux, 16 h	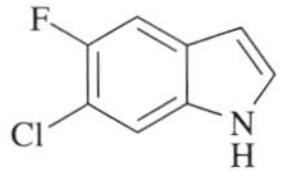 5-F, 6-Cl indole (50)	146
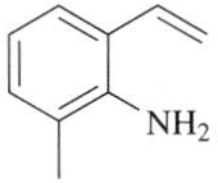 C₉ (3-Me-2-vinylaniline, NH₂)	PdCl₂(MeCN)₂, LiCl, benzoquinone, THF, reflux, 10 h	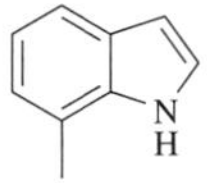 7-Me indole (82)	148
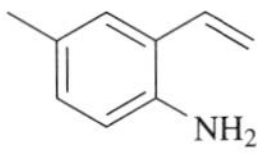 (4-Me-2-vinylaniline, NH₂)	PdCl₂(MeCN)₂, LiCl, benzoquinone, THF, reflux, 10 h	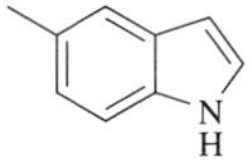 6-Me indole (83)	148
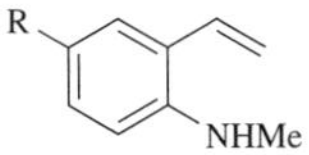 C₉₋₁₃ (R-substituted-2-vinyl-NHMe)	PdCl₂(MeCN)₂, LiCl, benzoquinone, THF, reflux, 18 h	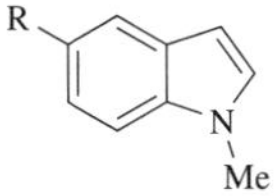 R-5-substituted-N-Me-indole: R = H (94), Cl (83), MeO (85), i-Pr (93), t-Bu (87)	148
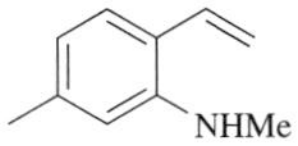 C₁₀ (4-Me-2-vinyl-NHMe)	PdCl₂(MeCN)₂, LiCl, benzoquinone, THF, reflux, 18 h	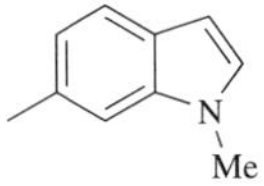 6-Me-N-Me-indole (81)	148
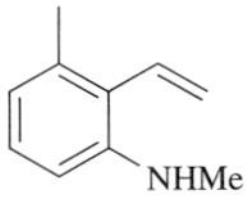 (3-Me-2-vinyl-NHMe)	PdCl₂(MeCN)₂, LiCl, benzoquinone, THF, reflux, 18 h	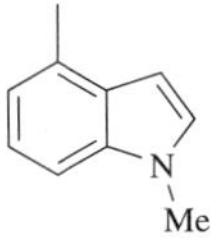 7-Me-N-Me-indole (79)	148

TABLE 4A. 2,3-UNSUBSTITUTED INDOLES FROM 2-VINYLANILINES AND -ANILIDES (*Continued*)

2-Vinylaniline or -anilide	Conditions	Product(s) and Yield(s) (%)	Refs.
C_{10-12}			
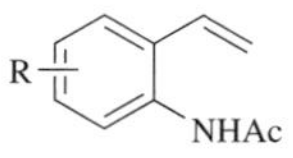 R-C₆H₃(CH=CH₂)NHAc	$PdCl_2$, CuCl, O_2, 1,3-propanediol, DME, 50–60°, 24 h	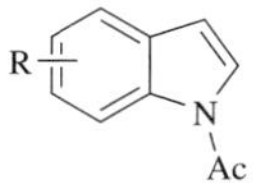 R-indole-N-Ac R: H (61), 6-Cl (64), 4-Me (49), 5-Me (55), 6-Me (70), 5-OMe (54), 6-OMe (35), 4-CO_2Me (67), 5-CO_2Me (65), 6-CO_2Me (57), 7-CO_2Me (62)	147
C_{12-15}			
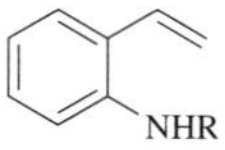 2-vinyl-NHR aniline	$PdCl_2(MeCN)_2$, LiCl, benzoquinone, THF, 18 h, reflux	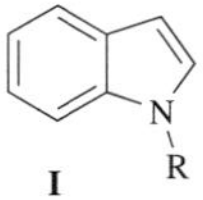 N-R indole (**I**) R: Bu (53), Bn (80)	148
C_{14-15}			
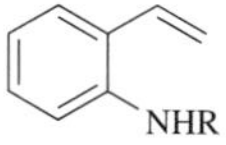 2-vinyl-NHR aniline	$Pd(OAc)_2$, $Cu(OAc)_2$, AcOH, DMF, 100°, 12 h	**I** R 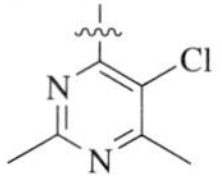 = 4-Cl-2,6-dimethylpyrimidin-5-yl (91) R = 2-ClC_6H_4 (98) R = 2-Cl-4-MeC$_6$H$_3$ (87)	378
C_{15}			
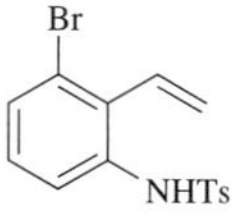 3-Br-2-vinyl-NHTs aniline	$PdCl_2(MeCN)_2$, LiCl, benzoquinone, THF, 18 h, reflux	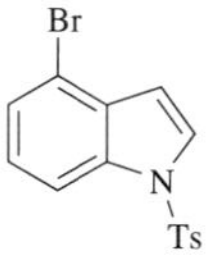 4-Br-N-Ts indole (77–80)	143, 144

C_{15-20}

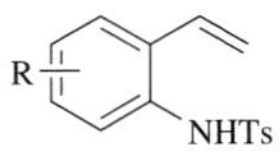

PdCl$_2$(MeCN)$_2$, LiCl, benzoquinone, DMF, 100°

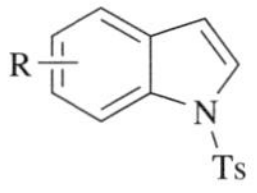

R	
H	(48)
5-Cl	(78)
4-Me	(87)
4-OMe	(68)
4-OTf	(88)
6-CH$_2$=CH	(47)
4-Ac	(58)
4-CO$_2$Me	(49)
4-CH(OAc)$_2$	(60)

145

TABLE 4B. 2,3-UNSUBSTITUTED INDOLES FROM 2-NITROSTYRENES

2-Nitrostyrene	Conditions	Product(s) and Yield(s) (%)	Refs.
C_8			
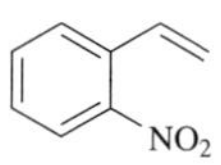 (2-vinyl nitrobenzene)	CO (20 atm), $PdCl_2(PPh_3)_2$, $SnCl_2$, 1,4-dioxane, 100°, 16 h	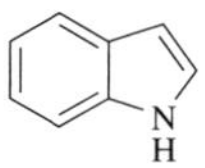 indole (50)	154, 155
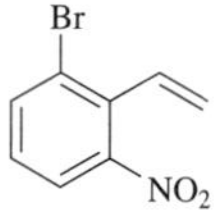 (3-bromo-2-vinyl nitrobenzene)	CO (4 atm), $Pd(OAc)_2$, PPh_3, Et_3N, DMF, MeOH, 60°	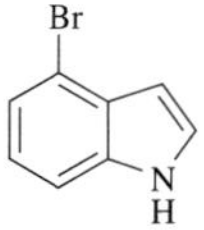 4-bromoindole (63)	156
C_{8-10}			
(R-substituted 2-vinyl nitrobenzene)	CO (4 atm), $Pd(OAc)_2$, PPh_3, MeCN, 70°	R-substituted indole	156

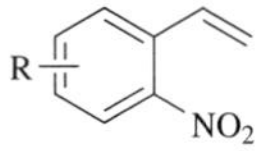

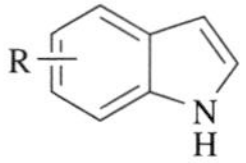

R	Time (h)	
H	15	(87)
4-OH	12	(96)
4-NO_2	26	(89)
5-Me	24	(51)
4-OMe	20	(89)
5-OMe	19	(63)
6-OMe	21	(40)
4-OTf	48	(40)
4-CO_2Me	23	(100)
5-CO_2Me	120	(47)
6-CO_2Me	21	(78)
7-CO_2Me	27	(71)

TABLE 5A. 2-SUBSTITUTED INDOLES FROM 2-ALLYLANILINES AND -ANILIDES

2-Allylaniline or -anilide	Conditions	Product(s) and Yield(s) (%)	Refs.
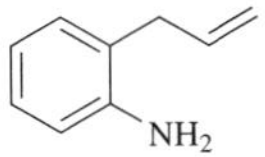 C₉ 2-allylaniline (NH₂)	PdCl₂(MeCN)₂, LiCl, benzoquinone, THF, reflux, 5 h	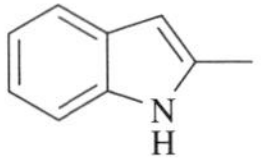 2-methylindole (86)	142
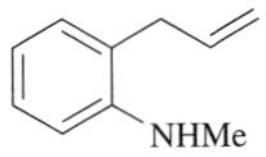 C₁₀ 2-allyl-N-methylaniline (NHMe)	PdCl₂(MeCN)₂, LiCl, benzoquinone, THF, reflux, 18 h	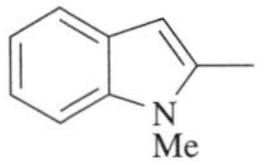 1,2-dimethylindole (89)	142
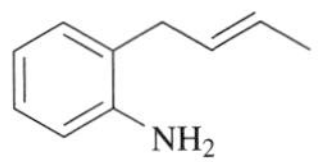 2-(but-2-enyl)aniline (NH₂)	PdCl₂(MeCN)₂, LiCl, benzoquinone, THF, reflux, 18 h	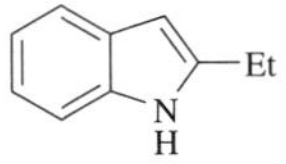 2-ethylindole (79)	142
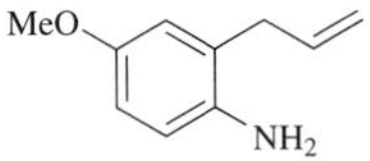 2-allyl-4-methoxyaniline (MeO, NH₂)	PdCl₂(MeCN)₂, LiCl, benzoquinone, THF, rt, 18 h	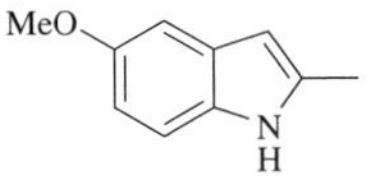 5-methoxy-2-methylindole (32)	142
C₁₁ 2-allyl-N-acetylaniline (NHAc)	PdCl₂(MeCN)₂, Cu(OAc)₂, Na₂CO₃, THF, reflux, 60 h	1-acetyl-2-methylindole (71)	142
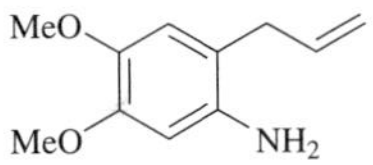 2-allyl-4,5-dimethoxyaniline (MeO, MeO, NH₂)	PdCl₂(MeCN)₂, LiCl, benzoquinone, THF, rt, 18 h	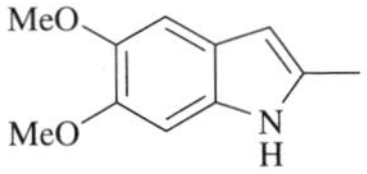 5,6-dimethoxy-2-methylindole (48)	142

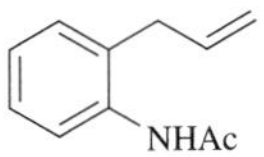

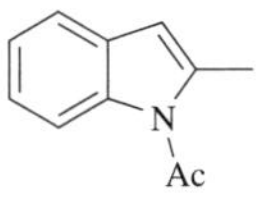

TABLE 5B. 2-SUBSTITUTED INDOLES FROM 2-HALOARYLENAMINES AND -IMINES

2-Haloarylenamine or -imine	Conditions	Product(s) and Yield(s) (%)	Refs.
C$_{15-19}$ 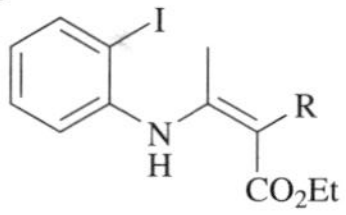 (enamine with R, CO$_2$Et, NH, I on aryl)	Pd(PPh$_3$)$_4$, Ag$_3$PO$_4$, DMSO, 100°	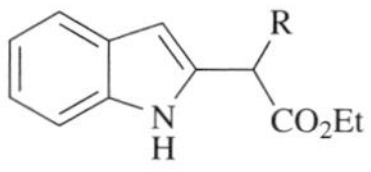 2-(CO$_2$Et-CHR)-indole R Time (h) Pr 3.0 (67) Bn 2.5 (82)	379
C$_{15-23}$ 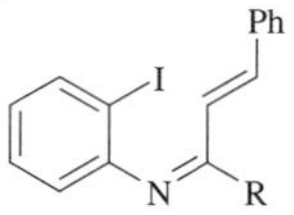 (Ph-CH=CH-C(R)=N-aryl-I)	Pd(OAc)$_2$, PPh$_3$, t-BuOK, DMSO, 120°, 12–16 h	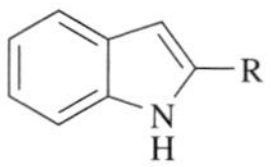 2-R-indole R H (71) 2-thienyl (71) Ph (63) 4-BrC$_6$H$_4$ (65) 4-ClC$_6$H$_4$ (75) R 4-FC$_6$H$_4$ (63) 4-O$_2$NC$_6$H$_4$ (51) 4-MeC$_6$H$_4$ (67) 4-CF$_3$C$_6$H$_4$ (52) 4-MeOC$_6$H$_4$ (70)	380
C$_{17-20}$ 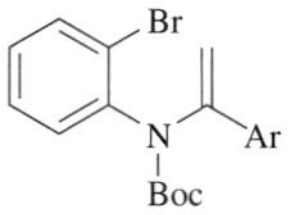 (2-Br-aryl-N(Boc)-C(=CH$_2$)-Ar)	Pd(PPh$_3$)$_4$, Et$_3$N, DMF, 100°, 20 h	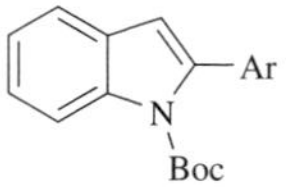 N-Boc 2-Ar-indole Ar 2-furyl (54) (44)^a 2-thienyl (63) (31)^a 4-ClC$_6$H$_4$ (59) 2-MeC$_6$H$_4$ (91) 4-MeOC$_6$H$_4$ (71)	162
C$_{24}$ 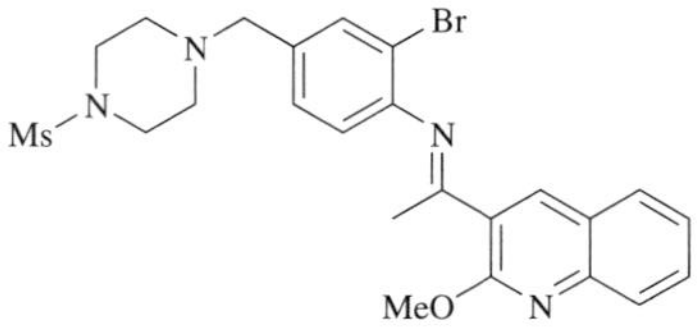 (Ms-piperazinyl-CH$_2$-aryl-Br with N=C(Me)-quinoline-OMe)	Pd(OAc)$_2$, P(o-tol)$_3$, Et$_3$N, DMF, reflux	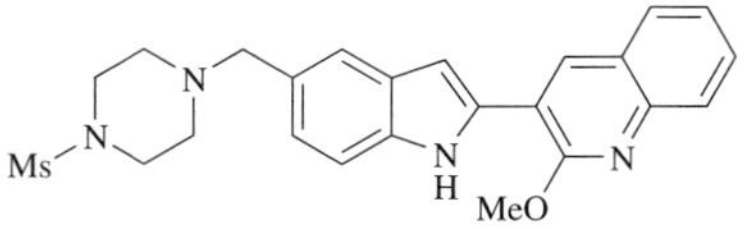 indole-quinoline-OMe with piperazine-Ms arm (55)	160

^{a} These yields refer to the corresponding N-deprotected indole.

TABLE 5C. 2-SUBSTITUTED INDOLES FROM 2-HALOARYLENAMINES AND -IMINES PREPARED IN SITU

Substrate	Reagent	Conditions	Product(s) and Yield(s) (%)	Refs.
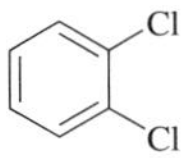 C₆ 1,2-dichlorobenzene	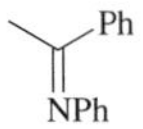 Ph-C(=NPh)Me	Pd₂(dba)₃, XPhos, t-BuONa, 1,4-dioxane, 110°, 14 h	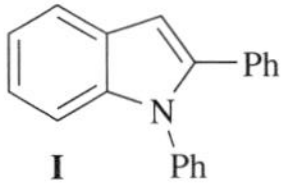 2-Ph, N-Ph indole (80)	196
	2-bromo-1-phenyl-1-propenyl, PhNH₂	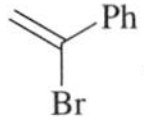 Pd₂(dba)₃, XPhos, t-BuONa, 1,4-dioxane, 110°, 72 h	**I** (77)	196
	Ph-C(=NR¹)Me with R²	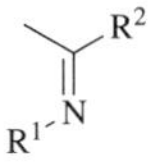 Pd₂(dba)₃, XPhos, t-BuONa, 1,4-dioxane, 110°, 14 h	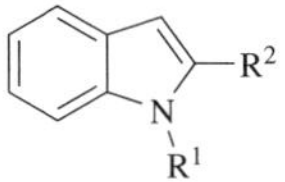 2-R², N-R¹ indole R¹ R² t-Bu i-Pr (55) Ph Ph (80) Ph (E)-CH=CHPh (75)	197
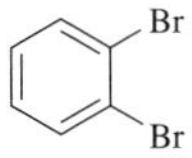 1,2-dibromobenzene	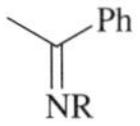 Ph-C(=NR)Me	Pd₂(dba)₃, XPhos, t-BuONa, 1,4-dioxane, 110°, 14 h	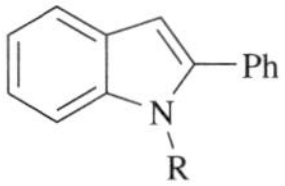 2-Ph, N-R indole R t-Bu (72) Ph (86) 2-ClC₆H₄ (77) 4-MeOC₆H₄ (80) Bn (56)	196
	C(=NPh)Me	 Pd₂(dba)₃, XPhos, t-BuONa, 1,4-dioxane, 110°, 14 h	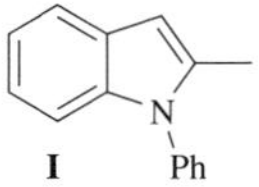 **I** (71)	196
	2-bromopropenyl, PhNH₂	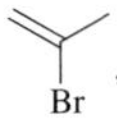 Pd₂(dba)₃, XPhos, t-BuONa, 1,4-dioxane, 50°, 3 h; 90°, 14 h	**I** (68)	196
	2-bromo-1-phenyl-1-propenyl, RNH₂	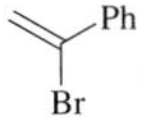 Pd₂(dba)₃, XPhos, t-BuONa, 1,4-dioxane, 110°, 14 h	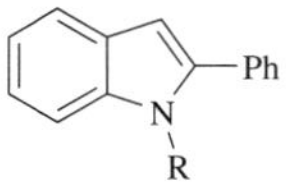 2-Ph, N-R indole R Ph (76) Bn (65)	196

TABLE 5C. 2-SUBSTITUTED INDOLES FROM 2-HALOARYLENAMINES AND -IMINES PREPARED IN SITU (Continued)

Substrate	Reagent	Conditions	Product(s) and Yield(s) (%)	Refs.
C_6				
2-Br, Br-C$_6$H$_4$	R^{2}C(Me)=NR^1	Pd$_2$(dba)$_3$, XPhos, t-BuONa, 1,4-dioxane, 110°, 14 h	2-R^2-N(R^1)-indole R^1 / R^2 / (%) — R^1 / R^2 / (%) Ph / Me (77) — c-C$_6$H$_{11}$ / Ph (80) t-Bu / CH=CMe$_2$ (78) — Ph / Ph (86) Bu / Ph (52) — 2-ClC$_6$H$_4$ / Ph (77) t-Bu / Ph (72) — 2-MeOC$_6$H$_4$ / Ph (80) t-Bu / 1-cyclohexenyl (63) — Bn / Ph (56) Ph / (N-Me-pyrrol-2-yl) (82) — Ph / (E)-CH=CHPh (83)	197
2-Cl,NH$_2$-C$_6$H$_4$	CH$_2$=C(R)Br	Pd$_2$(dba)$_3$, XPhos, toluene, t-BuONa, 110°, 20 h	2-R-indole (I) R / (%) Ph (65) 4-MeC$_6$H$_{14}$ (58) n-C$_8$H$_{17}$ (60) CH$_2$OBn (55)	165
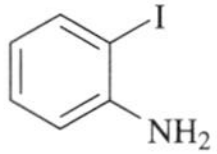 2-I,NH$_2$-C$_6$H$_4$	 RC(O)Me	Pd(OAc)$_2$, additive, DABCO, DMF, 105°, 3–12 h	2-R-indole (I) R / Additive / (%) CO$_2$H / — / (82) TMS / MgSO$_4$ / (64)^a	162
2-I,NH$_2$-C$_6$H$_4$	RC(O)Me	 Pd(dba)$_2$, dipf, t-BuONa, 1,4-dioxane, reflux, 20 h	2-R-indole (I) R / (%) MeCH(Me) (30) — 2-MeC$_6$H$_4$ (66) MeCH(Me)CH$_2$ (31) — 3-MeC$_6$H$_4$ (68) 2-thienyl (51) — 4-MeC$_6$H$_4$ (69) Me(CH$_2$)$_4$ (36) — 4-MeOC$_6$H$_4$ (60) Ph (71) — 3-CF$_3$C$_6$H$_4$ (75) 4-FC$_6$H$_4$ (68) — 2-Np (74)	381

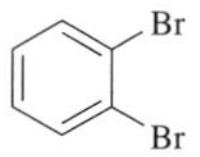

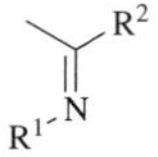

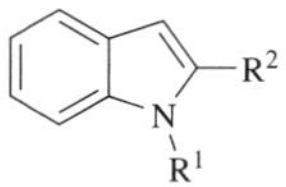

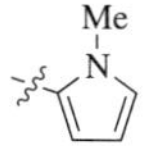

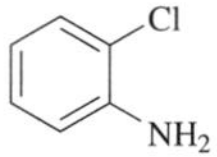

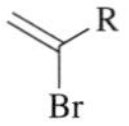

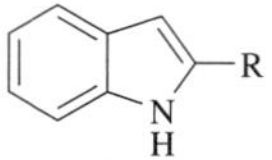

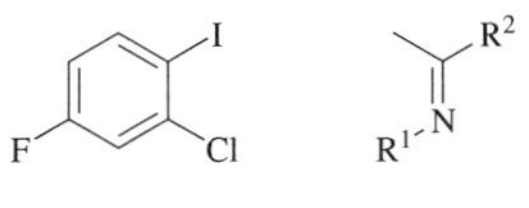

Substrate		Conditions	Product			Refs.
2-iodo-4-fluoro-1-chloro-benzene	ketimine (R¹-N=C(Me)R²)	Pd₂(dba)₃, XPhos, t-BuONa, 1,4-dioxane, 110°, 14 h	6-F-indole (R¹, R²)			197
			R¹	R²		
			t-Bu	n-C₅H₁₁	(60)	
			t-Bu	Ph	(83)	
			c-C₆H₁₁	Ph	(72)	
			Ph	Ph	(83)	

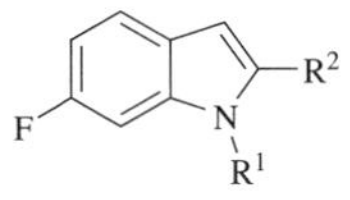

Substrate		Conditions	Product			Refs.
C₆₋₇ 2-X-4-R¹-aniline	alkyne (HC≡C-R²)	1. Ru₃(CO)₁₂, PF₆NH₄, toluene, 105°, 18 h; 2. Pd(OAc)₂, L2, t-BuOK, 105°, 24 h	5-R¹-2-R²-indole			382
			R¹	X	R²	
			H	Cl	Ph	(60)
			H	Cl	1-cyclohexenyl	(49)
			H	Br	Ph	(78)
			5-Cl	Br	Ph	(87)
			6-CF₃	Cl	1-cyclohexenyl	(35)
			5-Me	Br	Ph	(75)

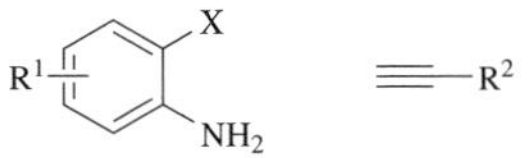

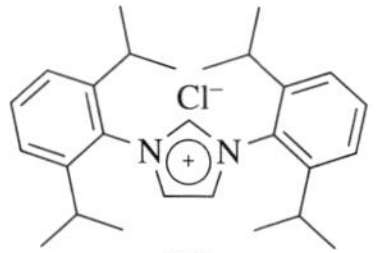

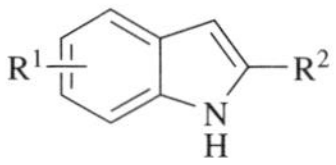

L2 = 1,3-bis(2,6-diisopropylphenyl)imidazolium chloride

Substrate		Conditions	Product		Refs.
2-bromo-4-R¹-aniline	1-bromo-alkene (CH₂=C(R²)Br)	Pd₂(dba)₃, DavePhos, t-BuONa, toluene, 100°, 20 h	5-R¹-2-R²-indole		165
			R¹	R²	
			H	Ph	(64)
			H	4-MeC₆H₄	(62)
			H	n-C₈H₁₇	(63)
			H	CH₂OBn	(61)
			Cl	Ph	(55)
			Cl	n-C₈H₁₇	(53)
			Me	Ph	(62)
			Me	4-MeC₆H₄	(61)
			Me	n-C₈H₁₇	(60)
			Me	CH₂OBn	(59)

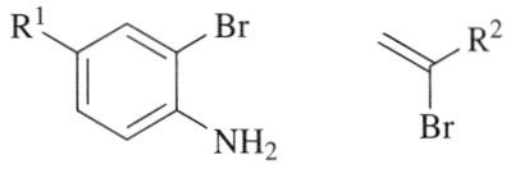

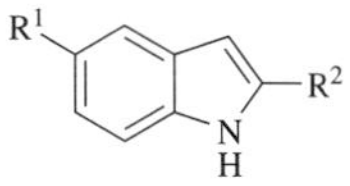

TABLE 5C. 2-SUBSTITUTED INDOLES FROM 2-HALOARYLENAMINES AND -IMINES PREPARED IN SITU (*Continued*)

	Substrate	Reagent	Conditions	Product(s) and Yield(s) (%)	Refs.
C$_{6-8}$	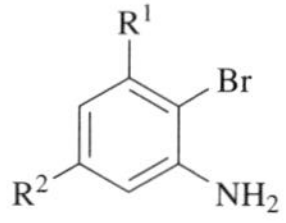 R^1, Br, NH$_2$, R^2 (aryl)	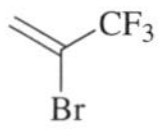 CF$_3$, Br (alkene)	Pd(OAc)$_2$, XPhos, Cs$_2$CO$_3$, toluene, 125°, 15 h	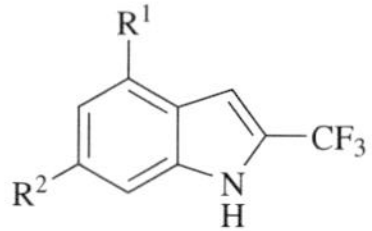 Indole with 2-CF$_3$, R^1, R^2 *b* R^1 R^2 H F (84) F F (66) H Cl (44) H Me (64) H CF$_3$ (81) H CF$_3$O (66) Me Me (17)	383
C$_{6-9}$	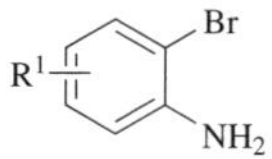 R^1, Br, NH$_2$	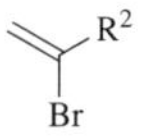 R^2, Br	Pd-PEPPSI-IP, Pd-coated 1200 mm capillary, *t*-BuONa, toluene, 75 atm back pressure, 15 μL min^{-1} flow rate, 215° 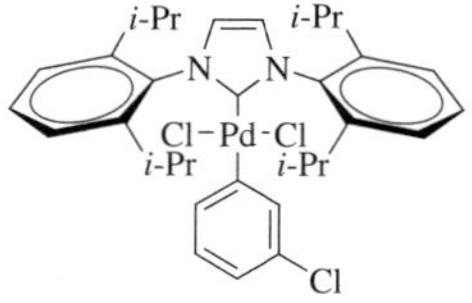 Pd-PEPPSI-IP (*i*-Pr, N, Cl-Pd-Cl, *i*-Pr, 3-Cl-C$_6$H$_4$)	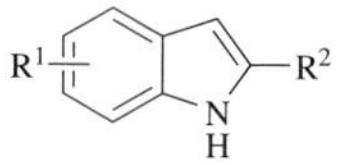 R^1 R^2 R^1 R^2 H Me (72) 5,7-F$_2$ Ph (72) H Et (81) 5-Me Me (82) H Ph (68) 5-Me Et (72) 5-F Me (73) 5-Me Ph (70) 5-F Et (76) 5,7-Me$_2$ Me (85) 5-F Ph (62) 5,7-Me$_2$ Et (72) 5-Cl Me (62) 5,7-Me$_2$ Ph (82) 5-Cl Et (70) 5-*i*-Pr Me (79) 5-Cl Ph (82) 5-*i*-Pr Et (83) 5,7-F$_2$ Me (65) 5-*i*-Pr Ph (80) 5,7-F$_2$ Et (64)	206
C$_{6-13}$	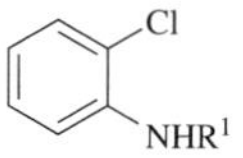 Cl, NHR^1	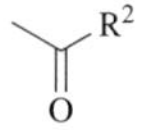 R^2, C=O	Pd[P(*c*-Bu)$_3$]$_2$, K$_3$PO$_4$, MgSO$_4$, DMA	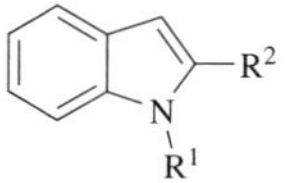 Indole with R^1, R^2 R^1 R^2 Time (h) Temp (°) H 4-MeOC$_6$H$_4$ 21 140 (80) H TMS 14 125 (80) Me CO$_2$H 14 140 (98) 4-MeOC$_6$H$_4$ Ph 14 140 (78)	164

Substrate	Reagent	Conditions	Product (Yield %)	Ref
C₇ 2-chloro-3-methylaniline	PhC(O)CH₃	Pd[P(t-Bu)₃]₂, K₃PO₄, MgSO₄, DMA, 90°, 2 h	4-methyl-2-phenylindole (81)	164
4-amino-3-chloro-benzonitrile	Et₂NC(O)C(O)CH₃	Pd[P(t-Bu)₃]₂, K₃PO₄, MgSO₄, DMA, 140°, 14 h	5-cyano-N,N-diethyl-1H-indole-2-carboxamide (65)	164
2,6-dichloro-4-(trifluoromethyl)aniline	PhC(O)CH₃	Pd[P(t-Bu)₃]₂, K₃PO₄, MgSO₄, DMA, 140°, 14 h	7-chloro-2-phenyl-5-(trifluoromethyl)-1H-indole (65)	164
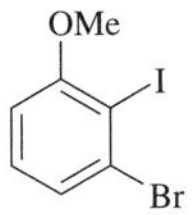 3-bromo-2-iodo-anisole	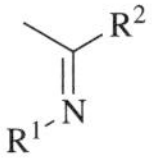 R²C(=NR¹)CH₃	Pd₂(dba)₃, XPhos, t-BuONa, 1,4-dioxane, 110°, 14 h	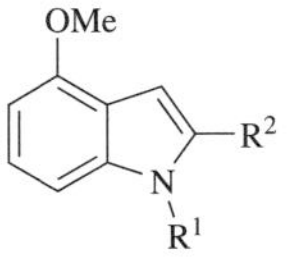 4-methoxy-indole (R¹/R²: Ph/Me (45); t-Bu/Ph (76); Ph/Ph (64))	197
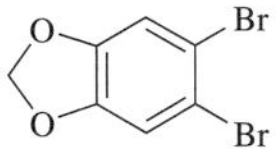 5,6-dibromo-benzo[d][1,3]dioxole	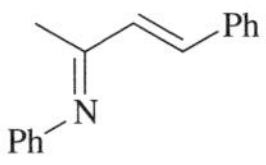 PhCH=CHC(=NPh)CH₃	Pd₂(dba)₃, XPhos, t-BuONa, 1,4-dioxane, 110°, 14 h	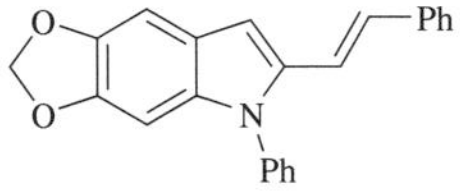 5-phenyl-7-(2-phenylvinyl)-[1,3]dioxolo-indole (71)	197
C₇₋₈ 4-chloro-aryl triflate	R²C(=NR³)CH₃	Pd₂dba₃, XPhos, t-BuOLi, 1,4-dioxane, 110°, 14–24 h	(see table below)	197

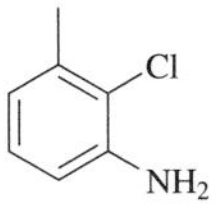

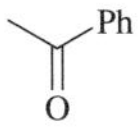

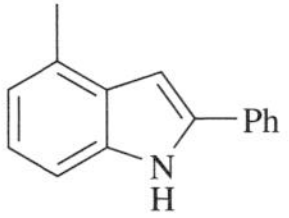

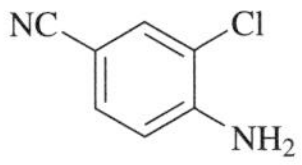

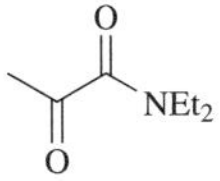

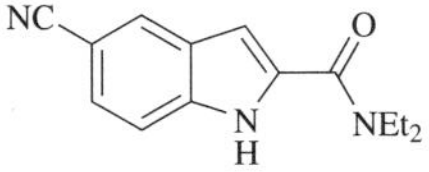

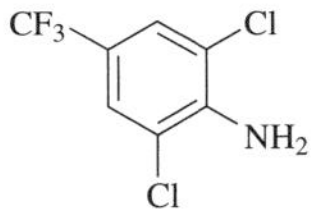

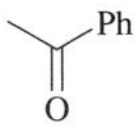

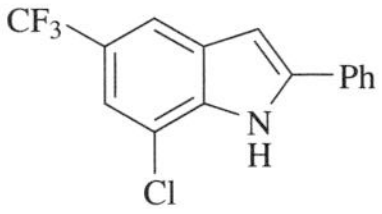

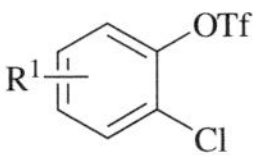

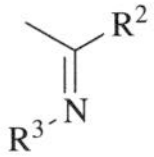

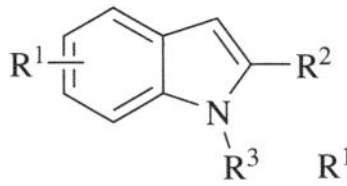

R¹	R²	R³	Addn Rate (mL/h)	(%)
6-F	Ph	Ph	0.1	(65)
5-Me	Me	Ph	0.1	(45)
5-Me	Ph	t-Bu	0.1	(70)
5-Me	Ph	Ph	0.16	(78)
6-MeO	Ph	Ph	0.12	(36)

TABLE 5C. 2-SUBSTITUTED INDOLES FROM 2-HALOARYLENAMINES AND -IMINES PREPARED IN SITU (*Continued*)

Substrate	Reagent	Conditions	Product(s) and Yield(s) (%)	Refs.
C$_{7-13}$				
R^1—Ar—Br, Br	R^{2}C(=NPh)-	Pd$_2$(dba)$_3$, XPhos, t-BuONa, 1,4-dioxane, 110°, 14 h	R^1 / R^2 : 5-MeO / (E)-CH=CHPh (74); 6-BnO / Ph (57); 6-BnO / (E)-CH=CHPh (60) [product: 2-R^2-N-Ph-indole with R^1]	197
C$_{10-16}$				
R^1—Ar—ONf, Cl	R^{3}N=C(R^2)-	Pd$_2$(dba)$_3$, XPhos, t-BuOLi, 1,4-dioxane, addition rate 0.1 mL/h,^{c} 110°, 10–20 h	R^1 / R^2 / R^3 : 6-F / 2-N-Me-pyrrolyl / Ph (78); 6-F / 4-pyridyl / Ph (83); 5-Me / (CH$_2$)$_4$Me / t-Bu (65); 6-CF$_3$ / 4-pyridyl / Ph (66); 5-Me / 4-pyridyl / Ph (88); 5-Me / 2-N-Me-pyrrolyl / Ph (88); 6-Ph / CH=CMe$_2$ / t-Bu (61); 6-Ph / 4-pyridyl / Ph (87)	197
C$_{10-17}$				
R^1—Ar—ONf, Cl	R^{2}N=C(Ph)-	Pd$_2$(dba)$_3$, XPhos, t-BuOLi, 1,4-dioxane, addition rate 0.1 mL/h,^{c} 110°, 10–20 h	R^1 / R^2 : H / Ph (92); 6-F / Ph (81); 6-F / 4-MeOC$_6$H$_4$ (64); 5-Me / t-Bu (93); 6-Me / t-Bu (79); 5-Me / Ph (86); 6-MeO / Ph (80); 6-NC / Ph (78); 5-Me / 4-MeOC$_6$H$_4$ (89); 6-MeO / 4-MeOC$_6$H$_4$ (63); 4,6-Me$_2$ / Ph (64); 6-t-BuO$_2$C / Ph (62); 6-Ph / c-C$_6$H$_{11}$ (75); Ph-CH=N- / Ph (80); Ph-CH=N- / Ph (72)	197

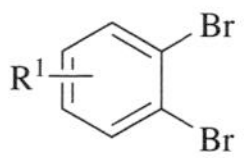

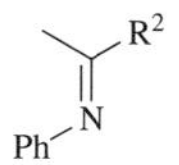

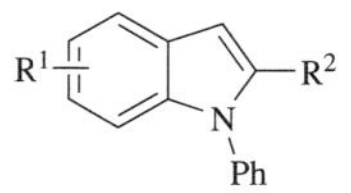

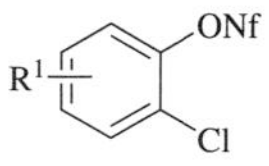

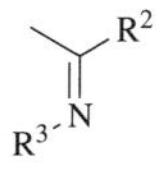

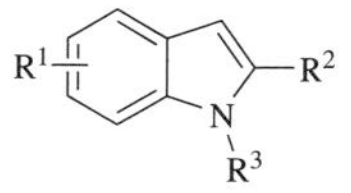

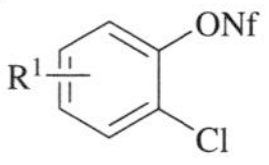

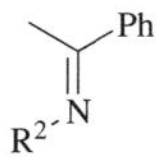

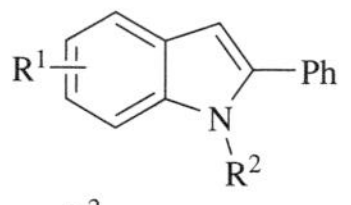

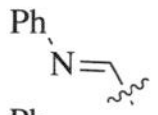

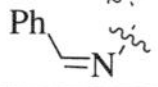

^{a} The number is the combined yield of the 2-silyl derivative and indole.
^{b} The yield was determined by NMR spectroscopy.
^{c} Refers to the addition rate of the halogenated arene.

TABLE 5D. 2-SUBSTITUTED INDOLES FROM 2-NITROSTYRENES

2-Nitrostyrene	Conditions	Product(s) and Yield(s) (%)	Refs.
C$_{9-14}$ 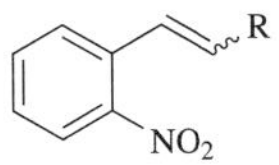 — styryl-R, o-NO$_2$; trans:cis = 50:50	Pd(OAc)$_2$, PPh$_3$, MeCN, CO (4 atm), 70°	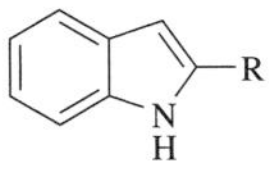 2-R-indole: R=Me, Time 24 h (96); R=Ph, 21 h (91)	156
C$_{10}$ 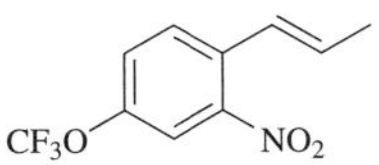 — CF$_3$O-substituted propenyl-nitrobenzene	Pd(OAc)$_2$, phen, DMF, CO (1 atm), 80°, 16 h	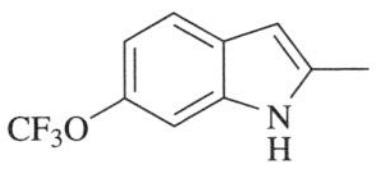 6-CF$_3$O-2-methylindole (84)^a	161
C$_{10-11}$ 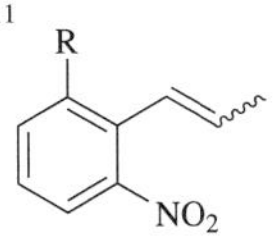 — 3-R-2-nitrostyrene	Pd(OAc)$_2$, MeCN, CO (4 atm), 70°	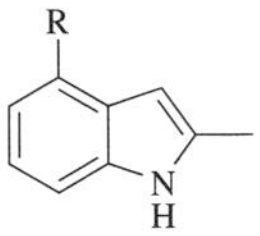 4-R-indole: R=Me, 17 h (66); CH$_2$OMe, 46 h (76); CO$_2$Me, 24 h (90)	158
C$_{10-14}$ 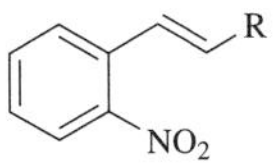 — 2-R-nitrostyrene	PdCl$_2$(PPh$_3$)$_2$, SnCl$_2$, 1,4-dioxane, CO (20 atm), 100°, 16 h	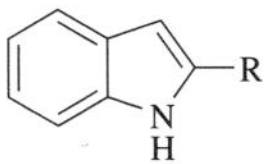 2-R-indole: R=CO$_2$Me (60); Ph (74)	154, 155
C$_{13}$ 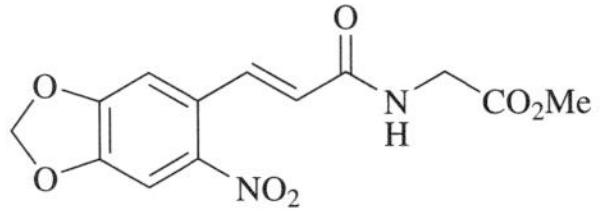 — methylenedioxy nitrocinnamoyl glycinate	Pd(TFA)$_2$, tmphen, DMF, CO (1 atm), 80°, 16 h	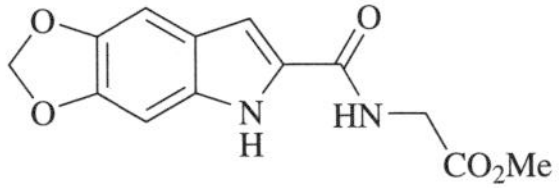 methylenedioxy-indole-2-carboxamide-CH$_2$CO$_2$Me (72)	161
C$_{14}$ 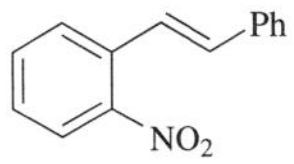 — 2-nitrostilbene	Pd(OAc)$_2$, PPh$_3$, MeCN, CO (4 atm), 70°, 18.5 h	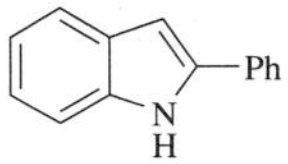 2-phenylindole (100)	156

TABLE 5D. 2-SUBSTITUTED INDOLES FROM 2-NITROSTYRENES (Continued)

2-Nitrostyrene	Conditions	Product(s) and Yield(s) (%)	Refs.
C_{14} 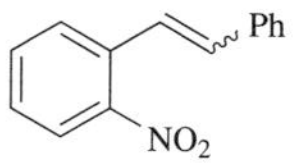 — styryl-NO$_2$-Ph, trans:cis = 50:50	Pd(OAc)$_2$, phen, DMF, 80°, 16 h	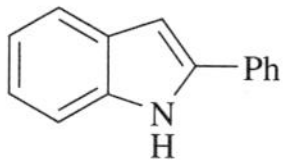 2-Ph-indole (86)	161
C_{14-15} 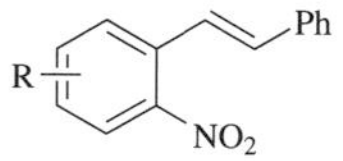 — R-styryl-NO$_2$-Ph	Pd(OAc)$_2$, phen, DMF, CO (1 atm), 80°, 16 h	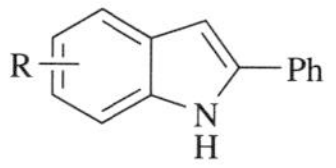 2-Ph-indole, R: H (87); 5-Cl (96); 5-Me (89)	161
C_{15} 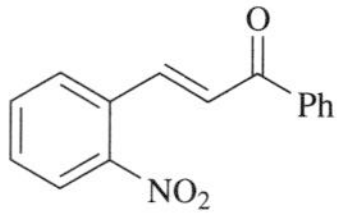 — PhC(O)-styryl-NO$_2$	Pd(OAc)$_2$, phen, DMF, CO (1 atm), 80°, 16 h	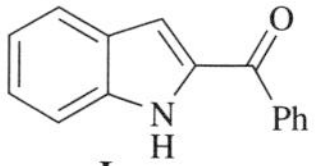 2-(PhCO)-indole **I** (84)	161
	PdCl$_2$(PPh$_3$)$_2$, SnCl$_2$, 1,4-dioxane, CO (20 atm), 100°, 16 h	**I** (52) 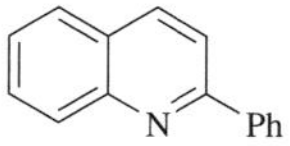 + 2-Ph-quinoline (34)	154, 155
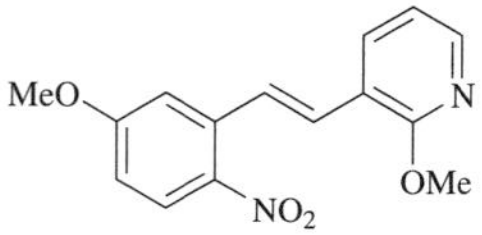 C_{15-16} — MeO-(2-OMe-pyridyl)-styryl-NO$_2$	Pd(OAc)$_2$, phen, DMF, CO (2 atm), 70°, 16 h	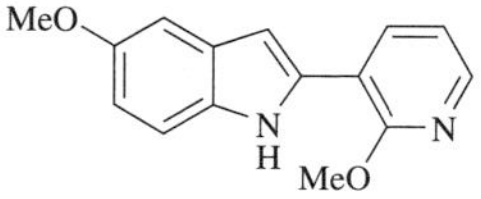 5-MeO-2-(2-OMe-pyridin-3-yl)-indole (72)	161
C_{15-16} 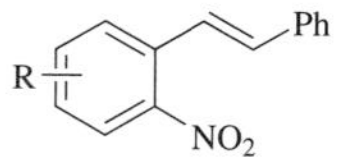 — R-styryl-NO$_2$-Ph	Pd(TFA)$_2$, tmphen, DMF, CO (1 atm), 80°, 16 h	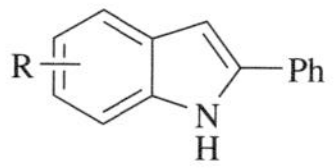 2-Ph-indole, R: 7-OMe (18); 5-CO$_2$Me (98); 6-NMe$_2$ (61)	161

C₁₅₋₂₅

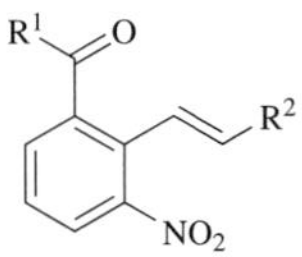

[Structure: R¹C(O)-aryl with NO₂ and CH=CH-R² substituents]

Pd(OAc)₂, phen, DMF,
CO (2 atm), 80°, 16 h

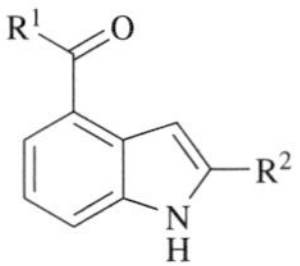

[Product: indole with R¹C(O) at 4-position and R² at 2-position]

R¹	R²	
NMe₂	2-furyl	(98)
N-pyrrolidino	4-FC₆H₄	(97)
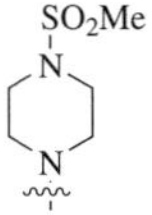 SO₂Me-piperazinyl	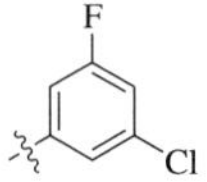 3-Cl-5-F-C₆H₃	(98)
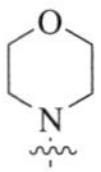 morpholino	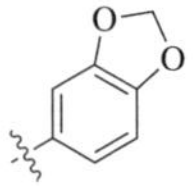 3,4-methylenedioxyphenyl	(98)
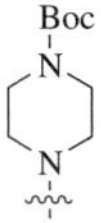 Boc-piperazinyl	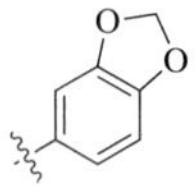 3,4-methylenedioxyphenyl	(99)

384

C₁₉₋₂₁

[Structure: 2-methoxynaphthalene with CH=CH linked to nitroarene bearing R]

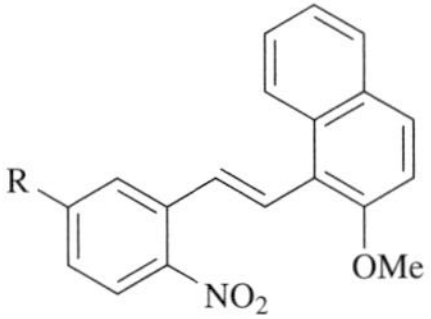

Cat, ligand, DMF,
CO (1 atm), 80°, 16 h

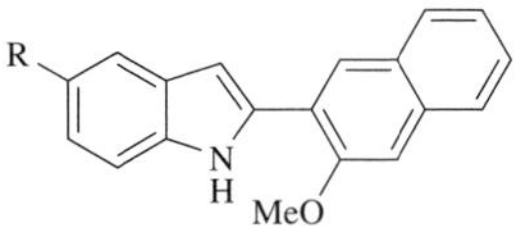

[Product: 2-(3-methoxynaphthalen-2-yl)-indole with R substituent]

R	Cat	Ligand	
Cl	Pd(OAc)₂	phen	(91)
CO₂Me	Pd(TFA)₂	tmphen	(78)

161

TABLE 5D. 2-SUBSTITUTED INDOLES FROM 2-NITROSTYRENES (*Continued*)

C_{24}

2-Nitrostyrene: MeO-quinoline-CH=CH-(NO2)-C6H3-CH2-N(piperazine)-Ms **I**

Conditions: See table

Product(s) and Yield(s) (%): indole-quinoline-MeO structure with CH2-piperazine-Ms substituent

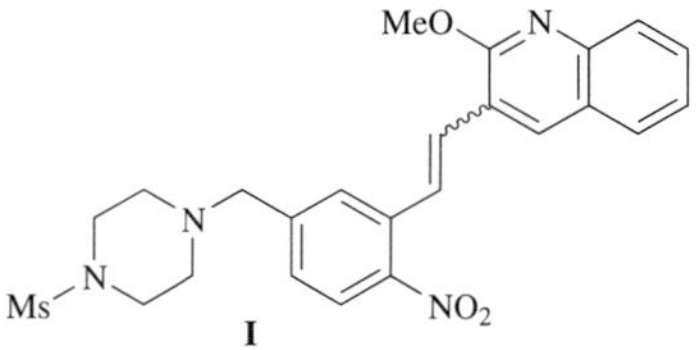

Refs.: 160

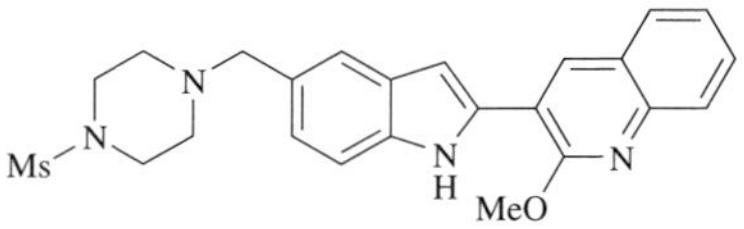

I config.	Catalyst	Ligand	Solvent	CO pressure (atm)	Temp (°)	Time (h)	
(E)	Pd(OAc)$_2$	PPh$_3$	MeCN	4	70	15	(95)
(E)	Pd(OAc)$_2$	phen	DMF	1	70	14	(94)
(E)	Pd(OCOCF$_3$)$_2$	tmphen	DMF	1	80	—	(95)
(E)	phen$_2$Pd(BF$_4$)$_2$	—	DMF	1	70	—	(99)
(E)	Pd(OCOCF$_3$)$_2$	tmphen	DMF	1	70	—	(100)
(E)	Pd(OAc)$_2$	tmphen	DMF	1	70	—	(92)
(Z)	Pd(OAc)$_2$	PPh$_3$	MeCN	4	70	15	(92)

[a] The yield was calculated by HPLC analysis.

TABLE 6A. 3-SUBSTITUTED INDOLES FROM 2-HALO- AND 2-PSEUDOHALO-N-ALLYLANILINES AND -ANILIDES

2-Halo-N-allylaniline or -anilide	Conditions	Product(s) and Yield(s) (%)	Refs.
C₉	Pd(OAc)₂, TPTTS, Et₃N, MeCN/H₂O (15:1), rt, 1.5 h	(97) **I**	169
	Pd(OAc)₂, (C₆F₁₃CH₂CH₂)₂PPh, Et₃N, scCO₂, 100°, 64 h	**I** (37)	170
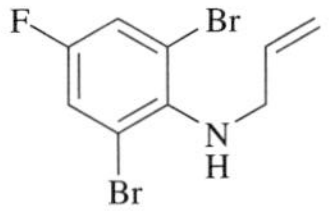	Pd(OAc)₂, P(o-tol)₃, Et₃N, MeCN, reflux, 5 h	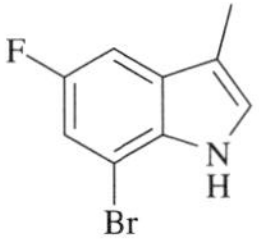 (65)	178
C₉₋₁₁	Pd(OAc)₂, base, Bu₄NCl, DMF	R | Base | Temp | Time (d) H | Na₂CO₃ | rt | 1 | (97) Me | Et₃N | rt | 2 | (81) Ac | NaOAc | 80° | 1 | (90)	168
	Pd(OAc)₂, Et₃N, MeCN, 110°, 72 h	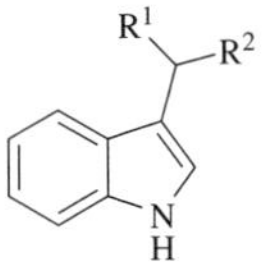 X | R¹ | R² I | H | H | (87) Br | H | H | (60) I | H | Me | (51) I | Me | Me | (73)	167
C₁₀	Pd(OAc)₂, Et₃N, Bu₄NCl, DMF, 80°, 1 d	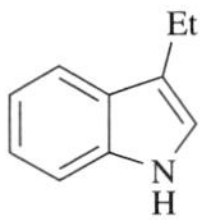 (73)	168

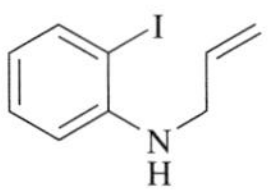

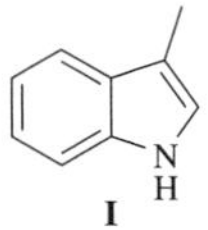

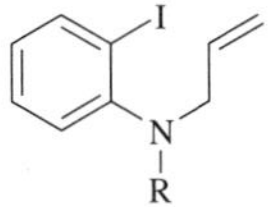

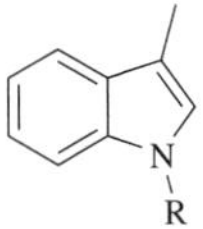

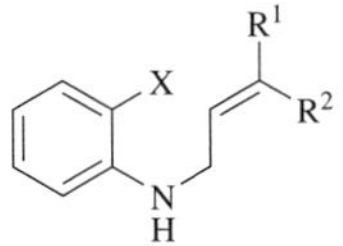

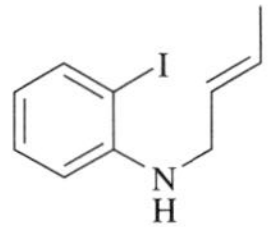

TABLE 6A. 3-SUBSTITUTED INDOLES FROM 2-HALO- AND 2-PSEUDOHALO-N-ALLYLANILINES AND -ANILIDES (Continued)

2-Halo-N-allylaniline or -anilide	Conditions	Product(s) and Yield(s) (%)	Refs.
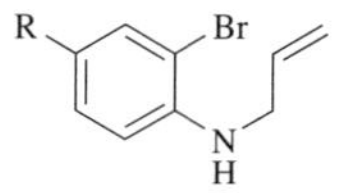 C10-12 (2-Br, N-allyl aniline with R substituent)	Pd(OAc)$_2$, ligand, Et$_3$N, MeCN, 110°, 72 h	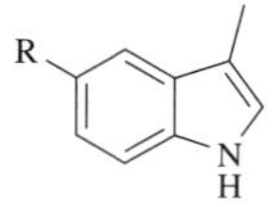 3-Me indole with R substituent; R = Me (77) ligand —; R = CO$_2$Et (50) ligand P(o-tol)$_3$	167
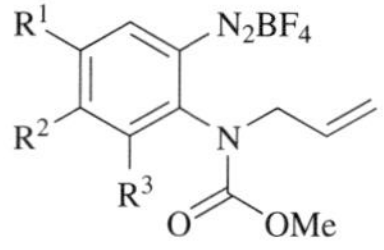 C11-13 (R^1, R^2, R^3 substituted N$_2$BF$_4$, N-allyl carbamate)	Pd(OAc)$_2$, NaOAc, MeOH, 50°, 0.5 h	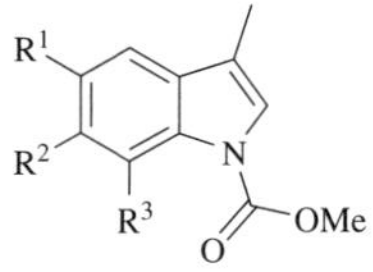 3-Me indole N-CO$_2$Me with R^1, R^2, R^3: R^1=H, R^2=H, R^3=H (85); R^1=Me, R^2=H, R^3=H (57); R^1=MeO, R^2=H, R^3=H (45); R^1=H, R^2=Me, R^3=Me (41); R^1=Me, R^2=Me, R^3=H (32)	385
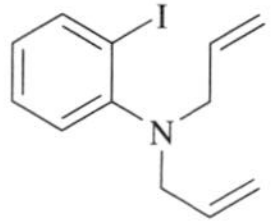 C12 (2-I, N,N-diallyl aniline)	Pd(OAc)$_2$, Et$_3$N, MeCN, 110°, 72 h	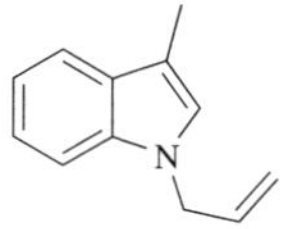 1-allyl-3-methylindole (87)	167
C12 (MeO$_2$C, Br, N-allyl, NH$_2$ aniline)	Pd(OAc)$_2$, Bu$_4$NCl, Na$_2$CO$_3$, DMF, 100°	3-Et-7-NH$_2$-5-CO$_2$Me indole (61)	179
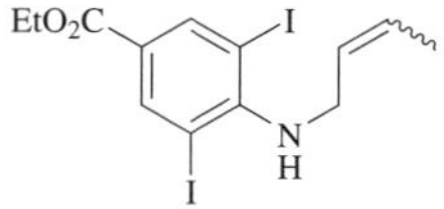 C13 (EtO$_2$C, I, N-allyl, I aniline)	Pd(OAc)$_2$, Bu$_4$NCl, NaHCO$_3$, DMF, 80°	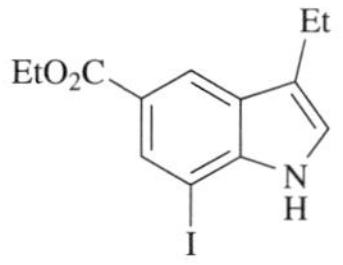 3-Et-7-I-5-CO$_2$Et indole (58)	179
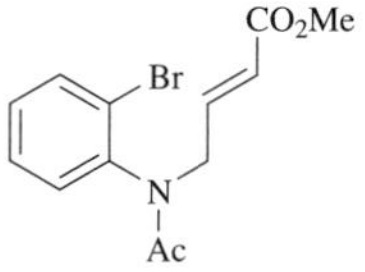 C13 (Br, CO$_2$Me-allyl, N-Ac aniline)	Pd(OAc)$_2$, PPh$_3$, NaHCO$_3$, DMF, 130°, 3 h	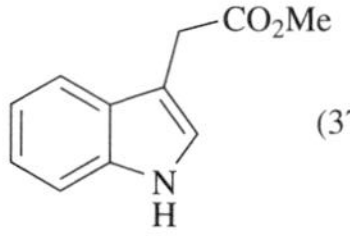 3-(CH$_2$CO$_2$Me)indole (37) + PhNHAc (6)	166

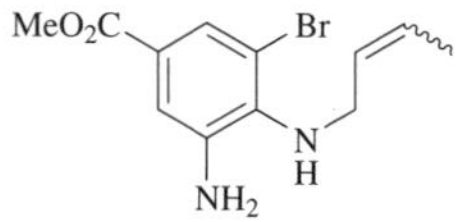

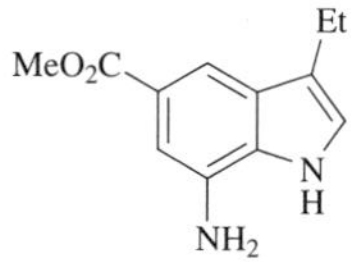

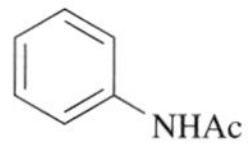

468

Substrate	Conditions	Product(s), Yield(s) (%)	Refs.
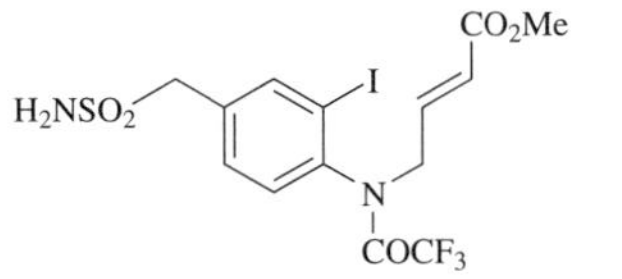 C_{14} substrate with H_2NSO_2, I, CO_2Me, $COCF_3$	$Pd(OAc)_2$, Bu_4NBr, Et_3N, DMF, 80°, 3 h	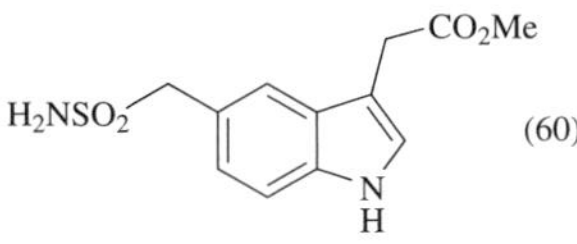 5-(H_2NSO_2)-3-(CH_2CO_2Me)-indole (60)	174
C_{16} 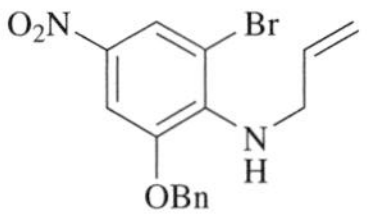 substrate (Br, O_2N, OBn, NH-allyl)	$Pd(OAc)_2$, Bu_4NBr, Et_3N, DMF, 24 h	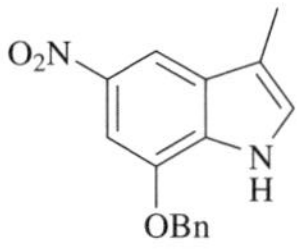 5-nitro-7-OBn-3-Me-indole (96)	171
C_{18} 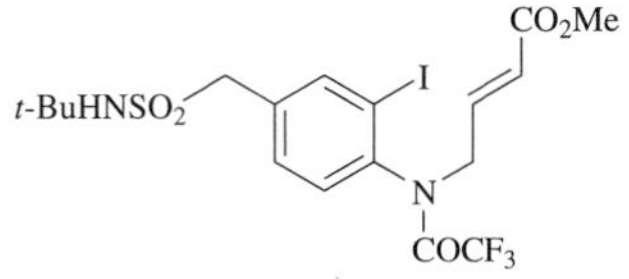 substrate (t-BuHNSO$_2$, I, CO_2Me, $COCF_3$)	$Pd(OAc)_2$, Bu_4NBr, Et_3N, DMF, 80°, 3 h	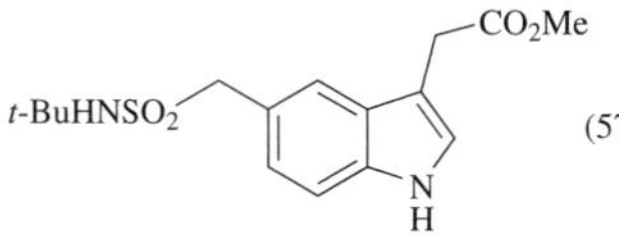 t-BuHNSO$_2$-indole-3-CH_2CO_2Me (57) + 5-O_2N-7-OBn-3-Me-indole (18)	174
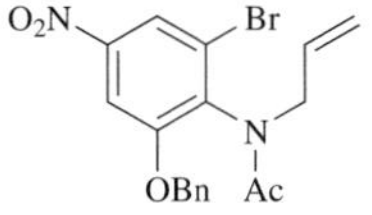 C_{20-22} substrate (O_2N, Br, N-Ac, OBn, allyl)	$Pd(OAc)_2$, $P(o\text{-tol})_3$, Bu_3N, toluene, 120°	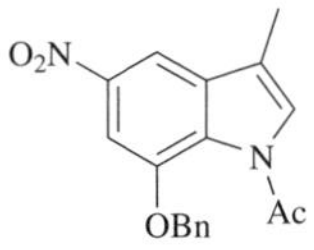 N-Ac-5-O_2N-7-OBn-3-Me-indole (74)	176
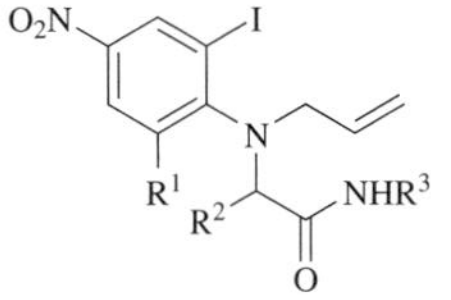 substrate with R^1, NHR^3, R^2	$Pd(OAc)_2$, PPh_3, Et_3N, toluene, 100°	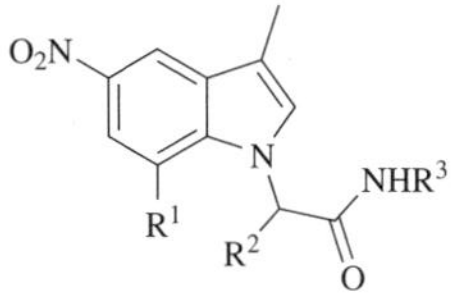 indole product	386

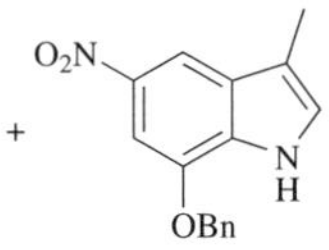

R^1	R^2	R^3	
H	Et	Bn	(64)
H	t-Bu	Bn	(70)
H	t-Bu	4-ClC$_6$H$_4$CH$_2$	(70)
Cl	t-Bu	4-ClC$_6$H$_4$CH$_2$	(74)

Substrate	Conditions	Product, Yield (%)	Refs.
C_{21} substrate (MeO$_2$C, Br, OBn, N-COCF$_3$, allyl)	$Pd(OAc)_2$, Bu_4NCl, Na_2CO_3, DMF, 100°, 2 h	5-(MeO$_2$C)-7-OBn-3-Et-indole (86)	179

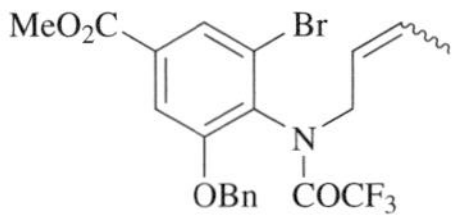

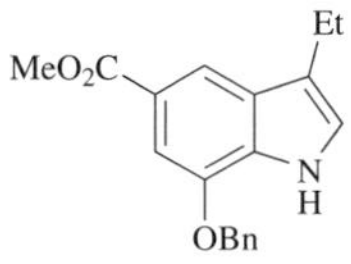

TABLE 6A. 3-SUBSTITUTED INDOLES FROM 2-HALO- AND 2-PSEUDOHALO-N-ALLYLANILINES AND -ANILIDES (Continued)

2-Halo-N-allylaniline or -anilide	Conditions	Product(s) and Yield(s) (%)	Refs.
C25			
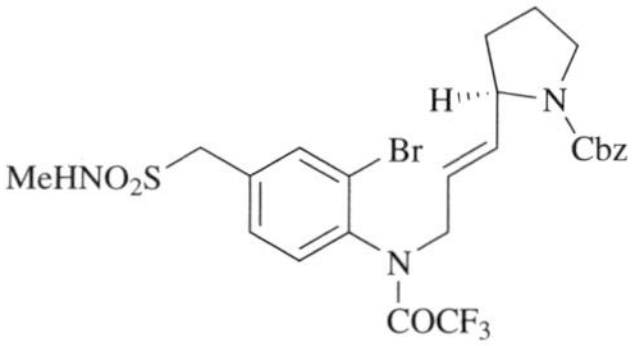 (structure: MeHNO2S-substituted 2-bromoaniline with N-Cbz pyrrolidine, N-COCF3, allyl)	Pd(OAc)2, Bu4NCl, Et3N, DMF/DME, 80°, 1 h	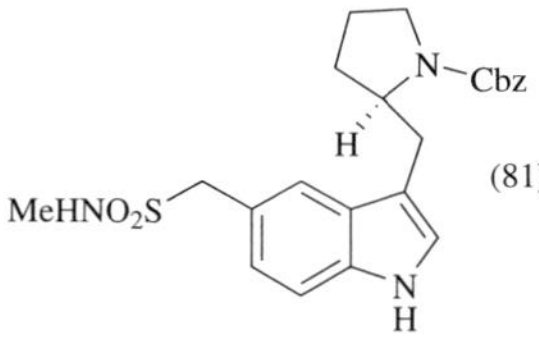 (indole product with N-Cbz pyrrolidine, MeHNO2S) (81)	175
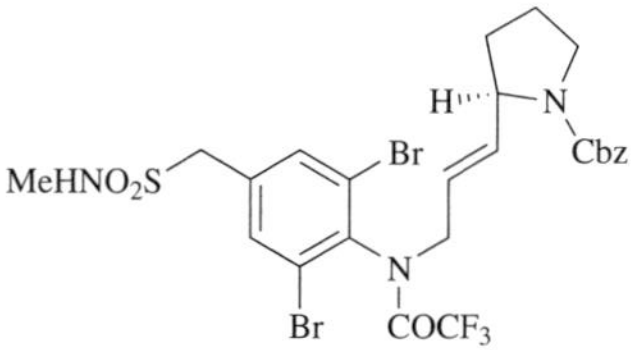 (structure: MeHNO2S-substituted 2,6-dibromoaniline with N-Cbz pyrrolidine, N-COCF3, allyl)	Pd(OAc)2, Bu4NCl, Et3N, DMF/DME, 80°, 1 h	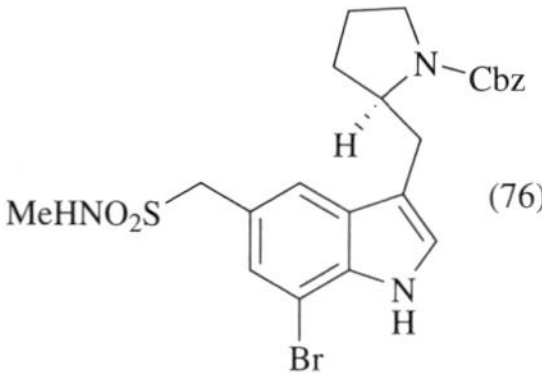 (7-Br indole product with N-Cbz pyrrolidine, MeHNO2S) (76)	175
C29			
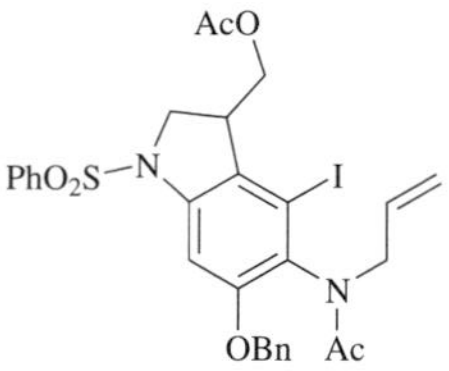 (structure: iodo-OBn-N-Ac-N-allyl aniline fused indoline with AcO, PhO2S-N)	Pd(OAc)2, Et3N, MeCN, 110°, 72 h	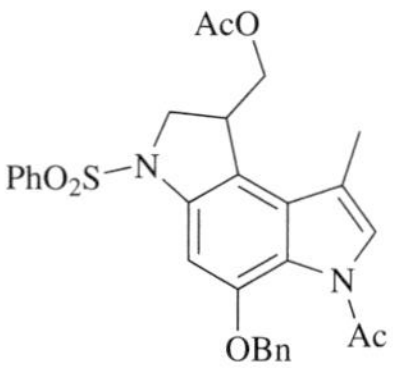 (tricyclic methyl-indole product, PhO2S-N, AcO, OBn, N-Ac) (70) 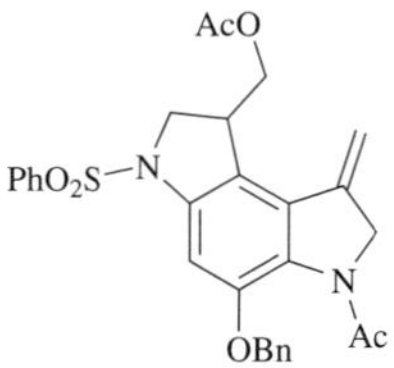 + (methylene tricyclic product, AcO, OBn, N-Ac, PhO2S-N) (10)	177

470

TABLE 6B. 3-SUBSTITUTED INDOLES FROM 2-HALO-N-ALLYLANILINES AND -ANILIDES PREPARED IN SITU

Substrate	Reagent	Conditions	Product(s) and Yield(s) (%)	Refs.
C_6	$R^2CHO, H_2N\diagup$; $\underset{Cl}{\diagdown}NC$	1. NH_4Cl, toluene/H_2O, rt 2. CF_3CO_2H 3. $Pd(OAc)_2$, PPh_3, Et_3N, toluene, 100°	$\begin{array}{ll} R^1 & R^2 \\ \hline H & Et \quad (51) \\ Cl & i\text{-}Bu \quad (74) \end{array}$	386
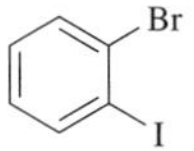	 $H_2N\diagup$	1. $Pd_2(dba)_3$, dppf, t-BuONa, toluene, rt to 140° 2. $X\!\!-\!\!\diagdown\!\!F$ 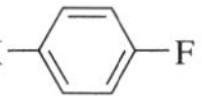	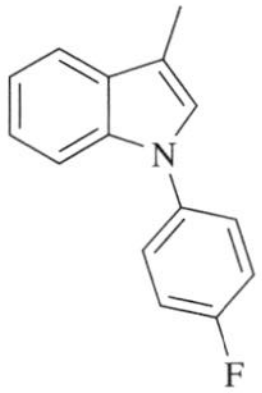 $\begin{array}{ll} X & \\ \hline I & (21) \\ Br & (30) \end{array}$	387
	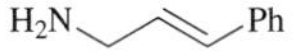 $H_2N\diagup\!\!\diagdown Ph$	$Pd_2(dba)_3$, dppf, t-BuONa, toluene, rt to 140°	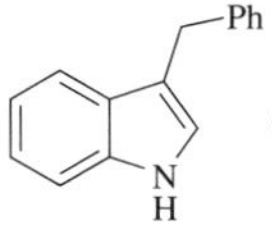 (59)	387
C_{6-7}	$H_2N\diagup$	$Pd_2(dba)_3$, dppf, t-BuONa, toluene, rt to 140°	$\begin{array}{ll} R & \\ \hline H & (85) \\ 4\text{-}F & (72) \\ 5\text{-}F & (65) \\ 6\text{-}F & (75) \\ 7\text{-}F & (63) \\ 5\text{-}Cl & (60) \\ \end{array}$ $\begin{array}{ll} R & \\ \hline 6\text{-}Cl & (56) \\ 5\text{-}Me & (73) \\ 6\text{-}CF_3 & (64) \\ 4\text{-}MeO & (71) \\ 6\text{-}MeO & (67) \\ 5\text{-}CF_3O & (71) \\ \end{array}$	387

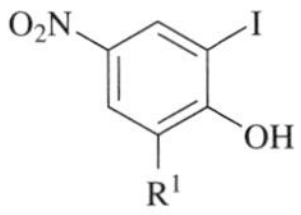

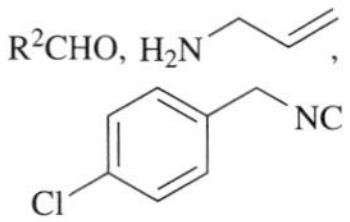

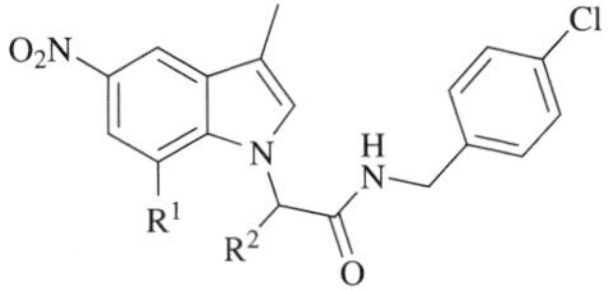

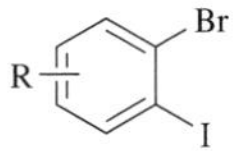

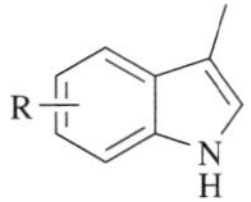

TABLE 6B. 3-SUBSTITUTED INDOLES FROM 2-HALO-N-ALLYLANILINES AND -ANILIDES PREPARED IN SITU (*Continued*)

Substrate	Reagent	Conditions	Product(s) and Yield(s) (%)	Refs.
C_{6-10} 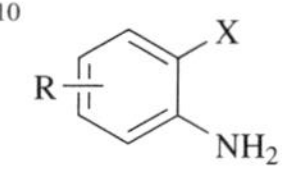 R⟨X, NH₂⟩	 Br⟨allyl⟩	Pd(OAc)₂, XPhos, K₂CO₃, DME, 80°	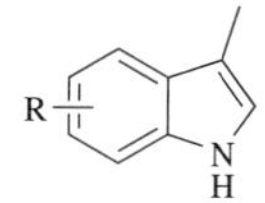 3-Me-indole R-substituted	388

X	R	Time (h)		X	R	Time (h)	
F	H	48	(—)	Cl	6-Me	48	(57)
Cl	H	24	(68)	Cl	7-Me	48	(46)
Cl	5-O₂N	22	(61)	Cl	4-NC	48	(15)
Cl	6-O₂N	48	(17)	Cl	5-NC	42	(32)
Cl	7-O₂N	22	(16)	Cl	4-MeO	48	(56)
F	5-Me	48	(—)	Cl	6-MeO	24	(23)
I	5-CF₃	48	(—)	Cl	6-NH₂CO	28	(50)
Cl	5-CF₃	28	(67)	Cl	6-MeO₂C	48	(15)
Cl	6-CF₃	48	(64)	Cl	5-t-Bu	20	(—)
Cl	6-Me	24	(14)	Cl	5-t-Bu	48	(67)

| C_{11-12} 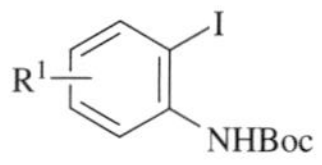 R¹⟨I, NHBoc⟩ | 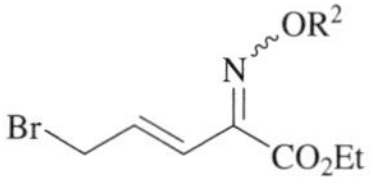 Br—CH=CH—C(CO₂Et)=N—OR² | 1. K₂CO₃, DMF, time 1 2. Pd(OAc)₂, PPh₃, additive, K₂CO₃, 60–65°, time 2 | indole-3-CH₂C(=N-OR²)CO₂Et, N-Boc, R¹ | 173 |

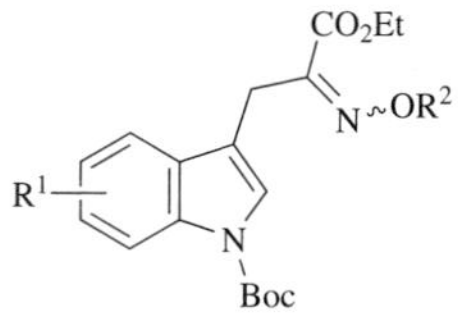

R¹	R²	Temp (°)	Time 1 (h)	Additive	Time 2 (h)	
H	Me	0; rt	—; 2	Bu₄NCl	17	(59)
H	Bn	0; rt	—; 2	Bu₄NCl	12	(62)
4-Br	Me	rt	30	—	4	(77)
3-O₂N	Me	rt	30	—	12	(65)
4-O₂N	Bn	rt	30	—	12	(77)
4-MeO	Me	0; rt	—; 2	—	12	(59)

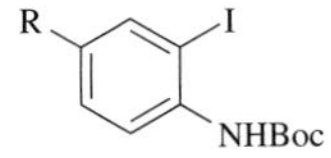

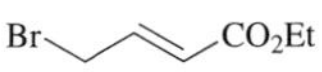

1. K_2CO_3, DMF, rt, time 1
2. $Pd(OAc)_2$, PPh_3, rt, 0.5 h; 60–65°, time 2

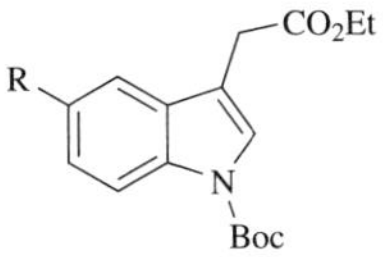

172

R	Time 1 (h)	Time 2 (h)	
Br	30	24	(67)
O_2N	3	19	(68)
MeO	94	5	(67)

TABLE 6C. 3-SUBSTITUTED INDOLES FROM 2-HALOARYLENAMINES

2-Haloarylenamine	Conditions	Product(s) and Yield(s) (%)	Refs.
C_{10} 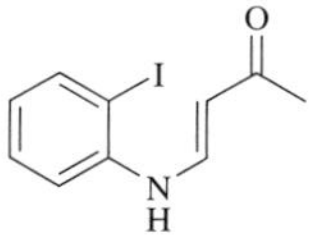	Pd(OAc)$_2$, Et$_3$N, DMF, 120° (sealed tube), 12 h	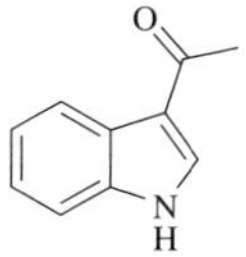 (70)	182
C_{10-12} 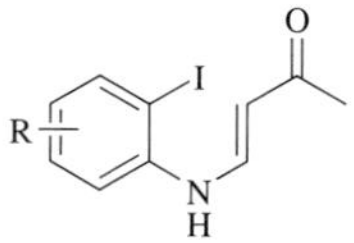	Pd(OAc)$_2$, P(o-tol)$_3$, Et$_3$N, MeCN, 100° (sealed tube), 20 h	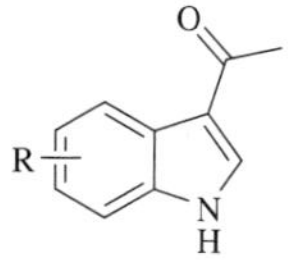 R / H (95) / 6-OMe (80) / 4-CO$_2$Me (95) / 5-CO$_2$Me (84) / 6-CO$_2$Me (86)	181
C_{11} 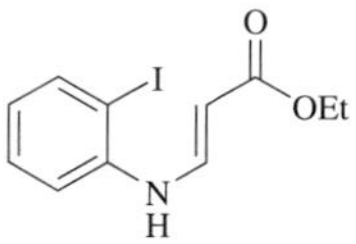	Pd(OAc)$_2$, Et$_3$N, DMF, 120° (sealed tube), 2 h	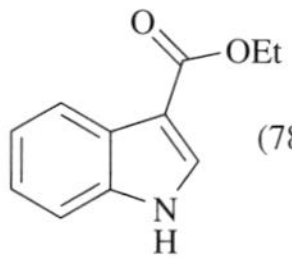 (78)	182
C_{11-13} 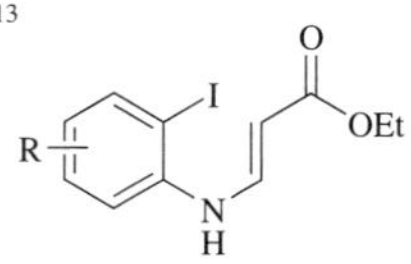	Pd(OAc)$_2$, P(o-tol)$_3$, Et$_3$N, MeCN, 100° (sealed tube), 20 h	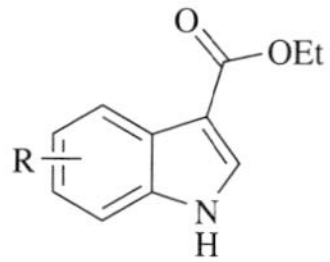 R / H (96) / 6-OMe (82) / 4-CO$_2$Me (93) / 5-CO$_2$Me (85) / 6-CO$_2$Me (90)	181

TABLE 6D. 3-SUBSTITUTED INDOLES FROM 2-HALOARYLENAMINES AND -IMINES PREPARED IN SITU

Substrate	Reagent	Conditions	Product(s) and Yield(s) (%)	Refs.
C_6				
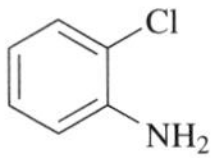 2-Cl-aniline	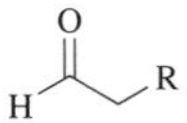 RCH₂CHO	Pd(dba)₂, XPhos, AcOK, DMA, 120°, 2–6 h	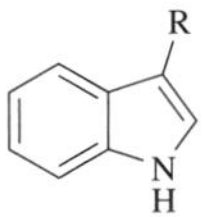 3-R-indole: R = HO(CH₂)₃ (63); Ph (70); 3,4,5-(MeO)₃C₆H₂CH₂ (85)	183
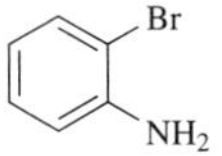 2-Br-aniline	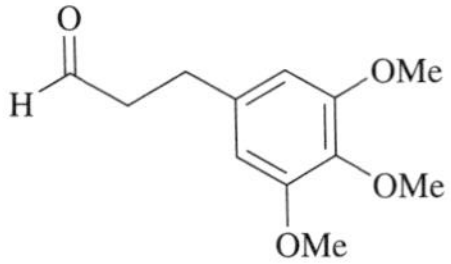 3,4,5-(MeO)₃C₆H₂CH₂CH₂CHO	Pd(dba)₂, XPhos, AcOK, DMA, 120°	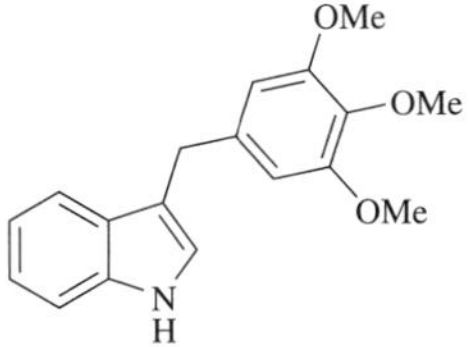 bis-indolylmethane with 3,4,5-(MeO)₃ (60)	183
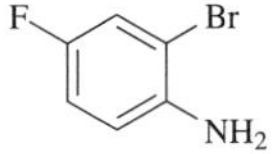 2-Br-4-F-aniline	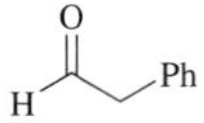 PhCH₂CHO	Pd(dba)₂, XPhos, AcOK, DMA, 120°	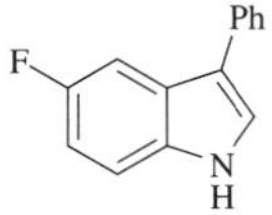 5-F-3-Ph-indole (81)	183
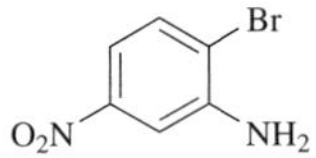 2-Br-4-NO₂-aniline (with NH₂)	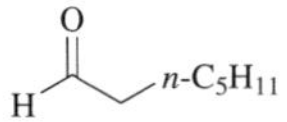 n-C₅H₁₁CH₂CHO	Pd(dba)₂, XPhos, AcOK, DMA, 120°	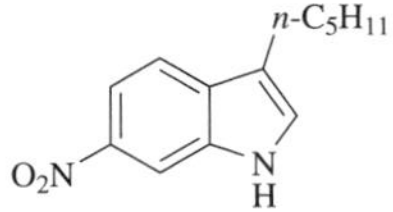 6-NO₂-3-n-C₅H₁₁-indole (15)	183
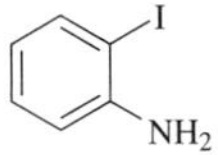 2-I-aniline	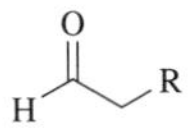 RCH₂CHO	Pd(OAc)₂, DABCO, DMF, 85°, 6–12 h	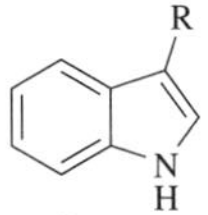 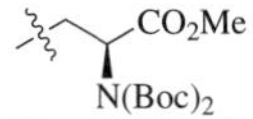 3-R-indole: R = (CH₂)₃OH (41); n-C₆H₁₃ (55); Ph (78); Bn (81); 3,4,5-(MeO)₃C₆H₂CH₂CH(CO₂Me)N(Boc)₂ (67); CH₂CH=C(Me)₂ with CH(N(Boc)₂) (81); 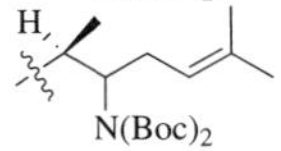 (43)	183

TABLE 6D. 3-SUBSTITUTED INDOLES FROM 2-HALOARYLENAMINES AND -IMINES PREPARED IN SITU (*Continued*)

Substrate	Reagent	Conditions	Product(s) and Yield(s) (%)	Refs.
C₆ (2-iodo-3-chloroaniline)	aldehyde with CO₂Me, N(Boc)₂	Pd(OAc)₂, DABCO, DMF, 80°, 12 h	4-Cl-indole with CH₂-CH(CO₂Me)N(Boc)₂ (80)	389
C₆₋₇ (R-substituted 2-iodoaniline)	aldehyde (CH₂)ₙN(Boc)₂	Pd(OAc)₂, DABCO, DMF, 85°, 12 h	indole with (CH₂)ₙN(Boc)₂ R / n: H 1 (56); H 2 (52); 5-F 1 (56); 5-O₂N 1 (50); 4-O₂N 1 (50)	390
	aldehyde CO₂Me, N(Boc)₂	Pd(OAc)₂, DABCO, DMF, 85°, 6–12 h	indole with CH₂-CH(CO₂Me)N(Boc)₂ R / n: 4-Cl 1 (16); 6-MeO 1 (49); 7-MeO 1 (79); 6-Me 1 (44); 6-Cl (74); 5-NO₂ (74); 6-NO₂ (64); 4-OMe (55); 5-OMe (51); 6-OMe (58); 7-OMe (71)	183
	aldehyde with phthalimide-(CH₂)ₙ	Pd(OAc)₂, DABCO, DMF, 85°, 12 h	indole with (CH₂)ₙ-phthalimide	390

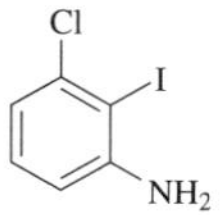

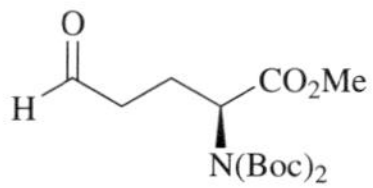

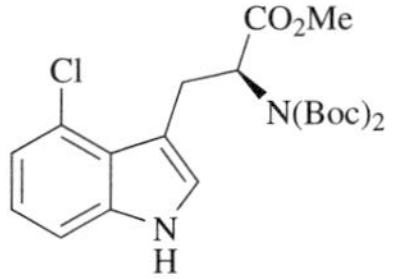

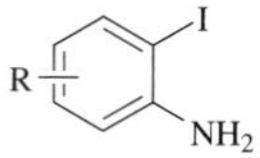

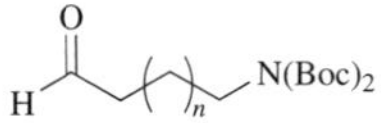

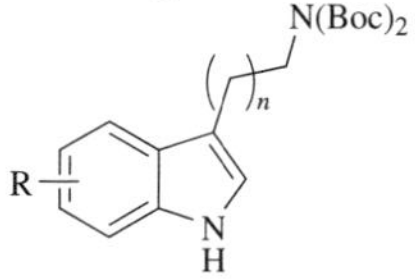

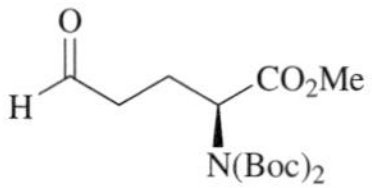

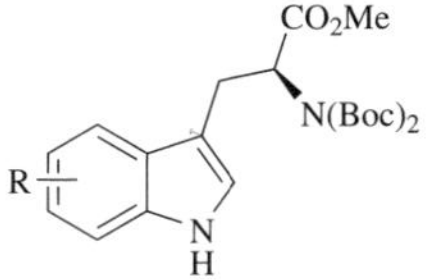

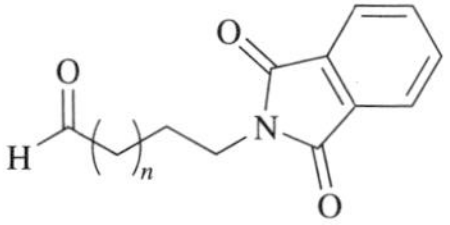

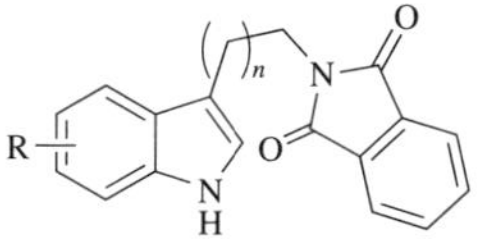

C_{6-8}

R¹—[benzene]—X, NH₂ + O=CH—CH₂R² → R²—[indole with R¹]—NH

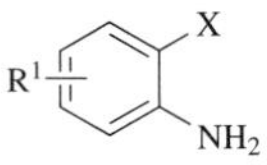

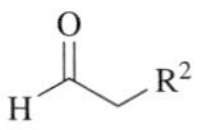

Pd₂(dba)₃, t-Bu₃P·HBF₄, KOAc, DMA, 120°, 9 h

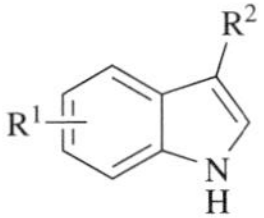

R	n		R	n
H	1	(52)	4-Cl	1	(23)
H	2	(77)	6-MeO	1	(50)
5-F	1	(49)	7-MeO	1	(66)
5-O₂N	1	(63)	4-MeO	1	(48)
4-O₂N	1	(60)	6-Me	1	(53)

391

R¹	R²	X	
5-O₂N	3,4,5-(MeO)₃C₆H₂CH₂	Br	(41)
6-O₂N	3,4,5-(MeO)₃C₆H₂CH₂	Br	(68)
6-O₂N	3,4,5-(MeO)₃C₆H₂CH₂	Cl	(58)
6-O₂N	Me(CH₂)₄	Cl	(65)
6-O₂N	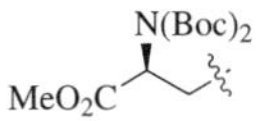 MeO₂C-CH(N(Boc)₂)-CH₂-	Cl	(65)
5-NC	Ph	Br	(61)
6-CF₃	3,4,5-(MeO)₃C₆H₂CH₂	Cl	(51)
6-MeCO	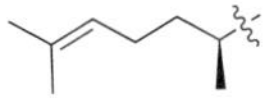 prenyl-type	Cl	(41)
4-MeO₂C	Me(CH₂)₄	Cl	(25)
6-MeO₂C	Me(CH₂)₅	Cl	(71)
7-MeO₂C	Me(CH₂)₅	Cl	(71)

C_7

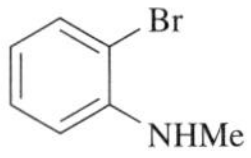

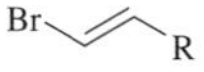

Br—[benzene]—NHMe + Br—CH=CH—R → R—[3-position of N-Me indole]

Pd₂(dba)₃, DavePhos, t-BuONa, toluene, 100°, 20 h

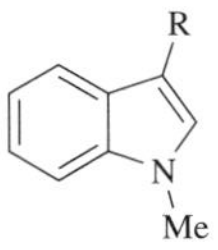

R	
Ph	(70)
4-MeC₆H₄	(69)
n-C₈H₁₇	(64)
CH₂OBn	(63)

165

TABLE 6D. 3-SUBSTITUTED INDOLES FROM 2-HALOARYLENAMINES AND -IMINES PREPARED IN SITU (*Continued*)

Substrate	Reagent	Conditions	Product(s) and Yield(s) (%)	Refs.
C₇				
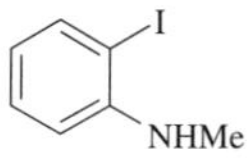 2-I-C₆H₄-NHMe	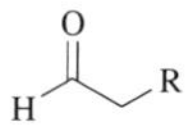 RCH₂CHO	Pd(OAc)₂, DABCO, DMF, 85°, 6–12 h	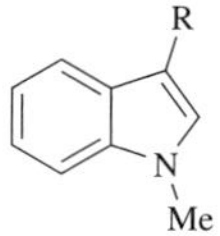 1-Me-3-R-indole; R = Bn (76), 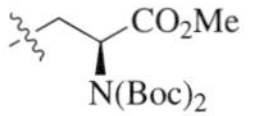 CH₂CO₂Me-CH-N(Boc)₂ (85)	183
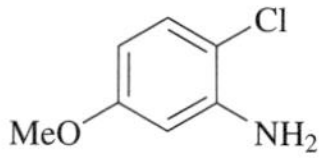 4-MeO-2-Cl-C₆H₃-NH₂	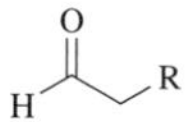 RCH₂CHO	Pd(dba)₂, XPhos, AcOK, DMA, 120°, 2–6 h	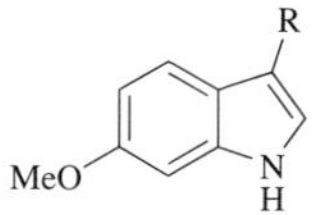 6-MeO-3-R-1H-indole; R = Bn (56), 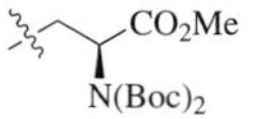 CH₂CO₂Me-CH-N(Boc)₂ (73)	183
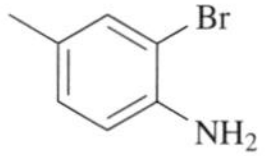 2-Br-4-Me-C₆H₃-NH₂	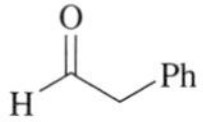 PhCH₂CHO	Pd(dba)₂, XPhos, AcOK, DMA, 120°	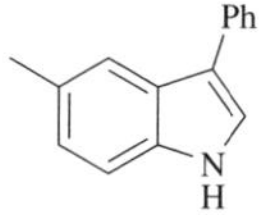 6-Me-3-Ph-1H-indole (46)	183
2-Br-4-Me-C₆H₃-NH₂	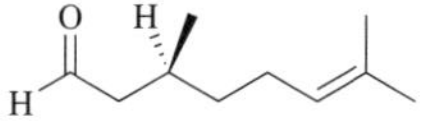 (R)-citronellal-type aldehyde	Pd(dba)₂, XPhos, AcOK, DMA, 120°	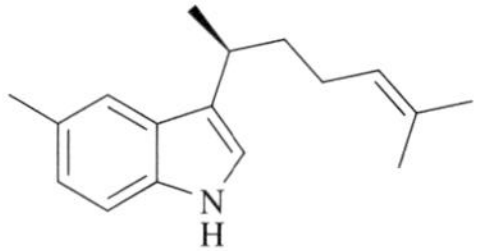 6-Me-3-(prenyl-methyl)-1H-indole (46)	183
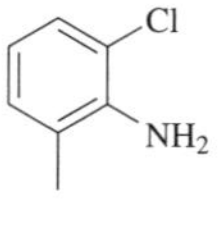 2-Cl-6-Me-C₆H₃-NH₂	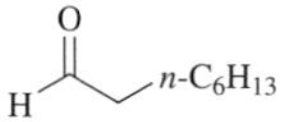 n-C₆H₁₃CH₂CHO	Pd(dba)₂, XPhos, AcOK, DMA, 120°	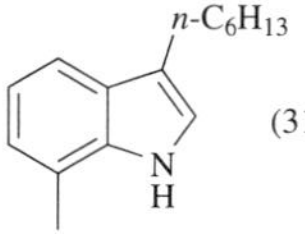 7-Me-3-(n-C₆H₁₃)-1H-indole (31)	183
C₈				
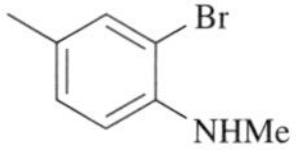 2-Br-4-Me-C₆H₃-NHMe	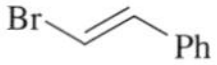 BrCH=CHPh	Pd₂(dba)₃, DavePhos, t-BuONa, toluene, 100°, 20 h	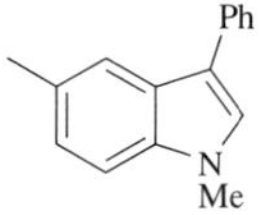 5-Me-1-Me-3-Ph-indole (72)	165

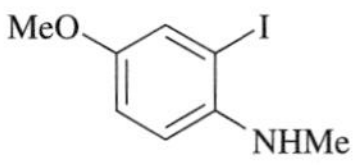 C9	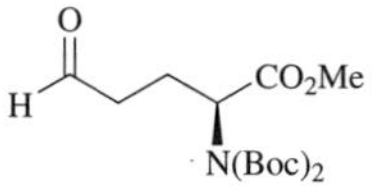 ![aldehyde with CO2Me N(Boc)2]	Pd(OAc)2, DABCO, DMF, 85°	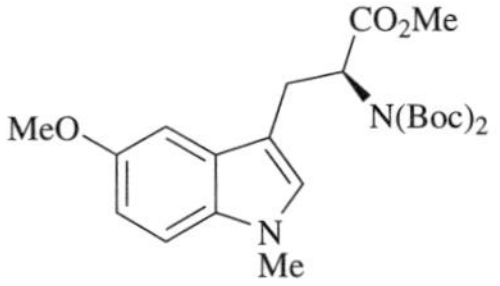 ![indole with MeO, N-Me, CH2CH(N(Boc)2)CO2Me] (84)	183
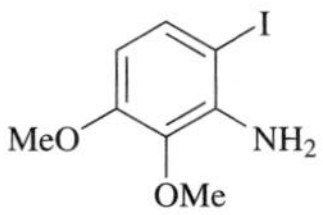	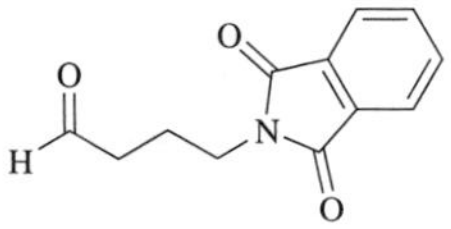 ![phthalimide aldehyde]	Pd(OAc)2, DABCO, DMF, 85°, 24 h	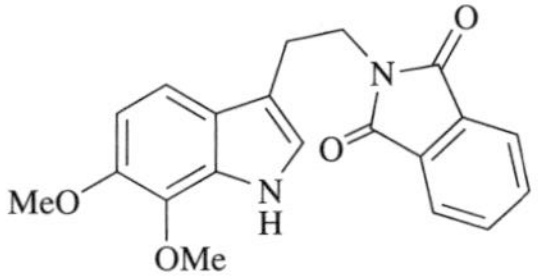 ![indole product with phthalimide, OMe, MeO] (60)	390
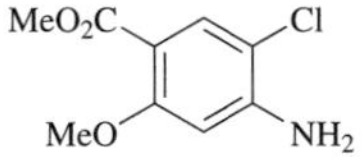	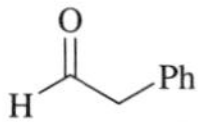 ![PhCH2CHO]	Pd(dba)2, XPhos, AcOK, DMA, 120°	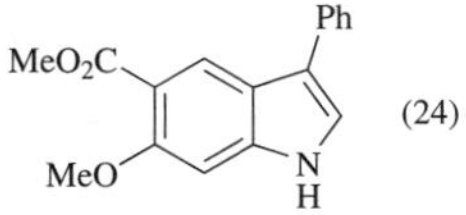 ![3-Ph indole with MeO2C, MeO] (24)	183
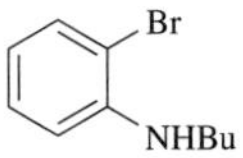 C10	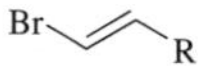 ![Br-CH=CH-R]	Pd2(dba)3, DavePhos, t-BuONa, toluene, 100°, 20 h	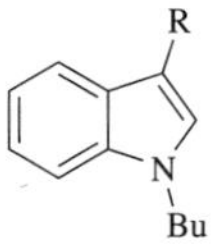 ![indole N-Bu with R] R Ph (61) 4-MeC6H4 (64) CH2OBn (49)	165
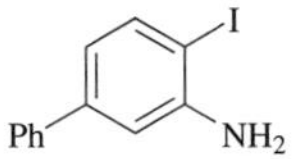 C12	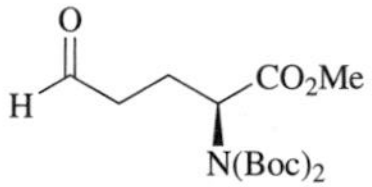 ![aldehyde with CO2Me N(Boc)2]	Pd(dba)2, XPhos, AcOK, DMA, 120°	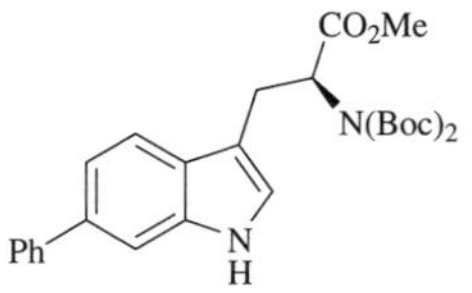 ![indole with Ph, CH2CH(N(Boc)2)CO2Me] (42)	183
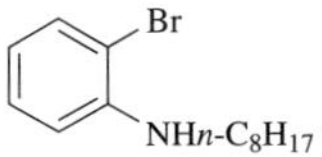 C14	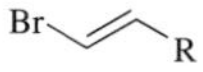 ![Br-CH=CH-R]	Pd2(dba)3, DavePhos, t-BuONa, toluene, 100°, 20 h	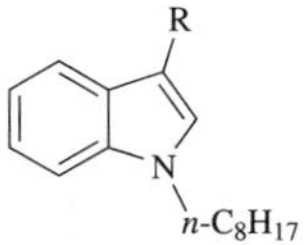 ![indole N-n-C8H17 with R] R Ph (68) CH2OBn (52)	165

TABLE 6E. 3-SUBSTITUTED INDOLES FROM ARYLENAMINES

Arylenamine	Conditions	Product(s) and Yield(s) (%)				Refs.
C_{21-27}	Pd(OAc)$_2$, Cu(OAc)$_2$, DMSO	R, R-substituted 3-aryl-1-tosylindole				208
		R	Temp (°)	Time (h)		
		4-F	80	24	(42)	
		4-F	120	6	(51)	
		4-MeO	80	17	(41)	
		4-MeO	120	5	(55)	
		3-MeO	80	22	(45)	
		3-MeO	120	6	(30)	
		4-NC	80	22	(29)	
		4-NC	150	6	(42)	
		4-EtO$_2$C	80	22	(60)	
		4-EtO$_2$C	120	6	(68)	
C_{22}	Pd(OAc)$_2$, Cu(OAc)$_2$, DMSO, 80°, 24 h	**I** (5-MeO-3-phenyl-1-tosylindole) + **II** (3-(3-methoxyphenyl)-1-tosylindole); **I** + **II** (38), **I:II** = 4.8:1				208

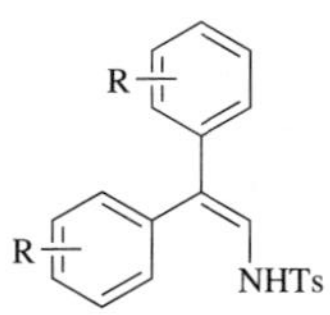

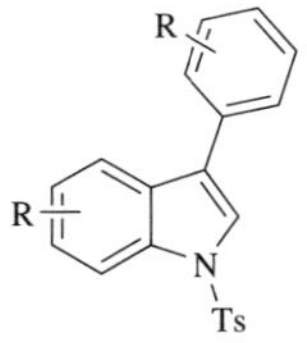

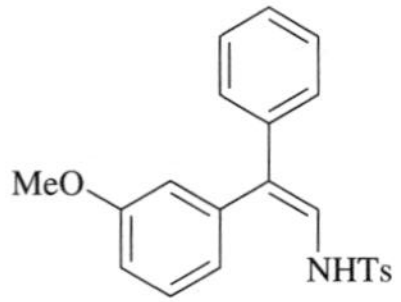

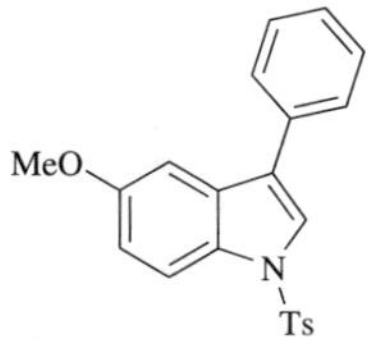

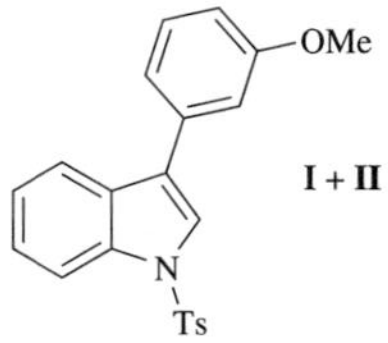

TABLE 6F. 3-SUBSTITUTED INDOLES FROM 2-NITROSTYRENES, NITROALKENES, AND NITROARENES

Substrate	Conditions	Product(s) and Yield(s) (%)	Refs.
C_{6-7} 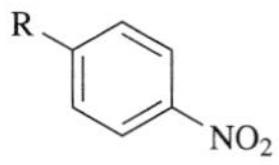 R–C_6H_4–NO_2 (≡–Ph)	[Pd(phen)_2][BF_4]_2, Ru_3(CO)_12, additive, DME, CO (59 atm), 170°	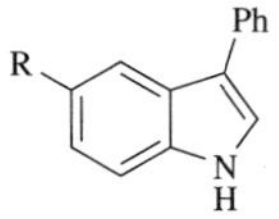 3-Ph-indole (R-substituted): R=H, additive —, 3 h (54); R=H, Me_2CO_3, 6 h (50); R=Me, —, 1.5 h (57); R=Me, Me_2CO_3, 6 h (51)	392
C_9 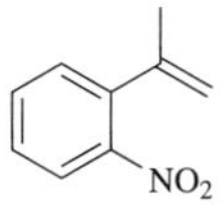 (2-nitro-isopropenylbenzene)	PdCl_2(PPh_3)_2, SnCl_2, 1,4-dioxane, CO (20 atm), 100°, 16 h	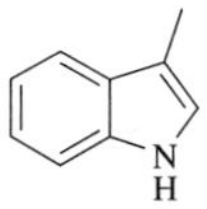 3-methylindole (57)	154, 155
C_10 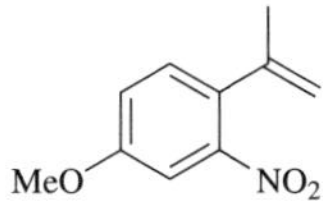 (4-MeO-2-NO_2-isopropenylbenzene)	Pd(OAc)_2, PPh_3, MeCN, CO (4 atm), 70°, 24 h	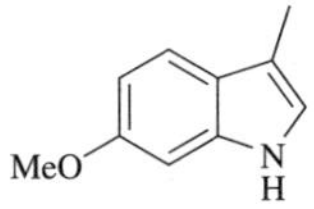 6-MeO-3-methylindole (81)	156
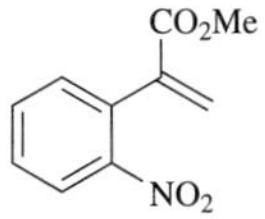 (2-nitro-α-CO_2Me-styrene)	Pd(OAc)_2, PPh_3, DMF, CO (6 atm), 110°, 72 h	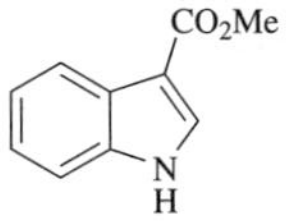 methyl indole-3-carboxylate (91)	393
C_{10-12} 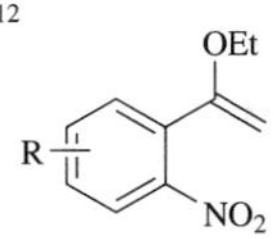 (R-substituted 2-nitro-α-OEt-styrene)	Pd(dba)_2, phen, dppp, DMF, CO (6 atm), 120°	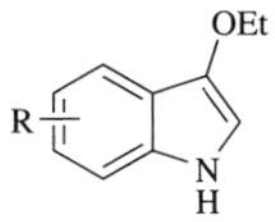 3-OEt-indole (R-substituted): R=H, 96 h (72); R=6-Cl, 72 h (63); R=6-NO_2, 72 h (85); R=4-Me, 48 h (84); R=6-OMe, 96 h (79); R=4-CO_2Me, 72 h (87)	180

TABLE 6F. 3-SUBSTITUTED INDOLES FROM 2-NITROSTYRENES, NITROALKENES, AND NITROARENES (Continued)

Substrate	Conditions	Product(s) and Yield(s) (%)	Refs.
C$_{12-15}$ 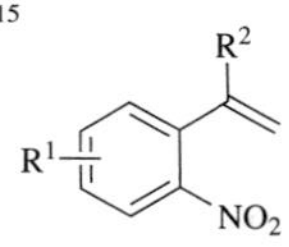 R^1–[aryl]–NO$_2$ with R^2 vinyl	Pd(dba)$_2$, dppp, phen, DMF, CO (6 atm), 72 h	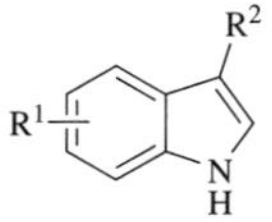 3-R^2-indole with R^1 substituent: R^1 / R^2 / Temp (°) 6-MeO / CO$_2$Et / 100 (81) 4-MeO$_2$C / CO$_2$Et / 120 (99) 5-Br / CO$_2$t-Bu / 100 (98) 5-Cl / CO$_2$t-Bu / 120 (92) 5-MeO / CO$_2$t-Bu / 120 (99) 5-MeO / SO$_2$Ph / 120 (74)	393
C$_{14-16}$ 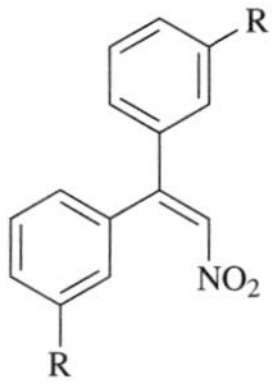 (m-R-C$_6$H$_4$)$_2$C=CH-NO$_2$	Pd(OAc)$_2$, phen, DMF, CO (1 atm), 110°	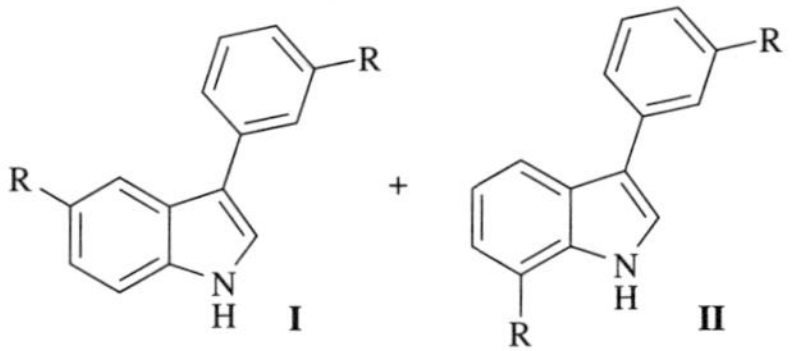 3-(m-R-phenyl)indoles I (5-R) + II (7-R) R / Time (h) / I+II / I:II Cl / 6 / (91) / 42:58 CF$_3$ / 8 / (86) / 51:49 MeO / 3 / (91) / 53:47	209
C$_{14-22}$ (p-R-C$_6$H$_4$)$_2$C=CH-NO$_2$ 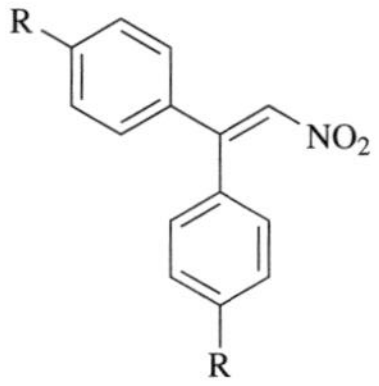	Pd(OAc)$_2$, phen, DMF, CO (1 atm), 110°	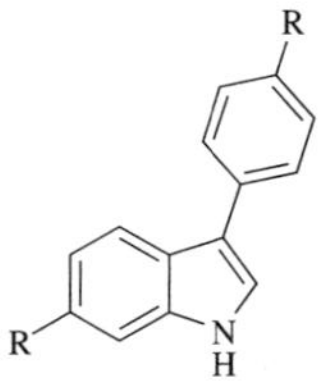 3-(p-R-phenyl)-6-R-indole R / Time (h) H / 3 (97) Cl / 6 (98) CF$_3$ / 16 (58) MeO / 3 (93) Me / 3 (87) t-Bu / 3 (92)	209

TABLE 7A. 2,3-DISUBSTITUTED INDOLES FROM 2-HALOARYLENAMINES AND -IMINES

2-Haloarylenamine or -imine	Conditions	Product(s) and Yield(s) (%)	Refs.
C_{12} 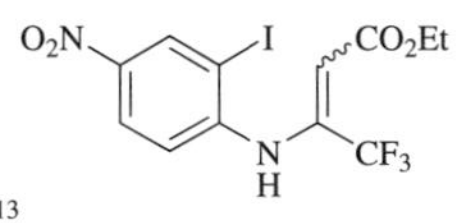 (4-O_2N-2-I-C$_6$H$_3$)NH-C(CF$_3$)=CH-CO$_2$Et	Pd(OAc)$_2$, PPh$_3$, NaHCO$_3$, DMF, 120°	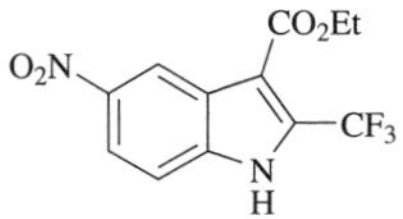 5-O_2N-2-CF$_3$-3-CO$_2$Et-indole (44)	45, 46
C_{12-13} 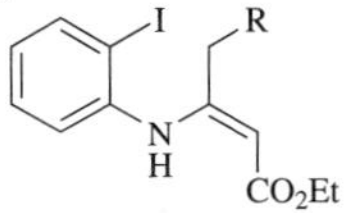 (2-I-C$_6$H$_4$)NH-C(CH$_2$R)=CH-CO$_2$Et	Pd(PPh$_3$)$_4$, Ag$_3$PO$_4$, DMSO, 100°	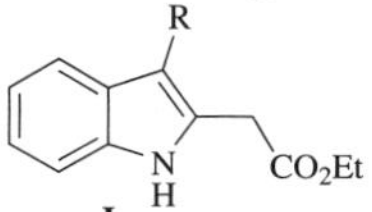 **I** 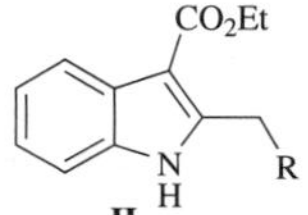 3-CO$_2$Et-2-CH$_2$R-indole + **II** 2-CH$_2$R-3-CO$_2$Et isomer R Time (h) **I** **II** H 3.5 (17) (79) Me 4.5 (6) (86)	379
C_{13-15} (4-R-2-Br-C$_6$H$_3$)NH-C(CF$_3$)=CH-CO$_2$Et 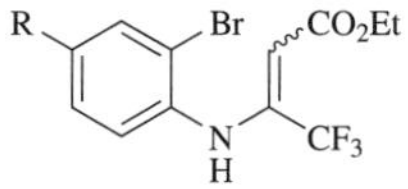	Pd(OAc)$_2$, PPh$_3$, Pr$_3$N, DMF, 120°	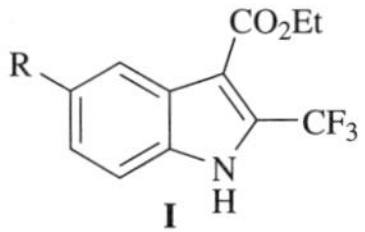 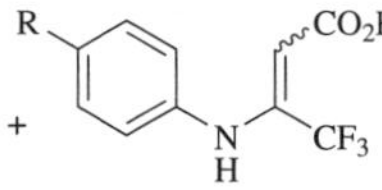 **I** 5-R-2-CF$_3$-3-CO$_2$Et-indole + **II** R-C$_6$H$_3$-NH-C(CF$_3$)=CH-CO$_2$Et R **I** **II** CN (54) (28) CO$_2$Et (68) (25)	45, 46
C_{14-20} (X, R^1-C$_6$H$_3$)N(OMe)=C(CF$_3$)-C≡C-R^2 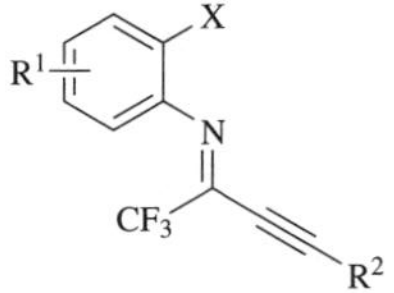	PdCl$_2$(PPh$_3$)$_2$, K$_3$PO$_4$, H$_2$O, DME, 60°	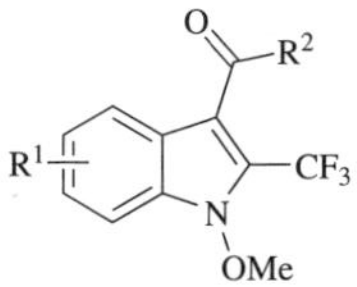 R^1-indole-2-CF$_3$-3-C(=O)R^2, N-OMe X R^1 R^2 Time (h) I H 2-thienyl 0.5 (85) I H t-Bu 0.5 (86) Br H Ph 0.5 (80) Cl H Ph 5 (13) I 4-F Ph 1 (76) I H 4-ClC$_6$H$_4$ 0.5 (70) I H 2-ClC$_6$H$_4$ 1 (70) I H 4-FC$_6$H$_4$ 1 (59) I H 4-MeC$_6$H$_4$ 0.5 (82) I H 4-MeOC$_6$H$_4$ 0.5 (85) I H 3-MeOC$_6$H$_4$ 0.5 (81) I 4-CF$_3$ Ph 3 (52) I 4-Me Ph 0.5 (82) I 5-MeO Ph 0.5 (78) I H 2-Np 1 (74)	394

TABLE 7A. 2,3-DISUBSTITUTED INDOLES FROM 2-HALOARYLENAMINES AND -IMINES (*Continued*)

2-Haloaryl[enamine or -imine]	Conditions	Product(s) and Yield(s) (%)	Refs.
C$_{15}$			
[EtO$_2$C-aryl-Br, C(CF$_3$)=C(CO$_2$Et)NH]	Pd(OAc)$_2$, PPh$_3$, NaHCO$_3$, DMF, 120°	[indole with 5-CO$_2$Et, 2-CF$_3$, 3-CO$_2$Et] (44)	45, 46
C$_{17}$			
[2-iodoaryl-N=C(CH$_2$Ph)-oxazolidinone]	Pd(dba)$_2$, XPhos, K$_3$PO$_4$, 1,4-dioxane, 80°, 23 h	[3-Ph-2-(oxazolidinon-3-yl)indole] (86)	395
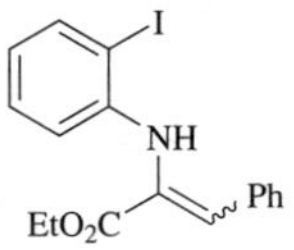 [2-iodoaryl-NH-C(CO$_2$Et)=CHPh]	Pd$_2$(dba)$_3$, P(*o*-tol)$_3$, Et$_3$N, DMF, 120°, 2 h	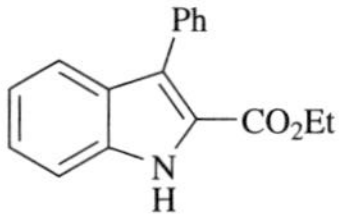 [3-Ph-2-CO$_2$Et-indole] (56)	396
C$_{17-19}$			
[2-iodoaryl-NH-C(Ph)=CH-C(O)R]	Pd$_2$(dba)$_3$, XPhos, K$_3$PO$_4$, 1,4-dioxane, 80°, 23 h	[2-Ph-3-C(O)R-indole] R = OEt (98) R = *N*-morpholino (91)	354

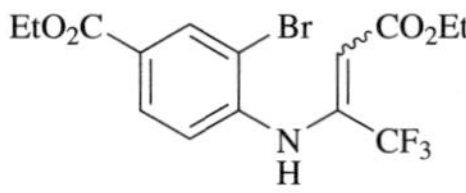

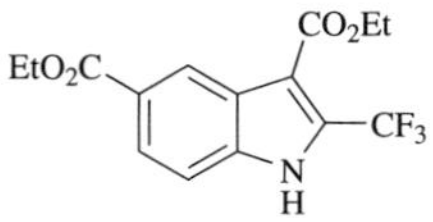

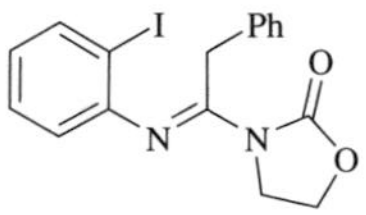

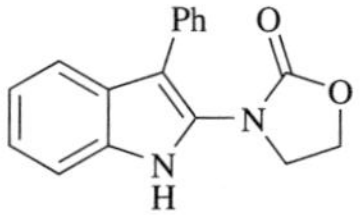

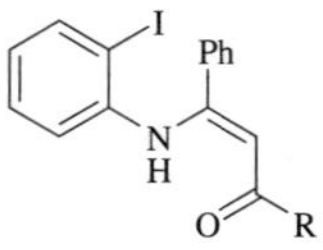

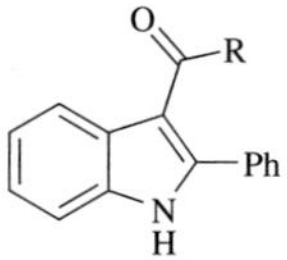

TABLE 7B. 2,3-DISUBSTITUTED INDOLES FROM 2-HALOARYLENAMINES AND -IMINES PREPARED IN SITU

Substrate	Reagent	Conditions	Product(s) and Yield(s) (%)	Refs.
C_6 2-X-aniline (NH_2)	R—≡—Ph	1. $Ru_3(CO)_{12}$, PF_6NH_4, toluene, 105°, 18 h; 2. $Pd(OAc)_2$, **L2**, t-BuOK, 105°, 24 h	2-Ph-3-R-indole: X=Cl, R=Et (—); X=Cl, R=Ph (89); X=Br, R=Ph (87)	382
2-Cl-aniline	R^1-C₆H₄-≡-R^2	1. $TiCl_4$, t-BuNH$_2$, toluene, 105°, 20 h; 2. $Pd(OAc)_2$, HIPrCl, t-BuOK, toluene, 105°, 20 h	**I** (2-Ar-3-R²) + **II** (3-Ar-2-R²): R^1 / R^2 / (I+II) / I:II — H / Bu / (81)[a] / 92:8; 3-CF₃ / Bu / (84)[b] / 97:3; H / Ph / (76) / —; H / n-C₆H₁₃ / (81)[a] / 92:8; 4-F / n-C₆H₁₃ / (74)[b] / >99:1; 4-Me / n-C₆H₁₃ / (81)[a] / 92:8; 4-OMe / n-C₆H₁₃ / (66) / >99:1; 3-CF₃ / n-C₆H₁₃ / (82)[b] / 97:3	184
2-Br-aniline	R^1-C₆H₄-≡-R^2	1. $TiCl_4$, t-BuNH$_2$, toluene, 105°, 20 h; 2. $Pd(OAc)_2$, HIPrCl, t-BuOK, toluene, 105°, 20 h	**I+II**: R^1 / R^2 / (I+II) / I:II — 4-Cl / n-C₆H₁₃ / (67) / >99:1; 2-Cl / n-C₆H₁₃ / (46) / >99:1	184

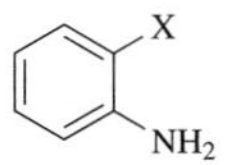

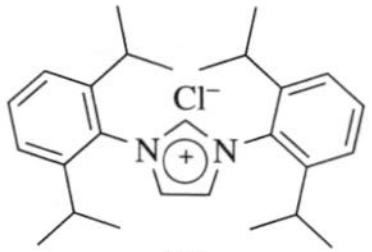

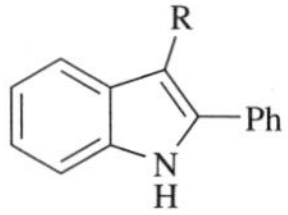

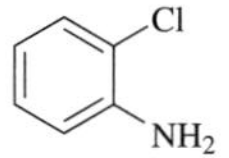

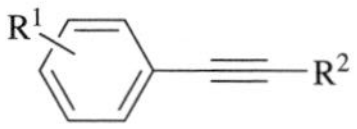

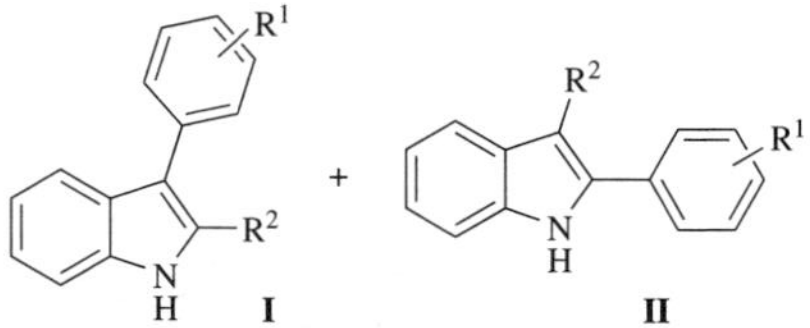

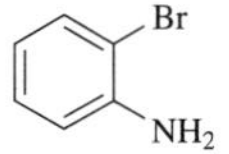

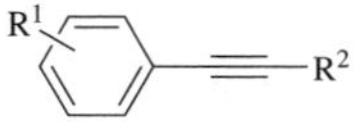

L2 = 1,3-bis(2,6-diisopropylphenyl)imidazolium chloride

TABLE 7B. 2,3-DISUBSTITUTED INDOLES FROM 2-HALOARYLENAMINES AND -IMINES PREPARED IN SITU (Continued)

Substrate	Reagent	Conditions	Product(s) and Yield(s) (%)	Refs.
C₆				
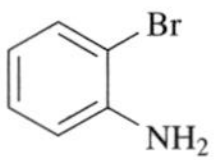 2-Br-aniline	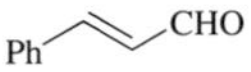 Ph-CH=CH-CHO	1. 2,2,2-trifluoroethanol, rt, 1 h 2. HCO₂H, RNC, rt, 1–3 d, solvent evaporated 3. MeCN, Pd(OAc)₂, PPh₃, 80°, 16–24 h	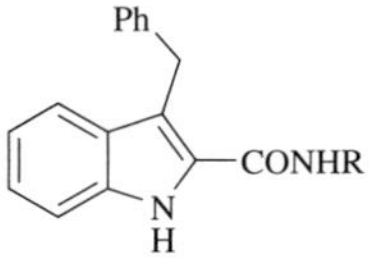 3-benzyl-2-CONHR-indole R CH₂CO₂Me (15) t-Bu (17) 1-cyclohexenyl (29) Bn (21)  4-Ph-cyclohexyl (32)	187
2-Br-aniline	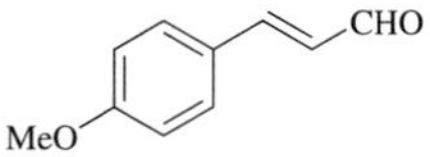 4-MeO-C₆H₄-CH=CH-CHO	1. 2,2,2-trifluoroethanol, rt, 1 h 2. HCO₂H, t-BuNC, rt, solvent evaporated 3. MeCN, Pd(OAc)₂, PPh₃, 80°	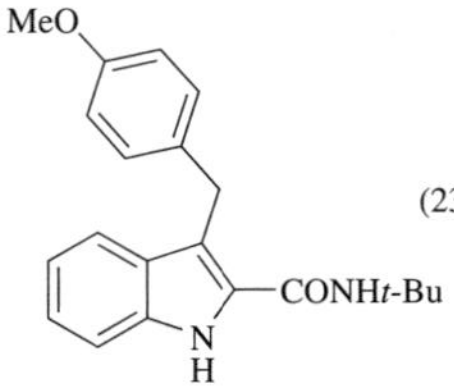 3-(4-MeO-benzyl)-2-CONH-t-Bu-indole (23)	187
2-Br-aniline	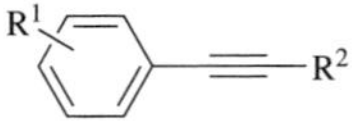 R¹-C₆H₄-C≡C-R²	1. TiCl₄, t-BuNH₂, toluene, 105°, 14 h 2. Pd(OAc)₂, PCy₃, t-BuOK, toluene, 105°, 22 h	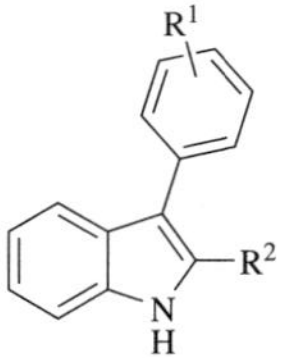 2-R²-3-(R¹-aryl)-indole R¹ R² H Ph (74) 4-OMe n-C₆H₁₃ (68) 3-OMe n-C₆H₁₃ (71)	184
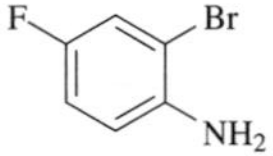 2-Br-4-F-aniline	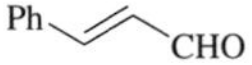 Ph-CH=CH-CHO	1. 2,2,2-trifluoroethanol, rt, 1 h 2. HCO₂H, BnNC, rt, solvent evaporated 3. MeCN, Pd(OAc)₂, PPh₃, 80°	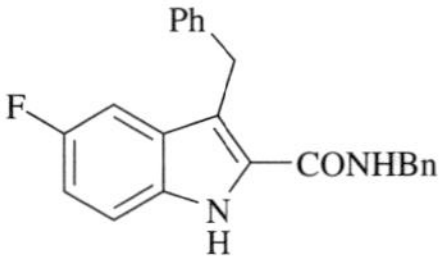 5-F-3-benzyl-2-CONHBn-indole (38)	187
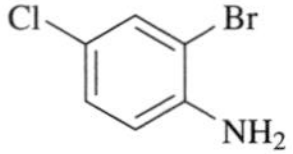 2-Br-4-Cl-aniline	 Ph-C≡C-Et	1. TiCl₄, t-BuNH₂, toluene, 105°, 14 h 2. Pd(OAc)₂, PCy₃, t-BuOK, toluene, 105°, 22 h	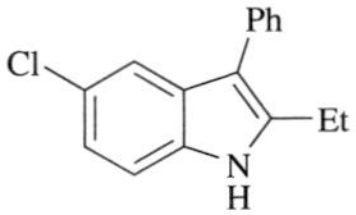 5-Cl-2-Et-3-Ph-indole (53)	184

486

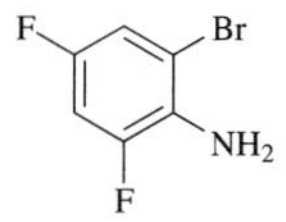

Substrate	Reagent	Conditions	Product (Yield %)	Refs.

Starting material (C6-13): 2-bromo-4-fluoro-6-fluoroaniline

Ph—≡—Et

1. TiCl$_4$, t-BuNH$_2$, toluene, 105°, 14 h
2. Pd(OAc)$_2$, PCy$_3$, t-BuOK, toluene, 105°, 22 h

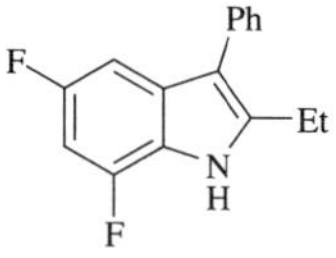

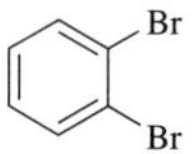

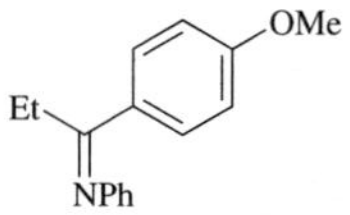

Product: 3-Ph, 2-Et, 5-F, 7-F indole (52)   184

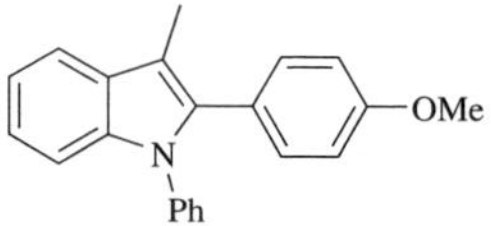

1,2-dibromobenzene

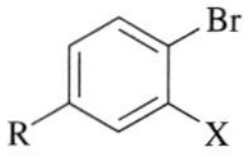

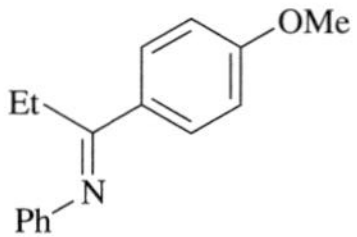

Ar(OMe)C(Et)=NPh

Pd$_2$(dba)$_3$, XPhos, t-BuONa, 1,4-dioxane, 110°, 14 h

Product: 2-(4-MeO-C$_6$H$_4$)-3-Et-1-Ph-indole (66)   196

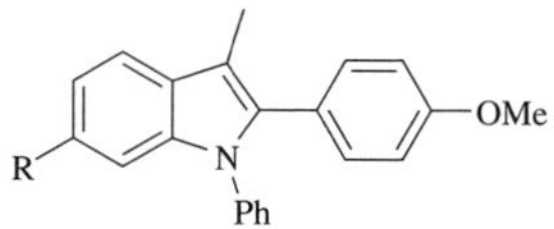

C$_{6-13}$: 2-bromo-X-R-benzene

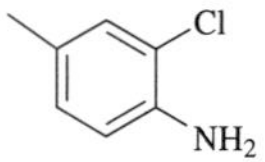

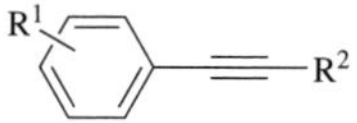

Ar(OMe)C(Et)=NPh

Pd$_2$(dba)$_3$, XPhos, t-BuONa, 1,4-dioxane, 110°, 14 h

X	R	Yield
Br	H	(66)
Cl	6-BnO	(76)

197

C$_7$: 2-chloro-4-methylaniline

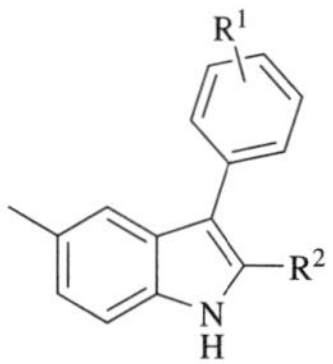

R^1-C$_6$H$_4$—≡—R^2

1. TiCl$_4$, t-BuNH$_2$, toluene, 105°, 14 h
2. Pd(OAc)$_2$, PCy$_3$, t-BuOK, toluene, 105°, 22 h

Product: 3-Ar(R^1)-2-R^2-5-Me-indole

R^1	R^2	Yield
H	Et	(71)
3-OMe	Bu	(55)[c]
4-OMe	Bu	(58)[c]
H	Ph	(71)
4-F	n-C$_6$H$_{13}$	(54)[c]
3-OMe	n-C$_6$H$_{13}$	(51)[c]
3-CF$_3$	n-C$_6$H$_{13}$	(71)[c]

184

C$_{13}$: 4-BnO-2-chloro-bromobenzene

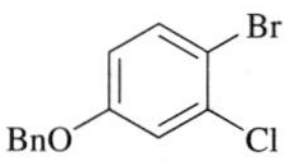

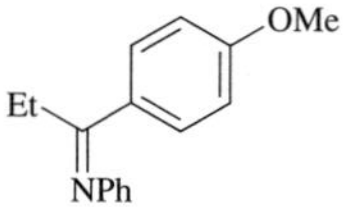

Ar(OMe)C(Et)=NPh

Pd$_2$(dba)$_3$, XPhos, t-BuONa, 1,4-dioxane, 110°, 14 h

Product: 2-(4-MeO-C$_6$H$_4$)-3-Me-6-BnO-1-Ph-indole (70)   196

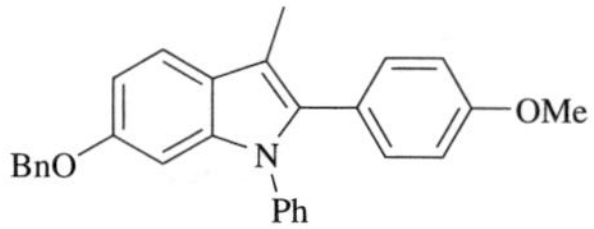

---

[a] Up to 8% of a regioisomer was isolated.
[b] Up to 5% of a regioisomer was isolated.
[c] Up to 4% of a regioisomer was isolated.

TABLE 7C. 2,3-DISUBSTITUTED INDOLES FROM ARYLENAMINES AND -IMINES

Arylenamine or -imine	Conditions	Product(s) and Yield(s) (%)	Refs.
C_{11-13} 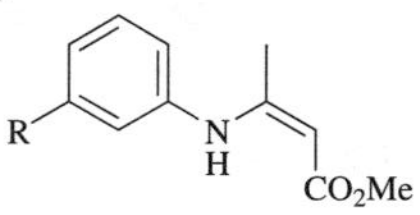 (structure with CO_2Me)	Pd(OAc)$_2$, Cu(OAc)$_2$, K$_2$CO$_3$, DMF, 12–16 h	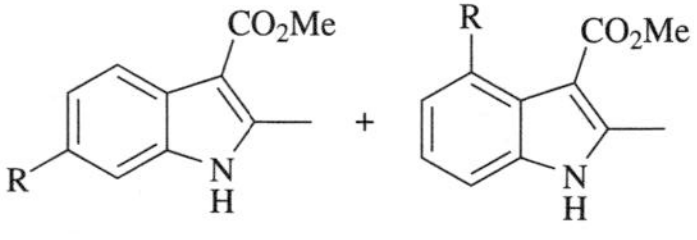 Structures **I** and **II** (both with CO_2Me); R / Temp (°) / **I** (yield) / **II** (yield) Cl / 80 / (72) / — F / 110 / (74) / — Me / 110 / (68) / — MeO / 140 / (68) / — MeCO / 80 / (54) / — R / Temp (°) / **I + II** / **I:II** Cl / 80 / (72) / 88:12 F / 110 / (74) / 53:47 Me / 110 / (68) / 92:8 MeO / 140 / (68) / >99:1 MeCO / 80 / (54) / >99:1	397
C_{11-14} (structure with CO_2Me) 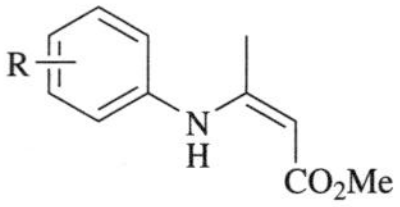	Pd(OAc)$_2$, Cu(OAc)$_2$, K$_2$CO$_3$, DMF, 12–16 h	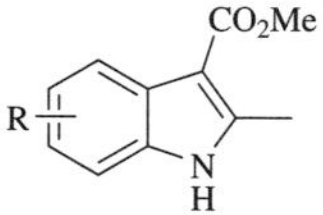 Structure with CO_2Me; R / Temp (°) / yield H / 80 / (72) 5-F / 110 / (74) 7-F / 140 / (78) 5-Cl / 80 / (64) 7-Cl / 80 / (53) 5-H$_2$NCO / 140 / (70) 5-NC / 140 / (65) R / Temp (°) / yield 7-MeO / 140 / (64) 5-Me / 80 / (72) 7-Me / 110 / (82) 5-MeCO / 140 / (52) 4,6-Me$_2$ / 140 / (62) 5-EtO$_2$C / 140 / (64)	397
C_{12-17} (structure with R^1, CO_2R^2) 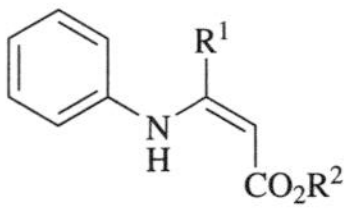	Pd(OAc)$_2$, Cu(OAc)$_2$, K$_2$CO$_3$, DMF, 12–16 h	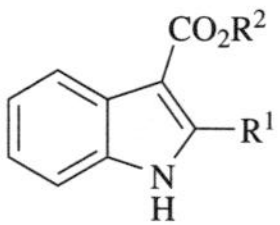 Structure with CO_2R^2, R^1 R^1 / R^2 / yield Me / Et / (71) Me / t-Bu / (79) Ph / Et / (68)	397
C_{19} (structure with NMe$_2$, N-OCOC$_6$F$_5$) 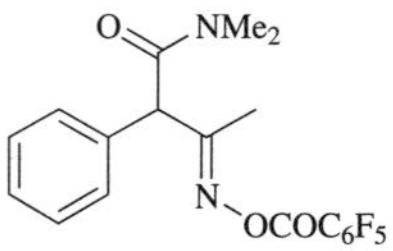	PdCl$_2$(MeCN)$_2$, MgO, DCE, reflux, 5 h	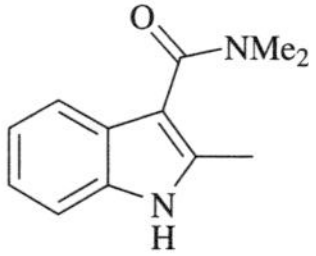 Structure with NMe$_2$ (68)	211

C_{21}

$PdCl_2(MeCN)_2$, MgO, DCE, reflux, 10 h

(68) 211

$PdCl_2(MeCN)_2$, MgO, DCE, reflux, 6 h

(59) +

(40) 211

C_{21-28}

$PdCl_2(MeCN)_2$, MgO, DCE, reflux

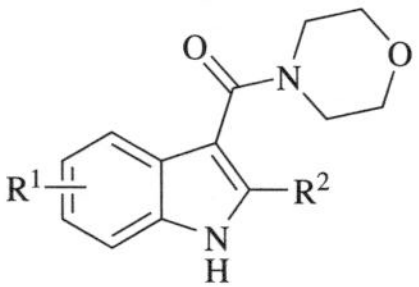

R^1	R^2	Time (h)	
6-Cl	Me	4	(70)
6-Br	Me	6	(75)
6-F	Me	4	(70)
H	Me	5	(73)
4-Me	Me	6	(46)
H	$3\text{-}BrC_6H_4$	10	(49)
6-Ph	Me	3.5	(72)
H	$CH_2\text{-}c\text{-}C_6H_{11}$	9	(91)
H	Bn	9	(63)
H	$(CH_2)_2Ph$	9	(55)

211

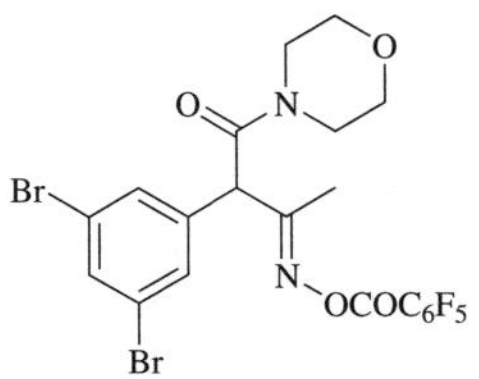

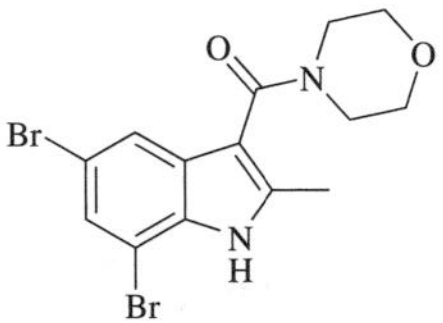

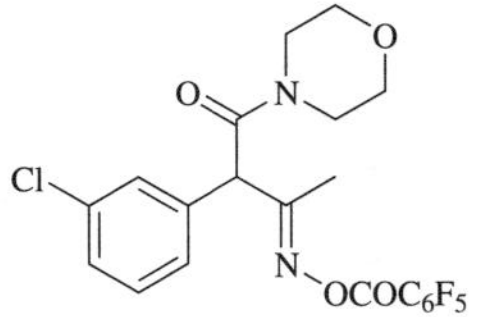

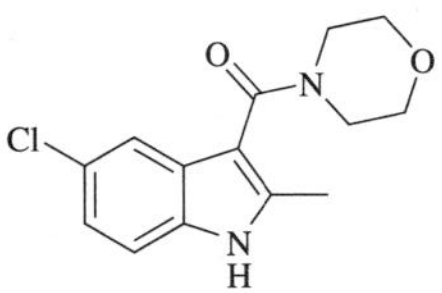

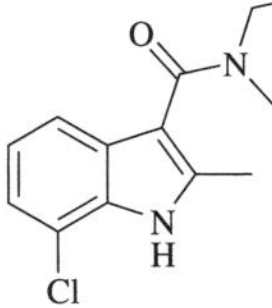

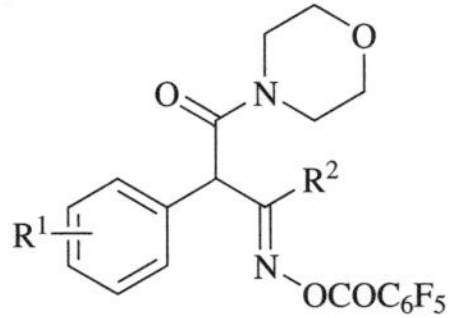

TABLE 7C. 2,3-DISUBSTITUTED INDOLES FROM ARYLENAMINES AND -IMINES (Continued)

Arylenamine or -imine	Conditions	Product(s) and Yield(s) (%)	Refs.
C_{26}			
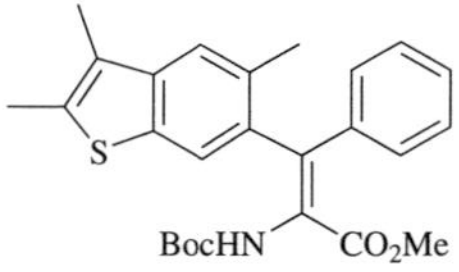	Pd(OAc)$_2$, Cu(OAc)$_2$·H$_2$O, DMF, 160°, 0.5 h	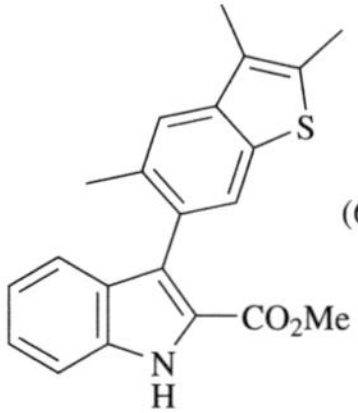 (60) 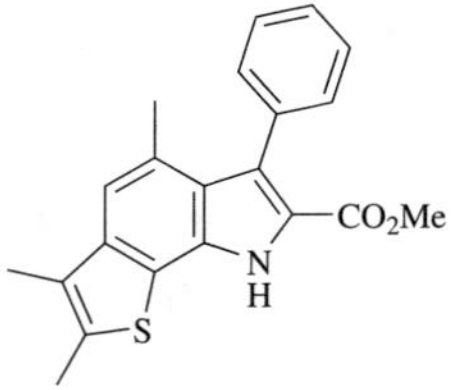 + (20)	212
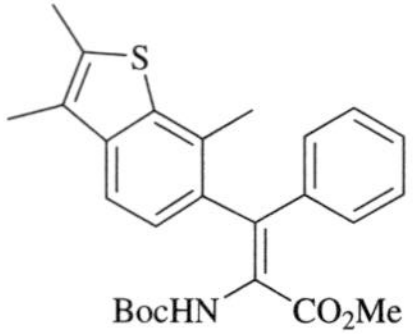	Pd(OAc)$_2$, Cu(OAc)$_2$·H$_2$O, DMF, 100°, 2 h	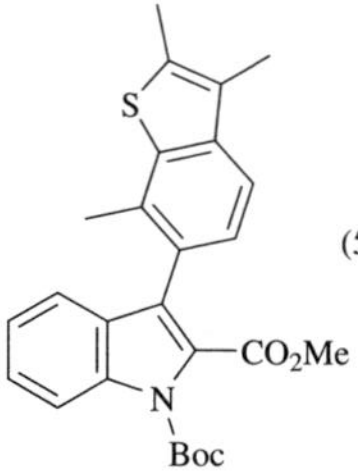 (50)	212
	Pd(OAc)$_2$, Cu(OAc)$_2$·H$_2$O, DMF, 100°, 2 h; 130°, 1 h	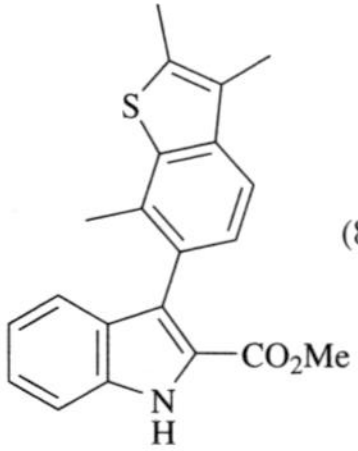 (85)	212

TABLE 7D. 2,3-DISUBSTITUTED INDOLES FROM 2-NITROSTYRENES, 2-ISOCYANOSTYRENE, AND 2-ALLYLANILINES

Substrate	Conditions	Product(s) and Yield(s) (%)	Refs.
C₉ (2-isocyanostyrene with R)	Pd(OAc)₂, dppp, Et₂NH, THF, 40°	2-aryl-3-(NEt₂-methyl)indole; R = H, 3.5 h (42)ᵃ; R = 4-O₂N, 2 h (24); R = 4-MeO, 21 h (39)	186
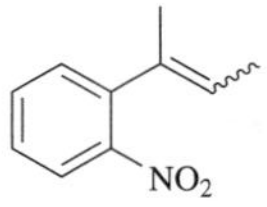 C₁₀ (2-nitrostyrene derivative, trans:cis = 1:2)	Pd(OAc)₂, PPh₃, MeCN, CO (4 atm), 70°, 21.5 h	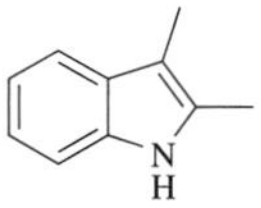 2,3-dimethylindole **I** (97)	156
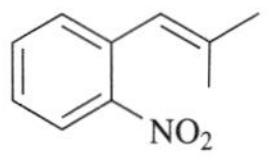 (2-methylallyl-nitrobenzene)	PdCl₂(PPh₃)₂, SnCl₂, 1,4-dioxane, CO (20 atm), 100°	**I** (52)	155
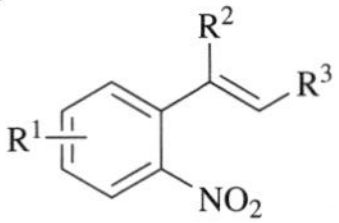 C₁₄₋₁₆ (substituted 2-nitrostyrene with R¹, R², R³)	Pd(dba)₂, dppp, phen, DMF, CO (6 atm), 72 h	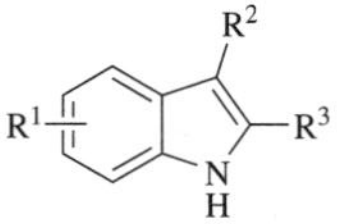 2,3-disubstituted indole; R¹=H, R²=EtO₂C, R³=Pr, 120° (84); R¹=5-MeO, R²=NC, R³=n-C₅H₁₁, 120° (45); R¹=H, R²=MeO₂C, R³=Ph, 100° (74)	393
C₁₆ (2-allylaniline with OTBS, NHMe)	PdCl₂(MeCN)₂, K₂CO₃, benzoquinone, THF, rt, 24 h	2-methyl-3-OTBS-N-methylindole (77)	157
C₁₇₋₂₂ (2-allylaniline with OTBS, R, NHMe)	PdCl₂(MeCN)₂, K₂CO₃, LiCl, benzoquinone, THF, 24 h	2-(CH₂R)-3-OTBS-N-methylindole; R = Me, 85° (50); R = Ph, rt (47)	157

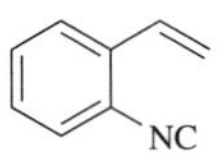

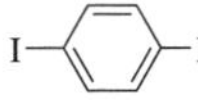

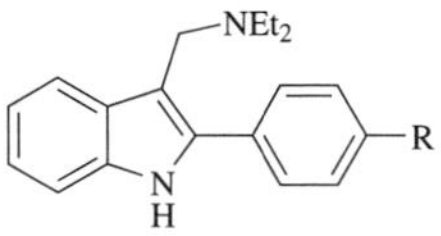

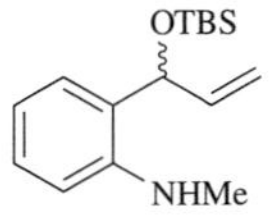

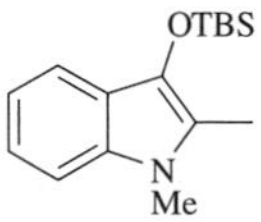

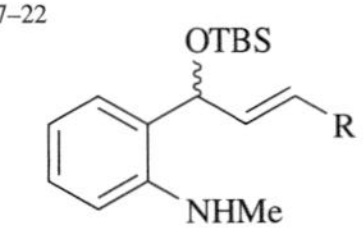

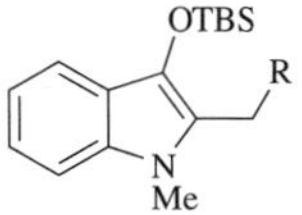

TABLE 7D. 2,3-DISUBSTITUTED INDOLES FROM 2-NITROSTYRENES, 2-ISOCYANOSTYRENE, AND 2-ALLYLANILINES (*Continued*)

Substrate	Conditions	Product(s) and Yield(s) (%)	Refs.
C_{18} 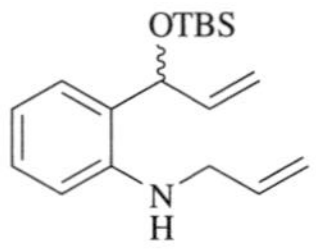	PdCl$_2$(MeCN)$_2$, K$_2$CO$_3$, LiCl, benzoquinone, THF, rt, 24 h	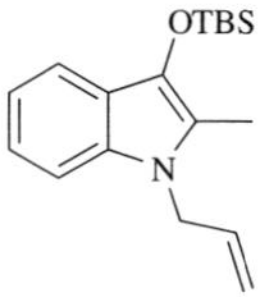 (54)	157
C_{23} 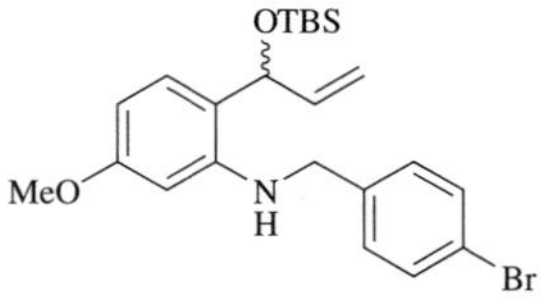	PdCl$_2$(MeCN)$_2$, K$_2$CO$_3$, benzoquinone, THF, rt, 24 h	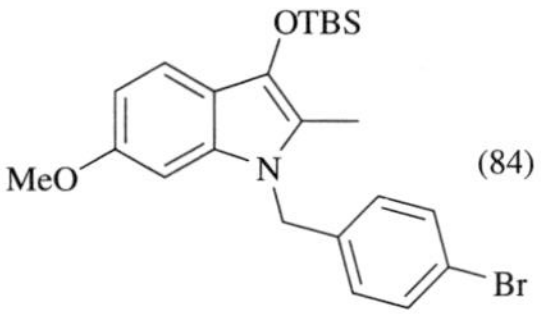 (84)	157

[a] The yield was determined by NMR spectroscopy.

TABLE 8. INDOLES VIA ARENE VINYLATION

Substrate	Conditions	Product(s) and Yield(s) (%)	Refs.
C_{11} 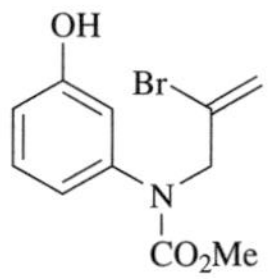 (2-bromoallyl-N-(2-hydroxyphenyl)-N-CO₂Me)	Herrmann's catalyst, Cs_2CO_3, DMA, 70°, 1 d 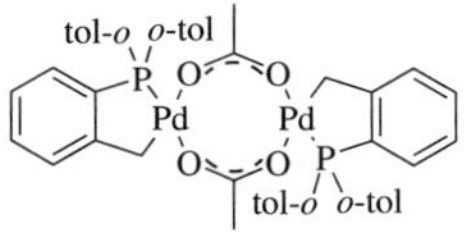 (Herrmann's catalyst structure shown)	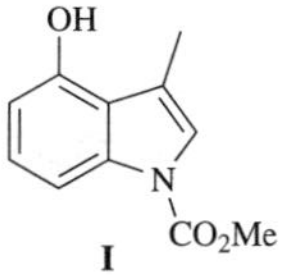 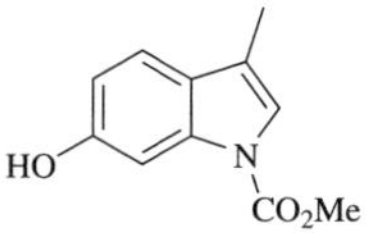 **I** (3-methyl-4-hydroxyindole-N-CO₂Me) + **II** (3-methyl-6-hydroxyindole-N-CO₂Me), **I + II** (79), **I:II** = 1:1	47
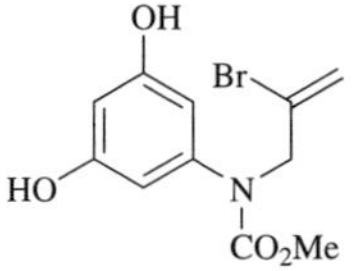 (2-bromoallyl-N-(3,5-dihydroxyphenyl)-N-CO₂Me)	Herrmann's catalyst, Cs_2CO_3, DMA, 70°, 1 d	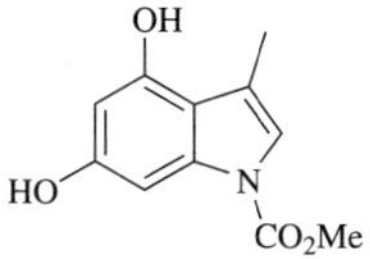 (3-methyl-4,6-dihydroxyindole-N-CO₂Me) (71)	47
C_{12} 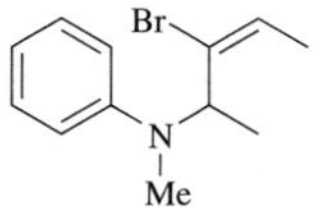 (bromo-butenyl-N-methyl-N-phenyl)	$Pd(OAc)_2$, XPhos, Cs_2CO_3, DME, 90°, 3 h	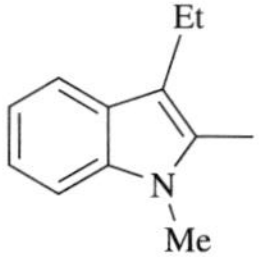 (3-ethyl-2-methyl-N-methylindole) (12)	398

TABLE 9. 2,3-UNSUBSTITUTED INDOLES VIA N-VINYLATION AND N-ARYLATION

Substrate	Conditions	Product(s) and Yield(s) (%)	Refs.
C₇ 3-iodo-2-methyl-nitrobenzene (O₂N, I, Me)	Br-CH₂CH₂-N(R)H; Pd(OAc)₂, (2-furyl)₃P, norbornene, Cs₂CO₃, MeCN, 135°, 20 h	7-methyl-6-nitroindole N-R R Ph (53) 4-O₂NC₆H₄ (70)	193
C₈ 2-bromo-β-chlorostyrene	RNH₂, Pd(OAc)₂, HP(t-Bu)₃BF₄, t-BuONa, toluene, 130°, 4 h	indole N-R R t-Bu (65) CMe₂Et (64) C(Me)₂CH=CH₂ (68) c-C₆H₁₁ (61) 2-MeC₆H₄ (85) CHMePh (77) 2,6-Me₂C₆H₃ (83) 1-adamantyl (66) 1-Np (78) 2,6-(i-Pr)₂C₆H₃ (73)	191
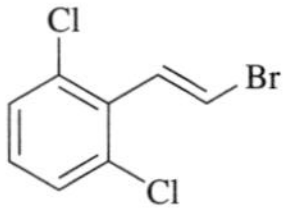 (E)-1,3-dichloro-2-(2-bromovinyl)benzene	PhNH₂, Pd₂(dba)₃, DavePhos, t-BuONa, toluene, 100°	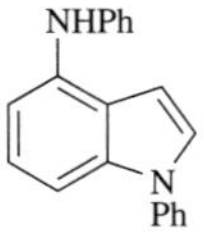 4-(phenylamino)-1-phenylindole (65)	190
same	1. 4-MeOC₆H₄NH₂, Pd₂(dba)₃, DavePhos, t-BuONa, toluene, 100°, 30 min 2. Morpholine	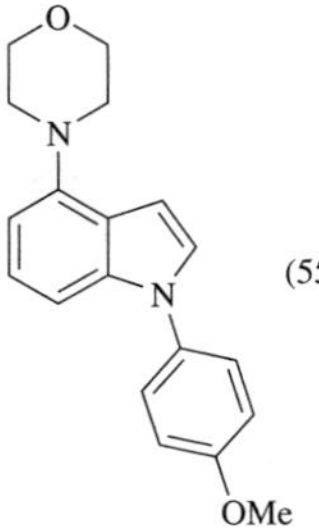 4-morpholino-1-(4-methoxyphenyl)indole (55)	190

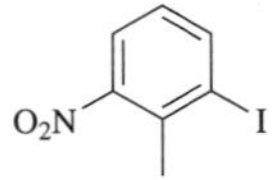

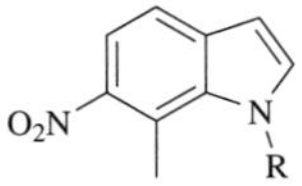

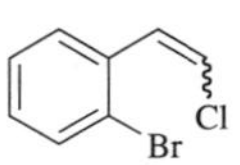

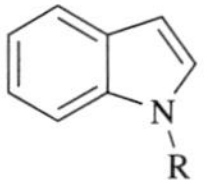

494

[Structure: 4-methoxyphenyl-N-methyl-N-(1-(4-methoxyphenyl)-1H-indol-4-yl)amine] (57)

1. 4-MeOC$_6$H$_4$NH$_2$, Pd$_2$(dba)$_3$, DavePhos, t-BuONa, toluene, 100°, 30 min
2. 4-MeOC$_6$H$_4$NHMe

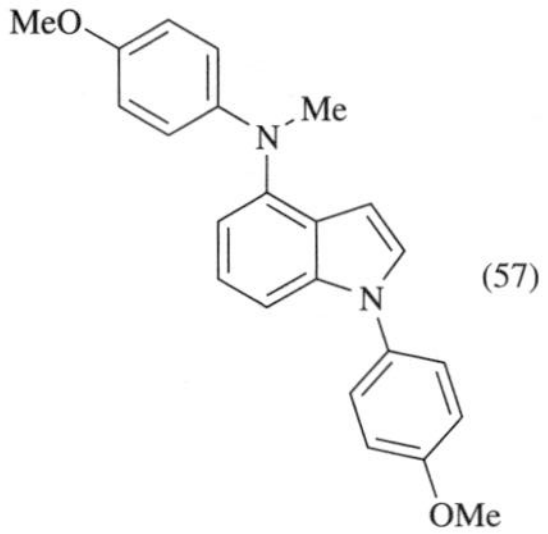

190

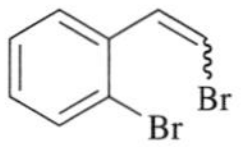

RNH$_2$, Pd$_2$(dba)$_3$, ligand, t-BuONa, toluene

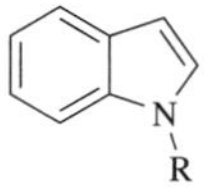

[Indole with N-R, structure I]

R	Ligand	Temp (°)	Time (h)	
Ph	SPhos	80	5	(85)
c-C$_6$H$_{11}$	DPEPhos	100	20	(65)
4-ClC$_6$H$_4$	SPhos	80	5	(77)
Bn	DPEPhos	100	20	(70)
4-MeOC$_6$H$_4$	SPhos	80	6	(75)
N=C(Ph)$_2$	SPhos	80	6	(62)

190

RNH$_2$, Pd$_2$(dba)$_3$, PhXPhos, Cs$_2$CO$_3$, 1,4-dioxane, 110°

[Indole with N-R, structure I]

R	Time (h)	
CO$_2$Et	24	(54)
CO$_2$t-Bu	10	(68)

190

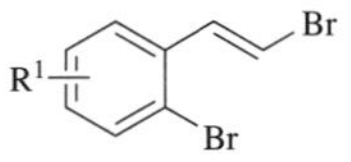

R^{2}NH$_2$, Pd$_2$(dba)$_3$, SPhos, Cs$_2$CO$_3$, toluene, 110°, 6 h

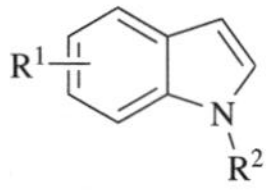

[Indole with R^1 and N-R^2]

R^1	R^2	
6-Cl	N-morpholino	(68)
5-Cl	Boc	(69)
6-Cl	Boc	(84)
7-Cl	Boc	(68)
6-Cl	4-MeOC$_6$H$_4$	(87)
6-Cl	4-MeOC$_6$H$_4$CH$_2$	(67)

192

TABLE 9. 2,3-UNSUBSTITUTED INDOLES VIA N-VINYLATION AND N-ARYLATION (Continued)

Substrate	Conditions	Product(s) and Yield(s) (%)	Refs.
C₈			
(2-Br, 6-Cl-phenyl vinyl bromide)	PhNH₂, Pd₂(dba)₃, DavePhos, t-BuONa, toluene, 100°	4-Cl-1-Ph-indole (80)	190
(dibromo-dichloro styrene)	BocNH₂, Pd₂(dba)₃, SPhos, Cs₂CO₃, toluene, 110°, 6 h	5-Cl-7-Cl-1-Boc-indole (39)	192
C₈₋₉			
(R-substituted 2-Br, vinyl chloride)	 H₂N-C(CH₃)₂Et, Pd(OAc)₂, HP(t-Bu)₃BF₄, t-BuONa, toluene, 130°, 4 h	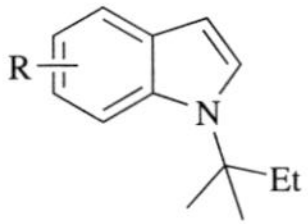 R: 5-F (65); 5-NO₂ (34); 6-Me (76) (1-t-amyl-indole)	191
	PhNH₂, Pd₂(dba)₃, SPhos, t-BuONa, toluene, 80°	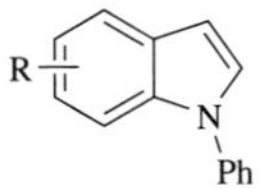 R: 5-F (59); 6-Me (73) (1-Ph-indole)	190
C₉			
(methylenedioxy bromo vinyl chloride)	 H₂N-C(CH₃)₂Et, Pd(OAc)₂, HP(t-Bu)₃BF₄, t-BuONa, toluene, 130°, 4 h	(methylenedioxy-1-t-amyl-indole) (64)	191
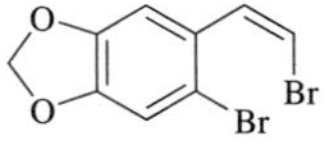 (methylenedioxy bromo vinyl bromide)	PhNH₂, Pd(OAc)₂, SPhos, t-BuONa, toluene, 80°	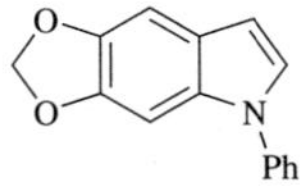 (methylenedioxy-1-Ph-indole) (67)	190

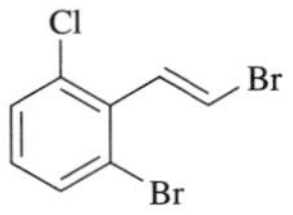

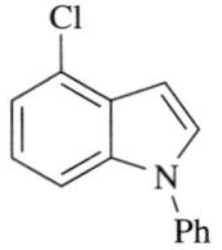

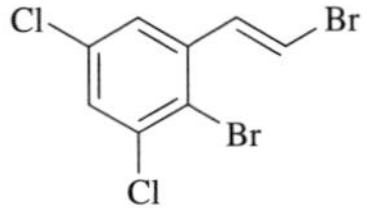

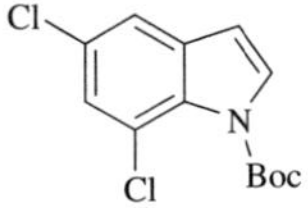

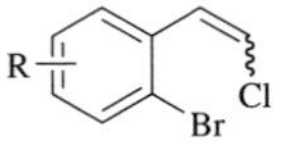

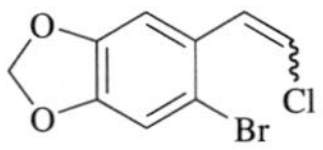

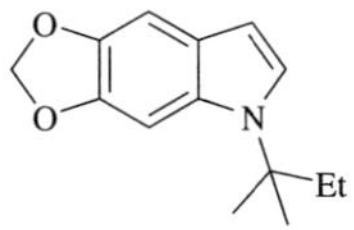

C$_{10}$	Pd(dba)$_2$, dmam-dtbpf, t-BuONa, o-xylene, 120°, 2–20 h	R	
		H	(73)
		4-F	(60)
		6-F	(74)
		4-Cl	(48)

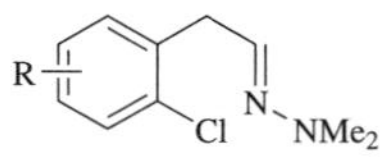

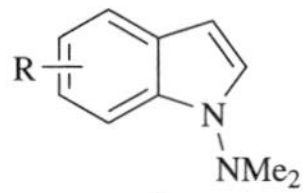

189

	Pd(dba)$_2$, (t-Bu)$_3$P, base, o-xylene, 120°, 2–20 h	R	Base	
		H	t-BuONa	(39)
		4-Cl	t-BuONa	(46)
		6-Cl	Rb$_2$CO$_3$	(18)

189

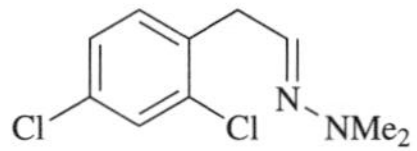

	Reagent, Pd(dba)$_2$, ligand, base, o-xylene, 120°, 24–48 h	Reagent	R	Ligand	Base	
		PhB(OH)$_2$	Ph	dmam-dtbpf	Cs$_2$CO$_3$	(29)
		pyrrole	1-pyrrolyl	(t-Bu)$_3$P	Rb$_2$CO$_3$	(24)

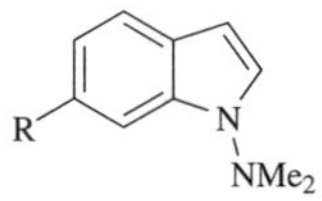

189

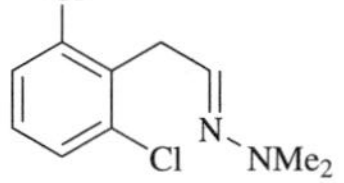

	Reagent, Pd(dba)$_2$, ligand, base, o-xylene, 120°, 24–48 h	Reagent	R	Ligand	Base	
		PhB(OH)$_2$	Ph	dmam-dtbpf	Cs$_2$CO$_3$	(56)
		PhB(OH)$_2$	Ph	(t-Bu)$_3$P	Cs$_2$CO$_3$	(40)
		pyrrole	1-pyrrolyl	(t-Bu)$_3$P	Rb$_2$CO$_3$	(54)
		indole	1-indolyl	(t-Bu)$_3$P	Rb$_2$CO$_3$	(40)
		piperazine	1-piperazinyl	dmam-dtbpf	t-BuONa	(30)
		PhMeNH	PhMeN	dmam-dtbpf	t-BuONa	(39)

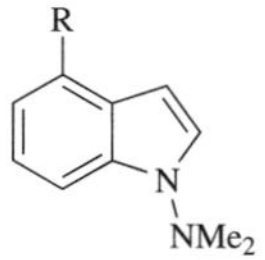

189

C$_{15}$	Piperazine, Pd(dba)$_2$, dmam-dtbpf, t-BuONa, o-xylene, 120°	(33)

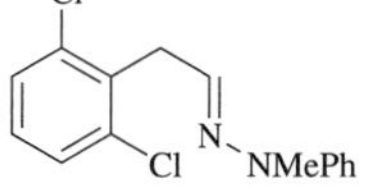

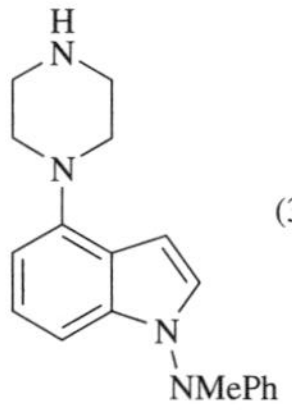

189

TABLE 9. 2,3-UNSUBSTITUTED INDOLES VIA *N*-VINYLATION AND *N*-ARYLATION (*Continued*)

Substrate	Conditions	Product(s) and Yield(s) (%)	Refs.
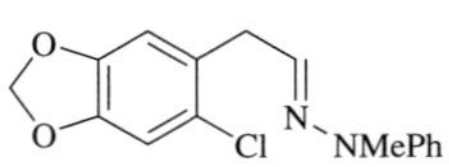 C₁₆	Pd(dba)₂, dmam-dtbpf, *t*-BuONa, *o*-xylene, 120°	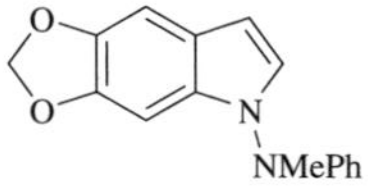 (33)	189

TABLE 10. 2-SUBSTITUTED INDOLES VIA N-VINYLATION AND N-ARYLATION

Substrate	Conditions	Product(s) and Yield(s) (%)	Refs.
C8 (2,2-dichlorovinyl aniline)	PhB(OH)$_2$, Pd(OAc)$_2$, SPhos, K$_3$PO$_4$•H$_2$O, toluene, 100°, 2 h	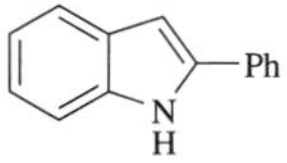 2-Ph indole (95)	200, 202
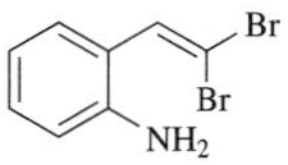 (2,2-dibromovinyl aniline)	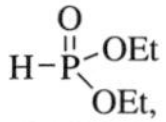 H–P(O)(OEt)$_2$, Pd(OAc)$_2$, dppf, Et$_3$N, toluene, 120°, 12 h	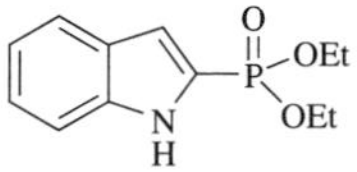 2-P(O)(OEt)$_2$ indole (63)	198
	CO$_2$-t-Bu vinyl, Pd(OAc)$_2$, P(o-tol)$_3$, K$_3$PO$_4$•H$_2$O, Et$_3$N, toluene, reflux	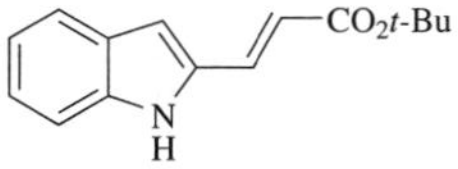 2-(CH=CH-CO$_2$-t-Bu) indole (50)	199
	RB(OH)$_2$, Pd(OAc)$_2$, SPhos, K$_3$PO$_4$•H$_2$O, toluene, 90°	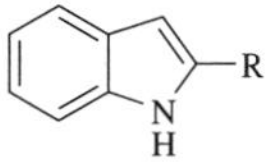 2-R indole R / Time (h) / (%) 3-thienyl / 12 / (86) Ph / 6 / (84) CH=CH-t-Bu / 5 / (80) 2-MeC$_6$H$_4$ / 4 / (82) 4-CF$_3$C$_6$H$_4$ / 7 / (83) 4-MeOC$_6$H$_4$ / 2 / (75)	200
	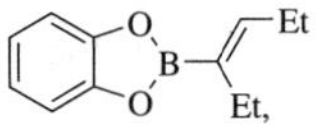 (Et-substituted vinyl boronate), Pd(OAc)$_2$, SPhos, K$_3$PO$_4$•H$_2$O, toluene, 90°, 6 h	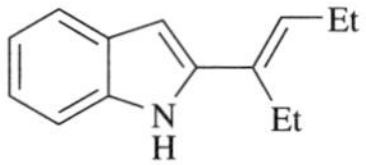 2-(C(Et)=CHEt) indole (73)	200
	Et$_3$B, Pd(OAc)$_2$, SPhos, K$_3$PO$_4$•H$_2$O, THF, 60°, 2 h	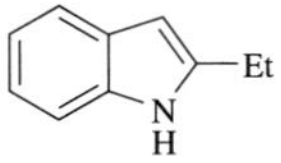 2-Et indole (77)	200

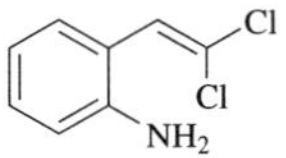

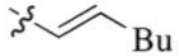

TABLE 10. 2-SUBSTITUTED INDOLES VIA N-VINYLATION AND N-ARYLATION (Continued)

Substrate	Conditions	Product(s) and Yield(s) (%)	Refs.
C8 (2-(1,2-dibromovinyl)aniline)	n-C6H13BBN, Pd(OAc)2, SPhos, K3PO4•H2O, THF, 60°, 3 h	2-(n-C6H13)-indole (79)	200
	BnO(CH2)4BBN, Pd(OAc)2, SPhos, K3PO4•H2O, THF, 60°, 4 h	2-((CH2)4OBn)-indole (78)	200
	≡—R, Pd/C, CuI, P(4-MeOC6H4)3, i-Pr2NH/toluene, 100°	2-(alkynyl)indole R CH2OH (40) (CH2)3OH (83) TMS (57) (CH2)3CN (50) (CH2)4Cl (70) R 3-pyridyl (81) Ph (85) n-C6H13 (71) (CH2)3OTHP (71)	205
	RB(OH)2, CO (12 atm), Pd(PPh3)4, K2CO3, 1,4-dioxane, 100°, 16 h	2-acyl indole R 3-thienyl (67) 4-ClC6H4 (70) 2-MeOC6H4 (40) 4-MeOC6H4 (61) 4-CF3C6H4 (73) 4-MeCONHC6H4 (29) (E)-PhCH=CH (67) 2-benzofuryl (58) R 3,4,5-(MeO)3C6H2 (63) 1-isoquinolinyl (40) 2-Np (61) 4-dibenzofuranyl (71)	203
	RB(OH)2, Pd(OAc)2, SPhos, K3PO4•H2O, toluene, 90°, 4 h	2-R-indole R 3-thienyl (86) Me(CH2)3CH=CH (80) Ph (84) 3-ClC6H4 (57) 4-ClC6H4 (60) 2-MeC6H4 (82) 4-MeC6H4 (88) R 4-MeOC6H4 (83) 4-CF3C6H4 (75) 4-MeCOC6H4 (23) (E)-PhCH=CH (68) 2-Me-3-MeOC6H3 (79) 2-Np (82)	202

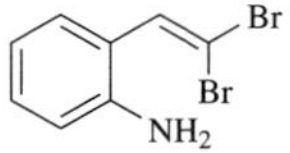

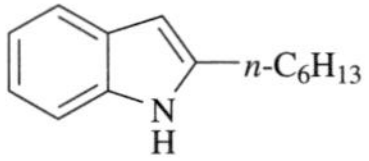

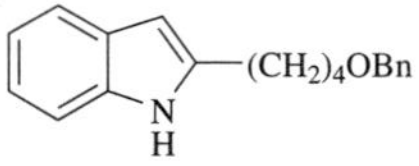

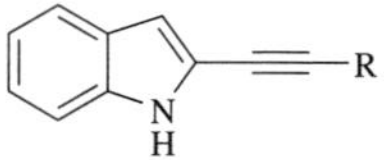

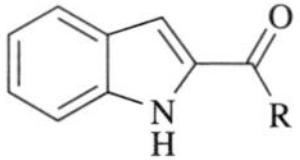

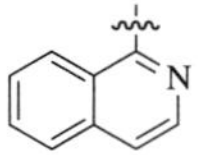

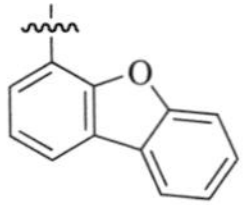

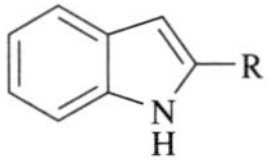

Starting material	Conditions	Product	R	Time (h)		Refs.
C_{8-15} (vinyl dibromide aniline, R substituted)	PhB(OH)$_2$, Pd(OAc)$_2$, SPhos, K$_3$PO$_4$•H$_2$O, toluene, 90°	2-Ph indole (I)	4-F 5-F 6-F 4-Me 6-CF$_3$ 4-CO$_2$Me 5-OBn	14 2 2.5 2 2.5 8.5 3	(88) (87) (80) (77) (90) (90) (86)	200
	PhB(OH)$_2$, Pd(OAc)$_2$, SPhos, K$_3$PO$_4$•H$_2$O, toluene, 90°, 4 h	I	4-F 5-F 6-F 4-Me 7-Me		(88) (87) (80) (77) (89)	202
			6-CF$_3$ 5-MeO$_2$C 6-MeO$_2$C 5-BnO 6-BnO		(90) (87) (90) (86) (72)	
	CO (10 atm), PdCl$_2$(PPh$_3$)$_2$, PPh$_3$, i-Pr$_2$NEt, THF/MeOH (1:1), 110°, 20 h	2-CO$_2$Me indole	R H 5-Cl 5-F 7-MeO		(70) (68) (78) (75)	204
			6-MeO$_2$C 5,6-(MeO)$_2$ 5-BnO		(73) (62) (72)	
	Pd(OAc)$_2$, t-Bu$_3$P•HBF$_4$, K$_2$CO$_3$, toluene, 100°	2-Br indole	R H 5-I 5-Br 6-F 7-MeO	Time (h) 14 24 24 14 14	(81) (68) (71) (82) (80)	207
			6-CF$_3$ 5-MeO$_2$C 6-MeO$_2$C 6-BnO	Time (h) 14 14 14 14	(72) (75) (84) (84)	
	PhB(OH)$_2$, CO (12 atm), Pd(PPh$_3$)$_4$, K$_2$CO$_3$, 1,4-dioxane, 85°, 24 h	2-benzoyl indole	R 4-Cl 5-MeO$_2$C 5-BnO		(68) (50) (73)	203

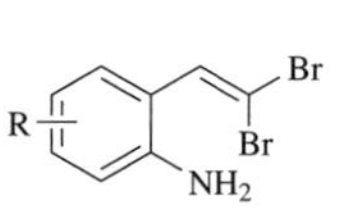

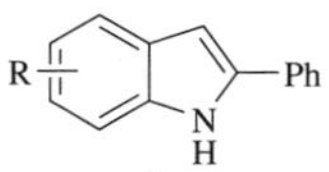

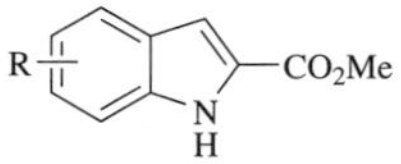

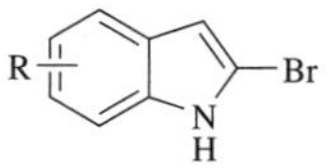

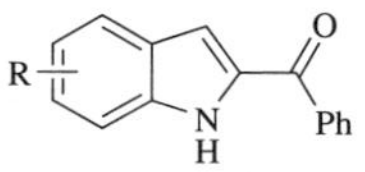

TABLE 10. 2-SUBSTITUTED INDOLES VIA N-VINYLATION AND N-ARYLATION (Continued)

Substrate	Conditions	Product(s) and Yield(s) (%)	Refs.

C_{8–16}

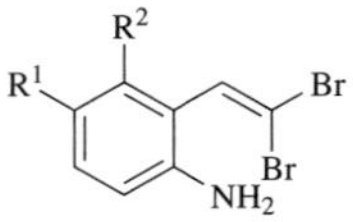

Substrate: (2-bromo-vinyl aniline with R¹, R² substituents, Br, NH₂)

Conditions: ≡≡—n-C₆H₁₃, Pd/C, CuI, P(4-MeOC₆H₄)₃, i-Pr₂NH/toluene, 100°

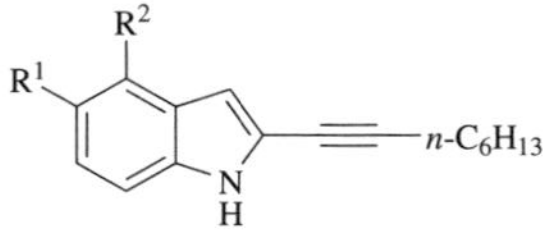

Product: 2-(n-C₆H₁₃)-indole with R¹, R²

R¹	R²	(yield)
H	F	(72)
MeO₂C	H	(84)
MeO	BnO	(81)

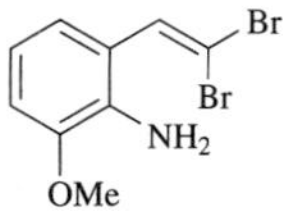

Refs.: 205

C₉

Substrate: (2-chlorostyrene with CF₃ group, X)

Conditions: RNH₂, Pd₂(dba)₃, DavePhos, t-BuONa, toluene, 140°

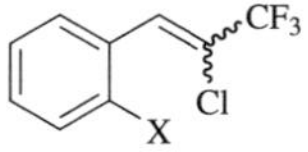

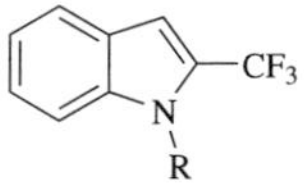

Product: 2-CF₃-N-R-indole

R	X	Time (h)	(%)
Bu	Br	26	(36)
Bu	Cl	14	(32)
4-pyridyl	Br	24	(24)
Ph	Cl	18	(83)
4-FC₆H₄	Br	14	(87)
4-FC₆H₄	Cl	12	(87)
4-ClC₆H₄	Br	32	(64)
4-O₂NC₆H₄	Br	36	(41)
2-MeC₆H₄	Br	12	(76)
2-MeC₆H₄	Cl	18	(72)
4-MeC₆H₄	Br	18	(94)

R	X	Time (h)	(%)
4-MeC₆H₄	Br	26	(96)
4-MeC₆H₄	Cl	16	(76)
4-MeOC₆H₄	Br	10	(84)
4-CF₃C₆H₄	Br	16	(68)
4-CF₃C₆H₄	Cl	24	(57)
Bn	Br	48	(—)
2,6-Me₂C₆H₃	Br	6	(62)
Ph(CH₂)₂	Br	10	(44)
1-Np	Br	16	(67)
1-Np	Cl	36	(71)

Refs.: 399

Substrate: (2-bromo-vinyl aniline with OMe, Br, NH₂)

Conditions: ≡≡—CO₂t-Bu, Pd(OAc)₂, P(o-tol)₃, K₃PO₄·H₂O, Et₃N, toluene, reflux

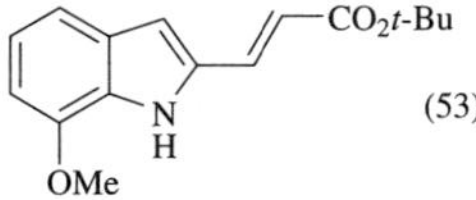

Product: 7-OMe-2-(CH=CH-CO₂t-Bu)-indole (53)

Refs.: 199

Substrate	Conditions	Product(s) and Yield(s) (%)	Refs.
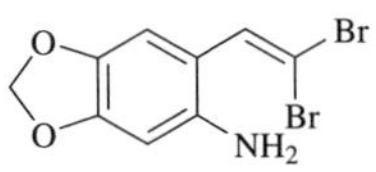 C9-15 ![vinyl dibromide aniline methylenedioxy]	PhB(OH)2, CO (12 atm), Pd(PPh3)4, K2CO3, 1,4-dioxane, 85°, 24 h	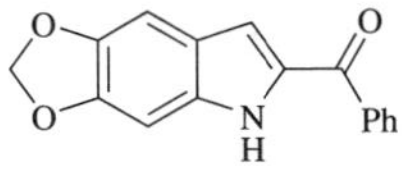 ![indole with C(O)Ph, methylenedioxy] (56)	203
	PdCl2(PPh3)2, PPh3, CO (10 atm), i-Pr2NEt, THF/MeOH (1:1), 110°, 20 h	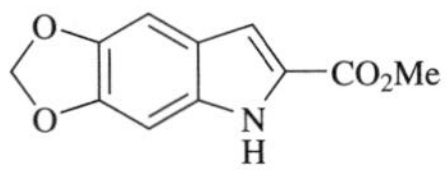 ![indole CO2Me, methylenedioxy] (67)	204
![styryl bromide with o-Br]	PhNH2, Pd2(dba)3, SPhos, t-BuONa, toluene	![N-Ph indole, R] R / Temp (°) Me / 80 (79) 4-ClC6H4 / 100 (51) 4-MeOC6H4 / 100 (81)	190
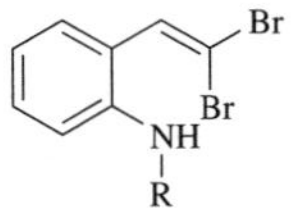 ![vinyl dibromide NHR aniline]	≡—n-C6H13, Pd/C, CuI, P(4-MeOC6H4)3, i-Pr2NH/toluene, 100°	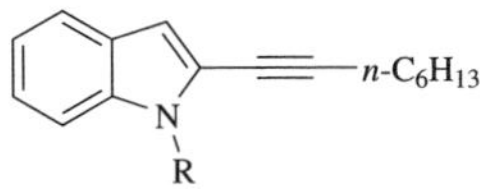 ![2-alkynyl N-R indole] R Me (61) i-Pr (60) Ph (65) Bn (51)	205
C10 ![MeO2C styryl bromide aniline]	![quinoline boronic acid with OMe], Pd(OAc)2, SPhos, K3PO4·H2O, toluene, rt, 2 min; 100°, 1.5 h	![2-(2-methoxyquinolin-3-yl)-6-CO2Me-indole] (86)	201
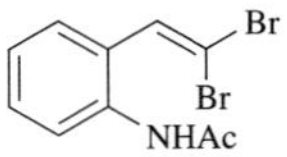 ![styryl bromide NHAc]	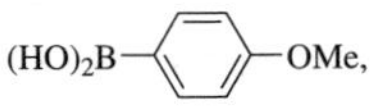 (HO)2B—C6H4—OMe, Pd2(dba)3, dppf, Et3N, toluene, 100°, 12 h	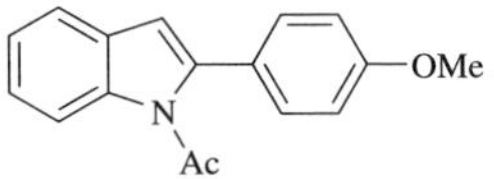 ![N-Ac 2-(4-methoxyphenyl)indole] (52)	198

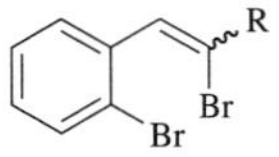

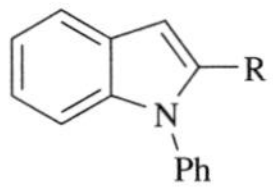

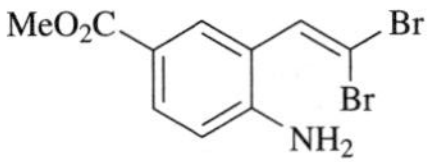

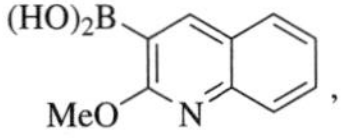

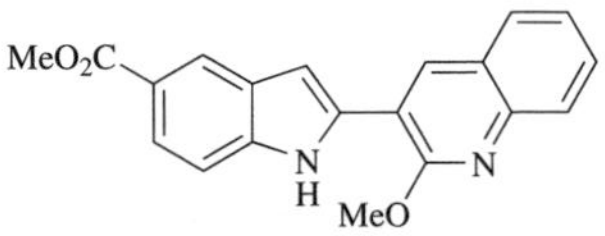

TABLE 10. 2-SUBSTITUTED INDOLES VIA N-VINYLATION AND N-ARYLATION (Continued)

Substrate	Conditions	Product(s) and Yield(s) (%)	Refs.
C_{10} 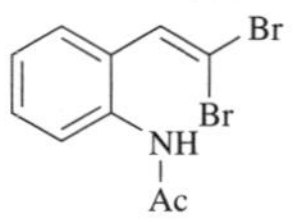 2-bromo-vinyl N-Ac aniline	$RB(OH)_2$, $Pd_2(dba)_3$, $P(o\text{-tol})_3$, K_2CO_3, toluene, 85°, 18 h	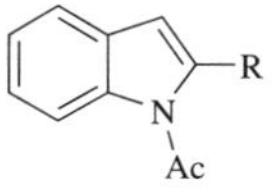 R: Ph (72); 2-MeC$_6$H$_4$ (17); 4-MeCOC$_6$H$_4$ (23)	202
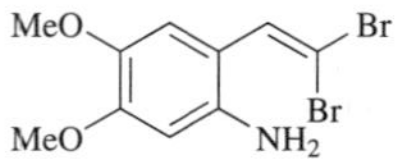 4,5-dimethoxy substrate	$PhB(OH)_2$, CO (12 atm), $Pd(PPh_3)_4$, K_2CO_3, 1,4-dioxane, 85°, 24 h	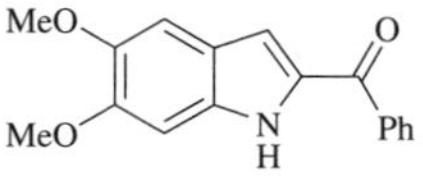 5,6-(MeO)$_2$-2-(PhCO)-indole (55)	203
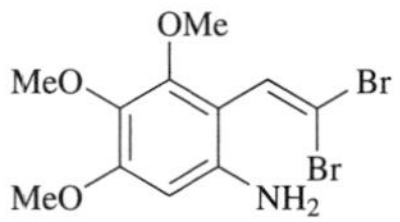 3,4-dimethoxy + OMe substrate	$PhB(OH)_2$, CO (12 atm), $Pd(PPh_3)_4$, K_2CO_3, 1,4-dioxane, 85°, 24 h	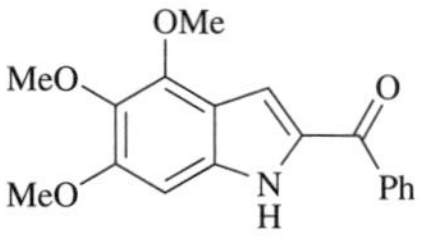 4,5,6-trimethoxy-2-(PhCO)-indole (65)	203
C_{11-15} 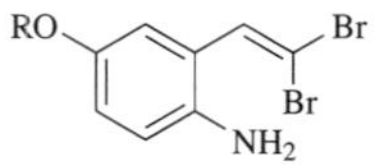 RO-substrate	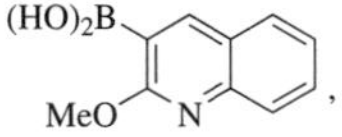 $(HO)_2B$-(2-MeO-quinolin-3-yl), $Pd(OAc)_2$, SPhos, $K_3PO_4 \cdot H_2O$, toluene, 100°	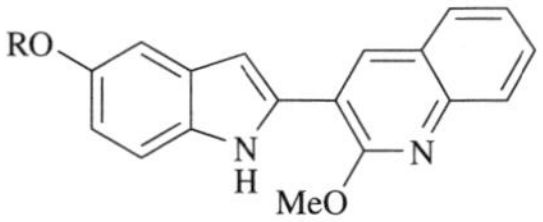  R: CH$_2$CH$_2$OMe (79); (CH$_2$)$_2$NMe(CH$_2$)$_2$OMe (88); CH$_2$CH$_2$N(piperidinyl) (94)	201
C_{11-16} 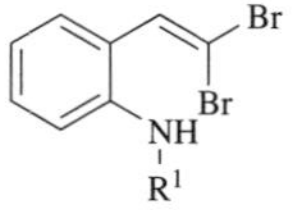 N-R^1 substrate	$R^2B(OH)_2$, $Pd(OAc)_2$, SPhos, $K_3PO_4 \cdot H_2O$, toluene, 90–100°, 4 h	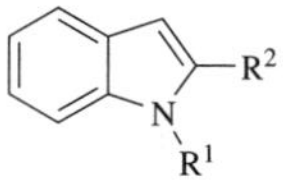 R^1 / R^2: i-Pr, Ph (71); Ph, Ph (92); Ph, 4-FC$_6$H$_4$ (86); Ph, 3,4-(MeO)$_2$C$_6$H$_3$ (60); 4-FC$_6$H$_4$, Ph (71); Bn, Ph (82); 4-CF$_3$C$_6$H$_4$, 2-FC$_6$H$_4$ (82); 3,4-(MeO)$_2$C$_6$H$_3$, 4-CF$_3$C$_6$H$_4$ (81)	202

C13	Br, PhNH2, Pd2(dba)3, XPhos, t-BuONa, 1,4-dioxane, 110°, 14 h	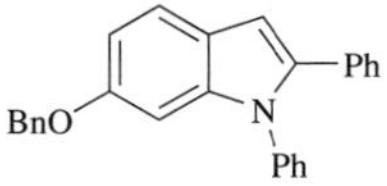 (57) 196
	CO2t-Bu, Pd2(dba)3, SPhos, K3PO4·H2O, Et3N, toluene, reflux	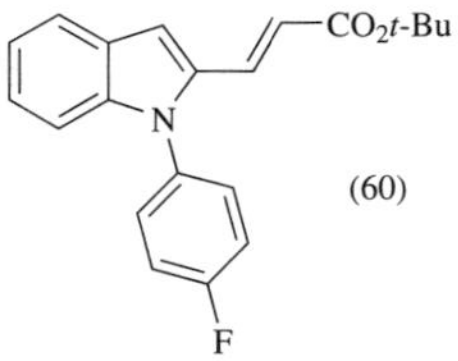 (60) 199
C14	PhB(OH)2, Pd(OAc)2, SPhos, K3PO4·H2O, toluene, 100°	(90) 200
	PhB(OH)2, Pd(OAc)2, SPhos, K3PO4·H2O, toluene, 100°	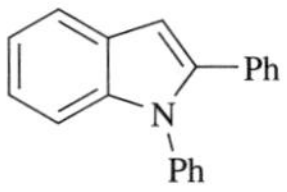 (90) 200
	(HO)2B—C6H4F, Pd(OAc)2, SPhos, K3PO4·H2O, toluene, 100°	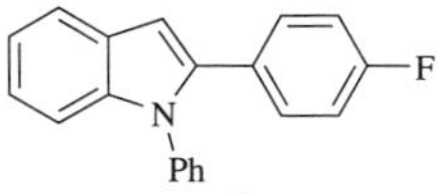 (86) 200
	Pd(OAc)2, (t-Bu)3P·HBF4, K2CO3, toluene, 100°, 14 h	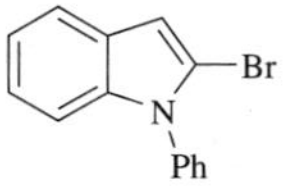 (72) 207

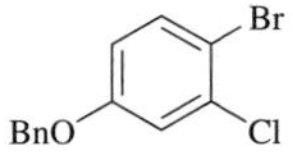

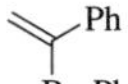

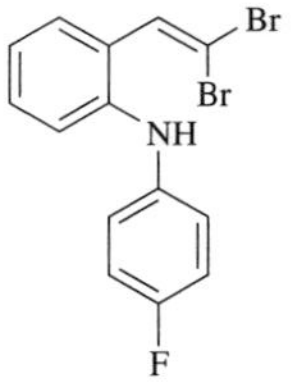

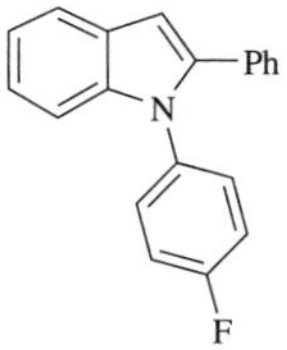

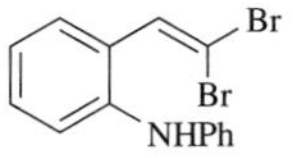

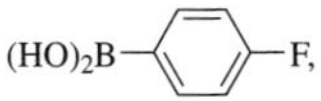

TABLE 10. 2-SUBSTITUTED INDOLES VIA N-VINYLATION AND N-ARYLATION (Continued)

Substrate	Conditions	Product(s) and Yield(s) (%)	Refs.
C₁₅	=R, Pd(OAc)₂, P(o-tol)₃, K₃PO₄•H₂O, Et₃N, toluene, reflux	R: CN (70); 4-MeOC₆H₄ (73)	199
	n-C₇H₁₅, OH, Pd(OAc)₂, Me₄NCl, K₃PO₄•H₂O, Et₃N, toluene, reflux	I + II (82), I:II = 75:25	199
	PhB(OH)₂, Pd(OAc)₂, SPhos, K₃PO₄•H₂O, toluene, 90°, 4 h	(82)	200
C₁₅₋₂₃	=CO₂t-Bu, Pd(OAc)₂, Me₄NCl, K₃PO₄•H₂O, Et₃N, toluene, reflux	R¹: H, H, 3-OBn; R²: 6-F (70), 6-CO₂Me (71), 4-OMe (39)	199
C₁₆	=CO₂t-Bu, Pd₂(dba)₃, SPhos, K₃PO₄•H₂O, Et₃N, toluene, reflux	(60)	199
	PhB(OH)₂, Pd(OAc)₂, SPhos, K₃PO₄•H₂O, toluene	Temp (°): 100, 90; Time (h): 2 (72), 4 (72)	200, 202

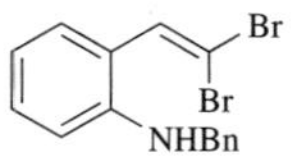

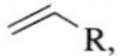

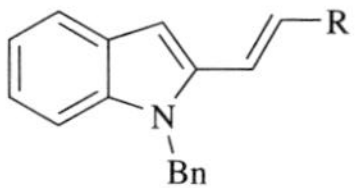

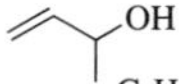

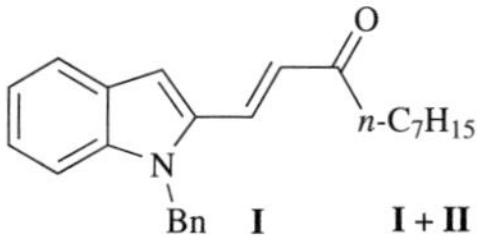

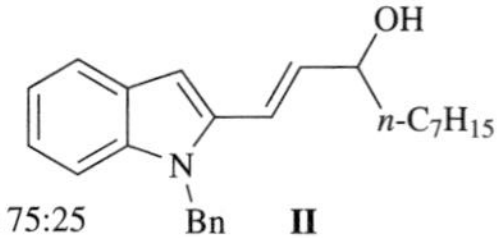

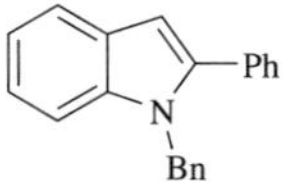

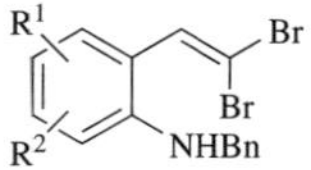

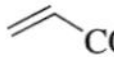

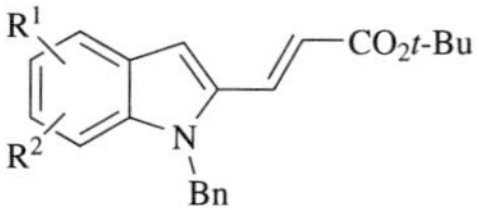

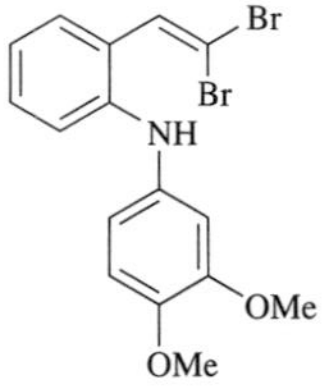

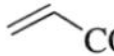

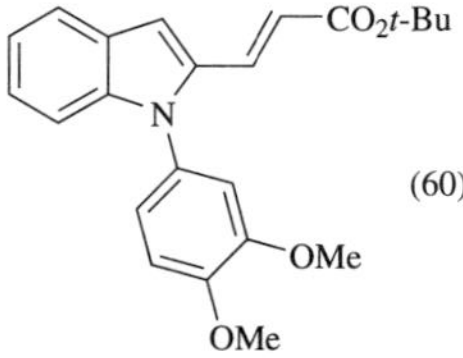

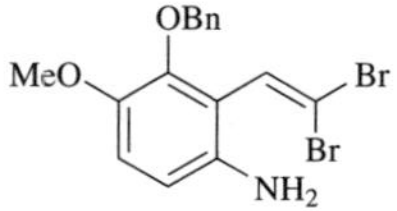

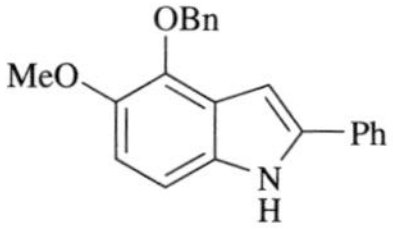

Substrate	Conditions	Product	Ref.
C$_{17-19}$![structure with MeO, CO$_2$Me, HN-C(O)R, I]	PdCl$_2$(dppf), KOAc, DMF; 90°, 4 h	![indole with MeO, CO$_2$Me, N-C(O)R] R: 2-Cl-3-pyridyl (91), Ph (45)	194
		OBn (85)	
C$_{17-21}$![structure with R^1, CO$_2$Et, HN-Ar(R^2), I]	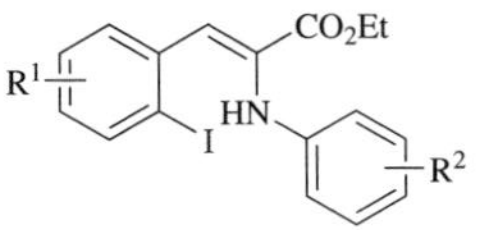 PdCl$_2$(dppf), KOAc, DMF; 90°, 4 h	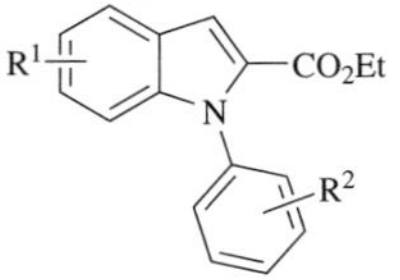 ![indole with R^1, CO$_2$Et, N-Ar(R^2)] R^1 / R^2 5-NO$_2$ / 2-Br (90) 6-NO$_2$ / 2-Br (83) H / 3,4-OCH$_2$O (92) 6-NO$_2$ / 3,4-OCH$_2$O (94) 5-OMe / 3,4-OCH$_2$O (90) H / 4-CO$_2$Et (89) 6-NO$_2$ / 4-CH$_2$CO$_2$Et (93) 5-OMe / 4-CO$_2$Et (94)	194
C$_{19}$![naphthalene with vinyl-Br, Br, NHBn]	∕=∕CO$_2$t-Bu, Pd(OAc)$_2$, P(o-tol)$_3$, K$_3$PO$_4$•H$_2$O, Et$_3$N, toluene, reflux	![benzoindole with CO$_2$t-Bu, N-Bn] (43)	199
C$_{22}$![BnO-aryl with vinyl-Br, Br, NHBn]	∕=∕CO$_2$t-Bu, Pd(OAc)$_2$, P(o-tol)$_3$, K$_3$PO$_4$•H$_2$O, Et$_3$N, toluene, reflux	![indole with CO$_2$t-Bu, BnO, N-Bn] (64)	199
![t-Bu aryl with vinyl-Br, Br, NH-Ph, t-Bu]	Pd(OAc)$_2$, (t-Bu)$_3$P•HBF$_4$, K$_2$CO$_3$, toluene, 100°, 14 h	![indole with Br, t-Bu, N-Ph, t-Bu] (81)	207

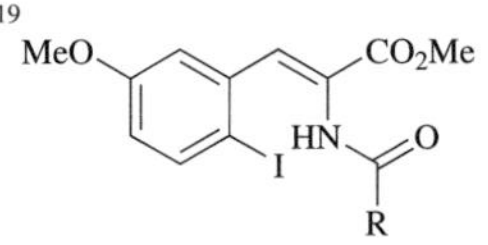

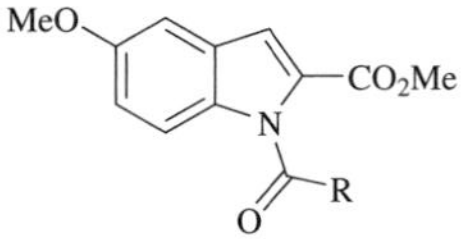

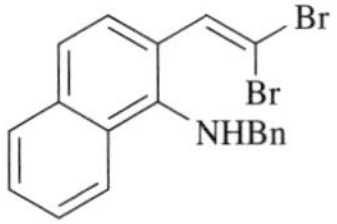

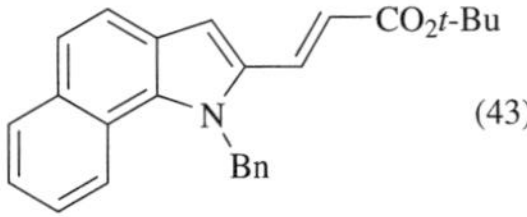

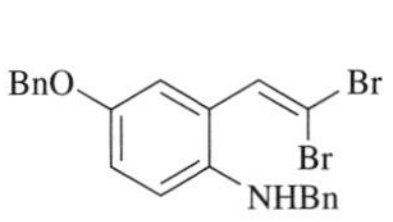

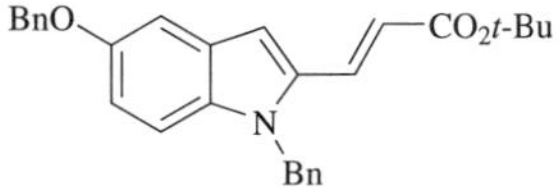

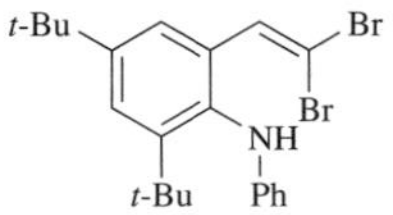

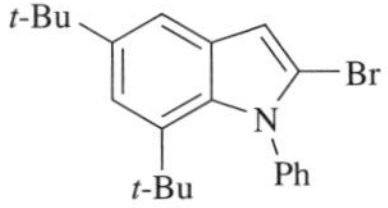

TABLE 10. 2-SUBSTITUTED INDOLES VIA *N*-VINYLATION AND *N*-ARYLATION (*Continued*)

Substrate	Conditions	Product(s) and Yield(s) (%)	Refs.
C_{22}			
MeO, OBn, Br, Br, NH, Ph substrate	Pd(OAc)$_2$, (*t*-Bu)$_3$P•HBF$_4$, K$_2$CO$_3$, toluene, 100°, 14 h	MeO, OBn, Br, N(Ph) indole (73)	207
BnO, BnO, Br, Br, NH$_2$ substrate	PhB(OH)$_2$, Pd(OAc)$_2$, SPhos, K$_3$PO$_4$•H$_2$O, toluene, 90°, 4 h	BnO, BnO, Ph, NH indole (57)	207

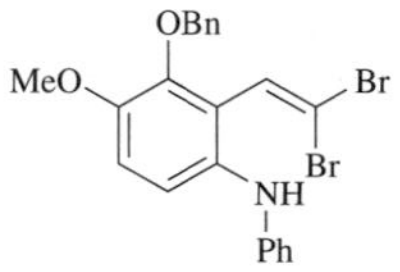

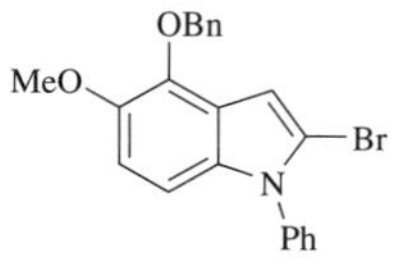

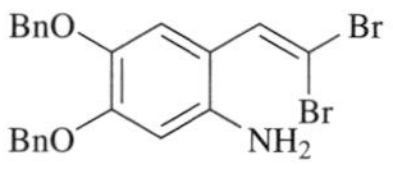

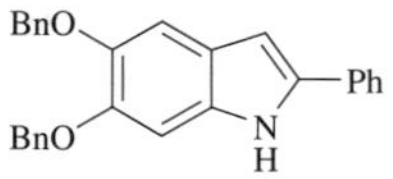

TABLE 11. 3-SUBSTITUTED INDOLES VIA N-VINYLATION AND N-ARYLATION

Substrate	Conditions	Product(s) and Yield(s) (%)	Refs.
C₉	PhNH₂, Pd₂(dba)₃, DavePhos, t-BuONa, toluene, 100°, 25 h	(73) I, N-Ph, 3-Me indole	190
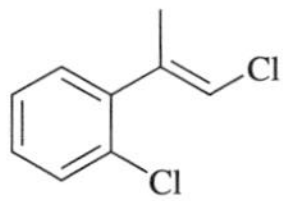	PhNH₂, Pd₂(dba)₃, DavePhos, t-BuONa, toluene, 100°, 25 h	I (82)	190
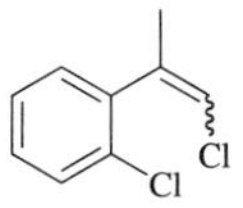	 H₂N–C(Et)(Me)₂, Pd(OAc)₂, HP(t-Bu)₃BF₄, t-BuONa, toluene, 130°, 4 h	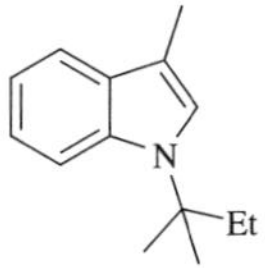 (66)	191
C₉₋₁₄ 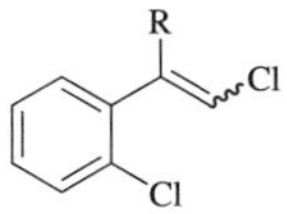	PhNH₂, Pd₂(dba)₃, DavePhos, t-BuONa, toluene, 100°	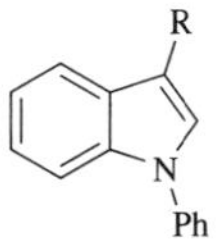 R / Me (78) / Ph (94)	190
C₁₁₋₁₅	BocNH₂, Pd₂(dba)₃, SPhos, Cs₂CO₃, toluene, 110°, 6 h	X / R¹ / R² : Br / 4-Cl / CO₂Et (61); Cl / 4-Cl / 3-furyl (61); Cl / 4-Cl / 4-MeOC₆H₄ (71); Br / 5-Cl / 4-MeOC₆H₄ (75); Br / 6-Cl / 4-MeOC₆H₄ (81); Br / 7-Cl / 4-MeOC₆H₄ (—)	192

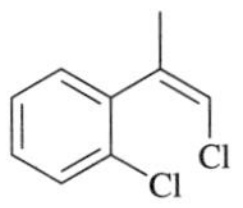

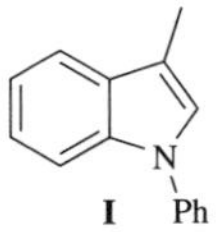

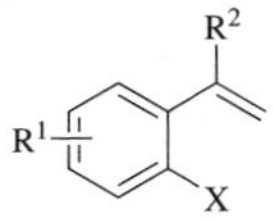

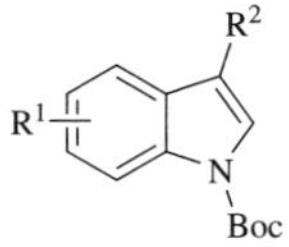

TABLE 12. 2,3-DISUBSTITUTED INDOLES VIA N-VINYLATION AND N-ARYLATION

Substrate	Conditions	Product(s) and Yield(s) (%)	Refs.
C$_9$ 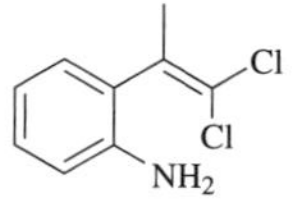 (2-chloro-propenyl aniline, Me, Cl)	PhB(OH)$_2$, Pd(OAc)$_2$, SPhos, K$_3$PO$_4$·H$_2$O, toluene, 100°, 2 h	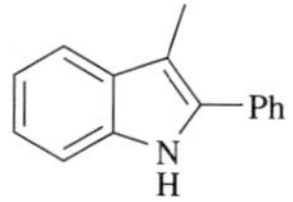 2-Ph-3-Me-indole (96)	200
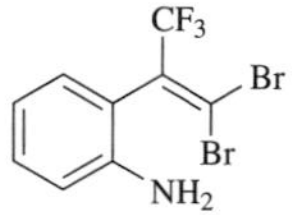 C$_{9-16}$ (2-bromo-CF$_3$-vinyl aniline)	PhB(OH)$_2$, Pd(OAc)$_2$, SPhos, K$_3$PO$_4$·H$_2$O, toluene, 100°, 1 h	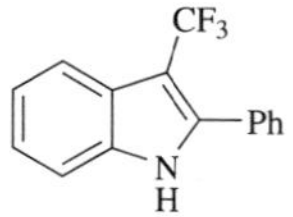 2-Ph-3-CF$_3$-indole (79)	200
C$_{10-17}$ (2-X-R-vinyl aniline)	PhB(OH)$_2$, Pd(OAc)$_2$, SPhos, K$_3$PO$_4$·H$_2$O, toluene, 90–100°, 4 h	2-Ph-3-R-indole: R / X CF$_3$ / Br (79) Me / Cl (96) 4-FC$_6$H$_4$ / Br (90) Ph—≡— / Br (77)	202
C$_{10-17}$ (3-Me-2-Cl-NH-R^2 substrate)	R^{3}B(OH)$_2$, Pd(OAc)$_2$, SPhos, K$_3$PO$_4$·H$_2$O, toluene, 90–100°, 4 h	R^1 / R^2 / R^3 5-O$_2$N / Me / Ph (90) H / Ph / 4-FC$_6$H$_4$ (96) H / 4-FC$_6$H$_4$ / Ph (94) H / 2-MeC$_6$H$_4$ / Ph (77) H / 4-CF$_3$C$_6$H$_4$ / 4-MeOC$_6$H$_4$ (79) H / 4-MeCOC$_6$H$_4$ / 2-MeC$_6$H$_4$ (75)	202
C$_{15}$ (2-chloro-propenyl-NHPh aniline) 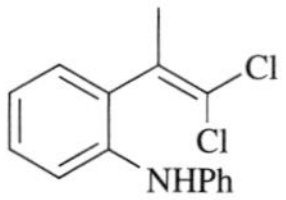	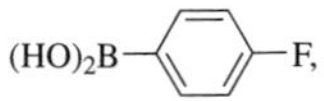 (HO)$_2$B–C$_6$H$_4$-F, Pd(OAc)$_2$, SPhos, K$_3$PO$_4$·H$_2$O, toluene, 100°	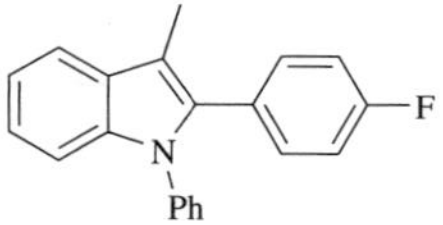 2-(4-FC$_6$H$_4$)-3-Me-1-Ph-indole (96)	200

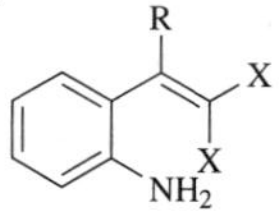

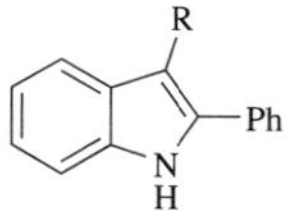

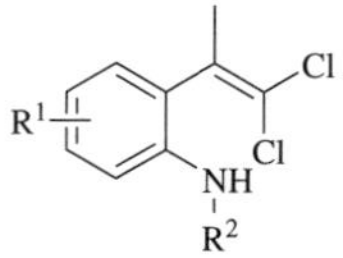

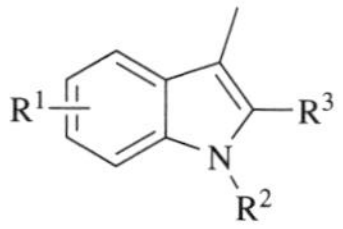

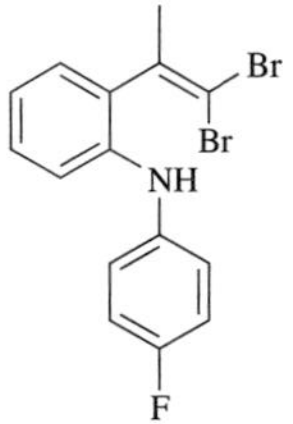

C₁₆	structure: 2,3-dibromo-N-(4-fluorophenyl)aniline derivative	PhB(OH)₂, Pd(OAc)₂, SPhos, K₃PO₄·H₂O, toluene, 100°	1-(4-fluorophenyl)-3-methyl-2-phenylindole (94)	200
	vinyl triflate with Ar and 2-bromophenyl	PhNH₂, Pd₂(dba)₃, DPEPhos, t-BuONa, toluene, 100°	2-Ar-3-methyl-1-phenylindole	210

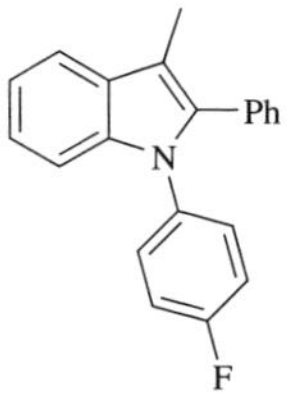

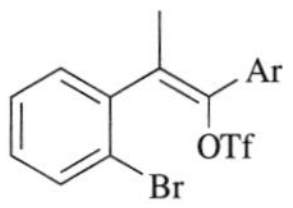

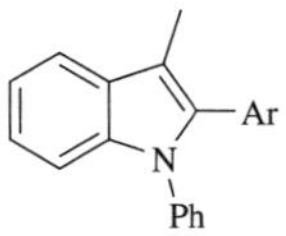

Ar	
Ph	(74)
4-FC₆H₄	(65)

TABLE 13. SOLID-PHASE SYNTHESIS OF INDOLES FROM ALKYNES

Substrate	Conditions	Product(s) and Yield(s) (%)[a]	Refs.
C₆ (substrate 1)	*Indole formation* TMS—≡—R, PdCl₂(PPh₃)₂, TMG, DMF, 110° *Cleavage* TFA, CH₂Cl₂, 15 min	(3-R-indole product) R % Mass Recovery CH₂CH₂OH 82 Ph 73	228
	Indole formation R¹—≡—R², PdCl₂(PPh₃)₂, TMG, DMF, 110° *Cleavage* TFA, CH₂Cl₂, 15 min	(2,3-disubstituted indole) R¹ R² % Mass Recovery t-Bu Me 55 Pr Pr 55 Ph Et 63[b] Ph Ph 97	228
C₆ (substrate 2)	*Indole formation* ≡—Ar¹, PdCl₂(PPh₃)₂, CuI, Et₃N, DMF, 70° *Bromination (at C3)* NBS, dioxane, 70°, 24 h *Suzuki cross-coupling* Ar²B(OH)₂, Pd(PPh₃)₄, K₂CO₃, DMF, 90°, 5–10 h *Cleavage* Bu₄NF, THF, 70°, 5 h	(2-Ar¹-3-Ar²-indole) Ar¹ Ar² % Purity 4-FC₆H₄ Ph (87) 91 4-FC₆H₄ 4-ClC₆H₄ (86) 92 4-FC₆H₄ 4-MeOC₆H₄ (86) 85 4-O₂NC₆H₄ 3-thienyl (93) 96 4-O₂NC₆H₄ 4-MeOC₆H₄ (87) 96 4-O₂NC₆H₄ 4-MeSC₆H₄ (99) 96 4-O₂NC₆H₄ 1-Np (91) 99 4-MeC₆H₄ 4-pyridyl (92) 84 4-MeC₆H₄ 4-MeOC₆H₄ (85) 93 4-MeC₆H₄ 4-MeO₂CC₆H₄ (94) 87	225

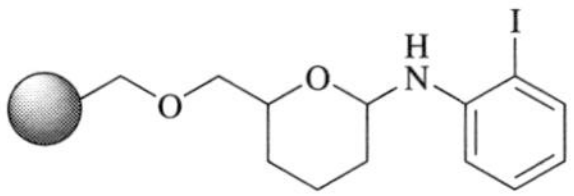

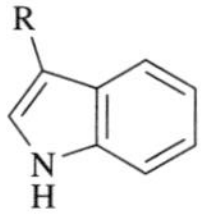

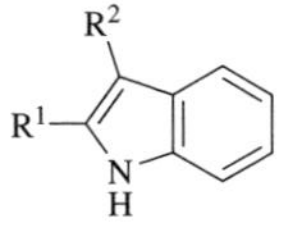

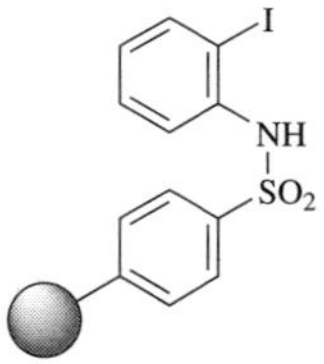

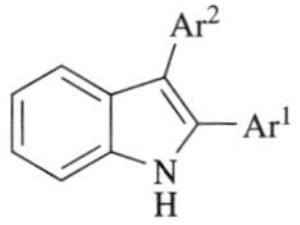

512

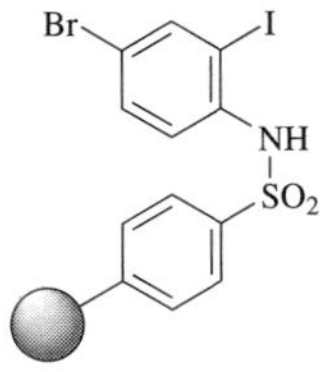

Indole formation
≡—TMS, PdCl$_2$(PPh$_3$)$_2$,
CuI, Et$_3$N, DMF, 70°
Bromination (at C2 and C3)
NBS, dioxane, 70°, 24 h
Suzuki cross-coupling
ArB(OH)$_2$, Pd(PPh$_3$)$_4$, K$_2$CO$_3$,
DMF, 90°, 5–10 h
Cleavage
Bu$_4$NF, THF, 70°, 5 h

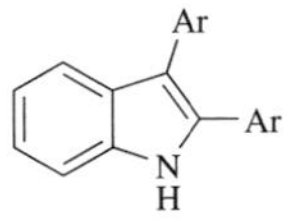

Ar		% Purity
4-MeC$_6$H$_4$	(81)	90
4-MeOC$_6$H$_4$	(84)	86

225

Indole formation
≡—R^1, PdCl$_2$(PPh$_3$)$_2$,
CuI, Et$_3$N, DMF, rt, 24 h
Acylation (at C3)
R^{2}COCl, AlCl$_3$, CH$_2$Cl$_2$, 12 h
Sonogashira cross-coupling (at C5)
≡—R^3, PdCl$_2$(PPh$_3$)$_2$,
CuI, Et$_3$N, DMF, 70°, 24 h
or *Suzuki cross-coupling*
ArB(OH)$_2$, PdCl$_2$(dppf), K$_2$CO$_3$,
dioxane, 90°, 24 h
Cleavage
t-BuOK, rt
Alkylation after cleavage
MeI

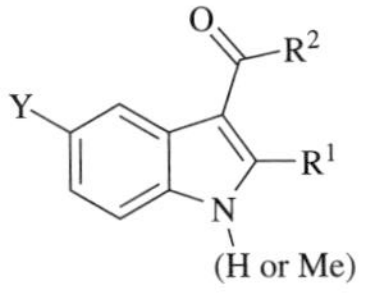

(10–20)

Y: R^3—≡—, Ar
R^1: Pr, Ph, 2-FC$_6$H$_4$, 4-FC$_6$H$_4$, 4-MeOC$_6$H$_4$, 4-MeC$_6$H$_4$
R^2: Me, *i*-Pr, *c*-Pr, Cy, Ph, 3-FC$_6$H$_4$, 4-FC$_6$H$_4$, 3-MeC$_6$H$_4$, 3-MeOC$_6$H$_4$,
4-MeOC$_6$H$_4$, 4-PhC$_6$H$_4$, 4-(*i*-Pr)C$_6$H$_4$, 1-Np
R^3: Pr, CH$_2$=C(Me), Ph, 2-FC$_6$H$_4$, 4-FC$_6$H$_4$, 4-MeOC$_6$H$_4$, 4-MeC$_6$H$_4$, Bn
Ar: Ph, 2-MeC$_6$H$_4$, 3,4-Cl$_2$C$_6$H$_3$, 3-F-4-MeC$_6$H$_3$, 2,3-Me$_2$C$_6$H$_3$,
4-(*i*-Pr)C$_6$H$_4$, 2-Cl-6-MeOC$_6$H$_3$, 2-NCC$_6$H$_4$, 4-PhOC$_6$H$_4$,
3,4-F$_2$C$_6$H$_3$, 2-Np,

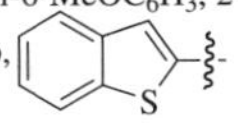

230

TABLE 13. SOLID-PHASE SYNTHESIS OF INDOLES FROM ALKYNES (Continued)

Substrate	Conditions	Product(s) and Yield(s) (%)[a]				Refs.
		R^1	R^2		% Purity	
C_{6-8}	*Indole formation*	H	Bu	(89)	92	
![substrate] R^1-C$_6$H$_3$(I)(NH-SO$_2$-C$_6$H$_4$-polymer)	≡≡—R^2, PdCl$_2$(PPh$_3$)$_2$, CuI, Et$_3$N, DMF, 60–70°, 5–16 h	H	MeOCH$_2$	(97)	93	229
		H	HOCH$_2$CH$_2$	(90)	—	
	Cleavage	H	(EtO)$_2$CH	(94)	—	
	Bu$_4$NF, THF, 70°, 5 h	H	Ph	(100)	95	
		H	4-MeC$_6$H$_4$	(97)	95	
		H	4-FC$_6$H$_4$	(100)	98	
		H	4-MeOC$_6$H$_4$	(95)	97	
		6-F	PhSCH$_2$	(85)	98	
		6-F	6-MeO-2-Np	(97)	95	
		6-OMe	MeOCH$_2$	(94)	86	
		6-OMe	4-O$_2$NC$_6$H$_4$	(90)	96	
		6-OMe	4-MeC$_6$H$_4$	(98)	85	
		6-OMe	4-MeOC$_6$H$_4$	(85)	91	
		5-CO$_2$Me	Bu	(87)	94	
		5-CO$_2$Me	2-pyridyl	(87)	100	
		5-CO$_2$Me	PhSCH$_2$	(85)	100	

		R	Ar		% Purity	
C_7	*Indole formation*	H	Ph	(87)	—	
![substrate with CONH-polymer, I, NH$_2$]	TMS—≡≡—, Pd(OAc)$_2$, PPh$_3$, Bu$_4$NI, Na$_2$CO$_3$, DMF, 80°, 5 h	H	4-MeC$_6$H$_4$	(65)	—	225
	Iodination (at C2)	H	4-MeOC$_6$H$_4$	(94)	85	
	NIS, CH$_2$Cl$_2$, rt, 2 h	H	1-Np	(96)	96	
	N-Alkylation	4-F-Bn	4-MeC$_6$H$_4$	(75)[c]	—	
	RBr, Cs$_2$CO$_3$, DMF	4-F-Bn	4-MeOC$_6$H$_4$	(74)[c]	—	
	Suzuki cross-coupling					
	ArB(OH)$_2$, Pd$_2$(dba)$_3$ or Pd(PPh$_3$)$_4$, K$_2$CO$_3$, DMF or DMF/H$_2$O, 80°, 6–20 h					
	Cleavage					
	TFA					

514

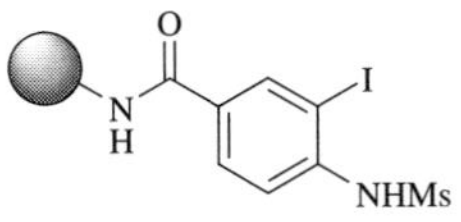

	Indole formation		
	R≡≡TMS, Pd(OAc)$_2$,		
	PPh$_3$, Bu$_4$NI, Na$_2$CO$_3$,		
	DMF, 80°, 6–16 h		
	Cleavage		
	TFA, CH$_2$Cl$_2$, 1 h		

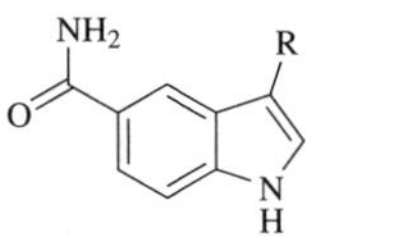

R		% Purity[c]	
Me	(77)	—	
CH$_2$CH$_2$OH	(95)	85	
CH$_2$CH$_2$Cl	(88)	96	226
Ph	(56)	—	
CH$_2$OCH$_2$-3-MeOC$_6$H$_4$	(90)	92	

Indole formation
R^1≡≡R^2, Pd(OAc)$_2$,
PPh$_3$, LiCl, K$_2$CO$_3$,
DMF, 80°, 15–22 h
Cleavage
TFA, CH$_2$Cl$_2$, 1 h

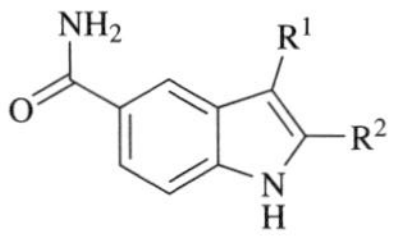

R^1	R^2		% Purity[c]	
Me	t-Bu	(87)	84	
Pr	Pr	(91)	82	226
Me	Ph	(86)	72	
CO$_2$Et	Ph	(38)[c]	—	
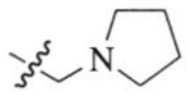 ⋯N(pyrrolidine)	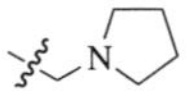 ⋯N(pyrrolidine)	(63)[d]	—	

Indole formation
≡≡R, PdCl$_2$(PPh$_3$)$_2$,
CuI, Et$_3$N,
DMF, 80°, 5–16 h
Cleavage
TFA, CH$_2$Cl$_2$, 1 h

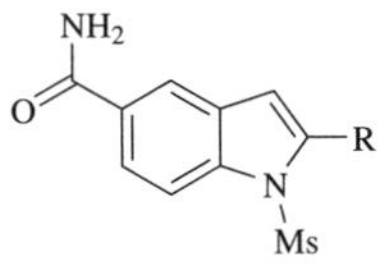

R		% Purity	
CH$_2$NMe$_2$	(86)	98	
n-C$_5$H$_{11}$	(90)	79	224
Ph	(87)	90	
Bn	(96)	90	

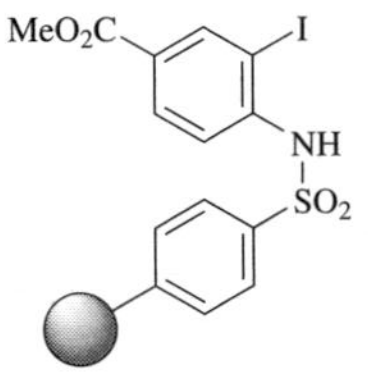

Indole formation
≡≡Ar^1, PdCl$_2$(PPh$_3$)$_2$,
CuI, Et$_3$N, DMF, 70°
Bromination (at C3)
NBS, dioxane, 70°, 24 h
Suzuki cross-coupling
Ar^{2}B(OH)$_2$, Pd(PPh$_3$)$_4$, K$_2$CO$_3$,
DMF, 90°, 5–10 h
Cleavage
Bu$_4$NF, THF, 70°

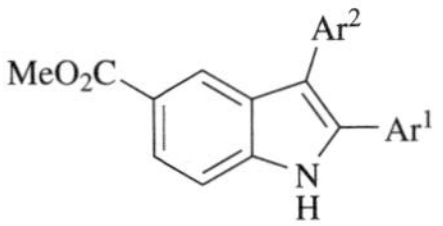

Ar^1	Ar^2		
Ph	3,4-OCH$_2$O-C$_6$H$_3$	(86)	225
Ph	2-Np	(85)	

TABLE 13. SOLID-PHASE SYNTHESIS OF INDOLES FROM ALKYNES (*Continued*)

Substrate	Conditions	Product(s) and Yield(s) (%)[a]	Refs.
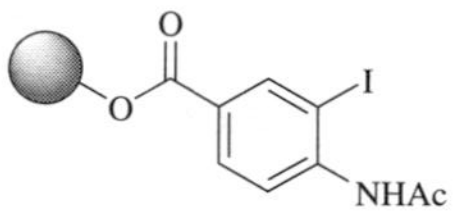 C₉ (resin-O-CO-C₆H₃(I)-NHAc)	*Indole formation* ≡≡—R, PdCl₂(PPh₃)₂, CuI, TMG, dioxane, 90°, 18 h *Cleavage* NaOH, *i*-PrOH, 50°, 5 h	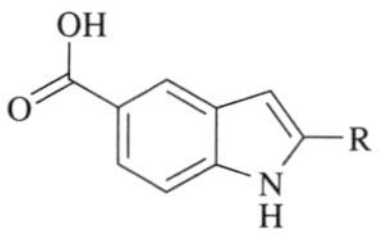 2-R-indole-5-carboxylic acid R *c* Pr (55) (CH₂)₃OH (95) *i*-Bu (82) Ph (72) 4-ClC₆H₄ (52) 4-MeOC₆H₄ (48) CH₂SPh (81)	222
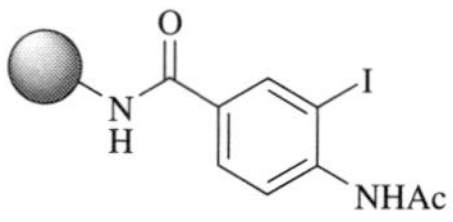 C₉ (resin-NH-CO-C₆H₃(I)-NHAc)	*Indole formation* Pr—≡≡—Pr, Pd(OAc)₂, PPh₃, Bu₄NCl, KOAc, DMF, 80°, 16 h *Cleavage* TFA, CH₂Cl₂, 1 h	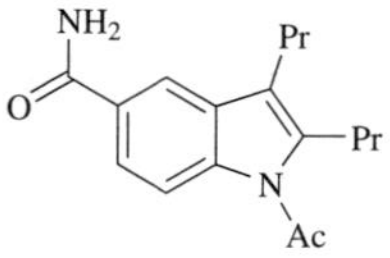 2,3-dipropyl-1-Ac-indole-5-carboxamide (95)	226
C₉₋₁₁ (resin-NH-CO-C₆H₃(I)-NHR)	*Indole formation* ≡≡—TMS, Pd(OAc)₂, PPh₃, Bu₄NCl, Na₂CO₃, DMF, 80°, 7–120 h *Cleavage* TFA, CH₂Cl₂, 1 h	3-methyl-1-R-indole-5-carboxamide R *c* Ac (93) COCHMe₂ (100)	226
C₁₁ (resin-NH-CO-C₆H₃(I)-NHCOCHMe₂)	*Indole formation* ≡≡—*t*-Bu, Pd(OAc)₂, PPh₃, LiCl, K₂CO₃, DMF, 80°, 120 h *Cleavage* TFA, CH₂Cl₂, 1 h	2-*t*-Bu-3-methyl-1-COCHMe₂-indole-5-carboxamide (75)	226

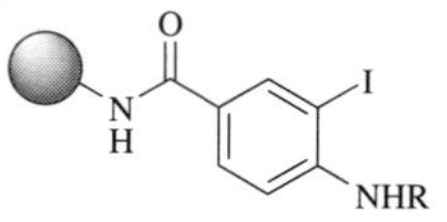

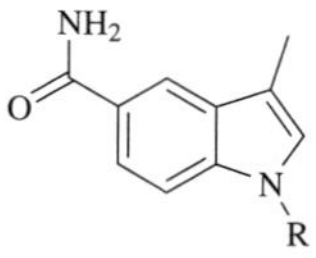

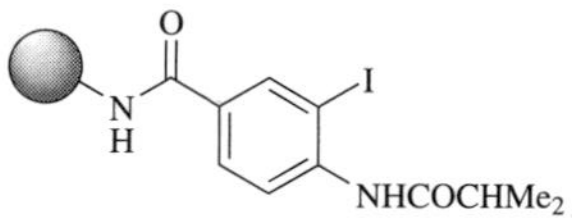

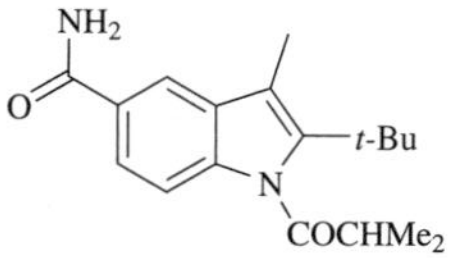

C_{15}

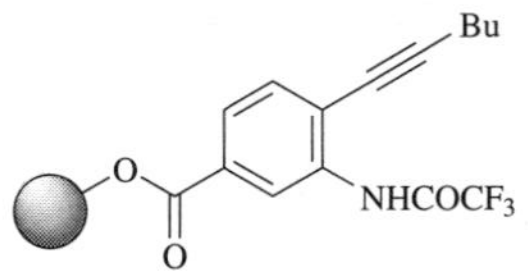

Indole formation 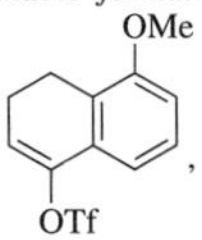
OMe, Pd(PPh$_3$)$_4$,
OTf
K$_2$CO$_3$, DMF, rt, 24 h
N-Alkylation
BnBr, NaH, DMF, rt, 4 h
Cleavage
TFA, CH$_2$Cl$_2$, 1 h

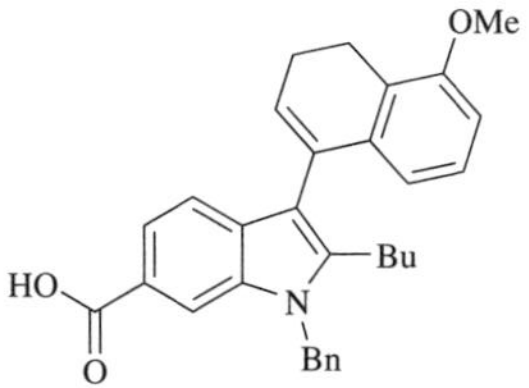

(81)   223

C_{15-22}

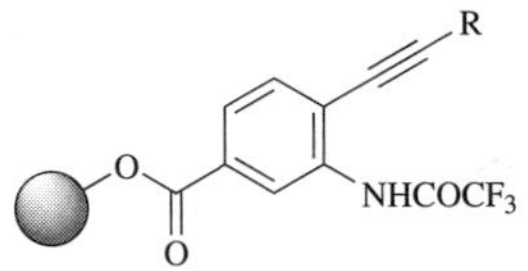

Indole formation 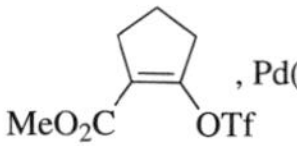
MeO$_2$C—OTf, Pd(PPh$_3$)$_4$,
K$_2$CO$_3$, DMF, rt, 24 h
Cleavage
TFA, CH$_2$Cl$_2$, 1 h

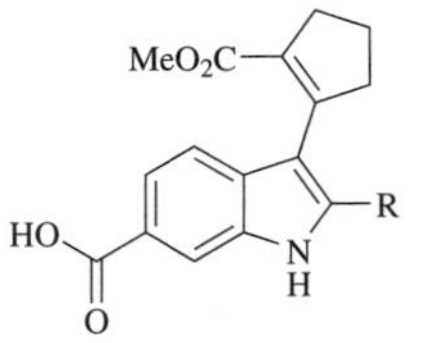

R	
Bu	(65)
c-C$_5$H$_9$	(75)
Ph	(60)
CH$_2$CH$_2$Ph	(60)
CH$_2$N(Me)COC$_6$H$_4$-4-OMe	(76)
CH$_2$CH$_2$OC$_6$H$_4$-3-CO$_2$Me	(68)

223

Indole formation 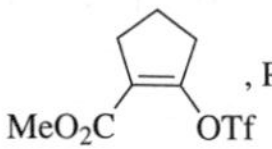
MeO$_2$C—OTf, Pd(PPh$_3$)$_4$,
K$_2$CO$_3$, DMF, rt, 24 h
N-Alkylation
BrCH$_2$CO$_2$Et, NaH, DMF, rt, 4 h
Cleavage
TFA, CH$_2$Cl$_2$, 2 h

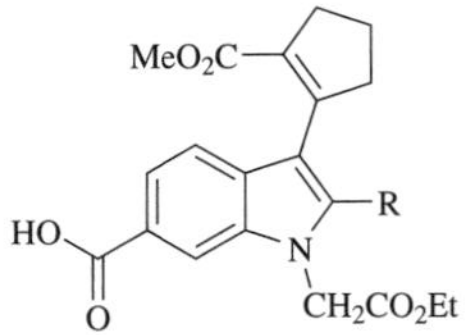

R	
Bu	(55)
c-C$_5$H$_9$	(38)
Ph	(71)
CH$_2$CH$_2$Ph	(67)
CH$_2$N(Me)COC$_6$H$_4$-4-OMe	(73)
CH$_2$CH$_2$OC$_6$H$_4$-3-CO$_2$Me	(40)

223

TABLE 13. SOLID-PHASE SYNTHESIS OF INDOLES FROM ALKYNES (*Continued*)

Substrate	Conditions	Product(s) and Yield(s) (%)[a]	Refs.
C$_{15-22}$ 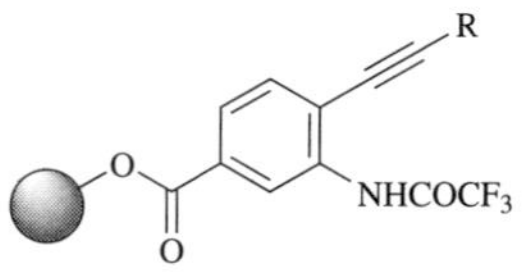	*Indole formation* 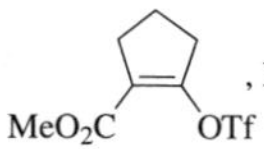 MeO$_2$C— OTf, Pd(PPh$_3$)$_4$, K$_2$CO$_3$, DMF, rt, 24 h *N-Alkylation* MeI, NaH, DMF, rt, 4 h *Cleavage* TFA, CH$_2$Cl$_2$, 2 h 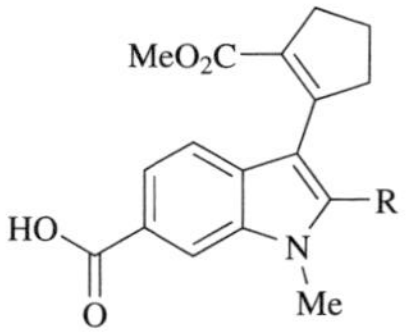	R: Bu (33); c-C$_5$H$_9$ (41); Ph (58); CH$_2$CH$_2$Ph (62); CH$_2$N(Me)COC$_6$H$_4$4-OMe (70); CH$_2$CH$_2$CH$_2$OC$_6$H$_4$3-CO$_2$Me (58)	223
C$_{21}$ 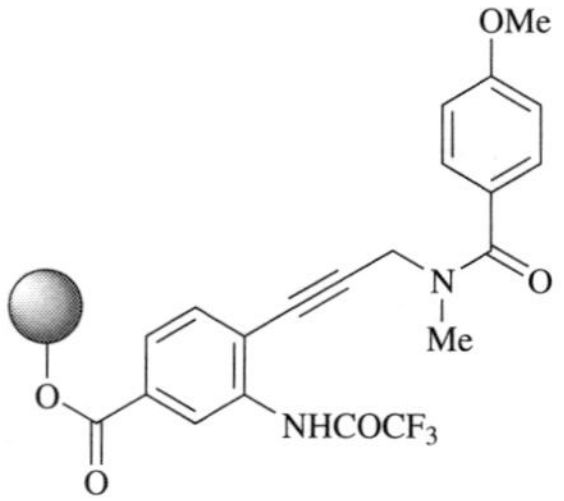	*Indole formation* 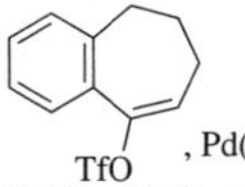 TfO— , Pd(PPh$_3$)$_4$, K$_2$CO$_3$, DMF, rt, 24 h *N-Alkylation* BrCH$_2$CO$_2$Et, NaH, DMF, rt, 4 h *Cleavage* TFA, CH$_2$Cl$_2$, rt, 1 h	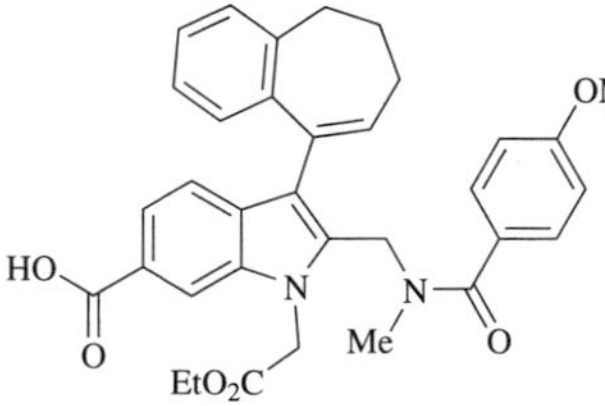 (55)	223
C$_{22-28}$ 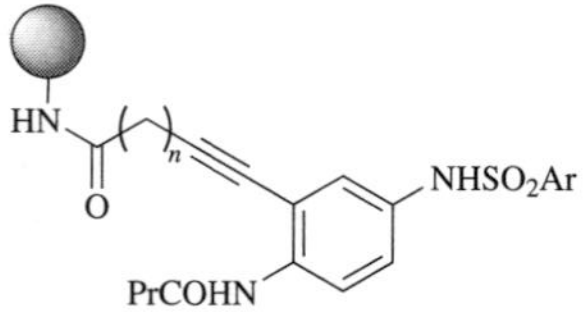	*Indole formation* PdCl$_2$(MeCN)$_2$, THF, MW, 160°, 10 min *Cleavage* TFA, CH$_2$Cl$_2$, rt, 2 h 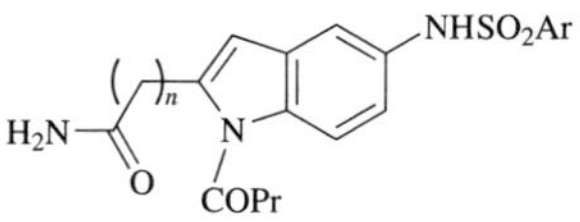	n / Ar 2 / 2-CF$_3$C$_6$H$_4$ (79) 4 / 2-CF$_3$C$_6$H$_4$ (71) 4 / 3-CF$_3$C$_6$H$_4$ (71) 8 / 2-CF$_3$C$_6$H$_4$ (65) 8 / 3-CF$_3$C$_6$H$_4$ (68) 8 / 4-CF$_3$C$_6$H$_4$ (75)	227

[a] The number is the crude yield given for all steps.
[b] The product was isolated as an 84:16 isomeric mixture.
[c] The number was determined by HPLC.
[d] The product was purified by preparative TLC.

TABLE 14. SOLID-PHASE SYNTHESIS OF INDOLES FROM ALKENES

Substrate	Conditions	Product(s) and Yield(s) (%)[a]	Refs.
C9 (2-iodoanilino acrylate on resin)	*Indole formation* Pd2(dba)3, ligand, Et3N, DMF, 110°, 15 h; *Cleavage* MeONa, MeOH/THF, 60°, 12 h	MeO2C-indole Ligand — %Yield (74) %Purity 83 P(o-tol)3 (78) 87	234
C9–11 (2-bromoanilino acrylate with R² on resin)	*Indole formation* Pd2(dba)3, P(o-tol)3, Et3N, DMF, 110°, 15 h; *Cleavage* MeONa, MeOH/THF, 60°, 12 h	MeO2C, R¹, R² indole R¹ / R² / (%) / %Purity H / H / (69) / 90 6-NO2 / H / (39) / 90 H / Me / (53) / 92 5-Me / H / (57) / — 5-CF3 / H / (65) / — 6-CF3 / H / (66) / 82 5-Me / Me / (43) / 93 6-CF3 / Me / (32) / 93	234
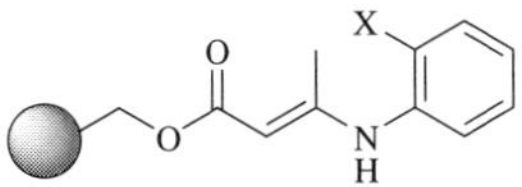 C10 (2-halo anilino methacrylate on resin)	*Indole formation* Pd(OAc)2, ligand, Et3N, DMF, 110°, 15 h; *Cleavage* MeONa, MeOH/THF, 60°, 12 h	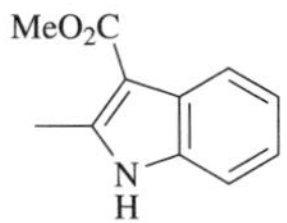 MeO2C, 2-methylindole X / Ligand / (%) / %Purity I / — / (27) / 91 I / P(o-tol)3 / (63) / 90 Br / P(o-tol)3 / (35) / 93	234
C13–16 (2-iodoanilino R-substituted acrylate on resin)	*Indole formation* Pd2(dba)3, P(o-tol)3, Et3N, DMF, 110°, 12 h; *Cleavage* MeONa, MeOH/THF, rt, 6–12 h	MeO2C, R indole R / (%) / %Purity 2-thienyl / (31) / 55 n-C5H11 / (60) / — c-C6H11 / (72) / — Ph / (48) / 70 4-MeOC6H4 / (40) / 52	234

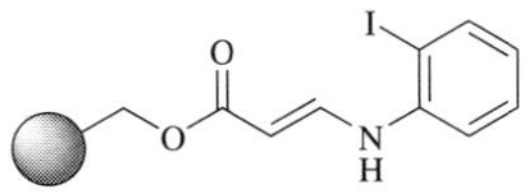

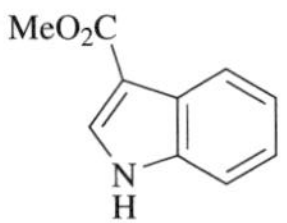

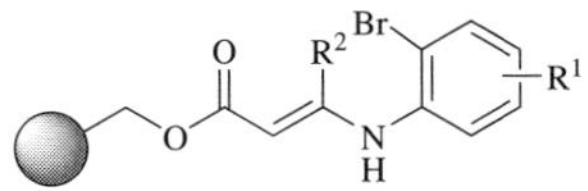

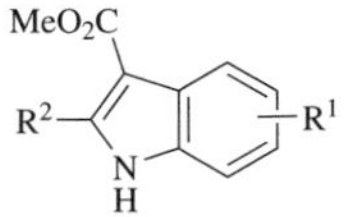

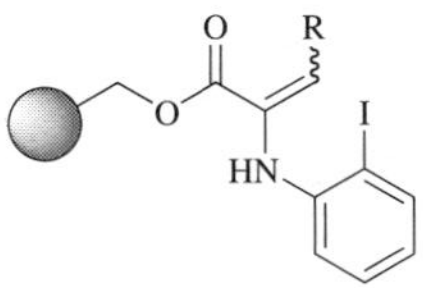

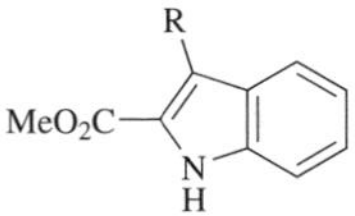

TABLE 14. SOLID-PHASE SYNTHESIS OF INDOLES FROM ALKENES (*Continued*)

Substrate	Conditions	Product(s) and Yield(s) (%)[a]	Refs.

C$_{13-16}$

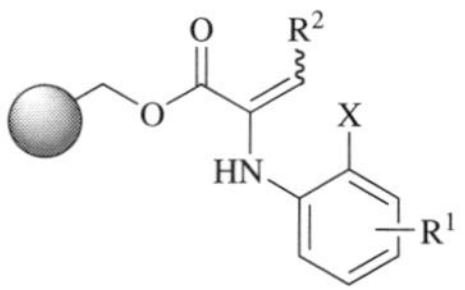

Substrate: (resin-bound acrylate with HN-aryl, X, R^1, R^2 substituents)

Conditions:
1. Pd$_2$dba$_3$, P(o-tol)$_3$, Et$_3$N, DMF, 120°, 2 h
2. MeONa, THF/MeOH

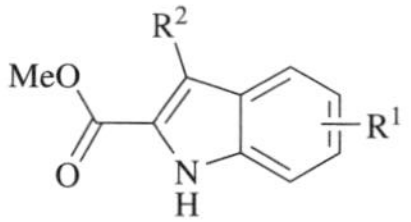

Products (indole with MeO$_2$C at 2-position, R^2 at 3-position, R^1 on ring):

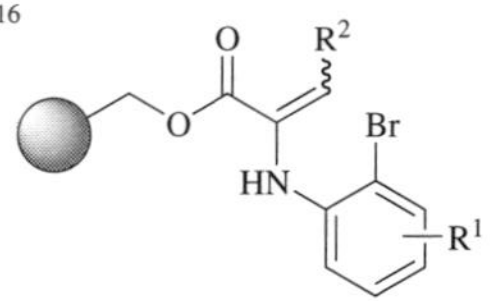

X	R^1	R^2	
I	H	2-thienyl	(31)
Br	H	2-pyridyl	(54)
Br	H	Ph	(52)
I	H	Ph	(48)
I	H	4-MeOC$_6$H$_4$	(40)
Br	4-Me	Ph	(62)

Ref. 396

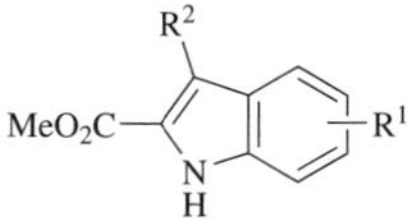

C$_{14-16}$

Substrate: (resin-bound with Br, HN, R^1, R^2)

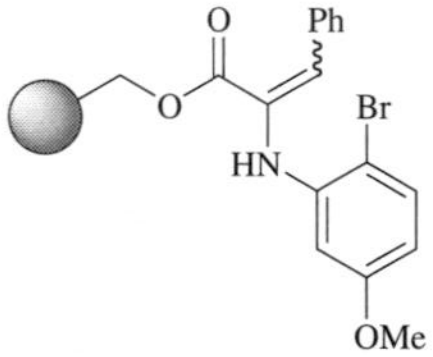

Conditions:
Indole formation
Pd$_2$(dba)$_3$, P(o-tol)$_3$, Et$_3$N, DMF, 110°, 12 h
Cleavage
MeONa, MeOH/THF, rt, 6–12 h

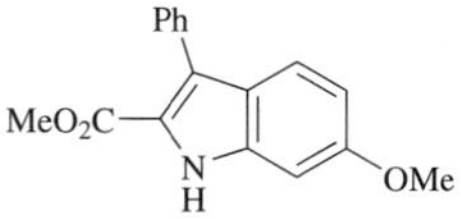

R^1	R^2		% Purity
H	2-pyridyl	(54)	65
H	Ph	(52)	82
5,7-F$_2$	Ph	(40)	69
6-NO$_2$	Ph	(60)	—
5-Me	Ph	(62)	88
5-CF$_3$	Ph	(56)	—
6-CF$_3$	Ph	(62)	—

Ref. 234

C$_{16}$

Substrate: (resin-bound with Br, OMe, HN, Ph)

Conditions:
Indole formation
Pd$_2$(dba)$_3$, ligand, base, DMF, 110°, 12 h
Cleavage
MeONa, MeOH/THF, rt, 6–12 h

Product: indole with MeO$_2$C at 2-position, Ph at 3-position, OMe on ring

Ligand	Base		% Purity
P(o-tol)$_3$	Et$_3$N	(25)	—
HP(t-Bu)$_3$BF$_4$	(c-C$_6$H$_{11}$)$_2$NMe	(67)	—

Ref. 234

[a] The number is the crude yield given for all steps.

TABLE 15. SOLID-PHASE SYNTHESIS VIA N-ARYLATION

Substrate	Conditions	Product(s) and Yield(s) (%)[a]	Refs.
C₅ (acrylate-NHAc on resin)	*Domino Heck reaction/Indole formation* 1-Br-2-XC₆H₄, Pd₂(dba)₃, (t-Bu)₃P, (c-C₆H₁₁)₂NMe, toluene, 100°, 24 h *Cleavage* MeONa, MeOH/THF, rt, 12 h	MeO₂C—indole X (yield) I (46) Br (78)	234 235
	Domino Heck reaction/Indole formation 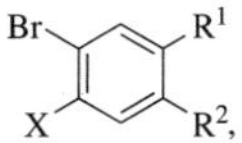 Br—(R¹, R²)—X, Pd₂(dba)₃, HP(t-Bu)₃BF₄, (c-C₆H₁₁)₂NMe *Cleavage* MeONa, MeOH/THF, rt, 12 h	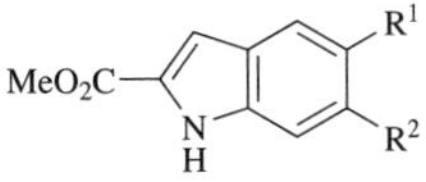 MeO₂C—indole(R¹,R²) X R¹ R² Solvent OTf H H DME (48) Br Me Me toluene (82)	234, 235
	Domino Heck reaction/Indole formation 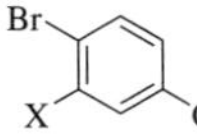 Br—(OMe)—X, Pd₂(dba)₃, HP(t-Bu)₃BF₄, (c-C₆H₁₁)₂NMe, toluene *Cleavage* MeONa, MeOH/THF, rt, 12 h	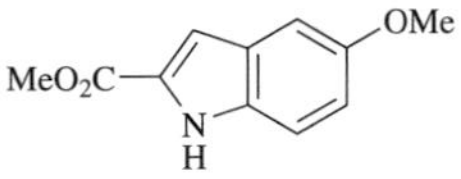 MeO₂C—indole-OMe (39) 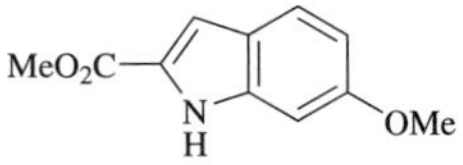 + MeO₂C—indole-OMe (31)	234, 235
C₁₀₋₁₁ (H₂N, TfO, R-aryl on resin)	*Indole formation* Pd₂(dba)₃, HP(t-Bu)₃BF₄, (c-C₆H₁₁)₂NMe, DME, 100°, 38 h *Cleavage* MeONa, MeOH/THF, rt, 12 h	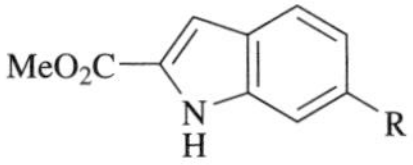 MeO₂C—indole-R R H (48) MeO (43)	234

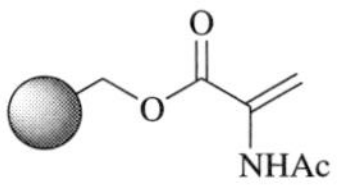

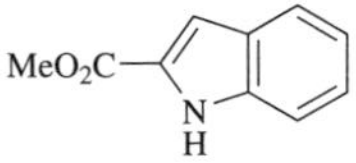

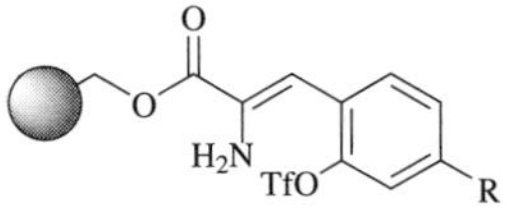

521

TABLE 15. SOLID-PHASE SYNTHESIS VIA N-ARYLATION (Continued)

Substrate	Conditions	Product(s) and Yield(s) (%)[a]	Refs.
C_{11} 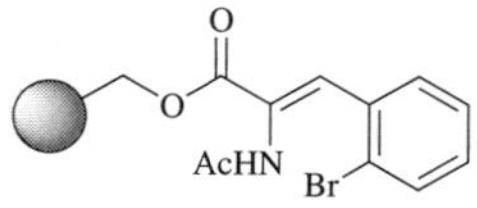 (AcHN substrate on resin)	*Indole formation* Pd$_2$(dba)$_3$, (t-Bu)$_3$P, (c-C$_6$H$_{11}$)$_2$NMe, toluene, 80°, 12 h *Cleavage* MeONa, MeOH/THF, rt, 12 h	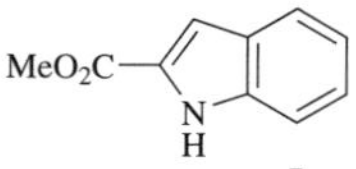 MeO$_2$C-indole-NH (99)	234
C_{16} 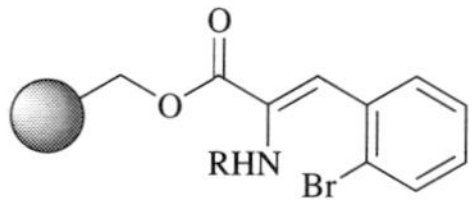 (RHN substrate on resin)	*Indole formation* Pd$_2$(dba)$_3$, (t-Bu)$_3$P, (c-C$_6$H$_{11}$)$_2$NMe, toluene, 80°, 12 h *Cleavage* MeONa, MeOH/THF, rt, 12 h	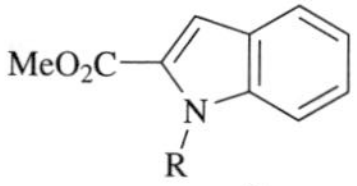 MeO$_2$C-indole-NR **I** R 2-MeC$_6$H$_4$ (44) 4-CF$_3$C$_6$H$_4$ (48)	234
C_{17} 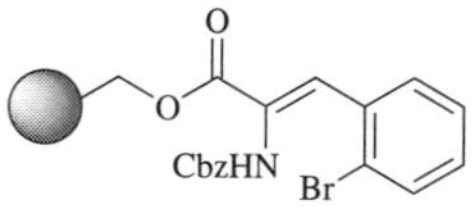 (CbzHN substrate on resin)	*Indole formation* Pd$_2$(dba)$_3$, (t-Bu)$_3$P, (c-C$_6$H$_{11}$)$_2$NMe, toluene, 80°, 12 h *Cleavage* MeONa, MeOH/THF, rt, 12 h	**I** (62)	234

[a] The number is the crude yield given for all steps.

TABLE 16. MISCELLANEOUS

Substrate	Conditions	Product(s) and Yield(s) (%)	Refs.
C₇₋₁₃ 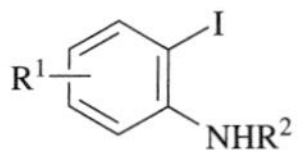 (2-iodoaniline derivative with NHR² and R¹)	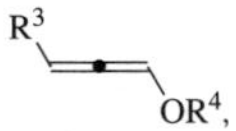 R³−CH=C=CH−OR⁴, Pd(OAc)₂, Bu₄NBr, NaOAc, DMSO, 110°	3-alkenyl indole product with R¹, R², R³, R⁵ and N-R² (table below)	400

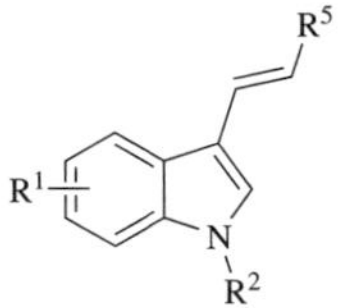

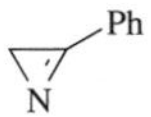

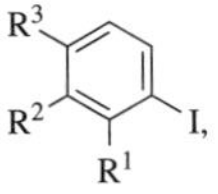

R¹	R²	R³	R⁴	R⁵	Time (h)		(E):(Z)
H	Me	Me	THP	H	1.5	(90)	—
H	Me	Pr	CH₂OEt	Et	1.5	(80)	>99:1
6-Cl	Me	Pr	MEM	Et	1.5	(77)	>99:1
H	Ac	Me	CH₂OEt	CO₂Me	3	(85)	—
5-NC	Me	Pr	MEM	Et	1.5	(87)	>99:1
H	Bn	Pr	MEM	Et	2.5	(84)	>99:1

| C₈ (aryl iodide with R¹, R², R³) | Pd(OAc)₂, (3-ClC₆H₄)₃P, norbornene, Cs₂CO₃, MeCN, reflux, 16 h^a | 2-phenylindole with R¹, R², R³ | 401 |

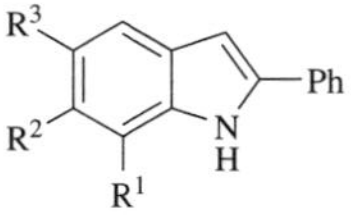

R¹	R²	R³	
Cl	H	H	(64)
MeO	H	H	(54)
Me	H	H	(55)
Me	F	H	(55)
Me	CF₃	H	(64)
Me	H	NHAc	(68)

| C₉ (5-chloro-2-iodo-N-methylaniline with CO₂Me) | Pr−CH=C=CH−OMEM, Pd(OAc)₂, Bu₄NBr, NaOAc, DMSO, 110°, 1.5 h | 3-(1-ethylpropenyl)indole with Cl, CO₂Me, N-Me, (76), (E):(Z) = 99:1 | 400 |

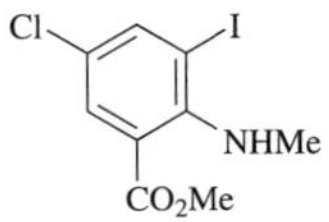

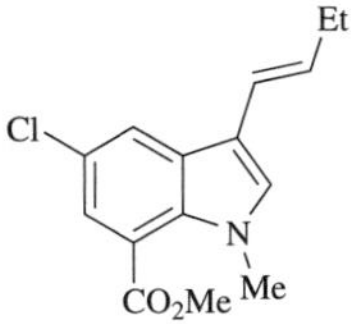

TABLE 16. MISCELLANEOUS (Continued)

Substrate	Conditions	Product(s) and Yield(s) (%)	Refs.
C_{13-19} 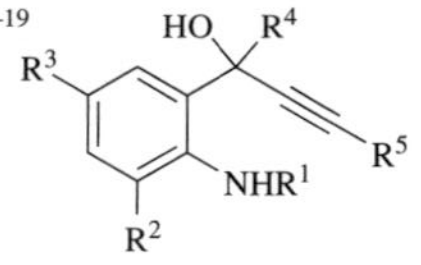 (structure with HO, R^4, R^5, NHR^1, R^2, R^3)	PdI$_2$, KI, CO (90 atm), MeOH, 100°, 2 h	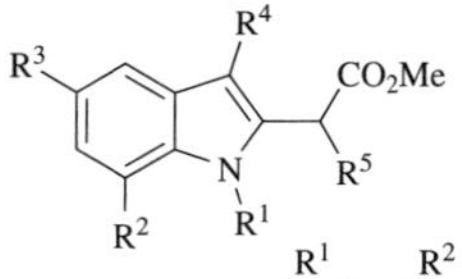 (indole product with CO$_2$Me, R^5, R^4, R^3, R^2, R^1)	402

R^1	R^2	R^3	R^4	R^5
H	H	H	Me	TMS (88)
H	H	H	Me	t-Bu (66)[b]
H	H	H	Me	t-Bu (75)
H	MeO	H	Me	TMS (42)
H	H	Cl	Me	t-Bu (68)
Me	H	H	Me	t-Bu (44)
H	MeO	H	Me	t-Bu (45)
H	H	H	Ph	TMS (63)
H	H	H	Ph	t-Bu (60)

| C_{13-21} (structure with R^1, R^2 alkyne, N–OH, R^3) | 1. cyanuric chloride, InCl$_3$, MeCN, reflux 2. PdCl$_2$(MeCN)$_2$, reflux, overnight | (N-acylindole with R^1, R^2, R^3C=O) | 403 |

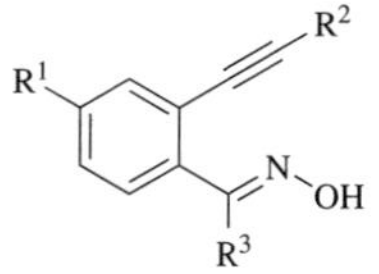

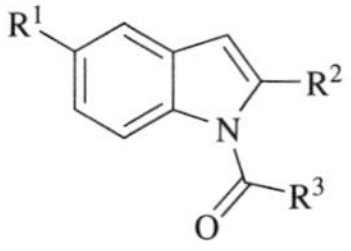

R^1	R^2	R^3		
H	c-C$_3$H$_5$	Me	(53)	
H	TMS	Me	(—)	
H	Bu	Me	(62)	
H	Bu	Et	(53)	
H	Ph	Me	(80)	
Cl	Ph	Me	(52)	
H	4-MeC$_6$H$_4$	Me	(74)	
H	4-MeOC$_6$H$_4$	Me	(62)	
H	Ph	Et	(74)	
Me	Ph	Me	(66)	
H	4-MeC$_6$H$_4$	Et	(70)	
H	4-MeOC$_6$H$_4$	Et	(60)	
Me	4-MeC$_6$H$_4$	Me	(70)	
H	Ph	Ph	(—)	

C$_{16-18}$

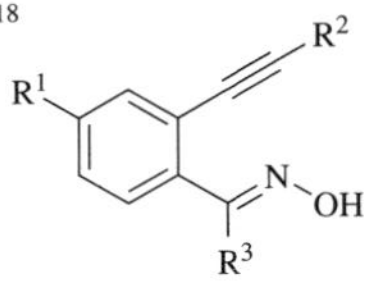

Starting material: R^1-substituted 2-alkynyl benzaldoxime (with R^2 on alkyne, R^3 on oxime carbon, N–OH)

1. cyanuric chloride, InCl$_3$, MeCN, reflux
2. PdCl$_2$(MeCN)$_2$, CuCl$_2$, reflux, overnight

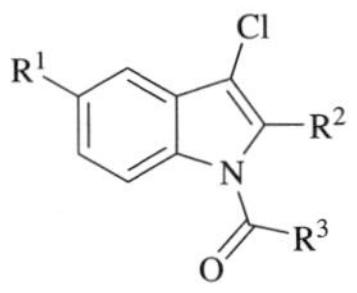

Product: 3-chloro-1-acyl indole (R^1 on benzene, R^2 at C2, N-C(O)R^3)

R^1	R^2	R^3
H	Ph	Me (66)
Cl	Ph	Me (50)
H	Ph	Et (70)
Me	Ph	Me (50)
H	4-MeC$_6$H$_4$	Et (61)
Me	4-MeC$_6$H$_4$	Me (63)

403

C$_{23-29}$

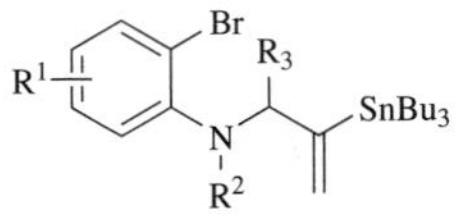

Starting material: 2-bromoaniline with N(R^2)-CH(R^3)-C(=CH$_2$)-SnBu$_3$ side chain, R^1 on ring

PdCl$_2$(allyl)$_2$, PPh$_3$, THF, 60°

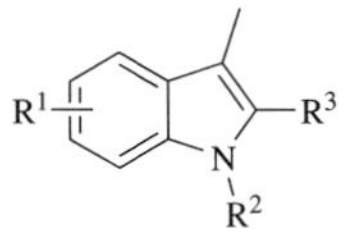

Product: 3-methyl indole (R^1 on benzene, Me at C3, N-R^2)

R^1	R^2	R^3	Time (h)	
5-Br	COMe	H	20	(82)
H	Bn	Me	48	(65)[c]

404

[a] 2H-Azirine was added at the rate of 0.26 mL/min.
[b] The reaction was performed under 60 atm of CO.
[c] The corresponding indolines were isolated in variable amounts.

REFERENCES

[1] Li, J. J.; Gribble, G. W. *Palladium in Heterocyclic Chemistry*; Pergamon: New York, 2000.
[2] Hegedus, L. S. *Angew. Chem., Int. Ed. Engl.* **1988**, *27*, 1113.
[3] Ackermann, L. *Synlett* **2007**, 507.
[4] Patil, S.; Buolamwini, J. K. *Curr. Org. Synth.* **2006**, *3*, 477.
[5] Gribble, G. W. *J. Chem. Soc., Perkin Trans. 1* **2000**, 1045.
[6] Minoru, I. *Trends in Heterocyclic Chemistry* **2001**, *7*, 75.
[7] Battistuzzi, G.; Cacchi, S.; Fabrizi, G. *Eur. J. Org. Chem.* **2002**, 2671.
[8] Undheim, K. In *Handbook of Organopalladium Chemistry for Organic Synthesis*; Negishi, E., Ed.; Wiley: New York, 2002; pp 409–492.
[9] Cacchi, S.; Fabrizi, G. *Chem. Rev.* **2005**, *105*, 2873.
[10] Humphrey, G. R.; Kuethe, J. T. *Chem. Rev.* **2006**, *106*, 2875.
[11] Ritleng, V.; Sirlin, C.; Pfeffer, M. *Chem. Rev.* **2002**, *102*, 1731.
[12] Kakiuchi, F.; Naoto Chatani, N. *Adv. Synth. Catal.* **2003**, *345*, 1077.
[13] Hassan, J.; Sévignon, M.; Gozzi, C.; Schulz, E.; Lemaire, M. *Chem. Rev.* **2002**, *102*, 1359.
[14] Alberico, D.; Scott, M. E.; Lautens, M. *Chem. Rev.* **2007**, *107*, 174.
[15] Baranano, D.; Mann, G.; Hartwig, J. F. *Curr. Org. Chem.* **1997**, *1*, 287.
[16] Hartwig, J. F. *Acc. Chem. Res.* **1998**, *31*, 852.
[17] Wolfe, J. P.; Wagaw, S.; Marcoux, J.-F.; Buchwald, S. L. *Acc. Chem. Res.* **1998**, *31*, 805.
[18] Hartwig, J. F. *Angew. Chem., Int. Ed.* **1998**, *37*, 2046.
[19] Hartwig, J. F. *Pure Appl. Chem.* **1999**, *71*, 1417.
[20] Yang, B. H.; Buchwald, S. L. *J. Organomet. Chem.* **1999**, *576*, 125.
[21] Prim, D.; Campagne, J.-M.; Joseph, D.; Andrioletti, B. *Tetrahedron* **2002**, *58*, 2041.
[22] Bjoern, S.; Ulrich, S. *Adv. Synth. Catal.* **2004**, *346*, 1599.
[23] Taylor, E. C.; Katz, A. H.; Salgado-Zamora, H.; McKillop, A. *Tetrahedron Lett.* **1985**, *26*, 5963.
[24] Hegedus, L. S.; Allen, G. F.; Waterman, E. L. *J. Am. Chem. Soc.* **1976**, *98*, 2674.
[25] Jutand, A. *Eur. J. Inorg. Chem.* **2003**, 2017.
[26] Kamijo, S.; Yamamoto, Y. *J. Am. Chem. Soc.* **2002**, *124*, 11940.
[27] Kamijo, S.; Yamamoto, Y. *Angew. Chem., Int. Ed.* **2002**, *41*, 3230.
[28] Kamijo, S.; Yamamoto, Y. *J. Org. Chem.* **2003**, *68*, 4764.
[29] Witulski, B.; Alayrac, C.; Tevzadze-Saeftel, L. *Angew. Chem., Int. Ed.* **2003**, *42*, 4257.
[30] Larock, R. C.; Yum, E. K. *J. Am. Chem. Soc.* **1991**, *113*, 6689.
[31] Larock, R. C.; Yum, E. K.; Refvik, M. D. *J. Org. Chem.* **1998**, *63*, 7652.
[32] Brown, D.; Grigg, R.; Sridharan, V.; Tambyrajah, V.; Thornton-Pett, M. *Tetrahedron* **1998**, *54*, 2595.
[33] Takeda, A.; Kamijo, S.; Yamamoto, Y. *J. Am. Chem. Soc.* **2000**, *122*, 5662.
[34] Ambrogio, I.; Cacchi, S.; Fabrizi, G. *Org. Lett.* **2006**, *8*, 2083.
[35] Tsutsumi, K.; Ogoshi, S.; Nishiguchi, S.; Kurosawa, H. *J. Am. Chem. Soc.* **1998**, *120*, 1938.
[36] Tsuji, J.; Mandai, T. *Angew. Chem., Int. Ed. Engl.* **1995**, *34*, 2589.
[37] Baize, M. W.; Blosser, P. W.; Plantevin, V.; Schimpff, D. G.; Gallucci, J. C.; Wojcicki, A. *Organometallics* **1996**, *15*, 164.
[38] Fournier-Nguefack, C.; Lhoste, P.; Sinou, D. *Synlett* **1996**, 553.
[39] Labrosse, J.-R.; Lhoste, P.; Sinou, D. *Tetrahedron Lett.* **1999**, *40*, 9025.
[40] Labrosse, J.-R.; Lhoste, P.; Sinou, D. *Org. Lett.* **2000**, *2*, 527.
[41] Yoshida, M.; Fujita, M.; Ishii, T.; Ihara, M. *J. Am. Chem. Soc.* **2003**, *125*, 4874.
[42] Yoshida, M.; Morishita, Y.; Fujita, M.; Ihara, M. *Tetrahedron* **2005**, *61*, 4381.
[43] Ambrogio, I.; Cacchi, S.; Fabrizi, G. *Tetrahedron Lett.* **2007**, *48*, 7721.
[44] Mori, M.; Chiba, K.; Ban, Y. *Tetrahedron Lett.* **1977**, *18*, 1037.
[45] Latham, E. J.; Stanfoth, S. P. *Chem. Commun.* **1996**, 2253.
[46] Latham, E. J.; Stanfoth, S. P. *J. Chem. Soc., Perkin Trans. 1* **1997**, 2059.
[47] Hennings, D. D.; Iwasa, S.; Rawal, V. H. *Tetrahedron Lett.* **1997**, *38*, 6379.
[48] Watanabe, M.; Yamamoto, T.; Nishiyama, M. *Angew. Chem., Int. Ed.* **2000**, *39*, 2501.
[49] Rudisill, D. E.; Stille, J. K. *J. Org. Chem.* **1989**, *54*, 5856.

[50] Sakamoto, T.; Kondo, Y.; Iwashita, S.; Nagano, T.; Yamanaka, H. *Chem. Pharm. Bull.* **1988**, *36*, 1305.
[51] Sonogashira, K.; Tohda, Y.; Hagihara, N. *Tetrahedron Lett.* **1975**, 4467.
[52] Sonogashira, K. In *Handbook of Organopalladium Chemistry for Organic Synthesis*; Negishi, E., Ed.; Wiley: New York, 2002; Vol. 1, pp 493–529.
[53] Shin, K.; Ogasawara, K. *Synlett* **1996**, 922.
[54] Kondo, Y.; Kojima, S.; Sakamoto, T. *Heterocycles* **1996**, *43*, 2741.
[55] Shin, K.; Ogasawara, K. *Synlett* **1996**, 922.
[56] Kondo, Y.; Kojima, S.; Sakamoto, T. *J. Org. Chem.* **1997**, *62*, 6507.
[57] Yasuhara, A.; Kanamori, Y.; Kaneko, M.; Numata, A.; Kondo, Y.; Sakamoto, T. *J. Chem. Soc., Perkin Trans. 1* **1999**, 529.
[58] Rodriguez, A. L.; Koradin, C.; Dohle, W.; Knochel, P. *Angew. Chem., Int. Ed.* **2000**, *39*, 2488.
[59] Dai, W.-M.; Sun, L.-P.; Guo, D.-S. *Tetrahedron Lett.* **2002**, *42*, 7699.
[60] Koradin, C.; Dohle, W.; Rodriguez, A. L.; Schmid, B.; Knochel, P. *Tetrahedron* **2003**, *59*, 1571.
[61] Suzuki, N.; Yasaki, S.; Yasuhara, A.; Sakamoto, T. *Chem. Pharm. Bull.* **2003**, *51*, 1170.
[62] Wattreson, S. H.; Dhar, T. G. M.; Ballentine, S. K.; Shen, Z.; Barrish, J. C.; Cheney, D.; Fleener, C. A.; Rouoleau, K. A.; Townsend, R.; Hollenbaugh, D. L.; Iwanowicz, E. J. *Bioorg. Med. Chem. Lett.* **2003**, *13*, 1273.
[63] Dai, W.-M.; Guo, D.-S.; Sun, L.-P. *Tetrahedron Lett.* **2001**, *42*, 5275.
[64] McLaughlin, M.; Palucki, M.; W. Davies, I. *Org. Lett.* **2006**, *8*, 3307.
[65] Messina, F.; Botta, M.; Corelli, F.; Villani, C. *Tetrahedron: Asymmetry* **2000**, *11*, 1681.
[66] Hiroya, K.; Itoh, S.; Ozawa, M.; Kanamori, Y.; Sakamoto, T. *Tetrahedron Lett.* **2002**, *43*, 1277.
[67] Cacchi, S.; Fabrizi, G.; Parisi, L. M. *Org. Lett.* **2003**, *5*, 3843.
[68] Chouzier, S.; Gruber, M.; Djakovitch, L. *J. Mol. Catal. A: Chem.* **2004**, *212*, 43.
[69] Hiroya, K.; Itoha, S.; Sakamoto, T. *Tetrahedron* **2005**, *61*, 10958.
[70] Iritani, K.; Matsubara, S.; Utimoto, K. *Tetrahedron Lett.* **1988**, *29*, 1799.
[71] Cacchi, S.; Carnicelli, V.; Marinelli, F. *J. Organomet. Chem.* **1994**, *475*, 289.
[72] Mahanty, J. S.; De, M.; Das, P.; Kundu, N. G. *Tetrahedron* **1997**, *53*, 13397.
[73] Van Esseveldt, B. C. J.; van Delft, F. L.; de Gelder, R.; Rutjes, F. P. J. T. *Org. Lett.* **2003**, *5*, 1717.
[74] Arcadi, A.; Cacchi, S.; Marinelli, F. *Tetrahedron Lett.* **1989**, *30*, 2581.
[75] Yu, M. S.; de Leon, L. L.; McGuire, M. A.; Botha, G. *Tetrahedron Lett.* **1998**, *39*, 9347.
[76] Torres, J. C.; Pilli, R. A.; Vargas, M. D.; Violante, F. A.; Garden, S. J.; Pinto, A. C. *Tetrahedron* **2002**, *58*, 4487.
[77] Terrasson, V.; Michaux, J.; Gaucher, A.; Wehbe, J.; Marque, S.; Prim, P.; Campagne, J.-M. *Eur. J. Org. Chem.* **2007**, 5332.
[78] Djakovitch, L.; Dufaud, V.; Zaidi, R. *Adv. Synth. Catal.* **2006**, *348*, 715.
[79] Oskooie, H. A.; Heravi, M. M.; Behbahani, F. K. *Molecules* **2007**, *12*, 1438.
[80] Sun, L.-P.; Huang, X.-H.; Dai, W.-M. *Tetrahedron* **2004**, *60*, 10983.
[81] McCarroll, A. J.; Bradshaw, T. D.; Westwell, A. D.; Matthews, C. S.; Stevens, M. F. G. *J. Med. Chem.* **2007**, *50*, 1707.
[82] Russo, O.; Messaoudi, S.; Hamze, A.; Olivi, N.; Peyrat, J.-F.; Brion, J.-D.; Sicsic, S.; Berque-Bestel, I.; Alami, M. *Tetrahedron* **2007**, *63*, 10671.
[83] Arcadi, A.; Cacchi, S.; Fabrizi, G.; Marinelli, F.; Parisi, L. M. *Heterocycles* **2004**, *64*, 475.
[84] Palimkar, S. S.; Kumar, P. H.; Lahoti, R. J.; Srinivasan, K. V. *Tetrahedron* **2006**, *62*, 5109.
[85] Beletskaya, I. P.; Kashin, A. N.; Litvinov, A. E.; Tyurin, V. S.; Valetsky, P. M; van Koten, G. *Organometallics* **2006**, *25*, 154.
[86] Ackermann, L. *Org. Lett.* **2005**, *7*, 439.
[87] Tang, Z.-Y.; Hu, Q.-S. *Adv. Synth. Catal.* **2006**, *348*, 846.
[88] Sanz, R.; Castroviejo, M. P.; Guilarte, V.; Pèrez, A.; Fanãnás, F. J. *J. Org. Chem.* **2007**, *72*, 5113.
[89] Kaspar, L. T.; Ackermann, L. *Tetrahedron* **2005**, *61*, 11311.
[90] Batsanov, A. S.; Collings, J. C.; Fairlamb, I. J. S.; Holland, J. P.; Howard, J. A. K.; Lin, Z.; Marder, T. B.; Parsons, A. C.; Ward, R. M.; Zhu, J. *J. Org. Chem.* **2005**, *70*, 703.
[91] Kálai, T.; Balog, M.; Jeko, J.; Hubbell, L.; Hideg, K. *Synthesis* **2002**, 2365.

[92] Amatore, C.; Blart, E.; Genêt, J. P.; Jutand, A.; Lemaire-Audoire, S.; Savignac, M. *J. Org. Chem.* **1995**, *60*, 6829.
[93] Gruber, M.; Chouzier, S.; Koehler, K.; Djakovitch, L. *Appl. Catal. A: Gen.* **2004**, *265*, 161.
[94] Pal, M.; Subramanian, V.; Batchu, V. R.; Dager, I. *Synlett* **2004**, 1965.
[95] Kabalka, G. W.; Wang, L.; Pagni, R. M. *Tetrahedron* **2001**, *57*, 8017.
[96] Hong, K. B.; Lee, C. W.; Yum, E. K. *Tetrahedron Lett.* **2004**, *45*, 693.
[97] Tsutsumi, K.; Kawase, T.; Kakiuchi, K.; Ogoshi, S.; Okada, Y.; Kurosawa, H. *Bull. Chem. Soc. Jpn.* **1999**, *72*, 2687.
[98] Korawa, Y.; Mori, M. *J. Org. Chem.* **2003**, *68*, 8068.
[99] Tsutsumi, K.; Yabukami, T.; Fujimoto, K.; Kawase, T.; Morimoto, T.; Kakiuchi, K. *Organometallics* **2003**, *22*, 2996.
[100] Cacchi, S.; Fabrizi, G.; Marinelli, F.; Moro, L.; Pace, P. *Synlett* **1997**, 1363.
[101] Grigg, R.; Teasdale, A.; Sridharan, V. *Tetrahedron Lett.* **1991**, *32*, 3859.
[102] Cacchi, S.; Fabrizi, G.; Pace, P. *J. Org. Chem.* **1998**, *63*, 1001.
[103] Shen, Z.; Lu, X. *Tetrahedron* **2006**, *62*, 10896.
[104] Zhao, J.; Larock, R. C. *J. Org. Chem.* **2006**, *71*, 5340.
[105] Cacchi, S.; Fabrizi, G.; Lamba, D.; Marinelli, F.; Parisi, L. M. *Synthesis* **2003**, 728.
[106] Cacchi, S.; Fabrizi, G.; Parisi, L. M. *Synthesis* **2004**, 1889.
[107] Cacchi, S.; Fabrizi, G.; Goggiamani, A. *Adv. Synth. Catal.* **2006**, *348*, 1301.
[108] Arcadi, A.; Cacchi, S.; Fabrizi, G.; Marinelli, F.; Parisi, L. M. *J. Org. Chem.* **2005**, *70*, 6213.
[109] Alonso, D. A.; Nájera, C.; Pacheco, M. C. *Adv. Synth. Catal.* **2002**, *344*, 172.
[110] Cacchi, S.; Fabrizi, G. In *Handbook of Organopalladium Chemistry for Organic Synthesis*; Negishi, E., Ed.; Wiley: New York, 2002; pp 1335–1359.
[111] Cacchi, S. *J. Organomet. Chem.* **1999**, *576*, 42.
[112] Ma, C.; Liu, X.; Li, X.; Flippen-Anderson, J.; Yu, S.; Cook, J. M. *J. Org. Chem.* **2001**, *66*, 4525.
[113] Nishikawa, T.; Wada, K.; Isobe, M. *Biosci., Biotechnol., Biochem.* **2002**, *66*, 2273.
[114] Walsh, T.; Toupence, R. B.; Ujjainwalla, F.; Young, J. R.; Goulet, M. T. *Tetrahedron* **2001**, *57*, 5233.
[115] Jeschke, T.; Wensbo, D.; Annby, U.; Gronowitz, S.; Cohen, L. A. *Tetrahedron Lett.* **1993**, *34*, 6471.
[116] Ma, C.; Liu, X.; Yu, S.; Zhao, S.; Cook, J. M. *Tetrahedron Lett.* **1999**, *40*, 657.
[117] Yu, J.; Wearing, X. Z.; Cook, J. M. *J. Org. Chem.* **2005**, *70*, 3963.
[118] Zhou, H.; Liao, X.; Yin, W.; Ma, J.; Cook, J. M. *J. Org. Chem.* **2006**, *71*, 251.
[119] Liu, X.; Deschamp, J. R.; Cook, J. M. *Org. Lett.* **2002**, *4*, 3339.
[120] Zhou, H.; Liao, X.; Cook, J. M. *Org. Lett.* **2004**, *6*, 249.
[121] Chen, C.-y.; Lieberman, D. R.; Larsen, R. D.; Reamer, R. A..; Verhoeven, T. R.; Reider, P. J.; Cottrell, I. F.; Houghton, P. G. *Tetrahedron Lett.* **1994**, *35*, 6981.
[122] Shen, M.; Li, G.; Lu, B. Z.; Hossain, A.; Roschangar, F.; Farina, V.; Senanayake, C. H. *Org. Lett.* **2004**, *6*, 4129.
[123] Arcadi, A.; Cacchi, S.; Marinelli, F. *Tetrahedron Lett.* **1992**, *33*, 3915.
[124] Arcadi, A.; Cacchi, S.; Carnicelli, V.; Marinelli, F. *Tetrahedron* **1994**, *50*, 437.
[125] Arcadi, A.; Cacchi, S.; Fabrizi, G.; Marinelli, F. *Synlett* **2000**, 394.
[126] Zhang, H.-C.; Brumfield, K. K.; Jaroskova, L.; Maryanoff, B. E. *Tetrahedron Lett.* **1998**, *39*, 4449.
[127] Dai, W.-M.; Guo, D.-S.; Sun, L.-P.; Huang, X.-H. *Org. Lett.* **2003**, *5*, 2919.
[128] Kondo, Y.; Sakamoto, T. Yamanaka, H. *Heterocycles* **1989**, *29*, 1013.
[129] Kondo, Y.; Shiga, F.; Murata, N.; Sakamoto, T.; Yamanaka, H. *Tetrahedron* **1994**, *50*, 11803.
[130] Yasuhara, A.; Kaneko, M.; Sakamoto, T. *Heterocycles* **1998**, *48*, 1793.
[131] Yasuhara, A.; Takeda, Y.; Suzuki, N.; Sakamoto, T. *Chem. Pharm. Bull.* **2002**, *50*, 235.
[132] Sashida, H.; Kawamukai, A. *Synthesis* **1999**, 1145.
[133] Netherton, M.; Fu, G. C. *Org. Lett.* **2001**, *3*, 4295.
[134] Old, D. W.; Wolfe, J. P.; Buchwald, S. L. *J. Am. Chem. Soc.* **1998**, *120*, 9722.
[135] Huang, X.; Anderson, K. W.; Zim, D.; Jiang, L.; Klapars, A.; Buchwald, S. L. *J. Am. Chem. Soc.* **2003**, *125*, 6653.
[136] Barder, T. E.; Walker, S. D.; Martinelli, J. R.; Buchwald, S. L. *J. Am. Chem. Soc.* **2005**, *127*, 4685.

[137] Chaplin, J. H.; Flynn, B. L. *Chem. Commun.* **2001**, 1594.
[138] Flynn, B. L.; Hamel, E.; Jung, M. K. *J. Med. Chem.* **2002**, *45*, 2670.
[139] Lu, B. Z.; Zhao, W.; Wei, H.-X.; Dufour, M.; Farina, V.; Senanayake, C. H. *Org. Lett.* **2006**, *8*, 3271.
[140] Nakamura, I.; Mizushima, Y.; Yamagishi, U.; Yamamoto, Y. *Tetrahedron* **2007**, *63*, 8670.
[141] Mukai, C.; Takahashi, Y. *Org. Lett.* **2005**, *7*, 5793.
[142] Hegedus, L. S.; Allen, G. F.; Bozell, J. J.; Waterman, E. L. *J. Am. Chem. Soc.* **1978**, *100*, 5800.
[143] Harrington, P. J.; Hegedus, L. S. *J. Org. Chem.* **1984**, *49*, 2657.
[144] Harrington, P. J.; Hegedus, L. S.; McDaniel, K. F. *J. Am. Chem. Soc.* **1987**, *109*, 4335.
[145] Krolski, M. E.; Renaldo, A. F.; Rudisill, D. E.; Stille, J. K. *J. Org. Chem.* **1988**, *53*, 1170.
[146] Adams, D. R.; Duncton, M. A. J.; Roffey, J. R. A.; Spencer, J. *Tetrahedron Lett.* **2002**, *43*, 7581.
[147] Kasahara, A.; Izumi, T.; Murakami, S.; Miyamoto, K.; Hino, T. *J. Heterocycl. Chem.* **1989**, *26*, 1405.
[148] Yamaguchi, M.; Arisawa, M.; Hirama, M. *Chem. Commun.* **1998**, 1399.
[149] Sato, T.; Ishida, S.; Ishibashi, H.; Ikeda, M. *J. Chem. Soc., Perkin Trans. 1* **1991**, 353.
[150] Hibino, S.; Sugino, E. *Heterocycles* **1987**, *26*, 1883.
[151] Cooper, M. K.; Yaniuk, D. W. *J. Organomet. Chem.* **1981**, *221*, 231.
[152] Subramanyam, C.; Noguchi, M.; Weinreb, S. M. *J. Org. Chem.* **1989**, *54*, 5580.
[153] Plevyak, J. E.; Heck, R. F. *J. Org. Chem.* **1978**, *43*, 2454.
[154] Akazome, M.; Kondo, T.; Watanabe, Y. *Chem. Lett.* **1992**, 769.
[155] Akazome, M.; Kondo, T.; Watanabe, Y. *J. Org. Chem.* **1994**, *59*, 3375.
[156] Söderberg, B. C.; Shriver, J. A. *J. Org. Chem.* **1997**, *62*, 5838.
[157] Gowan, M.; Caillé, A. S.; Lau, C. K. *Synlett* **1997**, 1312.
[158] Söderberg, B. C.; Chisnell, A. C.; O'Neil, S. N.; Shriver, J. A. *J. Org. Chem.* **1999**, *64*, 9731.
[159] Tollari, S.; Cenini, S.; Crotti, C.; Gianella, E. *J. Mol. Catal.* **1994**, *87*, 203.
[160] Kuethe, J. T.; Wong, A.; Qu, C.; Smitrovich, J.; Davies, I. W.; Hughes, D. L. *J. Org. Chem.* **2005**, *70*, 2555.
[161] Davies, I. W.; Smitrovich, J. H.; Sidler, R.; Qu, C.; Gresham, V.; Bazaral, C. *Tetrahedron* **2005**, *61*, 6425.
[162] Fuwa, H.; Sasaki, M. *Org. Lett.* **2007**, *9*, 3347.
[163] Che, C.-y.; Lieberman, D. R.; Larsen, R. D.; Verhoeven, T. R.; Reider, P. J. *J. Org. Chem.* **1997**, *62*, 2676.
[164] Nazaré, M.; Schneider, C.; Lindenschmidt, A.; Will, D. W. *Angew. Chem., Int. Ed.* **2004**, *43*, 4526.
[165] Barluenga, J.; Fernández, M. A.; Aznar, F.; Valdés, C. *Chem.—Eur. J.* **2005**, *11*, 2276.
[166] Mori, M.; Chiba, K.; Ban, Y. *Tetrahedron Lett.* **1977**, 1037.
[167] Odle, R.; Blevins, B.; Ratcliff, M.; Hegedus, L. S. *J. Org. Chem.* **1980**, *45*, 2709.
[168] Larock, R. C.; Babu, S. *Tetrahedron Lett.* **1987**, *28*, 5291.
[169] Genet, J. P.; Blart, E.; Savignac, M. *Synlett* **1992**, 715.
[170] Carrol, M. A.; Holmes, A. B. *Chem. Commun.* **1998**, 1395.
[171] Sundberg, R. J.; Pitts, W. J. *J. Org. Chem.* **1991**, *56*, 3048.
[172] Wensbo, D.; Annby, U.; Gronowitz, S. *Tetrahedron* **1995**, *51*, 10323.
[173] Wensbo, D.; Annby, U.; Gronowitz, S. *Tetrahedron* **1996**, *52*, 14975.
[174] Bosch, J.; Roca, T.; Armengol, M.; Fernández-Forner, D. *Tetrahedron* **2001**, *57*, 1041.
[175] Macor, J. E.; Ogilvie, R. J.; Wythes, M. J. *Tetrahedron Lett.* **1996**, *37*, 4289.
[176] Martin, P. *Helv. Chim. Acta* **1989**, *72*, 1554.
[177] Tietze, L. F.; Grote, T. *J. Org. Chem.* **1994**, *59*, 192.
[178] Zegar, S.; Tokar, C.; Enache, L. A.; Rajagopol, V.; Zeller, W.; O'Connell, M.; Singh, J.; Muellner, F. W.; Zembower, D. E. *Org. Process Res. Dev.* **2007**, *11*, 747.
[179] Charrier, N.; Demont, E.; Dunsdon, R.; Maile, G.; Naylor, A.; O'Brien, A.; Redshaw, S.; Theobald, P.; Vesey, D.; Walter, D. *Synlett* **2005**, 3071.
[180] Clawson, R. W., Jr.; Deavers, R. E., III; Akhmedov, N. G.; Söderberg, B. C. G. *Tetrahedron* **2006**, *62*, 10829.

[181] Kasahara, A.; Izumi, T.; Murakami, S.; Yanai, H.; Takatori, M. *Bull. Chem. Soc. Jpn.* **1986**, *59*, 927.
[182] Sakamoto, T.; Nagano, T.; Kondo, Y.; Yamanaka, H. *Synthesis* **1990**, 215.
[183] Jia, Y.; Zhu, J. *J. Org. Chem.* **2006**, *71*, 7826.
[184] Ackermann, L.; Sandmann, R.; Villar, A.; Ludwig T.; Kaspar, L. T. *Tetrahedron* **2008**, *64*, 769.
[185] Würtz, S.; Rakshit, S.: Neumann, J. J.; Dröge, T.; Glorius, F. *Angew. Chem. Int. Ed.* **2008**, *47*, 7230.
[186] Onitsuka, K.; Suzuki, S.; Takahashi, S. *Tetrahedron Lett.* **2002**, *43*, 6197.
[187] Kalinski, C.; Umkehrer, M.; Schmidt, J.; Ross. G.; Kolb, J.; Burdack, C.; Hiller, W.; Hoffmann, S. D. *Tetrahedron Lett.* **2006**, *47*, 4683.
[188] Yagoubi, M.; Cruz, A. C. F.; Nichols, P. L.; Elliott, R. L.; Willis, M. C. *Angew. Chem., Int. Ed.* **2010**, *49*, 7958.
[189] Watanabe, M.; Yamamoto, T.; Nishiyama, M. *Angew. Chem., Int. Ed.* **2000**, *39*, 2501.
[190] Willis, M. C.; Brace, G. N.; Findlay, T. J. K.; Holmes, I. P. *Adv. Synth. Catal.* **2006**, *348*, 851.
[191] Fletcher, A. J.; Matthew, N. B.; Willis, M. C. *Chem. Commun.* **2007**, 4764.
[192] Henderson, L. C.; Lindon, M. J.; Willis, M. C. *Tetrahedron* **2010**, *66*, 6632.
[193] Thansandote, P.; Raemy, M.; Rudolph, A.; Lautens, M. *Org. Lett.* **2007**, *9*, 5255.
[194] Brown, J. A. *Tetrahedron Lett.* **2000**, *41*, 1623.
[195] Siebeneicher, H.; Bytschkov, I.; Doye, S. *Angew. Chem., Int. Ed.* **2003**, *42*, 3042.
[196] Barluenga, J.; Jiménez-Aquino, A.; Valdés, C.; Aznar, F. *Angew. Chem., Int. Ed.* **2007**, *46*, 1529.
[197] Barluenga, J.; Jiménez-Aquino, A.; Aznar, F.; Valdés, C. *J. Am. Chem. Soc.* **2009**, *131*, 4031.
[198] Thielges, S.; Meddah, E.; Bisseret, P.; Eustache, J. *Tetrahedron Lett.* **2004**, *45*, 907.
[199] Fayol, A.; Fang, Y.-Q.; Lautens, M. *Org. Lett.* **2006**, *8*, 4203.
[200] Fang, Y.-Q.; Lautens, M. *Org. Lett.* **2005**, *7*, 3549.
[201] Fang, Y.-Q.; Karisch, R.; Lautens, M. *J. Org. Chem.* **2007**, *72*, 1341.
[202] Fang, Y.-Q.; Lautens, M. *J. Org. Chem.* **2008**, *73*, 538.
[203] Arthuis, M.; Pontikis, R.; Florent, J.-C. *Org. Lett.* **2009**, *11*, 4608.
[204] Vieira, T. O.; Meaney, L. A.; Shi, Y.-L.; Alper, H. *Org. Lett.* **2008**, *10*, 4899.
[205] Nagamochi, M.; Fang, Y.-Q.; Lautens, M. *Org. Lett.* **2007**, *9*, 2955.
[206] Shore, G.; Morin, S.; Mallik, D.; Organ, M. G. *Chem.—Eur. J.* **2008**, *14*, 1351.
[207] Newman, S. G.; Lautens, M. *J. Am. Chem. Soc.* **2010**, *132*, 11416.
[208] Inamoto, K.; Saito, T.; Hiroya, K.; Doi, T. *Synlett* **2008**, 3157.
[209] Hsieh, T. H. H.; Dong, V. M. *Tetrahedron* **2009**, *65*, 3062.
[210] Willis, M.; Brace, G. N.; Holmes, I. P. *Angew. Chem., Int. Ed.* **2005**, *44*, 403.
[211] Chiba, S.; Zhang, L.; Sanjaya, S.; Ang, G. Y. *Tetrahedron* **2010**, *66*, 5692.
[212] Abreu, A. S.; Ferreira, P. M. T.; Queiroz, M.-J. R. P.; Ferreira, I. C. F. R.; Calhelha, R. C.; Estevinho, L. M. *Eur. J. Org. Chem.* **2005**, 2951.
[213] Thompson, L. A.; Ellman, J. *Chem. Rev.* **1996**, *96*, 555.
[214] Tois, J.; Französén, R.; Koskinen, A. *Tetrahedron* **2003**, *59*, 5395.
[215] Bräse, S.; Kirchhoff, J. H.; Köbberling, J. *Tetrahedron* **2003**, *59*, 885.
[216] Bräse, S.; Gil, C.; Knepper, K. *Bioorg. Med. Chem.* **2002**, *10*, 2415.
[217] Scicinski, J. J.; Congreve, M. S.; Kay, C.; Ley, S. V. *Curr. Med. Chem.* **2002**, *9*, 2103.
[218] Katajisto, J.; Heinonen, P.; Loennberg, H. *Curr. Org. Chem.* **2004**, *8*, 977.
[219] Gil, C.; Bräse, S. *Curr. Opin. Chem. Biol.* **2004**, *8*, 230.
[220] Dolle, R. E. *J. Comb. Chem.* **2005**, *7*, 739.
[221] Patil, S. A.; Patil, R.; Miller, D. D. *Curr. Med. Chem.* **2009**, *16*, 2531.
[222] Fagnola, M. C.; Candiani, I.; Visentin, G.; Cabri, W.; Zarini, F.; Mongelli, N.; Bedeschi, A. *Tetrahedron Lett.* **1997**, *38*, 2307.
[223] Collini, M. D.; Ellingboe, J. W. *Tetrahedron Lett.* **1997**, *38*, 7963.
[224] Zhang, H.-C.; Brumfield, K. K.; Jaroskova, L.; Maryanoff, B. E. *Tetrahedron Lett.* **1998**, *39*, 4449.
[225] Zhang, H.-C.; Ye, H.; White, K. B.; Maryanoff, B. E. *Tetrahedron Lett.* **2001**, *42*, 4751.
[226] Zhang, H.-C.; Brumfield, K. K.; Maryanoff, B. E. *Tetrahedron Lett.* **1997**, *38*, 2439.
[227] Dai, W.-M.; Guo, D.-S.; Sun, L.-P.; Huang, X.-H. *Org. Lett.* **2003**, *5*, 2919.
[228] Smith, A. L.; Stevenson, G. I.; Swain, C. J.; Castro, J. L. *Tetrahedron Lett.* **1998**, *39*, 8317.

[229] Zhang, H.-C.; Ye, H.; Moretto, A. F.; Brumfield, K. K.; Maryanoff, B. E. *Org. Lett.* **2000**, *2*, 89.
[230] Wu, T. Y. H.; Ding, S.; Gray, N. S.; Schultz, P. G. *Org. Lett.* **2001**, *3*, 3827.
[231] Yun, W.; Mohan, R. *Tetrahedron Lett.* **1996**, *37*, 7189.
[232] Zhang, H.-C.; Maryanoff, B. E. *J. Org. Chem.* **1997**, *62*, 1804.
[233] Yamazaki, K.; Kondo, Y. *J. Comb. Chem.* **2002**, *4*, 191.
[234] Yamazaki, K.; Nakamura, Y.; Kondo, Y. *J. Org. Chem.* **2003**, *68*, 6011.
[235] Yamazaki, K.; Nakamura, Y.; Kondo, Y. *J. Chem. Soc., Perkin Trans. 1* **2002**, 2137.
[236] Bergman, J.; Koch, E.; Pelcman, B. *Tetrahedron Lett.* **1995**, *36*, 3945.
[237] Fiandanese, V.; Bottalico, D.; Marchese, G.; Punzi, A. *Tetrahedron* **2008**, *64*, 53.
[238] Hiroya, K.; Itoh, S.; Sakamoto, T. *J. Org. Chem.* **2004**, *69*, 1126.
[239] Cacchi, S; Fabrizi, G.; Parisi, L. M. *Org. Lett.* **2003**, *5*, 3843.
[240] Swamy, N. K.; Yazici, A.; Pyne, S. G. *J. Org. Chem.* **2010**, *75*, 3412.
[241] Dai, W.-M.; Guo, D.-S.; Sun, L.-P.; Huang, X.-H. *Org. Lett.* **2003**, *5*, 2919.
[242] Sun, L.-P.; Dai, W.-M. *Angew. Chem., Int. Ed.* **2006**, *45*, 7255.
[243] Slough, G. A.; Krchoňák, V.; Helquist, P.; Canham, S. M. *Org. Lett.* **2004**, *6*, 2909.
[244] Liu, F.; Ma, D. *J. Org. Chem.* **2007**, *72*, 4844.
[245] Kamijo, S.; Sasaki, Y.; Yamamoto, Y. *Tetrahedron Lett.* **2004**, *45*, 35.
[246] Ackermann, L.; Barfüßer, S.; Potukuchi, H. K. *Adv. Synth. Catal.* **2009**, *351*, 1064.
[247] Ohno, H.; Ohta, Y.; Oishi, S.; Fujii, N. *Angew. Chem., Int. Ed.* **2007**, *46*, 2295.
[248] Ohta, Y.; Chiba, H.; Oishi, S.; Fujii, N.; Ohno, H. *J. Org. Chem.* **2009**, *74*, 7052.
[249] Ohta, Y.; Chiba, H.; Oishi, S.; Fujii, N.; Ohno, H. *Org. Lett.* **2008**, *10*, 3535.
[250] Ohta, Y.; Oishi, S.; Fujii, N.; Ohno, H. *Org. Lett.* **2009**, *11*, 1979.
[251] Suzuki, Y.; Ohta, Y.; Oishi, S.; Fujii, N.; Ohno, H. *J. Org. Chem.* **2009**, *74*, 4246.
[252] Tobisu, M.; Fujihara, H.; Koh, K.; Chatani, N. *J. Org. Chem.* **2010**, *75*, 4841.
[253] Barberis, C.; Gordon, T. G.; Thomas, C.; Zhang, X.; Cusack, K. P. *Tetrahedron Lett.* **2005**, *46*, 8877.
[254] Melkonyan, F. S.; Karchava, A. V.; Yurovskaya, M. A. *J. Org. Chem.* **2008**, *73*, 4275.
[255] Melkonyan, F.; Topolyan, A.; Yurovskaya, M.; Karchava, A. *Eur. J. Org. Chem.*, **2008**, 5952.
[256] Cai, Q.; Li, Z.; Wei, J.; Ha, C.; Pei, D.; Ding, K. *Chem. Commun.* **2009**, 7581.
[257] Koenig, S. G.; Dankwardt, J. W.; Liu, Y.; Zhao, H.; Singh, S. P. *Tetrahedron Lett.* **2010**, *51*, 6549.
[258] Hodgkinson, R. C.; Schulz, J.; Willis, M. C. *Org. Biomol. Chem.* **2009**, *7*, 432.
[259] Willis, M. C.; Brace, G. N.; Findlay, T. J. K.; Holmes, I. P. *Adv. Synth. Catal.* **2006**, *348*, 851.
[260] Wang, Z.-J.; Yang, J.-G.; Yang, F.; Bao, W. *Org. Lett.* **2010**, *12*, 3034.
[261] Bernini, R.; Cacchi, S.; Fabrizi, G.; Filisti, E.; Sferrazza, A. *Synlett* **2009**, 1480.
[262] Gao, D.; Parvez, M.; Back, T. G. *Chem.—Eur. J.* **2010**, *16*, 14281.
[263] Bernini, R.; Fabrizi, G.; Sferrazza, A.; Cacchi, S. *Angew. Chem., Int. Ed.* **2009**, *48*, 8078.
[264] Würtz, S.; Rakshit, S.; Neumann, J. J.; Dröge, T.; Glorius, F. *Angew. Chem., Int. Ed.* **2008**, *47*, 7230.
[265] Shi, Z.; Zhang, C.; Delin Pan, S. L.; Ding, K.; Cui, Y.; Jiao, N. *Angew. Chem., Int. Ed.* **2009**, *48*, 4572.
[266] Arcadi, A.; Bianchi, G.; Marinelli, F. *Synthesis* **2004**, *4*, 610.
[267] Majumdar, K. C.; Samanta, S.; Chattopadhyay, B. *Tetrahedron Lett.* **2008**, 49, 7213.
[268] Miyazaki, Y.; Kobayashi, S. *J. Comb. Chem.* **2008**, *10*, 355.
[269] Ye, D.; Wang, J.; Zhang, X.; Zhou, Y.; Ding, F.; Feng, E.; Sun, H.; Liu, G.; Jiang, H.; Liu, H. *Green Chem.* **2009**, *11*, 1201.
[270] Ambrogio, I.; Arcadi, A.; Cacchi, S.; Fabrizi, G.; Marinelli, F. *Synlett* **2007**, 1775.
[271] Yamane, Y.; Liu, X.; Hamasaki, A.; Ishida, T.; Haruta, M.; Yokoyama, T.; Tokunaga, M. *Org. Lett.* **2009**, *11*, 5162.
[272] Alfonsi, M.; Arcadi, A.; Aschi, M.; Bianchi, G.; Marinelli, F. *J. Org. Chem.* **2005**, *70*, 2265.
[273] Praveen, C.; Karthikeyan, K.; Perumal, P. T. *Tetrahedron* **2009**, *65*, 9244.
[274] Patil, N. T.; Singh, V.; Konala, A.; Mutyala, A. K. *Tetrahedron Lett.* **2010**, *51*, 1493.
[275] Casado, R.; Contel, M.; Laguna, M.; Romero, P.; Sanz, S. *J. Am. Chem. Soc.* **2003**, *125*, 11925.
[276] Fukuda, Y.; Utimoto, K. *J. Org. Chem.* **1991**, *56*, 3729.
[277] Deetlefs, M.; Raubenheimer, H. G.; Esterhuysen, M. W. *Catal. Today* **2002**, *72*, 29.

[278] Arcadi, A.; Cerichelli, G.; Chiarini, M.; Di Giuseppe, S.; Marinelli, F. *Tetrahedron Lett.* **2000**, *41*, 9195.
[279] Mizushima, E.; Sato, K.; Hayashi, T.; Tanaka, M. *Angew. Chem., Int. Ed.* **2002**, *41*, 4563.
[280] Li, P.; Wang, L.; Wang, M.; You, F. *Eur. J. Org. Chem.* **2008**, 5946.
[281] Zhang, X.; Corma, A. *Angew. Chem., Int. Ed.* **2008**, *47*, 4358.
[282] Kothandaraman, P.; Rao, W.; Foo, S. J.; Chan, P. W. H. *Angew. Chem., Int. Ed.* **2010**, *49*, 4619.
[283] Shimada, T.; Nakamura, I.; Yamamoto, Y. *J. Am. Chem. Soc.* **2004**, *126*, 10546.
[284] Zeng, X.; Kinjo, R.; Donnadieu, B.; Bertrand, G. *Angew. Chem., Int. Ed.* **2010**, *49*, 942.
[285] Sakai, N.; Annaka, K.; Konakahara, T. *Org. Lett.* **2004**, *6*, 1527.
[286] Sakai, N.; Annaka, K.; Fujita, A.; Sato, A.; Konakahara, T. *J. Org. Chem.* **2008**, *73*, 4160.
[287] Lai, R.-Y.; Surekha, K.; Hayashi, A.; Ozawa, F.; Liu, Y.-H.; Peng, S.-M.; Liu, S.-T. *Organometallics* **2007**, *26*, 1062.
[288] Tsuchikama, K.; Hashimoto, Y.-k.; Endo, K.; Shibata, T *Adv. Synth. Catal.* **2009**, *351*, 2850.
[289] McDonald, F. E.; Chatterjee, A. K. *Tetrahedron Lett.* **1997**, *38*, 7687.
[290] Sanz, R.; Escribano, J.; Pedrosa, M. R.; Aguado, R.; Arnáiz, F. J. *Adv. Synth. Catal.* **2007**, *349*, 713.
[291] Li, X.; Wang, J.-Y.; Yu, W.; Wu L.-M. *Tetrahedron* **2009**, *65*, 1140.
[292] Okamoto, N.; Miwa, Y.; Minami, H.; Takeda, K.; Yanada, R. *Angew. Chem., Int. Ed.* **2009**, *48*, 9693.
[293] Okamoto, N.; Takeda, K.; Yanada, R. *J. Org. Chem.* **2010**, *75*, 7615.
[294] Cariou, K.; Ronan, B.; Mignani, S.; Fensterbank, L.; Malacria, M. *Angew. Chem., Int. Ed.* **2007**, *46*, 1881.
[295] Fürstner, A.; Davies, P. W. *J. Am. Chem. Soc.* **2005**, *127*, 15024.
[296] Fürstner, A.; Davies, P. W.; Gress, T. *J. Am. Chem. Soc.* **2005**, *127*, 8244.
[297] Cariou, K.; Ronan, B.; Mignani, S.; Fensterbank, L.; Malaria, M. *Angew. Chem., Int. Ed.* **2007**, *46*, 1881.
[298] Patil, S.; Patil, R. *Curr. Org. Synth.* **2007**, *4*, 201.
[299] Trost, B. M.; McClory, A. *Angew. Chem., Int. Ed.* **2007**, *46*, 2074.
[300] Isono, N.; Lautens, M. *Org. Lett.* **2009**, *11*, 1329.
[301] Boyer, A.; Isono, N.; Lackner, S.; Lautens, M. *Tetrahedron* **2010**, *66*, 6468.
[302] Burling, S.; Field, L. D.; Messerle, B. A. *Organometallics* **2000**, *19*, 87.
[303] Stuart, D. R.; Bertrand-Laperle, M.; Burgess, K. M. N.; Fagnou, K. *J. Am. Chem. Soc.* **2008**, *130*, 16474.
[304] Stuart, D. R.; Alsabeh, P.; Kuhn, M.; Fagnou, K. *J. Am. Chem. Soc.* **2010**, *132*, 18326.
[305] Chen, J.; Song, G.; Pan, C.-L.; Li, X. *Org. Lett.* **2010**, *12*, 5426.
[306] Saito, A.; Kanno, A.; Hanzawa, Y. *Angew. Chem., Int. Ed.* **2007**, *46*, 3931.
[307] Saito, A.; Oda, S.; Fukaya, H.; Hanzawa, Y. *J. Org. Chem.* **2009**, *74*, 1517.
[308] Doyle, M. P.; Shanklin, M. S.; Pho, H. Q.; Mahapatro, S. N. *J. Org. Chem.* **1988**, *53*, 1017.
[309] Etkin, N.; Babu, S. D.; Fooks, C. J.; Durst, T. *J. Org. Chem.* **1990**, *55*, 1093.
[310] Zaragoza, F. *Tetrahedron* **1995**, *51*, 8829.
[311] Chiba, S.; Hattori, G.; Narasaka, K. *Chem. Lett.* **2007**, *36*, 52.
[312] Kondo, T.; Okada, T.; Suzuki, T.; Mitsudo, T.-a. *J. Organomet. Chem.* **2001**, *622*, 149.
[313] Nair, R. N.; Lee, P. J.; Grotjahn, D. B. *Top. Catal.* **2010**, *53*, 1045.
[314] Nair, R. N.; Lee, P. J.; Rheingold, A. L.; Grotjahn, D. B. *Chem.—Eur. J.* **2010**, *16*, 7992.
[315] Hsu, G. C.; Kosar, W. P.; Jones, W. D. *Organometallics* **1994**, *13*, 385.
[316] Tsuji, Y.; Huh, K.-T.; Watanabe, Y. *J. Org. Chem.* **1987**, *52*, 1673.
[317] Tursky, M.; Lorentz-Petersen, L. L. R.; Olsen, L. B.; Madsen, R. *Org. Biomol. Chem.* **2010**, *8*, 5576.
[318] Cho, C. S.; Lim, H. K.; Shim, S. C.; Kim, T. J.; Choi, H. J. *Chem. Commun.* **1998**, 995.
[319] Cho, C. S.; Kim, J. H.; Shim, S. C. *Tetrahedron Lett.* **2000**, *41*, 1811.
[320] Cho, C. S.; Kim, J. H.; Kim, T.J.; Shim, S. C. *Tetrahedron* **2001**, *57*, 3321.
[321] Arisawa, M.; Terada, Y.; Nakagawa, M.; Nishida, A. *Angew. Chem., Int. Ed.* **2002**, *41*, 4732.
[322] Arisawa, M.; Terada, Y.; Takahashi, K.; Nakagawa, M.; Nishida, A. *J. Org. Chem.* **2006**, *71*, 4255.

323 Kasaya, Y.; Hoshi, K.; Terada, Y.; Nishida, A.; Shuto, S.; Arisawa, M. *Eur. J. Org. Chem.* **2009**, 4606.
324 Chiang, P.-Y.; Lin, Y.-C.; Wang, Y.; Liu, Y.-H. *Organometallics* **2010**, *29*, 5776.
325 Fürstner, A.; Hupperts, A. *J. Am. Chem. Soc.* **1995**, *117*, 4468.
326 Fürstner, A.; Hupperts, A.; Ptock, A.; Janssen, E. *J. Org. Chem.* **1994**, *59*, 5215.
327 Fürstner, A.; Ernst, A. *Tetrahedron* **1995**, *51*, 773.
328 Yin, Y.; Ma, W.; Chai, Z.; Zhao, G. *J. Org. Chem.* **2007**, *72*, 5731.
329 Okuma, K.; Seto, J.-i.; Sakaguchi, K.-i.; Ozaki, S.; Nagahora, N.; Shioji, K. *Tetrahedron Lett.* **2009**, *50*, 2943.
330 Kumar, M. P.; Liu, R.-S. *J. Org. Chem.* **2006**, *71*, 4951.
331 Velezheva, V. S.; Kornienko, A. G.; Topilin, S. V.; Turashev, A. D.; Peregudov, A. S.; Brennan, P. J. *J. Heterocycl. Chem.* **2006**, *43*, 873.
332 Velezheva, V. A.; Sokolov, A. I.; Kornienko, A. G.; Lyssenko, K. A.; Nelyubina, Y. V.; Godovikov, I. A.; Peregudov, A. S.; Mironov, A. F. *Tetrahedron Lett.* **2008**, *49*, 7106.
333 Lipińska, T. M.; Czarnocki, S. *J. Org. Lett.* **2006**, *8*, 367.
334 Amatore, C.; Jutand, A.; Suarez, A. *J. Am. Chem. Soc.* **1993**, *115*, 9531.
335 Amatore, C.; Jutand, A. *Acc. Chem. Res.* **2000**, *33*, 314.
336 Jia, Y.; Zhu, J. *Synlett* **2005**, 2469.
337 Dooleweerdt, K.; Ruhland, T.; Skrydstrup, T. *Org. Lett.* **2009**, *11*, 221.
338 Sanz, R.; Guilarte, V.; Pérez, A. *Tetrahedron Lett.* **2009**, *50*, 4423.
339 Sakai, H.; Tsutsumi, K.; Morimoto, T.; Kakiuchia, K. *Adv. Synth. Catal.* **2008**, *350*, 2498.
340 Layek, M.; Lakshmi, U.; Kalita, D.; Barange, D. K.; Islam, A.; Mukkanti, K.; Pal, M. *Beilstein J. Org. Chem.* **2009**, *5*, 46.
341 Monguchi, Y.; Mori, S.; Aoyagi, S.; Tsutsui, A.; Maegawa, T.; Sajiki, H. *Org. Biomol. Chem.* **2010**, *8*, 3338.
342 Ye, S.; Ding, Q.; Wang, Z.; Zhou, H.; Wu, J. *Org. Biomol. Chem.* **2008**, *6*, 4406.
343 Tyrrell, E.; Whiteman, L.; Williams, N. *Synthesis* **2009**, 829.
344 van Esseveldt, B. C. J.; van Delft, F. L.; Smits, J. M. M.; de Gelder, R.; Schoemaker, H. E.; Rutjesa, F. P. J. T. *Adv. Synth. Catal.* **2004**, *346*, 823.
345 Ototake, N.; Morimoto, Y.; Mokuya, A.; Fukaya, H.; Shida, Y.; Kitagawa, O. *Chem.—Eur. J.* **2010**, *16*, 6752.
346 Hiroya, K.; Matsumoto, S.; Ashikawa, M.; Kida, H.; Sakamoto, T. *Tetrahedron* **2005**, *61*, 12330.
347 Ambrogio, I.; Cacchi, S.; Fabrizi, G.; Prastaro, A. *Tetrahedron* **2009**, *65*, 8916.
348 Ackermann, L.; Sandmann, R.; Schinkel, M.; Kondrashov, M. V. *Tetrahedron* **2009**, *65*, 8930.
349 Yao, P. Y.; Zhang, Y.; Hsung, R. P.; Zhao, K. *Org. Lett.* **2008**, *10*, 4275.
350 Batail, N.; Bendjeriou, A.; Lomberget, T.; Barret, R.; Dufaud, V.; Djakovitcha, L. *Adv. Synth. Catal.* **2009**, *351*, 2055.
351 Solé, D.; Serrano, O. *J. Org. Chem.* **2008**, *73*, 9372.
352 Solé, D.; Serrano, O. *J. Org. Chem.* **2008**, *73*, 2476.
353 Fuwa, H.; Sasaki, M. *Org. Biomol. Chem.* **2007**, *5*, 2214.
354 Kramer, S.; Dooleweerdt, K.; Lindhardt, A. T.; Rottländer, M.; Skrydstrup, T. *Org. Lett.* **2009**, *11*, 4208.
355 Chae, J.; Konno, T.; Ishihara, T.; Yamanaka, H. *Chem. Lett.* **2004**, *33*, 314.
356 Konno, K.; Chae, J.; Ishihara, T.; Yamanaka, H. *J. Org. Chem.* **2004**, *69*, 8258.
357 Kondoh, A.; Yorimitsu, H.; Oshima, K. *Org. Lett.* **2010**, *12*, 1476.
358 Cui, X.; Li, J.; Fu, Y.; Liu, L.; Guo, Q. X. *Tetrahedron Lett.* **2008**, *49*, 3458.
359 Hwu, J. R.; Hsu, Y. C.; Josephrajan, T.; Tsay, S. C. *J. Mater. Chem.* **2009**, *19*, 3084.
360 Shi, Z.; Zhang, C.; Li, S.; Pan, D.; Ding, S.; Cui, Y.; Jiao, N. *Angew. Chem., Int. Ed.* **2009**, *48*, 4572.
361 Leogane, O.; Lebel, H. *Angew. Chem., Int. Ed.* **2008**, *47*, 350.
362 Parmentier, J. G.; Poissonnet, G.; Golstein, S. *Heterocycles* **2002**, *3*, 465.
363 Roschangar, F.; Liu, J.; Estanove, E.; Dufour, M.; Rodríguez, J.; Farina, V.; Hickey, E.; Hossain, A.; Jones, P. J.; Lee, H.; Lu, B. Z.; Varsolona, R.; Schröder, J.; Beaulieu, P.; Gillard, J.; Senanayakea, C. H. *Tetrahedron Lett.* **2008**, *49*, 363.

[364] Chen, Y.; Markina, N. A.; Larock, R. C. *Tetrahedron* **2009**, *65*, 8908.
[365] Shimamura, H.; Breazzano, S. P.; Garfunkle, J.; Kimball, F. S.; Trzupek, J. D.; Boger, D. L. *J. Am. Chem. Soc.* **2010**, *132*, 7776.
[366] Nakamura, I.; Nemoto, T.; Shiraiwa, N.; Terada, M. *Org. Lett.* **2009**, *11*, 1055.
[367] Sutou, N.; Kato, K.; Akita, H. *Tetrahedron: Asymmetry* **2008**, *19*, 1833.
[368] Gathergood, N.; Scammells, P. J. *Org. Lett.* **2003**, *5*, 921.
[369] Sugino, K.; Yoshimura, H.; Nishikawa, T.; Isobe, M. *Biosci., Biotechnol., Biochem.* **2008**, *72*, 2092.
[370] Álvarez, R.; Martínez, C.; Madich, Y.; Denis, J. G.; Aurrecoechea, J. M.; de Lera, Á. R. *Chem.-Eur. J.* **2010**, *16*, 12746.
[371] Guo, Y. J.; Tang, R. Y.; Li, J. H.; Zhong, P.; Zhanga, X. G. *Adv. Synth. Catal.* **2009**, *351*, 2615.
[372] Tang, S.; Xie, Y. X.; Li, J. H.; Wang, N. X. *Synthesis* **2007**, 1841.
[373] Arcadi, A.; Cianci, R.; Ferrara, G.; Marinelli, F. *Tetrahedron* **2010**, *66*, 2378.
[374] Cacchi, S.; Fabrizi, G.; Goggiamani, A.; Perboni, A.; Sferrazza, A.; Stabile, P. *Org. Lett.* **2010**, *12*, 3279.
[375] Han, X.; Lu, X. *Org. Lett.* **2010**, *12*, 3336.
[376] Bernini, R.; Cacchi, S.; Fabrizi, G.; Forte, G.; Petrucci, F.; Prastaro, A.; Niembro, S.; Shafir, A.; Vallribera, A. *Green Chem.* **2010**, *12*, 150.
[377] Mukai, C.; Takahashi, Y. *Org. Lett.* **2005**, *7*, 5793.
[378] Tsvelikhovsky, D.; Buchwald, S. *J. Am. Chem. Soc.* **2010**, *132*, 14048.
[379] Maruyama, J.; Yamashita, H.; Watanabe, T.; Arai, S.; Nishida, A. *Tetrahedron* **2009**, *65*, 1327.
[380] Mao, H.; Wan, J. P.; Pan, Y.; Sun, C. *Tetrahedron Lett.* **2010**, *51*, 1844.
[381] Choa, C. S.; Kimb, J. H.; Kimb, T. J.; Shimb, S. C. *J. Chem. Res.* **2004**, 630.
[382] Ackermann, L.; Althammer, A. *Synlett* **2006**, 3125.
[383] Kino, T.; Nagase, Y.; Horino, Y.; Yamakawa, T. *J. Mol. Catal. A: Chem.* **2008**, *282*, 34.
[384] Kuethe, J.; Davies, I. W. *Tetrahedron* **2006**, *62*, 11381.
[385] Siqueira, F. A.; Taylor, J. G.; Correia, C. R. D. *Tetrahedron Lett.* **2010**, *51*, 2102.
[386] El Kaim, L.; Gizzi, M.; Grimaud, L. *Org. Lett.* **2008**, *10*, 3417.
[387] Jensen, T.; Pedersen, H.; Bang-Andersen, B.; Madsen, R.; Jørgensen, M. *Angew. Chem., Int. Ed.* **2008**, *47*, 888.
[388] Weinrich, M. L.; Beck, H. P. *Tetrahedron Lett.* **2009**, *50*, 6968.
[389] Xu, Z.; Li, Q.; Zhang, L.; Jia, Y. *J. Org. Chem.* **2009**, *74*, 6859.
[390] Hu, C.; Qin, H.; Cui, Y.; Jia, Y. *Tetrahedron* **2009**, *65*, 9075.
[391] Xu, Z.; Hu, W.; Zhang, F.; Li, Q.; Lü, Z.; Zhang, L.; Jia, Y. *Synthesis* **2008**, 3981.
[392] Ragaini, F.; Ventriglia, F.; Hagar, M.; Fantauzzi, S.; Cenini, S. *Eur. J. Org. Chem.* **2009**, 2185.
[393] Söderberg, C. G.; Banini, S. R.; Turner, M. R.; Minter, A. R.; Arrington, A. K. *Synthesis* **2008**, 903.
[394] Chen, Z.; Zhu, J.; Xie, H.; Li, S.; Wu, Y.; Gong, Y. *Synlett* **2010**, 1418.
[395] Mantel, M. L. H.; Lindhardt, A. T.; Lupp, D.; Skrydstrup, T. *Chem.—Eur. J.* **2010**, *16*, 5437.
[396] Yamazaki, K.; Kondo, Y. *Chem. Commun.* **2002**, 210.
[397] Würtz, S.; Rakshit, S.; Neumann, J. J.; Dröge, T.; Glorius, F. *Angew. Chem., Int. Ed.* **2008**, *47*, 7230.
[398] Yagoubi, M.; Cruz, A. C. F.; Nichols, P. L.; Elliott, R. L.; Willis, M. C. *Angew. Chem., Int. Ed.* **2010**, *49*, 7958.
[399] Zhang, X. G.; Liu, Q.; Tang, R. Y.; Zhong, P.; Li, J. H. *Synthesis* **2010**, 1521.
[400] Boi, T.; Deagostino, A.; Prandi, C.; Tabasso, S.; Toppino, A.; Venturello, P. *Org. Biomol. Chem.* **2010**, *8*, 2020.
[401] Candito, D. A.; Lautens, M. *Org. Lett.* **2010**, *12*, 3312.
[402] Gabriele, B.; Mancuso, R.; Salerno, G.; Lupinacci, E.; Ruffolo, G.; Costa, M. *J. Org. Chem.* **2008**, *73*, 4971.
[403] Qiu, G.; Ding, Q.; Ren, H.; Peng, Y.; Wu, J. *Org. Lett.* **2010**, *12*, 3975.
[404] Lin, H.; Kazmaier, U. *Eur. J. Org. Chem.* **2009**, 1221.

CUMULATIVE CHAPTER TITLES BY VOLUME

Volume 1 (1942)

1. **The Reformatsky Reaction:** Ralph L. Shriner

2. **The Arndt-Eistert Reaction:** W. E. Bachmann and W. S. Struve

3. **Chloromethylation of Aromatic Compounds:** Reynold C. Fuson and C. H. McKeever

4. **The Amination of Heterocyclic Bases by Alkali Amides:** Marlin T. Leffler

5. **The Bucherer Reaction:** Nathan L. Drake

6. **The Elbs Reaction:** Louis F. Fieser

7. **The Clemmensen Reduction:** Elmore L. Martin

8. **The Perkin Reaction and Related Reactions:** John R. Johnson

9. **The Acetoacetic Ester Condensation and Certain Related Reactions:** Charles R. Hauser and Boyd E. Hudson, Jr.

10. **The Mannich Reaction:** F. F. Blicke

11. **The Fries Reaction:** A. H. Blatt

12. **The Jacobson Reaction:** Lee Irvin Smith

Volume 2 (1944)

1. **The Claisen Rearrangement:** D. Stanley Tarbell

2. **The Preparation of Aliphatic Fluorine Compounds:** Albert L. Henne

3. **The Cannizzaro Reaction:** T. A. Geissman

4. **The Formation of Cyclic Ketones by Intramolecular Acylation:** William S. Johnson

5. **Reduction with Aluminum Alkoxides (The Meerwein-Ponndorf-Verley Reduction):** A. L. Wilds

6. **The Preparation of Unsymmetrical Biaryls by the Diazo Reaction and the Nitrosoacetylamine Reaction:** Werner E. Bachmann and Roger A. Hoffman

7. **Replacement of the Aromatic Primary Amino Group by Hydrogen:** Nathan Kornblum

8. **Periodic Acid Oxidation:** Ernest L. Jackson

9. **The Resolution of Alcohols:** A. W. Ingersoll

10. **The Preparation of Aromatic Arsonic and Arsinic Acids by the Bart, Béchamp, and Rosenmund Reactions:** Cliff S. Hamilton and Jack F. Morgan

Volume 3 (1946)

1. **The Alkylation of Aromatic Compounds by the Friedel-Crafts Method:** Charles C. Price

2. **The Willgerodt Reaction:** Marvin Carmack and M. A. Spielman

3. **Preparation of Ketenes and Ketene Dimers:** W. E. Hanford and John C. Sauer

4. **Direct Sulfonation of Aromatic Hydrocarbons and Their Halogen Derivatives:** C. M. Suter and Arthur W. Weston

5. **Azlactones:** H. E. Carter

6. **Substitution and Addition Reactions of Thiocyanogen:** John L. Wood

7. **The Hofmann Reaction:** Everett L. Wallis and John F. Lane

8. **The Schmidt Reaction:** Hans Wolff

9. **The Curtius Reaction:** Peter A. S. Smith

Volume 4 (1948)

1. **The Diels-Alder Reaction with Maleic Anhydride:** Milton C. Kloetzel

2. **The Diels-Alder Reaction: Ethylenic and Acetylenic Dienophiles:** H. L. Holmes

3. **The Preparation of Amines by Reductive Alkylation:** William S. Emerson

4. **The Acyloins:** S. M. McElvain

5. **The Synthesis of Benzoins:** Walter S. Ide and Johannes S. Buck

6. **Synthesis of Benzoquinones by Oxidation:** James Cason

7. **The Rosenmund Reduction of Acid Chlorides to Aldehydes:** Erich Mosettig and Ralph Mozingo

8. **The Wolff-Kishner Reduction:** David Todd

Volume 5 (1949)

1. **The Synthesis of Acetylenes:** Thomas L. Jacobs

2. **Cyanoethylation:** Herman L. Bruson

3. **The Diels-Alder Reaction: Quinones and Other Cyclenones:** Lewis L. Butz and Anton W. Rytina

4. **Preparation of Aromatic Fluorine Compounds from Diazonium Fluoborates: The Schiemann Reaction:** Arthur Roe

5. **The Friedel and Crafts Reaction with Aliphatic Dibasic Acid Anhydrides:** Ernst Berliner

6. **The Gattermann-Koch Reaction:** Nathan N. Crounse

7. **The Leuckart Reaction:** Maurice L. Moore

8. **Selenium Dioxide Oxidation:** Norman Rabjohn

9. **The Hoesch Synthesis:** Paul E. Spoerri and Adrien S. DuBois

10. **The Darzens Glycidic Ester Condensation:** Melvin S. Newman and Barney J. Magerlein

Volume 6 (1951)

1. **The Stobbe Condensation:** William S. Johnson and Guido H. Daub

2. **The Preparation of 3,4-Dihydroisoquinolines and Related Compounds by the Bischler-Napieralski Reaction:** Wilson M. Whaley and Tutucorin R. Govindachari

3. **The Pictet-Spengler Synthesis of Tetrahydroisoquinolines and Related Compounds:** Wilson M. Whaley and Tutucorin R. Govindachari

4. **The Synthesis of Isoquinolines by the Pomeranz-Fritsch Reaction:** Walter J. Gensler

5. **The Oppenauer Oxidation:** Carl Djerassi

6. **The Synthesis of Phosphonic and Phosphinic Acids:** Gennady M. Kosolapoff

7. **The Halogen-Metal Interconversion Reaction with Organolithium Compounds:** Reuben G. Jones and Henry Gilman

8. **The Preparation of Thiazoles:** Richard H. Wiley, D. C. England, and Lyell C. Behr

9. **The Preparation of Thiophenes and Tetrahydrothiophenes:** Donald E. Wolf and Karl Folkers

10. **Reductions by Lithium Aluminum Hydride:** Weldon G. Brown

Volume 7 (1953)

1. **The Pechmann Reaction:** Suresh Sethna and Ragini Phadke

2. **The Skraup Synthesis of Quinolines:** R. H. F. Manske and Marshall Kulka

3. **Carbon–Carbon Alkylations with Amines and Ammonium Salts:** James H. Brewster and Ernest L. Eliel

4. **The von Braun Cyanogen Bromide Reaction:** Howard A. Hageman

5. **Hydrogenolysis of Benzyl Groups Attached to Oxygen, Nitrogen, or Sulfur:** Walter H. Hartung and Robert Simonoff

6. **The Nitrosation of Aliphatic Carbon Atoms:** Oscar Touster

7. **Epoxidation and Hydroxylation of Ethylenic Compounds with Organic Peracids:** Daniel Swern

Volume 8 (1954)

1. **Catalytic Hydrogenation of Esters to Alcohols:** Homer Adkins

2. **The Synthesis of Ketones from Acid Halides and Organometallic Compounds of Magnesium, Zinc, and Cadmium:** David A. Shirley

3. **The Acylation of Ketones to Form β-Diketones or β-Keto Aldehydes:** Charles R. Hauser, Frederic W. Swamer, and Joe T. Adams

4. **The Sommelet Reaction:** S. J. Angyal

5. **The Synthesis of Aldehydes from Carboxylic Acids:** Erich Mosettig

6. **The Metalation Reaction with Organolithium Compounds:** Henry Gilman and John W. Morton, Jr.

7. **β-Lactones:** Harold E. Zaugg

8. **The Reaction of Diazomethane and Its Derivatives with Aldehydes and Ketones:** C. David Gutsche

Volume 9 (1957)

1. **The Cleavage of Non-enolizable Ketones with Sodium Amide:** K. E. Hamlin and Arthur W. Weston

2. **The Gattermann Synthesis of Aldehydes:** William E. Truce

3. **The Baeyer-Villiger Oxidation of Aldehydes and Ketones:** C. H. Hassall

4. **The Alkylation of Esters and Nitriles:** Arthur C. Cope, H. L. Holmes, and Herbert O. House

5. **The Reaction of Halogens with Silver Salts of Carboxylic Acids:** C. V. Wilson

6. **The Synthesis of β-Lactams:** John C. Sheehan and Elias J. Corey

7. **The Pschorr Synthesis and Related Diazonium Ring Closure Reactions:** DeLos F. DeTar

Volume 10 (1959)

1. **The Coupling of Diazonium Salts with Aliphatic Carbon Atoms:** Stanley J. Parmerter

2. **The Japp-Klingemann Reaction:**   Robert R. Phillips

3. **The Michael Reaction:**   Ernst D. Bergmann, David Ginsburg, and Raphael Pappo

Volume 11 (1960)

1. **The Beckmann Rearrangement:**   L. Guy Donaruma and Walter Z. Heldt

2. **The Demjanov and Tiffeneau-Demjanov Ring Expansions:**   Peter A. S. Smith and Donald R. Baer

3. **Arylation of Unsaturated Compounds by Diazonium Salts:**   Christian S. Rondestvedt, Jr.

4. **The Favorskii Rearrangement of Haloketones:**   Andrew S. Kende

5. **Olefins from Amines: The Hofmann Elimination Reaction and Amine Oxide Pyrolysis:** Arthur C. Cope and Elmer R. Trumbull

Volume 12 (1962)

1. **Cyclobutane Derivatives from Thermal Cycloaddition Reactions:**   John D. Roberts and Clay M. Sharts

2. **The Preparation of Olefins by the Pyrolysis of Xanthates. The Chugaev Reaction:** Harold R. Nace

3. **The Synthesis of Aliphatic and Alicyclic Nitro Compounds:**   Nathan Kornblum

4. **Synthesis of Peptides with Mixed Anhydrides:**   Noel F. Albertson

5. **Desulfurization with Raney Nickel:**   George R. Pettit and Eugene E. van Tamelen

Volume 13 (1963)

1. **Hydration of Olefins, Dienes, and Acetylenes via Hydroboration:**   George Zweifel and Herbert C. Brown

2. **Halocyclopropanes from Halocarbenes:**   William E. Parham and Edward E. Schweizer

3. **Free Radical Addition to Olefins to Form Carbon–Carbon Bonds:**   Cheves Walling and Earl S. Huyser

4. **Formation of Carbon–Heteroatom Bonds by Free Radical Chain Additions to Carbon–Carbon Multiple Bonds:**   F. W. Stacey and J. F. Harris, Jr.

Volume 14 (1965)

1. **The Chapman Rearrangement:**   J. W. Schulenberg and S. Archer

2. **α-Amidoalkylations at Carbon:**   Harold E. Zaugg and William B. Martin

3. **The Wittig Reaction:**   Adalbert Maercker

Volume 15 (1967)

1. **The Dieckmann Condensation:** John P. Schaefer and Jordan J. Bloomfield

2. **The Knoevenagel Condensation:** G. Jones

Volume 16 (1968)

1. **The Aldol Condensation:** Arnold T. Nielsen and William J. Houlihan

Volume 17 (1969)

1. **The Synthesis of Substituted Ferrocenes and Other π-Cyclopentadienyl-Transition Metal Compounds:** Donald E. Bublitz and Kenneth L. Rinehart, Jr.

2. **The γ-Alkylation and γ-Arylation of Dianions of β-Dicarbonyl Compounds:** Thomas M. Harris and Constance M. Harris

3. **The Ritter Reaction:** L. I. Krimen and Donald J. Cota

Volume 18 (1970)

1. **Preparation of Ketones from the Reaction of Organolithium Reagents with Carboxylic Acids:** Margaret J. Jorgenson

2. **The Smiles and Related Rearrangements of Aromatic Systems:** W. E. Truce, Eunice M. Kreider, and William W. Brand

3. **The Reactions of Diazoacetic Esters with Alkenes, Alkynes, Heterocyclic, and Aromatic Compounds:** Vinod Dave and E. W. Warnhoff

4. **The Base-Promoted Rearrangements of Quaternary Ammonium Salts:** Stanley H. Pine

Volume 19 (1972)

1. **Conjugate Addition Reactions of Organocopper Reagents:** Gary H. Posner

2. **Formation of Carbon–Carbon Bonds via π-Allylnickel Compounds:** Martin F. Semmelhack

3. **The Thiele-Winter Acetoxylation of Quinones:** J. F. W. McOmie and J. M. Blatchly

4. **Oxidative Decarboxylation of Acids by Lead Tetraacetate:** Roger A. Sheldon and Jay K. Kochi

Volume 20 (1973)

1. **Cyclopropanes from Unsaturated Compounds, Methylene Iodide, and Zinc-Copper Couple:** H. E. Simmons, T. L. Cairns, Susan A. Vladuchick, and Connie M. Hoiness

2. **Sensitized Photooxygenation of Olefins:** R. W. Denny and A. Nickon

3. **The Synthesis of 5-Hydroxyindoles by the Nenitzescu Reaction:** George R. Allen, Jr.

4. **The Zinin Reaction of Nitroarenes:** H. K. Porter

Volume 21 (1974)

1. **Fluorination with Sulfur Tetrafluoride:** G. A. Boswell, Jr., W. C. Ripka, R. M. Scribner, and C. W. Tullock

2. **Modern Methods to Prepare Monofluoroaliphatic Compounds:** Clay M. Sharts and William A. Sheppard

Volume 22 (1975)

1. **The Claisen and Cope Rearrangements:** Sara Jane Rhoads and N. Rebecca Raulins

2. **Substitution Reactions Using Organocopper Reagents:** Gary H. Posner

3. **Clemmensen Reduction of Ketones in Anhydrous Organic Solvents:** E. Vedejs

4. **The Reformatsky Reaction:** Michael W. Rathke

Volume 23 (1976)

1. **Reduction and Related Reactions of α,β-Unsaturated Compounds with Metals in Liquid Ammonia:** Drury Caine

2. **The Acyloin Condensation:** Jordan J. Bloomfield, Dennis C. Owsley, and Janice M. Nelke

3. **Alkenes from Tosylhydrazones:** Robert H. Shapiro

Volume 24 (1976)

1. **Homogeneous Hydrogenation Catalysts in Organic Solvents:** Arthur J. Birch and David H. Williamson

2. **Ester Cleavages via S_N2-Type Dealkylation:** John E. McMurry

3. **Arylation of Unsaturated Compounds by Diazonium Salts (The Meerwein Arylation Reaction):** Christian S. Rondestvedt, Jr.

4. **Selenium Dioxide Oxidation:** Norman Rabjohn

Volume 25 (1977)

1. **The Ramberg-Bäcklund Rearrangement:** Leo A. Paquette

2. **Synthetic Applications of Phosphoryl-Stabilized Anions:** William S. Wadsworth, Jr.

3. **Hydrocyanation of Conjugated Carbonyl Compounds:** Wataru Nagata and Mitsuru Yoshioka

Volume 26 (1979)

1. **Heteroatom-Facilitated Lithiations:** Heinz W. Gschwend and Herman R. Rodriguez

2. **Intramolecular Reactions of Diazocarbonyl Compounds:** Steven D. Burke and Paul A. Grieco

Volume 27 (1982)

1. **Allylic and Benzylic Carbanions Substituted by Heteroatoms:** Jean-François Biellmann and Jean-Bernard Ducep

2. **Palladium-Catalyzed Vinylation of Organic Halides:** Richard F. Heck

Volume 28 (1982)

1. **The Reimer-Tiemann Reaction:** Hans Wynberg and Egbert W. Meijer

2. **The Friedländer Synthesis of Quinolines:** Chia-Chung Cheng and Shou-Jen Yan

3. **The Directed Aldol Reaction:** Teruaki Mukaiyama

Volume 29 (1983)

1. **Replacement of Alcoholic Hydroxy Groups by Halogens and Other Nucleophiles via Oxyphosphonium Intermediates:** Bertrand R. Castro

2. **Reductive Dehalogenation of Polyhalo Ketones with Low-Valent Metals and Related Reducing Agents:** Ryoji Noyori and Yoshihiro Hayakawa

3. **Base-Promoted Isomerizations of Epoxides:** Jack K. Crandall and Marcel Apparu

Volume 30 (1984)

1. **Photocyclization of Stilbenes and Related Molecules:** Frank B. Mallory and Clelia W. Mallory

2. **Olefin Synthesis via Deoxygenation of Vicinal Diols:** Eric Block

Volume 31 (1984)

1. **Addition and Substitution Reactions of Nitrile-Stabilized Carbanions:** Siméon Arseniyadis, Keith S. Kyler, and David S. Watt

Volume 32 (1984)

1. **The Intramolecular Diels-Alder Reaction:** Engelbert Ciganek

2. **Synthesis Using Alkyne-Derived Alkenyl- and Alkynylaluminum Compounds:** George Zweifel and Joseph A. Miller

Volume 33 (1985)

1. **Formation of Carbon–Carbon and Carbon–Heteroatom Bonds via Organoboranes and Organoborates:** Ei-Ichi Negishi and Michael J. Idacavage

2. **The Vinylcyclopropane-Cyclopentene Rearrangement:** Tomáš Hudlický, Toni M. Kutchan, and Saiyid M. Naqvi

Volume 34 (1985)

1. **Reductions by Metal Alkoxyaluminum Hydrides:**   Jaroslav Málek

2. **Fluorination by Sulfur Tetrafluoride:**   Chia-Lin J. Wang

Volume 35 (1988)

1. **The Beckmann Reactions: Rearrangements, Elimination-Additions, Fragmentations, and Rearrangement-Cyclizations:**   Robert E. Gawley

2. **The Persulfate Oxidation of Phenols and Arylamines (The Elbs and the Boyland-Sims Oxidations):**   E. J. Behrman

3. **Fluorination with Diethylaminosulfur Trifluoride and Related Aminofluorosulfuranes:**   Miloš Hudlický

Volume 36 (1988)

1. **The [3 + 2] Nitrone-Olefin Cycloaddition Reaction:**   Pat N. Confalone and Edward M. Huie

2. **Phosphorus Addition at sp^2 Carbon:**   Robert Engel

3. **Reduction by Metal Alkoxyaluminum Hydrides. Part II. Carboxylic Acids and Derivatives, Nitrogen Compounds, and Sulfur Compounds:**   Jaroslav Málek

Volume 37 (1989)

1. **Chiral Synthons by Ester Hydrolysis Catalyzed by Pig Liver Esterase:**   Masaji Ohno and Masami Otsuka

2. **The Electrophilic Substitution of Allylsilanes and Vinylsilanes:**   Ian Fleming, Jacques Dunoguès, and Roger Smithers

Volume 38 (1990)

1. **The Peterson Olefination Reaction:**   David J. Ager

2. **Tandem Vicinal Difunctionalization: β-Addition to α,β-Unsaturated Carbonyl Substrates Followed by α-Functionalization:**   Marc J. Chapdelaine and Martin Hulce

3. **The Nef Reaction:**   Harold W. Pinnick

Volume 39 (1990)

1. **Lithioalkenes from Arenesulfonylhydrazones:**   A. Richard Chamberlin and Steven H. Bloom

2. **The Polonovski Reaction:**   David Grierson

3. **Oxidation of Alcohols to Carbonyl Compounds via Alkoxysulfonium Ylides: The Moffatt, Swern, and Related Oxidations:**   Thomas T. Tidwell

Volume 40 (1991)

1. **The Pauson-Khand Cycloaddition Reaction for Synthesis of Cyclopentenones:** Neil E. Schore

2. **Reduction with Diimide:** Daniel J. Pasto and Richard T. Taylor

3. **The Pummerer Reaction of Sulfinyl Compounds:** Ottorino DeLucchi, Umberto Miotti, and Giorgio Modena

4. **The Catalyzed Nucleophilic Addition of Aldehydes to Electrophilic Double Bonds:** Hermann Stetter and Heinrich Kuhlmann

Volume 41 (1992)

1. **Divinylcyclopropane-Cycloheptadiene Rearrangement:** Tomáš Hudlický, Rulin Fan, Josephine W. Reed, and Kumar G. Gadamasetti

2. **Organocopper Reagents: Substitution, Conjugate Addition, Carbo/Metallo-cupration, and Other Reactions:** Bruce H. Lipshutz and Saumitra Sengupta

Volume 42 (1992)

1. **The Birch Reduction of Aromatic Compounds:** Peter W. Rabideau and Zbigniew Marcinow

2. **The Mitsunobu Reaction:** David L. Hughes

Volume 43 (1993)

1. **Carbonyl Methylenation and Alkylidenation Using Titanium-Based Reagents:** Stanley H. Pine

2. **Anion-Assisted Sigmatropic Rearrangements:** Stephen R. Wilson

3. **The Baeyer-Villiger Oxidation of Ketones and Aldehydes:** Grant R. Krow

Volume 44 (1993)

1. **Preparation of α,β-Unsaturated Carbonyl Compounds and Nitriles by Selenoxide Elimination:** Hans J. Reich and Susan Wollowitz

2. **Enone Olefin [2 + 2] Photochemical Cyclizations:** Michael T. Crimmins and Tracy L. Reinhold

Volume 45 (1994)

1. **The Nazarov Cyclization:** Karl L. Habermas, Scott E. Denmark, and Todd K. Jones

2. **Ketene Cycloadditions:** John Hyatt and Peter W. Raynolds

Volume 46 (1994)

1. **Tin(II) Enolates in the Aldol, Michael, and Related Reactions:** Teruaki Mukaiyama and Shū Kobayashi

2. **The [2,3]-Wittig Reaction:** Takeshi Nakai and Koichi Mikami

3. **Reductions with Samarium(II) Iodide:** Gary A. Molander

Volume 47 (1995)

1. **Lateral Lithiation Reactions Promoted by Heteroatomic Substituents:** Robin D. Clark and Alam Jahangir

2. **The Intramolecular Michael Reaction:** R. Daniel Little, Mohammad R. Masjedizadeh, Olof Wallquist (in part), and Jim I. McLoughlin (in part)

Volume 48 (1995)

1. **Asymmetric Epoxidation of Allylic Alcohols: The Katsuki-Sharpless Epoxidation Reaction:** Tsutomu Katsuki and Victor S. Martin

2. **Radical Cyclization Reactions:** B. Giese, B. Kopping, T. Göbel, J. Dickhaut, G. Thoma, K. J. Kulicke, and F. Trach

Volume 49 (1997)

1. **The Vilsmeier Reaction of Fully Conjugated Carbocycles and Heterocycles:** Gurnos Jones and Stephen P. Stanforth

2. **[6 + 4] Cycloaddition Reactions:** James H. Rigby

3. **Carbon–Carbon Bond-Forming Reactions Promoted by Trivalent Manganese:** Gagik G. Melikyan

Volume 50 (1997)

1. **The Stille Reaction:** Vittorio Farina, Venkat Krishnamurthy and William J. Scott

Volume 51 (1997)

1. **Asymmetric Aldol Reactions Using Boron Enolates:** Cameron J. Cowden and Ian Paterson

2. **The Catalyzed α-Hydroxyalkylation and α-Aminoalkylation of Activated Olefins (The Morita–Baylis–Hillman Reaction):** Engelbert Ciganek

3. **[4 + 3] Cycloaddition Reactions:** James H. Rigby and F. Christopher Pigge

Volume 52 (1998)

1. **The Retro–Diels–Alder Reaction. Part I. C–C Dienophiles:** Bruce Rickborn

2. **Enantioselective Reduction of Ketones:** Shinichi Itsuno

Volume 53 (1998)

1. **The Oxidation of Alcohols by Modified Oxochromium(VI)-Amine Complexes:** Frederick A. Luzzio

2. **The Retro-Diels-Alder Reaction. Part II. Dienophiles with One or More Heteroatoms:** Bruce Rickborn

Volume 54 (1999)

1. **Aromatic Substitution by the $S_{RN}1$ Reaction:** Roberto Rossi, Adriana B. Pierini, and Ana N. Santiago

2. **Oxidation of Carbonyl Compounds with Organohypervalent Iodine Reagents:** Robert M. Moriarty and Om Prakash

Volume 55 (1999)

1. **Synthesis of Nucleosides:** Helmut Vorbrüggen and Carmen Ruh-Pohlenz

Volume 56 (2000)

1. **The Hydroformylation Reaction:** Iwao Ojima, Chung-Ying Tsai, Maria Tzamarioudaki, and Dominique Bonafoux

2. **The Vilsmeier Reaction. 2. Reactions with Compounds Other Than Fully Conjugated Carbocycles and Heterocycles:** Gurnos Jones and Stephen P. Stanforth

Volume 57 (2001)

1. **Intermolecular Metal-Catalyzed Carbenoid Cyclopropanations:** Huw M. L. Davies and Evan G. Antoulinakis

2. **Oxidation of Phenolic Compounds with Organohypervalent Iodine Reagents:** Robert M. Moriarty and Om Prakash

3. **Synthetic Uses of Tosylmethyl Isocyanide (TosMIC):** Daan van Leusen and Albert M. van Leusen

Volume 58 (2001)

1. **Simmons-Smith Cyclopropanation Reaction:** André B. Charette and André Beauchemin

2. **Preparation and Applications of Functionalized Organozinc Compounds:** Paul Knochel, Nicolas Millot, Alain L. Rodriguez, and Charles E. Tucker

Volume 59 (2001)

1. **Reductive Aminations of Carbonyl Compounds with Borohydride and Borane Reducing Agents:** Ellen W. Baxter and Allen B. Reitz

Volume 60 (2002)

1. **Epoxide Migration (Payne Rearrangement) and Related Reactions:** Robert M. Hanson

2. **The Intramolecular Heck Reaction:** J. T. Link

Volume 61 (2002)

1. **[3 + 2] Cycloaddition of Trimethylenemethane and Its Synthetic Equivalents:** Shigeru Yamago and Eiichi Nakamura

2. **Dioxirane Epoxidation of Alkenes:** Waldemar Adam, Chantu R. Saha-Möller, and Cong-Gui Zhao

Volume 62 (2003)

1. **α-Hydroxylation of Enolates and Silyl Enol Ethers:** Bang-Chi Chen, Ping Zhou, Franklin A. Davis, and Engelbert Ciganek

2. **The Ramberg-Bäcklund Reaction:** Richard J. K. Taylor and Guy Casy

3. **The α-Hydroxy Ketone (α-Ketol) and Related Rearrangements:** Leo A. Paquette and John E. Hofferberth

4. **Transformation of Glycals into 2,3-Unsaturated Glycosyl Derivatives:** Robert J. Ferrier and Oleg A. Zubkov

Volume 63 (2004)

1. **The Biginelli Dihydropyrimidine Synthesis:** C. Oliver Kappe and Alexander Stadler

2. **Microbial Arene Oxidations:** Roy A. Johnson

3. **Cu, Ni, and Pd Mediated Homocoupling Reactions in Biaryl Syntheses: The Ullmann Reaction:** Todd D. Nelson and R. David Crouch

Volume 64 (2004)

1. **Additions of Allyl, Allenyl, and Propargylstannanes to Aldehydes and Imines:** Benjamin W. Gung

2. **Glycosylation with Sulfoxides and Sulfinates as Donors or Promoters:** David Crich and Linda B. L. Lim

3. **Addition of Organochromium Reagents to Carbonyl Compounds:** Kazuhiko Takai

Volume 65 (2005)

1. **The Passerini Reaction:** Luca Banfi and Renata Riva

2. **Diels-Alder Reactions of Imino Dienophiles:** Geoffrey R. Heintzelman, Ivona R. Meigh, Yogesh R. Mahajan, and Steven M. Weinreb

Volume 66 (2005)

1. **The Allylic Trihaloacetimidate Rearrangement:** Larry E. Overman and Nancy E. Carpenter

2. **Asymmetric Dihydroxylation of Alkenes:** Mark C. Noe, Michael A. Letavic, Sheri L. Snow, and Stuart McCombie

Volume 67 (2006)

1. **Catalytic Enantioselective Aldol Addition Reactions:** Erick M. Carreira, Alec Fettes, and Christiane Marti

2. **Benzylic Activation and Stereochemical Control in Reactions of Tricarbonyl(Arene)-Chromium Complexes:** Motokazu Uemura

Volume 68 (2007)

1. **Cotrimerizations of Acetylenic Compounds:** Nicolas Agenet, Olivier Buisine, Franck Slowinski, Vincent Gandon, Corinne Aubert, and Max Malacria

2. **Glycosylation on Polymer Supports:** Simone Bufali and Peter H. Seeberger

Volume 69 (2007)

1. **Dioxirane Oxidations of Compounds Other Than Alkenes:** Waldemar Adam, Cong-Gui Zhao, and Kavitha Jakka

2. **Electrophilic Fluorination with N–F Reagents:** Jérôme Baudoux and Dominique Cahard

Volume 70 (2008)

1. **The Catalytic Asymmetric Strecker Reaction:** Masakatsu Shibasaki, Motomu Kanai, and Tsuyoshi Mita

2. **The Synthesis of Phenols and Quinones via Fischer Carbene:** Marcey L. Waters and William D. Wulff

Volume 71 (2008)

1. **Ionic and Organometallic-Catalyzed Organosilane Reductions:** Gerald L. Larson and James L. Fry

Volume 72 (2008)

1. **Electrophilic Amination of Carbanions, Enolates, and Their Surrogates:** Engelbert Ciganek

2. **Desulfonylation Reactions:** Diego A. Alonso and Carmen Nájera

Volume 73 (2008)

1. **Allylboration of Carbonyl Compounds:** Hugo Lachance and Dennis G. Hall

Volume 74 (2009)

1. **Catalytic Asymmetric Hydrogenation of C=N Functions:** Hans-Ulrich Blaser and Felix Spindler

2. **Oxoammonium- and Nitroxide-Catalyzed Oxidations of Alcohols:** James M. Bobbitt, Christian Brückner, and Nabyl Merbouh

3. **Asymmetric Epoxidation of Electron-Deficient Alkenes:** Michael J. Porter and John Skidmore

Volume 75 (2011)

1. **Hydrocyanation of Alkenes and Alkynes:** T. V. RajanBabu

2. **Intermolecular C-H Insertions of Carbenoids:** Huw M. L. Davies and Phillip M. Pelphrey

3. **Cross-Coupling with Organosilicon Compounds:** Wen-Tau T. Chang, Russell C. Smith, Christopher S. Regens, Aaron D. Bailey, Nathan S. Werner, and Scott E. Denmark

4. **The Aza-Cope/Mannich Reaction:** Larry E. Overman, Philip G. Humphreys, and Gregory S. Welmaker

AUTHOR INDEX, VOLUMES 1–76

Volume number only is designated in this index

Adam, Waldemar, 61, 69
Adams, Joe T., 8
Adkins, Homer, 8
Agenet, Nicolas, 68
Ager, David J., 38
Albertson, Noel F., 12
Allen, George R., Jr., 20
Angyal, S. J., 8
Antoulinkis, Evan G., 57
Alonso, Diego A., 72
Apparu, Marcel, 29
Archer, S., 14
Arseniyadis, Siméon, 31
Aubert, Corinne, 68

Bachmann, W. E., 1, 2
Baer, Donald R., 11
Bailey, Aaron D., 75
Banfi, Luca, 65
Bataille, Carole J. R., 76
Baudoux, Jérôme, 69
Baxter, Ellen W., 59
Beauchemin, André, 58
Behr, Lyell C., 6
Behrman, E. J., 35
Bergmann, Ernst D., 10
Berliner, Ernst, 5
Biellmann, Jean-François, 27
Birch, Arthur J., 24
Blatchly, J. M., 19
Blatt, A. H., 1
Blaser, Hans-Ulrich, 74
Blicke, F. F., 1
Block, Eric, 30

Bloom, Steven H., 39
Bloomfield, Jordan J., 15, 23
Bobbitt, James M., 74
Bonafoux, Dominique, 56
Boswell, G. A., Jr., 21
Brand, William W., 18
Brewster, James H., 7
Brown, Herbert C., 13
Brown, Weldon G., 6
Brückner, Christian, 74
Bruson, Herman Alexander, 5
Bublitz, Donald E., 17
Buck, Johannes S., 4
Bufali, Simone, 68
Buisine, Olivier, 68
Burke, Steven D., 26
Butz, Lewis W., 5

Cacchi, Sandro, 76
Cahard, Dominique, 69
Caine, Drury, 23
Cairns, Theodore L., 20
Campagne, Jean-Marc, 76
Carmack, Marvin, 3
Carpenter, Nancy E., 66
Carreira, Eric M., 67
Carter, H. E., 3
Cason, James, 4
Castro, Bertrand R., 29
Casy, Guy, 62
Chamberlin, A. Richard, 39
Chang, Wen-Tau T., 75
Chapdelaine, Marc J., 38
Charette, André B., 58

Organic Reactions, Vol. 76, Edited by Scott E. Denmark et al.
© 2012 Organic Reactions, Inc. Published 2012 by John Wiley & Sons, Inc.

Chen, Bang-Chi, 62
Cheng, Chia-Chung, 28
Ciganek, Engelbert, 32, 51, 62, 72
Clark, Robin D., 47
Confalone, Pat N., 36
Cope, Arthur C., 9, 11
Corey, Elias J., 9
Cota, Donald J., 17
Cowden, Cameron J., 51
Crandall, Jack K., 29
Crich, David, 64
Crimmins, Michael T., 44
Crouch, R. David, 63
Crounse, Nathan N., 5

Daub, Guido H., 6
Dave, Vinod, 18
Davies, Huw M. L., 57, 75
Davis, Franklin A., 62
Denmark, Scott E., 45, 75
Denny, R. W., 20
DeLucchi, Ottorino, 40
DeTar, DeLos F., 9
Dickhaut, J., 48
Djerassi, Carl, 6
Donohoe, Timothy J., 76
Donaruma, L. Guy, 11
Drake, Nathan L., 1
DuBois, Adrien S., 5
Ducep, Jean-Bernard, 27
Dunoguès, Jacques, 37

Eliel, Ernest L., 7
Emerson, William S., 4
Engel, Robert, 36
England, D. C., 6

Fabrizi, Giancarlo, 76
Fan, Rulin, 41
Farina, Vittorio, 50
Ferrier, Robert J., 62
Fettes, Alec, 67
Fieser, Louis F., 1
Fleming, Ian, 37
Folkers, Karl, 6
Fry, James L., 71
Fuson, Reynold C., 1

Gadamasetti, Kumar G., 41
Gandon, Vincent, 68
Gaucher, Anne, 76
Gawley, Robert E., 35
Geissman, T. A., 2
Gensler, Walter J., 6
Giese, B., 48
Gilman, Henry, 6, 8
Ginsburg, David, 10
Göbel, T., 48
Goggiamani, Antonella, 76
Govindachari, Tuticorin R., 6
Grieco, Paul A., 26
Grierson, David, 39
Gschwend, Heinz W., 26
Gung, Benjamin W., 64
Gutsche, C. David, 8

Habermas, Karl L., 45
Hageman, Howard A., 7
Hall, Dennis G., 73
Hamilton, Cliff S., 2
Hamlin, K. E., 9
Hanford, W. E., 3
Hanson, Robert M., 60
Harris, Constance M., 17
Harris, J. F., Jr., 13
Harris, Thomas M., 17
Hartung, Walter H., 7
Hassall, C. H., 9
Hauser, Charles R., 1, 8
Hayakawa, Yoshihiro, 29
Heck, Richard F., 27
Heldt, Walter Z., 11
Heintzelman, Geoffrey R., 65
Henne, Albert L., 2
Hofferberth, John E., 62
Hoffman, Roger A., 2
Hoiness, Connie M., 20
Holmes, H. L., 4, 9
Houlihan, William J., 16
House, Herbert O., 9
Hudlický, Miloš, 35
Hudlický, Tomáš, 33, 41
Hudson, Boyd E., Jr., 1
Hughes, David L., 42
Huie, E. M., 36
Hulce, Martin, 38
Humphreys, Philip G., 75

Huyser, Earl S., 13
Hyatt, John A., 45

Idacavage, Michael J., 33
Ide, Walter S., 4
Ingersoll, A. W., 2
Innocenti, Paolo, 76
Itsuno, Shinichi, 52

Jackson, Ernest L., 2
Jacobs, Thomas L., 5
Jahangir, Alam, 47
Jakka, Kavitha, 69
Johnson, John R., 1
Johnson, Roy A., 63
Johnson, William S., 2, 6
Jones, Gurnos, 15, 49, 56
Jones, Reuben G., 6
Jones, Todd K., 45
Jorgenson, Margaret J., 18

Kanai, Motomu, 70
Kappe, C. Oliver, 63
Katsuki, Tsutomu, 48
Kende, Andrew S., 11
Kloetzel, Milton C., 4
Knochel, Paul, 58
Kobayashi, Shū, 46
Kochi, Jay K., 19
Kopping, B., 48
Kornblum, Nathan, 2, 12
Kosolapoff, Gennady M., 6
Kreider, Eunice M., 18
Krimen, L. I., 17
Krishnamurthy, Venkat, 50
Krow, Grant R., 43
Kuhlmann, Heinrich, 40
Kulicke, K. J., 48
Kulka, Marshall, 7
Kutchan, Toni M., 33
Kyler, Keith S., 31

Lachance, Hugo, 73
Lane, John F., 3
Larson, Gerald L., 71
Leffler, Marlin T., 1
Letavic, Michael A., 66

Lim, Linda B. L., 64
Link, J. T., 60
Little, R. Daniel, 47
Lipshutz, Bruce H., 41
Luzzio, Frederick A., 53

Malacria, Max, 68
McCombie, Stuart W., 66
McElvain, S. M., 4
McKeever, C. H., 1
McLoughlin, J. I., 47
McMurry, John E., 24
McOmie, J. F. W., 19
Maercker, Adalbert, 14
Magerlein, Barney J., 5
Mahajan, Yogesh R., 65
Málek, Jaroslav, 34, 36
Mallory, Clelia W., 30
Mallory, Frank B., 30
Manske, Richard H. F., 7
Marcinow, Zbigniew, 42
Marque, Sylvain, 76
Marti, Christiane, 67
Martin, Elmore L., 1
Martin, Victor S., 48
Martin, William B., 14
Masjedizadeh, Mohammad R., 47
Meigh, Ivona R., 65
Meijer, Egbert W., 28
Melikyan, G. G., 49
Merbouh, Nabyl, 74
Mikami, Koichi, 46
Miller, Joseph A., 32
Millot, Nicolas, 58
Miotti, Umberto, 40
Mita, Tsuyoshi, 70
Modena, Giorgio, 40
Molander, Gary, 46
Moore, Maurice L., 5
Morgan, Jack F., 2
Moriarty, Robert M., 54, 57
Morton, John W., Jr., 8
Mosettig, Erich, 4, 8
Mozingo, Ralph, 4
Mukaiyama, Teruaki, 28, 46

Nace, Harold R., 12
Nagata, Wataru, 25

Nájera, Carmen, 72
Nakai, Takeshi, 46
Nakamura, Eiichi, 61
Naqvi, Saiyid M., 33
Negishi, Ei-Ichi, 33
Nelke, Janice M., 23
Nelson, Todd D., 63
Newman, Melvin S., 5
Nickon, A., 20
Nielsen, Arnold T., 16
Noe, Mark C., 66
Noyori, Ryoji, 29

Ohno, Masaji, 37
Ojima, Iwao, 56
Otsuka, Masami, 37
Overman, Larry E., 66, 75
Owsley, Dennis C., 23

Pappo, Raphael, 10
Paquette, Leo A., 25, 62
Parham, William E., 13
Parmerter, Stanley M., 10
Pasto, Daniel J., 40
Paterson, Ian, 51
Pelphrey, Phillip M., 75
Pettit, George R., 12
Phadke, Ragini, 7
Phillips, Robert R., 10
Pierini, Adriana B., 54
Pigge, F. Christopher, 51
Pine, Stanley H., 18, 43
Pinnick, Harold W., 38
Porter, H. K., 20
Porter, Michael J., 74
Posner, Gary H., 19, 22
Prakash, Om, 54, 57
Price, Charles C., 3
Prim, Damien, 76

RajanBabu, T. V., 75
Rabideau, Peter W., 42
Rabjohn, Norman, 5, 24
Rathke, Michael W., 22
Raulins, N. Rebecca, 22
Raynolds, Peter W., 45
Reed, Josephine W., 41
Regens, Christopher S., 75

Reich, Hans J., 44
Reinhold, Tracy L., 44
Reitz, Allen B., 59
Rhoads, Sara Jane, 22
Rickborn, Bruce, 52, 53
Rigby, James H., 49, 51
Rinehart, Kenneth L., Jr., 17
Ripka, W. C., 21
Riva, Renata, 65
Roberts, John D., 12
Rodriguez, Alain L., 58
Rodriguez, Herman R., 26
Roe, Arthur, 5
Rondestvedt, Christian S., Jr., 11, 24
Rossi, Roberto, 54
Ruh-Polenz, Carmen, 55
Rytina, Anton W., 5

Saha-Möller, Chantu R., 61
Santiago, Ana N., 54
Sauer, John C., 3
Schaefer, John P., 15
Schore, Neil E., 40
Schulenberg, J. W., 14
Schweizer, Edward E., 13
Scott, William J., 50
Scribner, R. M., 21
Seeberger, Peter H., 68
Semmelhack, Martin F., 19
Sengupta, Saumitra, 41
Sethna, Suresh, 7
Shapiro, Robert H., 23
Sharts, Clay M., 12, 21
Sheehan, John C., 9
Sheldon, Roger A., 19
Sheppard, W. A., 21
Shibasaki, Masakatsu, 70
Shirley, David A., 8
Shriner, Ralph L., 1
Simmons, Howard E., 20
Simonoff, Robert, 7
Skidmore, John, 74
Slowinski, Franck, 68
Smith, Lee Irvin, 1
Smith, Peter A. S., 3, 11
Smith, Russell C., 75
Smithers, Roger, 37
Snow, Sheri L., 66
Spielman, M. A., 3

Spindler, Felix, 74
Spoerri, Paul E., 5
Stacey, F. W., 13
Stadler, Alexander, 63
Stanforth, Stephen P., 49, 56
Stetter, Hermann, 40
Struve, W. S., 1
Suter, C. M., 3
Swamer, Frederic W., 8
Swern, Daniel, 7

Takai, Kazuhiko, 64
Tarbell, D. Stanley, 2
Taylor, Richard J.K., 62
Taylor, Richard T., 40
Thoma, G., 48
Tidwell, Thomas T., 39
Todd, David, 4
Touster, Oscar, 7
Trach, F., 48
Truce, William E., 9, 18
Trumbull, Elmer R., 11
Tsai, Chung-Ying, 56
Tucker, Charles E., 58
Tullock, C. W., 21
Tzamarioudaki, Maria, 56

Uemura, Motokazu, 67

van Leusen, Albert M., 57
van Leusen, Daan, 57
van Tamelen, Eugene E., 12
Vedejs, E., 22
Vladuchick, Susan A., 20
Vorbrüggen, Helmut, 55

Wadsworth, William S., Jr., 25
Walling, Cheves, 13
Wallis, Everett S., 3
Wallquist, Olof, 47
Wang, Chia-Lin L., 34
Warnhoff, E. W., 18
Waters, Marcey L., 70
Watt, David S., 31
Weinreb, Steven M., 65
Welmaker, Gregory S., 75
Werner, Nathan S., 75
Weston, Arthur W., 3, 9
Whaley, Wilson M., 6
Wilds, A. L., 2
Wiley, Richard H., 6
Williamson, David H., 24
Wilson, C. V., 9
Wilson, Stephen R., 43
Wolf, Donald E., 6
Wolff, Hans, 3
Wollowitz, Susan, 44
Wood, John L., 3
Wulff, William D., 70
Wynberg, Hans, 28

Yamago, Shigeru, 61
Yan, Shou-Jen, 28
Yoshioka, Mitsuru, 25

Zaugg, Harold E., 8, 14
Zhao, Cong-Gui, 61, 69
Zhou, Ping, 62
Zubkov, Oleg A., 62
Zweifel, George, 13, 32

CHAPTER AND TOPIC INDEX, VOLUMES 1–76

Many chapters contain brief discussions of reactions and comparisons of alternative synthetic methods related to the reaction that is the subject of the chapter. These related reactions and alternative methods are not usually listed in this index. In this index, the volume number is in **boldface**, the chapter number is in ordinary type.

Acetoacetic ester condensation, **1**, 9
Acetylenes:
   cotrimerizations of, **68**, 1
   oxidation by dioxirane, **69**, 1
   reactions with Fischer carbene complexes, phenol and quinone formation, **70**, 2
   synthesis of, **5**, 1; **23**, 3; **32**, 2
Acid halides:
   reactions with esters, **1**, 9
   reactions with organometallic compounds, **8**, 2
α-Acylamino acid mixed anhydrides, **12**, 4
α-Acylamino acids, azlactonization of, **3**, 5
Acylation:
   of esters with acid chlorides, **1**, 9
   intramolecular, to form cyclic ketones, **2**, 4; **23**, 2
   of ketones to form diketones, **8**, 3
Acyl fluorides, synthesis of, **21**, 1; **34**, 2; **35**, 3
Acyl hypohalites, reactions of, **9**, 5
Acyloins, **4**, 4; **15**, 1; **23**, 2
Alcohols:
   conversion to fluorides, **21**, 1, 2; **34**, 2; **35**, 3
   conversion to olefins, **12**, 2
   oxidation of, **6**, 5; **39**, 3; **53**, 1; **74**, 2
   replacement of hydroxy group by nucleophiles, **29**, 1; **42**, 2

   resolution of, **2**, 9
Alcohols, synthesis:
   by allylstannane addition to aldehydes, **64**, 1
   by base-promoted isomerization of epoxides, **29**, 3
   by hydroboration, **13**, 1
   by hydroxylation of ethylenic compounds, **7**, 7
   by organochromium reagents to carbonyl compounds, **64**, 3
   by reduction, **6**, 10; **8**, 1; **71**, 1
   from organoboranes, **33**, 1; **73**, 1
Aldehydes, additions of allyl, allenyl, propargyl stannanes, **64**, 1
   addition of allylic boron compounds, **73**, 1
Aldehydes, catalyzed addition to double bonds, **40**, 4
Aldehydes, synthesis of, **4**, 7; **5**, 10; **8**, 4, 5; **9**, 2; **33**, 1
Aldol condensation, **16**; **67**, 1
   catalytic, enantioselective, **67**, 1
   directed, **28**, 3
   with boron enolates, **51**, 1
Aliphatic fluorides, **2**, 2; **21**, 1, 2; **34**, 2; **35**, 3
Alkanes: by reduction
   of alkyl halides with organochromium reagents, **64**, 3

Organic Reactions, Vol. 76, Edited by Scott E. Denmark et al.
© 2012 Organic Reactions, Inc. Published 2012 by John Wiley & Sons, Inc.

Alkanes: by reduction (*Continued*)
  of carbonyl groups with organosilanes, **71**, 1
  oxidation of, **69**, 1
Alkenes:
  arylation of, **11**, 3; **24**, 3; **27**, 2
  asymmetric dihydroxylation, **66**, 2
  cyclopropanes from, **20**, 1
  cyclization in intramolecular Heck reactions, **60**, 2
  from carbonyl compounds with organochromium reagents, **64**, 3
  dioxirane epoxidation of, **61**, 2
  epoxidation and hydroxylation of, **7**, 7
  epoxidation of electron-deficient, **74**, 3
  free-radical additions to, **13**, 3, 4
  hydroboration of, **13**, 1
  hydrocyanation of, **75**, 1
  hydrogenation with homogeneous catalysts, **24**, 1
  reactions with diazoacetic esters, **18**, 3
  reactions with nitrones, **36**, 1
  reduction by:
    alkoxyaluminum hydrides, **34**, 1
    diimides, **40**, 2
    organosilanes, **71**, 1
Alkenes, synthesis:
  from amines, **11**, 5
  from aryl and vinyl halides, **27**, 2
  by Bamford-Stevens reaction, **23**, 3
  by Claisen and Cope rearrangements, **22**, 1
  by dehydrocyanation of nitriles, **31**
  by deoxygenation of vicinal diols, **30**, 2
  from α-halosulfones, **25**, 1; **62**, 2
  by palladium-catalyzed vinylation, **27**, 2
  from phosphoryl-stabilized anions, **25**, 2
  by pyrolysis of xanthates, **12**, 2
  from silicon-stabilized anions, **38**, 1
  from tosylhydrazones, **23**, 3; **39**, 1
  by Wittig reaction, **14**, 3
Alkenyl- and alkynylaluminum reagents, **32**, 2
Alkenyl-
  lithiums, formation of, **39**, 1
  silanes, **75**, 3

Alkoxyaluminum hydride reductions, **34**, 1; **36**, 3
Alkoxyphosphonium cations, nucleophilic displacements on, **29**, 1
Alkoxysilanes, **75**, 3
Alkylation:
  of allylic and benzylic carbanions, **27**, 1
  with amines and ammonium salts, **7**, 3
  of aromatic compounds, **3**, 1
  of esters and nitriles, **9**, 4
  γ-, of dianions of β-dicarbonyl compounds, **17**, 2
  of metallic acetylides, **5**, 1
  of nitrile-stabilized carbanions, **31**
  with organopalladium complexes, **27**, 2
Alkylidenation by titanium-based reagents, **43**, 1
Alkylidenesuccinic acids, synthesis and reactions of, **6**, 1
Alkylidene triphenylphosphoranes, synthesis and reactions of, **14**, 3
Alkynes, hydrocyanation of, **75**, 1
Alkynylsilanes, **75**, 3
Allenylsilanes, electrophilic substitution reactions of, **37**, 2
Allylboration of carbonyl compounds, **73**, 1
Allylsilanes, **75**, 3
Allyl transfer reactions, **73**, 1
Allylic alcohols, synthesis:
  from epoxides, **29**, 3
  by Wittig rearrangement, **46**, 2
Allylic and benzylic carbanions, heteroatom-substituted, **27**, 1
Allylic hydroperoxides, in photooxygenations, **20**, 2
Allylic rearrangements, transformation of glycols into 2,3-unsaturated glycosyl derivatives, **62**, 4
Allylic rearrangements, trihaloacetimidate, **66**, 1
π-Allylnickel complexes, **19**, 2
Allylphenols, synthesis by Claisen rearrangement, **2**, 1; **22**, 1
Allylsilanes, electrophilic substitution reactions of, **37**, 2

Aluminum alkoxides:
  in Meerwein-Ponndorf-Verley reduction, **2**, 5
  in Oppenauer oxidation, **6**, 5
Amide formation by oxime rearrangement, **35**, 1
α-Amidoalkylations at carbon, **14**, 2
Amination:
  electrophilic, of carbanions and enolates, **72**, 1
  of heterocyclic bases by alkali amides, **1**, 4
  of hydroxy compounds by Bucherer reaction, **1**, 5
Amine oxides:
  Polonovski reaction of, **39**, 2
  pyrolysis of, **11**, 5
Amines:
  from allylstannane addition to imines, **64**, 1
  oxidation of, **69**, 1
  synthesis from organoboranes, **33**, 1
  synthesis by reductive alkylation, **4**, 3; **5**, 7
  synthesis by Zinin reaction, **20**, 4
  reactions with cyanogen bromide, **7**, 4
α-Aminoacid synthesis, via Strecker Reaction, **70**, 1
α-Aminoalkylation of activated olefins, **51**, 2
Aminophenols from anilines, **35**, 2
Anhydrides of aliphatic dibasic acids, Friedel-Crafts reaction with, **5**, 5
Anion-assisted sigmatropic rearrangements, **43**, 2
Anthracene homologs, synthesis of, **1**, 6
Anti-Markownikoff hydration of alkenes, **13**, 1
π-Arenechromium tricarbonyls, reaction with nitrile-stabilized carbanions, **31**
η^6-(Arene)chromium complexes, **67**, 2
Arndt-Eistert reaction, **1**, 2
Aromatic aldehydes, synthesis of, **5**, 6; **28**, 1
Aromatic compounds, chloromethylation of, **1**, 3
Aromatic fluorides, synthesis of, **5**, 4
Aromatic hydrocarbons, synthesis of, **1**, 6; **30**, 1

Aromatic substitution by the $S_{RN}1$ reaction, **54**, 1
Arsinic acids, **2**, 10
Arsonic acids, **2**, 10
Arylacetic acids, synthesis of, **1**, 2; **22**, 4
β-Arylacrylic acids, synthesis of, **1**, 8
Arylamines, synthesis and reactions of, **1**, 5
Arylation:
  by aryl halides, **27**, 2
  by diazonium salts, **11**, 3; **24**, 3
  γ-, of dianions of β-dicarbonyl compounds, **17**, 2
  of alkenes, **11**, 3; **24**, 3; **27**, 2
  of enolates, **76**, 2
  of ketones, **76**, 2
  of nitrile-stabilized carbanions, **31**
Aryldiazoacetates, **75**, 2
Arylglyoxals, condensation with aromatic hydrocarbons, **4**, 5
Arylsilanes, **75**, 3
Arylsulfonic acids, synthesis of, **3**, 4
Aryl halides, homocoupling of, **63**, 3
Aryl thiocyanates, **3**, 6
Asymmetric aldol reactions using boron enolates, **51**, 1
Asymmetric cyclopropanation, **57**, 1
Asymmetric dihydroxylation, **66**, 2
Asymmetric epoxidation, **48**, 1; **61**, 2; **74**, 3
Asymmetric hydrocyanation, **75**, 1
Asymmetric hydrogenation of C=N, **74**, 1
Asymmetric reduction, **71**, 1
Asymmetric Strecker reaction, **70**, 1
Atom transfer preparation of radicals, **48**, 2
Aza-Cope/Mannich reaction, **75**, 4
Aza-Payne rearrangements, **60**, 1
Azaphenanthrenes, synthesis by photocyclization, **30**, 1
Azides, synthesis and rearrangement of, **3**, 9
Azlactones, **3**, 5

Baeyer-Villiger reaction, **9**, 3; **43**, 3
Bamford-Stevens reaction, **23**, 3
Barbier Reaction, **58**, 2
Bart reaction, **2**, 10
Barton fragmentation reaction, **48**, 2

Béchamp reaction, **2**, 10
Beckmann rearrangement, **11**, 1; **35**, 1
Benzils, reduction of, **4**, 5
Benzoin condensation, **4**, 5
Benzoquinones:
　acetoxylation of, **19**, 3
　in Nenitzescu reaction, **20**, 3
　synthesis of, **4**, 6
Benzylic carbanions, **27**, 1; **67**, 2
Benzylsilanes, **75**, 3
Biaryls, synthesis of, **2**, 6; **63**, 3
Bicyclobutanes, from cyclopropenes, **18**, 3
Biginelli dihydropyrimidine synthesis, **63**, 1
Birch reaction, **23**, 1; **42**, 1
Bischler-Napieralski reaction, **6**, 2
Bis(chloromethyl) ether, **1**, 3; **19**, *warning*
Boron enolates, **51**, 1
Borane reagents, for allylic transfer, **73**, 1
Borohydride reduction, chiral, **52**, 2
　in reductive amination, **59**, 1
Boyland-Sims oxidation, **35**, 2
Bucherer reaction, **1**, 5

Cannizzaro reaction, **2**, 3
Carbanion, electrophilic amination, **72**, 1
Carbenes, **13**, 2; **26**, 2; **28**, 1
Carbene complexes in phenol and quinone synthesis, **70**, 2
Carbenoids,
　in cyclopropanation, **57**, 1; **58**, 1
　intermolecular C-H insertions of, **75**, 2
Carbohydrates, deoxy, synthesis of, **30**, 2
Carbo/metallocupration, **41**, 2
Carbon-carbon bond formation:
　by acetoacetic ester condensation, **1**, 9
　by acyloin condensation, **23**, 2
　by aldol condensation, **16**; **28**, 3; **46**, 1; **67**, 1
　by alkylation with amines and ammonium salts, **7**, 3
　by γ-alkylation and arylation, **17**, 2
　by allylic and benzylic carbanions, **27**, 1
　by amidoalkylation, **14**, 2
　by Cannizzaro reaction, **2**, 3
　by Claisen rearrangement, **2**, 1; **22**, 1
　by Cope rearrangement, **22**, 1
　by cyclopropanation reaction, **13**, 2; **20**, 1
　by Darzens condensation, **5**, 10
　by diazonium salt coupling, **10**, 1; **11**, 3; **24**, 3
　by Dieckmann condensation, **15**, 1
　by Diels-Alder reaction, **4**, 1, 2; **5**, 3; **32**, 1
　by free-radical additions to alkenes, **13**, 3
　by Friedel-Crafts reaction, **3**, 1; **5**, 5
　by Knoevenagel condensation, **15**, 2
　by Mannich reaction, **1**, 10; **7**, 3
　by Michael addition, **10**, 3
　by nitrile-stabilized carbanions, **31**
　by organoboranes and organoborates, **33**, 1
　by organocopper reagents, **19**, 1; **38**, 2; **41**, 2
　by organopalladium complexes, **27**, 2
　by organozinc reagents, **20**, 1
　by rearrangement of α-halosulfones, **25**, 1; **62**, 2
　by Reformatsky reaction, **1**, 1; **28**, 3
　by trivalent manganese, **49**, 3
　by Vilsmeier reaction, **49**, 1; **56**, 2
　by vinylcyclopropane-cyclopentene rearrangement, **33**, 2
Carbon-fluorine bond formation, **21**, 1; **34**, 2; **35**, 3; **69**, 2
Carbon-halogen bond formation,
　by replacement of hydroxy groups, **29**, 1
Carbon-heteroatom bond formation:
　by free-radical chain additions to carbon-carbon multiple bonds, **13**, 4
　by organoboranes and organoborates, **33**, 1
Carbon-nitrogen bond formation,
　by reductive amination, **59**, 1
Carbon-phosphorus bond formation, **36**, 2
Carbonyl compounds, addition of organochromium reagents, **64**, 3
Carbonyl compounds, α,β-unsaturated:
　formation by selenoxide elimination, **44**, 1
　vicinal difunctionalization of, **38**, 2

Carbonyl compounds, from nitro
  compounds, **38**, 3
  in the Passerini Reaction, **65**, 1
  oxidation with hypervalent iodine
    reagents, **54**, 2
  reactions with allylic boron
    compounds, **73**, 1
  reductive amination of, **59**, 1
Carbonylation as part of intramolecular
  Heck reaction, **60**, 2
Carboxylic acid derivatives, conversion to
  fluorides, **21**, 1, 2; **34**, 2; **35**, 3
Carboxylic acids:
  synthesis from organoboranes, **33**, 1
  reaction with organolithium reagents,
    18, 1
Catalytic asymmetric hydrogenation of
  C=N functions, **74**, 1
Catalytic enantioselective aldol addition,
  67, 1
C–H functionalization, **75**, 2
C–H insertions, intermolecular, with
  carbenoids, **75**, 2
Chapman rearrangement, **14**, 1; **18**, 2
Chloromethylation of aromatic
  compounds, **2**, 3; **9**, *warning*
Cholanthrenes, synthesis of, **1**, 6
Chromium reagents, **64**, 3; **67**, 2
Chugaev reaction, **12**, 2
Claisen condensation, **1**, 8
Claisen rearrangement, **2**, 1; **22**, 1
Cleavage:
  of benzyl-oxygen, benzyl-nitrogen, and
    benzyl-sulfur bonds, **7**, 5
  of carbon-carbon bonds by periodic
    acid, **2**, 8
  of esters via S_N2-type dealkylation, **24**,
    2
  of non-enolizable ketones with sodium
    amide, **9**, 1
  in sensitized photooxidation, **20**, 2
Clemmensen reduction, **1**, 7; **22**, 3
Collins reagent, **53**, 1
Condensation:
  acetoacetic ester, **1**, 9
  acyloin, **4**, 4; **23**, 2
  aldol, **16**
  benzoin, **4**, 5
  Biginelli, **63**, 1

Claisen, **1**, 8
Darzens, **5**, 10; **31**
Dieckmann, **1**, 9; **6**, 9; **15**, 1
directed aldol, **28**, 3
Knoevenagel, **1**, 8; **15**, 2
Stobbe, **6**, 1
Thorpe-Ziegler, **15**, 1; **31**
Conjugate addition:
  of hydrogen cyanide, **25**, 3; **75**, 1
  of organocopper reagents, **19**, 1; **41**, 2
Cope rearrangement, **22**, 1; **41**, 1; **43**, 2
Copper-catalyzed arylation of active
  methylenes, **76**, 2
Copper-catalyzed preparation of indoles
  by cyclization, **76**, 3
Copper-Grignard complexes, conjugate
  additions of, **19**, 1; **41**, 2
Corey-Winter reaction, **30**, 2
Coumarins, synthesis of, **7**, 1; **20**, 3
Coupling reaction of organostannanes, **50**,
  1
Cross-coupling of organosilicon
  compounds, **75**, 3
Cuprate reagents, **19**, 1; **38**, 2; **41**, 2
Curtius rearrangement, **3**, 7, 9
Cyanation, of *N*-heteroaromatic
  compounds, **70**, 1
Cyanoborohydride, in reductive
  aminations, **59**, 1
Cyanoethylation, **5**, 2
Cyanogen bromide, reactions with tertiary
  amines, **7**, 4
Cyclic ketones, formation by
  intramolecular acylation, **2**, 4; **23**, 2
Cyclization:
  of alkyl dihalides, **19**, 2
  of aryl-substituted aliphatic acids, acid
    chlorides, and anhydrides, **2**, 4;
    23, 2
  of α-carbonyl carbenes and carbenoids,
    26, 2
  cycloheptenones from α-bromoketones,
    29, 2
  of diesters and dinitriles, **15**, 1
  Fischer indole, **10**, 2
  intramolecular by acylation, **2**, 4
  intramolecular by acyloin
    condensation, **4**, 4

Cyclization: (*Continued*)
  intramolecular by Diels-Alder reaction, **32**, 1
  intramolecular by Heck reaction, **60**, 2
  intramolecular by Michael reaction, **47**, 2
  Nazarov, **45**, 1
  by radical reactions, **48**, 2
  of stilbenes, **30**, 1
  tandem cyclization by Heck reaction, **60**, 2
Cycloaddition reactions,
  of cyclenones and quinones, **5**, 3
  cyclobutanes, synthesis of, **12**, 1; **44**, 2
  cyclotrimerization of acetylenes, **68**, 1
  Diels-Alder, acetylenes and alkenes, **4**, 2
  Diels-Alder, imino dienophiles, **65**, 2
  Diels-Alder, intramolecular, **32**, 1
  Diels-Alder, maleic anhydride, **4**, 1
  [4 + 3], **51**, 3
  of enones, **44**, 2
  of ketenes, **45**, 2
  of nitrones and alkenes, **36**, 1
  Pauson-Khand, **40**, 1
  photochemical, **44**, 2
  retro-Diels-Alder reaction, **52**, 1; **53**, 2
  [6 + 4], **49**, 2
  [3 + 2], **61**, 1
Cyclobutanes, synthesis:
  from nitrile-stabilized carbanions, **31**
  by thermal cycloaddition reactions, **12**, 1
Cycloheptadienes, from
  divinylcyclopropanes, **41**, 1
  polyhalo ketones, **29**, 2
π-Cyclopentadienyl transition metal carbonyls, **17**, 1
Cyclopentenones:
  annulation, **45**, 1
  synthesis, **40**, 1; **45**, 1
Cyclopropane carboxylates, from diazoacetic esters, **18**, 3
Cyclopropanes:
  from α-diazocarbonyl compounds, **26**, 2
  from metal-catalyzed decomposition of diazo compounds, **57**, 1
  from nitrile-stabilized carbanions, **31**

  from tosylhydrazones, **23**, 3
  from unsaturated compounds, methylene iodide, and zinc-copper couple, **20**, 1; **58**, 1; **58**, 2
Cyclopropenes, synthesis of, **18**, 3

Darzens glycidic ester condensation, **5**, 10; **31**
DAST, **34**, 2; **35**, 3
Deamination of aromatic primary amines, **2**, 7
Debenzylation, **7**, 5; **18**, 4
Decarboxylation of acids, **9**, 5; **19**, 4
Dehalogenation of α-haloacyl halides, **3**, 3
Dehydrogenation:
  in synthesis of ketenes, **3**, 3
  in synthesis of acetylenes, **5**, 1
Demjanov reaction, **11**, 2
Deoxygenation of vicinal diols, **30**, 2
Desoxybenzoins, conversion to benzoins, **4**, 5
Dess-Martin Oxidation, **53**, 1
Desulfonylation reactions, **72**, 2
Desulfurization:
  of α-(alkylthio)nitriles, **31**
  in alkene synthesis, **30**, 2
  with Raney nickel, **12**, 5
Diazo compounds, carbenoids derived from, **57**, 1; **75**, 2
Diazoacetic esters, reactions with alkenes, alkynes, heterocyclic and aromatic compounds, **18**, 3; **26**, 2
α-Diazocarbonyl compounds, insertion and addition reactions, **26**, 2
Diazomethane:
  in Arndt-Eistert reaction, **1**, 2
  reactions with aldehydes and ketones, **8**, 8
Diazonium fluoroborates, synthesis and decomposition, **5**, 4
Diazonium salts:
  coupling with aliphatic compounds, **10**, 1, 2
  in deamination of aromatic primary amines, **2**, 7
  in Meerwein arylation reaction, **11**, 3; **24**, 3
  in ring closure reactions, **9**, 7

in synthesis of biaryls and aryl
quinones, **2**, 6
Dieckmann condensation, **1**, 9; **15**, 1
for synthesis of tetrahydrothiophenes,
6, 9
Diels-Alder reaction:
intramolecular, **32**, 1
retro-Diels-Alder reaction, **52**, 1; **53**, 2
with alkynyl and alkenyl dienophiles,
4, 2
with cyclenones and quinones, **5**, 3
with imines, **65**, 2
with maleic anhydride, **4**, 1
Dihydrodiols, **63**, 2
Dihydropyrimidine synthesis, **63**, 1
Dihydroxylation of alkenes, asymmetric,
66, 2
hydrogen-bond-mediated, **76**, 1
Diimide, **40**, 2
Diketones:
pyrolysis of diaryl, **1**, 6
reduction by acid in organic solvents,
22, 3
synthesis by acylation of ketones, **8**, 3
synthesis by alkylation of β-diketone
anions, **17**, 2
Dimethyl sulfide, in oxidation reactions,
39, 3
Dimethyl sulfoxide, in oxidation reactions,
39, 3
Diols:
deoxygenation of, **30**, 2
oxidation of, **2**, 8
Dioxetanes, **20**, 2
Dioxiranes, **61**, 2; **69**, 1
Dioxygenases, **63**, 2
Dirhodium catalysts, **75**, 2
Divinyl-aziridines, -cyclopropanes,
-oxiranes, and -thiiranes,
rearrangements of, **41**, 1
Doebner reaction, **1**, 8

Eastwood reaction, **30**, 2
Elbs reaction, **1**, 6; **35**, 2
Electrophilic
amination, **72**, 1
fluorination, **69**, 2
Enamines, reaction with quinones, **20**, 3
Enantioselective: aldol reactions, **67**, 1

allylation and crotylation, **73**, 1
Ene reaction, in photosensitized
oxygenation, **20**, 2
Enolates:
α-Arylation, **76**, 2
Fluorination of, **69**, 2
α-Hydroxylation of, **62**, 1
in directed aldol reactions, **28**, 3; **46**, 1;
51, 1
Enone cycloadditions, **44**, 2
Enzymatic reduction, **52**, 2
Enzymatic resolution, **37**, 1
Epoxidation:
of alkenes, **61**, 2; **74**, 3
of allylic alcohols, **48**, 1
with organic peracids, **7**, 7
Epoxide isomerizations, **29**, 3
Epoxide
formation, **48**, 1; **61**, 2; **74**, 3
migration, **60**, 1
Esters:
acylation with acid chlorides, **1**, 9
alkylation of, **9**, 4
alkylidenation of, **43**, 1
cleavage via S_N2-type dealkylation, **24**,
2
dimerization, **23**, 2
glycidic, synthesis of, **5**, 10
hydrolysis, catalyzed by pig liver
esterase, **37**, 1
β-hydroxy, synthesis of, **1**, 1; **22**, 4
β-keto, synthesis of, **15**, 1
reaction with organolithium reagents,
18, 1
reduction of, **8**, 1; **71**, 1
synthesis from diazoacetic esters, **18**, 3
synthesis by Mitsunobu reaction, **42**, 2
Ethers, synthesis by Mitsunobu reaction,
42, 2
Exhaustive methylation, Hofmann, **11**, 5

Favorskii rearrangement, **11**, 4
Ferrocenes, **17**, 1
Fischer carbene complexes, **70**, 2
Fischer indole cyclization, **10**, 2
Fluorinating agents, electrophilic, **69**, 2
Fluorination of aliphatic compounds, **2**, 2;
21, 1, 2; **34**, 2; **35**, 3; **69**, 2
of carbonyl compounds, **69**, 2

Fluorination of aliphatic compounds,
    (*Continued*)
    of heterocycles, **69**, 2
Fluorination:
    by DAST, **35**, 3
    by N-F reagents, **69**, 2
    by sulfur tetrafluoride, **21**, 1; **34**, 2
Formylation:
    by hydroformylation, **56**, 1
    of alkylphenols, **28**, 1
    of aromatic hydrocarbons, **5**, 6
    of aromatic compounds, **49**, 1
    of non-aromatic compounds, **56**, 2
Free radical additions:
    to alkenes and alkynes to form
        carbon-heteroatom bonds, **13**, 4
    to alkenes to form carbon-carbon
        bonds, **13**, 3
Freidel-Crafts catalysts, in nucleoside
    synthesis, **55**, 1
Friedel-Crafts reaction, **2**, 4; **3**, 1; **5**, 5; **18**, 1
Friedländer synthesis of quinolines, **28**, 2
Fries reaction, **1**, 11

Gattermann aldehyde synthesis, **9**, 2
Gattermann-Koch reaction, **5**, 6
Germanes, addition to alkenes and
    alkynes, **13**, 4
Glycals,
    fluorination of, **69**, 2
    transformation in glycosyl derivatives,
        62, 4
Glycosides, synthesis of, **64**, 2
Glycosylating agents, **68**, 2
Glycosylation on polymer supports, **68**, 2
Glycosylation, with sulfoxides and
    sulfinates, **64**, 2
Glycidic esters, synthesis and reactions of,
    5, 10
Gomberg-Bachmann reaction, **2**, 6; **9**, 7
Grundmann synthesis of aldehydes, **8**, 5

Halides, displacement reactions of, **22**, 2; **27**, 2
Halide-metal exchange, **58**, 2
Halides, synthesis:
    from alcohols, **34**, 2
    by chloromethylation, **1**, 3
    from organoboranes, **33**, 1
    from primary and secondary alcohols,
        29, 1
Haller-Bauer reaction, **9**, 1
Halocarbenes, synthesis and reactions of,
    13, 2
Halocyclopropanes, reactions of, **13**, 2
Halogen-metal interconversion reactions,
    6, 7
α-Haloketones, rearrangement of, **11**, 4
Halosilanes, **75**, 3
α-Halosulfones, synthesis and reactions of,
    25, 1; **62**, 2
Heck reaction, **27**, 2
    intramolecular, **60**, 2
Helicenes, synthesis by photocyclization,
    30, 1
Heteroarylsilanes, **75**, 3
Heterocyclic aromatic systems, lithiation
    of, **26**, 1
Heterocyclic bases, amination of, **1**, 4
    in nucleosides, **55**, 1
Heterodienophiles, **53**, 2
Hilbert-Johnson method, **55**, 1
Hoesch reaction, **5**, 9
Hofmann elimination reaction, **11**, 5; **18**, 4
Hofmann reaction of amides, **3**, 7, 9
Homocouplings mediated by Cu, Ni, and
    Pd, **63**, 3
Homogeneous hydrogenation catalysts, **24**, 1
Hunsdiecker reaction, **9**, 5; **19**, 4
Hydration of alkenes, dienes, and alkynes,
    13, 1
Hydrazoic acid, reactions and generation
    of, **3**, 8
Hydroboration, **13**, 1
Hydrocyanation,
    of alkenes and alkynes, **75**, 1
    of conjugated carbonyl compounds, **25**, 3
Hydroformylation, **56**, 1
Hydrogen cyanide, **25**, 3; **75**, 1
Hydrogenation catalysts, homogeneous,
    24, 1
Hydrogenation of C=N functions, **74**, 1
Hydrogenation of esters, with copper
    chromite and Raney nickel, **8**, 1

Hydrohalogenation, **13**, 4
Hydrosilylation, **75**, 3
Hydroxyaldehydes, aromatic, **28**, 1
α-Hydroxyalkylation of activated olefins, **51**, 2
α-Hydroxyketones:
　rearrangement, **62**, 3
　synthesis of, **23**, 2
Hydroxylation:
　of enolates, **62**, 1
　of ethylenic compounds with organic peracids, **7**, 7
Hypervalent iodine reagents, **54**, 2; **57**, 2

Imidates, rearrangement of, **14**, 1
Imines, additions of allyl, allenyl, propargyl stannanes, **64**, 1
　additions of cyanide, **70**, 1
　as dienophiles, **65**, 2
　catalytic asymmetric hydrogenation, **74**, 1
　synthesis, **70**, 1
Iminium ions, **39**, 2; **65**, 2; **75**, 4
Imino Diels-Alder reactions, **65**, 2
Indole synthesis,
　by catalyzed cyclization with alkenes, **76**, 3
　by catalyzed cyclization with alkynes, **76**, 3
　by Nenitzescu reaction, **20**, 3
　by reaction with TosMIC, **57**, 3
Ionic hydrogenation, **71**, 1
Isocyanides, in the Passerini reaction, **65**, 1
　sulfonylmethyl, reactions of, **57**, 3
Isoquinolines, synthesis of, **6**, 2, 3, 4; **20**, 3

Jacobsen reaction, **1**, 12
Japp-Klingemann reaction, **10**, 2

Katsuki-Sharpless epoxidation, **48**, 1
Ketene cycloadditions, **45**, 2
Ketenes and ketene dimers, synthesis of, **3**, 3; **45**, 2
α-Ketol rearrangement, **62**, 3
Ketones:
　acylation of, **8**, 3
　alkylidenation of, **43**, 1
　Baeyer-Villiger oxidation of, **9**, 3; **43**, 3
　cleavage of non-enolizable, **9**, 1
　comparison of synthetic methods, **18**, 1
　conversion to amides, **3**, 8; **11**, 1
　conversion to fluorides, **34**, 2; **35**, 3
　cyclic, synthesis of, **2**, 4; **23**, 2
　cyclization of divinyl ketones, **45**, 1
　reaction with diazomethane, **8**, 8
　reduction to aliphatic compounds, **4**, 8
　reduction by:
　　alkoxyaluminum hydrides, **34**, 1
　　organosilanes, **71**, 1
　reduction in anhydrous organic solvents, **22**, 3
　synthesis by oxidation of alcohols, **6**, 5; **39**, 3
　synthesis from acid chlorides and organo-metallic compounds, **8**, 2; **18**, 1
　synthesis from organoboranes, **33**, 1
　synthesis from organolithium reagents and carboxylic acids, **18**, 1
　synthesis from α,β-unsaturated carbonyl compounds and metals in liquid ammonia, **23**, 1
Kindler modification of Willgerodt reaction, **3**, 2
Knoevenagel condensation, **1**, 8; **15**, 2; **57**, 3
Koch-Haaf reaction, **17**, 3
Kornblum oxidation, **39**, 3
Kostaneki synthesis of chromanes, flavones, and isoflavones, **8**, 3

β-Lactams, synthesis of, **9**, 6; **26**, 2
β-Lactones, synthesis and reactions of, **8**, 7
Leuckart reaction, **5**, 7
Lithiation:
　of allylic and benzylic systems, **27**, 1
　by halogen-metal exchange, **6**, 7
　heteroatom facilitated, **26**, 1; **47**, 1
　of heterocyclic and olefinic compounds, **26**, 1
Lithioorganocuprates, **19**, 1; **22**, 2; **41**, 2
Lithium aluminum hydride reductions, **6**, 2
　chirally modified, **52**, 2
Lossen rearrangement, **3**, 7, 9

Mannich reaction, **1**, 10; **7**, 3; **75**, 4
Meerwein arylation reaction, **11**, 3; **24**, 3
Meerwein-Ponndorf-Verley reduction, **2**, 5
Mercury hydride method to prepare radicals, **48**, 2
Metal-catalyzed hydrocyanation, **75**, 1
Metalations with organolithium compounds, **8**, 6; **26**, 1; **27**, 1
Methylenation of carbonyl groups, **43**, 1
Methylenecyclopropane, in cycloaddition reactions, **61**, 1
Methylene-transfer reactions, **18**, 3; **20**, 1; **58**, 1
Michael reaction, **10**, 3; **15**, 1, 2; **19**, 1; **20**, 3; **46**, 1; **47**, 2
Microbiological oxygenations, **63**, 2
Mitsunobu reaction, **42**, 2
Moffatt oxidation, **39**, 3; **53**, 1
Morita-Baylis-Hillman reaction, **51**, 2

Nagata reaction, **25**, 3
Nazarov cyclization, **45**, 1
Nef reaction, **38**, 3
Nenitzescu reaction, **20**, 3
Nitriles:
 formation from:
  alkenes and alkynes, **75**, 1
  oximes, **35**, 2
 synthesis from organoboranes, **33**, 1
 α,β-unsaturated:
  by elimination of selenoxides, **44**, 1
Nitrile-stabilized carbanions:
 alkylation and arylation of, **31**
Nitroamines, **20**, 4
Nitro compounds, conversion to carbonyl compounds, **38**, 3
Nitro compounds, synthesis of, **12**, 3
Nitrone-olefin cycloadditions, **36**, 1
Nitrosation, **2**, 6; **7**, 6
Nitroxide-catalyzed oxidations, **74**, 2
Nucleosides, synthesis of, **55**, 1

Olefin formation, by reductive elimination of β-hydroxysulfones, **72**, 2
Olefins,
 hydrocyanation of, **75**, 1
 hydroformylation of, **56**, 1
Oligomerization of 1,3-dienes, **19**, 2

Oligosaccharide synthesis on polymer support, **68**, 2
Oppenauer oxidation, **6**, 5
Organoboranes:
 formation of carbon-carbon and carbon-heteroatom bonds from, **33**, 1
 in allylation of carbonyl compounds, **73**, 1
 isomerization and oxidation of, **13**, 1
 reaction with anions of α-chloronitriles, **31**, 1
Organochromium reagents:
 addition to carbonyl compounds, **64**, 3; **67**, 2
 addition to imines, **67**, 2
Organohypervalent iodine reagents, **54**, 2; **57**, 2
Organometallic compounds:
 of aluminum, **25**, 3
 of chromium, **64**, 3; **67**, 2
 of copper, **19**, 1; **22**, 2; **38**, 2; **41**, 2
 of lithium, **6**, 7; **8**, 6; **18**, 1; **27**, 1
 of magnesium, zinc, and cadmium, **8**, 2;
 of palladium, **27**, 2
 of silicon, **37**, 2
 of tin, **50**, 1; **64**, 1
 of zinc, **1**, 1; **20**, 1; **22**, 4; **58**, 2
Organonitriles, **75**, 1
Organosilanols, **75**, 3
Organosilicon hydride reductions, **71**, 1
Osmium tetroxide dihydroxylation
 asymmetric, **66**, 2
 hydrogen-bond directed, **76**, 1
Overman rearrangement of allylic imidates, **66**, 1
Oxidation:
 by dioxiranes, **61**, 2; **69**, 1
 by oxoammonium and nitroxide catalysts, **74**, 2
 of alcohols and polyhydroxy compounds, **6**, 5; **39**, 3; **53**, 1
 of aldehydes and ketones,
  Baeyer-Villiger reaction, **9**, 3; **43**, 3
 of amines, phenols, aminophenols, diamines, hydroquinones, and halophenols, **4**, 6; **35**, 2

of enolates and silyl enol ethers, **62**, 1
of α-glycols, α-amino alcohols, and polyhydroxy compounds by periodic acid, **2**, 8
with hypervalent iodine reagents, **54**, 2
of organoboranes, **13**, 1
of phenolic compounds, **57**, 2
with peracids, **7**, 7
by photooxygenation, **20**, 2
with selenium dioxide, **5**, 8; **24**, 4
Oxidative decarboxylation, **19**, 4
Oximes, formation by nitrosation, **7**, 6
Oxoammonium-catalyzed oxidation, **74**, 2
Oxochromium(VI)-amine complexes, **53**, 1
Oxo process, **56**, 1
Oxygenation of arenes by dioxygenases, **63**, 2

Palladium-catalyzed
arylation of enolates, **76**, 2
coupling of organostannanes, **50**, 1
indole synthesis by cyclization, **76**, 3
vinylic substitution, **27**, 2
Palladium intermediates in Heck reactions, **60**, 2
Passerini Reaction, **65**, 1
Pauson-Khand reaction to prepare cyclopentenones, **40**, 1
Payne rearrangement, **60**, 1
Pechmann reaction, **7**, 1
Peptides, synthesis of, **3**, 5; **12**, 4
Peracids, epoxidation and hydroxylation with, **7**, 7
in Baeyer-Villiger oxidation, **9**, 3; **43**, 3
Periodic acid oxidation, **2**, 8
Perkin reaction, **1**, 8
Persulfate oxidation, **35**, 2
Peterson olefination, **38**, 1
Phenanthrenes, synthesis by photocyclization, **30**, 1
Phenols, dihydric from phenols, **35**, 2
oxidation of, **57**, 2
synthesis from Fischer carbene complexes, **70**, 2
Phosphinic acids, synthesis of, **6**, 6
Phosphonic acids, synthesis of, **6**, 6
Phosphonium salts:
halide synthesis, use in, **29**, 1
synthesis and reactions of, **14**, 3

Phosphorus compounds, addition to carbonyl group, **6**, 6; **14**, 3; **25**, 2; **36**, 2
addition reactions at imine carbon, **36**, 2
Phosphoryl-stabilized anions, **25**, 2
Photochemical cycloadditions, **44**, 2
Photocyclization of stilbenes, **30**, 1
Photooxygenation of olefins, **20**, 2
Photosensitizers, **20**, 2
Pictet-Spengler reaction, **6**, 3
Pig liver esterase, **37**, 1
Polonovski reaction, **39**, 2
Polyalkylbenzenes, in Jacobsen reaction, **1**, 12
Polycyclic aromatic compounds, synthesis by photocyclization of stilbenes, **30**, 1
Polyhalo ketones, reductive dehalogenation of, **29**, 2
Pomeranz-Fritsch reaction, **6**, 4
Prévost reaction, **9**, 5
Pschorr synthesis, **2**, 6; **9**, 7
Pummerer reaction, **40**, 3
Pyrazolines, intermediates in diazoacetic ester reactions, **18**, 3
Pyridinium chlorochromate, **53**, 1
Pyrolysis:
of amine oxides, phosphates, and acyl derivatives, **11**, 5
of ketones and diketones, **1**, 6
for synthesis of ketenes, **3**, 3
of xanthates, **12**, 2
Pyrrolidines, by aza-Cope/Mannich reaction, **75**, 4

Quaternary ammonium
N-F reagents, **69**, 2
salts, rearrangements of, **18**, 4
Quinolines, synthesis of,
by Friedländer synthesis, **28**, 2
by Skraup synthesis, **7**, 2
Quinones:
acetoxylation of, **19**, 3
diene additions to, **5**, 3
synthesis of, **4**, 6
synthesis from Fischer carbene complexes, **70**, 2
in synthesis of 5-hydroxyindoles, **20**, 3

Ramberg-Bäcklund rearrangement, **25**, 1; **62**, 2
Radical formation and cyclization, **48**, 2
Rearrangements:
    allylic trihaloacetamidate, **66**, 1
    anion-assisted sigmatropic, **43**, 2
    Beckmann, **11**, 1; **35**, 1
    Chapman, **14**, 1; **18**, 2
    Claisen, **2**, 1; **22**, 1
    Cope, **22**, 1; **41**, 1, **43**, 2
    Curtius, **3**, 7, 9
    divinylcyclopropane, **41**, 1
    Favorskii, **11**, 4
    Lossen, **3**, 7, 9
    Ramberg-Bäcklund, **25**, 1; **62**, 2
    Smiles, **18**, 2
    Sommelet-Hauser, **18**, 4
    Stevens, **18**, 4
    [2,3] Wittig, **46**, 2
    vinylcyclopropane-cyclopentene, **33**, 2
Reduction:
    of acid chlorides to aldehydes, **4**, 7; **8**, 5
    of aromatic compounds, **42**, 1
    of benzils, **4**, 5
    of ketones, enantioselective, **52**, 2
    Clemmensen, **1**, 7; **22**, 3
    desulfurization, **12**, 5
    with diimide, **40**, 2
    by dissolving metal, **42**, 1
    by homogeneous hydrogenation catalysts, **24**, 1
    by hydrogenation of esters with copper chromite and Raney nickel, **8**, 1
    hydrogenolysis of benzyl groups, **7**, 5
    by lithium aluminum hydride, **6**, 10
    by Meerwein-Ponndorf-Verley reaction, **2**, 5
        chiral, **52**, 2
    by metal alkoxyaluminum hydrides, **34**, 1; **36**, 3
    by organosilanes, **71**, 1
    of mono- and polynitroarenes, **20**, 4
    of olefins by diimide, **40**, 2
    of α,β-unsaturated carbonyl compounds, **23**, 1
    by samarium(II) iodide, **46**, 3
    by Wolff-Kishner reaction, **4**, 8

Reductive alkylation, synthesis of amines, **4**, 3; **5**, 7
Reductive amination of carbonyl compounds, **59**, 1; **71**, 1
Reductive cyanation, **57**, 3
Redutive desulfonylation, **72**, 2
Reductive desulfurization of thiol esters, **8**, 5
Reformatsky reaction, **1**, 1; **22**, 4
Reimer-Tiemann reaction, **13**, 2; **28**, 1
Reissert reaction, **70**, 1
Resolution of alcohols, **2**, 9
Retro-Diels-Alder reaction, **52**, 1; **53**, 2
Ritter reaction, **17**, 3
Rosenmund reaction for synthesis of arsonic acids, **2**, 10
Rosenmund reduction, **4**, 7

Samarium(II) iodide, **46**, 3
Sandmeyer reaction, **2**, 7
Schiemann reaction, **5**, 4
Schmidt reaction, **3**, 8, 9
Selenium dioxide oxidation, **5**, 8; **24**, 4
Seleno-Pummerer reaction, **40**, 3
Selenoxide elimination, **44**, 1
Shapiro reaction, **23**, 3; **39**, 1
Silanes:
    addition to olefins and acetylenes, **13**, 4
    electrophilic substitution reactions, **37**, 2
    oxidation of, **69**, 1
    reduction with, **71**, 1
Silanolate salts, **75**, 3
Sila-Pummerer reaction, **40**, 3
Siliconates, **75**, 3
Silicon-based cross-coupling, **75**, 3
Silyl carbanions, **38**, 1
Silyl enol ether, α-hydroxylation, **62**, 1
Silyl compounds, cross-coupling of, **75**, 3
Simmons-Smith reaction, **20**, 1; **58**, 1
Simonini reaction, **9**, 5
Singlet oxygen, **20**, 2
Skraup synthesis, **7**, 2; **28**, 2
Smiles rearrangement, **18**, 2
Sommelet-Hauser rearrangement, **18**, 4
$S_{RN}1$ reactions of aromatic systems, **54**, 1
Solid-phase synthesis of indoles, **76**, 3
Sommelet reaction, **8**, 4
Stevens rearrangement, **18**, 4

Stetter reaction of aldehydes with olefins, **40**, 4
Strecker reaction, catalytic asymmetric, **70**, 1
Stilbenes, photocyclization of, **30**, 1
Stille reaction, **50**, 1
Stobbe condensation, **6**, 1
Substitution reactions using organocopper reagents, **22**, 2; **41**, 2
Sugars, synthesis by glycosylation with sulfoxides and sulfinates, **64**, 2
Sulfide reduction of nitroarenes, **20**, 4
Sulfonation of aromatic hydrocarbons and aryl halides, **3**, 4
Swern oxidation, **39**, 3; **53**, 1

Tetrahydroisoquinolines, synthesis of, **6**, 3
Tetrahydrothiophenes, synthesis of, **6**, 9
Thia-Payne rearrangement, **60**, 1
Thiazoles, synthesis of, **6**, 8
Thiele-Winter acetoxylation of quinones, **19**, 3
Thiocarbonates, synthesis of, **17**, 3
Thiocyanation of aromatic amines, phenols, and polynuclear hydrocarbons, **3**, 6
Thiophenes, synthesis of, **6**, 9
Thorpe-Ziegler condensation, **15**, 1; **31**
Tiemann reaction, **3**, 9
Tiffeneau-Demjanov reaction, **11**, 2
Tin(II) enolates, **46**, 1
Tin hydride method to prepare radicals, **48**, 2
Tipson-Cohen reaction, **30**, 2
Tosylhydrazones, **23**, 3; **39**, 1
Tosylmethyl isocyanide (TosMIC), **57**, 3
Transmetallation reactions, **58**, 2
Tricarbonyl(η^6-arene)chromium complexes, **67**, 2
Trihaloacetimidate, allylic rearrangements, **66**, 1
Trimerization, co-, acetylenic compounds, **68**, 1

Trimethylenemethane, [3 + 2] cycloaddtion of, **61**, 1
Trimethylsilyl cyanide, **75**, 1

Ullmann reaction:
  homocoupling mediated by Cu, Ni, and Pd, **63**, 3
  in synthesis of diphenylamines, **14**, 1
  in synthesis of unsymmetrical biaryls, **2**, 6
Unsaturated compounds, synthesis with alkenyl- and alkynylaluminum reagents, **32**, 2

Vilsmeier reaction, **49**, 1; **56**, 2
Vinylcyclopropanes, rearrangement to cyclopentenes, **33**, 2
Vinyldiazoacetates, **75**, 2
Vinyllithiums, from sulfonylhydrazones, **39**, 1
Vinylsilanes, electrophilic substitution reactions of, **37**, 2
Vinyl substitution, catalyzed by palladium complexes, **27**, 2
von Braun cyanogen bromide reaction, **7**, 4
Vorbrüggen reaction, **55**, 1
Willgerodt reaction, **3**, 2
Wittig reaction, **14**, 3; **31**
[2,3]-Wittig rearrangement, **46**, 2
Wolff-Kishner reaction, **4**, 8

Xanthates, synthesis and pyrolysis of, **12**, 2

Ylides:
  in Stevens rearrangement, **18**, 4
  in Wittig reaction, structure and properties, **14**, 3

Zinc-copper couple, **20**, 1; **58**, 1, 2
Zinin reduction of nitroarenes, **20**, 4